AF595089

Nutrition, Microbiota and Noncommunicable Diseases

Nutrition, Microbiota and Noncommunicable Diseases

Editor

Julio Plaza-Díaz

MDPI • Basel • Beijing • Wuhan • Barcelona • Belgrade • Manchester • Tokyo • Cluj • Tianjin

Editor
Julio Plaza-Díaz
Univ Granada
Biomed Res Ctr, Inst Nutr & Food Technol Jose Mataix
Spain

Editorial Office
MDPI
St. Alban-Anlage 66
4052 Basel, Switzerland

This is a reprint of articles from the Special Issue published online in the open access journal *Nutrients* (ISSN 2072-6643) (available at: https://www.mdpi.com/journal/nutrients/special_issues/Microbiota_Noncommunicable_Diseases).

For citation purposes, cite each article independently as indicated on the article page online and as indicated below:

LastName, A.A.; LastName, B.B.; LastName, C.C. Article Title. *Journal Name* **Year**, *Article Number*, Page Range.

ISBN 978-3-03936-916-4 (Hbk)
ISBN 978-3-03936-917-1 (PDF)

Contents

About the Editor

Julio Plaza-Díaz has a degree in pharmacy (2008) and PhD in biochemistry and molecular biology (2014) from the University of Granada, Spain. He has been a researcher at the University of Granada since 2011 and has participated in more than 10 funded research projects. He is the author or coauthor of 50 peer-reviewed publications in the field of identification, characterization, and evaluation of new probiotic/microbiome studies and its impact on health.

Editorial

Nutrition, Microbiota and Noncommunicable Diseases

Julio Plaza-Diaz [1,2,3]

1 Department of Biochemistry and Molecular Biology II, School of Pharmacy, University of Granada, 18,071 Granada, Spain; jrplaza@ugr.es; Tel.: +34-958-241-599
2 Instituto de Investigación Biosanitaria IBS.GRANADA, Complejo Hospitalario Universitario de Granada, 18,014 Granada, Spain
3 Children's Hospital of Eastern Ontario Research Institute, Ottawa, ON K1H 8L1, Canada

Received: 21 June 2020; Accepted: 29 June 2020; Published: 2 July 2020

The advent of new sequencing technologies has inspired the foundation of novel research to ascertain the connections between the microbial communities that reside in our gut and some physiological and pathological conditions. The microbiota, defined as the full collection of microbes (bacteria, fungi, and viruses, among others) that naturally exist within a particular biological niche, is estimated to contain 500–1000 species [1–4].

This Special Issue of *Nutrients*, "Nutrition, Microbiota, and Noncommunicable Diseases" contains 13 original publications and seven reviews investigating the contribution of intestinal microbiota on relevant health outcomes in a variety of populations, and animal studies which suggest the growing and extensive interests of research on this topic.

Seven studies were published examining the changes in intestinal microbiota in the human population. Two of these studies recruited patients with metabolic syndrome. Tenorio-Jimenez et al. [5] reported the anthropometric variables and biochemical and inflammatory biomarkers as well as the gastrointestinal microbiome composition changes in a randomized, crossover, placebo-controlled, single-center trial in adult patients newly diagnosed with metabolic syndrome treated either with *Lactobacillus reuteri* V3401 or a placebo during 12 weeks. *L. reuteri* V3401 administration improved selected inflammatory parameters and modified the gastrointestinal microbiome, especially *Verrucomicrobia* [5], and Bellikci-Koyu et al. [6] investigated the effects of regular kefir consumption on gut microbiota composition, and their relation with the components of metabolic syndrome in a parallel-group, randomized, controlled clinical trial for 12 weeks. Gut microbiota analysis showed that regular kefir consumption resulted in a significant increase only in *Actinobacteria* abundance [6].

In two more additional studies, one with healthy elderly women and another with patients with non-alcoholic fatty liver disease (NAFLD), Morita et al. [7] examined the effect of an exercise intervention (12 weeks, trunk muscle training or aerobic exercise training) on the composition of the intestinal microbiota in healthy elderly women. *Bacteroides* abundance was significantly increased only in the aerobic exercise group, particularly in subjects showing increases in the time spent in brisk walking [7], and Chong et al. [8] determined whether inulin supplementation after brief metronidazole therapy is effective in reducing alanine aminotransferase and maintaining weight loss achieved through a very-low-calorie diet among people with NAFLD. Treatment decreased the ratio of Firmicutes/Bacteroidetes [8].

Lau et al. [9] evaluated the association of probiotic ingestion with obesity, type 2 diabetes, hypertension, and dyslipidemia using data from the National Health and Nutrition Examination Survey, 1999–2014. Probiotic supplementation or yogurt consumption were associated with a lower prevalence of obesity and hypertension [9]. In another study with humans, Dalla Via et al. [10] verified whether trimethylamine-N-oxide urinary levels may be associated with the fecal relative abundance of specific bacterial taxa and the bacterial choline trimethylamine-lyase gene *cut*C in human fecal samples. Correlation analysis showed that the cut-Kp gene cluster was significantly associated with *Enterobacteriaceae* [10].

Finally, in one study with the pediatric population, Kong et al. [11] reported both oral and intestinal microbiota in patients with autism spectrum disorder and controls, with specific microbial patterns [11].

Regarding animal studies, six studies were published examining the changes in intestinal microbiota. Probiotic supplementation, high-fat diet, use of anorexic mice, fiber, and soy intake and antihypertensive effect in metabolomics profiles were analyzed in these studies. Valcarce et al. [12] reported the effect of a short-time probiotic supplementation consisting of a mixture of two probiotic bacteria with proven antioxidant and anti-inflammatory activities on zebrafish sperm quality and male behavior [12]. Hsu et al. [13] examined the alterations of gut microbiota, mediation of short-chain fatty acids (SCFAs) and their receptors, and downregulation of nutrient-sensing signals effects in rats that received a high-fat diet. Increased Firmicutes to Bacteroidetes ratio, *Akkermansia* and *Verrucomicrobia*, and reduced abundance in the genus *Lactobacillus* were associated with blood pressure elevation [13]. Dominique et al. [14] investigated the role of the microbiome and the ClpB protein in the deregulation and self-maintenance of anorexia pathology in mice. Plasma concentration of ClpB was increased in both limited food access and activity-based anorexia mice and it was correlated with the proportion of Enterobacteriaceae in the animal feces [14]. Sasaki et al. [15] investigated the effects of fiber intake timing on metabolism. Data have suggested that inulin is more easily digested by fecal microbiota during the active period than the inactive period. Inulin consumption at breakfast has a greater effect on the microbiota [15]. Tamura et al. [16] investigated soy protein intake effects on intestinal microbiota. Soy protein intake whether in the morning or evening led to a greater microbiota diversity and a decrease in cecal pH resulting from SCFA production compared with casein intake [16]. Finally, Ahn et al. [17] investigated the metabolomics changes in rats that received amlodipine. Serum levels of phosphatidylcholine, lysophosphatidylcholine, sphingomyelin, triglycerides with large numbers of double bonds, cholesterol, sterol derivatives, and cholesterol esters were increased. Amlodipine-induced compositional changes in the gut microbiota are a causal factor in inflammation [17].

Seven reviews investigating the impact of intestinal microbiota on relevant health outcomes in a variety of populations were published. Hills Jr. et al. [18] described a general vision about the gut microbiome and its important role in human health. Salli et al. [19] reported the health benefits of xylitol. The other reviews have described the intestinal microbiota changes in specific conditions, early infancy, hepatic ischemia-reperfusion and regeneration in liver surgery, vaginal microbiota, and cardiovascular diseases. Mesa et al. [20] reported the microbiome changes and how those modulate the inflammatory mechanisms related to physiological and pathological processes that are involved in the perinatal progress. Cornide-Petronio et al. [21] summarized the role of starvation, supplemented nutrition diet, nutritional status, and alterations in microbiota on hepatic ischemia/reperfusion and regeneration. Barrientos-Duran et al. [22] examined the most important aspect in the vaginal microbiota, with special emphasis in bacterial vaginosis, and the maintenance of eubiosis, and Sanchez-Rodriguez et al. [23] discussed how external factors such as dietary and physical activity habits influence host microbiota and atherogenesis, the potential mechanisms of the influence of gut microbiota in host blood pressure, and the alterations in the prevalence of those bacterial genera affecting vascular tone and the development of hypertension. Finally, Plaza-Diaz et al. [24] revisited the effects of sweeteners on gut microbiota.

The present Special Issue provides a summary of the progress on the topic of intestinal microbiota and its important role in human health in different populations, which will be of interest from a clinical and public health perspective. Nevertheless, more studies with more samples and comparable methods are necessary to understand the actual function of intestinal microbiota in disease development and health maintenance.

Funding: This research received no external funding.

Acknowledgments: Julio Plaza-Diaz is part of the University of Granada, Plan Propio de Investigación 2016, Excellence actions: Units of Excellence; Unit of Excellence on Exercise and Health (UCEES).

Conflicts of Interest: The author declares no conflict of interest.

References

1. Plaza-Diaz, J.; Ruiz-Ojeda, F.J.; Vilchez-Padial, L.M.; Gil, A. Evidence of the Anti-Inflammatory Effects of Probiotics and Synbiotics in Intestinal Chronic Diseases. *Nutrients* **2017**, *9*, 555. [CrossRef]
2. Plaza-Diaz, J.; Ruiz-Ojeda, F.J.; Gil-Campos, M.; Gil, A. Immune-Mediated Mechanisms of Action of Probiotics and Synbiotics in Treating Pediatric Intestinal Diseases. *Nutrients* **2018**, *10*, 42. [CrossRef] [PubMed]
3. Plaza-Diaz, J.; Gomez-Fernandez, A.; Chueca, N.; Torre-Aguilar, M.J.; Gil, A.; Perez-Navero, J.L.; Flores-Rojas, K.; Martin-Borreguero, P.; Solis-Urra, P.; Ruiz-Ojeda, F.J.; et al. Autism Spectrum Disorder (ASD) with and without Mental Regression is Associated with Changes in the Fecal Microbiota. *Nutrients* **2019**, *11*, 337. [CrossRef]
4. Álvarez-Mercado, A.I.; Navarro-Oliveros, M.; Robles-Sánchez, C.; Plaza-Díaz, J.; Sáez-Lara, M.J.; Muñoz-Quezada, S.; Fontana, L.; Abadía-Molina, F. Microbial population changes and their relationship with human health and disease. *Microorganisms* **2019**, *7*, 68. [CrossRef]
5. Tenorio-Jimenez, C.; Martinez-Ramirez, M.J.; Del Castillo-Codes, I.; Arraiza-Irigoyen, C.; Tercero-Lozano, M.; Camacho, J.; Chueca, N.; Garcia, F.; Olza, J.; Plaza-Diaz, J.; et al. Lactobacillus reuteri V3401 Reduces Inflammatory Biomarkers and Modifies the Gastrointestinal Microbiome in Adults with Metabolic Syndrome: The PROSIR Study. *Nutrients* **2019**, *11*, 1761. [CrossRef] [PubMed]
6. Bellikci-Koyu, E.; Sarer-Yurekli, B.P.; Akyon, Y.; Aydin-Kose, F.; Karagozlu, C.; Ozgen, A.G.; Brinkmann, A.; Nitsche, A.; Ergunay, K.; Yilmaz, E.; et al. Effects of Regular Kefir Consumption on Gut Microbiota in Patients with Metabolic Syndrome: A Parallel-Group, Randomized, Controlled Study. *Nutrients* **2019**, *11*, 2089. [CrossRef] [PubMed]
7. Morita, E.; Yokoyama, H.; Imai, D.; Takeda, R.; Ota, A.; Kawai, E.; Hisada, T.; Emoto, M.; Suzuki, Y.; Okazaki, K. Aerobic Exercise Training with Brisk Walking Increases Intestinal Bacteroides in Healthy Elderly Women. *Nutrients* **2019**, *11*, 868. [CrossRef] [PubMed]
8. Chong, C.Y.L.; Orr, D.; Plank, L.D.; Vatanen, T.; O'Sullivan, J.M.; Murphy, R. Randomised Double-Blind Placebo-Controlled Trial of Inulin with Metronidazole in Non-Alcoholic Fatty Liver Disease (NAFLD). *Nutrients* **2020**, *12*, 937. [CrossRef] [PubMed]
9. Lau, E.; Neves, J.S.; Ferreira-Magalhaes, M.; Carvalho, D.; Freitas, P. Probiotic Ingestion, Obesity, and Metabolic-Related Disorders: Results from NHANES, 1999–2014. *Nutrients* **2019**, *11*, 1482. [CrossRef]
10. Dalla Via, A.; Gargari, G.; Taverniti, V.; Rondini, G.; Velardi, I.; Gambaro, V.; Visconti, G.L.; De Vitis, V.; Gardana, C.; Ragg, E.; et al. Urinary TMAO Levels Are Associated with the Taxonomic Composition of the Gut Microbiota and with the Choline TMA-Lyase Gene (cutC) Harbored by Enterobacteriaceae. *Nutrients* **2019**, *12*, 62. [CrossRef]
11. Kong, X.; Liu, J.; Cetinbas, M.; Sadreyev, R.; Koh, M.; Huang, H.; Adeseye, A.; He, P.; Zhu, J.; Russell, H.; et al. New and Preliminary Evidence on Altered Oral and Gut Microbiota in Individuals with Autism Spectrum Disorder (ASD): Implications for ASD Diagnosis and Subtyping Based on Microbial Biomarkers. *Nutrients* **2019**, *11*, 2128. [CrossRef] [PubMed]
12. Valcarce, D.G.; Riesco, M.F.; Martinez-Vazquez, J.M.; Robles, V. Diet Supplemented with Antioxidant and Anti-Inflammatory Probiotics Improves Sperm Quality after Only One Spermatogenic Cycle in Zebrafish Model. *Nutrients* **2019**, *11*, 843. [CrossRef] [PubMed]
13. Hsu, C.N.; Hou, C.Y.; Lee, C.T.; Chan, J.Y.H.; Tain, Y.L. The Interplay between Maternal and Post-Weaning High-Fat Diet and Gut Microbiota in the Developmental Programming of Hypertension. *Nutrients* **2019**, *11*, 1982. [CrossRef] [PubMed]
14. Dominique, M.; Legrand, R.; Galmiche, M.; Azhar, S.; Deroissart, C.; Guerin, C.; do Rego, J.L.; Leon, F.; Nobis, S.; Lambert, G.; et al. Changes in Microbiota and Bacterial Protein Caseinolytic Peptidase B During Food Restriction in Mice: Relevance for the Onset and Perpetuation of Anorexia Nervosa. *Nutrients* **2019**, *11*, 2514. [CrossRef] [PubMed]

15. Sasaki, H.; Miyakawa, H.; Watanabe, A.; Nakayama, Y.; Lyu, Y.; Hama, K.; Shibata, S. Mice Microbiota Composition Changes by Inulin Feeding with a Long Fasting Period under a Two-Meals-Per-Day Schedule. *Nutrients* **2019**, *11*, 2802. [CrossRef]
16. Tamura, K.; Sasaki, H.; Shiga, K.; Miyakawa, H.; Shibata, S. The Timing Effects of Soy Protein Intake on Mice Gut Microbiota. *Nutrients* **2019**, *12*, 87. [CrossRef]
17. Ahn, Y.; Nam, M.H.; Kim, E. Relationship Between the Gastrointestinal Side Effects of an Anti-Hypertensive Medication and Changes in the Serum Lipid Metabolome. *Nutrients* **2020**, *12*, 205. [CrossRef]
18. Hills, R.D., Jr.; Pontefract, B.A.; Mishcon, H.R.; Black, C.A.; Sutton, S.C.; Theberge, C.R. Gut Microbiome: Profound Implications for Diet and Disease. *Nutrients* **2019**, *11*, 1613. [CrossRef]
19. Salli, K.; Lehtinen, M.J.; Tiihonen, K.; Ouwehand, A.C. Xylitol's Health Benefits beyond Dental Health: A Comprehensive Review. *Nutrients* **2019**, *11*, 1813. [CrossRef]
20. Mesa, M.D.; Loureiro, B.; Iglesia, I.; Fernandez Gonzalez, S.; Llurba Olive, E.; Garcia Algar, O.; Solana, M.J.; Cabero Perez, M.J.; Sainz, T.; Martinez, L.; et al. The Evolving Microbiome from Pregnancy to Early Infancy: A Comprehensive Review. *Nutrients* **2020**, *12*, 133. [CrossRef]
21. Cornide-Petronio, M.E.; Alvarez-Mercado, A.I.; Jimenez-Castro, M.B.; Peralta, C. Current Knowledge about the Effect of Nutritional Status, Supplemented Nutrition Diet, and Gut Microbiota on Hepatic Ischemia-Reperfusion and Regeneration in Liver Surgery. *Nutrients* **2020**, *12*, 284. [CrossRef] [PubMed]
22. Barrientos-Duran, A.; Fuentes-Lopez, A.; de Salazar, A.; Plaza-Diaz, J.; Garcia, F. Reviewing the Composition of Vaginal Microbiota: Inclusion of Nutrition and Probiotic Factors in the Maintenance of Eubiosis. *Nutrients* **2020**, *12*, 419. [CrossRef] [PubMed]
23. Sanchez-Rodriguez, E.; Egea-Zorrilla, A.; Plaza-Diaz, J.; Aragon-Vela, J.; Munoz-Quezada, S.; Tercedor-Sanchez, L.; Abadia-Molina, F. The Gut Microbiota and Its Implication in the Development of Atherosclerosis and Related Cardiovascular Diseases. *Nutrients* **2020**, *12*, 605. [CrossRef]
24. Plaza-Diaz, J.; Pastor-Villaescusa, B.; Rueda-Robles, A.; Abadia-Molina, F.; Ruiz-Ojeda, F.J. Plausible Biological Interactions of Low- and Non-Calorie Sweeteners with the Intestinal Microbiota: An Update of Recent Studies. *Nutrients* **2020**, *12*, 1153. [CrossRef] [PubMed]

Article

Randomised Double-Blind Placebo-Controlled Trial of Inulin with Metronidazole in Non-Alcoholic Fatty Liver Disease (NAFLD)

Clara Yieh Lin Chong [1], David Orr [2,*], Lindsay D. Plank [3], Tommi Vatanen [1], Justin M. O'Sullivan [1] and Rinki Murphy [4,*]

1 Liggins Institute, The University of Auckland, Auckland 1142, New Zealand; clara.chong@auckland.ac.nz (C.Y.L.C.); t.vatanen@auckland.ac.nz (T.V.); justin.osullivan@auckland.ac.nz (J.M.O.)
2 New Zealand Liver Transplant Unit, Auckland City Hospital, Auckland 1023, New Zealand
3 Department of Surgery, Faculty of Medical and Health Sciences, The University of Auckland, Auckland 1142, New Zealand; l.plank@auckland.ac.nz
4 Department of Medicine, Faculty of Medical and Health Sciences, The University of Auckland, Auckland 1142, New Zealand
* Correspondence: DOrr@adhb.govt.nz (D.O.); R.Murphy@auckland.ac.nz (R.M.); Tel.: +64-9-923-6313

Received: 5 March 2020; Accepted: 24 March 2020; Published: 27 March 2020

Abstract: *Background*: Non-alcoholic fatty liver disease (NAFLD) can be ameliorated by weight loss although difficult to maintain. Emerging evidence indicates that prebiotics and antibiotics improve NAFLD. *Aim*: To determine whether inulin supplementation after brief metronidazole therapy is effective in reducing alanine aminotransferase (ALT) and maintaining weight loss achieved through a very-low-calorie diet (VLCD) among people with NAFLD. *Methods*: Sixty-two people with NAFLD commenced 4-week VLCD using Optifast meal replacements (600 kcal/day). Sixty were then randomised into a 12-week double-blind, placebo-controlled, parallel three-arm trial: (1) 400 mg metronidazole twice daily in Week 1 then inulin 4 g twice daily OR (2) placebo twice daily in week one then inulin OR (3) placebo-placebo. Main outcomes were ALT and body weight at 12 weeks. Fecal microbiota changes were also evaluated. *Results*: Mean body mass index (BMI) and ALT reduced after VLCD by 2.4 kg/m^2 and 11 U/L, respectively. ALT further decreased after metronidazole-inulin compared to after placebo-placebo (mean ALT change −19.6 vs. −0.2 U/L, respectively; $p = 0.026$); however, weight loss maintenance did not differ. VLCD treatment decreased the ratio of *Firmicutes*/*Bacteroidetes* ($p = 0.002$). *Conclusion*: Brief metronidazole followed by inulin supplementation can reduce ALT beyond that achieved after VLCD in patients with NAFLD.

Keywords: prebiotics; alanine aminotransferase; antibiotic; Optifast; gut microbiome; inulin; metronidazole

1. Introduction

Non-alcoholic fatty liver disease (NAFLD) is defined by the pathological accumulation of fat in the liver and is now the leading cause of chronic liver disease [1]. NAFLD encompasses a spectrum of diseases ranging from simple fatty liver (steatosis) through to non-alcoholic steatohepatitis (NASH), which, in turn, leads to fibrosis, irreversible cirrhosis and, finally, hepatocellular carcinoma (HCC) in a small proportion of people [2,3]. The milder simple steatosis is characterised by the ectopic accumulation of fat in the liver, usually associated with energy-surplus-induced obesity. It is believed that multiple parallel factors (diet, insulin resistance, mitochondrial dysfunction and inflammation), acting synergistically in genetically predisposed individuals, are implicated in the development and progression of NAFLD.

An accumulating number of animal and human studies suggest a compelling role for gut microbiota in NAFLD, which is both transmitted by gut microbiota and reversed by a combination of ciprofloxacin and metronidazole antibiotics in animal models [4]. NAFLD is associated with dysbiosis of the gut microbiota, which is thought to lead to increased gut permeability, and abnormal choline and bile acid metabolism, leading to inflammation and increased hepatic fat accumulation [5]. An indication of the involvement of gut microbiota in NAFLD development was first apparent when hepatic steatosis developed in patients undergoing jejunal–ileal bypass surgery, coinciding with intestinal bacterial overgrowth in the blind loop. The hepatic steatosis regressed once patients were treated with the antibiotic metronidazole [6] which is commonly used for the treatment of small intestinal bacterial overgrowth [7]. While small intestinal bacterial overgrowth has been shown to be more prevalent in NAFLD [8–11], antibiotic treatment of NAFLD has not been investigated due to concerns about long-term use being associated with side effects, antimicrobial resistance and uncertain efficacy.

The cornerstone of NAFLD treatment currently is to offer lifestyle advice that targets 7% to 10% weight loss and is proven to be effective [12,13]. Recent evidence shows that very-low-calorie diets (VLCDs) [14] and bariatric surgery [15] are very effective in achieving weight loss and remission of associated comorbidities. Both these strategies alter gut microbiota, but to a lesser extent after dietary modification than after surgery [16–18]. However, the maintenance of weight loss remains a challenge and better alternatives to targeting specific mechanistic dysfunction are needed.

Prebiotics, which are nondigestible food ingredients that are fermented in the gut and modulate microbiota in a favourable way for the host, have shown promise in the treatment of NAFLD. A systematic review of 26 randomised controlled trials investigating the metabolic benefits of prebiotics concluded that prebiotics improve satiety, postprandial glucose and insulin in both healthy and obese individuals [19]. A meta-analysis of nine randomised controlled trials in NAFLD showed a reduction in body mass index (BMI) and an overall improvement in aminotransferase (ALT) with the use of prebiotics [20]. However, the use of a combination of strategies targeting gut microbiota dysbiosis of NAFLD such as VLCD, metronidazole and prebiotics in succession has not previously been investigated.

We hypothesised that the beneficial metabolic effects of short-term VLCD among adults with NAFLD could be enhanced by the brief use of metronidazole to target dysbiotic gut microbiota followed by a period of inulin supplementation to maintain this. We conducted a single centre, randomised, placebo-controlled, double-blind clinical three-arm trial of 12 weeks of inulin supplementation with or without an initial week of metronidazole cotreatment among adults with NAFLD who had all received four weeks of VLCDs.

2. Patients and Methods

2.1. Study Design

This study focused on adults with an established diagnosis of NAFLD attending Auckland City Hospital hepatology outpatient clinic. Patients either had histological evidence of NAFLD based on a liver biopsy, a phenotypic diagnosis based on the presence of BMI > 27 kg/m^2 and type 2 diabetes or metabolic syndrome (WHO criteria) with an elevated ALT (male > 40 U/L, female > 30 U/L) and age >18 years and <75 years. Exclusion criteria were alcohol consumption of more than 20 g per day for at least 3 consecutive months during the previous 5 years as assessed by a questionnaire. Participants were excluded if they had cirrhosis, hepatitis C or another liver disease, if they were awaiting or had previous bariatric surgery, had an allergy to eggs, nuts or metronidazole, a history of drug and alcohol abuse, a calculated eGFR less than 60 mL/min (MDRD formula) or current participation in other therapeutic trials. Ethics approval was from Health and Disability Ethics Committee NTX/12/05/040/AM02; ANZCTR registration number: 12613001002774, prospectively registered on 10 September 2013.

2.2. Randomisation and Treatment Groups

Sixty-two participants with NAFLD who met all eligibility criteria and provided written informed consent were provided with 3 Optifast meal replacements (600 kcal/day) per day for 4 weeks to initiate weight loss after which the 60 participants who attended the second study visit were randomly assigned to one of three parallel groups (1:1:1; Figure A1). The metronidazole and inulin group (Group MI) received metronidazole (dose of 400 mg twice daily for 7 days) along with inulin (at a dose of 4 g twice daily for 12 weeks); the placebo and inulin group (Group PI) received metronidazole-like placebo (twice daily for 7 days) along with inulin (at a dose of 4 g twice daily for 12 weeks); the placebo and inulin placebo group (Group PP) received metronidazole-like placebo (twice daily for 7 days) along with inulin-like placebo (containing maltodextrin at a dose of 4 g twice daily for 12 weeks).

The inulin dose was selected on the basis of previous prebiotic studies and was provided by Cargill Belgium. A metronidazole dose of 400 mg twice daily was selected as slightly lower than the standard dose of 400 mg three times daily used for various medical conditions, such as bacterial vaginosis, dental abscess and giardiasis, for increased adherence than three times daily. Metronidazole and matching placebo-containing maltodextrin were encapsulated by the Auckland Hospital Clinical Trials' Pharmacy department. All participants, their health care providers and assessment staff were blinded to treatment allocation. Participants were asked to take the inulin/matching placebo powder twice daily before breakfast and before dinner using a 4 g measuring spoon and two level spoonfuls dissolved into water. All participants were given a standardised set of recommendations about lifestyle changes and diet following the initial expected weight loss period during VLCD at time of study randomisation

In total, there were four time points in this study: baseline (study entry), Week 4 (after 4 weeks VLCD pre-randomisation), Week 16 (post-randomisation, at the end of treatment) and Week 28 (post-treatment follow up phase to evaluate whether there were any persistent effects detected beyond the treatment period) as shown in Figure 1. All participants underwent assessment for body weight, height, waist and hip circumference at each of these 4 timepoints. Blood samples for assessment of fasting lipids, glucose, insulin and liver Fibroscan CAP were obtained at baseline, Week 4 and then Week 16.

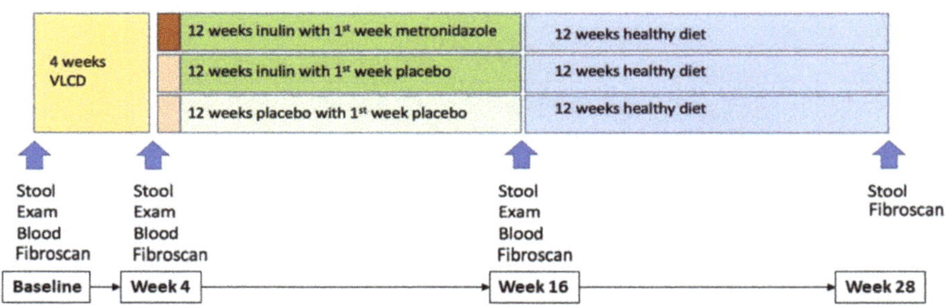

Figure 1. Assessment and sample collection timeline.

2.3. Stool Sample Collection

Stool samples were collected at each time point (Figure 1): baseline, Week 4, Week 16 and Week 28. Study participants collected the stool samples at home, using a sterile collection tube, prior to their hospital visits. Stool samples were stored at −70 °C from the beginning of the study (2013/2015) until DNA extraction was performed (2017).

DNA was extracted from stool samples using the QIAamp® Fast DNA Stool Mini Kit according to the manufacturer's protocol. Extracted DNA quality and quantity were measured using a NanoPhotometer N60 (IMPLEN, Germany; Table S1) and a Qubit (Invitrogen, US).

2.4. 16S rRNA Gene Amplicon Sequencing

Extracted DNA (mean yield = 6733.4 ng; mean 260/280 = 1.97; mean concentration = 33.7 ng/μL) was sent to the School of Biological Sciences (The University of Auckland, New Zealand) for 16S rRNA amplicon sequencing on an Illumina MiSeq sequencing platform. Sequences are available from SRA project number SUB5068044. Then, 16S rRNA gene amplicon sequencing (16S sequencing) libraries were prepared using the Nextera XT kit (Illumina). V3 and V4 regions were targeted for 16S sequencing by using the 16S Amplicon PCR Forward Primer (TCGT CGGCAGCGTCAGATGTGTATAAGAGACAGCCTACGGGNGGCWGCAG) and Reverse Primer (GTCTCGTGGGCTCGGAGATGTGTATAAGAGACAGGACTACHVGGGTATCTAATCC).

All amplicons were sequenced on the Illumina MiSeq 600 cycle run to generate an average of 121,346 sequence reads with paired-end (300 bp each) reads per sample.

2.5. 16S rRNA Amplicon Sequence Analyses

The 16S sequencing data were processed using QIIME 2 (v. 2018.4) [21]. Briefly, sequence quality control and denoising were performed using DADA2 [22]. The quality control step also included the filtering of PhiX reads and chimeric sequences. The sequences obtained after denoising were then classified using Greengenes 13_8 release data to identify amplicon sequencing variants (ASVs) for sequences with >99% sequence similarity. Samples that were included in downstream analyses had filtered sequence counts ranging from 12,123 to 109,977 (median 53,071). Three samples with less than 10,000 sequencing reads were removed.

2.6. Primary and Secondary Outcomes

The primary outcome was the proportion maintaining a ≥7% weight loss at the end of the 12-week variable treatment period compared to their baseline (before the fixed 4-week VLCD treatment period). Secondary outcomes measured at Week 16 (the end of the 12-week variable treatment period) included changes in ALT, glycaemia, lipids, Fibroscan® CAP from what was achieved at Week 4 (after VLCD treatment period) and the changes in gut microbial community from baseline to 28 weeks.

2.7. Statistical Analysis

The planned sample size for this pilot study was 60 subjects with an equal assignment to each of the three study groups (20 per group). We estimated that, with this sample size, the study would have 80% power to detect a difference in the proportion achieving a sustained weight loss of ≥7% at the end of the 12-week treatment period which we anticipated would be achieved by 50% of those receiving metronidazole and inulin supplemented diet, compared to 5% in the other two placebo-containing groups, with a two-sided type 1 error of 0.05. The primary outcome was assessed using Fisher's exact test. Pre-planned analyses for secondary outcomes were comparisons of the changes over the 12-week variable treatment period in the MI and PI groups with those in the PP group. Two-sample *t*-tests were used for these comparisons for normally distributed data and Mann–Whitney *U*-tests for non-normal data. Within-cohort changes over the VLCD period were analysed using paired *t*-test or Wilcoxon signed-rank test as appropriate. Data are presented as mean (SD) or median (Quartile 1, Quartile 3) for normally and non-normally distributed data, respectively.

2.8. Statistical Methods for Microbiota Analysis

Omnibus associations between microbial community structure and patient metadata were assessed using Permutational Analysis of Variance (PERMANOVA) (adonis function from the vegan package in R, 10,000 permutations) and Bray–Curtis dissimilarities. Wilcoxon matched-pairs signed-rank tests were used to assess relative abundance across timepoints. Associations between individual microbial taxa and patients' metadata were assessed using Multivariate Association with Linear Models (MaAsLin) [23], controlling for age as a possible confounding factor and repeated sampling per individual by a random effect. The Kruskal–Wallis test was used when more than two independent groups were compared. ASVs that were present in less than 20% of samples were filtered out. p-values were corrected for multiple testing using the Benjamini–Hochberg procedure [24,25] and FDR corrected p-values (q-values) were reported.

3. Results

3.1. Participants

Enrollment into this trial occurred between March 2013 and March 2015. Sixty-two participants entered the study and began VLCD, of whom 60 attended Visit 2 and were then randomised: 20 attended metronidazole and inulin (MI), 20 attended metronidazole placebo and inulin (PI) and 20 attended metronidazole placebo and inulin placebo (PP). Participant flow through the trial is shown in Figure A1. The mean age was 50 years (range 19–71), BMI 31.6 kg/m^2 (range 25.2–41.9) and ALT 66 U/L (range 30–141). The three groups were well matched with respect to demographic characteristics, clinical and laboratory data at study entry and after four weeks of VLCD (Table A1). Over the VLCD period, there were significant reductions in body weight, waist:hip ratio, blood pressure, ALT and gamma-glutamyl transferase (GGT), total and LDL cholesterol, triglycerides, fasting glucose, HbA1c, CRP and Fibroscan CAP score.

3.2. Primary and Secondary Outcomes

Of the 62 participants who were assessed at baseline, 60 were randomised and 56/60 (93.3%) participants completed the study. The clinical endpoint of achieving sustained weight loss of ≥7% at 16 weeks compared to baseline pre-VLCD was reached by 55% in group MI compared with 53% in group PI and 35% in group PP. These were not statistically significantly different ($p = 0.473$). At 28 weeks, a sustained weight loss of ≥7% was reached by 42% in group MI, 35% in group PI and 25% in group PP ($p = 0.584$).

Although there was no difference in BMI between the three treatment groups at 12 weeks, only the group receiving inulin with an initial one week of metronidazole, group MI, had a significant, further improvement in ALT (Table A2). However, this group had no significant change in Fibroscan® CAP score or in other markers of metabolic syndrome such as blood pressure, fasting lipids and glycaemia. No cases of adverse events requiring discontinuation of inulin were reported.

3.3. Gut Microbial Changes in Our Study Cohort

A total of 127 stool samples were obtained for analysis (Figure A2). Patient compliance in providing stool samples was highest at Week 4 with 38 (29.9%) stool samples and lowest at Week 28 with 26 (20.5%) stool samples (Figure A2). All four stool samples were obtained from 10 study participants.

After Optifast VLCD: VLCD treatment explained 5.3% (PERMANOVA, $p = 0.0024$) of the variance in microbial profiles. *Bacteroidetes* and *Firmicutes* were the two most highly represented bacterial phyla in our cohort (Figure 2). After four weeks of VLCD, the relative abundance of *Bacteroidetes* increased (Wilcoxon signed-rank test, $p = 0.047$) while *Firmicutes* decreased (Wilcoxon signed-rank test, $p = 0.01$) (Figure 2). Furthermore, the ratio of *Firmicutes/Bacteroidetes* decreased significantly after VLCD treatment (Wilcoxon signed-rank test, $p = 0.002$, $n = 30$) (Figure 3).

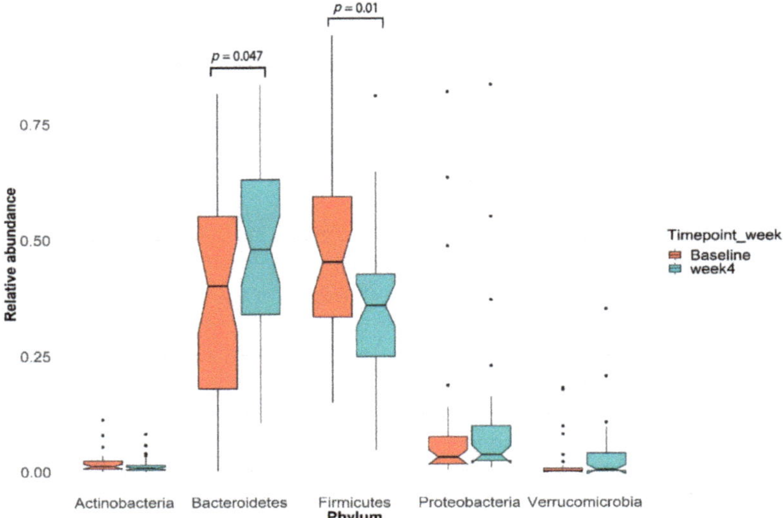

Figure 2. *Firmicutes* and *Bacteroidetes* are the dominant phyla in the subjects before and after very-low-calorie diet (VLCD) treatment. The figure shows boxplots of five typical human microbiota phyla. The boxes indicate the interquartile range (IQR) while the notch region shows the 95% confidence interval for the median and the whiskers extending from the boxes represent the distribution within 1.5 × IQR, with points beyond this range shown as outliers.

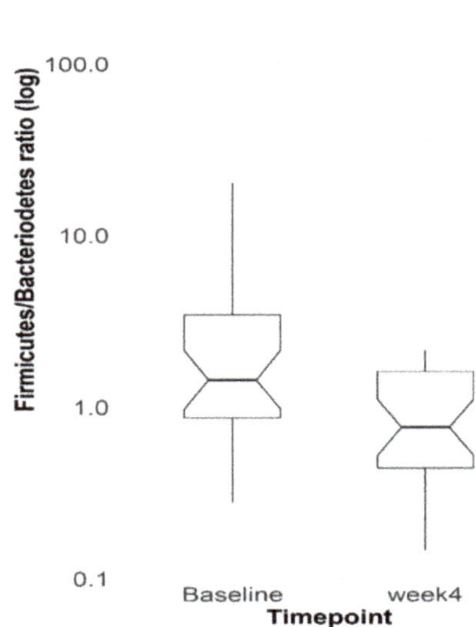

Figure 3. The ratio of *Firmicutes*/*Bacteroidetes* phyla decreased (Wilcoxon signed-rank test, $p = 0.002$, $n = 30$) from baseline to Week 4 after the VLCD diet. Boxplots as in Figure 2.

Linear modelling identified three statistically significant genera ($q < 0.1$, Table S2), all belonging to the phylum Firmicutes. *Roseburia*, *Streptococcus* and *Dialister* genera displayed an association with the VLCD treatment and were significantly lower after VLCD treatment compared to the other time points (Figure 4). However, the microbial alpha diversity metrics showed no significant change from baseline following a VLCD diet (Shannon, $p = 0.968$; Wilcoxon signed-rank test, $n = 30$).

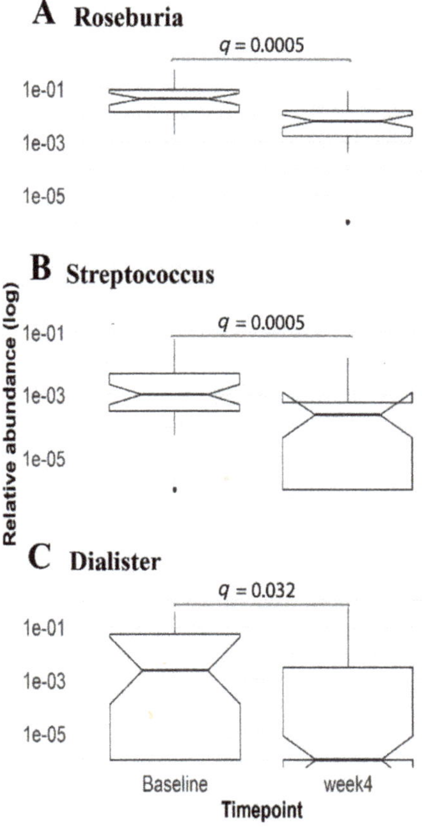

Figure 4. The relative abundance of genera *Roseburia* (**A**), *Streptococcus* (**B**) and *Dialister* (**C**) were lower (linear mixed-effects model, q-value = 0.0005, 0.0005 and 0.03, respectively) after VLCD treatment compared to baseline. Baseline, $n = 35$; Week 4, $n = 38$. Boxplots as in Figure 2.

3.4. Enrichment of Distinct Gut Microbial Profile in Our Study Cohort

There were no significant differences in alpha diversity (Shannon's diversity index) at Week 16 between intervention groups ($p = 0.755$, Kruskal–Wallis test). Similarly, comparison between Week 4 and Week 16 found no significant difference between groups ($p = 0.949$, Kruskal–Wallis test).

Linear modelling identified three taxa, genera *Roseburia*, *Anaerotruncus* and family *Lachnospiraceae*, all belonging to the phylum Firmicute, were associated to the antibiotic/prebiotic treatment period ($q = 0.026$) (Table S3). However, comparison between groups found no differences ($p = 0.097$, Kruskal–Wallis test).

Linear modelling revealed a suggestive association between genus *Turicibacter* and plasma ALT levels (q-value = 0.086, Table S4) when corrected for patient age. However, neither linear modelling nor PERMANOVA identified any significant associations between microbial taxa and plasma ALT in the cross-sectional model at Week 16 ($n = 28$ samples).

4. Discussion

There are currently few approved treatment options for NAFLD patients beyond dietary measures to lose weight. In this study, four weeks of VLCD resulted in a significant weight loss. Subsequent inulin supplementation for 12 weeks, with or without an initial one week of metronidazole), did not improve weight loss maintenance. The transition to a real food diet after a period of VLCD meal replacements is usually associated with weight regain and recurrence of NAFLD. Indeed, only 48% of participants were able to maintain a ≥7% weight loss after 12 weeks of transition to a real food diet. Despite similar weight loss maintenance, the group who received metronidazole and inulin (after the initial VLCD period) achieved a further significant reduction in ALT. The reduction in ALT suggests reduced steatohepatitis, which, surprisingly, in our study—in contrast to other prebiotic studies that have shown an improvement in ALT commensurate with weight loss [20]—occurred without further weight loss. This finding supports the potential role of metronidazole in improving steatohepatitis through the treatment of intestinal bacterial overgrowth and/or through altering gut microbial functions that enhance efflux of free fatty acids and de novo lipogenesis in the liver. These mechanisms can occur without weight loss as the cause and are not necessarily apparent by simple characterisation of microbial abundances in the faeces. This is because there is substantial inter-individual variability in gut microbiota among patients with NAFLD and bacterial abundance in faecal samples do not directly demonstrate activity or metabolite production of the taxa present in the small intestine.

Nonetheless, four weeks of VLCD (Optifast) had a major effect on decreasing the ratio of *Firmicutes/Bacteroidetes* in faeces, as well as decreasing the abundance of genera *Roseburia, Streptococcus* and *Dialister*. The genus *Roseburia*, a member of clostridial cluster XIVa [26], consists of obligate Gram-positive anaerobic bacteria and is an important butyrate-producing colonic bacterium [27–30] and suggested to be able to alleviate inflammation by stimulating Treg cell differentiation [31,32]. Butyrate is a short-chain fatty acid produced mainly by the enteric microbiome [33,34]. It is a crucial element in the normal development of colonic epithelial cells [35] and preferred energy source in the colonic mucosa [36]. A previous study has shown that a butyrate-producing probiotic MIYAIRI 588 strain of *Clostridium butyricum* effectively improved hepatic indexes in an animal model [33]. Butyrate has also been suggested to confer various beneficial metabolic effects such as enhancing mitochondrial activity [34], increasing insulin sensitivity [37], conveying anti-inflammatory potential [38] as well as increasing intestinal barrier function [39]. However, the role of butyrate in NAFLD is controversial as patients with NASH were shown to have higher faecal butyrate compared to healthy subjects [40]. *Roseburia* has also been detected as significantly elevated in NAFLD patients compared to healthy controls [41]. This genus is suggested to be one of the gut microbiota biomarkers that is shared by obese patients with metabolic disease and is negatively associated with body mass index (BMI) [32]. Ironically, a depletion of *Roseburia* was observed after four weeks of Optifast VLCD treatment, with a significant reduction in BMI in our study. In fact, our observation is similar to Duncan et al. [42] and, subsequently, Alemán et al. [43] who observed a significant reduction of genus *Roseburia* after VLCD intervention. Since *Roseburia* are predominantly polysaccharide-degrading bacteria [44], we postulate that the observed reduction in the genus *Roseburia* is actually due to reduced dietary carbohydrates from VLCD and not directly linked to BMI.

Similarly, the genus *Streptococcus*, a possible biomarker of NAFLD [45] reduced significantly after VLCD treatment compared to baseline (Figure 4). This corroborates previous work where *Streptococcus* was enriched in NAFLD and NAFLD-cirrhosis patients [46,47] compared to both healthy subjects [48,49] and obese individuals [45]. Future studies could investigate metabolic activities and molecular mechanisms linking *Streptococcus* and NAFLD aetiology.

Optifast-based VLCD reduced the alpha diversity (Shannon's diversity index) of gut microbiota seen between baseline to Week 4 (Figure S1). Altered diversity has also been shown to occur in a similar Optifast-based VLCD study of three months duration in 18 obese participants, although changes regressed during the subsequent weight maintenance phase and return to a real food diet [18]. It is noticeable that upon transition to a food diet, metronidazole-inulin and placebo-inulin groups both shared a similar fluctuation pattern in Shannon's diversity index for all time points compared to the placebo-placebo group which demonstrated a relatively stable pattern.

Due to limited stool sample collection after VLCD, lack of statistical power precluded the assessment of any gut microbial differences in abundance between metronidazole treated and non-metronidazole treated groups. Further, we cannot rule out the possibility that low compliance in the stool sample collection may have added further bias in microbiome analysis. However, linear modelling revealed that the genus *Turicibacter* was associated with the plasma ALT levels within placebo-inulin and placebo-placebo groups. *Turicibacter* has been suggested to be responsive to the cholesterol level in the diet [50]. This is in line with the fact that hepatic free cholesterol accumulation and altered cholesterol homeostasis will lead to liver injury and eventually contribute to the pathogenesis of NAFLD/NASH [51]. We suggest an association between *Turicibacter* and plasma ALT levels which clearly need further research.

Finally, maltodextrin may not have been as inert as a placebo should be, given there is some evidence that maltodextrin detrimentally impacts the intestinal environment [52–54]. However, most of these studies used much higher doses of maltodextrin than was used in the form of placebo to match inulin in this study.

5. Conclusions

In conclusion, this is the first clinical trial evidence that supplementation with prebiotic inulin following brief metronidazole therapy can further reduce ALT after four weeks of VLCD therapy in patients with NAFLD. A prominent shift in phyla Firmicutes, Bacteroidetes, genera *Roseburia, Streptococcus* and *Dialister* were seen after four weeks of Optifast treatment. Unfortunately, limited stool samples collection after VLCD treatment resulted in insufficient power to detect a significant difference in gut microbiota with additional metronidazole or inulin. Nevertheless, a potential role of metronidazole, together with inulin in altering the gut microbial function (e.g., metabolites production), is suggested to alleviate steatohepatitis, as evidenced by the reduction of ALT in the MI group. Future studies are recommended to examine the effect on microbial metabolites which does not manifest in measuring the diversity of microbiota. Furthermore, this clinical therapeutic approach requires validation in larger clinical studies with the possibility of low compliance on stool samples collection is accommodated.

Supplementary Materials: The following are available online at http://www.mdpi.com/2072-6643/12/4/937/s1, Figure S1: Violin plot on the Shannon's diversity index for each group on each time point, Table S1: DNA quality and quantity measured using a NanoPhotometer N60, Table S2: Linear modelling results showing genera associated with the VLCD treatment, Table S3: Linear modelling results showing genera associated with the antibiotic/prebiotic treatment period, Table S4: Linear modelling results showing suggestive association between genus *Turicibacter* and plasma ALT levels.

Author Contributions: Conceptualisation, D.O. and R.M.; data curation, C.Y.L.C. and T.V.; formal analysis, C.Y.L.C., D.O., L.D.P., T.V. and J.M.O.; funding acquisition, D.O. and R.M.; investigation, C.Y.L.C., D.O., L.D.P., J.M.O. and R.M.; methodology, D.O., L.D.P., T.V., J.M.O. and R.M.; project administration, D.O.; resources, J.M.O. and R.M.; Supervision, D.O., T.V., J.M.O. and R.M.; visualisation, C.Y.L.C., T.V. and J.M.O.; writing—original draft, C.Y.L.C. and R.M.; writing—review and editing, C.Y.L.C., D.O., L.D.P., T.V., J.M.O. and R.M. All authors have read and agreed to the published version of the manuscript.

Funding: This research was funded by A+ Trust, New Zealand Society of Gastroenterology.

Acknowledgments: The inulin used in this work was provided gratis by Cargill Belgium. We are grateful to Gil Hardy for overseeing the packaging of inulin and maltodextrin performed by AnQual Laboratories. We thank Auckland Pharmacy Clinical Trials for the encapsulation of the metronidazole and placebo tablets.

Conflicts of Interest: The authors declare no conflicts of interest. The funders had no role in the design of the study; in the collection, analyses, or interpretation of data; in the writing of the manuscript, or in the decision to publish the results.

Abbreviations

VLCD	Very low calorie diet
NAFLD	Non-alcoholic fatty liver disease
ALT	Alanine aminotransferase
ASVs	Amplicon Sequence Variants

Appendix A

Table A1. Baseline and week 4 characteristics of study participants.

Clinical Characteristics	Metronidazole-Inulin (n = 20)	Placebo-Inulin (n = 20)	Placebo-Placebo (n = 20)	Total (n = 60)
Age (y)	50.6 (10.4)	51.5 (13.5)	46.7 (11.2)	49.6 (11.8)
Sex (M:F)	10:10	8:12	13:7	31:29
Body mass index (kg/m^2)				
Baseline	31.4 (3.4)	30.9 (3.5)	32.6 (4.3)	31.6 (3.8)
Week 4	29.0 (3.4)	27.1 (7.3)	29.9 (3.8)	29.1 (3.6) ***
Weight (kg)				
Baseline	89.8 (12.4)	85.6 (15.5)	94.2 (15.9)	89.9 (14.9)
Week 4	83.0 (12.0)	73.7 (21.6)	86.4 (13.4)	82.7 (13.3) ***
Waist:hip ratio				
Baseline	0.98 (0.07)	0.99 (0.04)	0.98 (0.06)	0.98 (0.05)
Week 4	0.93 (0.06)	0.94 (0.05)	0.92 (0.08)	0.96 (0.06) ***
Blood pressure (mm Hg)				
Systolic at Baseline	132 (16)	129 (14)	125 (17)	129 (16)
Systolic at Week 4	121 (13)	121 (16)	121 (21)	121 (17) ***
Diastolic at Baseline	83 (10)	77 (10)	77 (11)	79 (11)
Diastolic at Week 4	76 (10)	74 (10)	74 (14)	75 (11) ***
ALT (U/L)				
Baseline	64.0 (28.6)	70.0 (26.1)	60.2 (28.1)	64.7 (27.4)
Week 4	59.8 (25.0)	53.3 (22.0)	46.2 (21.1)	53.2 (23.1) **
AST (U/L)				
Baseline	34 (25, 46)	41 (31, 47)	33 (27, 40)	36 (29, 45)
Week 4	41 (31, 55)	36 (30, 47)	30 (27, 37)	36(29, 43)
GGT (U/L)				
Baseline	101 (61, 141)	164 (76, 246)	71 (42, 134)	93 (57, 163)
Week 4	49 (29, 73)	83 (35, 124)	35 (22, 65)	46(27, 88) ***
Bilirubin (mol/L)				
Baseline	9.5 (7.5, 13.0)	9.5 (6.5, 3.5)	9.5 (7.5, 12.0)	9.5 (7, 13)
Week 4	11.0 (8.5,12.0)	10.0 (6.0,15.0)	11.0 (8.0,15.0)	10 (8, 14) *
Albumin (g/L)				
Baseline	46.0 (2.5)	47.4 (2.8)	45.3 (2.1)	46.2 (2.6)
Week 4	45.5 (2.7)	47.3 (2.6)	46.1 (2.2)	46.3 (2.6)
Ferritin (µg/L)				
Baseline	228 (172)	214 (190)	254 (222)	232 (194)
Week 4	295 (255)	221 (166)	241 (204)	251(208)
Cholesterol (mmol/L)				
Total (Baseline)	4.77 (0.98)	5.16 (1.29)	5.01 (0.86)	4.98 (1.06)
Total (Week 4)	3.77 (1.05)	3.78 (1.00)	3.64 (0.76)	3.73 (0.93) ***
High-density (Baseline)	1.35 (0.43)	1.41 (0.36)	1.26 (0.27	1.34 (0.36)
High-density (Week 4)	1.32 (0.40)	1.37 (0.32)	1.18 (0.21)	1.29 (0.32) *
Low-density (Baseline)	2.54 (0.74)	2.88 (1.28)	2.96 (0.76)	2.79 (0.97)
Low–density (Week 4)	1.85 (0.72)	1.94 (0.91)	1.96 (0.72)	1.92 (0.78) ***

Table A1. *Cont.*

Triglycerides (mmol/L)				
Baseline Week 4	2.01 (1.07) 0.98 (0.49)	1.96 (0.80) 1.06 (0.46)	1.74 (0.79) 1.07 (0.54)	1.91 (0.89) 1.04 (0.49) ***
Type 2 diabetes	7/20	7/20	7/20	21/60
Fasting glucose (mmol/L)				
Baseline Week 4	6.32 (1.80) 5.70 (1.46)	6.02 (1.70) 5.50 (1.39)	6.07 (1.85) 5.19 (0.77)	6.17 (1.77) 5.46 (1.23) ***
HbA1c (mmol/mol)				
Baseline Week 4	44.3 (11.5) 40.4 (8.3)	48.1 (12.9) 42.5 (8.6)	44.6 (11.5) 39.3 (7.4)	45.6 (11.9) 40.7 (8.1) ***
CRP (mg/L)				
Baseline Week 4	2.0 (0.5, 4.0) 1.0 (0.5, 4.0)	2.0 (1.0, 5.0) 0.8 (0.5, 3.0)	2.5 (0.5, 6.0) 1.5 (0.5, 3.0)	2.0 (0.5, 5.0) 1.0 (0.5, 3.0) **
Fibroscan® CAP (dB/m)				
Baseline Week 4	307 (51) 267 (55)	299 (58) 246 (51)	296 (63) 258 (69)	301 (57) 257 (58) ***

Data are mean (SD), median (Q1, Q3) or number of patients. * $p < 0.05$, ** $p < 0.01$, *** $p < 0.001$ (paired t-test or Wilcoxon signed rank test).

Appendix B

Table A2. Changes in characteristics at week 16, after 12 weeks of treatment following randomisation to metronidazole/inulin, placebo/inulin or placebo/placebo.

Clinical Characteristics	Group MI: Metronidazole-Inulin (n = 19)	Group PI: Placebo-Inulin (n = 19)	Group PP: Placebo-Placebo (n = 18)	p-Value Group MI vs. PP	p-Value Group PI vs. PP
Body mass index (kg/m^2)	0.10 (1.20)	0.12 (1.80)	0.21 (1.36)	0.792	0.859
Weight (kg)	0.35 (3.38)	0.29 (5.02)	0.47 (4.10)	0.922	0.910
Waist:hip ratio	−0.00 (0.04)	−0.01 (0.04)	−0.00 (0.06)	0.893	0.604
Blood pressure (mm Hg)					
Systolic Diastolic	3.4 (11.3) −1.4 (10.2)	2.2 (16.7) 0.4 (12.0)	−7.2 (18.4) 0.5 (13.8)	0.043 0.641	0.117 0.981
ALT (U/L)	−19.6 (25.5)	−2.2 (16.7)	−0.2 (24.5)	**0.026**	0.856
AST (U/L)	−14 (−32, −7)	−3 (−22, 2)	−4 (−9, 5)	**0.006**	0.849
GGT (U/L)	13 (−2, 29)	26 (−1, 91)	17 (8, 62)	0.274	0.693
Bilirubin (µmol/L)	−1 (−3, 1)	−2 (−3, 0)	1 (−2, 2)	0.115	0.033
Albumin (g/L)	0.13 (1.81)	−1.17 (2.46)	−1.14 (2.57)	0.131	0.979
Ferritin (µg/L)	−56.3 (60.1)	−27.6 (63.5)	−1.3 (72.7)	0.031	0.291
Cholesterol (mmol/L)					
Total HDL LDL	1.09 (1.10) 0.16 (0.15) 2.28 (5.56)	1.32 (1.13) 0.20 (0.24) 0.83 (0.77)	0.91 (0.57) 0.13 (0.17) 0.55 (0.49)	0.560 0.638 0.181	0.208 0.344 0.246
Triglycerides (mmol/L)	0.52 (0.32)	0.50 (0.79)	0.56 (0.64)	0.831	0.814
Fasting glucose (mmol/L)	0.39 (1.32)	0.18 (0.70)	1.14 (2.05)	0.228	0.077
HbA1c (mmol/mol)	1.94 (4.25)	0 (5.79)	2.06 (6.02)	0.947	0.316
CRP (mg/L)	0.0 (−1.0, 0.75)	1.0 (0.0, 1.5)	0.0 (0.0, 1.0)	0.884	0.235
Fibroscan CAP (dB/m)	14.5 (65.6)	22.4 (71.6)	10.8 (42.6)	0.859	0.605

Data are mean (SD) or median (Q1, Q3).

Appendix C

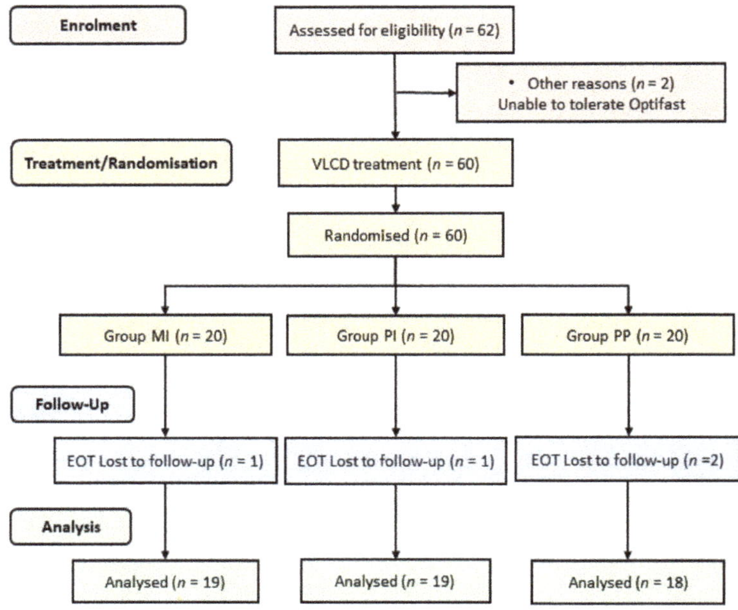

Figure A1. Flow diagram showing different phases of the study and number of participants at each stage. EOT: end of treatment; MI: metronidazole-inulin; PI: placebo-inulin; PP: placebo-placebo.

Appendix D

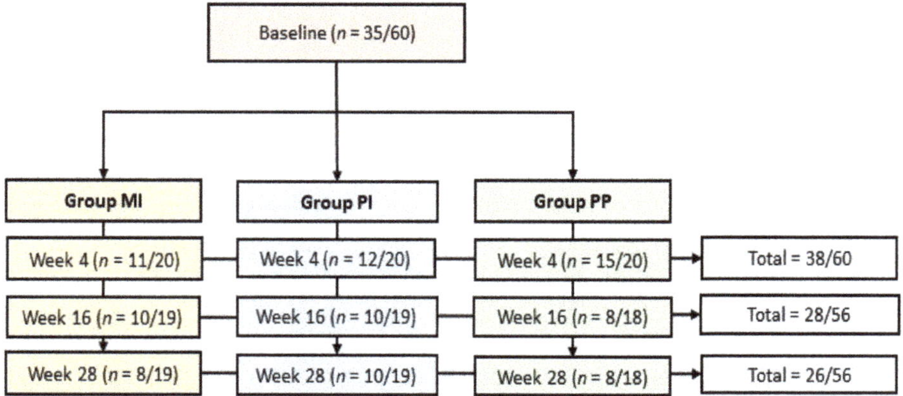

Figure A2. Flow diagram showing total stool sample collection at different time points from different groups. MI: metronidazole-inulin; PI: placebo-inulin; PP: placebo-placebo.

References

1. Shouhed, D.; Steggerda, J.; Burch, M.; Noureddin, M. The role of bariatric surgery in nonalcoholic fatty liver disease and nonalcoholic steatohepatitis. *Expert Rev. Gastroenterol. Hepatol.* **2017**, *11*, 797–811. [CrossRef]
2. Calzadilla Bertot, L.; Adams, L. The Natural Course of Non-Alcoholic Fatty Liver Disease. *Int. J. Mol. Sci.* **2016**, *17*, 774. [CrossRef] [PubMed]

3. Ekstedt, M.; Nasr, P.; Kechagias, S. Natural History of NAFLD/NASH. *Curr. Hepatol. Rep.* **2017**, *16*, 391–397. [CrossRef] [PubMed]
4. Henao-Mejia, J.; Elinav, E.; Jin, C.; Hao, L.; Mehal, W.Z.; Strowig, T.; Thaiss, C.A.; Kau, A.L.; Eisenbarth, S.C.; Jurczak, M.J.; et al. Inflammasome-mediated dysbiosis regulates progression of NAFLD and obesity. *Nature* **2012**, *482*, 179–185. [CrossRef]
5. Aron-Wisnewsky, J.; Gaborit, B.; Dutour, A.; Clement, K. Gut microbiota and non-alcoholic fatty liver disease: New insights. *Clin. Microbiol. Infect.* **2013**, *19*, 338–348. [CrossRef] [PubMed]
6. Drenick, E.J.; Fisler, J.; Johnson, D. Hepatic Steatosis After Intestinal Bypass—Prevention and Reversal by Metronidazole, Irrespective of Protein-Calorie Malnutrition. *Gastroenterology* **1982**, *82*, 535–548. [CrossRef]
7. Di Stefano, M.; Miceli, E.; Missanelli, A.; Mazzocchi, S.; Corazza, G.R. Absorbable vs. non-absorbable antibiotics in the treatment of small intestine bacterial overgrowth in patients with blind-loop syndrome. *Aliment. Pharmacol. Ther.* **2005**, *21*, 985–992. [CrossRef]
8. Wigg, A.J. The role of small intestinal bacterial overgrowth, intestinal permeability, endotoxaemia, and tumour necrosis factor alpha in the pathogenesis of non-alcoholic steatohepatitis. *Gut* **2001**, *48*, 206–211. [CrossRef]
9. Shanab, A.A.; Scully, P.; Crosbie, O.; Buckley, M.; O'Mahony, L.; Shanahan, F.; Gazareen, S.; Murphy, E.; Quigley, E.M.M. Small Intestinal Bacterial Overgrowth in Nonalcoholic Steatohepatitis: Association with Toll-Like Receptor 4 Expression and Plasma Levels of Interleukin 8. *Dig. Dis. Sci.* **2011**, *56*, 1524–1534. [CrossRef]
10. Rafiei, R.; Bemanian, M.; Rafiei, F.; Bahrami, M.; Fooladi, L.; Ebrahimi, G.; Hemmat, A.; Torabi, Z. Liver disease symptoms in non-alcoholic fatty liver disease and small intestinal bacterial overgrowth. *Rom. J. Intern. Med.* **2018**, *56*, 85–89. [CrossRef]
11. Kapil, S.; Duseja, A.; Sharma, B.K.; Singla, B.; Chakraborti, A.; Das, A.; Ray, P.; Dhiman, R.K.; Chawla, Y. Small intestinal bacterial overgrowth and toll-like receptor signaling in patients with non-alcoholic fatty liver disease. *J. Gastroenterol. Hepatol.* **2016**, *31*, 213–221. [CrossRef] [PubMed]
12. European Association for the Study of the Liver (EASL); European Association for the Study of Diabetes (EASD); European Association for the Study of Obesity (EASO). EASL–EASD–EASO Clinical Practice Guidelines for the management of non-alcoholic fatty liver disease. *J. Hepatol.* **2016**, *64*, 1388–1402. [CrossRef] [PubMed]
13. Chalasani, N.; Younossi, Z.; Lavine, J.E.; Charlton, M.; Cusi, K.; Rinella, M.; Harrison, S.A.; Brunt, E.M.; Sanyal, A.J. The diagnosis and management of nonalcoholic fatty liver disease: Practice guidance from the American Association for the Study of Liver Diseases. *Hepatology* **2018**, *67*, 328–357. [CrossRef] [PubMed]
14. Lean, M.E.; Leslie, W.S.; Barnes, A.C.; Brosnahan, N.; Thom, G.; McCombie, L.; Peters, C.; Zhyzhneuskaya, S.; Al-Mrabeh, A.; Hollingsworth, K.G.; et al. Primary care-led weight management for remission of type 2 diabetes (DiRECT): An open-label, cluster-randomised trial. *Lancet* **2018**, *391*, 541–551. [CrossRef]
15. Knop, F.K.; Taylor, R. Mechanism of Metabolic Advantages After Bariatric Surgery: It's all gastrointestinal factors versus it's all food restriction. *Diabetes Care* **2013**, *36*, S287–S291. [CrossRef] [PubMed]
16. Davies, N.K.; O'Sullivan, J.M.; Plank, L.D.; Murphy, R. Altered gut microbiome after bariatric surgery and its association with metabolic benefits: A systematic review. *Surg. Obes. Relat. Dis.* **2019**, *15*, 656–665. [CrossRef]
17. Damms-Machado, A.; Mitra, S.; Schollenberger, A.E.; Kramer, K.M.; Meile, T.; Königsrainer, A.; Huson, D.H.; Bischoff, S.C. Effects of Surgical and Dietary Weight Loss Therapy for Obesity on Gut Microbiota Composition and Nutrient Absorption. *Biomed. Res. Int.* **2015**, 1–12. [CrossRef]
18. Heinsen, F.A.; Fangmann, D.; Müller, N.; Schulte, D.M.; Rühlemann, M.C.; Türk, K.; Settgast, U.; Lieb, W.; Baines, J.F.; Schreiber, S.; et al. Beneficial Effects of a Dietary Weight Loss Intervention on Human Gut Microbiome Diversity and Metabolism Are Not Sustained during Weight Maintenance. *Obes. Facts* **2016**, *9*, 379–391. [CrossRef]
19. Kellow, N.J.; Coughlan, M.T.; Reid, C.M. Metabolic benefits of dietary prebiotics in human subjects: A systematic review of randomised controlled trials. *Br. J. Nutr.* **2014**, *111*, 1147–1161. [CrossRef]
20. Loman, B.R.; Hernández-Saavedra, D.; An, R.; Rector, R.S. Prebiotic and probiotic treatment of nonalcoholic fatty liver disease: A systematic review and meta-analysis. *Nutr. Rev.* **2018**, *76*, 822–839. [CrossRef]
21. Bolyen, E.; Rideout, J.R.; Dillon, M.R.; Bokulich, N.A.; Abnet, C.C.; Al-Ghalith, G.A.; Alexander, H.; Alm, E.J.; Arumugam, M.; Asnicar, F.; et al. Reproducible, interactive, scalable and extensible microbiome data science using QIIME 2. *Nat. Biotechnol.* **2019**, *37*, 852–857. [CrossRef] [PubMed]

22. Callahan, B.J.; McMurdie, P.J.; Rosen, M.J.; Han, A.W.; Johnson, A.J.A.; Holmes, S.P. DADA2: High-resolution sample inference from Illumina amplicon data. *Nat. Methods* **2016**, *13*, 581–583. [CrossRef] [PubMed]
23. Morgan, X.C.; Tickle, T.L.; Sokol, H.; Gevers, D.; Devaney, K.L.; Ward, D.V.; Reyes, J.A.; Shah, S.A.; LeLeiko, N.; Snapper, S.B.; et al. Dysfunction of the intestinal microbiome in inflammatory bowel disease and treatment. *Genome Biol.* **2012**, *13*, R79. [CrossRef] [PubMed]
24. Chen, S.Y.; Feng, Z.; Yi, X. A general introduction to adjustment for multiple comparisons. *J. Thorac. Dis.* **2017**, *9*, 1725–1729. [CrossRef]
25. Benjamini, Y.; Hochberg, Y. Controlling the False Discovery Rate—A Practical and Powerful Approach to Multiple Testing. *J. R. Stat. Soc. B* **1995**, *57*, 289–300. [CrossRef]
26. Neyrinck, A.M.; Possemiers, S.; Verstraete, W.; De Backer, F.; Cani, P.D.; Delzenne, N.M. Dietary modulation of clostridial cluster XIVa gut bacteria (*Roseburia* spp.) by chitin–glucan fiber improves host metabolic alterations induced by high-fat diet in mice. *J. Nutr. Biochem.* **2012**, *23*, 51–59. [CrossRef]
27. Jost, T.; Lacroix, C.; Braegger, C.; Chassard, C. Assessment of bacterial diversity in breast milk using culture-dependent and culture-independent approaches. *Br. J. Nutr.* **2013**, *110*, 1253–1262. [CrossRef]
28. Rivière, A.; Selak, M.; Lantin, D.; Leroy, F.; De Vuyst, L. Bifidobacteria and Butyrate-Producing Colon Bacteria: Importance and Strategies for Their Stimulation in the Human Gut. *Front. Microbiol.* **2016**, *7*, 979. [CrossRef]
29. Tamanai-Shacoori, Z.; Smida, I.; Bousarghin, L.; Loreal, O.; Meuric, V.; Fong, S.B.; Bonnaure-Mallet, M.; Jolivet-Gougeon, A. *Roseburia* spp.: A marker of health? *Future Microbiol.* **2017**, *12*, 157–170. [CrossRef]
30. Hold, G.L.; Schwiertz, A.; Aminov, R.I.; Blaut, M.; Flint, H.J. Oligonucleotide Probes That Detect Quantitatively Significant Groups of Butyrate-Producing Bacteria in Human Feces. *Appl. Environ. Microbiol.* **2003**, *69*, 4320–4324. [CrossRef]
31. Machiels, K.; Joossens, M.; Sabino, J.; De Preter, V.; Arijs, I.; Eeckhaut, V.; Ballet, V.; Claes, K.; Van Immerseel, F.; Verbeke, K.; et al. A decrease of the butyrate-producing species *Roseburia hominis* and *Faecalibacterium prausnitzii* defines dysbiosis in patients with ulcerative colitis. *Gut* **2014**, *63*, 1275–1283. [CrossRef]
32. Zeng, Q.; Li, D.; He, Y.; Li, Y.; Yang, Z.; Zhao, X.; Liu, Y.; Wang, Y.; Sun, J.; Feng, X.; et al. Discrepant gut microbiota markers for the classification of obesity-related metabolic abnormalities. *Sci. Rep.* **2019**, *9*, 13424. [CrossRef] [PubMed]
33. Endo, H.; Niioka, M.; Kobayashi, N.; Tanaka, M.; Watanabe, T. Butyrate-Producing Probiotics Reduce Nonalcoholic Fatty Liver Disease Progression in Rats: New Insight into the Probiotics for the Gut-Liver Axis. *PLoS ONE* **2013**, *8*, e63388. [CrossRef] [PubMed]
34. Rose, S.; Bennuri, S.C.; Davis, J.E.; Wynne, R.; Slattery, J.C.; Tippett, M.; Delhey, L.; Melnyk, S.; Kahler, S.G.; MacFabe, D.F.; et al. Butyrate enhances mitochondrial function during oxidative stress in cell lines from boys with autism. *Transl. Psychiatry* **2018**, *8*, 42. [CrossRef] [PubMed]
35. Pryde, S.E.; Duncan, S.H.; Hold, G.L.; Stewart, C.S.; Flint, H.J. The microbiology of butyrate formation in the human colon. *FEMS Microbiol. Lett.* **2002**, *217*, 133–139. [CrossRef]
36. Roediger, W.E. The colonic epithelium in ulcerative colitis: An energy-deficiency disease? *Lancet* **1980**, *316*, 712–715. [CrossRef]
37. Vrieze, A.; Van Nood, E.; Holleman, F.; Salojärvi, J.; Kootte, R.S.; Bartelsman, J.F.W.M.; Dallinga–Thie, G.M.; Ackermans, M.T.; Serlie, M.J.; Oozeer, R.; et al. Transfer of Intestinal Microbiota From Lean Donors Increases Insulin Sensitivity in Individuals With Metabolic Syndrome. *Gastroenterology* **2012**, *143*, 913–916.e7. [CrossRef]
38. Lührs, H.; Gerke, T.; Schauber, J.; Dusel, G.; Melcher, R.; Scheppach, W.; Menzel, T. Cytokine-activated degradation of inhibitory κB protein α is inhibited by the short-chain fatty acid butyrate. *Int. J. Colorectal Dis.* **2001**, *16*, 195–201. [CrossRef]
39. Peng, L.; Li, Z.R.; Green, R.S.; Holzman, I.R.; Lin, J. Butyrate Enhances the Intestinal Barrier by Facilitating Tight Junction Assembly via Activation of AMP-Activated Protein Kinase in Caco-2 Cell Monolayers. *J. Nutr.* **2009**, *139*, 1619–1625. [CrossRef]
40. Rau, M.; Rehman, A.; Dittrich, M.; Groen, A.K.; Hermanns, H.M.; Seyfried, F.; Beyersdorf, N.; Dandekar, T.; Rosenstiel, P.; Geier, A. Fecal SCFAs and SCFA-producing bacteria in gut microbiome of human NAFLD as a putative link to systemic T-cell activation and advanced disease. *United Eur. Gastroenterol. J.* **2018**, *6*, 1496–1507. [CrossRef]

41. Raman, M.; Ahmed, I.; Gillevet, P.M.; Probert, C.S.; Ratcliffe, N.M.; Smith, S.; Greenwood, R.; Sikaroodi, M.; Lam, V.; Crotty, P.; et al. Fecal Microbiome and Volatile Organic Compound Metabolome in Obese Humans With Nonalcoholic Fatty Liver Disease. *Clin. Gastroenterol. Hepatol.* **2013**, *11*, 868–875.e3. [CrossRef] [PubMed]
42. Duncan, S.H.; Belenguer, A.; Holtrop, G.; Johnstone, A.M.; Flint, H.J.; Lobley, G.E. Reduced Dietary Intake of Carbohydrates by Obese Subjects Results in Decreased Concentrations of Butyrate and Butyrate-Producing Bacteria in Feces. *Appl. Environ. Microbiol.* **2007**, *73*, 1073–1078. [CrossRef]
43. Alemán, J.O.; Bokulich, N.A.; Swann, J.R.; Walker, J.M.; De Rosa, J.C.; Battaglia, T.; Costabile, A.; Pechlivanis, A.; Liang, Y.; Breslow, J.L.; et al. Fecal microbiota and bile acid interactions with systemic and adipose tissue metabolism in diet-induced weight loss of obese postmenopausal women. *J. Transl. Med.* **2018**, *16*, 244. [CrossRef] [PubMed]
44. Chassard, C.; Goumy, V.; Leclerc, M.; Del'homme, C.; Bernalier-Donadille, A. Characterization of the xylan-degrading microbial community from human faeces. *FEMS Microbiol. Ecol.* **2007**, *61*, 121–131. [CrossRef] [PubMed]
45. Nistal, E.; Sáenz de Miera, L.E.; Ballesteros Pomar, M.; Sánchez Campos, S.; García Mediavilla, M.V.; Álvarez Cuenllas, B.; Linares, P.; Olcoz, J.L.; Arias Loste, M.T.; García Lobo, J.M.; et al. An altered fecal microbiota profile in patients with non-alcoholic fatty liver disease (NAFLD) associated with obesity. *Rev. Esp. Enferm. Dig.* **2019**, *111*, 275–282. [CrossRef]
46. Caussy, C.; Tripathi, A.; Humphrey, G.; Bassirian, S.; Singh, S.; Faulkner, C.; Bettencourt, R.; Rizo, E.; Richards, L.; Xu, Z.Z.; et al. A gut microbiome signature for cirrhosis due to nonalcoholic fatty liver disease. *Nat. Commun.* **2019**, *10*, 1406. [CrossRef]
47. Ponziani, F.R.; Bhoori, S.; Castelli, C.; Putignani, L.; Rivoltini, L.; Del Chierico, F.; Sanguinetti, M.; Morelli, D.; Paroni Sterbini, F.; Petito, V.; et al. Hepatocellular Carcinoma Is Associated With Gut Microbiota Profile and Inflammation in Nonalcoholic Fatty Liver Disease. *Hepatology* **2019**, *69*, 107–120. [CrossRef]
48. Shen, F.; Zheng, R.D.; Sun, X.Q.; Ding, W.J.; Wang, X.Y.; Fan, J.G. Gut microbiota dysbiosis in patients with non-alcoholic fatty liver disease. *Hepatobiliary Pancreat. Dis. Int.* **2017**, *16*, 375–381. [CrossRef]
49. Jiang, W.; Wu, N.; Wang, X.; Chi, Y.; Zhang, Y.; Qiu, X.; Hu, Y.; Li, J.; Liu, Y. Dysbiosis gut microbiota associated with inflammation and impaired mucosal immune function in intestine of humans with non-alcoholic fatty liver disease. *Sci. Rep.* **2015**, *5*, 8096. [CrossRef]
50. Dimova, L.G.; Zlatkov, N.; Verkade, H.J.; Uhlin, B.E.; Tietge, U.J.F. High-cholesterol diet does not alter gut microbiota composition in mice. *Nutr. Metab.* **2017**, *14*, 15. [CrossRef]
51. Arguello, G.; Balboa, E.; Arrese, M.; Zanlungo, S. Recent insights on the role of cholesterol in non-alcoholic fatty liver disease. *Biochim. Biophys. Acta Mol. Basis Dis.* **2015**, *1852*, 1765–1778. [CrossRef] [PubMed]
52. Nickerson, K.P.; Chanin, R.; McDonald, C. Deregulation of intestinal anti-microbial defense by the dietary additive, maltodextrin. *Gut Microbes.* **2015**, *6*, 78–83. [CrossRef] [PubMed]
53. Baer, D.J.; Stote, K.S.; Henderson, T.; Paul, D.R.; Okuma, K.; Tagami, H.; Kanahori, S.; Gordon, D.T.; Rumpler, W.V.; Ukhanova, M.; et al. The metabolizable energy of dietary resistant maltodextrin is variable and alters fecal microbiota composition in adult men. *J. Nutr.* **2014**, *177*, 1023–1029. [CrossRef] [PubMed]
54. Arnold, A.R.; Chassaing, B. Maltodextrin, modern stressor of the intestinal environment. *Cell Mol. Gastroenterol. Hepatol.* **2019**, *7*, 475–476. [CrossRef] [PubMed]

Article

Relationship Between the Gastrointestinal Side Effects of an Anti-Hypertensive Medication and Changes in the Serum Lipid Metabolome

Yoomin Ahn [1], Myung Hee Nam [2] and Eungbin Kim [1,*]

1 Department of Systems Biology, Yonsei University, Seoul 03722, Korea; dbals12345@yonsei.ac.kr
2 Environmental Risk and Welfare Research Team, Korea Basic Science (KBSI), Seoul 02855, Korea; nammh@kbsi.re.kr
* Correspondence: eungbin@yonsei.ac.kr; Tel.: +82-2-2123-2651

Received: 28 November 2019; Accepted: 6 January 2020; Published: 13 January 2020

Abstract: An earlier study using a rat model system indicated that the active ingredients contained in the anti-hypertensive medication amlodipine (AMD) appeared to induce various bowel problems, including constipation and inflammation. A probiotic blend was found to alleviate intestinal complications caused by the medicine. To gain more extensive insight into the beneficial effects of the probiotic blend, we investigated the changes in metabolite levels using a non-targeted metabolic approach with ultra-performance liquid chromatography-quadrupole/time-of-fligh (UPLC-q/TOF) mass spectrometry. Analysis of lipid metabolites revealed that rats that received AMD had a different metabolome profile compared with control rats and rats that received AMD plus the probiotic blend. In the AMD-administered group, serum levels of phosphatidylcholines, lysophosphatidylcholines, sphingomyelins, triglycerides with large numbers of double bonds, cholesterols, sterol derivatives, and cholesterol esters (all $p < 0.05$) were increased compared with those of the control group and the group that received AMD plus the probiotic blend. The AMD-administered group also exhibited significantly decreased levels of triglycerides with small numbers of double bonds (all $p < 0.05$). These results support our hypothesis that AMD-induced compositional changes in the gut microbiota are a causal factor in inflammation.

Keywords: lipid metabolome; amlodipine; probiotics; corticosterone; ACTH; gut bacteriome

---

1. Introduction

Many metabolic diseases (e.g., obesity, hypertension, and diabetes) have emerged as serious health problems in developed countries, mainly as a result of changes in eating habits and developments in the food industry. For example, according to a report published in 2017 by the Organization for Economic Co-operation and Development (OECD), the obesity rate in the U.S. was 30.9% in 2000 and increased to 38.2% in 2014. In the case of hypertension, although prevalence decreased from 1999 to 2016, the absolute burden caused by hypertension has increased [1]. Unlike infectious diseases, metabolic disorders are typically chronic and manageable rather than remediable, forcing patients to take medications almost ad infinitum. Along with their beneficial effects, long-term medications may cause some unwanted effects. There is increasing evidence that some side effects of long-term medications, including gastrointestinal (GI) disorders (such as constipation, diarrhea, and irritable bowel syndrome (IBS)), are related to disruption of the gut microbial population, referred to as dysbiosis [2–5]. Many medications cause gut microbiota dysbiosis even though they are not considered antibiotics [6,7].

The human gut microbiota is a complex ecosystem, consisting of approximately 1–4 $\times 10^{15}$ microbial cells. The gut microbiota establishes a close relationship with the host through interactions among themselves and with host cells in the GI tract [5,8–10]. Hence, it seems logical rather than surprising that maintenance of a well-balanced gut microbial community is a prerequisite for healthy functioning of the whole system. Indeed, the gut microbiota is proposed to be an essential "organ" that functions to maintain nutrient metabolism, immune function, and metabolic homeostasis [11–14]. Recent studies show that the gut microbiota affects neurodevelopment and diverse brain functions by regulating the gut–brain axis, the bidirectional communication between the brain and the gut [15–18]. Many of these studies reported only correlative or associative findings; however, efforts have been undertaken to examine causality and mechanism in the microbiome.

We previously reported that amlodipine (AMD), the active ingredient in a hypertension medicine, is an aggravating factor in various bowel problems, including constipation and inflammation. This is because it induces compositional changes in the gut microbiota, since normalization of the gut microbiota alleviates intestinal complications caused by AMD [19]. To investigate the effects of the gut microbiome on the host, we performed a comparative analysis of lipid metabolome in serum samples from rats that received saline (null control), AMD, or AMD plus a probiotic blend (AMD+PB). We chose to examine lipid metabolites because they are strongly associated with high blood pressure [20], and AMD is used as a treatment for hypertension.

2. Materials and Methods

2.1. Experimental Rats

A total of 18 six-week-old male Sprague Dawley rats were randomly divided into three groups (n = 6/group) to receive saline (null control), AMD, or AMD+PB. The probiotic blend (PB) was obtained in powder form and consisted of *Bifidobacterium lactis* CBT BL3 (KCTC 11904BP), *Bifidobacterium longum* CBT BG7 (KCTC 12200BP), *Bifidobacterium bifidum* CBT BF3 (KCTC 12199BP), *Lactobacillus acidophilus* CBT LA1 (KCTC 11906BP), *Lactobacillus rhamnosus* CBT LR5 (KCTC 12202BP), and *Streptococcus thermophilus* CBT ST3 (KCTC 11870BP) (Cell Biotech Co., Ltd., Seoul, Korea). The PB also contained the excipients fructooligosaccharide, lactose, galactooligosaccharide, orange flavor powder, milk flavor powder, Mg-stearate, L-ascorbic acid, vitamin E, dry-formed vitamin A, vitamin B6 hydrochloride, and vitamin B1 hydrochloride. There were approximately equal numbers (ca. 1.67×10^9 CFUs/g) of viable cells of each of the six bacterial strains in the PB. The total number of viable cells in the powdered form of the product was determined by measurement to be 1×10^{10} CFUs/g, which was diluted in water for oral administration of 1×10^7 CFUs/day.

Three rats were housed in a single cage, so two cages were used for each treatment group. After a one-week acclimation period, oral gavage of PB was administered each day in a dose of ~1 × 10^7 CFUs. Starting in the third week, AMD was administered to the rats daily for 2 weeks by oral gavage (2 mg/kg/day). The daily dose of AMD was determined as previously described [21]. All rats were housed under the following conditions: temperature 23 ± 1 °C, relative humidity 55–65%, and a 12 h light cycle. Metabolic data (weight, food intake, and water intake) were collected every day. Weight data were measured individually for each animal, but food and water intake were measured for each cage rather than for each animal. The use and care of the animals were reviewed and approved by the Institutional Animals Care and Use Committee at the Cell Biotech R&D Centre (CBTJ-15-02). All animal procedures were in accordance with the Guide for the Care and Use of Laboratory Animals issued by the Laboratory Animal Resources Commission of Cell Biotech R&D Centre.

2.2. Serum Collection and Serum Lipid Metabolite Analysis

On day 28, the rats were sacrificed by CO_2 asphyxiation. It should be noted that in the AMD group one rat died before scarification. Blood samples were collected from the heart in micro tubes, kept at 4 °C for 1 h, and then centrifuged at 2200× *g* for 15 min. The supernatant was stored at −80 °C until use. Each serum sample was prepared by adding 180 μL of isopropyl alcohol to 45 μL of serum (serum:IPA, 1:4) and then vortexing for 1 min. The mixture was incubated at −20 °C for 3 h. Then, the samples were centrifuged at 14,000× rpm at 4 °C for 20 min. The supernatant was then diluted with an equal volume of deionized water and injected into an ultra-performance liquid chromatography/quadrupole time-of-flight mass spectrometry (UPLCQ/TOF–MS) machine (Waters Corporation, Milford, MA, USA). The lipid metabolites in the serum were separated using an Acquity UPLC CSH C18 column (2.1 × 100 mm, 1.7 µL particle size; Waters Corporation). The column temperature was 55 °C. The mobile phase consisted of acetonitrile:water (60:40) with 10 mM ammonium formate in 0.1% formic acid (A) and isopropanol:acetonitrile (90:10) with 10 mM ammonium formate in 0.1% formic acid (B). The flow rate was set at 0.4 mL/min. The samples were eluted using the following conditions: initial 40% B to 53% at 2 min, to 50% A at 2.1 min, to 54% B at 12 min, to 70% B at 12.1 min, to 1% B at 18 min, to 40% B at 18.1 min, followed by equilibration for an additional 2 min. Mass acquisition was performed in positive and negative electrospray ionization modes. Mass data were collected in the range of m/z 60–1400 for 20 min with a scan time of 0.25 s and an inter-scan time of 0.02 s. The source and desolvation temperatures were 120 and 550 °C, respectively.

2.3. *Processing and Analysis of Mass Spectrometry Data*

The Progenesis QI software (Waters Corporation) was used for data processing, including mass ion alignment, normalization, and peak picking. The intensities of the mass peaks for each sample were normalized according to the total ion intensity and Pareto scaled using SIMCA-P+ 12 software (Umetrics, San Jose, CA, USA).

To differentiate among the intensities of the mass peaks in each treatment group, principal component analysis (PCA) was performed. In addition, orthogonal partial least-square discriminant analysis (OPLS-DA) was used for the selection of metabolites.

Metabolites were identified by matching the measured mass spectra with references in the Human Metabolomics Database (http://www.hmdb.ca/) and METLIN (http://metlin.scripps.edu/). Lipids identified in the samples were validated on the basis of isotope similarity and fragmentation patterns. Hierarchical clustering analysis was performed using PermutMatrix (version 1.9.3, ATGC team, LIRMM, Montpellier, France) with the Pearson distance and Ward's aggregation method.

Statistical analysis of stress hormone data and lipid metabolomic data was performed using GraphPad Prism (version 7.03; GraphPad Software, Inc., San Diego, CA, USA). Data are expressed as the mean ± SEM. The significance of differences among the data were measured by one-way ANOVA followed by Tukey's post-hoc test, or by the Kruskal–Wallis test followed by Dunn's post-hoc test for data that did not follow the normal distribution.

2.4. *The Criteria for Metabolite Selection*

Metabolites were selected on the basis of the following criteria: a) all differences between groups were significant ($p < 0.05$), b) the metabolite level was at least twice as high in the AMD group than in the control group and similar between the control group and the AMD+PB group, and c) the highest relative level of the metabolite was greater than 10.

3. Results and Discussion

3.1. *Statistical Analysis of the Serum Metabolome*

Multivariate statistical analysis of the metabolome data was performed to identify statistically significant endogenous metabolites. First, PCA was conducted to determine the inherent similarities in the spectral profiles of the treatment groups. As shown in Figure 1, the control group and the AMD group were clearly divided into two clusters on the PCA score plot, whereas the AMD+PB group displayed a pattern almost identical to that of the control group. This result is in good agreement with the previous finding that the PB alleviated intestinal complications caused by AMD [19].

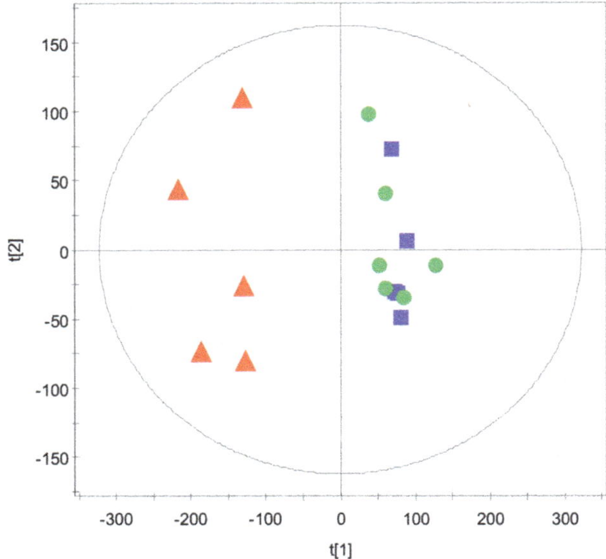

Figure 1. Principal component analysis (PCA) score plot of the metabolome analysis of the three treatment groups. Red triangles (▲): anti-hypertensive medication amlodipine (AMD) group. Blue squares (■): AMD plus a probiotic blend (AMD+PB) group. Green circles (●): control group.

3.2. Screening and Identification of Candidate Markers for Lipid Metabolites

To initially distinguish the differences among serum metabolites, hierarchical clustering analysis was performed to identify metabolites that were significantly increased or decreased among the treatment groups (Figure 2). Table 1 summarizes a detailed subgroup analysis of the metabolites. The G1a subgroup included cholesterol esters (CEs [22:6] and [20:4]), sterol derivatives, sphingomyelins (SMs), lysophosphatidylcholines (LysoPCs [18:0] and [16:0]), several phosphatidylcholines (PCs), and cholesterol. The G1b subgroup included PCs (18:0/22:6) and several triglycerides (TGs) with more than 10 double bonds (e.g., TG [60:12]). G1 metabolite levels overall were strongly increased in the AMD group but recovered in the AMD+PB group to the same level as those in the control group. On the other hand, the G2 metabolites included several TGs and diglycerides (DGs) with less than five double bonds (e.g., DG [34:1]). In addition, the TGs in G2 included monosaturated species (TG [48:1], TG [52:1], and TG [50:1]).

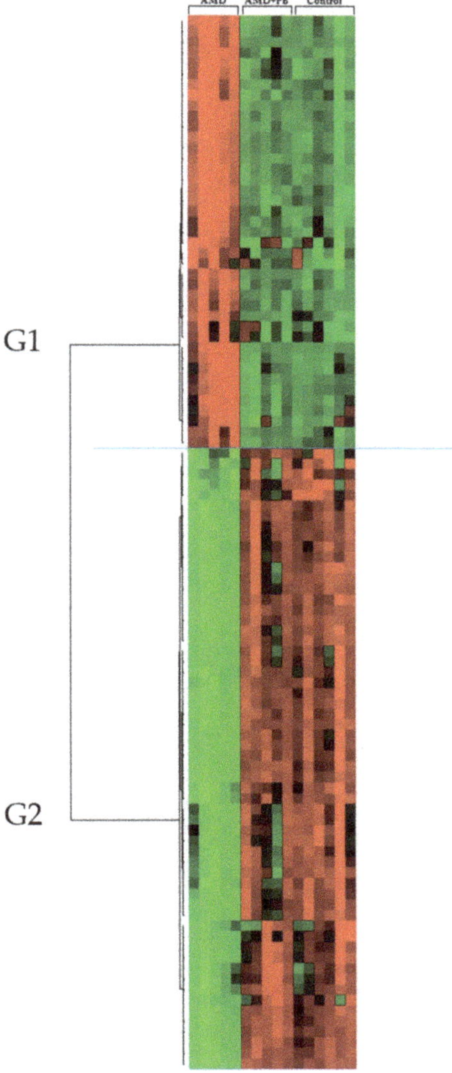

Figure 2. Hierarchical clustering analysis of the UPLC-HDMS metabolomics results. The rows display the metabolites, and the columns display the samples. Metabolites that significantly decreased relative to the average level across the samples are displayed in green, while those that significantly increased are displayed in red. The brightness of each color corresponds to the intensity of the difference compared with the average value.

Table 1. List of identified metabolites.

Group	Identification	Mean			p Value		Fold Change	
		C	A	A+P	A/C	A/A+P	A/C	A/A+P
G1	CE (22:6)	6 ± 3	57 ± 20	7 ± 1	0.004	0.005	8.97	8.38
	CE (20:4)	12 ± 3	43 ± 7	14 ± 2	0.000	0.001	3.69	3.08
	Sterol derivatives	42 ± 13	135 ± 37	49 ± 8	0.003	0.005	3.22	2.77
	SM (d18:1/24:1)	120 ± 27	343 ± 66	122 ± 17	0.001	0.001	2.86	2.82
	SM (d16:1/18:0)	207 ± 44	563 ± 110	223 ± 44	0.001	0.001	2.72	2.52
	LysoPC (18:0)	894 ± 160	2136 ± 356	840 ± 50	0.001	0.001	2.39	2.54
	PC (18:0/20:4)	1561 ± 302	3524 ± 573	1830 ± 339	0.001	0.001	2.26	1.93
	PC (16:0/18:0)	66 ± 11	148 ± 27	73 ± 12	0.001	0.002	2.24	2.04
	PC (16:0/16:0)	54 ± 8	118 ± 14	62 ± 10	0.000	0.000	2.17	1.91
	Cholesterol	217 ± 53	445 ± 112	229 ± 17	0.007	0.012	2.05	1.94
	LysoPC (16:0)	1484 ± 205	2893 ± 284	1385 ± 131	0.000	0.000	1.95	2.09
	TG (60:12)	25 ± 9	201 ± 64	22 ± 13	0.003	0.003	8.20	9.32
	TG (60:11)	56 ± 15	221 ± 75	43 ± 20	0.007	0.005	3.92	5.20
	TG (58:10)	88 ± 26	287 ± 71	83 ± 25	0.002	0.002	3.28	3.46
	TG (60:10)	39 ± 6	108 ± 33	30 ± 10	0.009	0.005	2.74	3.64
	PC (18:0/22:6)	279 ± 75	687 ± 244	285 ± 23	0.018	0.021	2.46	2.41
G2	TG (48:2)	159 ± 18	16 ± 6	175 ± 51	0.000	0.002	0.10	0.09
	TG (48:1)	112 ± 16	15 ± 8	124 ± 51	0.000	0.008	0.14	0.12
	TG (51:2)	148 ± 15	21 ± 7	155 ± 20	0.000	0.000	0.14	0.14
	TG (51:3)	144 ± 22	20 ± 7	156 ± 12	0.000	0.000	0.14	0.13
	TG (53:3)	133 ± 24	24 ± 8	134 ± 15	0.000	0.000	0.18	0.18
	TG (54:2)	443 ± 51	116 ± 34	410 ± 89	0.000	0.001	0.26	0.28
	TG (50:2)	1177 ± 140	356 ± 102	1290 ± 210	0.000	0.000	0.30	0.28
	TG (54:3)	1296 ± 143	430 ± 77	1288 ± 131	0.000	0.000	0.33	0.33
	TG (52:1)	231 ± 25	84 ± 23	203 ± 48	0.000	0.003	0.36	0.41
	TG (50:1)	562 ± 72	209 ± 63	579 ± 161	0.000	0.004	0.37	0.36
	TG (56:3)	163 ± 27	42 ± 22	148 ± 36	0.000	0.001	0.26	0.28
	TG (50:3)	830 ± 83	179 ± 67	945 ± 112	0.000	0.000	0.22	0.19
	DG (34:1)	184 ± 18	61 ± 18	191 ± 15	0.000	0.000	0.33	0.32
	TG (52:5)	489 ± 101	179 ± 50	549 ± 118	0.000	0.001	0.37	0.33
	TG (52:2)	3124 ± 323	1189 ± 347	3363 ± 272	0.000	0.000	0.38	0.35

Abbreviations: C = null control group, A = AMD-administered group, A+P = AMD plus probiotic blend-administered group. Metabolites are arranged in order of the magnitude of the A/C fold change.

PCs and LysoPCs regulate immune function. PCs inhibit the TNF-α-induced upregulation of pro-inflammatory cytokines [22,23] and stimulate universal anti-inflammatory effects in the liver [24]. In contrast to the PCs, research on the immunomodulatory functions of LysoPCs shows conflicting results. Some studies show that LysoPCs contribute to the progression of inflammation by upregulating IL-1β-induced inducible nitric oxide synthase (NOS) [25] and also act as a death effector in the lipo-apoptosis of hepatocytes [26], which are key cells in innate immunity [27]. In addition, LysoPCs are involved in cardiovascular complications related to diabetes, rheumatoid arthritis, and atherosclerosis [28,29], as well as the activation of inflammatory responses via the acceleration of endothelial chemokine secretion [29,30]. However, other studies suggest that LysoPCs regulate inflammatory responses by inhibiting the secretion of pro-inflammatory cytokines such as TNF-α [31]. LysoPCs were evaluated as a biomarker because PC is converted to LysoPC by phospholipase A2 under inflammatory conditions [32–34]. Some studies suggest that LysoPCs are immunoregulatory lipid messengers under normal and pathogen-induced physiological conditions [35] because they can mediate signaling through G-protein-coupled receptors and be recognized as autoantigens [36]. Notwithstanding ambiguous results concerning pathways and mechanisms, it is certain that LysoPCs are involved in inflammation. Accordingly, it is apparent that increases of phospholipids such as PC and LysoPC in AMD-administered rats are associated with inflammation.

TGs are associated with the immune system. An excess of TGs causes diseases like hypertriglyceridemia [37], which is related to systemic inflammation [38]. Sterols play an essential role in countless biological processes including reproduction, metabolism, development, and immunity [39].

Cholesterols contribute to protection against infection by amplifying the inflammatory response and are the precursors of steroid hormones (including sex hormones, growth hormones, and glucocorticoids like corticosterone) [40]. However, excessive or prolonged cholesterol-induced immune responses can cause chronic inflammatory diseases like atherosclerosis [41]. Therefore, we performed additional analysis to examine and compare the levels of two hormones, adrenocorticotropic hormone (ACTH) and corticosterone, which are both representative stress hormones associated with immune reaction in rats [42–44].

3.3. Identification and Comparison of Corticosterone and Adrenocorticotropic Hormone (ACTH)

The probiotic blend used in this study was previously shown to have beneficial effects on human subjects with irritable bowel syndrome [45] and on animals with indomethacin-induced small intestine injury [46]. In our previous experiment, it was found to bring down increased levels of inflammatory cytokines in AMD-administered rats [19]. Glucocorticoids, including corticosterone in rodents and cortisol in humans, are anti-inflammatory steroid hormones [47,48]. In this context, we hypothesized that the probiotic blend could normalize potential anomalies in the level of corticosterone.

As shown in Figure 3, corticosterone levels were much higher in the AMD group than in the other two groups. The stress related to handling by the investigators was almost the same among the groups. Like the AMD group and the AMD+PB group, the control group was also subjected to oral gavage. Because the stress from oral gavage was the same among the groups, it is reasonable to hypothesize that higher corticosterone levels in the AMD group were caused by AMD-induced activation of the hypothalamic-pituitary-adrenocortical (HPA) axis or by a direct effect of AMD on the adrenal cortex, either or both of which were blocked by the PB co-treatment. In contrast, the ACTH levels decreased slightly more in the AMD group than in the control group (Figure 4). Considering that ACTH has a short half-life in plasma [49] and corticosterone itself is a negative regulator of ACTH secretion, the observed reduction of ACTH is likely a reflection of feedback inhibition of the HPA axis by corticosterone [47].

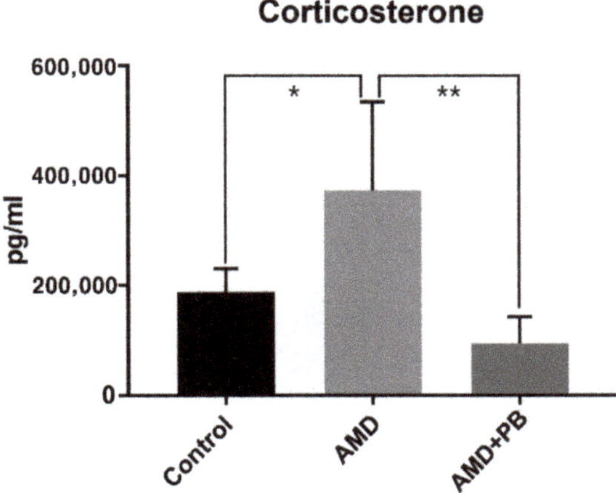

Figure 3. Corticosterone levels in the rat sera (*; p value < 0.05, **; p value < 0.01).

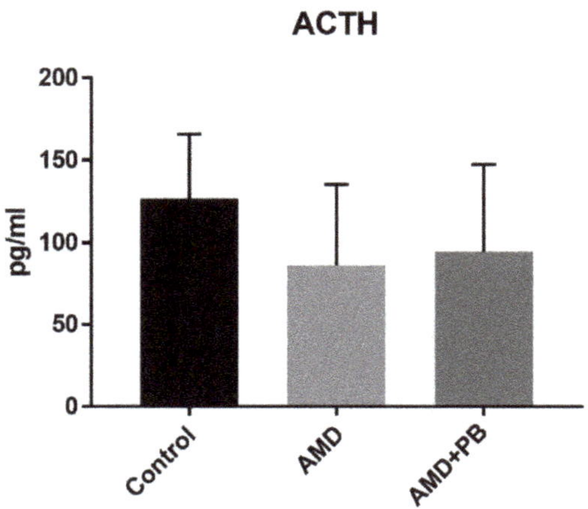

Figure 4. Adrenocorticotropic hormone (ACTH) levels in the rat sera.

4. Conclusions

Composition and stability of the gut microbiome is known to be affected by nutrition and disease, as well as antibiotics or medication [50]. Gut microbes influence lipid processing of hosts by engaging in gene expression related to the host's cholesterol and TG metabolism [51]. In this study, we revealed the change of lipid profiles in the serum of AMD-administered rats. Considering that impairment of the fine balance between gut microbes and the host's immune system leads to systemic inflammation [52], it can be postulated that the change of serum lipid profiles by AMD may reflect the disturbance of the gut microbial environment by AMD. Combined with these facts, our results suggest that AMD-induced dysbiosis leads to inflammation and changes in metabolic pathways, which in turn promotes the secretion of corticosterone to relieve the symptoms (Figure 5).

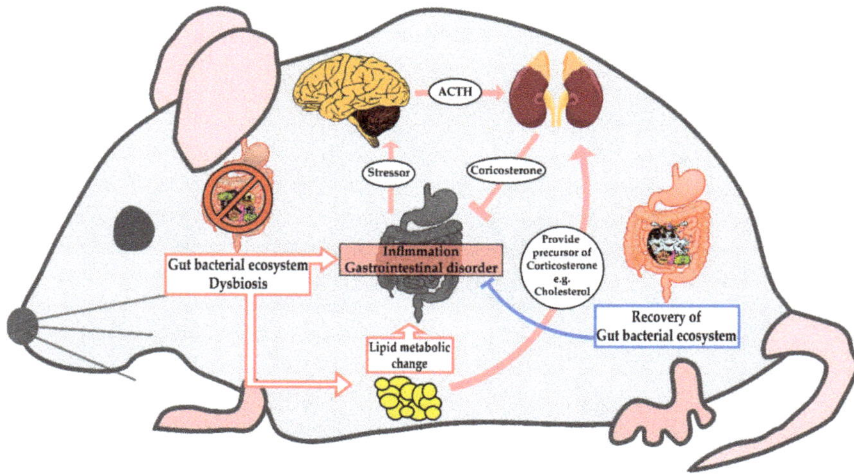

Figure 5. A cartoon summarizing the hypothesized effects of AMD-induced dysbiosis on lipid metabolism.

Author Contributions: Conceptualization, E.K. and Y.A.; methodology, M.H.N.; formal analysis, M.H.N.; investigation, Y.A.; resources, M.H.N. and Y.A.; data curation, Y.A. and M.H.N.; writing—original draft preparation, Y.A..; writing—review and editing, E.K.; visualization, Y.A.; supervision, E.K.; project administration, E.K.; funding acquisition, E.K. All authors have read and agreed to the published version of the manuscript.

Funding: This research was financially supported in part by Cell Biotech Co., Ltd. through the Cell Biotech R&D Centre and in part by the Strategic Initiative for Microbiomes in Agriculture and Food (918011-4) funded by the Ministry of Agriculture, Food and Rural Affairs. M.N. acknowledges the financial support from the National Research Foundation of Korea (NRF-2018M3A9F3056901).

Acknowledgments: Thanks to Kyeong-Hoon Jeong for advice on the stress hormones of mice.

Conflicts of Interest: The authors declare no conflicts of interest.

References

1. Dorans, K.S.; Mills, K.T.; Liu, Y.; He, J. Trends in Prevalence and Control of Hypertension According to the 2017 American College of Cardiology/American Heart Association (ACC/AHA) Guideline. *J. Am. Heart Assoc.* **2018**, 7. [CrossRef]
2. Carding, S.; Verbeke, K.; Vipond, D.T.; Corfe, B.M.; Owen, L.J. Dysbiosis of the Gut Microbiota in Disease. *Microb. Ecol. Health Dis.* **2015**, *26*, 26191. [CrossRef] [PubMed]
3. Chassard, C.; Dapoigny, M.; Scott, K.P.; Crouzet, L.; Del'homme, C.; Marquet, P.; Martin, J.C.; Pickering, G.; Ardid, D.; Eschalier, A.; et al. Functional Dysbiosis within the Gut Microbiota of Patients with Constipated-Irritable Bowel Syndrome. *Aliment. Pharm.* **2012**, *35*, 828–838. [CrossRef] [PubMed]
4. Duboc, H.; Rainteau, D.; Rajca, S.; Humbert, L.; Farabos, D.; Maubert, M.; Grondin, V.; Jouet, P.; Bouhassira, D.; Seksik, P.; et al. Increase in Fecal Primary Bile Acids and Dysbiosis in Patients with Diarrhea-Predominant Irritable Bowel Syndrome. *Neurogastroenterol. Motil.* **2012**, *24*, 513–e247. [CrossRef] [PubMed]
5. Belizário, J.E.; Faintuch, J. Microbiome and Gut Dysbiosis. *Exp. Suppl.* **2018**, *109*, 459–476. [CrossRef] [PubMed]
6. Brüssow, H. Problems with the Concept of Gut Microbiota Dysbiosis. *Microb. Biotechnol.* **2019**, *1751–7915*, 13479. [CrossRef]
7. Le Bastard, Q.; Al-Ghalith, G.A.; Grégoire, M.; Chapelet, G.; Javaudin, F.; Dailly, E.; Batard, E.; Knights, D.; Montassier, E. Systematic Review: Human Gut Dysbiosis Induced by Non-Antibiotic Prescription Medications. *Aliment. Pharm.* **2018**, *47*, 332–345. [CrossRef]
8. Althani, A.A.; Marei, H.E.; Hamdi, W.S.; Nasrallah, G.K.; El Zowalaty, M.E.; Al Khodor, S.; Al-Asmakh, M.; Abdel-Aziz, H.; Cenciarelli, C. Human Microbiome and Its Association with Health and Diseases. *J. Cell. Physiol.* **2016**, *231*, 1688–1694. [CrossRef]
9. Guinane, C.M.; Cotter, P.D. Role of the Gut Microbiota in Health and Chronic Gastrointestinal Disease: Understanding a Hidden Metabolic Organ. *Ther. Adv. Gastroenterol.* **2013**, *6*, 295–308. [CrossRef]
10. Gill, S.R.; Pop, M.; Deboy, R.T.; Eckburg, P.B.; Turnbaugh, P.J.; Samuel, B.S.; Gordon, J.I.; Relman, D.A.; Fraser-Liggett, C.M.; Nelson, K.E. Metagenomic Analysis of the Human Distal Gut Microbiome. *Science* **2006**, *312*, 1355–1359. [CrossRef]
11. Mazmanian, S.K.; Liu, C.H.; Tzianabos, A.O.; Kasper, D.L. An Immunomodulatory Molecule of Symbiotic Bacteria Directs Maturation of the Host Immune System. *Cell* **2005**, *122*, 107–118. [CrossRef] [PubMed]
12. Belizário, J.E.; Faintuch, J.; Garay-Malpartida, M. Gut Microbiome Dysbiosis and Immunometabolism: New Frontiers for Treatment of Metabolic Diseases. *Mediat. Inflamm.* **2018**, *2018*, 2037838. [CrossRef] [PubMed]
13. Thursby, E.; Juge, N. Introduction to the Human Gut Microbiota. *Biochem. J.* **2017**, *474*, 1823–1836. [CrossRef] [PubMed]
14. Tremaroli, V.; Bäckhed, F. Functional Interactions between the Gut Microbiota and Host Metabolism. *Nature* **2012**, *489*, 242–249. [CrossRef]
15. Liang, S.; Wu, X.; Jin, F. Gut-Brain Psychology: Rethinking Psychology from the Microbiota–Gut–Brain Axis. *Front. Integr. Neurosci.* **2018**, *12*, 33. [CrossRef]
16. Erny, D.; Hrabě de Angelis, A.L.; Prinz, M. Communicating Systems in the Body: How Microbiota and Microglia Cooperate. *Immunology* **2017**, *150*, 7–15. [CrossRef]
17. Carabotti, M.; Scirocco, A.; Maselli, M.A.; Severi, C. The Gut-Brain Axis: Interactions between Enteric Microbiota, Central and Enteric Nervous Systems. *Ann. Gastroenterol.* **2015**, *28*, 203–209.

18. Erny, D.; Hrabě de Angelis, A.L.; Jaitin, D.; Wieghofer, P.; Staszewski, O.; David, E.; Keren-Shaul, H.; Mahlakoiv, T.; Jakobshagen, K.; Buch, T.; et al. Host Microbiota Constantly Control Maturation and Function of Microglia in the CNS. *Nat. Neurosci.* **2015**, *18*, 965–977. [CrossRef]
19. Song, S.-C.; An, Y.-M.; Shin, J.-H.; Chung, M.-J.; Seo, J.-G.; Kim, E. Beneficial Effects of a Probiotic Blend on Gastrointestinal Side Effects Induced by Leflunomide and Amlodipine in a Rat Model. *Benef. Microbes* **2017**, *8*, 801–808. [CrossRef]
20. Zicha, J.; Kunes, J.; Devynck, M.A. Abnormalities of Membrane Function and Lipid Metabolism in Hypertension: A Review. *Am. J. Hypertens.* **1999**, *12*, 315–331. [CrossRef]
21. Sevilla, M.A.; Voces, F.; Carrón, R.; Guerrero, E.I.; Ardanaz, N.; San Román, L.; Arévalo, M.A.; Montero, M.J. Amlodipine Decreases Fibrosis and Cardiac Hypertrophy in Spontaneously Hypertensive Rats: Persistent Effects after Withdrawal. *Life Sci.* **2004**, *75*, 881–891. [CrossRef] [PubMed]
22. Eros, G.; Varga, G.; Vairadi, R.; Czobel, M.; Kaszaki, J.; Ghyczy, M.; Boros, M. Anti-Inflammatory Action of a Phosphatidylcholine, Phosphatidylethanolamine and N-Acylphosphatidylethanolamine-Enriched Diet in Carrageenan-Induced Pleurisy. *Eur. Surg. Res.* **2009**, *42*, 40–48. [CrossRef]
23. Treede, I.; Braun, A.; Sparla, R.; Kühnel, M.; Giese, T.; Turner, J.R.; Anes, E.; Kulaksiz, H.; Füllekrug, J.; Stremmel, W.; et al. Anti-Inflammatory Effects of Phosphatidylcholine. *J. Biol. Chem.* **2007**, *282*, 27155–27164. [CrossRef] [PubMed]
24. Oneta, C.M.; Mak, K.M.; Lieber, C.S. Dilinoleoylphosphatidylcholine Selectively Modulates Lipopolysaccharide-Induced Kupffer Cell Activation. *J. Lab. Clin. Med.* **1999**, *134*, 466–470. [CrossRef]
25. Taniuchi, M.; Otani, H.; Kodama, N.; Tone, Y.; Sakagashira, M.; Yamada, Y.; Mune, M.; Yukawa, S. Lysophosphatidylcholine Up-Regulates IL-1 Beta-Induced INOS Expression in Rat Mesangial Cells. *Kidney Int. Suppl.* **1999**, *71*, S156–S158. [CrossRef]
26. Han, M.S.; Park, S.Y.; Shinzawa, K.; Kim, S.; Chung, K.W.; Lee, J.-H.; Kwon, C.H.; Lee, K.-W.; Lee, J.-H.; Park, C.K.; et al. Lysophosphatidylcholine as a Death Effector in the Lipoapoptosis of Hepatocytes. *J. Lipid Res.* **2008**, *49*, 84–97. [CrossRef]
27. Zhou, Z.; Xu, M.-J.; Gao, B. Hepatocytes: A Key Cell Type for Innate Immunity. *Cellular and Molecular Immunology* **2015**, *13*, 301. [CrossRef]
28. Choi, J.; Zhang, W.; Gu, X.; Chen, X.; Hong, L.; Laird, J.M.; Salomon, R.G. Lysophosphatidylcholine Is Generated by Spontaneous Deacylation of Oxidized Phospholipids. *Chem. Res. Toxicol.* **2011**, *24*, 111–118. [CrossRef]
29. Wi, S.J.; Seo, S.Y.; Cho, K.; Nam, M.H.; Park, K.Y. Lysophosphatidylcholine Enhances Susceptibility in Signaling Pathway against Pathogen Infection through Biphasic Production of Reactive Oxygen Species and Ethylene in Tobacco Plants. *Phytochemistry* **2014**, *104*, 48–59. [CrossRef]
30. Riederer, M.; Lechleitner, M.; Hrzenjak, A.; Koefeler, H.; Desoye, G.; Heinemann, A.; Frank, S. Endothelial Lipase (EL) and EL-Generated Lysophosphatidylcholines Promote IL-8 Expression in Endothelial Cells. *Atherosclerosis* **2011**, *214*, 338–344. [CrossRef]
31. Xu, T.; Feng, G.; Zhao, B.; Zhao, J.; Pi, Z.; Liu, S.; Song, F.; Liu, Z. A Non-Target Urinary and Serum Metabolomics Strategy Reveals Therapeutical Mechanism of Radix Astragali on Adjuvant-Induced Arthritis Rats. *J. Chromatogr. B* **2017**, *1048*, 94–101. [CrossRef] [PubMed]
32. Shin, J.-H.; Nam, M.H.; Lee, H.; Lee, J.-S.; Kim, H.; Chung, M.-J.; Seo, J.-G. Amelioration of Obesity-Related Characteristics by a Probiotic Formulation in a High-Fat Diet-Induced Obese Rat Model. *Eur. J. Nutr.* **2018**, *57*, 2081–2090. [CrossRef] [PubMed]
33. Yaligar, J.; Gopalan, V.; Kiat, O.W.; Sugii, S.; Shui, G.; Lam, B.D.; Henry, C.J.; Wenk, M.R.; Tai, E.S.; Velan, S.S. Evaluation of Dietary Effects on Hepatic Lipids in High Fat and Placebo Diet Fed Rats by In Vivo MRS and LC-MS Techniques. *Plos ONE* **2014**, *9*, e91436. [CrossRef] [PubMed]
34. Abbott, M.J.; Tang, T.; Sul, H.S. The Role of Phospholipase A2-Derived Mediators in Obesity. *Drug Discov. Today Dis. Mech.* **2010**, *7*, e213–e218. [CrossRef]
35. Fox, L.M.; Cox, D.G.; Lockridge, J.L.; Wang, X.; Chen, X.; Scharf, L.; Trott, D.L.; Ndonye, R.M.; Veerapen, N.; Besra, G.S.; et al. Recognition of Lyso-Phospholipids by Human Natural Killer T Lymphocytes. *PLoS Biol.* **2009**, *7*, e1000228. [CrossRef]
36. Frasch, S.C.; Zemski-Berry, K.; Murphy, R.C.; Borregaard, N.; Henson, P.M.; Bratton, D.L. Lysophospholipids of Different Classes Mobilize Neutrophil Secretory Vesicles and Induce Redundant Signaling through G2A. *J. Immunol.* **2007**, *178*, 6540–6548. [CrossRef]

37. AbouRjaili, G.; Shtaynberg, N.; Wetz, R.; Costantino, T.; Abela, G.S. Current Concepts in Triglyceride Metabolism, Pathophysiology, and Treatment. *Metabolism* **2010**, *59*, 1210–1220. [CrossRef]
38. Jonkers, I.J.; Mohrschladt, M.F.; Westendorp, R.G.; van der Laarse, A.; Smelt, A.H. Severe Hypertriglyceridemia with Insulin Resistance Is Associated with Systemic Inflammation: Reversal with Bezafibrate Therapy in a Randomized Controlled Trial. *Am. J. Med.* **2002**, *112*, 275–280. [CrossRef]
39. Wollam, J.; Antebi, A. Sterol Regulation of Metabolism, Homeostasis, and Development. *Annu. Rev. Biochem.* **2011**, *80*, 885–916. [CrossRef]
40. Mao, Z.; Li, J.; Zhang, W. Hormonal Regulation of Cholesterol Homeostasis. In *Cholesterol—Good, Bad and the Heart*; Nagpal, M.L., Ed.; InTech: London, UK, 2018. [CrossRef]
41. Tall, A.R.; Yvan-Charvet, L. Cholesterol, Inflammation and Innate Immunity. *Nat. Rev. Immunol.* **2015**, *15*, 104–116. [CrossRef]
42. Milot, M.R.; James, J.S.; Merali, Z.; Plamondon, H. A Refined Blood Collection Method for Quantifying Corticosterone. *Lab. Anim.* **2012**, *41*, 77–83. [CrossRef] [PubMed]
43. Vahl, T.P.; Ulrich-Lai, Y.M.; Ostrander, M.M.; Dolgas, C.M.; Elfers, E.E.; Seeley, R.J.; D'Alessio, D.A.; Herman, J.P. Comparative Analysis of ACTH and Corticosterone Sampling Methods in Rats. *Am. J. Physiol. Endocrinol. Metab.* **2005**, *289*, E823–E828. [CrossRef] [PubMed]
44. Djordjević, J.; Cvijić, G.; Davidović, V. Different Activation of ACTH and Corticosterone Release in Response to Various Stressors in Rats. *Physiol. Res.* **2003**, *52*, 67–72. [PubMed]
45. Yoon, J.S.; Sohn, W.; Lee, O.Y.; Lee, S.P.; Lee, K.N.; Jun, D.W.; Lee, H.L.; Yoon, B.C.; Choi, H.S.; Chung, W.-S.; et al. Effect of Multispecies Probiotics on Irritable Bowel Syndrome: A Randomized, Double-Blind, Placebo-Controlled Trial: Probiotics in Irritable Bowel Syndrome. *J. Gastroenterol. Hepatol.* **2014**, *29*, 52–59. [CrossRef] [PubMed]
46. Byun, S.J.; Lim, T.J.; Lim, Y.J.; Seo, J.G.; Chung, M.J. *In Vivo* Effects of s-Pantoprazole, Polaprenzinc, and Probiotic Blend on Chronic Small Intestinal Injury Induced by Indomethacin. *Benef. Microbes* **2016**, *7*, 731–737. [CrossRef] [PubMed]
47. Jeong, K.H.; Jacobson, L.; Pacák, K.; Widmaier, E.P.; Goldstein, D.S.; Majzoub, J.A. Impaired Basal and Restraint-Induced Epinephrine Secretion in Corticotropin-Releasing Hormone-Deficient Mice. *Endocrinology* **2000**, *141*, 1142–1150. [CrossRef]
48. Chio, S.L.; Sin, Y.M. Changes in Corticosterone Levels under Different Degrees of Acute Inflammation in Mice. *Agents Actions* **1992**, *36*, 93–98. [CrossRef]
49. Kant, G.J.; Leu, J.R.; Anderson, S.M.; Mougey, E.H. Effects of Chronic Stress on Plasma Corticosterone, ACTH and Prolactin. *Physiol. Behav.* **1987**, *40*, 775–779. [CrossRef]
50. Hur, K.Y.; Lee, M.-S. Gut Microbiota and Metabolic Disorders. *Diabetes Metab. J.* **2015**, *39*, 198. [CrossRef]
51. Falcinelli, S.; Picchietti, S.; Rodiles, A.; Cossignani, L.; Merrifield, D.L.; Taddei, A.R.; Maradonna, F.; Olivotto, I.; Gioacchini, G.; Carnevali, O. *Lactobacillus Rhamnosus* Lowers Zebrafish Lipid Content by Changing Gut Microbiota and Host Transcription of Genes Involved in Lipid Metabolism. *Sci. Rep.* **2015**, *5*, 9336. [CrossRef]
52. He, M.; Shi, B. Gut Microbiota as a Potential Target of Metabolic Syndrome: The Role of Probiotics and Prebiotics. *Cell Biosci.* **2017**, *7*, 54. [CrossRef] [PubMed]

Article

The Timing Effects of Soy Protein Intake on Mice Gut Microbiota

Konomi Tamura [1,†], Hiroyuki Sasaki [1,2,†], Kazuto Shiga [1], Hiroki Miyakawa [1] and Shigenobu Shibata [1,*]

1 Laboratory of Physiology and Pharmacology, School of Advanced Science and Engineering, Waseda University, Shinjuku-ku, Tokyo 162-8480, Japan; tmr-knm@akane.waseda.jp (K.T.); hiroyuki-sasaki@asagi.waseda.jp (H.S.); k_shiga@fuji.waseda.jp (K.S.); hgbbst-hiroki@toki.waseda.jp (H.M.)

2 National Institute of Advanced Industrial Science and Technology, AIST-Waseda University Computational Bio Big-Data Open Innovation Laboratory (CBBD-OIL), Shinjuku-ku, Tokyo 169-8555, Japan

* Correspondence: shibatas@waseda.jp; Tel.: +81-3-5369-7318

† These authors contributed equally to this work.

Received: 22 November 2019; Accepted: 25 December 2019; Published: 27 December 2019

Abstract: Soy protein intake is known to cause microbiota changes. While there are some reports about the effect of soy protein intake on gut microbiota and lipid metabolism, effective timing of soy protein intake has not been investigated. In this study, we examined the effect of soy protein intake timing on microbiota. Mice were fed twice a day, in the morning and evening, to compare the effect of soy protein intake in the morning with that in the evening. Mice were divided into three groups: mice fed only casein protein, mice fed soy protein in the morning, and mice fed soy protein in the evening under high-fat diet conditions. They were kept under the experimental condition for two weeks and were sacrificed afterward. We measured cecal pH and collected cecal contents and feces. Short-chain fatty acids (SCFAs) from cecal contents were measured by gas chromatography. The microbiota was analyzed by sequencing 16S rRNA genes from feces. Soy protein intake whether in the morning or evening led to a greater microbiota diversity and a decrease in cecal pH resulting from SCFA production compared to casein intake. In addition, these effects were relatively stronger by morning soy protein intake. Therefore, soy protein intake in the morning may have relatively stronger effects on microbiota than that in the evening.

Keywords: soy protein; microbiota; lipid metabolism; circadian; chrono-nutrition

1. Introduction

Mammals have approximately 100 trillion bacteria in their gut that comprise the microbiota. Gut microbiota has profound influences on the host's physiological conditions such as nutrient absorption, metabolism, and immunity [1]. Microbial alterations cause inflammatory bowel diseases and metabolic disorders. For example, concerning the microbiota of an obese person, the relative abundance of *Firmicutes*, which is a factor of obesity, is increased and *Bacteroidetes*, which prevents fat accumulation, is decreased [2]. In addition, an altered microbiota causes obesity because germ-free mice show an increase in body fat when injected with the microbiota of an obese mice [3].

Intestinal bacteria digest non-digestible food components such as dietary fibers, oligosaccharides, resistant starches, and resistant proteins, and produce short-chain fatty acids (SCFAs) [4,5]. SCFAs include acetic acid, propionic acid, lactic acid, and butyric acid. SCFAs are used as an energy source for colonic epithelial cells [6]. SCFAs maintain gut acidic conditions and prevent the growth of harmful bacteria such as *Enterobacteriaceae* and *Clostridia* [4,7]. SCFAs also have beneficial effects on mammalian energy metabolism and regulate the metabolism of fatty acid, glucose, and cholesterol [4].

The alteration of microbiota depends on various factors such as age, stress, disease, drugs, and diet [8]. There are many reports about the relationship between diet and microbiota. The microbiota can be rapidly affected by dietary changes [9]. Some studies evaluated the impact of protein on the microbiota. Many proteins are absorbed by the small intestine. However, segments of some proteins pass through the small intestine to reach the large intestine [10]. These resistant proteins and amino acids are metabolized by intestinal microbiota to SCFAs [11]. For example, soy protein intake causes higher microbial diversity and SCFA levels [12,13].

The circadian clock system plays an important role in maintaining physiological conditions such as the sleep-wake cycle, body temperature, and metabolism [14]. In the mammalian circadian system, there are two clocks: the main central oscillator in the suprachiasmatic nuclei (SCN), and the peripheral oscillator in peripheral organs. The SCN clock is mainly entrained by light-dark stimuli, and it regulates the peripheral clocks. The peripheral clocks are entrained by pharmacological agents, food nutrients, and mental or physical stress [15–19]. The SCFAs produced by the microbiota also entrain the circadian clock [20]. Microbiota exhibits diurnal oscillations in composition and function in both mice and humans. In addition, jet lag induced circadian disruption changes microbiota, and when feces from jet lag mice were transferred to germ-free mice, they became obese [21]. Thus, the activity of microbiota is strongly associated with circadian rhythm.

Chrono-nutrition is the science of nutrition, which is based on chronobiology. Hormonal secretion and the metabolism and absorption of nutrients have circadian variations [16]. Therefore, some food components have the most effective intake timing [22,23]. For example, fish oil intake in the morning rather than the evening is more effective to reduce lipids in mice [23]. Intake of water-soluble dietary fiber in the morning has a greater effect on microbiota diversity rather than in the evening [24]. In this study, we examined the effective timing of soy protein intake. Soy protein, especially β-conglycinin, is known to have beneficial effects on hepatic lipid metabolism, prevention of hepatic steatosis, and reduction of body fat in both rodents and humans [25–29]. It is also reported that soy protein intake has a superior effect on microbiota [12,26,27]. As hormonal secretion as well as microbial composition and function exhibit diurnal oscillations, the effective timing to alter microbiota can be different according to the food components [14,21]. However, the effective timing of soy protein intake has not been investigated. Therefore, in this study, we examined the effect of soy protein intake on mice gut microbiota based on chrono-nutrition, such as morning intake or evening intake.

2. Materials and Methods

2.1. Animals and Diets

We used 105 of ICR 8-week-old male mice (Tokyo Laboratory Animals, Tokyo, Japan) in this study. The mice were kept under 12 h light/12 h dark condition. Lights-on time was defined as zeitgeber time 0 (ZT0) and lights-off time as ZT12. Each mouse was housed in a plastic cage individually, at a temperature of 22 ± 2 °C, humidity of 60% ± 5%, and light intensity of 100-150 lux. We prepared two kinds of diet, a high-fat diet (HFD) with casein and HFD with soy protein (Fujipro F, Fuji Oil Co., Osaka, Japan) (Table 1).

To produce metabolic syndrome models with obesity, high inflammation, and abnormal microbiota, the mice were fed a HFD with casein and water ad libitum for one week before commencing the experiments. Thereafter, mice were fed HFD with casein or soy protein according to the experimental protocols. The Committee for Animal Experimentation at Waseda University approved all experimental protocols (permission protocol 2018-A030).

Table 1. Nutrition components (g) in each diet (100 g).

	Casein Diet	Soy Diet
Casein	22.86	-
Soy protein	-	23.78
L-cysteine	0.18	0.18
βcorn starch	13.71	12.79
αcorn starch	15.5	15.5
Sucrose	10	10
Soybean oil	4	4
Lard	24	24
Cellulose	5	5
Mineral mixture	3.5	3.5
Vitamin mixture	1	1
Choline bitartrate	0.25	0.25
Total	100	100

2.2. Experimental Design

In this study, we prepared two kinds of diets, HFD with casein (casein diet) and HFD with soy protein (soy diet), as described previously. For microbiota deterioration, mice were fed a casein diet for one week before commencing the experiments.

In experiment 1, we examined the effects of soy protein intake in a short period. Soy protein is known to have anti-obesity effects. Obesity causes a change in microbiota composition. Therefore, to eliminate the effect of body weight differences, we conducted the experiments over a short period, before body weight change. Mice were given free access to the casein diet (Casein group) or soy diet (Soy group) for 10 days, and they were sacrificed at ZT12, ZT20, or ZT4. Ten mice were prepared for each time point and group. We measured the cecal pH, and we collected the cecal contents, feces, blood, and liver samples (Figure 1a).

In experiment 2, we examined the timing effects of soy protein intake. To compare the effect of soy protein intake in the morning to that in the evening, mice were fed 1.8 g diets twice a day in the morning (ZT12) and evening (ZT20). When mice were given 4 h of access to food in the morning and evening, the amount of food consumption was different between the morning and evening [24]. Therefore, we fed mice 1.8 g diets twice a day, so that morning and evening food consumption is the same. It is reported that mice were able to consume all of 1.8 g diet within 4 h [30,31]. The experiment period was set to 14 days since it takes a few days for mice to adapt to the 2-meals-per-day feeding pattern. The mice were fed only casein diet (Casein group), soy diet in the morning and casein diet in the evening (M-Soy group), or casein diet in the morning and soy diet in the evening (E-Soy group). The mice were kept under experimental conditions for two weeks, and then they were sacrificed at ZT12, 20, or 4. Five mice were prepared for each time point and group. We measured the cecal pH and collected the cecal contents, feces, blood, and liver samples (Figure 1b).

2.3. Cholesterol and Triglyceride Measurement

Serum cholesterol and triglyceride (TG) levels were measured using cholesterol and triglyceride kit (FUJIFILM Wako Pure Chemical Co., Osaka, Japan). The assay was performed according to the manufacturer's instructions.

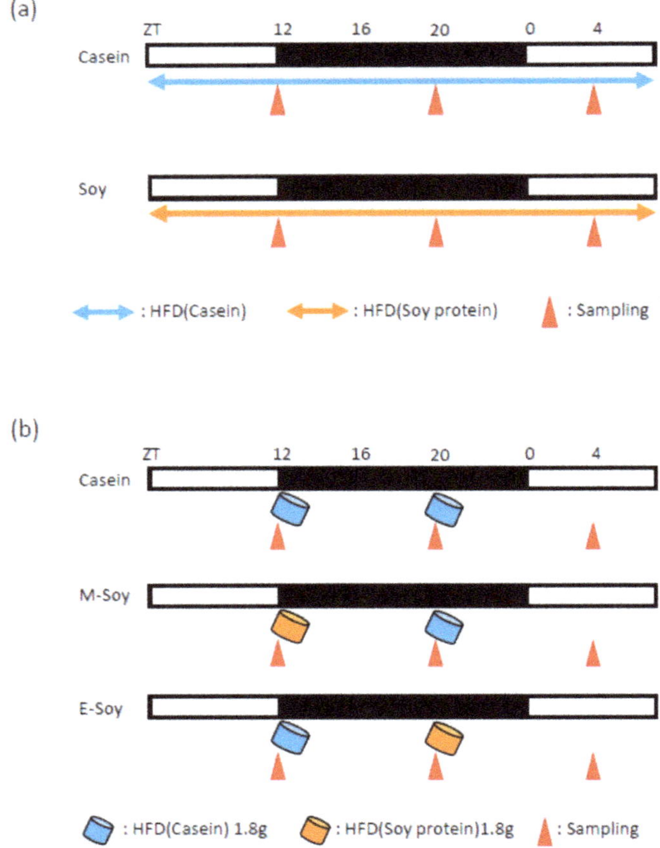

Figure 1. Experimental design. (**a**) Experimental protocol to examine the effect of soy protein intake. (**b**) Experimental protocol to examine the effect of soy protein intake timing. The white and black bars indicate environmental 12 h light and dark conditions, respectively. The horizontal blue arrow indicates free access to a high-fat diet (HFD) with casein. The horizontal orange arrow indicates free access to HFD with soy protein. The blue cylinder indicates the feeding timing of 1.8 g of HFD with casein. The orange cylinder indicates the feeding timing of 1.8 g of HFD with soy protein. The red triangles indicate the sampling time. High-fat diet, casein diet feeding, soy protein diet feeding, morning soy protein diet feeding, evening soy protein diet feeding: HFD, Casein, Soy, M-Soy, E-Soy, respectively.

2.4. Real-Time RT-PCR

Relative liver mRNA levels were measured by real-time RT-PCR. The mice were anesthetized with isoflurane and sacrificed at ZT 12, ZT 20, or ZT4. We collected livers to measure mRNA levels at each time point. Total liver RNA was extracted using RNA-*Solv* Reagent (Omega Bio-Tek Inc., Norcross, GA, USA). RNA concentration of each sample was adjusted using a spectrophotometer (GE Healthcare Japan Co., Tokyo, Japan). The RNA was reverse-transcribed and amplified using One-Step SYBR RT-PCR kit (Takara Bio Inc., Shiga, Japan) with specific primer pairs (Table 2) on Piko Real PCR system (Thermo Fisher Scientific, Waltham, MA, USA). The relative expression levels of target genes were normalized with *GAPDH*. The data were analyzed using the ΔΔCt method.

Table 2. Sequences of Primers for Real-time RT-PCR.

Gene	Forward	Reverse
Acc1	GCACTCCCGATTCATAATG	CCCAAATCAGAAAGTGTATC
Cyp7α1	AGACCGCACATAAAGCCCGG	CTTTCATTGCTTCAGGGCTC
Fas	TGGGTTCTAGCCAGCAGAGT	ACCACCAGAGACCGTTATGC
Gapdh	TGGTGAAGGTCGGTGTGAAC	AATGAAGGGGTCGTTGATGG
Hmgcr	GATCATCCAGTTGGTCAATGC	GCAAGCTTTGTGGAGAGGAG
Srebp1c	CGCTACCGGTCTTCTATCAATG	CAAGAAGCGGATGTAG

2.5. Cecal pH Measurement

Cecal pH was measured using pH meter (Euthech Instruments, Vernon Hills, IL, USA). The electrode of the pH meter was inserted directly into the cecum, immediately after collection.

2.6. Short-Chain Fatty Acid (SCFA) Measurement

Short-chain fatty acid (SCFA) in cecal contents was measured through gas chromatography (Shimadzu Co., Kyoto, Japan) as described in a previous report [32]. Cecal contents were acidified with sulfuric acid and SCFAs were extracted from 50 mg of cecal contents by shaking in 50 μL of sulfuric acid, 400 μL of diethyl ether, and 200 μL of ethanol (FUJIFILM Wako Pure Chemical Co., Osaka, Japan). The mixture was centrifuged at 18700× *g* for 30 s. The supernatant (1 μL) was injected into the capillary column (InertCap Pure WAX (30 m × 0.25 mm, df = 0.5 μm), GL Science, Tokyo, Japan) of gas chromatography coupled to a flame ionization detector. The initial temperature was 80 °C and the final temperature was 200 °C. Helium was used as carrier gas and quantification of the samples was performed using calibration curves for acetic, lactic, propionic, and butyric acids. A standard curve plotted for the quantitation of each acid was in the samples.

2.7. Fecal DNA Extraction

Fecal DNA was extracted according to the previous report with modifications [33]. We collected feces from the rectum, when we sacrificed the mice at each time point. Approximately 0.2 g fecal sample was suspended in a 50 mL tube containing 20 mL PBS. The suspension was filtered through a 100 μm nylon filter (Corning Inc., New York, NY, USA). The tube was washed with 10 mL PBS and then filtered through the filter. The filtrates were centrifuged at 9000× *g* for 20 min at 4 °C, and the supernatants were removed. Each precipitate was suspended in 1.5 mL TE 10 buffer (10 mM Tris-HCl (FUJIFILM Wako Pure Chemical Co., Osaka, Japan)/10 mM EDTA (DOJINDO, Tokyo, Japan)), and the suspension was transferred to 2 mL microtube. The suspensions were centrifuged at 9560× *g* for 5 min at 4 °C, and the supernatants were removed. Each precipitate was suspended in 800 μL TE 10 buffer. The suspensions were added 100 μL lysozyme (150 mg/mL) (FUJIFILM Wako Pure Chemical Co., Osaka, Japan) and then incubated for 1 h at 37 °C. Achromopeptidase (20 μL, 100 units/μL, FUJIFILM Wako Pure Chemical Co., Osaka, Japan) was added to the suspension and then incubated for 30 min at 37 °C. The suspension was treated with 50 μL of 20% sodium dodecyl sulfate and proteinase K (Promega Co., Madison, WI, USA) and then incubated for 1 h at 55 °C. To extract DNA, 980 μL PCI (phenol/chloroform/isoamyl alcohol) (Invitrogen, Carlsbad, CA, USA) was added and centrifuged at 6000× *g* for 10 min at 20 °C. The supernatant was transferred to a new 2 mL microtube and then suspended with 100 μL of 3 M sodium acetate and 900 μL isopropanol (FUJIFILM Wako Pure Chemical Co., Osaka, Japan). The suspensions were centrifuged at 6000× *g* for 10 min at 20 °C, and the supernatants were removed. The DNA pellet was rinsed with 1 mL of 70% ethanol and dried. The DNA was purified by treatment with 99 μL TE buffer and 1 μL RNase (10 μg/mL) (FUJIFILM Wako Pure Chemical Co., Osaka, Japan), then precipitated with 100 μL of 20% PEG solution (TOKYO Chemical Industry Co., Tokyo, Japan). The DNA was pelleted by centrifugation at 10,000× *g* at 4 °C, rinsed with 500 μL of 70% ethanol, and dissolved in 50 μL TE buffer.

2.8. 16S rRNA Gene Sequencing

16S rRNA gene sequencing was performed according to Illumina instructions (16S Metagenomic Sequencing Library Preparation). V3-V4 variable regions of the 16S rRNA gene were amplified using forward primer (5′-TCGTCGGCAGCGTCAGATG TGTATAAGAGACAGCCTACGGGNGGCWGCAG-3′) and reverse primer (5′-GTCTCGTGGG CTCGGAGATGTGTATAAGAGACAGGACTACHVGGGTATCTAATCC-3′). Amplicon PCR was performed with 2.5 μL microbial DNA (5 ng/μL), 5 μL of each amplicon PCR primer (1 μM), and 12.5 μL of 2 × KAPA HiFi HotStart Ready Mix (Kapa Biosystems, Wilmington, MA, USA) under conditions of 3 min at 95 °C, 25 cycles of 95 °C for 30 s, 55 °C for 30 s, and 72 °C for 30 s, and a final extension of 72 °C for 5 min. The PCR products were purified using AMPure XP beads (Beckman Coulter, Inc., Brea, CA, USA).

The Nextera XT Index Kit v2 (Illumina Inc., San Diego, CA, USA) was used to join dual indices and Illumina sequencing adapters. Index PCR was performed in 5 μL PCR production, 5 μL of each Nextera XT Index primer, 25 μL of 2 × KAPA HiFi HotStart Ready Mix, and 10 μL of PCR Grade water under conditions of 3 min at 95 °C, 8 cycles of 95 °C for 30 s, 55 °C for 30 s, and 72 °C for 30 s, and a final extension of 72 °C for 5 min. The PCR products were purified using AMPure XP beads. The quality of the purification was checked using Agilent 2100 Bioanalyzer with DNA 1000 kit (Agilent Technologies, Santa Clara, CA, USA). Finally, the DNA library concentration was diluted to 4 nM.

The DNA library was sequenced using Miseq reagent kit v3 (Illumina Inc., San Diego, CA, USA) on Illumina Miseq 2 × 300 bp platform. This sequencing was performed following manufacturer instructions.

2.9. Analysis of 16S rRNA Gene Sequences

16S rRNA gene sequence reads were processed through quantitative insights into microbial ecology (QIIME) pipeline version 1.9.1 [34]. Quality-filtered sequence reads were assigned to operational taxonomic units (OTUs) at 97% identity with the UCLUST algorithm [35]. These reads were then compared to the reference sequence collections in the Greengenes database (August 2013 version). A total of 4,034,110 reads were obtained from 105 samples. On average, 38,420 ± 2427 reads were obtained per sample. Taxonomy summary, alpha-diversity (within-sample), beta-diversity (between-sample dissimilarity), and principal coordinate analysis (PCoA) were calculated and generated by QIIME. PCoA analysis was also calculated using unweighted UniFrac distances.

2.10. Metagenome Prediction

The functional profiles of microbial communities were predicted through phylogenetic investigation of communities by reconstruction of unobserved states (PICRUSt) [36]. The functional predictions were assigned to almost all Kyoto Encyclopedia of Genes and Genomes (KEGG) ortholog (KO) functional profiles of microbial communities via 16S sequences. We selected and examined categories related to "Amino Acid Metabolism" and "Energy Metabolism" for analysis of simplification and clarity.

2.11. Statistical Analysis

Data are expressed as mean ± standard error of the mean (SEM). In this study, we compared the feeding condition at each time point, because we focused on the difference in feeding condition rather than time point. Statistical analysis was performed using GraphPad Prism version 6.03 (GraphPad Software, San Diego, CA, USA). The data were tested for normality and equality of variances using a D'Agostino-Pearson test/Kolmogorov-Smirnov test and Bartlett's test, respectively. Parametric analysis was conducted using one-way ANOVA with Tukey test or Student's *t*-test for post-hoc analysis, and non-parametric analysis was conducted using the Kruskal-Wallis test with Dunn's test or the Mann-Whitney test for post-hoc analysis. The differences in microbiota composition were tested using the permutational multivariate analysis of variance (PERMANOVA). PERMANOVA was analyzed by QIIME.

3. Results

3.1. Soy Protein Intake Affected Lipid Metabolism and the Gut Microbiota

It has already been reported that soy protein not only reduces serum cholesterol and triglycerides, but also changes the microbiota composition, leading to considerable microbial diversity [12,25,26]. To examine whether the results of our study are similar to previous reports, we considered the effect of soy protein feeding in the free-feeding condition.

First, we examined the effect of soy protein on lipid metabolism. The food consumption (Casein group: 3.88 ± 0.15 g/day, Soy group: 3.92 ± 0.11 g/day) and final body weight (Casein group: 42.69 ± 0.59 g, Soy group: 42.15 ± 0.71 g) showed no differences between the groups. We showed the data of each time point and the average of a 3-time point (AVE). The serum cholesterol of the Soy group was significantly lower than that of the Casein group at ZT20 and AVE. Serum TG level showed no significant difference between both groups (Figure 2a). We measured the mRNA expression levels of fatty acid and cholesterol metabolism-related genes from liver samples. *Acc1*(ZT12, ZT20, and AVE), *Fasn* (ZT12, ZT20, and AVE), and *Srebp1c* (ZT12, ZT4, and AVE) expression levels were significantly lower in the Soy group than those in the Casein group. *Cyp7α1* expression level tended to be higher in the Soy group than that in the Casein group at ZT20 (Figure 2b).

To examine the effect of soy protein on microbiota, we measured cecal pH and SCFA production. Cecal pH was significantly lower in the Soy group than that in the Casein group (Figure 3a). Acetic acid (ZT12, ZT4, and AVE), propionic acid (ZT12 and ZT4), lactic acid, and butyric acid levels were significantly or tended to be higher in the Soy group than those in the Casein group (Figure 3b).

As cecal pH was decreased and SCFA production was increased, soy protein intake may alter the microbiota. Therefore, we analyzed the microbiota from feces. The Soy group showed significantly higher alpha-diversity for the Simpson index than that in the Casein group at ZT20 (Figure 4a). The PCoA of unweighted UniFrac distance showed that the beta-diversity of microbiota composition was significantly different between the Soy group and the Casein group (Figure 4b). Concerning the relative abundances of microbes at the phylum level, *Bacteroidetes* (ZT20) and *Proteobacteria* (ZT12, ZT20, and AVE) in the Soy group were significantly higher than those in the Casein group. *Firmicutes* (ZT12, ZT20, and AVE) in the Soy group were significantly lower than those in the Casein group (Figure 4c). At the genus level, *Bifidobacterium* (ZT12, ZT4, and AVE), *Enterococcus* (ZT20 and AVE), *[Ruminococcus]* (ZT20 and AVE), and *Desulfovibrio* (ZT20 and AVE) in the Soy group were significantly higher than those in the Casein group. *Lactococcus* in the Soy group was significantly lower than that in the Casein group (Figure 4d).

To infer the metagenome functional content based on the microbial community profiles obtained from 16S rRNA gene sequences, we used PICRUSt. The microbial communities could be distinguished based on their functions. The KEGG pathways associated with amino acid and energy metabolisms were significantly upregulated in the Soy group. The pathways associated with glycine, serine, and threonine metabolisms (ZT12, ZT20, and AVE) and lysine biosynthesis (ZT20 and AVE) in the Soy group were significantly upregulated compared to those in the Casein group (Figure 5a). The pathways associated with methane metabolism (ZT12, ZT20, and AVE) and nitrogen metabolism (ZT12, ZT20, and AVE) in the Soy group were significantly upregulated compared to those in the Casein group (Figure 5b).

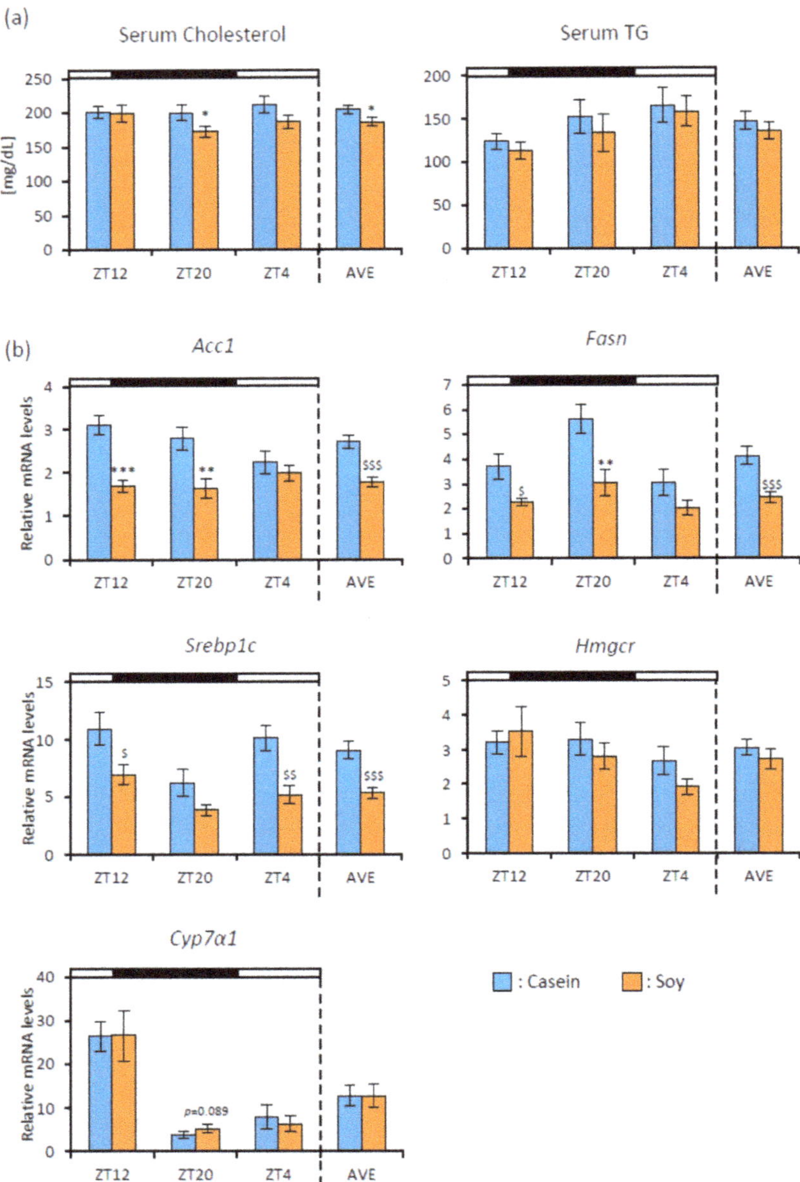

Figure 2. Serum lipid and gene expression levels in the liver. (**a**) Serum cholesterol and triglyceride levels (ZT12, ZT20, ZT4, and an average of three points) of mice that were fed each diet for 10 days. (**b**) Relative RNA expression levels of fatty acid and cholesterol metabolism-related genes in the liver (ZT12, ZT20, ZT4, and an average of three points) of mice that were fed each diet for 10 days. Data are represented as mean ± SEM ($n = 10$). * $p < 0.05$, ** $p < 0.01$, *** $p < 0.001$ versus Casein, evaluated using Student's t-test. \$ $p < 0.05$, \$\$ $p < 0.01$, \$\$\$ $p < 0.001$ versus Casein, evaluated using the Mann-Whitney test. High-fat diet, casein diet feeding, soy protein diet feeding, an average value of three points: HFD, Casein, Soy, AVE, respectively.

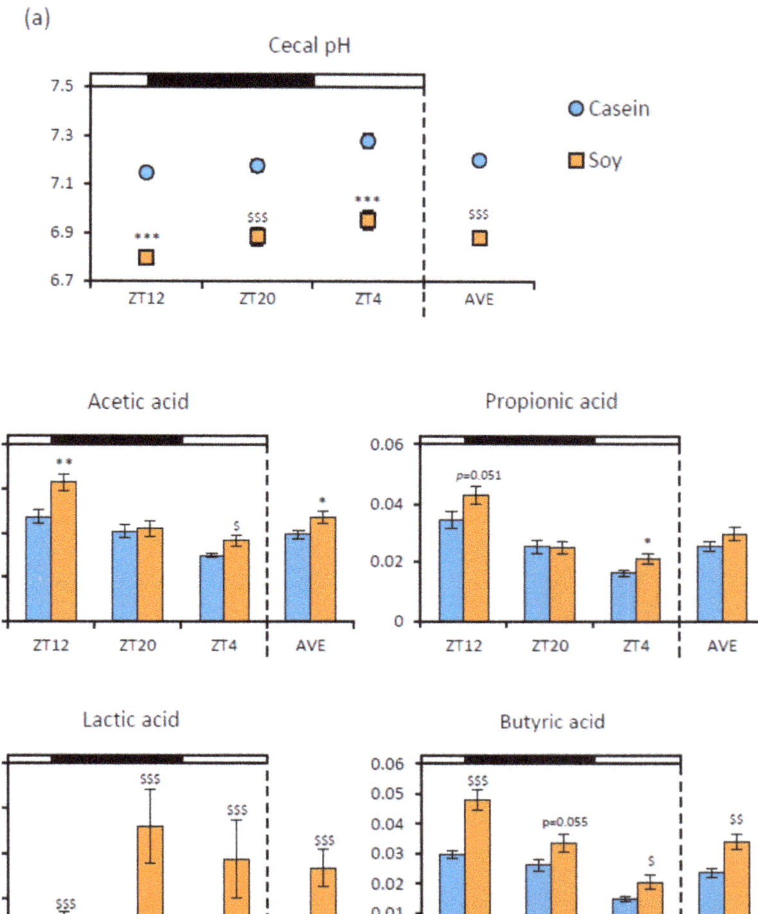

Figure 3. The effect of soy protein intake on cecal pH and SCFA levels. (**a**) Cecal pH levels (ZT12, ZT20, ZT4, and an average of three points) of mice that were fed each diet for 10 days. (**b**) Cecal short-chain fatty acids (SCFA) levels (ZT12, ZT20, ZT4, and an average of three points) of mice that were fed each diet for 10 days. Data are represented as mean ± SEM ($n = 10$). * $p < 0.05$, ** $p < 0.01$, *** $p < 0.001$ versus Casein, evaluated using Student's t-test. \$ $p < 0.05$, \$\$ $p < 0.01$, \$\$\$ $p < 0.001$ versus Casein, evaluated using the Mann-Whitney test.

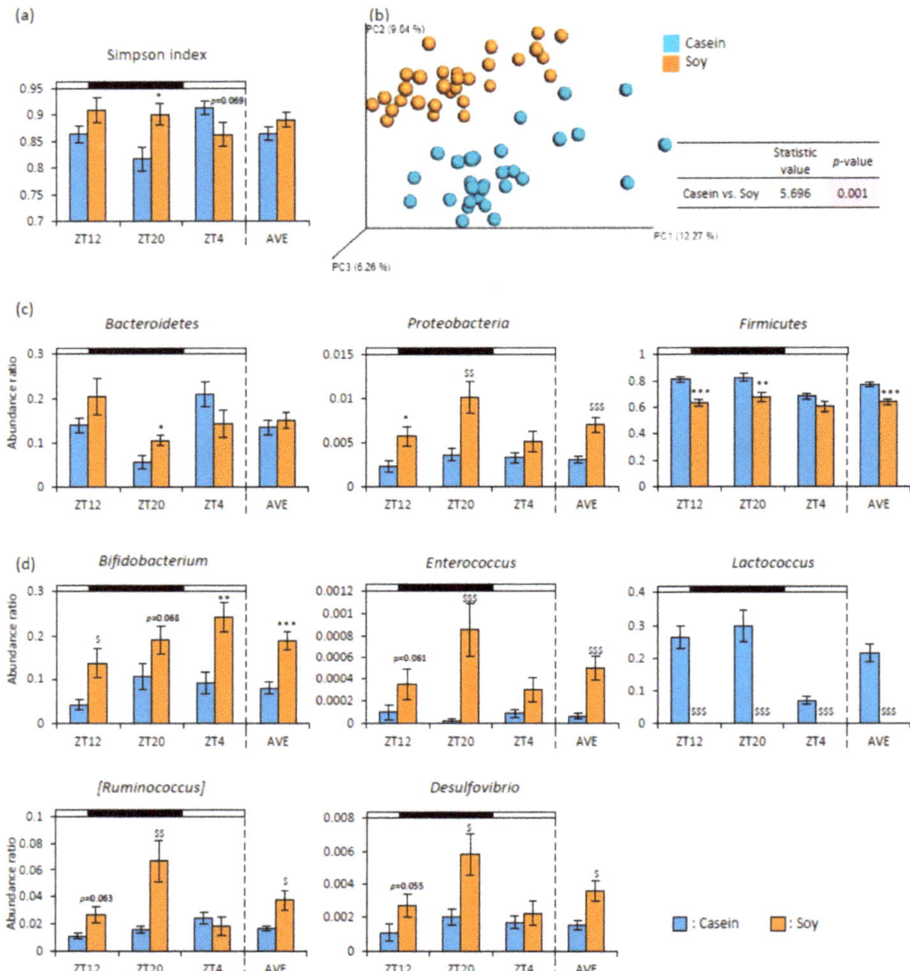

Figure 4. The effect of soy protein intake on microbiota. (**a**) Alpha-diversity about Simpson index (ZT12, ZT20, ZT4, and an average of three points) of mice that were fed each diet for 10 days. (**b**) Beta-diversity in comparison of each diet. The PCoA plots of unweighted UniFrac distance metrics obtained from sequencing the 16S rRNA gene in feces (n = 30). (**c**) The relative abundance of microbes at the Phylum level, and (**d**) at the Genus level of mice that were fed each diet for 10 days (ZT12, ZT20, ZT4, and an average of three points). Data (**a**,**c**,**d**) are represented as mean ± SEM (n = 10). * $p < 0.05$, ** $p < 0.01$, *** $p < 0.001$ versus Casein, evaluated using Student's t-test. \$ $p < 0.05$, \$\$ $p < 0.01$, \$\$\$ $p < 0.001$ versus Casein, evaluated using the Mann-Whitney test. The table in (**b**) indicates the result using PERMANOVA.

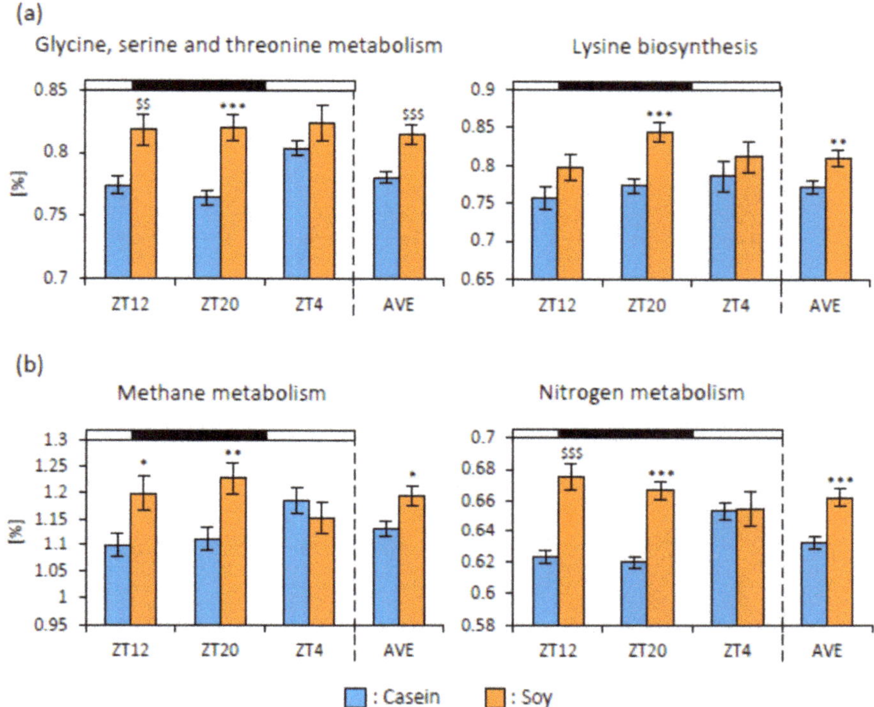

Figure 5. The functional predictions of microbial communities. (**a**) The functional predictions about categories related to "Amino Acid Metabolism" and (**b**) "Energy Metabolism" of microbial communities in mice that were fed each diet for 10 days (ZT12, ZT20, ZT4, and an average of three points). Data are represented as mean ± SEM ($n = 10$). * $p < 0.05$, ** $p < 0.01$, *** $p < 0.001$ versus Casein, evaluated using Student's *t*-test. $$ $p < 0.01$, $$$ $p < 0.001$ versus Casein, evaluated using the Mann-Whitney test.

3.2. Soy Protein Intake in the Morning Affected the Gut Microbiota More Than That in the Evening

Since soy protein intake improved microbiota, we examined the effect of soy protein intake timing on microbiota. To compare the effect of soy protein intake in the morning and evening, mice were fed 1.8 g diets twice a day in the morning (ZT12) and evening (ZT20). Mice were fed only casein diet (Casein group), soy diet in the morning and casein diet in the evening (M-Soy group), or casein diet in the morning and soy diet in the evening (E-Soy group). Mice were kept under the experimental condition for two weeks and were then sacrificed at ZT12, 20 or 4 (Figure 1b). We measured cecal pH and collected cecal contents, feces, blood, and liver sample.

First, we examined the effect of soy protein intake timing on lipid metabolism. The final body weight (Casein group: 43.57 ± 0.76 g, M-Soy group: 43.86 ± 0.51 g, E-Soy group: 42.50 ± 0.77 g) showed no significant difference among the groups. We showed the data of each time point and the average of a 3-time point (AVE). Serum cholesterol in the M-Soy group was significantly higher than that in the Casein and the E-Soy groups at ZT12. The serum TG level showed no significant difference (Figure 6a). We measured the mRNA expression levels of fatty acid and cholesterol metabolism-related genes in the liver sample. *Acc1* expression level was significantly lower in the E-Soy group than that in the Casein group at ZT12 and AVE. *Fasn* expression level in the E-Soy group at ZT12 was significantly lower than that in the M-Soy group and tended to be lower than that in the Casein group. *Cyp7α1* expression level in the Casein group was significantly higher than that in the M-Soy group at ZT20 and ZT4, and that in the E-Soy group at ZT4 (Figure 6b).

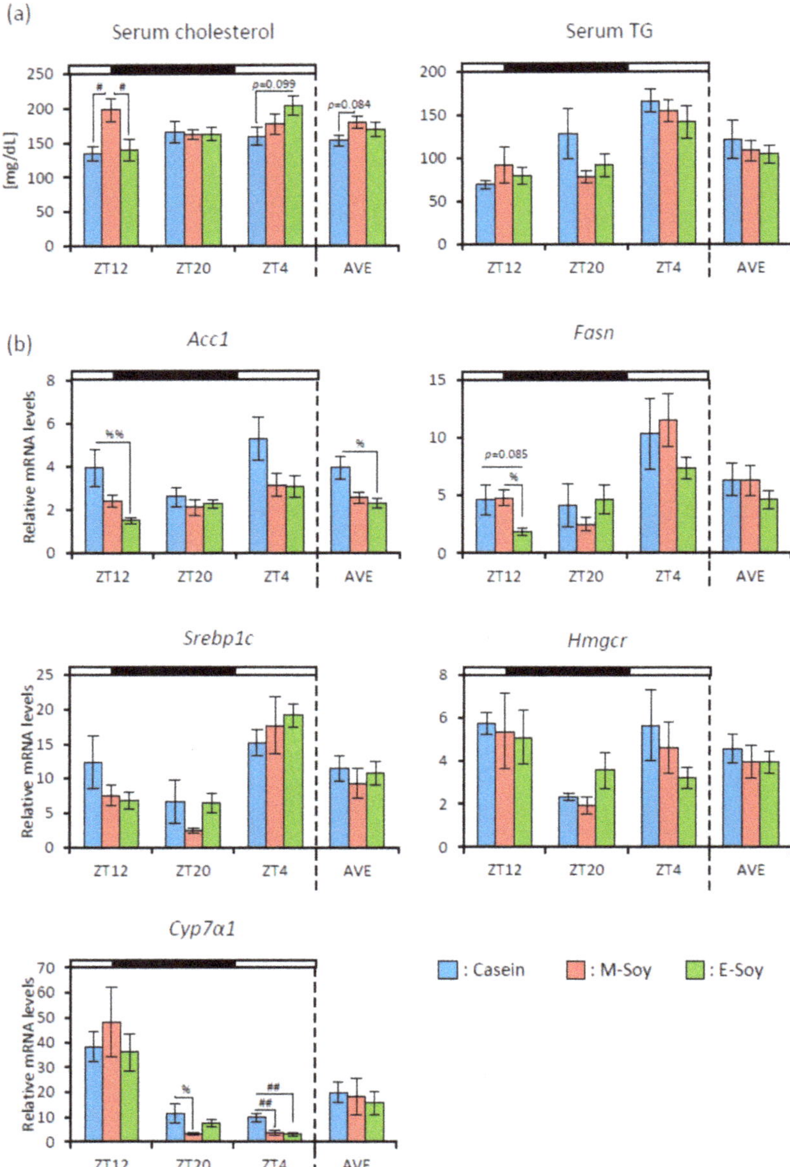

Figure 6. Serum lipid levels and gene expression levels in the liver. (**a**) Serum cholesterol and triglyceride levels (ZT12, ZT20, ZT4, and an average of three points) of mice that were kept in each feeding condition for two weeks. (**b**) Relative RNA expression levels of fatty acid and cholesterol metabolism-related genes in the liver (ZT12, ZT20, ZT4, and an average of three points) of mice that were kept in each feeding condition for two weeks. Data are represented as mean ± SEM (n = 5). # $p < 0.05$, ## $p < 0.01$ evaluated using one-way ANOVA with Tukey's post-hoc test. % $p < 0.05$, %% $p < 0.01$ the Kruskal-Wallis test with Dunn's post-hoc test. High-fat diet, casein diet feeding, morning soy protein diet feeding, evening soy protein diet feeding, the average value of three points: HFD, Casein, M-Soy, E-Soy, AVE, respectively.

To examine the effect of soy protein intake timing on microbiota, we measured cecal pH and SCFA production. Cecal pH in the M-Soy group tended to be lower than that in the Casein group at ZT12, and significantly lower than that in the other groups at ZT20. The E-Soy group showed a significantly lower pH than those in the other groups at ZT4. The M-Soy and the E-Soy groups showed significantly lower cecal pH than that in the Casein group on AVE (Figure 7a). The lactic acid in the M-Soy group was significantly higher than that in the Casein group and tended to be higher than that in the E-Soy group at ZT20. The lactic acid in the E-Soy group was significantly higher than that in the Casein group at ZT4. Only the M-Soy group showed a significantly higher level of lactic acid than that in the Casein group on AVE. Butyric acid in the M-Soy group tended to be higher than that in the Casein group at ZT20 and ZT4, and significantly higher than that in the E-Soy group at ZT20. Butyric acid in the E-Soy group was significantly higher than that in the other groups at ZT4. Only the M-Soy group showed a significantly higher level of butyric acid than that in the Casein group on AVE (Figure 7b).

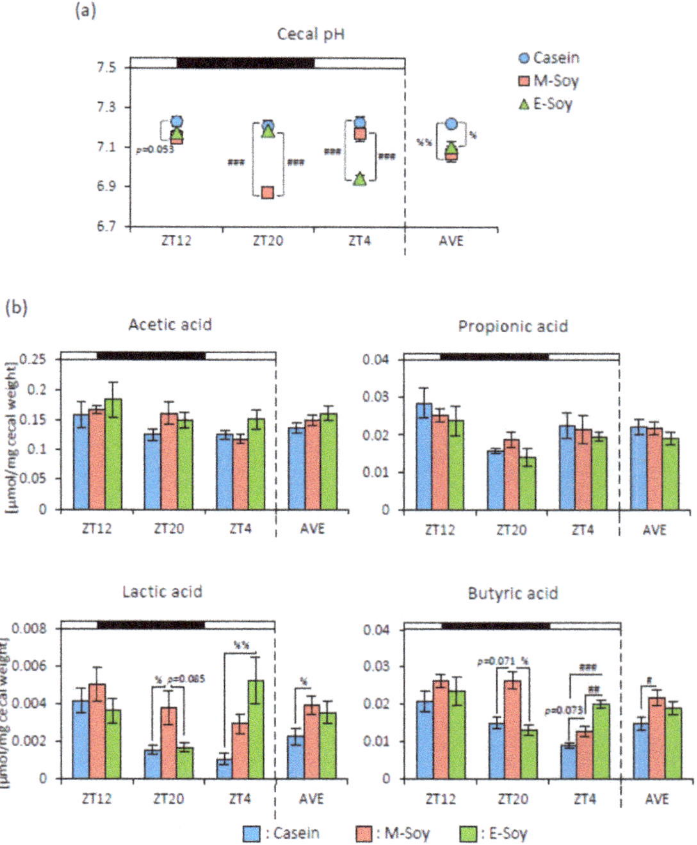

Figure 7. The effect of soy protein intake timing on cecal pH and SCFA levels. (**a**) Cecal pH levels (ZT12, ZT20, ZT4, and an average of three points) of mice that were kept in each feeding condition for two weeks. (**b**) Cecal SCFA levels (ZT12, ZT20, ZT4 and average of three points) of mice that were kept in each feeding condition for two weeks. Data are represented as mean ± SEM ($n = 5$). # $p < 0.05$, ## $p < 0.01$, ### $p < 0.001$ evaluated using one-way ANOVA with Tukey's post-hoc test. % $p < 0.05$, %% $p < 0.01$ the Kruskal-Wallis test with Dunn's post-hoc test.

As the cecal pH was decreased and SCFA production was increased, soy protein intake in the morning may strongly alter the microbiota. Therefore, we analyzed the microbiota from feces. The M-Soy group showed tend to higher alpha-diversity for Simpson index than that shown by the other groups on AVE (Figure 8a). The PCoA of unweighted UniFrac distance showed that the beta-diversity of microbiota composition was significantly different between the Casein and the M-Soy group at ZT20. On the other hand, between the Casein and the E-Soy group, the beta-diversity of microbiota composition didn't show a significant difference at ZT4 (Figure 8b). The PCoA of unweighted UniFrac distance of all-time points was significantly different between the Casein and the M-Soy groups (statistic value = 2.478, $p = 0.002$), and relatively different between the Casein and the E-Soy groups (statistic value = 1.460, $p = 0.048$) (Figure 8c). For the relative abundance of microbes at the phylum level, *Bacteroidetes* was significantly higher in the M-Soy group than that in the Casein group on AVE. The relative abundance of *Firmicutes* was significantly lower in the E-Soy group than that in the Casein group at ZT20 and AVE (Figure 8d). In the genus level, the relative abundance of *Lactococcus* in the M-Soy group was significantly lower than that in the other groups at ZT20, and the relative abundance of *Lactococcus* in E-Soy group was significantly lower than that in the other groups at ZT4. On AVE, the relative abundance of *Lactococcus* in the M-Soy group was significantly lower and that in the E-Soy group tended to be lower than that in the Casein group. The relative abundance of *[Ruminococcus]* in the E-Soy group was significantly lower than that in the other groups at ZT20 (Figure 8e).

To infer the metagenome functional content based on microbial community profiles obtained from 16S rRNA gene sequences, we used PICRUSt. The microbial communities could be distinguished based on their functions. The KEGG pathways associated with glycine, serine, and threonine metabolism in the M-Soy group were significantly or tended to be upregulated at ZT12 and AVE compared to those in the Casein group (Figure 9a). The pathways associated with methane metabolism in the M-Soy group were significantly or tended to be upregulated compared to those in the Casein group at ZT12 and AVE, and compared to those in the E-Soy group at ZT20. The pathways associated with methane metabolism in the E-Soy group were significantly upregulated compared to those in the Casein group at ZT12. The pathways associated with nitrogen metabolism in the M-Soy group tended to be upregulated compared to those in the other groups on AVE (Figure 9b).

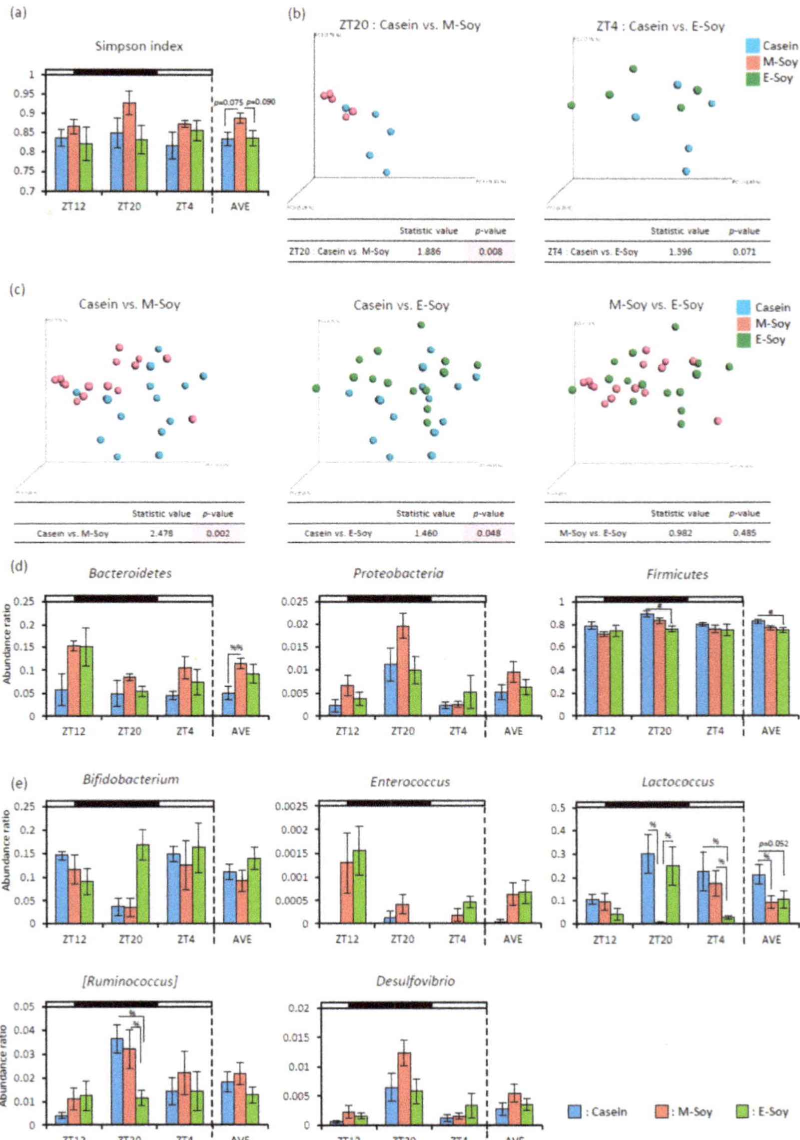

Figure 8. The effect of soy protein intake timing on microbiota. (**a**) Alpha-diversity about Simpson index (ZT12, ZT20, ZT4, and an average of three points) of mice that were kept in each feeding condition for two weeks. (**b**) Beta-diversity in comparison of Casein and M-Soy at ZT20, or Casein and E-Soy at ZT4, shortly after soy protein intake ($n = 5$). (**c**) Beta-diversity in comparison of Casein and M-Soy, Casein and E-Soy, or M-Soy and E-Soy. The PCoA plots of unweighted UniFrac distance metrics obtained from sequencing the 16S rRNA gene in feces ($n = 15$). (**d**) The relative abundance of microbes at the Phylum level, and (**e**) at the Genus level of mice that were kept in each feeding condition for two weeks. Data (**a**,**d**,**e**) are represented as mean ± SEM ($n = 5$). # $p < 0.05$ evaluated using one-way ANOVA with Tukey's post-hoc test. % $p < 0.05$, %% $p < 0.01$ the Kruskal-Wallis test with Dunn's post-hoc test. The tables in (**b**,**c**) indicate the result using PERMANOVA.

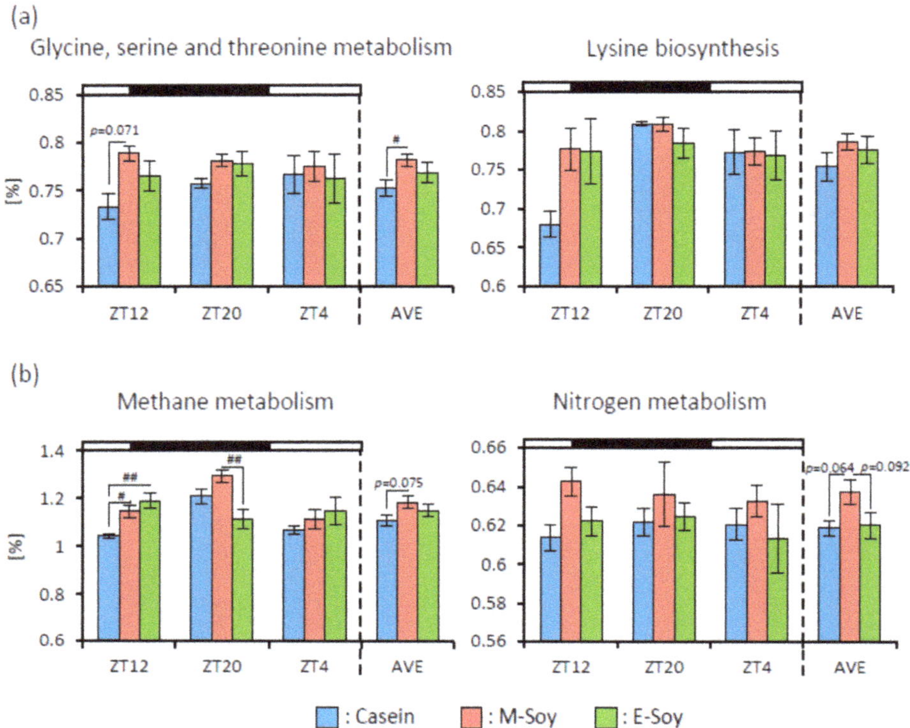

Figure 9. The functional predictions of microbial communities. (**a**) The functional predictions about categories related to "Amino Acid Metabolism" and (**b**) "Energy Metabolism" of microbial communities in mice that were kept under each feeding condition for two weeks (ZT12, ZT20, ZT4, and an average of three points). Data are represented as mean ± SEM (n = 5). # $p < 0.05$, ## $p < 0.01$ evaluated using one-way ANOVA with Tukey's post-hoc test.

4. Discussion

In this study, 10 days of soy protein intake reduced serum cholesterol and fatty acid synthesis related genes expression levels were experimented with in mice. In addition, soy protein changed microbial conditions and decreased cecal pH caused by SCFA production. Two weeks of soy protein feeding in the morning or evening resulted in a decrease in cecal pH and an increase in SCFA and microbiota diversity change after soy protein intake. In addition, soy protein intake in the morning may have a longer effect on SCFA production and cecal pH reduction than that of soy protein intake in the evening. It was suggested that soy protein might attenuate abnormality in gut microbiota effectively when taken in the morning rather than in the evening.

In experiment 1, soy protein reduced serum cholesterol level and fatty acid synthesis related genes such as *Acc1*, *Fasn*, and *Srebp1c* expression levels (Figure 2). It has already been reported that soy protein reduces serum cholesterol, TG, and fatty acid synthesis related genes expression levels [12,26]. In these reports, the effects of long-term soy protein intake were examined. In our study, mice were fed soy protein only for 10 days. However, serum cholesterol and fatty acid synthesis related genes expression levels were decreased. It was reported that SCFAs produced by the microbiota upregulated the expression of GLP-1 via activation of the MAPK pathway. GLP-1 induced reduction in mRNA expression of the fatty acid synthesis related genes [37–39]. Therefore, it is suggested that SCFA production by soy protein intake may be related to a reduction in fatty acid synthesis related gene expression. Since the current study focused on the effects of soy protein on microbiota, we measured only mRNA levels but not protein levels of the fatty acid synthesis related genes. To support these mRNA data, it might be necessary to measure the protein levels of enzyme activity.

Soy protein also changed the microbiota composition (Figure 4). Soy protein intake enhanced the production of SCFA, especially lactic and butyric acids, and decreased cecal pH (Figure 3). Previous studies also reported that soy protein causes greater diversity of microbiota than milk protein does, and their microbiota showed different compositions [12,25–27]. However, these previous studies examined the effects of long-term soy protein intake with bodyweight changes. Therefore, these reports could not exclude the possibility that soy protein provides anti-obesity effects and then improves the microbiota. In the current study, only 10 days of soy protein intake changed the microbiota without body weight changes. Our previous study reported that water-soluble dietary fiber changed microbiota in 10 days [24]. It is suggested that soy protein intake itself changes the microbiota and even short-term intake is effective in changing the microbiota.

In the present study, the relative abundances of various bacteria were changed by soy protein intake both at the phylum and genus levels (Figure 4c,d). These results were similar to previous reports [27,40]. *Firmicutes* are known as the obese factor and its relative abundance is higher in obese people [2]. In this experiment, *Firmicutes* was decreased by soy protein intake. This result suggests that soy protein may have an anti-obesity effect by decreasing *Firmicutes*. *Bifidobacterium* and *Enterococcus* are known to produce acetic and lactic acids through fermentative metabolism [41–43]. An increase in the relative abundance of *Bifidobacterium* and *Enterococcus* might cause an increase in acetic and lactic acids. It has also been reported that oral administration of *Bifidobacterium breve* to infants may prevent digestive disease [44]. Thus, Soy protein intake may prevent diseases of the intestines by increasing the relative abundance of *Bifidobacterium*. These changes in microbiota may be related to a decrease in serum lipid and hepatic fatty acid synthesis related gene expression. *Lactococcus* is known to produce lactic acid. However, *Lactococcus* was decreased by soy protein intake in this study. The reason may be because *Lactococcus* is the bacteria commonly found in raw milk, cheese, and other dairy products [45]. In this experiment, *Proteobacteria* and *Desulfovibrio* were increased by soy protein intake. It was reported that an increase in the abundance of *Proteobacteria* can change microbiota [46]. Further, *Desulfovibrio* in *Proteobacteria* can induce barrier dysfunction [47]. In the present study, we observed a greater diversity in the Soy group. However, some negative bacteria were increased. The reason underlying this increase in bacterial number and the associated mechanism under soy protein feeding condition remain unclear. In the future, this may be clarified by examining in detail the relationship between diet and microbiota.

In experiment 2, we examined the effect of soy protein intake timing (morning or evening) on microbiota. There were smaller effects of the diets on serum lipid, hepatic fatty acid, and cholesterol metabolism-related gene expression levels in experiment 2 than those in experiment 1 (Figure 2; Figure 6), because the amount of soy protein intake was smaller in experiment 2 than in experiment 1. In addition, it has been reported that mice under time-restricted feeding of 8 h per day were protected against obesity and hepatic steatosis, with improved energy expenditure [48]. In this experiment, mice were restricted in not only the amount but also the timing of feeding. Therefore, the feeding schedule itself may have powerfully reduced the serum lipid levels and hepatic gene expression of the fatty acid synthesis related genes, and then this protocol may mask the effects of soy protein.

At first, we compared the effects of soy protein on the microbiota at 8 h after feeding initiation in both the groups (ZT20 for the morning intake and ZT4 for the evening intake), because microbiota was altered shortly after water-soluble dietary fiber intake under two-meals-per-day schedule [24] and rapidly affected by dietary changes [9]. The cecal pH and the amount of lactic acid and butyric acid showed similar effects in both morning and evening groups (Figure 7). These data suggest that soy protein has beneficial effects on the microbiota in the morning intake or evening intake as compared to the casein intake. In addition, the previous feeding has a strong effect on microbial conditions. Since we collected samples every 8 h in this study, we can discuss the effects throughout the day by taking the average of 3-time points. The morning soy protein intake also showed lower cecal pH before soy protein intake (ZT12) and higher levels of lactic and butyric acids at a one day average

(Figure 7). Overall, it is suggested that soy protein intake in the morning may have long term effects on SCFA production and cecal pH reduction than that of the soy protein intake in the evening. Regarding microbiota, soy protein intake in the morning, not in the evening caused greater diversity on one day average (Figure 8a). In addition, microbiota varied to a greater extent by soy protein intake in the morning than that in the evening, shortly after soy protein intake (Figure 8b), and throughout the day (Figure 8c).

We do not know the detailed mechanism of such different effects on microbiota between morning intake and evening intake groups. The difference between soy protein intake in the morning and evening may be caused by a difference in fasting time before each diet. Two-meal-per-day feeding conditions were set close to a human feeding pattern in this study. In general, the fasting period before breakfast is the longest compared to other meal times in human eating habits. It has been reported that food signal after a long fasting strongly determines the peripheral clock phase [49]. It is also reported that consumption of water-soluble dietary fiber at breakfast which is after a longer fasting period, had greater effects on the microbiota [24]. Soy proteins contain proteins that are resistant to digestion [50], therefore resistant protein becomes the good food for microbiota just like water-soluble dietary fiber. In this study, fasting time was longer before the morning diet than before the evening diet. The intake of soy protein including resistant protein after long fasting may also have a greater impact on the microbiota. Furthermore, it has been reported that microbiota composition has circadian dynamics [21]. Even in mice that were fed the same diet, different reactions may occur if the microbiota is different in the morning or evening. The abdominal temperature, bowel movement, and endocrine system may influence the microbiota diversity [51–54], and these factors show circadian rhythm [55–57]. Thus, the different microbial reactions that were observed based on the soy protein intake in the morning or evening may be explained by the differences in the fasting time before each diet, microbiota circadian oscillations, and gut functional rhythm.

We used PICRUSt analysis to infer the functional capabilities of microbial communities. Soy protein intake upregulated the KEGG pathways associated with amino acid metabolism, especially glycine, serine, and threonine metabolism and lysine biosynthesis (Figure 5a). It has been reported that downregulating the pathway associated with amino acid metabolism has been observed in diarrheic calves and dogs and may be a feature of microbiota-associated diseases [58,59]. It is suggested that soy protein may improve microbiota. The pathways associated with methane and nitrogen metabolisms in the Soy group were significantly upregulated compared to those in the Casein group (Figure 5b). The upregulation of the nitrogen metabolism pathway by soy protein intake suggested that the indigestible component of soy protein might be metabolized by microbiota. The pathways associated with glycine, serine, and threonine metabolism, methane metabolism, and nitrogen metabolism were significantly upregulated in the M-Soy group compared to those in the Casein group (Figure 9). It is suggested that soy protein intake in the morning may have a stronger effect on upregulating these pathways than that in the evening. However, PICRUSt is only a predictor of metagenomic function. Therefore, metabolomic approaches are preferred in identifying factual changes in the metabolic function of microbiota by soy protein intake and its timing and identifying biomarkers for unstable gut microbiota.

5. Conclusions

In summary, the present experiments showed that soy protein intake and its timing affected the microbiota. The change in microbiota caused SCFA production and a decrease in cecal pH. In particular, soy protein may be effective in improving lipid metabolism and changing microbiota even with short-term intake. In addition, with respect to the timing of soy protein intake, morning intake may have relatively stronger effects on microbiota than evening intake would. This study provides evidence that soy protein intake and its timing are important factors that affect microbiota composition. To our knowledge, this is the first study to examine the effect of protein intake timing on microbiota and predict the functional profiles of microbial communities affected by soy protein. Therefore, our results are expected to be useful in designing future studies that may focus on the effects of foods or beverages in improving microbiota composition at different mealtimes and in providing important information for chrono-nutrition research.

Author Contributions: Research design and data analysis, K.T., H.S., S.S.; writing manuscript, K.T., H.S., S.S.; performing experiment, K.T., H.S., K.S., H.M. All authors have read and agreed to the published version of the manuscript.

Funding: This work was partially supported by the Council for Science, Technology, and Innovation, SIP, "Technologies for creating next-generation agriculture, forestry, and fisheries" (funding agency: Bio-oriented Technology Research Advancement Institution, NARO) (Shibata. S.), Japan Society for the Promotion of Science (JSPS) KAKENHI (A and Houga) (Shibata. S), and Fuji Foundation for Protein Research (Sasaki. H).

Conflicts of Interest: The authors declare no conflict of interest.

References

1. Marchesi, J.R.; Adams, D.H.; Fava, F.; Hermes, G.D.A.; Hirschfield, G.M.; Hold, G.; Quraishi, M.N.; Kinross, J.; Smidt, H.; Tuohy, K.M.; et al. The gut microbiota and host health: A new clinical frontier. *Gut* **2016**, *65*, 330–339. [CrossRef] [PubMed]
2. Ley, R.E.; Turnbaugh, P.J.; Klein, S.; Gordon, J.I. Human gut microbes associated with obesity. *Nature* **2006**, *444*, 1022–1023. [CrossRef] [PubMed]
3. Turnbaugh, P.J.; Ley, R.E.; Mahowald, M.A.; Magrini, V.; Mardis, E.R.; Gordon, J.I. An obesity-associated gut microbiome with increased capacity for energy harvest. *Nature* **2006**, *444*, 1027–1031. [CrossRef] [PubMed]
4. den Besten, G.; van Eunen, K.; Groen, A.K.; Venema, K.; Reijngoud, D.J.; Bakker, B.M. The role of short-chain fatty acids in the interplay between diet, gut microbiota, and host energy metabolism. *J. Lipid Res.* **2013**, *54*, 2325–2340. [CrossRef]
5. Morita, T.; Kasaoka, S.; Oh-hashi, A.; Ikai, M.; Numasaki, Y.; Kiriyama, S. Resistant proteins alter cecal short-chain fatty acid profiles in rats fed high amylose cornstarch. *J. Nutr.* **1998**, *128*, 1156–1164. [CrossRef]
6. Cook, S. Review article: Short chain fatty acids in health and disease. *Aliment. Pharmacol. Ther.* **1998**, *12*, 499–507. [CrossRef]
7. Duncan, S.H.; Louis, P.; Thomson, J.M.; Flint, H.J. The role of ph in determining the species composition of the human colonic microbiota. *Environ. Microbiol.* **2009**, *11*, 2112–2122. [CrossRef]
8. Sekirov, I.; Russell, S.L.; Antunes, L.C.M.; Finlay, B.B. Gut microbiota in health and disease. *Physiol. Rev.* **2010**, *90*, 859–904. [CrossRef]
9. David, L.A.; Maurice, C.F.; Carmody, R.N.; Gootenberg, D.B.; Button, J.E.; Wolfe, B.E.; Ling, A.V.; Devlin, A.S.; Varma, Y.; Fischbach, M.A.; et al. Diet rapidly and reproducibly alters the human gut microbiome. *Nature* **2014**, *505*, 559–563. [CrossRef]
10. Davila, A.-M.; Blachier, F.; Gotteland, M.; Andriamihaja, M.; Benetti, P.-H.; Sanz, Y.; Tomé, D. Intestinal luminal nitrogen metabolism: Role of the gut microbiota and consequences for the host. *Pharmacol. Res.* **2013**, *68*, 95–107. [CrossRef]
11. Hamer, H.M.; Preter, V.D.; Windey, K.; Verbeke, K. Functional analysis of colonic bacterial metabolism: Relevant to health? *Am. J. Physiol.—Gastrointest. Liver Physiol.* **2012**, *302*, G1–G9. [CrossRef] [PubMed]
12. Butteiger, D.N.; Hibberd, A.A.; McGraw, N.J.; Napawan, N.; Hall-Porter, J.M.; Krul, E.S. Soy protein compared with milk protein in a western diet increases gut microbial diversity and reduces serum lipids in golden syrian hamsters. *J. Nutr.* **2016**, *146*, 697–705. [CrossRef] [PubMed]
13. An, C.; Kuda, T.; Yazaki, T.; Takahashi, H.; Kimura, B. Caecal fermentation, putrefaction and microbiotas in rats fed milk casein, soy protein or fish meal. *Appl. Microbiol. Biotechnol.* **2014**, *98*, 2779–2787. [CrossRef] [PubMed]
14. Gachon, F.; Nagoshi, E.; Brown, S.A.; Ripperger, J.; Schibler, U. The mammalian circadian timing system: From gene expression to physiology. *Chromosoma* **2004**, *113*, 103–112. [CrossRef]
15. Narishige, S.; Kuwahara, M.; Shinozaki, A.; Okada, S.; Ikeda, Y.; Kamagata, M.; Tahara, Y.; Shibata, S. Effects of caffeine on circadian phase, amplitude and period evaluated in cells in vitro and peripheral organs in vivo in per2::Luciferase mice. *Br. J. Pharmacol.* **2014**, *171*, 5858–5869. [CrossRef]
16. Tahara, Y.; Shibata, S. Chrono-biology, chrono-pharmacology, and chrono-nutrition. *J. Pharmacol. Sci.* **2014**, *124*, 320–335. [CrossRef] [PubMed]

17. Furutani, A.; Ikeda, Y.; Itokawa, M.; Nagahama, H.; Ohtsu, T.; Furutani, N.; Kamagata, M.; Yang, Z.-H.; Hirasawa, A.; Tahara, Y.; et al. Fish oil accelerates diet-induced entrainment of the mouse peripheral clock via gpr120. *PLoS ONE* **2015**, *10*, e0132472. [CrossRef]
18. Tahara, Y.; Shiraishi, T.; Kikuchi, Y.; Haraguchi, A.; Kuriki, D.; Sasaki, H.; Motohashi, H.; Sakai, T.; Shibata, S. Entrainment of the mouse circadian clock by sub-acute physical and psychological stress. *Sci. Rep.* **2015**, *5*, 11417. [CrossRef]
19. Tahara, Y.; Shibata, S. Circadian rhythms of liver physiology and disease: Experimental and clinical evidence. *Nat. Rev. Gastroenterol. Hepatol.* **2016**, *13*, 217–226. [CrossRef] [PubMed]
20. Tahara, Y.; Yamazaki, M.; Sukigara, H.; Motohashi, H.; Sasaki, H.; Miyakawa, H.; Haraguchi, A.; Ikeda, Y.; Fukuda, S.; Shibata, S. Gut microbiota-derived short chain fatty acids induce circadian clock entrainment in mouse peripheral tissue. *Sci. Rep.* **2018**, *8*, 1395. [CrossRef]
21. Thaiss Christoph, A.; Zeevi, D.; Levy, M.; Zilberman-Schapira, G.; Suez, J.; Tengeler Anouk, C.; Abramson, L.; Katz Meirav, N.; Korem, T.; Zmora, N.; et al. Transkingdom control of microbiota diurnal oscillations promotes metabolic homeostasis. *Cell* **2014**, *159*, 514–529. [CrossRef]
22. Takahashi, M.; Ozaki, M.; Miyashita, M.; Fukazawa, M.; Nakaoka, T.; Wakisaka, T.; Matsui, Y.; Hibi, M.; Osaki, N.; Shibata, S. Effects of timing of acute catechin-rich green tea ingestion on postprandial glucose metabolism in healthy men. *J. Nutr. Biochem.* **2019**, *73*, 108221. [CrossRef]
23. Oishi, K.; Konishi, T.; Hashimoto, C.; Yamamoto, S.; Takahashi, Y.; Shiina, Y. Dietary fish oil differentially ameliorates high-fructose diet-induced hepatic steatosis and hyperlipidemia in mice depending on time of feeding. *J. Nutr. Biochem.* **2018**, *52*, 45–53. [CrossRef] [PubMed]
24. Sasaki, H.; Miyakawa, H.; Watanabe, A.; Nakayama, Y.; Lyu, Y.; Hama, K.; Shibata, S. Mice microbiota composition changes by inulin feeding with a long fasting period under a two-meals-per-day schedule. *Nutrients* **2019**, *11*, 2802. [CrossRef] [PubMed]
25. Aoyama, T.; Fukui, K.; Takamatsu, K.; Hashimoto, Y.; Yamamoto, T. Soy protein isolate and its hydrolysate reduce body fat of dietary obese rats and genetically obese mice (yellow kk). *Nutrition* **2000**, *16*, 349–354. [CrossRef]
26. Panasevich, M.R.; Schuster, C.M.; Phillips, K.E.; Meers, G.M.; Chintapalli, S.V.; Wankhade, U.D.; Shankar, K.; Butteiger, D.N.; Krul, E.S.; Thyfault, J.P.; et al. Soy compared with milk protein in a western diet changes fecal microbiota and decreases hepatic steatosis in obese oletf rats. *J. Nutr. Biochem.* **2017**, *46*, 125–136. [CrossRef]
27. Watanabe, K.; Igarashi, M.; Li, X.; Nakatani, A.; Miyamoto, J.; Inaba, Y.; Sutou, A.; Saito, T.; Sato, T.; Tachibana, N.; et al. Dietary soybean protein ameliorates high-fat diet-induced obesity by modifying the gut microbiota-dependent biotransformation of bile acids. *PLoS ONE* **2018**, *13*, e0202083. [CrossRef]
28. Kohno, M.; Hirotsuka, M.; Kito, M.; Matsuzawa, Y. Decreases in serum triacylglycerol and visceral fat mediated by dietary soybean & beta-conglycinin. *J. Atheroscler. Thromb.* **2006**, *13*, 247–255.
29. Yamazaki, T.; Kishimoto, K.; Miura, S.; Ezaki, O. Dietary β-conglycinin prevents fatty liver induced by a high-fat diet by a decrease in peroxisome proliferator-activated receptor γ2 protein. *J. Nutr. Biochem.* **2012**, *23*, 123–132. [CrossRef]
30. Feillet, C.A.; Ripperger, J.A.; Magnone, M.C.; Dulloo, A.; Albrecht, U.; Challet, E. Lack of food anticipation in per2 mutant mice. *Curr. Biol.* **2006**, *16*, 2016–2022. [CrossRef]
31. Hirao, A.; Nagahama, H.; Tsuboi, T.; Hirao, M.; Tahara, Y.; Shibata, S. Combination of starvation interval and food volume determines the phase of liver circadian rhythm in per2: Luc knock-in mice under two meals per day feeding. *Am. J. Physiol. -Gastrointest. Liver Physiol.* **2010**, *299*, G1045–G1053. [CrossRef] [PubMed]
32. Huazano, A.; López, M.G. Metabolism of short chain fatty acids in colon and faeces of mice after a supplementation of diets with agave fructans. *Lipid Metab.* **2013**, *8*, 163–182.
33. Nishijima, S.; Suda, W.; Oshima, K.; Kim, S.-W.; Hirose, Y.; Morita, H.; Hattori, M. The gut microbiome of healthy japanese and its microbial and functional uniqueness. *DNA Res.* **2016**, *23*, 125–133. [CrossRef] [PubMed]
34. Caporaso, J.G.; Kuczynski, J.; Stombaugh, J.; Bittinger, K.; Bushman, F.D.; Costello, E.K.; Fierer, N.; Peña, A.G.; Goodrich, J.K.; Gordon, J.I.; et al. Qiime allows analysis of high-throughput community sequencing data. *Nat. Methods* **2010**, *7*, 335–336. [CrossRef]
35. Edgar, R.C. Search and clustering orders of magnitude faster than blast. *Bioinformatics* **2010**, *26*, 2460–2461. [CrossRef]

36. Langille, M.G.; Zaneveld, J.; Caporaso, J.G.; McDonald, D.; Knights, D.; Reyes, J.A.; Clemente, J.C.; Burkepile, D.E.; Vega Thurber, R.L.; Knight, R.; et al. Predictive functional profiling of microbial communities using 16s rrna marker gene sequences. *Nat. Biotechnol.* **2013**, *31*, 814–821. [CrossRef]
37. den Besten, G.; Gerding, A.; van Dijk, T.H.; Ciapaite, J.; Bleeker, A.; van Eunen, K.; Havinga, R.; Groen, A.K.; Reijngoud, D.-J.; Bakker, B.M. Protection against the metabolic syndrome by guar gum-derived short-chain fatty acids depends on peroxisome proliferator-activated receptor γ and glucagon-like peptide-1. *PLoS ONE* **2015**, *10*, e0136364. [CrossRef]
38. Ding, X.; Saxena, N.K.; Lin, S.; Gupta, N.A.; Anania, F.A. Exendin-4, a glucagon-like protein-1 (glp-1) receptor agonist, reverses hepatic steatosis in ob/ob mice. *Hepatology* **2006**, *43*, 173–181. [CrossRef]
39. ZHANG, J.; Sun, Y.s.; Zhao, L.; Chen, T.; Fan, M.; Jiao, H.C.; Zhao, J.; Wang, X.; Li, F.c.; Li, H.; et al. Scfas-induced glp-1 secretion links the regulation of gut microbiome on hepatic lipogenesis in chickens. *BioRxiv* **2019**. [CrossRef]
40. Huang, H.; Krishnan, H.B.; Pham, Q.; Yu, L.L.; Wang, T.T.Y. Soy and gut microbiota: Interaction and implication for human health. *J. Agric. Food Chem.* **2016**, *64*, 8695–8709. [CrossRef]
41. Lee, K.-Y.; So, J.-S.; Heo, T.-R. Thin layer chromatographic determination of organic acids for rapid identification of bifidobacteria at genus level. *J. Microbiol. Methods* **2001**, *45*, 1–6. [CrossRef]
42. O'Callaghan, A.; van Sinderen, D. Bifidobacteria and their role as members of the human gut microbiota. *Front. Microbiol.* **2016**, *7*, 925. [CrossRef] [PubMed]
43. Hanchi, H.; Mottawea, W.; Sebei, K.; Hammami, R. The genus enterococcus: Between probiotic potential and safety concerns-an update. *Front. Microbiol.* **2018**, *9*, 1791. [CrossRef] [PubMed]
44. Wang, C.; Shoji, H.; Sato, H.; Nagata, S.; Ohtsuka, Y.; Shimizu, T.; Yamashiro, Y. Effects of oral administration of bifidobacterium breve on fecal lactic acid and short-chain fatty acids in low birth weight infants. *J. Pediatric Gastroenterol. Nutr.* **2007**, *44*, 252–257. [CrossRef] [PubMed]
45. Casalta, E.; Montel, M.-Ç. Safety assessment of dairy microorganisms: The lactococcus genus. *Int. J. Food Microbiol.* **2008**, *126*, 271–273. [CrossRef] [PubMed]
46. Shin, N.-R.; Whon, T.W.; Bae, J.-W. Proteobacteria: Microbial signature of dysbiosis in gut microbiota. *Trends Biotechnol.* **2015**, *33*, 496–503. [CrossRef] [PubMed]
47. Earley, H.; Lennon, G.; Balfe, A.; Kilcoyne, M.; Clyne, M.; Joshi, L.; Carrington, S.; Martin, S.T.; Coffey, J.C.; Winter, D.C.; et al. A preliminary study examining the binding capacity of akkermansia muciniphila and desulfovibrio spp., to colonic mucin in health and ulcerative colitis. *PLoS ONE* **2015**, *10*, e0135280. [CrossRef]
48. Hatori, M.; Vollmers, C.; Zarrinpar, A.; DiTacchio, L.; Bushong Eric, A.; Gill, S.; Leblanc, M.; Chaix, A.; Joens, M.; Fitzpatrick James, A.J.; et al. Time-restricted feeding without reducing caloric intake prevents metabolic diseases in mice fed a high-fat diet. *Cell Metab.* **2012**, *15*, 848–860. [CrossRef]
49. Kuroda, H.; Tahara, Y.; Saito, K.; Ohnishi, N.; Kubo, Y.; Seo, Y.; Otsuka, M.; Fuse, Y.; Ohura, Y.; Hirao, A.; et al. Meal frequency patterns determine the phase of mouse peripheral circadian clocks. *Sci. Rep.* **2012**, *2*, 711. [CrossRef]
50. Higaki, N.; Sato, K.; Suda, H.; Suzuka, T.; Komori, T.; Saeki, T.; Nakamura, Y.; Ohtsuki, K.; Iwami, K.; Kanamoto, R. Evidence for the existence of a soybean resistant protein that captures bile acid and stimulates its fecal excretion. *Biosci. Biotechnol. Biochem.* **2006**, *70*, 2844–2852. [CrossRef]
51. Worthmann, A.; John, C.; Rühlemann, M.C.; Baguhl, M.; Heinsen, F.-A.; Schaltenberg, N.; Heine, M.; Schlein, C.; Evangelakos, I.; Mineo, C.; et al. Cold-induced conversion of cholesterol to bile acids in mice shapes the gut microbiome and promotes adaptive thermogenesis. *Nat. Med.* **2017**, *23*, 839–849. [CrossRef] [PubMed]
52. Roager, H.M.; Hansen, L.B.S.; Bahl, M.I.; Frandsen, H.L.; Carvalho, V.; Gøbel, R.J.; Dalgaard, M.D.; Plichta, D.R.; Sparholt, M.H.; Vestergaard, H.; et al. Colonic transit time is related to bacterial metabolism and mucosal turnover in the gut. *Nat. Microbiol.* **2016**, *1*, 16093. [CrossRef] [PubMed]
53. Neuman, H.; Debelius, J.W.; Knight, R.; Koren, O. Microbial endocrinology: The interplay between the microbiota and the endocrine system. *FEMS Microbiol. Rev.* **2015**, *39*, 509–521. [CrossRef] [PubMed]
54. Galley, J.D.; Bailey, M.T. Impact of stressor exposure on the interplay between commensal microbiota and host inflammation. *Gut Microbes* **2014**, *5*, 390–396. [CrossRef] [PubMed]
55. Ohnishi, N.; Tahara, Y.; Kuriki, D.; Haraguchi, A.; Shibata, S. Warm water bath stimulates phase-shifts of the peripheral circadian clocks in per2: Luciferase mouse. *PLoS ONE* **2014**, *9*, e100272. [CrossRef]

56. Hoogerwerf, W.A. Role of clock genes in gastrointestinal motility. *Am. J. Physiol. Gastrointest. Liver Physiol.* **2010**, *299*, G549–G555. [CrossRef]
57. Pilorz, V.; Helfrich-Förster, C.; Oster, H. The role of the circadian clock system in physiology. *Pflügers Arch. Eur. J. Physiol.* **2018**, *470*, 227–239. [CrossRef]
58. Gomez, D.E.; Arroyo, L.G.; Costa, M.C.; Viel, L.; Weese, J.S. Characterization of the fecal bacterial microbiota of healthy and diarrheic dairy calves. *J. Vet. Intern. Med.* **2017**, *31*, 928–939. [CrossRef]
59. Minamoto, Y.; Otoni, C.C.; Steelman, S.M.; Büyükleblebici, O.; Steiner, J.M.; Jergens, A.E.; Suchodolski, J.S. Alteration of the fecal microbiota and serum metabolite profiles in dogs with idiopathic inflammatory bowel disease. *Gut Microbes* **2015**, *6*, 33–47. [CrossRef]

Article

Urinary TMAO Levels Are Associated with the Taxonomic Composition of the Gut Microbiota and with the Choline TMA-Lyase Gene (*cutC*) Harbored by Enterobacteriaceae

Alessandro Dalla Via [1], Giorgio Gargari [1], Valentina Taverniti [1], Greta Rondini [1], Ilaria Velardi [1], Veniero Gambaro [2], Giacomo Luca Visconti [2], Valerio De Vitis [1], Claudio Gardana [1], Enzio Ragg [1], Andrea Pinto [1], Patrizia Riso [1] and Simone Guglielmetti [2,*]

[1] Department of Food, Environmental and Nutritional Sciences (DeFENS), University of Milan, 20122 Milan, Italy; alessandro.dallavia@unimi.it (A.D.V.); gargari.g@gmail.com (G.G.); valentina.taverniti@unimi.it (V.T.); gremary8687@gmail.com (G.R.); i.velardi@hotmail.com (I.V.); valeriodevitis88@gmail.com (V.D.V.); claudio.gardana@unimi.it (C.G.); enzio.ragg@unimi.it (E.R.); andrea.pinto@unimi.it (A.P.); patrizia.riso@unimi.it (P.R.)

[2] Department of Pharmaceutical Sciences, University of Milan, 20122 Milan, Italy; veniero.gambaro@unimi.it (V.G.); giacomo.visconti@unimi.it (G.L.V.)

* Correspondence: simone.guglielmetti@unimi.it

Received: 7 December 2019; Accepted: 19 December 2019; Published: 25 December 2019

Abstract: Gut microbiota metabolization of dietary choline may promote atherosclerosis through trimethylamine (TMA), which is rapidly absorbed and converted in the liver to proatherogenic trimethylamine-N-oxide (TMAO). The aim of this study was to verify whether TMAO urinary levels may be associated with the fecal relative abundance of specific bacterial taxa and the bacterial choline TMA-lyase gene *cutC*. The analysis of sequences available in GenBank grouped the *cutC* gene into two main clusters, cut-Dd and cut-Kp. A quantitative real-time polymerase chain reaction (qPCR) protocol was developed to quantify *cutC* and was used with DNA isolated from three fecal samples collected weekly over the course of three consecutive weeks from 16 healthy adults. The same DNA was used for 16S rRNA gene profiling. Concomitantly, urine was used to quantify TMAO by ultra-performance liquid chromatography coupled with tandem mass spectrometry (UPLC-MS/MS). All samples were positive for *cutC* and TMAO. Correlation analysis showed that the cut-Kp gene cluster was significantly associated with *Enterobacteriaceae*. Linear mixed models revealed that urinary TMAO levels may be predicted by fecal cut-Kp and by 23 operational taxonomic units (OTUs). Most of the OTUs significantly associated with TMAO were also significantly associated with cut-Kp, confirming the possible relationship between these two factors. In conclusion, this preliminary method-development study suggests the existence of a relationship between TMAO excreted in urine, specific fecal bacterial OTUs, and a *cutC* subgroup ascribable to the choline-TMA conversion enzymes of *Enterobacteriaceae*.

Keywords: choline; trimethylamine; trimethylamine n-oxide; 16S rRNA gene profiling; qPCR; linear mixed models

---

1. Introduction

From infancy, the microorganisms colonizing the human gastrointestinal tract (GIT), collectively known as GIT microbiota, act as a "hidden" metabolic organ that exerts indispensable functions for the development and physiology of the human organism, such as the production of vitamins, modulation of the immune system, competitive exclusion toward exogenous pathogenic bacteria, xenobiotic detoxification, and production of short-chain fatty acids [1]. Nonetheless, detrimental activities have also been associated with gut commensal microorganisms, such as the production of carcinogens by the bacterial nitroreductases and azoreductases [2], or the conversion of primary bile acids to toxic compounds by the microbiota-associated enzyme cholesterol dehydrogenase and 7-α-dehydroxylase [3]. In addition, it was proposed that the intestinal bacterial enzymatic activities that produce trimethylamine (TMA) may promote atherosclerosis. TMA, in fact, is readily absorbed from the intestinal tract and, once in the liver, is converted into trimethylamine-N-oxide (TMAO) [4], whose plasma level has been identified as a metabolite strongly associated with atherosclerosis in a large case-control cohort for cardiovascular disease [5]. In particular, TMAO was proposed to promote atherogenesis by increasing cholesterol in macrophages and enhancing the accumulation of foam cells in artery walls [4,5]. Nonetheless, the literature has contradicted the role of TMAO, and recent studies have questioned its deleterious role in the cardiovascular system [6], suggesting, on the contrary, that TMAO could have protective functions [7,8].

Reportedly, a dominant contribution to the production of TMA in the gut comes from the microbial metabolism of diet-derived substrates such as carnitine- and choline-containing molecules [4,5,9]. Choline is an essential nutrient that is used by cells to synthesize membrane phospholipids. Furthermore, choline is the precursor of the neurotransmitter acetylcholine and a major source for methyl groups via its metabolite, trimethylglycine (betaine) [10]. The main dietary sources of the choline moiety, which is mostly present in food as lecithin (i.e., phosphatidylcholine), were reported to be eggs, liver, soybeans, and pork [11]. Although they are also present in numerous other foods [12], recent surveys in the USA indicated that choline may be underconsumed in specific populations (e.g., pregnant women and vegans) [13]. Based on the average observed choline intake in healthy European populations, a panel of the European Food Safety Authority set the adequate intake of choline at 400 mg/day [14].

Recent literature has suggested that the enhanced abundance of choline utilization genes in the intestinal microbiome is associated with increased TMA levels in the gut and, subsequently, with a higher hepatic production of TMAO. Proof of the importance of choline-derived TMA in the context of TMAO toxicity was recently provided by the study of Craciun and Balskus, in which the specific inhibition in mouse intestine of the microbial choline TMA-lyase (the primary enzymatic activity involved in the production of TMA from choline [15]) resulted in a significant reduction in plasma TMAO levels and recovery from dietary-induced platelet aggregation and thrombus formation [16].

Choline TMA-lyase is discontinuously distributed in bacterial taxa. Consequently, it was speculated that the phylogenetic composition of the microbiota is plausibly a poor predictor of the intestinal potential to convert choline into TMA [15,17,18]. However, in another study, the taxonomic structure of the gut microbiota was used to predict genes involved in choline metabolism [19] by means of PICRUSt, a bioinformatic tool used to infer the functional profiles of the microbial communities from 16S rRNA gene profiling data [20]. Although the toxicity of TMAO has been extensively investigated in the last 10 years, the association potentially existing among host TMAO levels, gut microbiota composition, and the intestinal microbial metabolization of choline has been only marginally considered. In this context, we developed a molecular protocol for the targeted quantification in the fecal microbiome of the bacterial gene *cutC* coding for the glycyl radical enzyme homolog choline TMA-lyase [15,21]. This protocol was applied to quantify the *cutC* gene abundance in the fecal samples collected at different time points from a group of healthy adults. Then, the obtained results were analyzed in comparison with the bacterial taxonomic composition and the urinary levels of TMAO concomitantly determined in the same population to deduce the potential association of excreted TMAO with gut microbial taxa and/or specific choline TMA-lyase enzymes.

2. Materials and Methods

2.1. Design and Use of Primers Targeting the cutC Gene

The primers used in polymerase chain reaction (PCR) for the amplification of the *cutC* gene were designed as follows. The GenBank database and Conserved Domain Database (CDD) at the National Center for Biotechnology Information (NCBI) were queried to select 52 nonredundant representative bacterial proteins of the choline trimethylamine-lyase protein family TIGR04394 (choline_CutC; EC Number 4.3.99.4), including the CutC enzymes of *Desulfovibrio desulfuricans* [4], and *Klebsiella pneumoniae* [22]. Then, the corresponding CDS nucleotide sequences of selected proteins were used to build a UPMGA tree upon ClustalW multiple alignments. According to the obtained dendrogram, sequences were clustered in two groups: One including the *cutC* sequence of *K. pneumoniae*, named cut-Kp, and one including the *cutC* sequence of *D. desulfuricans*, named cut-Dd (Supplementary Figure S1). Finally, a pair of primers was designed in the most conserved regions of each group of sequences: cut-Dd-F, 5′-CGTGTTGACCAGTACATGTA-3′ and cut-Dd-R 5′-GCTGGTAACCTGCGAAGAA-3′ (expected amplicon of 185 bp); cut-Kp-F, 5′-GATCTGACCTATCTGATTATGG-3′, and cut-Kp-R, 5′-TTGTGGAGCATCATCTTGAT-3′ (expected amplicon of 190 bp).

2.2. PCR Detection of cutC Gene in Single Strains

The two primer pairs designed as described above were used in endpoint PCR with the genomic DNA extracted from 64 bacterial strains (Table S1). Reaction mix was prepared in 25 μL, including 0.5 units of DreamTaq Polymerase (ThermoFisher, Fermentas, Waltham, MA, USA), 1× concentration of DreamTaq Polymerase Buffer (ThermoFisher, Fermentas,), 0.25 μM of each primer, 200 μM of deoxyribonucleotide triphosphate (dNTPs), and 0.5 mM of MgCl2. The PCR cycle program used was the following: Initial denaturation at 95 °C for 2 min, followed by 35 cycles of denaturation at 94 °C for 45 s, annealing at 58 °C for 45 s for the cut-Dd couple and 56 °C for 45 s for the cut-Kp couple, and extension at 72 °C for 20 s. A final extension of 7 min at 72 °C was then applied.

2.3. Detection of Choline-Utilization Activity in Single Strains

Bacterial strains were grown in the respective culture medium (reported in Table S1) for 48 h. Afterward, the biomasses were collected by centrifugation at 9500 g for 10 min. The cell pellets were then washed with sterile PBS and resuspended in fresh medium with the addition of 0.2% filter-sterilized choline. Bacteria were incubated at 37 °C for 48 h in glass tubes with screw cap. Afterward, supernatants were collected and used for mass spectrometry (MS) and nuclear magnetic resonance (NMR) analyses. The MS analyzes were performed by directly injecting 5 μL of diluted broth cultures after the removal of the bacterial cells by centrifugation and subsequent filtration with a 0.45-μm syringe filter. In detail, the broth cultures were analyzed in full scan in the range from 50 u to 400 u on an HR-MS Orbitrap model Exactive with a HESI-II probe for electrospray ionization (Thermo Scientific, San Jose, CA, USA). The resolution, gain control, mass tolerance, and maximum ion injection time was set to 50 K, 1E6, 2 ppm, and 100 ms, respectively. The MS data were processed using Xcalibur software (Thermo Scientific). Choline and TMA were used as reference standard. Choline and TMA were also directly detected in broth cultures by ^{1}H-NMR with a 60 MHz benchtop NMR spectrometer Spinsolve 60 Carbon Ultra, Magritek GmbH (Aachen, Germany).

2.4. Study Population

Study participants were recruited within the University campus. In total, four females and 12 males aged 21–45 (mean: 29.8 years) were enrolled (Table S2). The inclusion criteria were as follows: Healthy adult volunteers of both sexes who provided signed informed consent of their participation in the study. The exclusion criteria were as follows: Antibiotic consumption in the month preceding the start of the study, consumption of antacids or prokinetic gastrointestinal drugs, episodes of viral or bacterial enteritis in the two months prior to the study, episodes of gastric or duodenal ulcers in the previous five years, pregnancy or breastfeeding, recent history of alcohol abuse or suspected drug use, and any severe disease that may interfere with treatment. Ethical permission was granted by the University of Milan Ethics Committee (ref: opinion no. 37/16, 15 December 2016).

2.5. Collection of Fecal and Urine Samples

Three fecal sample were collected weekly over the course of three consecutive weeks from each volunteer. All the participants were asked to follow their regular diet during the three weeks. Concomitantly to the fecal sample, the volunteers provided 24-h urine collection.

Urine samples were collected over 24 h in sterile tanks and on the same days that fecal samples were been collected. The volume of collected urine was recorded in order to calculate the daily excretion of trimethylamine oxide (TMAO). Immediately after delivery, part of the urine samples was transferred in 10-mL sterile tubes and stored at −80 °C until analysis.

2.6. Analysis of cutC *Gene by Quantitative Real-Time PCR*

The cutC gene was quantified in fecal DNA with quantitative real-time PCR (qPCR) with both primer pairs, cut-Dd and cut-Kp. To this aim, DNA was extracted from feces using the kit PowerLyzer® PowerFecal® DNA Isolation Kit (MO BIO Laboratories, Inc.), starting from 0.25 ± 0.02 mg of sample according to the manufacturer's instructions. Primer pairs were tested with a gradient qPCR in a range of eight temperatures in order to find the most efficient annealing temperature using DNA of Streptococcus dysgalactiae 485 and Klebsiella sp. A1.2 as reference DNA. In addition, the amplification efficiency of the two pairs of primers was tested in qPCR experiments with six serial 1:3 dilutions of genomic DNA isolated from *Streptococcus dysgalactiae* 485, *Klebsiella* sp. A1.2, and human fecal metagenomic DNA. All DNA (bacterial and metagenomic) serial dilutions were tested with primer concentrations of 0.5 μM, 0.4 μM, and 0.3 μM. Efficiency curves were obtained with Bio-Rad software by setting samples as "standard" and obtaining a curve with efficiency (E) parameter and R2 value. Based on the results of these setup experiments, primers were then used at a final concentration of 0.5 μM, as with this concentration, we obtained an R2 value of 0.98. In addition, two randomly selected fecal DNA samples were tested at the different concentration by adding 70 ng, 50 ng, 25 ng, and 10 ng in qPCR reactions. Based on Ct value comparison between the different DNA concentrations, the *cutC* gene quantification was subsequently performed using 50 ng of total DNA. The reaction mix contained the SsoFast TM Eva-SuperGreen Supermix 2× (Bio-Rad Laboratories), deionized Milli-Q water (Millipore), and primers. All DNA samples (5 μL in each well) were tested in technical duplicate. The qPCR cycles employed were the following: Initial denaturation at 95 °C for 3 min, followed by 44 cycles of denaturation at 95 °C for 30 s, annealing at 58 °C (for cut-Dd primers) or 58.5 °C (for cut-Kp primers) for 30 s, and elongation at 72 °C for 5 s. A final denaturation ramp between 65 °C and 95 °C for 5 s was performed for the melting curve analysis. Moreover, specificity of qPCR reaction was confirmed by checking the presence of only one amplification and of the expected size in electrophoresis on a 2% agarose gel. A total of 48 fecal samples were analyzed. Each sample was analyzed with each primer set in duplicate. The $2^{-\Delta\Delta Ct}$ method was used for the relative quantification of *cutC* gene, using the EUB panbacterial primers [Muyzer] targeting the 16S rRNA gene as reference. Data were reported as relative increase of *cutC* copy number compared to the level of the sample that showed the highest significant Ct in qPCR set as 1.

2.7. Analysis of the Bacterial Taxonomic Composition of Fecal Samples

The bacterial community structure of the fecal microbiota was analyzed as described elsewhere [23,24], with DNA extracted from feces as described in Section 2.2. In brief, extracted DNA was analyzed through 16S rRNA gene profiling. Sequencing reads were generated at the Institute for Genome Sciences (University of Maryland, School of Medicine, Baltimore, MD, USA) with Illumina HiSeq 2500 rapid run sequencing of the V3–V4 variable region. Sequencing reads were equally distributed among the samples. Sequences were filtered and trimmed based on their quality.

We obtained a sequence length of 301 bp for both R1 and R2 sequences with an average quality score (Phred score) higher than 35. Sequencing reads were rarefied at 5000 per sample. Subsequently, sequence reads were analyzed through the bioinformatic pipeline Quantitative Insights into Microbial Ecology (QIIME) version 1.9.1 [25] with the GreenGenes database updated to version 13.5. The relative abundance of bacteria in each fecal sample was reported at the taxonomic levels of phylum, class, order, family, genus, and operational taxonomic units (OTUs). Sequence were deposited in the European Nucleotide Archive (ENA) of the European Bioinformatics Institute under accession code PRJEB34169.

2.8. TMAO Quantification in Urine Samples

TMAO levels in urine samples were determined by ultra-performance liquid chromatography coupled to tandem mass spectrometry (UPLC-MS/MS) (Waters Acquity UPLC system). The analysis method involved the use of a totally porous column with stationary C8 stable bond (Agilent Poroshell C8-SB) and a mobile phase consisting of a gradient acetonitrile and formate buffer (3 mM of ammonium formate and 0.1% formic acid). The UPLC system was equipped with a triple quadrupole detector, which allowed the development of a "multiple reaction monitoring" (MRM) method for the analysis of TMAO. In detail, once thawed at room temperature and after centrifugation at 6000 rpm for 5 min, 25 μL of urine sample were diluted in 950 μL of UPLC mobile phase (1/1 (*v*/*v*) acetonitrile/ultra-pure sterile water + 0.025% of formic acid), and 25 μL of deuterated internal standard solution (1 ppm, TMAO-d9, Spectra 2000) were used for the normalization of results [26]. The UPLC samples were prepared mixing 950 μL of mobile phase [1/1 (*v*/*v*) acetonitrile/ultra-pure sterile water + 0.025% formic acid), 25 μL of urine sample, and 25 μL of deuterium-labeled methyl d9-TMAO solution (1 ppm; Spectra 2000 S.r.l., Roma, Italy). Mobile phase: 1/1 (*v*/*v*) acetonitrile/ultra-pure sterile water + 0.025% of formic acid. The run time per sample was 8 min. Sample freezing and thawing or their prolonged storage at room temperature did not have an impact on the TMAO quantification. A triple set of working standards of TMAO (trimethylamine N-oxide dihydrate, Fluka) at concentrations of 5 ppm, 50 ppm, 100 ppm was prepared according to the method described above, replacing the 25 μL of urine sample with 25 μL of standard solution. The average response factor was used for calculation.

2.9. Statistical Analysis

Statistical analyses of data were carried out using R statistic software (version 3.4.2). Concerning *cutC* gene and TMAO data, intrasubject variability was defined "high" when variance among the three replicates results were higher than twice the median of all variances. Correlation analyses were performed using the Kendall and Spearman formula with the items specified in the text as predictors and dependent variables. Significance was set at $p \leq 0.05$, and mean differences in the range $0.05 < p < 0.10$ were accepted as trends. To find associations among TMAO levels, bacterial taxa relative abundance, and *cutC* gene abundance, the machine learning supervised linear mixed model (LMM) algorithm was used. In brief, the LMM was performed using "lmer" function in the "lme4" library [27]. All samples were used in the LMM analysis (n = 48), considering that three measurements were available for each subject. The Akaike's Information Criterion (AIC) was used to test the goodness of fit of the LMM. The AIC index/value depends on the ANOVA test results between two models: The model that considered the effect of the predictors and the null model.

3. Results

3.1. Distribution of the cutC Gene among Bacterial Taxa

According to the literature, the ability of intestinal bacteria to convert the choline moiety to TMA is primarily associated with a recently discovered choline utilization (cut) genetic region harboring the *cutC* gene, which encodes a glycyl radical enzyme catalyzing C–N bond cleavage [15,18]. For this reason, we designed primers specifically targeting the *cutC* gene. These primers were intended for quantitative PCR (qPCR) experiments, and we avoided the use of degenerations in their sequence.

In contrast, to target all putative *cutC* sequences identified in GenBank, we clustered the putative *cutC* genes into two groups (named Dd and Kp) according to sequence similarity (Supplementary Figure S1) and designed a pair of primers for each group in the most conserved sequence regions. Group Dd included putative *cutC* genes from *Firmicutes* (*Anaerococcus, Clostridium, Enterococcus, Streptococcus*), *Proteobacteria* (*Desulfotalea, Desulfovibrio, Enterobacter*), and *Actinobacteria* (*Olsenella*). Group Kp comprised putative *cutC* gene sequences from *Proteobacteria* (*Aeromonas, Enterobacter, Erwinia, Escherichia, Klebsiella, Pectobacterium, Pelobacter, Proteus, Providencia, Raoultella, Serratia*) and *Firmicutes* (*Desulfosporosinus, Enterococcus*).

Subsequently, the two primer sets were used in endpoint PCR reactions to test the presence of putative *cutC* genes within the genomic DNA isolated from the pure cultures of 64 bacterial strains. We obtained an amplicon of the expected size from seven strains. Specifically, strains *Streptococcus dysgalactiae* 485, *S. dysgalactiae* 486, and *S. dysgalactiae* A1.3 gave a band of the expected size with primers cut-Dd. In addition, strains *Enterococcus gilvus* MD179, *Enterococcus hirae* MD160, *Klebsiella oxytoca* MIMgr, and *Klebsiella* sp. MIMgr were positive with primers cut-Kp (Figure 1A,B). MS and NMR analyses revealed the ability to metabolize choline and produce TMA only for the same seven strains that resulted in positive PCR experiments (Figure 1C and Supplementary Figure S2).

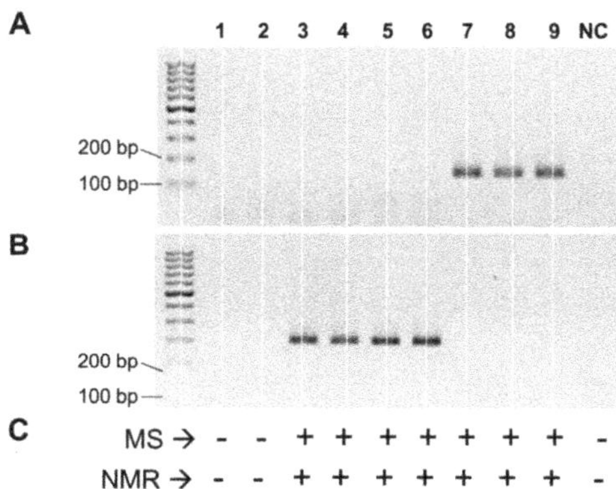

Figure 1. Detection of the choline-utilization activity in pure bacterial cultures. Panels (**A**,**B**) represent agarose gel resulting from end-point PCR with primers cut-Dd (**A**) and cut-Kp (**B**). Panel (**C**) summarizes the detection of TMA in cell-free broth by mass spectrometry (MS) and nuclear magnetic resonance (NMR); +, TMA detected; -, TMA not detected. Lanes: **1**, *Escherichia coli* 3.1; **2**, *Lactococcus garvieae* FMBgr; **3**, *Enterococcus gilvus* MD160; **4**, *Enterococcus hirae* MD179; **5**, *Klebsiella oxytoca* MIMgr; **6**, *Klebsiella* sp. A1.2; **7**, *Streptococcus dysgalactiae* 485; **8**, *Streptococcus dysgalactiae* 486; **9**, *Streptococcus dysgalactiae* A 1.2; NC, negative control (i.e., M17 broth incubated without bacteria).

3.2. Bacterial Taxonomic Structure of the Fecal Microbiota

The metagenomic DNA isolated from the feces collected at three time points from 16 healthy adults (n = 48) was used in 16S rRNA gene profiling experiments. A total of 12,588,795 filtered high-quality sequence reads were generated with an average of 13,340 ± 8677 (mean ± standard deviation; max-min 11,594–4570) per sample.

We failed to stratify samples according to the 16S rRNA gene profiling data, indicating that fecal bacterial community structure was homogeneous among samples and among subjects (Supplementary Figure S3). In addition, we also observed that the overall composition of the fecal microbiota in each subject remained mostly stable over the three collection time points (Supplementary Figure S3). Globally, 182 bacterial genera were estimated, with a minimum of 36 and a maximum of 98 genera per fecal sample. *Bacteroides* was the most prevalent genus, followed by four genera of the order *Clostridiales* (undefined *Ruminococcaceae*, undefined *Lachnospiraceae*, *Ruminococcus*, and *Faecalibacterium*)

(Supplementary Figure S4A). At the family level, most of the reads were ascribed to only three families, i.e., *Ruminococcaceae*, *Bacteroidaceae*, and *Lachnospiraceae* (Supplementary Figure S4B).

3.3. *Putative* cutC *Genes in Human Fecal Metagenomic DNA*

In order to investigate the presence of *cutC* genes in the human gut microbiome, the cut-Dd and cut-Kp primer sets were used in qPCR experiments using the same fecal metagenomic DNA as a template from healthy adults used for microbiota profiling. All analyzed fecal samples gave a positive signal in qPCR with both primer pairs (Figure 2). In general, cut-Kp was detected at a higher relative concentration than cut-Dd (median ΔΔCt of 5.33 and 0.85 for cut-Kp and cut-Dd, respectively) (Figure 2A,B). In addition, with both cut-Kp and cut-Dd, six volunteers out of 16 showed a variance among the three replicates that was higher than twice the median of all variances, indicating a higher intrasubject variability (Figure 2A,B).

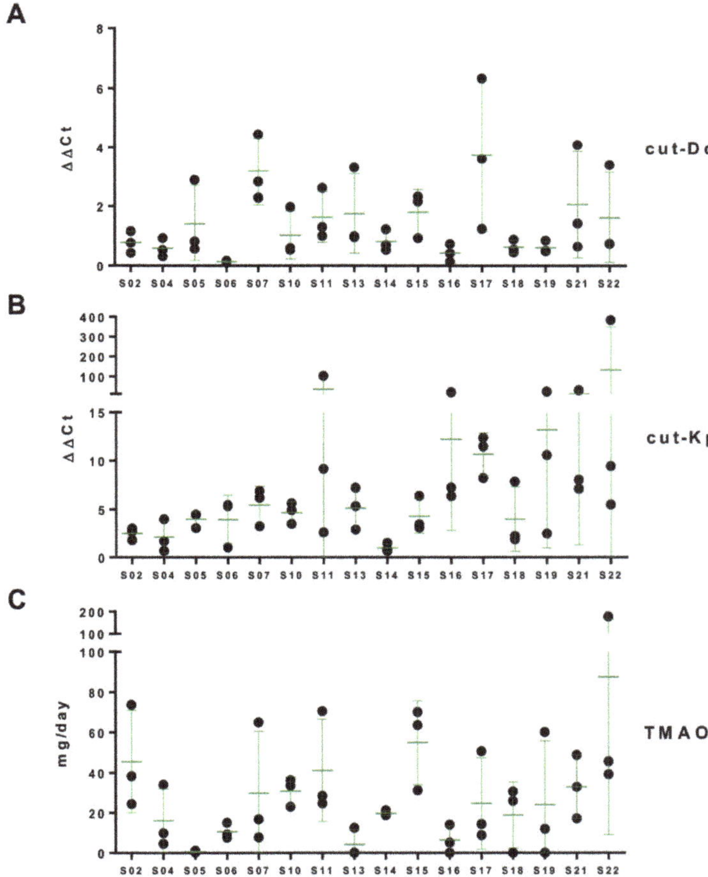

Figure 2. Fecal levels of the *cutC* gene and daily urinary excretion of trimethylamine-N-oxide (TMAO). The relative abundance of *cutC* was determined by quantitative real-time polymerase chain reaction (qPCR) with the primer pair cut-Dd-F/R (panel **A**) and cut-Kp-F/R (**B**). The TMAO concentration was determined by ultra-performance liquid chromatography coupled with tandem mass spectrometry (UPLC-MS/MS) in urine collected over 24 h (**C**). Green bars represent the mean ± standard deviation of three measurements per subject.

Subsequently, we performed correlation analyses between the *cutC* abundances determined with qPCR and the 16S rRNA gene profiling data to find potential relationships between the choline TMA-lyase genes and specific bacterial taxa of the fecal microbiota. To this end, we used the median relative abundance of bacterial taxa in fecal samples as predictors, whereas the dependent variables considered were the median abundances of cut-Dd and cut-Kp determined by qPCR per subject. We found that cut-Dd was positively correlated with taxa belonging to the phylum *Firmicutes*, including an undefined *Mogibacteriaceae* genus, *Oscillospira*, and the family *Christensenellaceae*. On the contrary, cut-Dd was negatively correlated with the *Firmicutes* order *Bacillales*, the *Firmicutes* genus *Streptococcus*, and the *Proteobacteria* genus *Haemophilus* (Supplementary Figure S5). Conversely, cut-Kp was positively associated with *Proteobacteria*. In particular, inside this phylum, a significant correlation was found with the family *Enterobacteriaceae* (Supplementary Figure S5).

3.4. Daily Urinary Excretion of TMAO

Subjects were asked to collect 24-h urine specimens the same days when the fecal samples were taken. Then, the levels of TMAO were quantified by UPLC-MS in all urine samples, revealing wide variability among the investigated healthy adults, with levels of urinary TMAO excretion ranging from less than 1 mg to more than 175 mg per day (Figure 2). We also observed an evident intrasubject variability in five volunteers whose TMAO excretion showed a variance among the three replicates that was higher than twice the median of all variances (Figure 2C). In particular, four out of the five volunteers with wide intrasubject variability (i.e., S07, S11, S19, and S22) were found to possess high intrasubject variability for *cutC* gene levels determined in qPCR experiments (Figure 2).

3.5. Associations among Urinary TMAO, Fecal cutC, and Fecal Bacterial Taxa

A linear mixed model was used to infer potential significant relationships among the datasets collected from volunteers at the three time points considered (Figure 3). TMAO was significantly associated with the cut-Kp/cut-Dd synergy ($p < 0.001$). Furthermore, studying the association of the single *cut* gene types, we observed that the relationship with TMAO was mainly determined by cut-Kp (Figure 3). In addition, we found a significant association between TMAO and 23 operational taxonomic units (OTUs). Conversely, cut-Kp and cut-Dd were significantly associated with 18 and eight OTUs, respectively. Notably, most of the OTUs that were significantly associated with cut-Kp (i.e., 15 out of 18) were also associated with TMAO, confirming the relationship between these two variables. Nine of the identified OTUs belonged to the phylum *Bacteroidetes*, while the remaining 21 were ascribed to *Firmicutes*. In addition, 80% of the OTUs (n = 24) belonged to only three families: *Bacteroidaceae*, *Lachnospiraceae*, and *Ruminococcaceae*. In particular, the most significant association (i.e., $p < 0.001$) referred to *Bacteroides caccae*, an undefined *Lachnospiraceae* genus, and several undefined *Ruminoccaceae* species (for TMAO and cut-Kp), *Bacteroides fragilis*, and an undefined *Clostridiales* species (for cut-Kp only) and an *Oscillospira* species (for cut-Dd) (Figure 3).

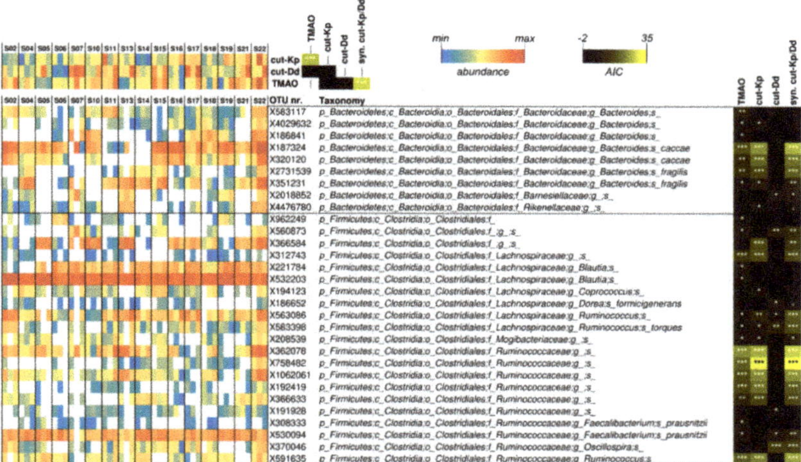

Figure 3. Analysis of the associations among fecal *cutC* gene abundances, fecal bacterial operational taxonomic units (OTUs), and urinary excreted TMAO carried out through a linear mixed model (LMM). Only OTUs that showed a significant association with *cutC* or TMAO are reported. The heatmap on the right represents TMAO levels, and *cutC* gene and OTU relative abundances. White boxes in the blue-yellow-red heatmap indicate that the OTU was not detected in that specific sample. The taxonomic lineage of each taxon is shown: p, phylum; c, class; o, order; f, family; g, genus; s, species. The black-yellow heatmap represents the Akaike's information criterion (AIC) values of the LMM analysis. Asterisks indicate significant associations: * $p < 0.05$; ** $p < 0.01$; *** $p < 0.001$; $^{+}$, $p < 0.1$. syn. cut-Kp/Dd = synergy between cut-Kp and cut-Dd in LMM analysis.

4. Discussion

A growing number of studies have linked host TMAO levels to different diseases or prepathological metabolic states [28,29]. Conversely, TMAO has also been proposed as a beneficial factor that may promote protein stabilization and protect cells from osmotic and hydrostatic stresses according to a compensatory response mechanism [30]. The biological role of TMAO is therefore still debated. Nonetheless, a growing number of scientific studies have suggested that this molecule may play an important role in health and diseases [6].

It has been suggested that an important contribution to the hepatic production of TMAO is given by the TMA produced in the gut by microbial degradation of TMA-containing dietary molecules [31]. In particular, TMAO levels and their physiological consequences were shown to be significantly affected by the TMA derived from choline [15]. In this context, we studied the levels of TMAO excreted daily with urine, the composition of the intestinal microbiota, and the abundance of the choline TMA-lyase gene *cutC* in a group of healthy adult subjects with an Italian dietary pattern. The aim of this observational study was to verify whether TMAO levels excreted with the urine might be associated with the relative abundance of specific bacterial taxa and the bacterial gene *cutC* in feces. Literature focusing on the relationship among these three elements, particularly in non-diseased populations, is limited and partially contradicting [15,17–19].

The gene *cutC*, encoding the lyase enzyme essential for the conversion of choline into TMA [32], is not evenly distributed across bacterial taxa due to gene loss and horizontal gene transfer events that differently involve strains within the same species [15,18,33]. Therefore, predicting the choline degradation potential of a microbial ecosystem solely based on the taxonomic composition has many intrinsic limitations. The use of primers selectively targeting a specific enzymatic conserved domain may overcome this problem, permitting the selective quantification of the abundance of a gene coding for a specific enzymatic activity in the metagenomic DNA. A similar approach was used by Martinez-Del Campo et al., who designed degenerate primers for the PCR amplification of the *cutC* gene from fecal metagenomic DNA and single strains [18]. The use of degenerate primers was necessitated by the fact that the CutC protein possesses sequence heterogeneity. In particular, Martinez-Del Campo et al. showed that the amino acid sequences deduced from the predicted bacterial *cutC* genes can be clustered into two groups (clades 1 and 2, [18]), which correspond to the CutC types I and II identified by Jameson et al. within a neighbor-joining phylogenetic tree constructed from amino acid sequences of glycyl radical enzymes [32]. The same result was found in our study by generating a distance tree based on the nucleotide sequences of putative *cutC* genes (Supplementary Figure S1). In particular, cluster cut-Dd corresponded to clade 1 and CutC type I, whereas cut-Kp included sequences coding for putative proteins found in clade 2 and CutC type II reported by the authors of [18] and [32], respectively.

For this reason, we developed two nondegenerate primer pairs located at the level of the catalytic site of the encoded enzyme that were useful for the amplification in (q)PCR experiments of the two clusters of the gene *cutC*.

When the two primer sets were used with the DNA of single strains, the only positive amplification signals were obtained with the bacteria that demonstrated the ability to metabolize choline in the biotransformation assay and produce TMA, confirming the suitability of these molecular probes to target choline-TMA-converting bacteria. Specifically, the bacterial strains identified here as able to degrade choline to TMA include species previously confirmed to exert this conversion, such as *Streptococcus dysgalactiae* [18]. In addition, we found *Klebsiella oxytoca*, which was reported to harbor a putative cut gene cluster [34], but has never been confirmed phenotypically. We also identified two positive *Enterococcus* strains. Reportedly, TMA production from choline has also been described for some enterococci, but not for the species *E. gilvus*, which is often isolated from food matrices, including meat, milk, and cheeses [35,36], and for the zoonotic pathogen *E. hirae* [37].

The qPCR experiments conducted showed that putative bacterial *cutC* genes were present in the fecal samples of all healthy adult subjects investigated. The high prevalence of this bacterial gene in the human gut microbiome was reported in a previous study, in which the presence of *cutC* homologs was observed in 96.6% of the assembled stool metagenomes of healthy individuals from the Human Microbiome Project (HMP) [18].

Reportedly, most of the TMA produced in the gut is absorbed into the portal circulation by passive diffusion [38]. Then, approximately 95% of the absorbed TMA is oxidized in the liver by flavin monooxygenases and excreted in the urine within 24 h [31,39]. Therefore, in this study, we performed a quantification of TMAO levels in urine samples obtained by 24-h collection.

The data presented here revealed a marked variability of both *cutC* and TMAO levels over the three time points considered in approximately 40% of volunteers. This instability was plausibly due to the variability of the daily food consumption of each subject. In this study, volunteers were free to follow their usual diet. Therefore, the analysis of multiple time points at approximately one-week intervals was useful to address the observed temporal instability of these parameters. To the best of our knowledge, this is the first work to report the stability of intestinal *cutC* and urinary TMAO levels over time.

This study has several limitations:

1. First, we quantified the abundance of a gene of the intestinal microbiome without considering if and how much this gene was expressed. This could therefore limit the possibility of associating the abundance of this gene with its product.
2. Furthermore, the production of TMA, in addition to the presence of the bacterial gene that allows its production (*cutC*), depends on the availability of the choline substrate, which mainly comes from the diet.
3. Nonetheless, the contribution to the TMA produced in the intestine and, consequently, to the TMAO generated in the liver, derives from different chemical moieties (mainly choline, betaine, and carnitine) and includes different microbial metabolic pathways, such as those involving the

carnitine monooxygenase CntAB and the glycine betaine reductase GrdH, in addition to the choline TMA-lyase CutC [40].

4. In addition, TMAO urinary levels may also depend on host factors that may largely vary from subject to subject, such as (i) the gut-to-blood barrier permeability to TMA [41], (ii) the oxidation of TMA in the liver by flavin monooxygenase [5], and the kidney function [42].
5. Finally, TMAO can also be ingested directly from foods such as fish and seafood, which are naturally rich in this molecule [43].

However, despite the limitations described above, this study showed that changes in urine TMAO levels are associated with changes in the fecal abundance of the *cutC* gene and variations in the relative abundance of several bacterial taxonomic units of the fecal microbiota. In particular, TMAO was significantly associated with the levels of a specific subcategory of the *cutC* gene, which we named cut-Kp here. This result could be explained by the relative abundance of cut-Kp, which, by qPCR results, was approximately six-times higher than that of cut-Dd. According to correlation analysis, the most important contribution to cut-Kp gene abundance is provided by *Proteobacteria*, particularly by *Enterobacteriaceae*. This result is supported by the fact that cut-Kp has been quantified with primers designed on a cluster of gene sequences having the *cutC* of the *Enterobacteriaceae* species *K. pneumoniae* as a reference. Reportedly, the analysis of human gut metagenomes revealed a high proportion of the genera *Klebsiella* and *Escherichia*, which harbor three potential TMA-producing pathways, suggesting the importance of these bacteria for TMA cycling in the human gut [44].

Most OTUs that were found to be significantly associated with TMAO also had cut-Kp, confirming the relationship between TMAO and cut-Kp levels. A few OTUs were also associated with cut-Dd. All the taxonomic units associated with TMAO and *cutC* belong to only two taxonomic orders, *Bacteroidales* and *Clostridiales*. In particular, almost all the OTUs are attributable to only three families: *Bacteroidaceae*, *Lachnospiraceae*, and *Ruminococcaceae*. Notably, these families have been identified as the most metabolically active bacteria of the human microbiota and play a dominant role in the colonic fermentation of dietary fibers [45,46]. Reportedly, many of these bacteria do not display choline-utilization activities (e.g., cut genes have never been identified in *Bacteroidetes* and *Faecalibacterium*). Nonetheless, we can hypothesize an indirect association of these bacteria with *cutC* and TMAO based on the speculation that the higher presence of these bacteria might determine a greater utilization of the available nutritional sources in the colon, reducing substrates for the remaining bacterial communities. The latter may then receive selective pressure for the expansion of the activities related to the metabolization of the residual energy and carbon sources such as choline, resulting in increased TMA production.

5. Conclusions

Here, we described the results of a preliminary method-development study, which suggests the existence of a relationship between the levels of TMAO excreted in urine, some intestinal taxonomic groups belonging to the most active bacterial families of the colonic microbiota, and a subgroup of the *cutC* gene ascribable to the choline-TMA conversion enzymes of *Enterobacteriaceae*, named cut-Kp, whose relative abundance can be determined with the qPCR protocol developed in this study. Nonetheless, considering the limitations listed above, particularly concerning dietary intake, it is plausible to hypothesize that the results of this study may vary in other populations.

Supplementary Materials: The following are available online at http://www.mdpi.com/2072-6643/12/1/62/s1, Figure S1: UPGMA hierarchical clustering based on ClustalW alignment of amino acid sequences of the choline trimethylamine lyase CutC. Figure S2. Verification of choline utilization and TMA production by single bacterial strains. Figure S3: Bacterial community structure of fecal samples. Figure S4: Tukey box and whisker plots representing the most abundant genera (A) and families (B) detected by 16S rRNA gene profiling in fecal samples collected from the adult volunteers participating in this study. Figure S5: Correlations among the fecal relative abundances of the choline TMA-lyase gene *cutC* and bacterial taxa. Table S1: Bacterial strains used for the screening of choline utilization activity. Table S2: Basic characteristics of the study participants.

Author Contributions: Conceptualization, S.G. and P.R.; methodology, S.G., A.D.V., V.T., I.V., V.G., G.R., E.R., C.G., V.D.V. and A.P.; formal analysis, S.G. and G.G; investigation, A.D.V., V.T., G.R., I.V., G.L.V. and A.P.; resources, S.G., V.G. and A.P.; data curation, S.G., A.D.V., V.T., G.G., and A.P.; writing—original draft preparation, S.G.; writing—review and editing, all authors; visualization, S.G., G.G. and A.D.V.; supervision, S.G, V.T. and P.R.; project administration, S.G.; funding acquisition, S.G., P.R. and V.G. All authors have read and agreed to the published version of the manuscript.

Funding: This research was partially funded by the University of Milan Funding "Linea 2-2016", PRISM Project. We acknowledge the European Joint Programming Initiative "A Healthy Diet for a Healthy Life" (JPI-HDHL—http://www.healthydietforhealthylife.eu/) and Mipaaft (Italy; D.M. 8245/7303/2016) for contributing to the grant awarded to G.G.

Acknowledgments: We thank Renata Piccinini for kindly providing the *S. dysgalactiae* strains.

Conflicts of Interest: The authors declare no conflict of interest.

References

1. Mohajeri, M.H.; Brummer, R.J.M.; Rastall, R.A.; Weersma, R.K.; Harmsen, H.J.M.; Faas, M.; Eggersdorfer, M. The role of the microbiome for human health: From basic science to clinical applications. *Eur. J. Nutr.* **2018**, *57*, 1–14. [CrossRef] [PubMed]
2. Claus, S.P.; Guillou, H.; Ellero-Simatos, S. The gut microbiota: A major player in the toxicity of environmental pollutants? *NPJ Biofilms Microbiomes* **2016**, *2*, 16003. [CrossRef] [PubMed]
3. Ridlon, J.M.; Harris, S.C.; Bhowmik, S.; Kang, D.J.; Hylemon, P.B. Consequences of bile salt biotransformations by intestinal bacteria. *Gut Microbes* **2016**, *7*, 22–39. [CrossRef] [PubMed]
4. Tang, W.H.; Wang, Z.; Levison, B.S.; Koeth, R.A.; Britt, E.B.; Fu, X.; Wu, Y.; Hazen, S.L. Intestinal microbial metabolism of phosphatidylcholine and cardiovascular risk. *N. Engl. J. Med.* **2013**, *368*, 1575–1584. [CrossRef] [PubMed]
5. Wang, Z.; Klipfell, E.; Bennett, B.J.; Koeth, R.; Levison, B.S.; Dugar, B.; Feldstein, A.E.; Britt, E.B.; Fu, X.; Chung, Y.M.; et al. Gut flora metabolism of phosphatidylcholine promotes cardiovascular disease. *Nature* **2011**, *472*, 57–63. [CrossRef]
6. Nowinski, A.; Ufnal, M. Trimethylamine N-oxide: A harmful, protective or diagnostic marker in lifestyle diseases? *Nutrition* **2018**, *46*, 7–12. [CrossRef]
7. Huc, T.; Drapala, A.; Gawrys, M.; Konop, M.; Bielinska, K.; Zaorska, E.; Samborowska, E.; Wyczalkowska-Tomasik, A.; Paczek, L.; Dadlez, M.; et al. Chronic, low-dose TMAO treatment reduces diastolic dysfunction and heart fibrosis in hypertensive rats. *Am. J. Physiol. Heart Circ. Physiol.* **2018**, *315*, H1805–H1820. [CrossRef]
8. Collins, H.L.; Drazul-Schrader, D.; Sulpizio, A.C.; Koster, P.D.; Williamson, Y.; Adelman, S.J.; Owen, K.; Sanli, T.; Bellamine, A. L-Carnitine intake and high trimethylamine N-oxide plasma levels correlate with low aortic lesions in ApoE(-/-) transgenic mice expressing CETP. *Atherosclerosis* **2016**, *244*, 29–37. [CrossRef]
9. Koeth, R.A.; Wang, Z.; Levison, B.S.; Buffa, J.A.; Org, E.; Sheehy, B.T.; Britt, E.B.; Fu, X.; Wu, Y.; Li, L.; et al. Intestinal microbiota metabolism of L-carnitine, a nutrient in red meat, promotes atherosclerosis. *Nat. Med.* **2013**, *19*, 576–585. [CrossRef]
10. Zeisel, S.H.; Blusztajn, J.K. Choline and human nutrition. *Annu. Rev. Nutr.* **1994**, *14*, 269–296. [CrossRef]
11. Zeisel, S.H.; Mar, M.H.; Howe, J.C.; Holden, J.M. Concentrations of choline-containing compounds and betaine in common foods. *J. Nutr.* **2003**, *133*, 1302–1307. [CrossRef] [PubMed]
12. Wiedeman, A.M.; Barr, S.I.; Green, T.J.; Xu, Z.; Innis, S.M.; Kitts, D.D. Dietary choline intake: Current state of knowledge across the life cycle. *Nutrients* **2018**, *10*, 1513. [CrossRef] [PubMed]
13. Wallace, T.C.; Blusztajn, J.K.; Caudill, M.A.; Klatt, K.C.; Natker, E.; Zeisel, S.H.; Zelman, K.M. Choline: The underconsumed and underappreciated essential nutrient. *Nutr. Today* **2018**, *53*, 240–253. [CrossRef] [PubMed]
14. EFSA Panel on Dietetic Products, Nutrition and Allergies (NDA). Dietary reference values for choline. *EFSA J.* **2016**, *14*, 70.
15. Craciun, S.; Balskus, E.P. Microbial conversion of choline to trimethylamine requires a glycyl radical enzyme. *Proc. Natl. Acad. Sci. USA* **2012**, *109*, 21307–21312. [CrossRef] [PubMed]

16. Roberts, A.B.; Gu, X.; Buffa, J.A.; Hurd, A.G.; Wang, Z.; Zhu, W.; Gupta, N.; Skye, S.M.; Cody, D.B.; Levison, B.S.; et al. Development of a gut microbe-targeted nonlethal therapeutic to inhibit thrombosis potential. *Nat. Med.* **2018**, *24*, 1407–1417. [CrossRef]
17. Rath, S.; Heidrich, B.; Pieper, D.H.; Vital, M. Uncovering the trimethylamine-producing bacteria of the human gut microbiota. *Microbiome* **2017**, *5*, 54. [CrossRef]
18. Martinez-del Campo, A.; Bodea, S.; Hamer, H.A.; Marks, J.A.; Haiser, H.J.; Turnbaugh, P.J.; Balskus, E.P. Characterization and detection of a widely distributed gene cluster that predicts anaerobic choline utilization by human gut bacteria. *mBio* **2015**, *6*, e00042-15. [CrossRef]
19. Xu, K.Y.; Xia, G.H.; Lu, J.Q.; Chen, M.X.; Zhen, X.; Wang, S.; You, C.; Nie, J.; Zhou, H.W.; Yin, J. Impaired renal function and dysbiosis of gut microbiota contribute to increased trimethylamine-N-oxide in chronic kidney disease patients. *Sci. Rep.* **2017**, *7*, 1445. [CrossRef]
20. Langille, M.G.; Zaneveld, J.; Caporaso, J.G.; McDonald, D.; Knights, D.; Reyes, J.A.; Clemente, J.C.; Burkepile, D.E.; Vega Thurber, R.L.; Knight, R.; et al. Predictive functional profiling of microbial communities using 16S rRNA marker gene sequences. *Nat. Biotechnol.* **2013**, *31*, 814–821. [CrossRef]
21. Craciun, S.; Marks, J.A.; Balskus, E.P. Characterization of choline trimethylamine-lyase expands the chemistry of glycyl radical enzymes. *ACS Chem. Biol.* **2014**, *9*, 1408–1413. [CrossRef] [PubMed]
22. Kalnins, G.; Kuka, J.; Grinberga, S.; Makrecka-Kuka, M.; Liepinsh, E.; Dambrova, M.; Tars, K. Structure and function of CutC choline lyase from human microbiota bacterium klebsiella pneumoniae. *J. Biol. Chem.* **2015**, *290*, 21732–21740. [CrossRef] [PubMed]
23. Cattaneo, C.; Gargari, G.; Koirala, R.; Laureati, M.; Riso, P.; Guglielmetti, S.; Pagliarini, E. New insights into the relationship between taste perception and oral microbiota composition. *Sci. Rep.* **2019**, *9*, 3549. [CrossRef] [PubMed]
24. Gargari, G.; Taverniti, V.; Gardana, C.; Cremon, C.; Canducci, F.; Pagano, I.; Barbaro, M.R.; Bellacosa, L.; Castellazzi, A.M.; Valsecchi, C.; et al. Fecal clostridiales distribution and short-chain fatty acids reflect bowel habits in irritable bowel syndrome. *Environ. Microbiol.* **2018**, *20*, 3201–3213. [CrossRef] [PubMed]
25. Caporaso, J.G.; Kuczynski, J.; Stombaugh, J.; Bittinger, K.; Bushman, F.D.; Costello, E.K.; Fierer, N.; Pena, A.G.; Goodrich, J.K.; Gordon, J.I.; et al. QIIME allows analysis of high-throughput community sequencing data. *Nat. Methods* **2010**, *7*, 335–336. [CrossRef] [PubMed]
26. Johnson, D.W. A flow injection electrospray ionization tandem mass spectrometric method for the simultaneous measurement of trimethylamine and trimethylamine N-oxide in urine. *J. Mass Spectrom.* **2008**, *43*, 495–499. [CrossRef]
27. Bates, D.; Mächler, M.; Bolker, B.; Walker, S. Fitting linear mixed-effects models using lme4. *J. Stat. Softw. Artic.* **2015**, *67*, 1–48. [CrossRef]
28. Chhibber-Goel, J.; Singhal, V.; Parakh, N.; Bhargava, B.; Sharma, A. The metabolite Trimethylamine-N-Oxide is an emergent biomarker of human health. *Curr. Med. Chem.* **2017**, *24*, 3942–3953. [CrossRef]
29. Janeiro, M.H.; Ramirez, M.J.; Milagro, F.I.; Martinez, J.A.; Solas, M. Implication of Trimethylamine N-Oxide (TMAO) in disease: Potential biomarker or new therapeutic target. *Nutrients* **2018**, *10*, 1398. [CrossRef]
30. Ufnal, M.; Nowinski, A. Is increased plasma TMAO a compensatory response to hydrostatic and osmotic stress in cardiovascular diseases? *Med. Hypotheses* **2019**, *130*, 109271. [CrossRef]
31. Zeisel, S.H.; Warrier, M. Trimethylamine N-Oxide, the microbiome, and heart and kidney disease. *Annu. Rev. Nutr.* **2017**, *37*, 157–181. [CrossRef] [PubMed]
32. Jameson, E.; Fu, T.; Brown, I.R.; Paszkiewicz, K.; Purdy, K.J.; Frank, S.; Chen, Y. Anaerobic choline metabolism in microcompartments promotes growth and swarming of Proteus mirabilis. *Environ. Microbiol.* **2016**, *18*, 2886–2898. [CrossRef] [PubMed]
33. Falony, G.; Vieira-Silva, S.; Raes, J. Microbiology meets big data: The case of gut microbiota-derived Trimethylamine. *Annu. Rev. Microbiol.* **2015**, *69*, 305–321. [CrossRef] [PubMed]
34. Zarzycki, J.; Erbilgin, O.; Kerfeld, C.A. Bioinformatic characterization of glycyl radical enzyme-associated bacterial microcompartments. *Appl. Environ. Microbiol.* **2015**, *81*, 8315–8329. [CrossRef] [PubMed]
35. Fracalanzza, S.A.; Scheidegger, E.M.; Santos, P.F.; Leite, P.C.; Teixeira, L.M. Antimicrobial resistance profiles of enterococci isolated from poultry meat and pasteurized milk in Rio de Janeiro, Brazil. *Mem. Inst. Oswaldo Cruz* **2007**, *102*, 853–859. [CrossRef] [PubMed]
36. Colombo, E.; Franzetti, L.; Frusca, M.; Scarpellini, M. Phenotypic and genotypic characterization of lactic acid bacteria isolated from Artisanal Italian goat cheese. *J. Food Prot.* **2010**, *73*, 657–662. [CrossRef] [PubMed]

37. De Jong, A.; Simjee, S.; Rose, M.; Moyaert, H.; El Garch, F.; Youala, M.; Group, E.S. Antimicrobial resistance monitoring in commensal enterococci from healthy cattle, pigs and chickens across Europe during 2004-14 (EASSA Study). *J. Antimicrob. Chemother.* **2019**, *74*, 921–930. [CrossRef]
38. Bennett, B.J.; de Aguiar Vallim, T.Q.; Wang, Z.; Shih, D.M.; Meng, Y.; Gregory, J.; Allayee, H.; Lee, R.; Graham, M.; Crooke, R.; et al. Trimethylamine-N-oxide, a metabolite associated with atherosclerosis, exhibits complex genetic and dietary regulation. *Cell Metab.* **2013**, *17*, 49–60. [CrossRef]
39. Zhang, A.Q.; Mitchell, S.; Smith, R. Fish odour syndrome: Verification of carrier detection test. *J. Inherit. Metab. Dis.* **1995**, *18*, 669–674. [CrossRef]
40. Jameson, E.; Quareshy, M.; Chen, Y. Methodological considerations for the identification of choline and carnitine-degrading bacteria in the gut. *Methods* **2018**, *149*, 42–48. [CrossRef]
41. Ufnal, M.; Pham, K. The gut-blood barrier permeability—A new marker in cardiovascular and metabolic diseases? *Med. Hypotheses* **2017**, *98*, 35–37. [CrossRef] [PubMed]
42. Johnson, C.; Prokopienko, A.J.; West, R.E., 3rd; Nolin, T.D.; Stubbs, J.R. Decreased kidney function is associated with enhanced hepatic flavin monooxygenase activity and increased circulating Trimethylamine N-Oxide concentrations in mice. *Drug Metab. Dispos. Biol. Fate Chem.* **2018**, *46*, 1304–1309. [CrossRef] [PubMed]
43. Seibel, B.A.; Walsh, P.J. Trimethylamine oxide accumulation in marine animals: Relationship to acylglycerol storage. *J. Exp. Biol.* **2002**, *205*, 297–306. [PubMed]
44. Jameson, E.; Doxey, A.C.; Airs, R.; Purdy, K.J.; Murrell, J.C.; Chen, Y. Metagenomic data-mining reveals contrasting microbial populations responsible for trimethylamine formation in human gut and marine ecosystems. *Microb. Genom.* **2016**, *2*, e000080. [CrossRef]
45. Wilson, M. The indigenous microbiota of the gastrointestinal tract. In *The Human Microbiota in Health and Disease: An Ecological and Community-Based Approach*, 1st ed.; CRC Press, Taylor & Francis Group: Boca Raton, FL, USA, 2018; 472p.
46. Gosalbes, M.J.; Durban, A.; Pignatelli, M.; Abellan, J.J.; Jimenez-Hernandez, N.; Perez-Cobas, A.E.; Latorre, A.; Moya, A. Metatranscriptomic approach to analyze the functional human gut microbiota. *PLoS ONE* **2011**, *6*, e17447. [CrossRef]

Article

Mice Microbiota Composition Changes by Inulin Feeding with a Long Fasting Period under a Two-Meals-Per-Day Schedule

Hiroyuki Sasaki [1,2,†], Hiroki Miyakawa [1,†], Aya Watanabe [1], Yuki Nakayama [1], Yijin Lyu [1], Koki Hama [1] and Shigenobu Shibata [1,*]

1. Laboratory of Physiology and Pharmacology, School of Advanced Science and Engineering, Waseda University, Shinjuku-ku, Tokyo 162-8480, Japan; hiroyuki-sasaki@asagi.waseda.jp (H.S.); hgbbst-hiroki@toki.waseda.jp (H.M.); aya_watanabe7115@suou.waseda.jp (A.W.); yukibecky-6991@akane.waseda.jp (Y.N.); ikin@fuji.waseda.jp (Y.L.); koby_frekoi418@ruri.waseda.jp (K.H.)
2. AIST-Waseda University Computational Bio Big-Data Open Innovation Laboratory (CBBD-OIL), National Institute of Advanced Industrial Science and Technology, Shinjuku-ku, Tokyo 169-8555, Japan

* Correspondence: shibatas@waseda.jp; Tel.: +81-3-5369-7318

† These authors contributed equally to this work.

Received: 23 October 2019; Accepted: 14 November 2019; Published: 16 November 2019

Abstract: Water-soluble dietary fiber is known to modulate fecal microbiota. Although there are a few reports investigating the effects of fiber intake timing on metabolism, there are none on the effect of intake timing on microbiota. Therefore, in this study, we examined the timing effects of inulin-containing food on fecal microbiota. Mice were housed under conditions with a two-meals-per-day schedule, with a long fasting period in the morning and a short fasting period in the evening. Then, 10–14 days after inulin intake, cecal content and feces were collected, and cecal pH and short-chain fatty acids (SCFAs) were measured. The microbiome was determined using 16S rDNA sequencing. Inulin feeding in the morning rather than the evening decreased the cecal pH, increased SCFAs, and changed the microbiome composition. These data suggest that inulin is more easily digested by fecal microbiota during the active period than the inactive period. Furthermore, to confirm the effect of fasting length, mice were housed under a one-meal-per-day schedule. When the duration of fasting was equal, the difference between morning and evening nearly disappeared. Thus, our study demonstrates that consuming inulin at breakfast, which is generally after a longer fasting period, has a greater effect on the microbiota.

Keywords: microbiota; inulin; circadian rhythm; feeding timing

1. Introduction

In the gut of mammals, the microbiota includes 100 trillion bacteria. Disordered microbiota alteration is involved in the development of various diseases [1]. *Firmicutes* are bacteria related to obesity, while *Bacteroidetes* suppress fat accumulation in mice fed a high-fat diet (HFD) [2]. When the feces of obese mice are transplanted into germ-free mice, obesity develops [3]. Moreover, *Fusobacterium*, including *Fusobacterium nucleatum*, are increased in patients with colorectal cancer compared with healthy subjects [4,5]. In addition to physical illnesses, a relationship of microbiota with psychological illness has also been reported. In patients with major depression, *Bacteroidetes*, *Proteobacteria*, and *Actinobacteria* are significantly increased compared with healthy subjects [6]. These results suggest that intestinal bacteria are related to the development of diseases and that maintaining homeostasis of the microbiota is important for the mental and physical health of the host.

Short-chain fatty acids (SCFAs) are produced when the microbiota ferments and degrades non-digestible food components [7]. The SCFAs lower intestinal pH, suppress the growth of pathogenic bacteria in the gut, and function as a regulator of metabolism and immunity [8]. Among SCFAs, acetic acid is a liver energy substrate used for fat synthesis, and propionic acid is used as a material for gluconeogenesis in the liver. Butyrate promotes the induction of regulatory T cells in the large intestine [9,10]. Furthermore, SCFAs also increase insulin sensitivity in the liver and muscles through GPR43, a receptor for SCFAs in white fat, as well as increase energy efficiency [11].

The circadian rhythm, controlled by clock genes, plays an important role in daily locomotor activity rhythms and physiological events, such as the sleep–wake cycle, hormone secretion, and the sympathetic nervous system [12,13]. Clocks in peripheral tissues are regulated by the central clock in the suprachiasmatic nucleus and external cues such as food, temperature, and exercise [14–16]. It has been reported that circadian rhythms are also present in the intestinal flora and controlled by dietary composition [17–20]. Furthermore, disturbance of the circadian clock due to jet lag alters microbial populations. For example, when the stool of jet-lagged mice is transplanted into germ-free mice, the recipient mice become obese [19]. Recently, however, it has been reported that SCFAs produced by gut microbiota can synchronize the circadian clock [21].

The microbiota composition changes depending on food components. In particular, foods rich in dietary fiber have a strong effect on the microbiota and are known as prebiotics [22]. According to Gibson et al., prebiotics are defined as "nondigestible food ingredients that beneficially affect the host by selectively stimulating the growth and/or activity of one or a limited number of bacterial species already resident in the colon, and thus attempt to improve host health" [23]. Inulin is a water-soluble dietary fiber and, thus, a prebiotic. It is particularly involved in the growth of bacteria that produce lactic acid [24] and promote the absorption of minerals such as calcium and magnesium [25,26].

It has been suggested that meal timing and daily eating habits may affect the development and prevention of lifestyle-related diseases such as obesity. A study by Hatori et al. demonstrates that restricted feeding in an activity period for mice without reducing calorific intake prevents metabolic diseases in mice fed a HFD [27]. Mice consuming milk fat late in the activity period have elevated hepatic fat and increased serum triglycerides and free fatty acids [28]. In addition, scheduled access to a HFD during the inactivity period increases body weight in mice compared with access during the activity period [29,30]. Moreover, in human experiments, the combination of a late dinner with a short sleep duration is associated with the risk of obesity [31]. In addition, the risk of obesity has been related to eating supper after 20:00 in the evening [32]. In recent years, it has been suggested that the influence of food on lipid metabolism is different depending on the time of food intake. In mice fed a high-fructose diet, fish oil given earlier in the activity period rather than later more effectively lowered lipids [33].

There are many reports indicating that time of food intake affects energy metabolism, but there are still relatively few reports describing the effect of eating time on microbiota. Furthermore, there are few reports on the dual effect of food type, particularly dietary fiber, and intake time on microbiota. Therefore, in the current study, we investigated whether inulin intake during the morning has a stronger effect on the microbiota than inulin intake during the evening with a two-meals-per-day schedule in mice.

2. Materials and Methods

2.1. Mice

In this study, we used eight-week-old male ICR mice (Tokyo Laboratory Animals, Tokyo, Japan). The mice were kept in a room maintained on a 12 h light/12 h dark (LD) cycle (lights on from 08:00 to 20:00). Zeitgeber time 0 (ZT0) was designated as lights-on time and ZT12 as lights-off time under the LD cycle. The mice were housed either in groups (five mice per cage; experiments 1 and 2) or individually (experiments 3 and 4) in plastic cages. The cages were maintained at a temperature of 22 ± 2 °C, humidity of 60 ± 5%, and light intensity of 100–150 lux. The mice were provided with a HFD containing 45% kcal of fat (Diet 12451; Research Diets Inc., New Brunswick, NJ, USA) with cellulose

(Oriental Yeast Co., Ltd., Tokyo, Japan) or inulin (Fuji FF; Fuji Nihon Seito Co., Tokyo, Japan) [34,35] and water ad libitum. This HFD is a diet used as a model for obesity, diabetes, and fatty liver in rodents [36,37]. Inulin has been reported to attenuate HFD-induced lipid metabolism and microbiota change [38]. In addition, the metabolic syndrome caused by obesity and abnormal lipid metabolism in the liver are related to microbiota change [39–41]. Therefore, we used an HFD to enhance the attenuating effects of inulin with the condition of microbiota change. The animal experiment was conducted with permission from the Committee for Animal Experimentation of the School of Science and Engineering at Waseda University (permission # 09A11, 10A11) and in accordance with the law (No. 105) and notification (No. 6) of the Japanese government.

2.2. Scheduled Feeding

We prepared two types of feeding conditions. In type 1 (experiments 1 and 2), only the feeding time was controlled, while in type 2 (experiments 3 and 4), both the start time of feeding and the amount of food were controlled.

In type 1 feeding, all of the mice could approach the feed box during the permitted time. We defined the morning as ZT12–20 and the evening as ZT20–4. The mice had free access to the feed box for predetermined four-hour periods (morning meal as ZT12–16 and evening meal as ZT20–0). Throughout the remaining time, the feed box was locked. Food intake was calculated by measuring the weight of the food in the feed box at the start and end of the experiment. The total consumed food was divided by the number of mice and the number of days in the experiment. In the type 1 experiments, we housed the mice as a group to avoid the stress induced by individual housing.

In type 2 feeding, all of the mice were housed in cages containing food dispensers that released food pellets under the regulation of a timer. The mice were fed 90% of the amount of food that was consumed in experiment 1 (Figure 1c). In experiment 3, the mice were fed two meals per day at ZT12 (defined as morning) or ZT20 (defined as evening); the meal size was 1.8 g. In experiment 4, the mice were fed one meal per day at ZT12 (morning) or ZT20 (evening); the meal size was 3.6 g.

We adjusted the concentration of dietary fiber so that the amount of inulin was approximately equal between experiments.

2.3. Cecal pH Measurement

The cecal pH was measured by inserting the glass tip of an electrode of a pH meter (pH Spear; Eutech Instruments, Vernon Hills, IL, USA) directly into the cecum.

2.4. Measurement of SCFAs

The SCFAs were measured via gas chromatography and flame ionization detection (Shimadzu Corp., Kyoto, Japan) as described by a previous report [42] with some modifications. A total of 0.05 g of cecal content was acidified with 0.05 mL sulfuric acid (FUJIFILM Wako Pure Chemical Corp., Osaka, Japan). Then, the SCFAs were extracted by shaking with 0.4 mL of diethyl ether (FUJIFILM Wako Pure Chemical Corp., Osaka, Japan) and 0.2 mL of ethanol (FUJIFILM Wako Pure Chemical Corp., Osaka, Japan), which was then centrifuged at 14,000 rpm at room temperature for 30 s. A total of 1 μL of the organic phase was injected into the capillary column (InertCap Pure WAX (30 m × 0.25 mm, df = 0.5 μm); GL Science Inc., Tokyo, Japan). The initial temperature was 80 °C, and the final temperature was 200 °C. Helium was used as a carrier gas. Quantification of the samples was performed using calibration curves for acetic, lactic, propionic, and butyric acids. A standard curve for each acid was conducted for their quantitation in the samples.

2.5. Fecal DNA Extraction

The fecal DNA was extracted as previously described, with some modifications [43]. About 0.2 g of the fecal sample was suspended in a 50 mL Falcon tube containing 20 mL PBS. The suspension was filtered through a 100-μm mesh nylon filter (Corning Inc., New York NY, USA). The debris on the filter was washed with 10 mL of Phosphate buffered salts (PBS). The filtrates were centrifuged at 4000 rpm for 20 min at 4 °C, and each precipitate was then suspended with 1.5 mL of TE10 buffer (10 mM Tris-HCl (FUJIFILM Wako Pure Chemical Corp., Osaka, Japan)/10 mM ethylenediaminetetraacetic acid (EDTA) (DOJINDO, Tokyo, Japan)). The suspensions were transferred to 2-mL microtubes before being centrifuged at 10,000 rpm for five minutes at 4 °C. Following this, each precipitate was suspended again with 0.8 mL of TE10 buffer. The DNA was extracted using 1 mL of PCI (Invitrogen, Carlsbad, CA, USA) and isolated with 0.1 mL of lysozyme (FUJIFILM Wako Pure Chemical Corp., Osaka, Japan) and 0.02 mL of achromopeptidase (FUJIFILM Wako Pure Chemical Corp., Osaka, Japan). The DNA was purified via treatment with RNase (Promega Corp., Madison, WI, USA), followed by precipitation with 20% PEG solution (Tokyo Chemical Industry Co., Ltd., Tokyo, Japan). Finally, the DNA was rinsed with 70% ethanol and dissolved in 50 μL TE buffer.

2.6. 16 S rDNA Gene Sequencing

The 16S rDNA gene sequencing was performed according to the instructions of Illumina. V3–V4 variable regions of the 16S rDNA gene were amplified by PCR using the following primers:

forward primer = 5′-TCGTCGGCAGCGTCAGATGTGTATAAGAGACAGCCTA

CGGGNGGCWGCAG-3′;
reverse primer = 5′-GTCTCGTGGGCTCGGAGATGTGTATAAGA

GACAGGACTACHVGGGTATCTAATCC-3′.

The PCR amplification was performed with 2.5 μL microbial DNA (5 ng/μL), 5 μL of each primer (1 μmol/L), and 12.5 μL 2 × KAPA HiFi HotStart Ready Mix (Kapa Biosystems Inc., Wilmington, MA, USA). The following PCR procedure was used: 95 °C for three minutes, followed by 25 cycles of 95 °C for 30 s, 55 °C for 30 s, and 72 °C for 30 s. Finally, an extension was performed at 72 °C for five minutes. The Amplicon PCR products were purified using AMPure XP beads (Beckman Coulter, Inc., Brea, CA, USA), according to the manufacturer's instructions. A Nextera XT Index Kit v2 (Illumina Inc., Hayward, CA, USA) was used for the Illumina sequencing adapters and attachment of the dual indices. An index PCR was performed with 5.0 μL PCR product, 5.0 μL of each of the Nextera XT Index Primers, 25 μL 2× KAPA HiFi HotStart Ready Mix, and 10 μL PCR-Grade Water. The PCR was performed via the following procedure: 95 °C for three minutes, followed by eight cycles of 95 °C for 30 s, 55 °C for 30 s, and 72 °C for 30 s. Finally, an extension was performed at 72 °C for five minutes. The index PCR products were purified using AMPure XP beads (Beckman Coulter, Inc., Brea, CA, USA). The quality of the purifications was checked using the Agilent 2100 Bioanalyzer with a DNA1000 Kit (Agilent Technologies Inc., Santa Clara, CA, USA). Finally, the DNA library was diluted to 4 nmol/L.

Then, the DNA library was sequenced using the Miseq Reagent Kit v3 (Illumina Inc.) in the Illumina Miseq 2 × 300 bp platform, according to the manufacturer's instructions.

2.7. Analysis of 16S rDNA Gene Sequences

The 16S rDNA sequence reads were processed by the Quantitative Insights into Microbial Ecology (QIIME) pipeline version 1.9.1 [44]. The quality-filtered sequence reads were assigned to operational taxonomic units at 97% identity with the UCLUST algorithm [45]; these reads were then compared with reference sequence collections in the Greengenes database (August 2013 version). A total of 6,680,549 reads were obtained from the 91 samples. On average, 73,412 ± 4606 reads were obtained per sample. The taxonomy summary at the phylum to genus levels, alpha diversity such as the Simpson diversity index, beta diversity, and principal coordinate analysis (PCoA) were calculated and generated using QIIME. A PCoA analysis was also calculated using weighted UniFrac distances.

2.8. Predicted Metagenomes

In experiments 3 and 4, the functional profiles of microbial communities were predicted by the Phylogenetic Investigation of Communities by Reconstruction of Unobserved States (PICRUSt) [46]. The functional predictions were assigned to the Kyoto Encyclopedia of Genes and Genomes (KEGG) ortholog functional profiles of microbial communities via 16S sequences. We selected and examined categories related to "carbohydrate metabolism" for simplification and clarity of the analysis.

2.9. Statistical Analysis

The data were expressed as means ± standard error of the mean (SEM). All statistical analyses were performed using GraphPad Prism (version 6.03, GraphPad Software Inc., San Diego, CA, USA). We checked whether the data showed a normal or non-normal distribution and equal or biased variation via the D'Agostino-Pearson test/Kolmogorov–Smirnov test and F-value test/Bartlett's test, respectively. If the data showed a normal distribution and equal variation, the statistical significance was determined by the Student's t-test or one-way ANOVA with a Tukey's test or two-way ANOVA with a Tukey's post-hoc analysis if the interaction was significant. If the interaction was not significant but the main effect was, Sidak's post-hoc analysis was used. If the data showed a non-normal distribution or biased variation, the statistical significance was determined by the Mann–Whitney test or Kruskal–Wallis test with a Dunn post-hoc analysis and a two-stage linear step-up procedure of the Benjamini, Krieger, and Yekutieli test for multiple comparisons. The permutational multivariate analysis of variance (PERMANOVA) was used to assess the change of the microbiota composition. The PERMANOVA was analyzed by QIIME.

3. Results

3.1. Inulin Intake Changed Microbiota Composition under Both Morning and Evening Timings

In this experiment, cellulose, an insoluble dietary fiber, was added to the HFD as a control for inulin because the dietary fiber contained in this HFD was cellulose. The amount of dietary fiber in the food was kept the same between each group. Indeed, when comparing the HFD with and without cellulose, there was no significant difference in the body weight, food intake, cecal pH, amount of SCFAs, or microbiota composition of the mice (Figure S1). Therefore, this concentration of cellulose did not appear to affect these factors.

Some reports have suggested that inulin consumption induces changes in the microbiota composition [47–50]. Here, we divided the mice into two groups. Group 1 received cellulose and was fed an HFD with 2.5% cellulose in the morning and evening. Group 2 received inulin and was fed HFD with 2.5% inulin in the morning and evening. The mice were housed under each condition for 10 days, after which they were sacrificed at ZT20 (four hours after the morning intake) or ZT4 (four hours after the evening intake) on days 10–11 (Figure 1a). We sampled the cecal content and feces from the rectum and measured the cecal pH. There was no significant difference in the body weight between the two groups before sacrifice (Figure 1b), nor was there was a difference in the food intake between them (Figure 1c). There were no standard error bars in the food intake volume because of the group housing.

The cecal pH was significantly lower in the inulin group than in the cellulose group at both ZT20 and ZT4 (Figure 1d). The propionic acid level was significantly higher in the inulin group than in the cellulose group at ZT4, and the lactic acid level was significantly higher in the inulin group than in the cellulose group at ZT20. At ZT4, there was only a slight increase in the inulin group compared with the cellulose group. In the cellulose group, the lactic acid level was significantly different between ZT20 and ZT4. There were no significant differences in the acetic acid, propionic acid, and total SCFA levels between the cellulose and inulin group at either ZT20 or ZT4 (Figure 1e–i).

As the propionic and lactic acid increased and the cecal pH decreased, the microbiota may have changed due to inulin feeding. Therefore, we extracted 16S rDNA from the mice feces and analyzed the microbiota. In the cellulose group, the values of alpha-diversity as described by the Simpson index were significantly higher at ZT4 than at ZT20. The Simpson index was significantly higher in the inulin group than in the cellulose group at ZT20, but there was no significant difference observed at ZT4 (Figure 1j). Next, we examined the differences in the changes of the relative abundance of taxa between the inulin group and cellulose group. Some of the detected bacteria are shown in Figure 2. At the phylum level, the relative abundance of *Firmicutes* was significantly lower in the inulin group than in the cellulose group at ZT20. However, there was no significant difference in the relative abundance of *Bacteroidetes* between the inulin and cellulose groups, though the levels were increased slightly in the former (Figure 2a). At the genus level, the relative abundance of *Lactococcus* and *Streptococcus* significantly decreased in the inulin group at ZT20, and the relative abundance of *Oscillospira* significantly decreased in the inulin group at ZT4 (Figure 2b). We analyzed the PCoA of the weighted UniFrac distances and determined the beta-diversity of the microbiota composition (Figure S2a). In this experiment, we focused on the influence of inulin on the microbiota; thus, we primarily compared cellulose and inulin feeding. The beta-diversity of the microbiota composition was significantly different between the cellulose and inulin groups at ZT20 but not at ZT4 (Figure S2b,c).

These results suggest that inulin consumption changes microbiota composition. In addition, the inulin feeding time may have different effects on the microbiota because changes in the microbiota were more prominent at ZT20 (morning) than at ZT4 (evening), which showed significant and non-significant differences, respectively, compared with the cellulose group.

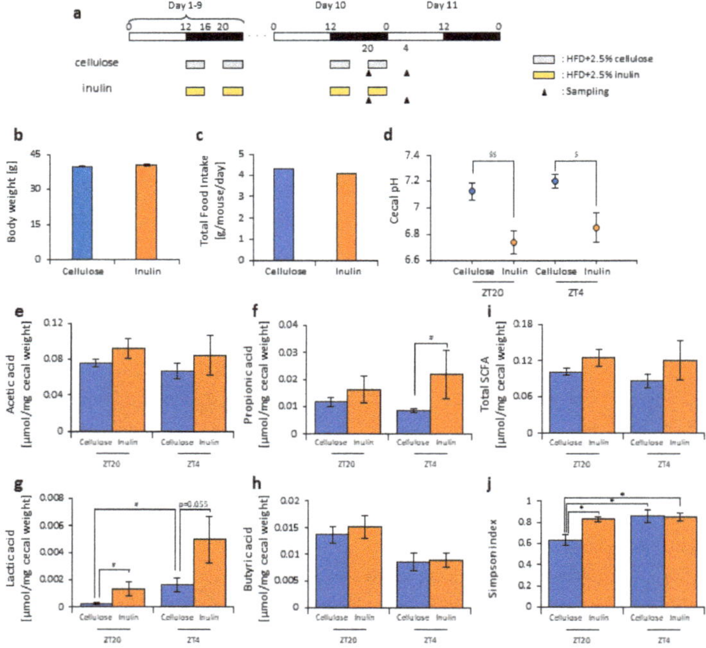

Figure 1. Inulin feeding decreased cecal pH and increased short-chain-fatty-acids. (**a**) Experimental schedule, where the white and black bars indicate environmental 12 h light and dark conditions, respectively. The gray bar indicates feeding with a high-fat-diet (HFD) and 2.5% cellulose. The yellow bar indicates feeding with HFD and 2.5% inulin. The black arrowhead indicates the sampling time. (**b**) Body

weight before sampling. (**c**) Average daily food intake. (**d**) Cecal pH of mice housed for 10 days for each group. (**e–i**) The short-chain fatty acids (SCFAs) of mice, including (**e**) acetic acid, (**f**) propionic acid, (**g**) lactic acid, (**h**) butyric acid, and (**i**) total SCFAs. (**j**) Bacterial alpha diversity. Comparison of the Simpson index estimation of the 16S rDNA gene libraries at 97% similarity from the sequencing analysis. All values except (**c**) are represented as mean ± SEM (cellulose at ZT20 (n = 5) and 4 (n = 5); inulin at ZT20 (n = 5) and 4 (n = 5)). * $p < 0.05$, evaluated using the two-way ANOVA with Tukey's post hoc test. $$ $p < 0.01$, $ $p < 0.05$, evaluated using the two-way ANOVA with Sidak's post hoc test. # $p < 0.05$, evaluated using the Mann–Whitney test with a two-stage linear step-up procedure of the Benjamini, Krieger, and Yekutieli test for multiple comparisons.

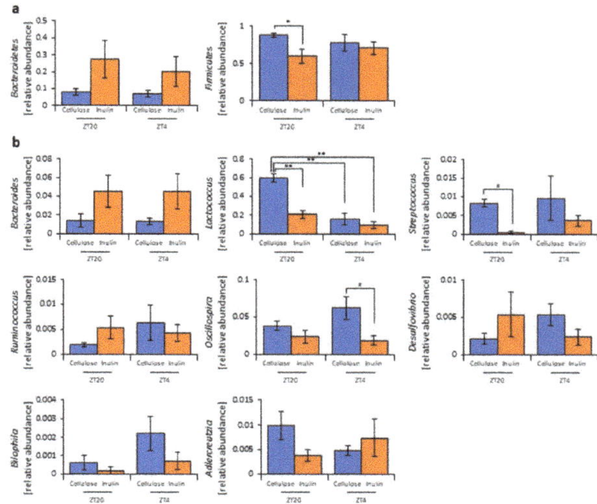

Figure 2. Inulin feeding changed the relative abundance of some bacteria. (**a**) Phylum level. (**b**) Genus level. All values are represented as mean ± SEM (cellulose at ZT20 (n = 5) and 4 (n = 5); inulin at ZT20 (n = 5) and 4 (n = 5)). ** $p < 0.01$, * $p < 0.05$, evaluated using the two-way ANOVA with Tukey's post hoc test. # $p < 0.05$, evaluated using the Mann–Whitney test with a two-stage linear step-up procedure of the Benjamini, Krieger, and Yekutieli test for multiple comparisons.

3.2. Inulin Intake in the Morning Rather than the Evening Strongly Affected the Microbiota Composition under Time-Restricted Feeding Conditions

In this study, inulin may have had different effects on the microbiota depending on the feeding times. However, in experiment 1, we did not measure the effect of the feeding pattern. It is possible that the effect of inulin was increased at ZT20 (four hours after the morning intake) due to the high consumption in the morning. In the next experiment, we examined whether morning or evening inulin feeding affected the microbiota under the two meals-per-day schedule. The mice were divided into three groups. Group 1 received cellulose and was fed an HFD with 5% cellulose in the morning and evening. Group 2 received inulin in the morning and was fed an HFD with 5% inulin in the morning and an HFD with 5% cellulose in the evening. Group 3 received inulin in the evening and was fed an HFD with 5% cellulose in the morning and an HFD with 5% inulin in the evening. The mice were housed under each condition for 10 days, after which they were sacrificed at ZT20 and ZT4 on days 10–11 (Figure 3a). We sampled cecal contents and feces and measured the cecal pH. There was no significant difference in body weight between any group before sacrifice (Figure 3b), nor was there a large difference in total food intake between them. However, the total food intake was slightly higher if inulin intake was in the morning rather than in the evening (Figure 3c). The cecal pH was significantly lower in the morning inulin group than in the morning cellulose or evening inulin groups at ZT20. On the contrary, the pH was significantly lower in the evening inulin group than in the evening cellulose and morning inulin groups at ZT4 (Figure 3d). The acetic acid, propionic acid, lactic acid, butyric acid, and total SCFA levels were significantly higher in the morning inulin group than in the morning cellulose or evening inulin groups at ZT20. However, the acetic acid, propionic acid, lactic acid, butyric acid, and total SCFA levels were significantly higher in the evening inulin group than in the evening cellulose or morning inulin groups at ZT4 (Figure 3e–i).

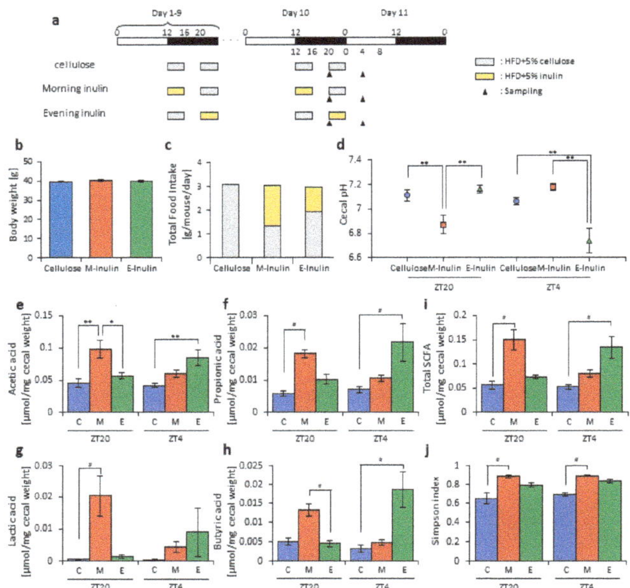

Figure 3. Morning inulin feeding decreased cecal pH and increased short-chain-fatty-acids more than evening inulin feeding. (**a**) Experimental schedule, where white and black bars indicate environmental 12 h light and dark conditions, respectively. The gray bar indicates feeding with a high-fat-diet (HFD) and 5% cellulose. The yellow bar indicates feeding with HFD and 5% inulin. The black arrowhead indicates the sampling time. (**b**) Body weight before sampling. (**c**) Average daily food intake. The gray bar indicates the average daily food intake of cellulose, and the yellow bar indicates the average daily food intake of inulin. (**d**) Cecal pH of mice housed for 10 days for each group. (**e–i**) SCFAs of mice, including (**e**) acetic acid, (**f**) propionic acid, (**g**) lactic acid, (**h**) butyric acid, and (**i**) total SCFAs. (**j**) Bacterial alpha diversity. Comparison of the Simpson index estimation of the 16S rDNA gene libraries at 97% similarity from the sequencing analysis. All values except (**c**) are represented as mean ± SEM (cellulose at ZT20 (n = 5) and 4 (n = 5); morning inulin at ZT20 (n = 5) and 4 (n = 5); evening inulin at ZT20 (n = 5) and 4 (n = 5)). ** $p < 0.01$, * $p < 0.05$, evaluated using the two-way ANOVA with Tukey's post hoc test. # $p < 0.05$, evaluated using the Kruskal–Wallis test with Dunn post hoc test with a two-stage linear step-up procedure of the Benjamini, Krieger, and Yekutieli test for multiple comparisons. Cellulose, morning inulin, and evening inulin are C, M, or E, respectively.

Next, we extracted 16S rDNA from the mice feces and analyzed the microbiota. The value of alpha-diversity as determined by the Simpson index was significantly higher in the morning inulin group than in the morning cellulose or evening inulin groups at both ZT20 and ZT4 (Figure 3j). We also examined the differences between the changes of the relative abundance of taxa between the inulin and cellulose groups. Some of the detected bacteria are shown in Figure 4. At the phylum level, the relative abundance of *Bacteroidetes* was significantly higher in the morning inulin group than in the morning cellulose group at ZT20. Meanwhile, the relative abundance of *Firmicutes* was significantly lower in the morning inulin group than in the morning cellulose and evening inulin groups at ZT20 as well as significantly lower in the evening inulin group than in the evening cellulose group at ZT4

(Figure 4a). At the genus level, the relative abundance of *Lactococcus* was significantly decreased in the morning inulin group at both ZT20 and ZT4 and in the evening inulin group at ZT4, while the relative abundance of *Dorea* and *Allobaculum* was significantly increased in the morning inulin group (Figure 4b). We analyzed the PCoA of the weighted UniFrac distances and determined the beta-diversity of the microbiota composition (Figure S3a). At ZT20, the beta-diversity of the microbiota was significantly different between the cellulose and morning inulin groups and the morning and evening inulin groups (Figure S3b). At ZT4, the beta-diversity of the microbiota composition was significantly different among all of the groups (Figure S3c).

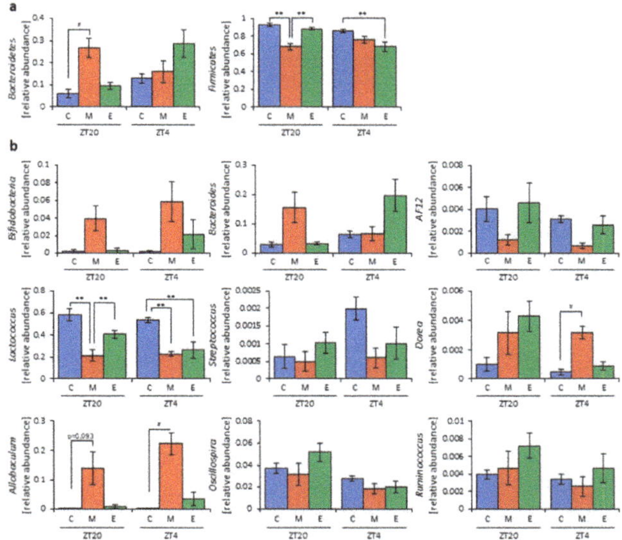

Figure 4. Morning inulin feeding changed the relative abundance of some bacteria. (**a**) Phylum level. (**b**) Genus level. All values are represented as mean ± SEM (cellulose at ZT20 (n = 5) and 4 (n = 5); morning inulin at ZT20 (n = 5) and 4 (n = 5); evening inulin at ZT20 (n = 5) and 4 (n = 5)). ** $p < 0.01$, evaluated using the two-way ANOVA with Tukey's post hoc test. # $p < 0.05$, evaluated using the Kruskal–Wallis test with Dunn post hoc test with a two-stage linear step-up procedure of the Benjamini, Krieger, and Yekutieli test for multiple comparisons. Cellulose, morning inulin, and evening inulin are C, M, or E, respectively.

These results suggest that morning inulin feeding affected the microbiota more than evening inulin feeding. However, the inulin intake was higher in the morning inulin group. Therefore, the increased consumption of the morning inulin group may have had more of an impact on the microbiota. To eliminate the effects of different food intakes, we prepared an apparatus to supply equal food amounts at two meals per day in the next experiment.

3.3. *Inulin Feeding in the Morning Affected the Microbiota Composition More than that in the Evening under Restricted Food Amount Conditions*

In this experiment, we provided the mice with two meals per day of 1.8 g of food at ZT12 (morning) and ZT20 (evening) to achieve equal food intake. The mice were divided into three groups. Group 1 received cellulose and was fed 1.8 g of an HFD with 5% cellulose in both the morning and evening. Group 2 received inulin in the morning and was fed 1.8 g of an HFD with 5% inulin in the morning and 1.8 g of an HFD with 5% cellulose in the evening. Group 3 received inulin in the evening and was fed 1.8 g of an HFD with 5% cellulose in the morning and 1.8 g of an HFD with 5% inulin in the evening. The mice were housed under each condition for 14 days, after which they were sacrificed at

ZT20 and ZT4 on days 14–15 (Figure 5a). We sampled cecal content and feces and measured the cecal pH. There was no significant difference in body weight between any group before sacrifice (Figure 5b). The cecal pH was significantly lower in the morning inulin group than in the morning cellulose group and significantly lower in the evening inulin group than in the evening cellulose group. Moreover, the cecal pH was significantly lower in the morning inulin group than in the evening inulin group (Figure 5c). The propionic acid, lactic acid, butyric acid, and total SCFA levels were significantly higher in the morning inulin group than in the morning cellulose group, while the propionic acid level was significantly higher in the evening inulin group than in the evening cellulose group (Figure 5d–h).

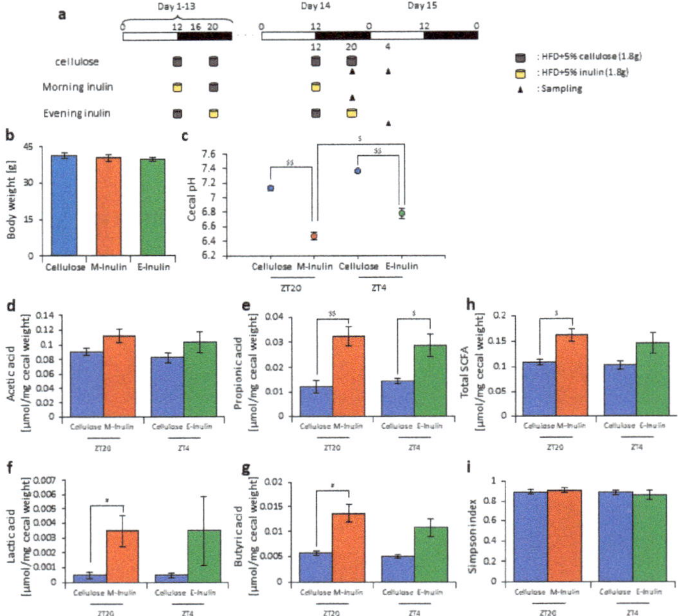

Figure 5. Morning inulin feeding decreased cecal pH and increased short-chain-fatty-acids more than evening inulin feeding under equivalent feeding conditions. (**a**) Experimental schedule, where white and black bars indicate environmental 12 h light and dark conditions, respectively. The gray cylinder indicates the 1.8 g of high-fat-diet (HFD) with 5% cellulose. The yellow cylinder indicates the 1.8 g of HFD with 5% inulin. The black arrowhead indicates the sampling time. (**b**) Body weight before sampling. (**c**) Cecal pH of mice housed for 14 days for each group. (**d–h**) SCFAs of mice, including (**d**) acetic acid, (**e**) propionic acid, (**f**) lactic acid, (**g**) butyric acid, and (**h**) total SCFAs. (**i**) Bacterial alpha diversity. Comparison of the Simpson index estimation of the 16S rDNA gene libraries at 97% similarity from the sequencing analysis. All values are represented as mean ± SEM (cellulose at ZT20 (n = 4) and 4 (n = 4); morning inulin (n = 5); evening inulin (n = 5)). $$ $p < 0.01$, $ $p < 0.05$, evaluated using the two-way ANOVA with Sidak's post hoc test. # $p < 0.05$, evaluated using the Mann–Whitney test with a two-stage linear step-up procedure of the Benjamini, Krieger, and Yekutieli test for multiple comparisons.

Next, we extracted 16S rDNA from the mice feces and analyzed the microbiota. The value of alpha-diversity as determined by the Simpson index showed no significant difference between any group (Figure 5i). We also examined the differences in the changes of the relative abundance of taxa. Bacteria detected in over half of all samples are shown in Table 1. At the phylum level, the relative abundance of *Proteobacteria* was significantly increased in the morning inulin group, while the relative abundance of *TM7* was significantly decreased in the morning and evening inulin groups (Table 1a). At the genus level, the relative abundance of *Butyricimonas* was significantly increased in the morning inulin group, while the relative abundance of *AF12*, *Staphylococcus*, *Lactococcus*, *Oscillospira*, *Bilophila*, and *Desulfovibrio* was significantly decreased in the morning inulin group. Meanwhile, the relative abundance of *AF12*, *Odoribacter*, and *Oscillospira* was significantly decreased in the evening inulin group (Table 1b). The number of bacteria changed by inulin feeding in the morning was higher than that changed by inulin feeding in the evening. We analyzed the PCoA of the weighted UniFrac distances and determined the beta-diversity of the microbiota composition (Figure S4a). The beta-diversity of the microbiota was significantly different between the cellulose group and morning inulin group (Figure S4b), but no significant difference was observed between the cellulose group and evening inulin group (Figure S4c). We predicted the functional profiles from sequencing data with PICRUSt. Among the categories related to "carbohydrate metabolism", the relative abundance of fructose and mannose metabolism was significantly increased in the morning inulin group but not in the evening inulin group (Figure S4d).

These results suggest that inulin feeding in the morning may affect the microbiota, even if the food intake amount is the same in the morning and evening.

Table 1. The relative abundance of some bacteria under the condition of two meals per day. (**a**). Phylum level. (**b**). Genus level.

a. Phylum Level

Bacterial	ZT20			ZT4		
	Cellulose	M-Inulin	*p*-Value	Cellulose	E-Inulin	*p*-Value
Actinobacteria	0.0072 ± 0.0039	0.0103 ± 0.0081	0.7143	0.0690 ± 0.0647	0.0907 ± 0.0440	0.8254
Bacteroidetes	0.1535 ± 0.0463	0.2499 ± 0.0654	0.669	0.1647 ± 0.0467	0.3644 ± 0.1203	0.2085
Deferribacteres	0.0011 ± 0.0010	0.0005 ± 0.0004	0.9999	0.0002 ± 0.0002	0.0004 ± 0.0003	0.8413
Firmicutes	0.8221 ± 0.0467	0.6863 ± 0.0713	0.3232	0.7486 ± 0.0228	0.5294 ± 0.0865	0.0743
Proteobacteria	0.0111 ± 0.0033	0.0019 ± 0.0004	0.0159 #	0.0152 ± 0.0030	0.0054 ± 0.0017	0.1905
TM7	0.0043 ± 0.0025	0.0004 ± 0.0003	0.0159 #	0.0017 ± 0.0002	0.0001 ± 0.0001	0.0159 #
Verrucomicrobia	0.0006 ± 0.0003	0.0473 ± 0.0219	0.1905	0.0003 ± 0.003	0.0132 ± 0.0117	0.1032

b. Genus Level

Bacterial	ZT20			ZT4		
	Cellulose	M-Inulin	*p*-Value	Cellulose	E-Inulin	*p*-Value
Bifidobacterium	0.0022 ± 0.0018	0.0080 ± 0.0077	0.5556	0.0063 ± 0.0062	0.0088 ± 0.0044	0.6825
Adlercreutzia	0.0046 ± 0.0019	0.0022 ± 0.0004	0.2344	0.0047 ± 0.0014	0.0026 ± 0.0006	0.6428
Bacteroides	0.0463 ± 0.0175	0.1100 ± 0.0262	0.0993	0.0701 ± 0.0238	0.1953 ± 0.0627	0.1352
Parabacteroides	0.0010 ± 0.0003	0.0009 ± 0.0002	0.8247	0.0015 ± 0.0006	0.0010 ± 0.0003	0.6229
AF12	0.0075 ± 0.0023	0.0011 ± 0.0006	0.0435 #	0.0052 ± 0.0013	0.0009 ± 0.0002	0.0317 #
Butyricimonas	0.0004 ± 0.0001	0.0008 ± 0.0004	0.0317 #	0.0005 ± 0.0003	0.0002 ± 0.0001	0.7937
Odoribacter	0.0017 ± 0.0003	0.0006 ± 0.0002	0.1795	0.0020 ± 0.0008	0.0003 ± 0.0001	0.0308 $
[Prevotella]	0.0144 ± 0.0085	0.0303 ± 0.0098	0.2857	0.0118 ± 0.0064	0.0375 ± 0.0279	0.9762
Staphylococcus	0.0013 ± 0.0006	0.0002 ± 0.0001	0.0397 #	0.0009 ± 0.0006	0.0005 ± 0.0002	0.8254
Lactobacillus	0.0294 ± 0.0230	0.0136 ± 0.0036	0.6825	0.1152 ± 0.0811	0.0631 ± 0.0323	0.873
Lactococcus	0.2862 ± 0.0453	0.0957 ± 0.0128	0.0159 #	0.1691 ± 0.0446	0.0856 ± 0.0136	0.2857
Streptococcus	0.0034 ± 0.0016	0.0011 ± 0.0002	0.1545	0.0037 ± 0.0016	0.0019 ± 0.0010	0.357
Clostridium	0.0002 ± 0.0001	0.0001 ± 0.00005	0.3889	0.0014 ± 0.0012	0.0003 ± 0.0002	0.3889
Dehalobacterium	0.0018 ± 0.0002	0.0017 ± 0.0006	0.873	0.0010 ± 0.0001	0.0010 ± 0.0007	0.1746
Coprococcus	0.0049 ± 0.0009	0.0081 ± 0.0032	0.9999	0.0035 ± 0.0007	0.0019 ± 0.0005	0.1905
Dorea	0.0032 ± 0.0011	0.0032 ± 0.0012	0.9966	0.0041 ± 0.0031	0.0020 ± 0.0012	0.5089
Roseburia	0.0018 ± 0.0010	0.0053 ± 0.0038	0.9762	0.0005 ± 0.0004	0.0004 ± 0.0001	0.5635
[Ruminococcus]	0.0305 ± 0.0113	0.0384 ± 0.0105	0.6271	0.0327 ± 0.0129	0.0240 ± 0.0094	0.5957
Oscillospira	0.0803 ± 0.0131	0.0290 ± 0.0088	0.0148 $	0.0600 ± 0.0173	0.0172 ± 0.0074	0.0404 $
Ruminococcus	0.0096 ± 0.0019	0.0067 ± 0.0032	0.4961	0.0056 ± 0.0017	0.0025 ± 0.0010	0.161
Allobaculum	0.0017 ± 0.0007	0.1764 ± 0.0544	0.1905	0.0662 ± 0.0649	0.1669 ± 0.1094	0.5238
Bilophila	0.0013 ± 0.0003	0.0003 ± 0.0002	0.0159 #	0.0011 ± 0.0005	0.0002 ± 0.0001	0.1111

Table 1. *Cont.*

b. Genus Level						
Bacterial	**ZT20**			**ZT4**		
	Cellulose	**M-Inulin**	***p*-Value**	**Cellulose**	**E-Inulin**	***p*-Value**
Desulfovibrio	0.0037 ± 0.0011	0.0006 ± 0.0004	0.047 #	0.0021 ± 0.0007	0.0019 ± 0.0002	0.371
Akkermansia	0.0006 ± 0.0002	0.0473 ± 0.0219	0.1905	0.0003 ± 0.0003	0.0132 ± 0.0117	0.1032

(**a**) Number of bacteria significantly changed by M-inulin/all number of bacteria well-detected = 2/7. Number of bacteria significantly changed by E-inulin/all number of bacteria well-detected = 1/7. # $p < 0.05$, evaluated using the Mann–Whitney test with a two-stage linear step-up procedure of the Benjamini, Krieger, and Yekutieli test for multiple comparisons. (**b**) Number of bacteria significantly changed by M-inulin/all number of bacteria well-detected = 7/24. Number of bacteria significantly changed by E-inulin/all number of bacteria well-detected = 3/24. $ $p < 0.05$, evaluated using the two-way ANOVA with Sidak post hoc test. # $p < 0.05$, evaluated using the Mann–Whitney test with a two-stage linear step-up procedure of the Benjamini, Krieger, and Yekutieli test for multiple comparisons.

3.4. A Relationship Was Observed between the Length of Fasting Time and Inulin Feeding Stimulation

In experiment 3, it was observed that inulin intake in the morning may have an effect on the microbiota and that the fasting time factored into this effect in the morning. The morning inulin group fasted for 16 h after the previous feeding, while the evening inulin group fasted for 8 h after the previous feeding, meaning that the time until breakfast was longer than the time until dinner. Therefore, the difference in the length of fasting time may have changed the effect on the microbiota. To test this hypothesis, we prepared an experiment with equal fasting times based on one meal a day, in which 3.6 g of food was given to the mice at either ZT12 (morning) or ZT20 (evening). The mice were divided into four groups. Group 1 received cellulose in the morning and was fed 3.6 g of an HFD with 5% cellulose in the morning. Group 2 received inulin in the morning and was fed 3.6 g of an HFD with 5% inulin in the morning. Group 3 received cellulose in the evening and was fed 3.6 g of an HFD with 5% cellulose in the evening. Group 4 received inulin in the evening and was fed 3.6 g of an HFD with 5% inulin in the evening. The mice were housed under each condition for 14 days, after which they were sacrificed at ZT20 and ZT4 on days 14–15 (Figure 6a). We sampled cecal content and feces and measured the cecal pH. The body weight was significantly increased in the evening cellulose and inulin groups compared with the morning cellulose and inulin groups. (Figure 6b). The cecal pH was significantly lower in the morning and evening inulin groups than in the morning and evening cellulose groups (Figure 6c). The propionic and lactic acid levels were significantly higher in the morning inulin group than in the morning cellulose group. In addition, the butyric acid level was higher, albeit not significantly, in the morning inulin group than in the morning cellulose group. Meanwhile, the lactic and butyric acid levels were significantly higher in the evening inulin group than in the evening cellulose group, and the propionic acid level was higher, albeit not significantly, in the evening inulin group than in the evening cellulose group (Figure 6d–h).

Next, we extracted 16S rDNA from the mice feces and analyzed the microbiota. The value of alpha-diversity as determined by the Simpson index was significantly higher in the morning cellulose group than in the evening cellulose group (Figure 6i). We also examined the differences in the changes of the relative abundance of taxa. Bacteria detected in over half of all samples are shown in Table 2. At the phylum level, the relative abundance of *Actinobacteria* was increased in the morning inulin group, but there was no significant difference in the relative abundance in the evening inulin group (Table 2a). At the genus level, the relative abundance of *Bifidobacterium* and *Allobaculum* was significantly increased in the morning inulin group, while the relative abundance of *Streptococcus*, *Oscillospira*, and *Ruminococcus* was significantly decreased in the morning inulin group. Meanwhile, the relative abundance of *Dorea* and *Allobaculum* was significantly increased in the evening inulin group, and the relative abundance of *Staphylococcus* and *Lactococcus* was significantly decreased in the evening inulin group (Table 2b). The number of bacteria changed by inulin feeding in either the morning or the evening was similar. We analyzed the PCoA of the weighted UniFrac distances and determined the beta-diversity of the microbiota composition (Figure S5a). The beta-diversity of the microbiota was not significantly different between the cellulose and inulin groups in either the morning or evening (Figure S5b,c). We predicted the functional profiles from sequencing data with PICRUSt. Among the categories related to "carbohydrate metabolism", the relative abundance of fructose and mannose metabolism was not significantly different between the cellulose and inulin groups in either the morning or evening (Figure S5d). These results suggest that inulin intake in either the morning or evening with equal fasting periods does not change microbiota beta-diversity.

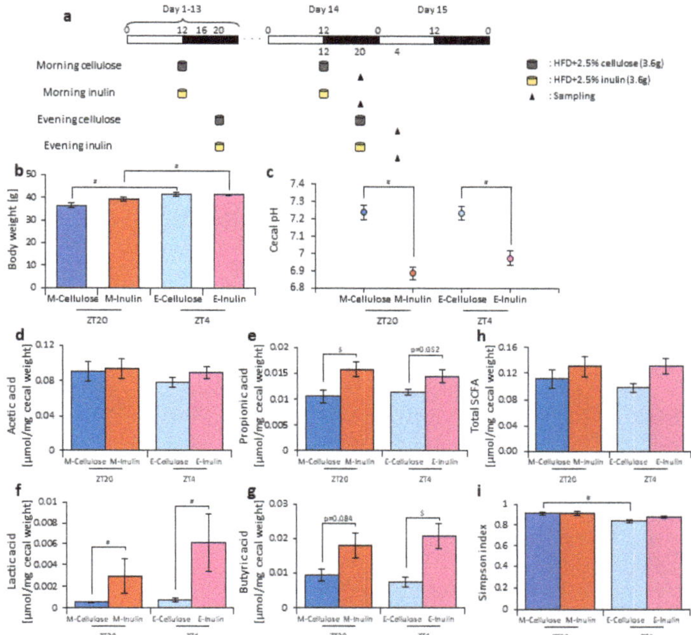

Figure 6. When fasting times are equal, the difference between morning and evening inulin feeding disappears. (**a**) Experimental schedule, where white and black bars indicate environmental 12 h light and dark conditions, respectively. The gray cylinder indicates the 3.6 g high-fat-diet (HFD) with 2.5% cellulose. The yellow cylinder indicates the 3.6 g of HFD with 2.5% inulin. The black arrowhead indicates the sampling time. (**b**) Body weight before sampling. (**c**) Cecal pH of mice housed for 14 days for each group. (**d**–**h**) SCFAs of mice, including (**d**) acetic acid, (**e**) propionic acid, (**f**) lactic acid, (**g**) butyric acid, and (**h**) total SCFAs. (**i**) Bacterial alpha diversity. Comparison of the Simpson index estimation of the 16S rDNA gene libraries at 97% similarity from the sequencing analysis. All values are represented as mean ± SEM (morning cellulose (n = 6); morning inulin (n = 6); evening cellulose (n = 6); evening inulin (n = 6)). $\$ p < 0.05$, evaluated using the two-way ANOVA with Sidak's post hoc test. $\# p < 0.05$, evaluated using the Mann–Whitney test with a two-stage linear step-up procedure of the Benjamini, Krieger, and Yekutieli test for multiple comparisons. The table in (**j**) indicates the results using permutational multivariate analysis of variance (PERMANOVA). Morning cellulose, morning inulin, evening cellulose, or evening inulin are represented as M-cellulose, M-inulin, E-cellulose or E-inulin, respectively.

Table 2. The relative abundance of some bacteria under the condition of one meal per day. (**a**). Phylum level. (**b**). Genus level.

a. Phylum Level						
Bacterial	**ZT20**			**ZT4**		
	Cellulose	**M-Inulin**	***p*-Value**	**Cellulose**	**E-Inulin**	***p*-Value**
Actinobacteria	0.0041 ± 0.0017	0.0364 ± 0.0199	0.0174 #	0.0321 ± 0.0154	0.0871 ± 0.0380	0.1797
Bacteroidetes	0.0759 ± 0.0296	0.1372 ± 0.0337	0.3748	0.1024 ± 0.0275	0.1609 ± 0.0411	0.4069
Deferribacteres	0.0037 ± 0.0016	0.0018 ± 0.0006	0.5714	0.0007 ± 0.0004	0.0002 ± 0.00006	0.4459
Firmicutes	0.80427 ± 0.0240	0.7513 ± 0.0306	0.1797	0.7832 ± 0.0207	0.6906 ± 0.0373	0.1298
Proteobacteria	0.1119 ± 0.0161	0.0706 ± 0.0120	0.0799	0.0790 ± 0.0141	0.0568 ± 0.0102	0.4449
Verrucomicrobia	0.0001 ± 0.0001	0.0025 ± 0.0011	0.0606	0.0025 ± 0.0022	0.0042 ± 0.0036	0.5455

b. Genus Level						
Bacterial	**ZT20**			**ZT4**		
	Cellulose	**M-Inulin**	***p*-Value**	**Cellulose**	**E-Inulin**	***p*-Value**
Bifidobacterium	0.0004 ± 0.0003	0.0305 ± 0.0192	0.0043 ##	0.0266 ± 0.0151	0.0785 ± 0.0367	0.1775
Adlercreutzia	0.0036 ± 0.0017	0.0037 ± 0.0010	0.5714	0.0051 ± 0.0008	0.0041 ± 0.0008	0.3874
Bacteroides	0.0222 ± 0.0103	0.0327 ± 0.0078	0.4395	0.0149 ± 0.0051	0.0313 ± 0.0101	0.1801
Parabacteroides	0.0059 ± 0.0027	0.0032 ± 0.0010	0.8983	0.0048 ± 0.0012	0.0025 ± 0.0005	0.3874
Butyricimonas	0.0002 ± 0.00008	0.0001 ± 0.00004	0.6623	0.0003 ± 0.0001	0.0002 ± 0.0001	0.6591
Odoribacter	0.0015 ± 0.0003	0.0022 ± 0.0012	0.8983	0.0023 ± 0.0004	0.0031 ± 0.0007	0.3874
[Prevotella]	0.0066 ± 0.0031	0.0151 ± 0.0102	0.5541	0.0033 ± 0.0018	0.0056 ± 0.0045	0.9805
Mucispirillum	0.0037 ± 0.0016	0.0018 ± 0.0006	0.5714	0.0007 ± 0.0004	0.0001 ± 0.00006	0.4459
Staphylococcus	0.0005 ± 0.0002	0.00005 ± 0.00003	0.145	0.0009 ± 0.0004	0.00007 ± 0.00002	0.0022 ##
Lactobacillus	0.0080 ± 0.0031	0.0088 ± 0.0027	0.8983	0.1050 ± 0.0530	0.01225 ± 0.0046	0.1797
Lactococcus	0.2490 ± 0.0332	0.1468 ± 0.0205	0.094	0.3408 ± 0.0271	0.1824 ± 0.0494	0.0078 $$
Streptococcus	0.0048 ± 0.0011	0.0013 ± 0.0002	0.0087 ##	0.0037 ± 0.0005	0.0023 ± 0.0006	0.0931
SMB53	0.0102 ± 0.0079	0.0130 ± 0.0078	0.9394	0.0467 ± 0.0224	0.0132 ± 0.0076	0.1688
Dehalobacterium	0.0017 ± 0.0004	0.0024 ± 0.0006	0.5628	0.0007 ± 0.0002	0.0021 ± 0.0008	0.132
Blautia	0.0004 ± 0.0001	0.0003 ± 0.00008	0.7381	0.0002 ± 0.0001	0.0003 ± 0.0001	0.1991
Coprococcus	0.0093 ± 0.0009	0.0116 ± 0.0026	0.7879	0.0028 ± 0.0005	0.0048 ± 0.0016	0.3874
Dorea	0.0017 ± 0.0004	0.0034 ± 0.0007	0.077	0.0011 ± 0.0003	0.0042 ± 0.0006	0.0016 $$
Roseburia	0.00006 ± 0.00002	0.00005 ± 0.00003	0.9242	0.00007 ± 0.00002	0.0001 ± 0.00007	0.3398
[Ruminococcus]	0.0652 ± 0.0085	0.0509 ± 0.0082	0.3544	0.0299 ± 0.0059	0.0347 ± 0.0073	0.882
Anaerotruncus	0.0004 ± 0.0001	0.0001 ± 0.00005	0.1797	0.0001 ± 0.00004	0.0003 ± 0.0001	0.8312
Oscillospira	0.0473 ± 0.0076	0.0213 ± 0.0052	0.0043 $$	0.0217 ± 0.0042	0.0158 ± 0.0022	0.6797
Ruminococcus	0.0082 ± 0.0009	0.0034 ± 0.0007	0.0012 $$	0.0035 ± 0.0008	0.0020 ± 0.0003	0.7285
Allobaculum	0.0011 ± 0.0006	0.1122 ± 0.0562	0.0022 ##	0.0327 ± 0.0157	0.1703 ± 0.0513	0.0449 #
Catenibacterium	0.0005 ± 0.0001	0.0003 ± 0.0001	0.3874	0.0004 ± 0.00009	0.0003 ± 0.0001	0.474
Desulfovibrio	0.0010 ± 0.0005	0.0011 ± 0.0007	0.9073	0.0006 ± 0.0002	0.0023 ± 0.0009	0.1001
Citrobacter	0.0030 ± 0.0011	0.0026 ± 0.0004	0.7879	0.0026 ± 0.0004	0.0015 ± 0.0003	0.0931
Klebsiella	0.0417 ± 0.0147	0.0328 ± 0.0138	0.5714	0.0412 ± 0.0105	0.0291 ± 0.0124	0.1797
Akkermansia	0.0001 ± 0.0001	0.0025 ± 0.0011	0.4545	0.0024 ± 0.0022	0.0041 ± 0.0036	0.5455

(**a**) Number of bacteria significantly changed by M-inulin/all number of bacteria well-detected = 1/6. Number of bacteria significantly changed by E-inulin/all number of bacteria well-detected = 0/6. # $p < 0.05$, evaluated using the Mann–Whitney test with a two-stage linear step-up procedure of the Benjamini, Krieger, and Yekutieli test for multiple comparisons. (**b**) Number of bacteria significantly changed by M-inulin/all number of bacteria well-detected = 5/28. Number of bacteria significantly changed by E-inulin/all number of bacteria well-detected = 4/28. $$ $p < 0.01$, evaluated using the two-way ANOVA with Sidak post hoc test. ## $p < 0.01$, # $p < 0.05$, evaluated using the Mann–Whitney test with a two-stage linear step-up procedure of the Benjamini, Krieger, and Yekutieli test for multiple comparisons.

4. Discussion

In this study, inulin intake changed the composition and profile of the gut microbiota, increased SCFAs, and decreased the cecal pH (Figure 1, Figure 2 and Figure S2). SCFAs are important for health because they improve energy metabolism in the liver and muscles and immune function in the large intestine [9–11]. In addition, the effect of inulin on the microbiota was dependent on the timing of inulin intake. Therefore, we gave inulin to the mice in either the morning or evening. The microbiota was more affected by inulin feeding in the morning than in the evening (Figure 5 and Figure S4) because the fasting period was longer for the latter. There has been previous research on fasting time and dietary effects. Previous studies examining postprandial glucose metabolism have shown that breakfast, rather than dinner, can suppress postprandial hyperglycemia and that one of the primary factors is the difference in fasting time [51]. Additionally, in a previous study examining the circadian clock, a meal after a long fasting period strongly synchronized the peripheral clock [52,53]. Under a two-meals-per-day schedule in mice, the same amount of chow after 16 h of fasting could reset the *Per2* gene expression rhythm in the liver clock compared with the same amount of chow after 6 h of fasting; in the two-meal experiments presented here, we used exactly the same protocol. In the current experiment, there was no difference in the cecal pH or SCFAs measurements between morning and evening with the same fasting duration (Figure 6c,h). Considering actual human life, the fasting time until breakfast is generally the longest among the three meals. Thus, these results, along with those of the previous study [53], support that inulin intake in the morning is most effective at attenuating HFD-induced changes of the gut microbiota. However, since the gut microbiota is also related to the circadian clock, there may be a difference between morning and evening in the gut microbiota composition, regardless of the fasting time. In addition to the daily feeding model used in this study, a feeding model for equalizing fasting time has been considered [52,54,55]. By using these feeding models, the relationship between fasting time and the effects of foods may be clarified. Moreover, the feeding model of this study has too long a starvation period compared with actual human life. Therefore, a feeding model that mimics the actual human lifestyle of three meals a day, as reported by Kuroda et al., may be considered for future experiments [52].

In this study, we first regulated the access time to inulin-containing food under a two-meals-per-day schedule because mice access food in the morning rather than in the evening under ad-lib food conditions [54,56]. Under these feeding conditions, we found clear effects of inulin in the morning. Therefore, in ad-lib feeding conditions, functional food intake at an earlier time during the active period may be a considerable factor in microbiota changes. Next, we regulated the food volume under a two-meals-per-day schedule. Once again, inulin in the morning had a clear effect on the microbiota, clarifying the importance of inulin intake in the morning on the beta-diversity and profile of the microbiota. However, in these experimental conditions, we did not control feeding and/or digestive speed; therefore, volume- and speed-controlled feeding systems may be required to determine the effect of feeding time.

The first meal after a long fast, most often breakfast, resets the phase of peripheral clock [52,53]. We recently demonstrated that cellobiose, a water-soluble dietary fiber, produces SCFAs, allowing them to reset the peripheral clock [21]. Taken together, these results suggest that the intake of inulin-containing foods in the morning may help reset the peripheral clock through SCFAs production.

Previous studies have reported that inulin consumption increases *Bifidobacteria* and *Akkermancia muciniphila* and decreases gram-positive cocci in humans and mice [47–50]. In this study, the gram-positive cocci *Streptococcus* and *Staphylococcus* decreased, but the *A. muciniphila* was not significantly changed. We considered that the degree of polymerization of inulin is one of the reasons that the results of this study differ from previous studies. In the structure of inulin, fructose is a monomer linked by 2–60 molecules with β-glycosidic bonds. The inulin used in this study had 16 fructose bonds (a degree of polymerization of 16) [34]. It has been reported that the influence on the microbiota is different depending on the degree of polymerization of inulin [48]. Therefore, the results may have been different with other degrees of polymerization.

Streptococcus is known to produce lactic acid [57,58], and *Streptococcus mutans* increases in the intestines of type 2 diabetes patients and is induced by a high-calorie diet [59]. Furthermore, *Staphylococcus aureus* is increased in obese patients, and *Staphylococcus* has a positive correlation with energy intake [60,61]. The SCFAs produced by inulin feeding increase the concentration of GLP-1 in the blood and promote insulin secretion [62]. Furthermore, SCFAs regulate insulin activity in adipose tissue through the GPR43 receptor [11]. Therefore, it has been suggested that inulin may be an effective food against diabetes. In fact, in rats and humans, inulin consumption inhibits blood glucose levels and lowers blood triglyceride levels and total cholesterol levels [35,63,64]. In this study, the species level was not fully detected, and blood glucose levels and triglycerides were not measured. If these factors were measured, we may have been able to clarify the relationship between the gut microbiota and glucose metabolism.

It has been reported that SCFAs produced by ingestion of water-soluble dietary fiber prevent fat accumulation in adipose tissue via GPR43 [11]. However, it has also been reported that water-soluble dietary fiber does not involve SCFAs and suppresses fatty acid accumulation itself. For example, water-soluble dietary fiber may form a highly viscous matrix in the small intestine, increase the viscosity of the small intestine, and then physically suppress fat absorption [65,66]. These reports should be considered when investigating the association between gut microbiota and lipid metabolism.

The analysis of the carbohydrate metabolism identified a significant association with the fructose and mannose metabolism in the morning inulin group under two meals per day but not under one meal per day. Inulin is a fructan polymerized with fructose. Therefore, it may be possible that fructose metabolism is more activated by inulin in the morning than in the evening. Furthermore, the production of SCFAs may be increased because fructose is metabolized in the morning. In addition, fructose metabolism may also be related to fasting time. PICRUSt is only a predictive tool. To determine accurate functional information of the related bacteria, metagenomic studies should be conducted. Additionally, the number of mice in each group should be increased to provide more accurate explanations regarding the microbiota and PICRUSt analysis.

5. Conclusions

In summary, inulin intake in the morning rather than in the evening affected the gut microbiota, promoted SCFAs production, and lowered the cecal pH. The difference between the morning and evening results was related to the fasting duration, suggesting that there may be a relationship between fasting duration and meal stimulation regarding control of the microbiota.

Supplementary Materials: The following are available online at http://www.mdpi.com/2072-6643/11/11/2802/s1, Figure S1: the addition of cellulose does not affect the microbiota. Figure S2: inulin feeding in the morning changed the microbiota composition more than in the evening. Figure S3: morning inulin feeding changed the microbiota composition more than evening inulin feeding. Figure S4: morning inulin feeding changed the microbiota composition and the relative abundance of inferred functional profile more than evening inulin feeding under equivalent feeding conditions. Figure S5: the microbiota composition and the relative abundance of inferred functional profile was not significantly changed by inulin feeding, even under the condition of one meal per day.

Author Contributions: H.S., H.M. and S.S.; designed the research and analyzed the data. H.S. and S.S.; wrote the manuscript. H.S., H.M., A.W., Y.N., Y.L. and K.H.; performed the experiments.

Funding: This work was partially supported by the Council for Science, Technology, and Innovation, SIP, "Technologies for creating next-generation agriculture, forestry, and fisheries" (funding agency: Bio-oriented Technology Research Advancement Institution, NARO) (Shibata. S.) and Japan Society for the Promotion of Science (JSPS) KAKENHI (A and Houga) (Shibata. S).

Conflicts of Interest: We declare that there are no conflicts of interest related to this study.

References

1. Hooks, K.B.; O'Malley, M.A. Dysbiosis and its discontents. *mBio* **2017**, *8*, e01492–e01517. [CrossRef]
2. Ley, R.E.; Turnbaugh, P.J.; Klein, S.; Gordon, J.I. Microbial ecology: Human gut microbes associated with obesity. *Nature* **2006**, *444*, 1022–1023. [CrossRef] [PubMed]
3. Turnbaugh, P.J.; Ley, R.E.; Mahowald, M.A.; Magrini, V.; Mardis, E.R.; Gordon, J.I. An obesity-associated gut microbiome with increased capacity for energy harvest. *Nature* **2006**, *444*, 1027–1031. [CrossRef] [PubMed]

4. Jahani-Sherafat, S.; Alebouyeh, M.; Moghim, S.; Ahmadi Amoli, H.; Ghasemian-Safaei, H. Role of gut microbiota in the pathogenesis of colorectal cancer; a review article. *Gastroenterol. Hepatol. Bed Bench* **2018**, *11*, 101–109.
5. Mai, V.; Morris, J.G., Jr. Need for prospective cohort studies to establish human gut microbiome contributions to disease risk. *J. Natl. Cancer Inst.* **2013**, *105*, 1850–1851. [CrossRef] [PubMed]
6. Jiang, H.; Ling, Z.; Zhang, Y.; Mao, H.; Ma, Z.; Yin, Y.; Wang, W.; Tang, W.; Tan, Z.; Shi, J.; et al. Altered fecal microbiota composition in patients with major depressive disorder. *Brain Behav. Immun.* **2015**, *48*, 186–194. [CrossRef] [PubMed]
7. Koh, A.; De Vadder, F.; Kovatcheva-Datchary, P.; Backhed, F. From dietary fiber to host physiology: Short-chain fatty acids as key bacterial metabolites. *Cell* **2016**, *165*, 1332–1345. [CrossRef] [PubMed]
8. den Besten, G.; van Eunen, K.; Groen, A.K.; Venema, K.; Reijngoud, D.J.; Bakker, B.M. The role of short-chain fatty acids in the interplay between diet, gut microbiota, and host energy metabolism. *J. Lipid Res.* **2013**, *54*, 2325–2340. [CrossRef]
9. Cummings, J.H.; Pomare, E.W.; Branch, W.J.; Naylor, C.P.; Macfarlane, G.T. Short chain fatty acids in human large intestine, portal, hepatic and venous blood. *Gut* **1987**, *28*, 1221–1227. [CrossRef]
10. Okada, T.; Fukuda, S.; Hase, K.; Nishiumi, S.; Izumi, Y.; Yoshida, M.; Hagiwara, T.; Kawashima, R.; Yamazaki, M.; Oshio, T.; et al. Microbiota-derived lactate accelerates colon epithelial cell turnover in starvation-refed mice. *Nat. Commun.* **2013**, *4*, 1654. [CrossRef]
11. Kimura, I.; Ozawa, K.; Inoue, D.; Imamura, T.; Kimura, K.; Maeda, T.; Terasawa, K.; Kashihara, D.; Hirano, K.; Tani, T.; et al. The gut microbiota suppresses insulin-mediated fat accumulation via the short-chain fatty acid receptor gpr43. *Nat. Commun.* **2013**, *4*, 1829. [CrossRef] [PubMed]
12. Bass, J.; Takahashi, J.S. Circadian integration of metabolism and energetics. *Science* **2010**, *330*, 1349–1354. [CrossRef] [PubMed]
13. Shibata, S.; Tahara, Y.; Hirao, A. The adjustment and manipulation of biological rhythms by light, nutrition, and abused drugs. *Adv. Drug Deliv. Rev.* **2010**, *62*, 918–927. [CrossRef] [PubMed]
14. Tahara, Y.; Aoyama, S.; Shibata, S. The mammalian circadian clock and its entrainment by stress and exercise. *J. Physiol. Sci. JPS* **2017**, *67*, 1–10. [CrossRef] [PubMed]
15. Tahara, Y.; Shibata, S. Entrainment of the mouse circadian clock: Effects of stress, exercise, and nutrition. *Free Radic. Biol. Med.* **2018**, *119*, 129–138. [CrossRef] [PubMed]
16. Sasaki, H.; Hattori, Y.; Ikeda, Y.; Kamagata, M.; Iwami, S.; Yasuda, S.; Tahara, Y.; Shibata, S. Forced rather than voluntary exercise entrains peripheral clocks via a corticosterone/noradrenaline increase in per2::Luc mice. *Sci. Rep.* **2016**, *6*, 27607. [CrossRef]
17. Leone, V.; Gibbons, S.M.; Martinez, K.; Hutchison, A.L.; Huang, E.Y.; Cham, C.M.; Pierre, J.F.; Heneghan, A.F.; Nadimpalli, A.; Hubert, N.; et al. Effects of diurnal variation of gut microbes and high-fat feeding on host circadian clock function and metabolism. *Cell Host Microbe* **2015**, *17*, 681–689. [CrossRef]
18. Liang, X.; Bushman, F.D.; FitzGerald, G.A. Rhythmicity of the intestinal microbiota is regulated by gender and the host circadian clock. *Proc. Natl. Acad. Sci. USA* **2015**, *112*, 10479–10484. [CrossRef]
19. Thaiss, C.A.; Zeevi, D.; Levy, M.; Zilberman-Schapira, G.; Suez, J.; Tengeler, A.C.; Abramson, L.; Katz, M.N.; Korem, T.; Zmora, N.; et al. Transkingdom control of microbiota diurnal oscillations promotes metabolic homeostasis. *Cell* **2014**, *159*, 514–529. [CrossRef]
20. Zarrinpar, A.; Chaix, A.; Yooseph, S.; Panda, S. Diet and feeding pattern affect the diurnal dynamics of the gut microbiome. *Cell Metab.* **2014**, *20*, 1006–1017. [CrossRef]
21. Tahara, Y.; Yamazaki, M.; Sukigara, H.; Motohashi, H.; Sasaki, H.; Miyakawa, H.; Haraguchi, A.; Ikeda, Y.; Fukuda, S.; Shibata, S. Gut microbiota-derived short chain fatty acids induce circadian clock entrainment in mouse peripheral tissue. *Sci. Rep.* **2018**, *8*, 1395. [CrossRef] [PubMed]
22. David, L.A.; Materna, A.C.; Friedman, J.; Campos-Baptista, M.I.; Blackburn, M.C.; Perrotta, A.; Erdman, S.E.; Alm, E.J. Host lifestyle affects human microbiota on daily timescales. *Genome Biol.* **2014**, *15*, R89. [CrossRef] [PubMed]
23. Gibson, G.R.; Roberfroid, M.B. Dietary modulation of the human colonic microbiota: Introducing the concept of prebiotics. *J. Nutr.* **1995**, *125*, 1401–1412. [CrossRef] [PubMed]
24. Valcheva, R.; Dieleman, L.A. Prebiotics: Definition and protective mechanisms. *Best Pract. Res. Clin. Gastroenterol.* **2016**, *30*, 27–37. [CrossRef] [PubMed]

25. Hoving, L.R.; Katiraei, S.; Pronk, A.; Heijink, M.; Vonk, K.K.D.; Amghar-El Bouazzaoui, F.; Vermeulen, R.; Drinkwaard, L.; Giera, M.; van Harmelen, V.; et al. The prebiotic inulin modulates gut microbiota but does not ameliorate atherosclerosis in hypercholesterolemic apoe*3-leiden.Cetp mice. *Sci. Rep.* **2018**, *8*, 16515. [CrossRef] [PubMed]
26. Demigne, C.; Jacobs, H.; Moundras, C.; Davicco, M.J.; Horcajada, M.N.; Bernalier, A.; Coxam, V. Comparison of native or reformulated chicory fructans, or non-purified chicory, on rat cecal fermentation and mineral metabolism. *Eur. J. Nutr.* **2008**, *47*, 366–374. [CrossRef] [PubMed]
27. Hatori, M.; Vollmers, C.; Zarrinpar, A.; DiTacchio, L.; Bushong, E.A.; Gill, S.; Leblanc, M.; Chaix, A.; Joens, M.; Fitzpatrick, J.A.; et al. Time-restricted feeding without reducing caloric intake prevents metabolic diseases in mice fed a high-fat diet. *Cell Metab.* **2012**, *15*, 848–860. [CrossRef]
28. Wang, X.; Xue, J.; Yang, J.; Xie, M. Timed high-fat diet in the evening affects the hepatic circadian clock and pparalpha-mediated lipogenic gene expressions in mice. *Genes Nutr.* **2013**, *8*, 457–463. [CrossRef]
29. Arble, D.M.; Bass, J.; Laposky, A.D.; Vitaterna, M.H.; Turek, F.W. Circadian timing of food intake contributes to weight gain. *Obesity* **2009**, *17*, 2100–2102. [CrossRef]
30. Haraguchi, A.; Aoki, N.; Ohtsu, T.; Ikeda, Y.; Tahara, Y.; Shibata, S. Controlling access time to a high-fat diet during the inactive period protects against obesity in mice. *Chronobiol. Int.* **2014**, *31*, 935–944. [CrossRef]
31. Hsieh, S.D.; Muto, T.; Murase, T.; Tsuji, H.; Arase, Y. Association of short sleep duration with obesity, diabetes, fatty liver and behavioral factors in japanese men. *Intern. Med.* **2011**, *50*, 2499–2502. [CrossRef] [PubMed]
32. Baron, K.G.; Reid, K.J.; Kern, A.S.; Zee, P.C. Role of sleep timing in caloric intake and bmi. *Obesity* **2011**, *19*, 1374–1381. [CrossRef] [PubMed]
33. Oishi, K.; Konishi, T.; Hashimoto, C.; Yamamoto, S.; Takahashi, Y.; Shiina, Y. Dietary fish oil differentially ameliorates high-fructose diet-induced hepatic steatosis and hyperlipidemia in mice depending on time of feeding. *J. Nutr. Biochem.* **2018**, *52*, 45–53. [CrossRef] [PubMed]
34. Wada, T.; Ohguchi, M.; Iwai, Y. A novel enzyme of bacillus sp. 217c-11 that produces inulin from sucrose. *Biosci. Biotechnol. Biochem.* **2003**, *67*, 1327–1334. [CrossRef]
35. Wada, T.; Sugatani, J.; Terada, E.; Ohguchi, M.; Miwa, M. Physicochemical characterization and biological effects of inulin enzymatically synthesized from sucrose. *J. Agric. Food Chem.* **2005**, *53*, 1246–1253. [CrossRef]
36. Kowalski, T.J.; Farley, C.; Cohen-Williams, M.E.; Varty, G.; Spar, B.D. Melanin-concentrating hormone-1 receptor antagonism decreases feeding by reducing meal size. *Eur. J. Pharmacol.* **2004**, *497*, 41–47. [CrossRef]
37. Oh, J.; Lee, S.R.; Hwang, K.T.; Ji, G.E. The anti-obesity effects of the dietary combination of fermented red ginseng with levan in high fat diet mouse model. *Phytother. Res. PTR* **2014**, *28*, 617–622. [CrossRef]
38. Weitkunat, K.; Schumann, S.; Petzke, K.J.; Blaut, M.; Loh, G.; Klaus, S. Effects of dietary inulin on bacterial growth, short-chain fatty acid production and hepatic lipid metabolism in gnotobiotic mice. *J. Nutr. Biochem.* **2015**, *26*, 929–937. [CrossRef]
39. Ponziani, F.R.; Pecere, S.; Gasbarrini, A.; Ojetti, V. Physiology and pathophysiology of liver lipid metabolism. *Expert Rev. Gastroenterol. Hepatol.* **2015**, *9*, 1055–1067. [CrossRef]
40. Federico, A.; Dallio, M.; Di Sarno, R.; Giorgio, V.; Miele, L. Gut microbiota, obesity and metabolic disorders. *Minerva Gastroenterol. E Dietol.* **2017**, *63*, 337–344.
41. Pascale, A.; Marchesi, N.; Marelli, C.; Coppola, A.; Luzi, L.; Govoni, S.; Giustina, A.; Gazzaruso, C. Microbiota and metabolic diseases. *Endocrine* **2018**, *61*, 357–371. [CrossRef]
42. Huazano, A.; López, M.G. Metabolism of short chain fatty acids in colon and faeces of mice after a supplementation of diets with agave fructans. *Lipid Metab.* **2013**, *8*, 163–182.
43. Nishijima, S.; Suda, W.; Oshima, K.; Kim, S.W.; Hirose, Y.; Morita, H.; Hattori, M. The gut microbiome of healthy japanese and its microbial and functional uniqueness. *DNA Res. Int. J. Rapid Publ. Rep. Genes Genomes* **2016**, *23*, 125–133. [CrossRef]
44. Caporaso, J.G.; Kuczynski, J.; Stombaugh, J.; Bittinger, K.; Bushman, F.D.; Costello, E.K.; Fierer, N.; Pena, A.G.; Goodrich, J.K.; Gordon, J.I.; et al. Qiime allows analysis of high-throughput community sequencing data. *Nat. Methods* **2010**, *7*, 335–336. [CrossRef]
45. Edgar, R.C. Search and clustering orders of magnitude faster than blast. *Bioinformatics* **2010**, *26*, 2460–2461. [CrossRef]

46. Langille, M.G.; Zaneveld, J.; Caporaso, J.G.; McDonald, D.; Knights, D.; Reyes, J.A.; Clemente, J.C.; Burkepile, D.E.; Vega Thurber, R.L.; Knight, R.; et al. Predictive functional profiling of microbial communities using 16s rrna marker gene sequences. *Nat. Biotechnol.* **2013**, *31*, 814–821. [CrossRef]
47. Gibson, G.R.; Beatty, E.R.; Wang, X.; Cummings, J.H. Selective stimulation of bifidobacteria in the human colon by oligofructose and inulin. *Gastroenterology* **1995**, *108*, 975–982. [CrossRef]
48. Zhu, L.; Qin, S.; Zhai, S.; Gao, Y.; Li, L. Inulin with different degrees of polymerization modulates composition of intestinal microbiota in mice. *FEMS Microbiol. Lett.* **2017**, *364*, 364. [CrossRef]
49. Zhang, S.; Yang, J.; Henning, S.M.; Lee, R.; Hsu, M.; Grojean, E.; Pisegna, R.; Ly, A.; Heber, D.; Li, Z. Dietary pomegranate extract and inulin affect gut microbiome differentially in mice fed an obesogenic diet. *Anaerobe* **2017**, *48*, 184–193. [CrossRef]
50. Brighenti, F.; Casiraghi, M.C.; Canzi, E.; Ferrari, A. Effect of consumption of a ready-to-eat breakfast cereal containing inulin on the intestinal milieu and blood lipids in healthy male volunteers. *Eur. J. Clin. Nutr.* **1999**, *53*, 726–733. [CrossRef]
51. Takahashi, M.; Ozaki, M.; Kang, M.I.; Sasaki, H.; Fukazawa, M.; Iwakami, T.; Lim, P.J.; Kim, H.K.; Aoyama, S.; Shibata, S. Effects of meal timing on postprandial glucose metabolism and blood metabolites in healthy adults. *Nutrients* **2018**, *10*, 1763. [CrossRef]
52. Kuroda, H.; Tahara, Y.; Saito, K.; Ohnishi, N.; Kubo, Y.; Seo, Y.; Otsuka, M.; Fuse, Y.; Ohura, Y.; Hirao, A.; et al. Meal frequency patterns determine the phase of mouse peripheral circadian clocks. *Sci. Rep.* **2012**, *2*, 711. [CrossRef]
53. Hirao, A.; Nagahama, H.; Tsuboi, T.; Hirao, M.; Tahara, Y.; Shibata, S. Combination of starvation interval and food volume determines the phase of liver circadian rhythm in per2::Luc knock-in mice under two meals per day feeding. *Am. J. Physiol. Gastrointest. Liver Physiol.* **2010**, *299*, G1045–G1053. [CrossRef]
54. Ikeda, Y.; Sasaki, H.; Ohtsu, T.; Shiraishi, T.; Tahara, Y.; Shibata, S. Feeding and adrenal entrainment stimuli are both necessary for normal circadian oscillation of peripheral clocks in mice housed under different photoperiods. *Chronobiol. Int.* **2015**, *32*, 195–210. [CrossRef]
55. Sasaki, H.; Hattori, Y.; Ikeda, Y.; Kamagata, M.; Iwami, S.; Yasuda, S.; Shibata, S. Phase shifts in circadian peripheral clocks caused by exercise are dependent on the feeding schedule in per2::Luc mice. *Chronobiol. Int.* **2016**, *33*, 849–862. [CrossRef]
56. Sasaki, H.; Ohtsu, T.; Ikeda, Y.; Tsubosaka, M.; Shibata, S. Combination of meal and exercise timing with a high-fat diet influences energy expenditure and obesity in mice. *Chronobiol. Int.* **2014**, *31*, 959–975. [CrossRef]
57. Kilic, A.O.; Pavlova, S.I.; Ma, W.G.; Tao, L. Analysis of lactobacillus phages and bacteriocins in american dairy products and characterization of a phage isolated from yogurt. *Appl. Environ. Microbiol.* **1996**, *62*, 2111–2116.
58. Gilliland, S.E. Health and nutritional benefits from lactic acid bacteria. *FEMS Microbiol. Rev.* **1990**, *7*, 175–188. [CrossRef]
59. Koh, A.; Molinaro, A.; Stahlman, M.; Khan, M.T.; Schmidt, C.; Manneras-Holm, L.; Wu, H.; Carreras, A.; Jeong, H.; Olofsson, L.E.; et al. Microbially produced imidazole propionate impairs insulin signaling through mtorc1. *Cell* **2018**, *175*, 947–961.e917. [CrossRef]
60. Gomes, A.C.; Hoffmann, C.; Mota, J.F. The human gut microbiota: Metabolism and perspective in obesity. *Gut Microbes* **2018**, *9*, 308–325. [CrossRef]
61. Bervoets, L.; Van Hoorenbeeck, K.; Kortleven, I.; Van Noten, C.; Hens, N.; Vael, C.; Goossens, H.; Desager, K.N.; Vankerckhoven, V. Differences in gut microbiota composition between obese and lean children: A cross-sectional study. *Gut Pathog.* **2013**, *5*, 10. [CrossRef]
62. Delzenne, N.M.; Cani, P.D.; Neyrinck, A.M. Modulation of glucagon-like peptide 1 and energy metabolism by inulin and oligofructose: Experimental data. *J. Nutr.* **2007**, *137*, 2547s–2551s. [CrossRef]
63. Russo, F.; Chimienti, G.; Riezzo, G.; Pepe, G.; Petrosillo, G.; Chiloiro, M.; Marconi, E. Inulin-enriched pasta affects lipid profile and lp(a) concentrations in italian young healthy male volunteers. *Eur. J. Nutr.* **2008**, *47*, 453–459. [CrossRef]
64. Letexier, D.; Diraison, F.; Beylot, M. Addition of inulin to a moderately high-carbohydrate diet reduces hepatic lipogenesis and plasma triacylglycerol concentrations in humans. *Am. J. Clin. Nutr.* **2003**, *77*, 559–564. [CrossRef]

65. Shirouchi, B.; Kawamura, S.; Matsuoka, R.; Baba, S.; Nagata, K.; Shiratake, S.; Tomoyori, H.; Imaizumi, K.; Sato, M. Dietary guar gum reduces lymph flow and diminishes lipid transport in thoracic duct-cannulated rats. *Lipids* **2011**, *46*, 789–793. [CrossRef]
66. Eastwood, M.A.; Morris, E.R. Physical properties of dietary fiber that influence physiological function: A model for polymers along the gastrointestinal tract. *Am. J. Clin. Nutr.* **1992**, *55*, 436–442. [CrossRef]

 *nutrients*

Article

Changes in Microbiota and Bacterial Protein Caseinolytic Peptidase B During Food Restriction in Mice: Relevance for the Onset and Perpetuation of Anorexia Nervosa

Manon Dominique [1,2,3], Romain Legrand [1], Marie Galmiche [1,2,3], Saïda Azhar [1], Camille Deroissart [1], Charlène Guérin [2,3], Jean-Luc do Rego [3,4], Fatima Leon [3,4], Séverine Nobis [3,4], Grégory Lambert [1], Nicolas Lucas [1] and Pierre Déchelotte [1,2,3,5,*]

1. TargEDys SA, University of Rouen Normandy, 76183 Rouen, France; mdominique@targedys.com (M.D.); legrandromain@hotmail.fr (R.L.); mgalmiche@targedys.com (M.G.); saida-az@hotmail.fr (S.A.); camillederoissart@gmail.com (C.D.); glambert@targedys.com (G.L.); lucasnicolas@hotmail.fr (N.L.)
2. Inserm UMR1073, Nutrition, Gut and Brain Laboratory, University of Rouen Normandy, Unirouen, 76183 Rouen, France; charlene.guerin@univ-rouen.fr
3. Institute for Research and Innovation in Biomedicine (IRIB), University of Rouen Normandy, Unirouen, 76183 Rouen, France; jean-luc.do-rego@univ-rouen.fr (J.-L.d.R.); Fati.jpl@free.fr (F.L.); severine-nobis@orange.fr (S.N.)
4. Animal Behavior Platform, Service Commun d'Analyse Comportementale (SCAC), University of Rouen Normandy, 76183 Rouen, France
5. Rouen University Hospital, CHU Charles Nicolle, 76183 Rouen, France

* Correspondence: Pierre.Dechelotte@chu-rouen.fr; Tel.: +2-35-14-82-20

Received: 9 September 2019; Accepted: 15 October 2019; Published: 18 October 2019

Abstract: Microbiota contributes to the regulation of eating behavior and might be implicated in the pathophysiology of anorexia nervosa. ClpB (Caseinolytic peptidase B) protein produced mainly by the *Enterobacteriaceae* family has been identified as a conformational mimetic of α-MSH, which could result in similar anorexigenic effects. The aim of this study was to highlight the role of the microbiome and the ClpB protein in deregulation and self-maintenance of anorexia pathology. Male C57Bl/6 mice were undergone to the ABA (Activity-Based Anorexia) protocol: after 5 days of acclimatization, both ABA and LFA (Limited Food Access) mice had progressively limited access to food until D17. At the end of protocol, the plasma ClpB concentration and *Enterobacteriaceae* DNA in colonic content were measured. As expected, dietary restriction induced lost weight in LFA and ABA mice. At D10, colonic permeability and plasma concentration of the ClpB protein were significantly increased in LFA and ABA mice vs. controls. At D17, plasma concentration of ClpB was increased in LFA and ABA mice and, it was correlated with proportion of *Enterobacteriaceae* in the faeces. These abnormally high ClpB concentrations and all associated factors, and therefore might contribute to the initiation and/or perpetuation of anorexia nervosa by interfering with satiety signaling.

Keywords: anorexia; food restriction; ClpB; microbiota; *Enterobacteriaceae*

1. Introduction

Eating Disorders (ED) are public health problems that have continued to worsen in recent years with a prevalence of 3.5% from 2000–2006 to 7.8% in 2013–2018 [1]. Among these disorders, anorexia nervosa (AN) is characterized by a difficulty in maintaining a minimum weight and an obsession with weight and body shape [2], the pathophysiology of which is multifactorial and remains partially debated [3].

Among the proposed mechanisms of AN, the role of the gut microbiota in regulating the physiology of AN is increasingly recognized [4,5]. Indeed, studies have shown that intestinal microbial composition is influenced directly by food, in the short and long term [6–8]. Conversely, behavior [9] and appetite [10] are modulated at least in part by several gut-microbiota derived signals, among which bacterial products (e.g., peptides, neurotransmitters) have been shown to influence peripheral and central mechanisms of satiety, reward [11,12] and anxiety [13]. Finally, microbiota composition is implicated in the regulation of body composition: dysbiosis has been reported both in obese individuals [14] and in patients with AN [15]. Moreover, increased *Escherichia coli*, a leading representative of *Enterobacteriaceae* in gut microbiota was also observed in anorexic patients [16]. Altogether, these data strongly suggest that dysfunction of the microbiota-intestine-brain axis in response to exogenous triggering factors might be a key factor in the onset and/or perpetuation of ED [10,17]. Communication between microbiota, gut and brain may rely on various microbiota-derived signals, such as proteins, peptides, monoamines, metabolites, or even gut-produced immunoglobulins gaining access to the brain or modulating afferent neuronal or hormonal regulations generated in the splanchnic area [17]. Among bacterial proteins, ClpB (Caseinolytic peptidase B), a heat shock protein produced by *Enterobacteriaceae* [18] including *E. coli* is of particular relevance to the control of satiety [19] since it holds in common a six amino acid discontinuous epitope sharing molecular mimicry with α-melanocyte-stimulating hormone (α-MSH), the main central neuropeptide signaling satiety in the hypothalamus [20,21]. In addition, other studies have shown that α-MSH could also be found at peripheral level [22]. Moreover, α-MSH could induce the activation of MC4R present on intestinal enteroendocrine L cells [22,23]. Through this specificity, ClpB could stimulate the secretion by enteroendocrine L cells of the satiating hormones GLP-1 or PYY and activate vagal and hormonal pathways leading to hypothalamic activation of the POMC neurons releasing α-MSH [10,11]. In accordance with a role of this protein in the physiological and pathological regulation of eating behavior, ClpB was found naturally in the plasma of healthy subjects and at a higher level in patients with eating disorders [24].

In addition to the direct effect of ClpB mentioned previously, the hypothesis that microbial proteins may also modulate eating behavior through the intestinal production of specific immunoglobulins (Ig) can be suggested. Indeed, previous reports have detected Ig which react with α-MSH, in the sera of both healthy individuals and rats [25]. The levels of these Ig correlate with psychological traits characteristic of eating disorders [25]. This suggests that α-MSH reactive Ig may interfere with melanocortin signaling in both normal and pathological conditions. Moreover, a recent study showed that the levels of α-MSH-reactive IgG, the binding of melanocortin 4 receptor (MC4R) and the cellular internalization rate of MC4R-expressing cells were all lower in obese subjects [26]. Inverse results were found in anorexic and bulimic patients [26]. Other studies also confirmed the implication of α-MSH reactive Ig in the physiological regulation of feeding and mood [27]. In patients with eating disorders, increasing ClpB plasma levels correlated with plasma levels of anti-ClpB and anti-α-MSH Ig [19]. These factors emphasize the physiological involvement of anti-α-MSH Ig in the regulation of food intake.

Thus, bacterial ClpB protein appears as a candidate for interfering with endogenous pathway of satiety regulation. To get further insights in its involvement during food restriction, we performed the present study in a well-established model of food restriction in rodents, the Activity-Based Anorexia (ABA) model, and evaluated the impact of food restriction on the plasma ClpB protein and its related Ig and on the proportion of *Enterobacteriaceae*.

2. Materils and Methods

2.1. Animal Experimentation

Animal experimentation procedures were approved by the Local Ethical Committee of Normandy (approval CENOMEXA n°1112–05). Male C57Bl/6 mice (Janvier Labs, Genest-Saint-Isle, France), at 7 weeks old were kept in holding cages (four mice per cage) at environmental conditions 22 °C ± 3 °C and relative humidity of 40 ± 20% on a 12 h light-dark cycle with lights on at 10:00 a.m. During acclimatization period, all mice were given *ad libitum* access to water and standard food (Kliba Nafag, Germany).

At D1 of the protocol, all mice were randomized individually into 3 groups: An *ad libitum* group (Control, $n = 16$), a limited-food access group (LFA, $n = 16$) and an activity-based anorexia group (ABA, $n = 16$). ABA mice were placed individually in cages with an activity wheel connected to Running Wheel ® software (Intellibio, Seichamps, France).

Food access was progressively limited in ABA and LFA groups from 6 h per day at D6, to 3 h at D9 and until the end of the experiment. Mice always had free access to water. Body weight, water and food intake were measured at 9:00 a.m. each day.

At D10, 8 mice of each group were chosen according to their weight and were anaesthetized by ketamine/xylazine (Imalgene® 1000, Murial/Xylazine Rompun 2%, Bayer) intraperitoneally and were euthanized by decapitation. Blood samples were taken from the mesenteric artery before decapitation. The hypothalamus was taken to perform qPCR to analyze the anorexigenic (POMC) and orexigenic (AgRP) neuronal populations. Intracolonic faeces were taken to perform qPCR to analyze the *Enterobacteriaceae* DNA. The plasma was recovered after centrifugation (3000× *g*, 20 min, 4 °C). Samples were taken and stored at −80 °C if their analysis was not done immediately.

At D17, the end of the experiment, remaining mice underwent the same procedures as D10.

2.2. Permeability

Colon permeability was assessed by measured FITC-dextran (4 kDa) (Sigma) by Ussing chambers. FITC-dextran (5 mg/mL) was placed on the mucosal side. After 3 h at 37 °C, medium from the serosal side was removed and stored at −80 °C. The fluorescence level of FITC-dextran (excitation at 485 nm, emission at 535 nm) was measured in a 96-well black plate with spectrometer Chameleon V (Hidex, Turku, Finland). Values were converted to concentration (mg/mL) using a concentration standard curve.

2.3. ClpB Concentration

The presence of the protein ClpB was measured by the technique of enzyme linked immunosorbent assay (ELISA) previously described by Breton et al., 2016 [9]. For this, two antibodies were used: rabbit polyclonal anti-ClpB (Delphi Genetics, Brussels, BEL) and a mouse monoclonal antibody anti-ClpB (Delphi Genetics, Brussels, BEL). The optical density was determined at 405 nm using a microplate reader Infinite F50 (Tecan Life Sciences, Switzerland). Each determination was performed in duplicate.

2.4. ClpB and α-MSH Ig Assay

Plasma levels of Ig reacting with ClpB or α-MSH were measured using enzyme-linked immunosorbent assay according to a published protocol [28]. For this, a concentration of 2 μg/mL of ClpB protein (Delphi Genetics, Brussels, BEL) or α-MSH peptides (Bachem, Budendorf, Swiss) were used to coat 96-well Maxisorp plates (Nunc, Rochester, NY, USA). Mice plasma samples were diluted at 1:200 in dissociative buffer (3 M NaCl and 1.5 M glycine buffer, pH 8.9) to determine the total Ig levels. Two antibodies were used for detection: Alkaline phosphatase (AP)-conjugated goat anti-mouse IgG or anti-mouse IgM (1:2000) (Jackson ImmunoResearch Laboratories, St. Thomas Place, Ely, UK). The optical density was determined at 405 nm using an Infinite F50 microplate reader (Tecan Life Sciences, Switzerland). Blank optical density values (without the addition of plasma samples) were subtracted from the sample optical density values. Each sample was done in duplicate.

2.5. qPCR Assay for Faecal Enterobacteriaceae DNA

Enterobacteriaceae DNA in faeces were extracted with the ZymoBIOMICS Kit according to the protocol given by the supplier (ZymoResearch, Irvine, CA, USA). After extraction, the total DNA was quantified using a NanoDrop spectrophotometer (ThermoScientific, Waltham, MA, USA). qPCR was performed on 1 ng/μL of DNA and with Light Cycler®480 SYBR® Green I Master (Roche, Swiss). The primers for detection of *Enterobacteriaceae* were: 5′-TGTGCCCAGATGGGATTAGC-3′ and 3′-TTAACCTTGCGGCCGTACTC-5′. The relative quantity of each DNA was calculated using standard curves normalized to a reference *16s DNA* gene.

2.6. RT-qPCR Assay for Hypothalamus Neuronal Populations mRNA

Hypothalamic total RNA was extracted within cold TRIZOL reagent (Invitrogen, Carlsbad, CA, USA). After extraction, the total RNA was quantified using a NanoDrop spectrophotometer (ThermoScientific, Waltham, MA, USA). cDNA was generated by reverse transcription with 1 μg of total RNA using M-MLV Reverse Transcriptase (200 U/μL) (ThermoFisher, Waltham, MA, USA). RT-qPCR was performed on all samples using a BioRad CFX96 Real Time PCR System (BioRad, Hercules, CA, USA) and SYBR Green Master Mix (Life Technologies, Carlsbad, CA, USA). The primers for detection of *pomc* were: 5′-CCTCCTGCTTCAGACCTCCA-3′ and 5′-GGCTGTTCATCTCCGTTGC-3′; for *agrp*, 5′-GCAGACCGAGCAGAAGAT-3′ and 5′-CTGTTGTCCCAAGCAGGA-3′. The relative quantity of each mRNA was calculated from standard curves, normalized to a reference *gapdh* gene.

2.7. Statistical Analysis

Data are shown as means +/− standard error of means (SEM). Before statistical analysis the normality was evaluated by the Kolmogorov-Smirnov test. Then, statistical significance was calculated by the unpaired *t*-test, one-way ANOVA or two-way ANOVA, as appropriate. All statistical calculations were performed using Prism 6.0 software (GraphPad Software, Inc., San Diego, CA, USA) and $p < 0.05$ was considered as significant.

3. Results

3.1. Body Weight and Food Intake

During the adaptation phase (D1–D6), the animals had the weight between 20 g and25 g. From the beginning of the dietary restriction (D6), the animals started to lose weight. The ABA and LFA mice lost significant weight compared to the control (** $p < 0.01$, D7, D8). This difference in weight loss continues until the end of the experiment (D17) (*** $p < 0.001$, D9 to D17). From D10 until the end of endurance, ABA mice lost significantly more weight than the LFA mice (* $p < 0.05$, D10, D12; ** $p < 0.01$, D13 to D17) (Supplementary data, Figure S1A).

Food intake (Supplementary data, Figure S1B) of ABA mice increased between D4 and D6 as compared to LFA and the control mice (*** $p < 0.001$) (Supplementary data, Figure S1B), (*** $p < 0.001$, D6) (Supplementary data, Figure S1C). However, since the beginning of the limited access to food (D7), food intake decreased significantly in all groups (*** $p < 0.001$) (Supplementary data, Figure S1B,C). At D10, when access time to food is shortest, food intake was significantly reduced compared to D6 (before restriction) (*** $p < 0.001$) with a reduction of 34% in the LFA group and 58% in the ABA group (Supplementary data, Figure S1C).

From D11, restriction was even intensified for the ABA group compared to the LFA group (* $p < 0.05$, D12; ** $p < 0.01$, D13; *** $p < 0.001$, D15) (Supplementary data, Figure S1B) until the end of this experiment (reduction of 16.4%) (Supplementary data, Figure S1C).

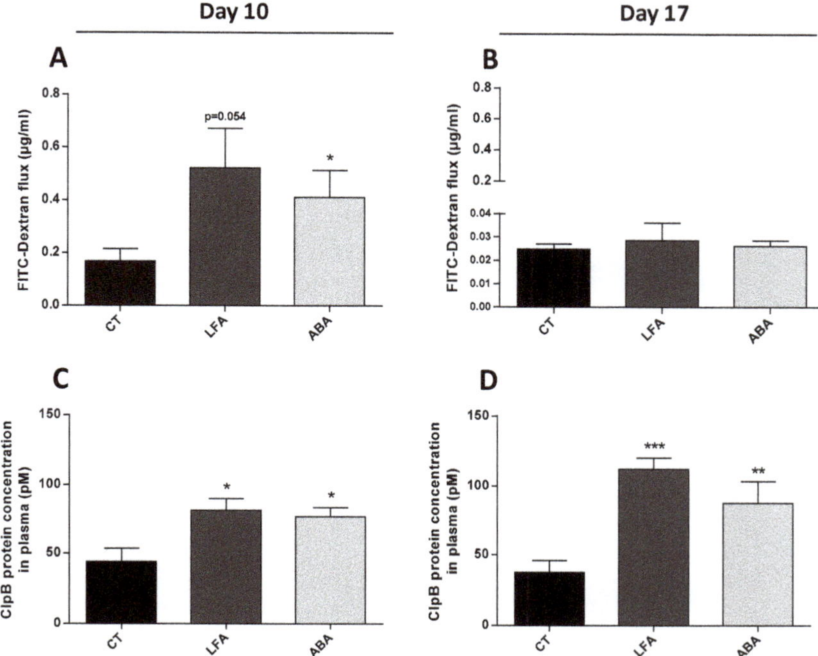

Figure 1. Intestinal permeability measure and ClpB concentration in plasma. The intestinal permeability was measured by an ELISA assay after FITC-dextran passage in the ussing chamber (**A**) at D10 and (**B**) at D17. The ClpB concentration was measured in plasma in pM by an ELISA assay (**C**) at D10 and (**D**) at D17. Data are means ± SEM. Unpaired Mann-Whitney test (**A**, $p = 0.0541$) or ne-way ANOVA test with Holm-Sidak's post-tests (**C**,**D**); *** $p < 0.001$, ** $p < 0.01$, * $p < 0.05$.

3.2. Wheel Activity

Total wheel activity increased during the restriction phase (D6–D10) of ABA mice. (Supplementary data, Figure S2A,B) as compared to the adaptation phase (D2–D5) which resulted mainly from an increased activity during the dark phase. (Supplementary data, Figure S2C,D). From D9, wheel activity decreased during the dark phase (Supplementary data, Figure S2C,D), while it increased during the light phase (Supplementary data, Figure S2E,F) vs. D6–D10.

3.3. Intestinal Permeability, ClpB and Immunoglobulins Plasma Levels

Intestinal permeability assessed in vitro by FITC-dextran flux increased at D10 (Figure 1A) while no difference was observed at D17 (Figure 1B).

ClpB protein concentration in plasma increased significantly at D10 (Figure 1C) and D17 (Figure 1D) in ABA and LFA groups vs. controls.

The plasma levels of anti-α-MSH IgG were increased at D10 and D17 in LFA vs. controls and the ABA group (Figure 2A,B). IgM anti-α-MSH levels were increased at D10 and D17 in LFA and ABA vs. controls (Figure 2E,F). The anti-ClpB IgG were increased at D17, but not at D10 in LFA vs. controls and ABA group (Figure 2C,D). The anti-ClpB IgM were increased at D10 but not at D17 in LFA vs. controls and vs. ABA group at D10 (Figure 2G,H).

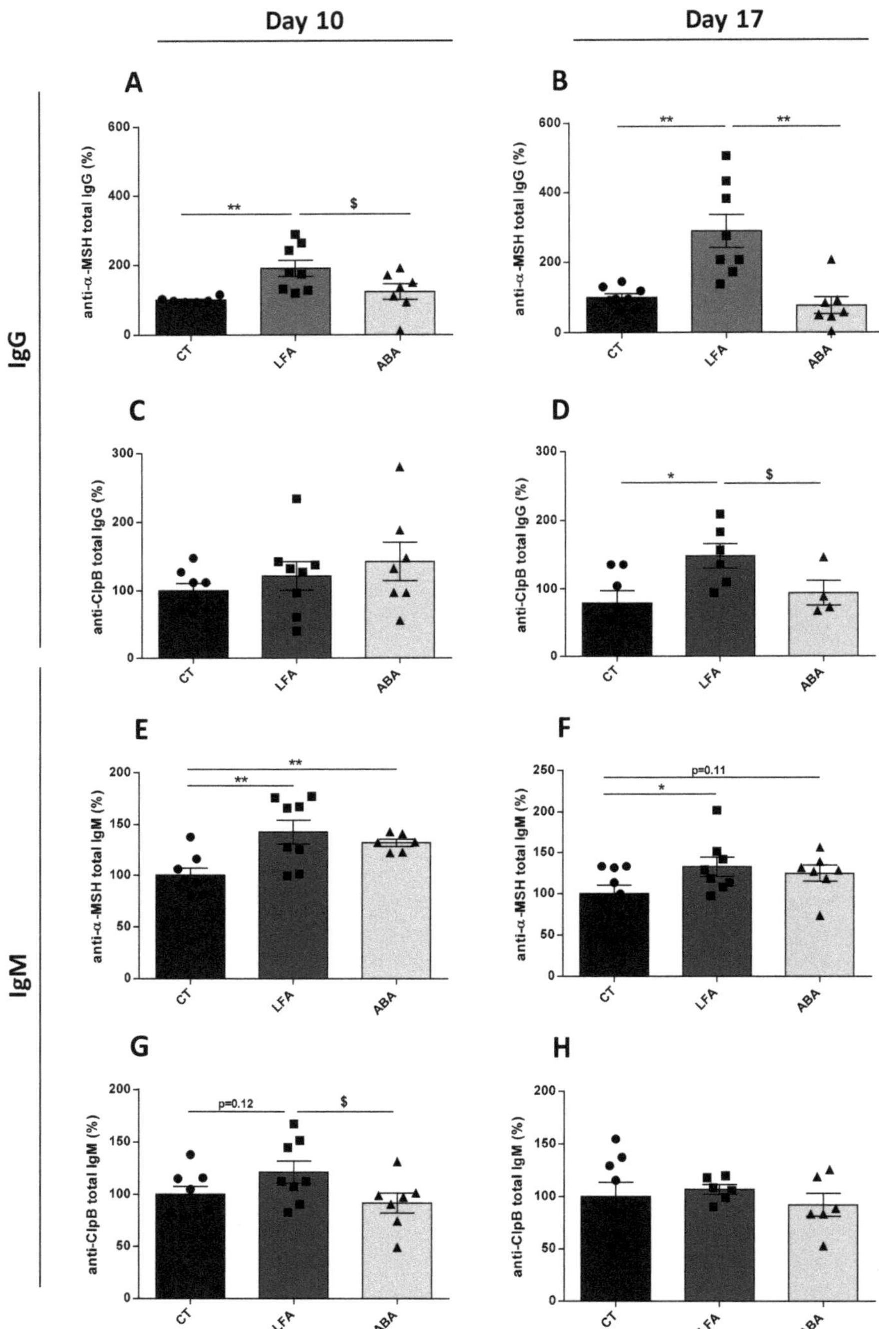

Figure 2. Impact of food restriction and physical activity on modulation of anti-α-MSH and anti-ClpB IgG and IgM. Anti-α-MSH and anti-ClpB IgG antibodies (%) (**A**,**C**) at D10 and (**B**,**D**) at D17 were measured in plasma. Anti-α-MSH and anti-ClpB IgM were measured in the same way at (**E**,**G**) at D10 and (**F**,**H**) at D17 in plasma. Data are means ± SEM. Unpaired Mann-Whitney test (**D**,**F**) or unpaired *t*-test (**A**,**B**,**E**,**G**,**H**); $^{**}\ p < 0.01$, $^{*}\ p < 0.05$, $^{\$}\ p < 0.10$.

3.4. Faecal *Enterobacteriaceae* DNA

Relative quantitative amount of *Enterobacteriaceae* DNA in faeces was not different at D10 (Figure 3A) but increased at D17 (Figure 3B). This increased in *Enterobacteriaceae* DNA is positively correlated with the ClpB plasma concentration (Figure 3C).

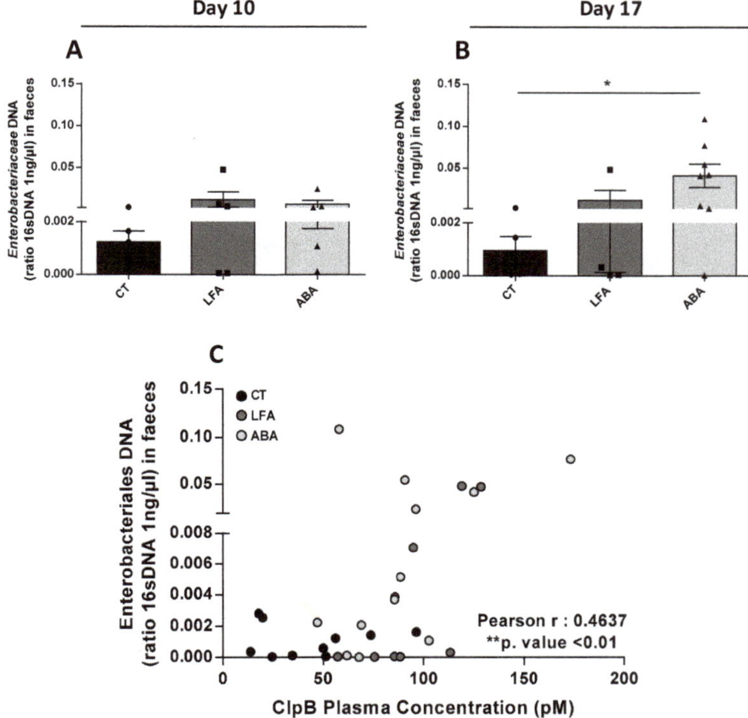

Figure 3. *Enterobacteriaceae* DNA in faeces and correlation with ClpB Plasma concentration. Relative quantitative expression of *Enterobacteriaceae* DNA in faeces by qPCR (**A**) at D10 and (**B**) at D17. The relative expression was calculated with a 1 ng/µL *Enterobacteriaceae* concentration normalized by *16sDNA* gene. (**C**) Correlation between *Enterobacteriaceae* DNA in faeces and ClpB plasma concentration (pM). Data are means ± SEM. Unpaired Mann-Whitney test (**A**,**B**) or Pearson correlation (**C**); ** $p < 0.01$, * $p < 0.05$.

3.5. Hypothalamic Neuropeptides

Hypothalamic POMC mRNA relative expression increased in LFA and ABA groups at D10 (Figure 4A) and D17 (Figure 4B). AgRP mRNA relative expression was not altered at either D10 (Figure 4C) or D17 (Figure 4D).

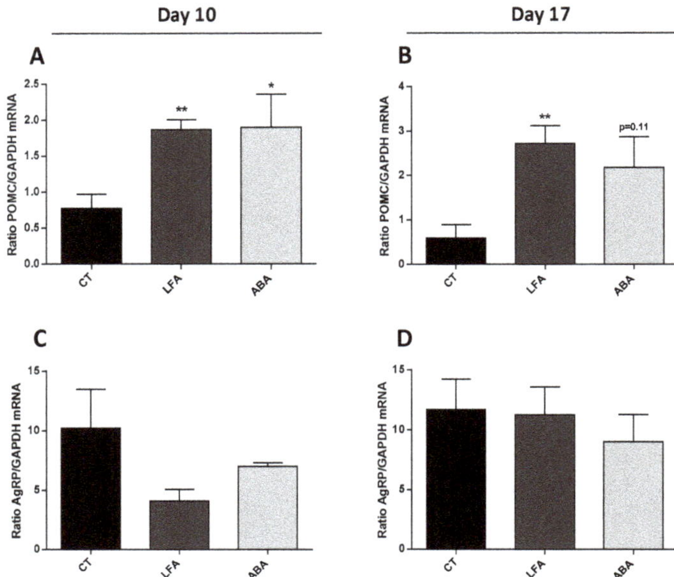

Figure 4. Impact of food restriction on neuronal population gene expression. Relative quantitative expression of (**A**,**B**) POMC and (**C**,**D**) AgRP mRNA in the hypothalamus by qPCR. The relative abundance of mRNA was calculated as the ratio of the normalized level (SQ of gene of interest mRNA/SQ of GAPDH mRNA). Data are means ± SEM. Unpaired t-test (**A**,**B**); ** $p < 0.01$, * $p < 0.05$.

4. Discussion

In this study, we highlighted that an increase of ClpB plasma concentration correlated with the relative amount of *Enterobacteriaceae* in faeces of food restricted mice. As expected in the ABA model, mice significantly lost weight which was amplified by the wheel activity of ABA mice [29,30]. Reduced food intake in ABA was not a consequence of wheel activity, since activity ceased when food was again available, which rather suggests that the hyperactivity was a consequence of food restriction.

Several studies have previously reported an increase permeability in the colon of ABA mice [31,32] which suggests that a dysfunction of the intestinal barrier may occur during anorexia nervosa. In the present study, the increased intestinal permeability observed at D10 was associated with an increase in the plasma concentration of the ClpB protein. Accordingly, a previous study from our group [33] reported alterations of the colonic mucosa proteome in ABA mice, suggesting that a decreased energy supply to the colonic mucosa may compromise its functional integrity metabolism [33].

No difference in intestinal permeability or ClpB levels were observed between the LFA and ABA groups. This suggests that physical activity alone has no significant effect on colonic barrier function in this model, and that food restriction induces the increase of ClpB protein plasma level. This increased ClpB may result from an increased transcellular passage of this protein across the enterocytes. In fact, the enterocyte endocytosis of intact proteins is a well-established process. Milk proteins such as β-lactoglobulin (18.36 kDa) and α-lactalbumin (14.2 kDa) can cross the enterocytes via a non-specific liquid phase endocytosis mechanism and reach the basolateral side by a transcytosis mechanism [34]. Even larger proteins such as the 44 kDa glycoprotein Horseradish Peroxidase (HRP) can enter the intestinal absorptive cells by apical endocytosis [35]. Thus, endocytosis of the whole 96 kDa ClpB looks plausible.

Alternatively, paracellular passage may be possible for lower molecular weight proteins or fragments. An increased paracellular passage may be allowed by a degradation of the intercellular tight junction proteins network (e.g., occludin, claudin-1) [36], as already reported in inflammatory bowel diseases [37,38], irritable bowel syndrome [39,40], obesity [41] and malnutrition states [42] including the ABA model [31], and several intestinal diseases [43]. This paracellular pathway may be of relevance for the fragments of ClpB. Indeed, the ClpB protein has a capacity to fragment naturally as shown in vitro (Mogk et al., 1999), and smaller fragments may access the basolateral space before finally reaching the plasma compartment. This hypothesis is consistent with the increased ex vivo colonic passage of the FITC Dextran molecule (4 kDa) observed in food restricted mice. Our home-made ELISA test probably identified both the whole ClpB protein and several of its fragments.

In the present study, we also observed that food restriction induced changes in the plasma levels of anti-α-MSH IgG and IgM. The most consistent finding was an increase of anti-α-MSH IgG and IgM in the LFA group at D10 and D17. This immune activation may result from the activation of the hypothalamo-pituitary axis with the increased release of CRF and related-peptides such as α-MSH [44], and consequently an increase in the corresponding Ig. Our results suggest a higher increase of anti-α-MSH Ig as compared to anti-ClpB Ig. This may reflect the fact that the primary antibodies used for the ELISA assay may recognize other epitopes in the α-MSH structure in addition to those in common with ClpB. Furthermore, this increase could be a consequence of food restriction, with or without activity, which is stressful for mice. A previous study showed that repeated exposure of rats to mild stress induced by food restriction and repeated blood sampling increased the levels and affinity of α-MSH reactive IgG Ig [27]; passive transfer of these Ig purified from the blood of stressed to naïve animals induced acute food intake and suppressed anxiety. This suggests that the production of these Ig might be an adaptive response to stress aiming to counteract its effects by blunting the satiating effect of α-MSH. The moderate increase of anti-α-MSH Ig in the ABA group raises the question of a possible immunosuppressive effect caused by intense physical activity, which remains debated [45]. Although other studies need to be done, these results confirm the hypothesis proposed by Fetissov et al. that the anorexia physiopathology performed from altered signaling between the gut microbiota, the immune system and the neuropeptides involved in feeding behavior regulation [46].

In our study, colonic content analysis showed an increase proportion of *Enterobacteriaceae* DNA in the ABA group at D17 compared to controls, with intermediate values in the LFA group. This is in accordance with the increased ClpB plasma level at D17, since *Enterobacteriaceae* are known to produce the ClpB protein [18,24]. Already at D10 an increased production and/or release of ClpB by *Enterobacteriaceae* may have occurred before a significant growth of this family. Accordingly, the *Enterobacteriaceae* DNA correlated with ClpB plasma concentration across groups. Moreover, the significant correlation between the ClpB protein plasma levels and the relative amount of *Enterobacteriaceae* seen only in the anorexic mice suggests that dietary restriction impairs microbiota composition. The increased *Enterobacteriaceae* and ClpB protein production is in accordance with previous papers reporting increased *Enterobacteriaceae* in patients with anorexia and in malnourished animals [15]. Under critical dietary restriction conditions, an increased production of ClpB may also be an adaptive process to support the survival of the microorganism since ClpB is a chaperone protein [47]. Accordingly, Breton et al. reported that the production of ClpB by in vitro *E. coli* was increased in the stationary growth phase, after the disposal of added nutrients during the exponential phase [11]. The food restriction led to an increase in *Enterobacteriaceae* population, combined with an increase in both ClpB production and colonic permeability, together these factors have led to abnormally high plasma levels of ClpB.

The increased plasma concentration of ClpB or its fragments allows a direct central effect at the hypothalamic level of stimulating the POMC-related satiating pathways [11] which in turn contribute to either the onset or the perpetuation of anorexia and hyperactivity which could be explain by a satiating and anxiogenic effects of ClpB mimicking α-MSH [19].

It is important to emphasize that this biological approach of anorexia nervosa does not come in contradiction with the well-established triggering role of psychological stress. Indeed, stress might reduce food intake at the hypothalamic level or via the mesocorticolimbic system, but also at the peripheral level by increasing intestinal permeability [48,49] and altering microbiota virulence, proliferation and release of pro-inflammatory and anorexigenic signals acting on neuronal afferents [50,

51]. Using the data from this article and others, we can propose anintegrative perspective (Figure 5) which links dietary restriction, stress, microbiota-gut-brain axis dysregulation (including increased ClpB signaling) and the on-going self-maintenance of anorexia nervosa. When fully confirmed, this approach may open innovative therapeutic perspectives via modulation of gut microbiota by different nutritional, microbial or pharmacological approaches [52,53].

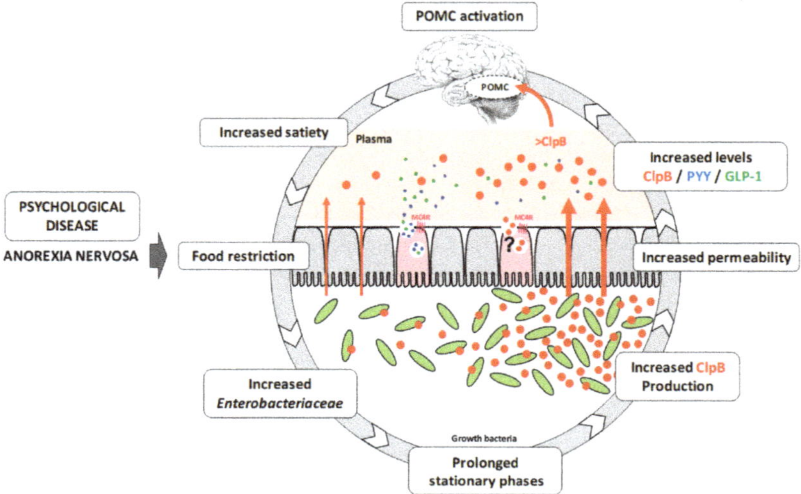

Figure 5. Vicious circle of the physiopathology of anorexia nervosa. Anorexia is characterized by psychological disorders (deformation of the self-image, obsessive fear of gaining weight) which are the cause of a restriction of food intake. This limited dietary intake leads to dysbiosis, characterized by an increase in *Enterobacteriaceae* within the microbiota. This increase generates an increased production of the ClpB protein, resulting from the prolongation of the stationary growth phase of these bacteria. In parallel, dietary restriction also causes an increase in intestinal permeability, which explains the increased passage of this protein through the intestinal mucosa. This protein is then found in the bloodstream with the other satietogenic peptides (GLP-1, PYY) released via the activation of the MC4R receptor present on the L cells. The mechanism of passage of this protein through the mucosa remains unknown, but hypotheses suggest that it may pass through the mucosa in fragments or through a mechanism of endocytosis. Finally, because of its anorectic action, the ClpB protein can activate anorexigenic neuronal populations such as POMC, whose response will lead to an increase in satiety. As well, the vicious cycle of the physiology of anorexia nervosa will can continue...

5. Conclusions

In conclusion, we have shown here that bacterial ClpB plasma levels increase during dietary restriction in mice, regardless of physical activity, and correlates with amount of *Enterobacteriaceae* in feces. This brings additional arguments for the role of the gut microbiota in the mechanisms of eating disorders and so, suggests its impact in the perpetuation and self-maintenance of the anorexia. These data suggest that nutritional or probiotic interventions aiming to restore gut microbiota may be useful in the therapeutic strategy of eating disorders.

Supplementary Materials: The following are available online at http://www.mdpi.com/2072-6643/11/10/2514/s1, Figure S1: Food restriction model confirmation–Body weight and Food intake, Figure S2: ABA model confirmation–Wheel activity.

Author Contributions: Conceptualization, R.L., N.L., G.L., P.D.; methodology, R.L., N.L.; formal analysis, M.D.; investigation, M.D., M.G., S.A., C.D., C.G., F.L., S.N.; resources, J.-L.d.R., P.D., G.L.; writing—original draft preparation, M.D. and P.D.; writing—review & editing, R.L., N.L., G.L., P.D.; visualization, M.D.; supervision, P.D.; project administration, P.D.; funding acquisition, P.D. and G.L.

Funding: This research received no external funding

Acknowledgments: Supported by European Union and Normandie Regional Council. Europe gets involved in Normandie with European Regional Development Fund (ERDF). The support of the TargEDys SA company and of the Ministry of Industry and Technology for the funding of M.D.'s PhD CIFRE thesis contract is acknowledged. The authors would like to thank Antonia Trower, Marketing Manager at TargEDys SA, for her help in manuscript proofreading.

Conflicts of Interest: M.D. and M.G. are PhD employee of TargEDys SA; N.L., R.L., S.A., C.D. were employees of TargEDys SA. P.D. is consultant for TargEDys SA, shareholder and cofounder of the company. G.L. is the CEO of TargEDys SA. Other co-authors declare no conflict of interest.

References

1. Galmiche, M.; Déchelotte, P.; Lambert, G.; Tavolacci, M.P. Prevalence of eating disorders over the 2000–2018 period: A systematic literature review. *Am. J. Clin. Nutr.* **2019**, *109*, 1402–1413. [CrossRef]
2. American Psychiatric Association. *DSM–5*, 5th ed.; University of Michigan: Ann Arbor, MI, USA, 2013.
3. Gorwood, P.; Blanchet-Collet, C.; Chartrel, N.; Duclos, J.; Dechelotte, P.; Hanachi, M.; Fetissov, S.; Godart, N.; Melchior, J.-C.; Ramoz, N.; et al. New Insights in Anorexia Nervosa. *Front. Neurosci.* **2016**, *10*, 256. [CrossRef]
4. Kleiman, S.C.; Carroll, I.M.; Tarantino, L.M.; Bulik, C.M. Gut feelings: A role for the intestinal microbiota in anorexia nervosa? *Int. J. Eat. Disord.* **2015**, *48*, 449–451. [CrossRef]
5. Roubalová, R.; Procházková, P.; Papežová, H.; Smitka, K.; Bilej, M.; Tlaskalová-Hogenová, H. Anorexia nervosa: Gut microbiota-immune-brain interactions. *Clin. Nutr. Edinb. Scotl.* **2019**. [CrossRef]
6. David, L.A.; Maurice, C.F.; Carmody, R.N.; Gootenberg, D.B.; Button, J.E.; Wolfe, B.E.; Ling, A.V.; Devlin, A.S.; Varma, Y.; Fischbach, M.A.; et al. Diet rapidly and reproducibly alters the human gut microbiome. *Nature* **2014**, *505*, 559–563. [CrossRef]
7. Wu, G.D.; Chen, J.; Hoffmann, C.; Bittinger, K.; Chen, Y.-Y.; Keilbaugh, S.A.; Bewtra, M.; Knights, D.; Walters, W.A.; Knight, R.; et al. Linking long-term dietary patterns with gut microbial enterotypes. *Science* **2011**, *334*, 105–108. [CrossRef]
8. Singh, R.K.; Chang, H.-W.; Yan, D.; Lee, K.M.; Ucmak, D.; Wong, K.; Abrouk, M.; Farahnik, B.; Nakamura, M.; Zhu, T.H.; et al. Influence of diet on the gut microbiome and implications for human health. *J. Transl. Med.* **2017**, *15*, 73. [CrossRef]
9. Dinan, T.G.; Cryan, J.F. Microbes, Immunity, and Behavior: Psychoneuroimmunology Meets the Microbiome. *Neuropsychopharmacology* **2017**, *42*, 178–192. [CrossRef]
10. Van de Wouw, M.; Schellekens, H.; Dinan, T.G.; Cryan, J.F. Microbiota-Gut-Brain Axis: Modulator of Host Metabolism and Appetite. *J. Nutr.* **2017**, *147*, 727–745. [CrossRef]
11. Breton, J.; Tennoune, N.; Lucas, N.; Francois, M.; Legrand, R.; Jacquemot, J.; Goichon, A.; Guérin, C.; Peltier, J.; Pestel-Caron, M.; et al. Gut Commensal E. coli. Proteins Activate Host Satiety Pathways following Nutrient-Induced Bacterial Growth. *Cell Metab.* **2016**, *23*, 324–334. [CrossRef]
12. Strandwitz, P. Neurotransmitter modulation by the gut microbiota. *Brain Res.* **2018**, *1693*, 128–133. [CrossRef]
13. Lach, G.; Schellekens, H.; Dinan, T.G.; Cryan, J.F. Anxiety, Depression, and the Microbiome: A Role for Gut Peptides. *Neurother. J. Am. Soc. Exp. Neurother.* **2018**, *15*, 36–59. [CrossRef]
14. Castaner, O.; Goday, A.; Park, Y.-M.; Lee, S.-H.; Magkos, F.; Shiow, S.-A.T.E.; Schröder, H. The Gut Microbiome Profile in Obesity: A Systematic Review. *Int. J. Endocrinol.* **2018**, *2018*, 9. [CrossRef]
15. Breton, J.; Déchelotte, P.; Ribet, D. Intestinal microbiota and Anorexia Nervosa. *Clin. Nutr. Exp.* **2019**, in press. [CrossRef]
16. Million, M.; Angelakis, E.; Maraninchi, M.; Henry, M.; Giorgi, R.; Valero, R.; Vialettes, B.; Raoult, D. Correlation between body mass index and gut concentrations of Lactobacillus reuteri, Bifidobacterium animalis, Methanobrevibacter smithii and Escherichia coli. *Int. J. Obes.* **2013**, *37*, 1460–1466. [CrossRef]
17. Fetissov, S.O. Role of the gut microbiota in host appetite control: Bacterial growth to animal feeding behaviour. *Nat. Rev. Endocrinol.* **2017**, *13*, 11–25. [CrossRef]
18. Legrand, R.; Lucas, N.; Dominique, M.; Azhar, S.; Deroissart, C.; Le Solliec, M.-A.; Rondeaux, J.; Nobis, S.; Guérin, C.; Léon, F.; et al. Commensal Hafnia alvei strain reduces food intake and fat mass in obese mice-a new potential probiotic for appetite and body weight management. *Int. J. Obes.* **2019**, in press.

19. Tennoune, N.; Chan, P.; Breton, J.; Legrand, R.; Chabane, Y.N.; Akkermann, K.; Järv, A.; Ouelaa, W.; Takagi, K.; Ghouzali, I.; et al. Bacterial ClpB heat-shock protein, an antigen-mimetic of the anorexigenic peptide α-MSH, at the origin of eating disorders. *Transl. Psychiatry* **2014**, *4*, e458. [CrossRef]
20. Kim, M.S.; Rossi, M.; Abusnana, S.; Sunter, D.; Morgan, D.G.; Small, C.J.; Edwards, C.M.; Heath, M.M.; Stanley, S.A.; Seal, L.J.; et al. Hypothalamic localization of the feeding effect of agouti-related peptide and alpha-melanocyte-stimulating hormone. *Diabetes* **2000**, *49*, 177–182. [CrossRef]
21. Guan, H.-Z.; Dong, J.; Jiang, Z.-Y.; Chen, X. α-MSH Influences the Excitability of Feeding-Related Neurons in the Hypothalamus and Dorsal Vagal Complex of Rats. *BioMed Res. Int.* **2017**, *2017*, 2034691. [CrossRef]
22. Manning, S.; Batterham, R.L. Enteroendocrine MC4R and energy balance: Linking the long and the short of it. *Cell Metab.* **2014**, *20*, 929–931. [CrossRef]
23. Panaro, B.L.; Tough, I.R.; Engelstoft, M.S.; Matthews, R.T.; Digby, G.J.; Møller, C.L.; Svendsen, B.; Gribble, F.; Reimann, F.; Holst, J.J.; et al. The melanocortin-4 receptor is expressed in enteroendocrine L cells and regulates the release of peptide YY and glucagon-like peptide 1 in vivo. *Cell Metab.* **2014**, *20*, 1018–1029. [CrossRef]
24. Breton, J.; Legrand, R.; Akkermann, K.; Järv, A.; Harro, J.; Déchelotte, P.; Fetissov, S.O. Elevated plasma concentrations of bacterial ClpB protein in patients with eating disorders. *Int. J. Eat. Disord.* **2016**, *49*, 805–808. [CrossRef]
25. Fetissov, S.O.; Harro, J.; Jaanisk, M.; Järv, A.; Podar, I.; Allik, J.; Nilsson, I.; Sakthivel, P.; Lefvert, A.K.; Hökfelt, T. Autoantibodies against neuropeptides are associated with psychological traits in eating disorders. *Proc. Natl. Acad. Sci. USA* **2005**, *102*, 14865–14870. [CrossRef]
26. Lucas, N.; Legrand, R.; Bôle-Feysot, C.; Breton, J.; Coëffier, M.; Akkermann, K.; Järv, A.; Harro, J.; Déchelotte, P.; Fetissov, S.O. Immunoglobulin G modulation of the melanocortin 4 receptor signaling in obesity and eating disorders. *Transl. Psychiatry* **2019**, *9*, 87. [CrossRef]
27. Sinno, M.H.; Do Rego, J.C.; Coëffier, M.; Bole-Feysot, C.; Ducrotté, P.; Gilbert, D.; Tron, F.; Costentin, J.; Hökfelt, T.; Déchelotte, P.; et al. Regulation of feeding and anxiety by alpha-MSH reactive autoantibodies. *Psychoneuroendocrinology* **2009**, *34*, 140–149. [CrossRef]
28. Fetissov, S.O. Neuropeptide autoantibodies assay. *Methods Mol. Biol.* **2011**, *789*, 295–302.
29. Lewis, D.Y.; Brett, R.R. Activity-based anorexia in C57/BL6 mice: Effects of the phytocannabinoid, Delta9-tetrahydrocannabinol (THC) and the anandamide analogue, OMDM-2. *Eur Neuropsychopharmacol.* **2010**, *20*, 622–631. [CrossRef]
30. Nobis, S.; Goichon, A.; Achamrah, N.; Guérin, C.; Azhar, S.; Chan, P.; Morin, A.; Bôle-Feysot, C.; do Rego, J.C.; Vaudry, D.; et al. Alterations of proteome, mitochondrial dynamic and autophagy in the hypothalamus during activity-based anorexia. *Sci. Rep.* **2018**, *8*, 7233. [CrossRef]
31. Jésus, P.; Ouelaa, W.; François, M.; Riachy, L.; Guérin, C.; Aziz, M.; Do Rego, J.-C.; Déchelotte, P.; Fetissov, S.O.; Coëffier, M. Alteration of intestinal barrier function during activity-based anorexia in mice. *Clin. Nutr. Edinb. Scotl.* **2014**, *33*, 1046–1053. [CrossRef]
32. Achamrah, N.; Nobis, S.; Breton, J.; Jésus, P.; Belmonte, L.; Maurer, B.; Legrand, R.; Bôle-Feysot, C.; do Rego, J.L.; Goichon, A.; et al. Maintaining physical activity during refeeding improves body composition, intestinal hyperpermeability and behavior in anorectic mice. *Sci. Rep.* **2016**, *6*, 21887. [CrossRef]
33. Nobis, S.; Achamrah, N.; Goichon, A.; L'Huillier, C.; Morin, A.; Guérin, C.; Chan, P.; do Rego, J.L.; do Rego, J.C.; Vaudry, D.; et al. Colonic Mucosal Proteome Signature Reveals Reduced Energy Metabolism and Protein Synthesis but Activated Autophagy during Anorexia-Induced Malnutrition in Mice. *Proteomics* **2018**, *18*, 1700395. [CrossRef]
34. Caillard, I.; Tomé, D. Transport of beta-lactoglobulin and alpha-lactalbumin in enterocyte-like Caco-2 cells. *Reprod. Nutr. Dev.* **1995**, *35*, 179–188. [CrossRef]
35. Blok, J.; Mulder-Stapel, A.A.; Ginsel, L.A.; Daems, W.T. Endocytosis in absorptive cells of cultured human small-intestinal tissue: Horseradish peroxidase, lactoperoxidase, and ferritin as markers. *Cell Tissue Res.* **1981**, *216*, 1–13. [CrossRef]
36. Buckley, A.; Turner, J.R. Cell Biology of Tight Junction Barrier Regulation and Mucosal Disease. *Cold Spring Harb. Perspect. Biol.* **2018**, *10*, a029314. [CrossRef]
37. Poritz, L.S.; Garver, K.I.; Green, C.; Fitzpatrick, L.; Ruggiero, F.; Koltun, W.A. Loss of the tight junction protein ZO-1 in dextran sulfate sodium induced colitis. *J. Surg. Res.* **2007**, *140*, 12–19. [CrossRef]

38. Xu, P.; Elamin, E.; Elizalde, M.; Bours, P.P.H.A.; Pierik, M.J.; Masclee, A.A.M.; Jonkers, D.M.A.E. Modulation of Intestinal Epithelial Permeability by Plasma from Patients with Crohn's Disease in a Three-dimensional Cell Culture Model. *Sci. Rep.* **2019**, *9*, 2030. [CrossRef]
39. Bertiaux-Vandaële, N.; Youmba, S.B.; Belmonte, L.; Lecleire, S.; Antonietti, M.; Gourcerol, G.; Leroi, A.-M.; Déchelotte, P.; Ménard, J.-F.; Ducrotté, P.; et al. The expression and the cellular distribution of the tight junction proteins are altered in irritable bowel syndrome patients with differences according to the disease subtype. *Am. J. Gastroenterol.* **2011**, *106*, 2165–2173. [CrossRef]
40. Cheng, P.; Yao, J.; Wang, C.; Zhang, L.; Kong, W. Molecular and cellular mechanisms of tight junction dysfunction in the irritable bowel syndrome. *Mol. Med. Rep.* **2015**, *12*, 3257–3264. [CrossRef]
41. Brun, P.; Castagliuolo, I.; Leo, V.D.; Buda, A.; Pinzani, M.; Palù, G.; Martines, D. Increased intestinal permeability in obese mice: New evidence in the pathogenesis of nonalcoholic steatohepatitis. *Am. J. Physiol.Gastrointest. Liver Physiol.* **2007**, *292*, G518–G525. [CrossRef]
42. Genton, L.; Cani, P.D.; Schrenzel, J. Alterations of gut barrier and gut microbiota in food restriction, food deprivation and protein-energy wasting. *Clin. Nutr.* **2015**, *34*, 341–349. [CrossRef]
43. Bischoff, S.C.; Barbara, G.; Buurman, W.; Ockhuizen, T.; Schulzke, J.-D.; Serino, M.; Tilg, H.; Watson, A.; Wells, J.M. Intestinal permeability-a new target for disease prevention and therapy. *BMC Gastroenterol.* **2014**, *14*, 189. [CrossRef]
44. Dhillo, W.S.; Small, C.J.; Seal, L.J.; Kim, M.-S.; Stanley, S.A.; Murphy, K.G.; Ghatei, M.A.; Bloom, S.R. The Hypothalamic Melanocortin System Stimulates the Hypothalamo-Pituitary-Adrenal Axis in vitro and in vivo in Male Rats. *Neuroendocrinology* **2002**, *75*, 209–216. [CrossRef]
45. Nieman, D.C.; Nehlsen-Cannarella, S.L. The Effects of Acute and Chronic Exercise on Immunoglobulins. *Sports Med.* **1991**, *11*, 183–201. [CrossRef]
46. Fetissov, S.O.; Hökfelt, T. On the origin of eating disorders: Altered signaling between gut microbiota, adaptive immunity and the brain melanocortin system regulating feeding behavior. *Curr. Opin. Pharmacol.* **2019**, *48*, 82–91. [CrossRef]
47. Lee, S.; Sowa, M.E.; Watanabe, Y.; Sigler, P.B.; Chiu, W.; Yoshida, M.; Tsai, F.T.F. The structure of ClpB: A molecular chaperone that rescues proteins from an aggregated state. *Cell* **2003**, *115*, 229–240. [CrossRef]
48. Vanuytsel, T.; van Wanrooy, S.; Vanheel, H.; Vanormelingen, C.; Verschueren, S.; Houben, E.; Salim Rasoel, S.; Tóth, J.; Holvoet, L.; Farré, R.; et al. Psychological stress and corticotropin-releasing hormone increase intestinal permeability in humans by a mast cell-dependent mechanism. *Gut* **2014**, *63*, 1293–1299. [CrossRef]
49. Wallon, C.; Yang, P.-C.; Keita, A.V.; Ericson, A.-C.; McKay, D.M.; Sherman, P.M.; Perdue, M.H.; Söderholm, J.D. Corticotropin-releasing hormone (CRH) regulates macromolecular permeability via mast cells in normal human colonic biopsies in vitro. *Gut* **2008**, *57*, 50–58. [CrossRef]
50. Biaggini, K.; Borrel, V.; Szunerits, S.; Boukherroub, R.; N'Diaye, A.; Zébré, A.; Bonnin-Jusserand, M.; Duflos, G.; Feuilloley, M.; Drider, D.; et al. Substance P enhances lactic acid and tyramine production in Enterococcus faecalis V583 and promotes its cytotoxic effect on intestinal Caco-2/TC7 cells. *Gut Pathog.* **2017**, *9*, 20. [CrossRef]
51. Biaggini, K.; Barbey, C.; Borrel, V.; Feuilloley, M.; Déchelotte, P.; Connil, N. The pathogenic potential of Pseudomonas fluorescens MFN1032 on enterocytes can be modulated by serotonin, substance P and epinephrine. *Arch. Microbiol.* **2015**, *197*, 983–990. [CrossRef]
52. Lam, Y.Y.; Maguire, S.; Palacios, T.; Caterson, I.D. Are the Gut Bacteria Telling Us to Eat or Not to Eat? Reviewing the Role of Gut Microbiota in the Etiology, Disease Progression and Treatment of Eating Disorders. *Nutrients* **2017**, *9*, 602. [CrossRef]
53. Dominique, M.; Breton, J.; Guérin, C.; Bole-Feysot, C.; Lambert, G.; Déchelotte, P.; Fetissov, S. Effects of Macronutrients on the In Vitro Production of ClpB, a Bacterial Mimetic Protein of α-MSH and Its Possible Role in Satiety Signaling. *Nutrients* **2019**, *11*, 2115. [CrossRef]

Article

New and Preliminary Evidence on Altered Oral and Gut Microbiota in Individuals with Autism Spectrum Disorder (ASD): Implications for ASD Diagnosis and Subtyping Based on Microbial Biomarkers

Xuejun Kong [1,2,*,†], Jun Liu [1,2,†], Murat Cetinbas [3], Ruslan Sadreyev [3], Madelyn Koh [1], Hui Huang [4], Adetaye Adeseye [4], Puhan He [4], Junli Zhu [5], Hugh Russell [6], Clara Hobbie [1], Kevi Liu [1] and Andrew B. Onderdonk [7]

1. Athinoula A. Martinos Center for Biomedical Imaging, Massachusetts General Hospital, 149 13th Street Charlestown, Boston, MA 02129, USA
2. Harvard Medical School, Boston, 25 Shattuck Street, Boston, MA 02115, USA
3. Department of Molecular Biology, Massachusetts General Hospital, 185 Cambridge Street, Boston, MA 02114, USA
4. Harvard School of Dental Medicine, 188 Longwood Avenue, Boston, MA 02115, USA
5. Fisher college, 1 Arlington Street, Boston, MA 02116, USA
6. Precidiag, Inc., 27 Strathmore Road, Natick, MA 01760, USA
7. Brigham and Women's Hospital, 75 Francis Street, Boston, MA 02115, USA

* Correspondence: xkong1@mgh.harvard.edu

† These authors contributed equally to this paper.

Received: 3 August 2019; Accepted: 23 August 2019; Published: 6 September 2019

Abstract: Autism Spectrum Disorder (ASD) is a complex neurological and developmental disorder characterized by behavioral and social impairments as well as multiple co-occurring conditions, such as gastrointestinal abnormalities, dental/periodontal diseases, and allergies. The etiology of ASD likely involves interaction between genetic and environmental factors. Recent studies suggest that oral and gut microbiome play important roles in the pathogenesis of inflammation, immune dysfunction, and disruption of the gut–brain axis, which may contribute to ASD pathophysiology. The majority of previous studies used unrelated neurotypical individuals as controls, and they focused on the gut microbiome, with little attention paid to the oral flora. In this pilot study, we used a first degree-relative matched design combined with high fidelity 16S rRNA (ribosomal RNA) gene amplicon sequencing in order to characterize the oral and gut microbiotas of patients with ASD compared to neurotypical individuals, and explored the utility of microbiome markers for ASD diagnosis and subtyping of clinical comorbid conditions. Additionally, we aimed to develop microbiome biomarkers to monitor responses to a subsequent clinical trial using probiotics supplementation. We identified distinct features of gut and salivary microbiota that differed between ASD patients and neurotypical controls. We next explored the utility of some differentially enriched markers for ASD diagnosis and examined the association between the oral and gut microbiomes using network analysis. Due to the tremendous clinical heterogeneity of the ASD population, we explored the relationship between microbiome and clinical indices as an attempt to extract microbiome signatures assocociated with clinical subtypes, including allergies, abdominal pain, and abnormal dietary habits. The diagnosis of ASD currently relies on psychological testing with potentially high subjectivity. Given the emerging role that the oral and gut microbiome plays in systemic diseases, our study will provide preliminary evidence for developing microbial markers that can be used to diagnose or guide treatment of ASD and comorbid conditions. These preliminary results also serve as a starting point to test whether altering the oral and gut microbiome could improve co-morbid conditions in patients with ASD and further modify the core symptoms of ASD.

Keywords: autism spectrum disorders; gut microbiota; oral microbiota; dysbiosis; co-occurring conditions; allergy; abdominal pain; biomarker discovery

---

1. Introduction

Autism Spectrum Disorder (ASD) is a complex neurological and developmental disorder with a rapidly increasing prevalence on a global scale [1]. The etiology of ASD likely involves an interplay between genetic and environmental factors, as well as both systemic inflammation and inflammation of the central nervous system (CNS) [2–4]. Recent studies suggest that microbiome dysregulation plays an important role in the pathogenesis of inflammation [5–8], which may contribute to the manifestation of ASD symptoms [9–12]. Evidence from animal studies supports a link between microbiome dysregulation, inflammation in the body, and development of ASD [13,14]. Patients with autism often have difficulties maintaining a balanced diet, due to multiple factors such as highly selective food preference, organic gastrointestinal (GI) diseases, and oral motor difficulties, and they show high rates of gut dysbiosis compared to neurotypical individuals [9,12]. Notably, some studies demonstrated a correlation between the severity of GI dysfunction and the severity of behavioral symptoms [15]. Gut dysbiosis may affect the CNS via the vagus nerve, microbial metabolites and neuroinflammation [16–18].

While most studies agree that the microbiome composition is different between autistic and neurotypical populations, these studies have yielded inconsistent results as to the nature or extent of these GI bacterial community differences [12,19]. Environmental factors are the dominant determinants for gut microbiome composition [20–22], yet most previous studies using age and sex matched controls have not adequately controlled for environmental influences [12,23,24]. In addition, compared to the gut, the oral microbiome is understudied, despite dental plaque and saliva samples being easier to obtain than stool samples. Alterations of the oral microbiota are associated with not only periodontal diseases [25], but also the upper GI tract flora [26], systemic diseases such as Rheumatoid Arthritis [27] and neurological conditions such as Alzheimer's disease [28]. Epidemiological studies have demonstrated a higher prevalence of oral health issues among patients with ASD, as compared to neurotypical individuals [29]. Only two studies to date have explored differences in oral microbiota between children with autism and controls [30,31]. Results from these studies have low degrees of concordance, likely due to the different sequencing methodologies and study designs.

Here, we have designed a pilot study to investigate the oral and gut microbiome simultaneously in patients with ASD and their first-degree family members. This would control for genetic and lifestyle factors while investigating the existence of ASD-microbiome signatures and whether these signatures hold any diagnostic value. Furthermore, to explore the poorly understood oral microbiome, we have directly compared oral and gut microbiome to explore their relationship in ASD and their association with systemic clinical indices. These questions are important to address in order to detail the roles of the human microbiome in ASD, and its utility in guiding diagnosis of ASD, clinical subtypes, and potential targeted interventions.

Given the multitude of factors that influence microbiome-host interactions, a secondary goal of the study attempts to characterize the potential relationships between the gut and oral microbiome and relevant clinical indices, including allergy, abdominal pain and dietary habits. Previously, Plaza-Diaz investigated gut microbiome in ASD patients with or without mental regression and found microbiome signatures associated with different psychiatric subtypes [32]. However, the association between medical subtypes and microbiome has been poorly explored in ASD patients.

Research on high impact diseases such as Rheumatoid Arthritis has revealed fascinating associations between oral and gut microbiomes [27]. Our study will serve as a starting point to address the complex interplay between the oral microbiome and the gut microbiome in the phenotypic presentation and pathophysiology of ASD. We believe that this study will open new horizons and opportunities in disease investigation and management. As a pre-probiotics clinical trial pilot project, we hope that this study and its continuation will provide insight for whether this new methodology with combined oral and fecal data can be used to (1) screen, diagnose, and determine subtypes of ASD, (2) stratify patients who may respond to probiotics therapy, (3) provide guidance on treatment strategies and develop targeted probiotic formulation, and (4) help to monitor treatment efficacy.

2. Materials and Methods

2.1. Study Participants

We recruited 20 patients diagnosed with ASD (autism spectrum disorder) and compared them with 19 family members (parent or sibling) as neurotypical controls. Patients had been diagnosed with ASD according to DSM-5 (Diagnostic and Statistical Manual of Mental Disorders) criteria [33]. Individuals with ASD between 7–25 years old with a disease duration of at least 6 weeks were enlisted. Exclusion criteria for all subjects included known genetic conditions, clinically evident serious infections or inflammatory conditions, history of cancer, severe dental/periodontal diseases or possession of dental braces. Subjects who had received probiotic treatment were asked to stop treatment at least one week prior to sample collection and subjects were excluded if they had taken antibiotics in the preceding month. Neurotypical controls had to meet the following criteria: biological sibling or biological parent of autistic subjects with IQ equal to or greater than 80 who do not have a diagnosis of ASD, attention deficit hyperactivity disorder, other intellectual developmental disorders, or psychiatric conditions. For recruitment of control subjects, siblings of the same gender and comparable age (+/− 5 years apart) received the highest priority, but an opposite-gender sibling was recruited for a control as needed. If the subject with ASD had no siblings, a parent acting as primary caretaker was recruited. Demographics and characteristics of study subjects are available in Table S1 and summarized in Table 1. Visual dental inspections were performed to determine oral health status for all subjects. Lifestyle questionnaires were distributed to assess factors that could affect microbiome status and create a GI clinical indices (GSI) score (Table 1, Table S1) [34,35].

Table 1. Characteristics of study participants and microbiome lifestyle factors.

	Autistic	*Neurotypical*
Subjects	20	19
Age (1st–3rd quartile)	15 (13–18)	29 (11–50)
Gender (n)		
Female	25% (5)	58% (11)
Male	75% (15)	42% (8)
Neighborhood in last 5 years (n)		
Cities	10% (2)	11% (2)
Suburbs	90% (18)	89% (17)
Countryside	0% (0)	0% (0)
Pets (n)		
Yes	10% (2)	11% (2)
No	85% (17)	84% (16)
n/a	5% (1)	5% (1)
Abdominal tenderness during exam (n)		
Yes	0% (0)	0% (0)
No	95% (19)	95% (18)
n/a	5% (1)	5% (1)

Table 1. *Cont.*

	Autistic	*Neurotypical*
Allergies (n)		
Yes	60% (12)	37% (7)
No	40% (8)	63% (12)
Drink alcohol (n)		
Yes	0% (0)	11% (2)
No	95% (19)	84% (16)
n/a	5% (1)	5% (1)
Recreational drugs (n)		
Yes	0% (0)	0% (0)
No	95% (19)	19% (18)
n/a	5% (1)	5% (1)
Tobacco products (n)		
Yes	0% (0)	0% (0)
No	95% (19)	95% (18)
n/a	5% (1)	5% (1)
First 6 months of life		
Breast Fed	70% (14)	74% (14)
Bottle Fed	15% (3)	5% (1)
Both	25% (5)	16% (3)
n/a	5% (1)	5% (1)
Picky Eater		
Yes	20% (4)	11% (2)
No	80% (16)	84% (16)
n/a	0% (0)	5% (1)
Servings of vegetables and fruits per day (n)		
Less than three	65% (13)	74% (14)
Three	30% (6)	21% (4)
More than three	5% (1)	5% (1)

ASD patients were recruited from clinics at Massachusetts General Hospital (MGH), Beth Israel Deaconess Medical Center, community ASD education events, and charity ASD programs in Boston. The study was approved by institutional review board of MGH (Boston, MA, USA, IRB protocol number: 2017P000573). Informed consents were obtained from subjects or the legal guardians of the subjects. All methods were performed in accordance with the relevant guidelines and regulations.

2.2. Sample Handling and Collection

To obtain oral microbiome samples, participants were asked to produce 1–3 mL of saliva after refraining from eating, drinking and oral hygiene practice for 1 h. Samples were collected with sterile DNA- and RNA-free 15 mL Falcon tubes and immediately frozen at −80 °C. De-identified and coded samples were shipped to Precidiag Inc. (Natick, MA, USA) for DNA extraction and sequencing on dry ice. Stool samples were collected by the participants at home under the supervision of trained parents with a HR-Easy Stool Collection Kit (Precidiag, Inc.) and stored at room temperature, followed by de-identification and shipment to a Precidiag CLIA-certified laboratory for DNA extraction and sequencing analysis. The HR-Easy Stool Collection Kit provides a superior method for collection, storage and stabilizing stool samples for microbiome study at ambient temperature for up to a month with minimal alterations when compared with freshly-collected samples (Yu et al., manuscript in preparation). Microbial DNA was then extracted using a HR-Easy Fecal DNA Kit (Precidiag, Inc.) according to the manufacturer's instructions and DNA samples were carefully quantified with a nanodrop spectrophotometer. A260/A280 ratios were also measured to confirm high-purity DNA yield. DNA samples were frozen at −20 °C until use.

2.3. 16S rRNA Gene Amplicon Sequencing

Microbial 16S rRNA V3-V4 genomic regions from total oral and gut DNA samples were amplified with the following primers 341F: 5′AATGATACGGCGACCACCGAGATCTA-CACTCTTTCCCTAC ACGACGCTCTTCCGATCTCCTACGGGAGGCAGCAGCCTACGGGNBGCASCAG3′ and 805R: 5′CAAGCAGAAGACGGCATACGAGATNNNNNNG-TGACTGGAGTTCAGACGTGTGCTCTTT CCGATCTGACTACNVGGGTATCTAATCC3′ via polymerase chain reaction (PCR) (95 °C for 2 min, followed by 25 cycles at 95 °C for 30 s, 55 °C for 30 s, and 72 °C for 30 s, and a final extension at 72 °C for 5 min). PCR products were purified and analyzed using a Bioanalyzer DNA kit, followed by quantification with real-time PCR. Serially diluted PhiX control library (Illumina, San Diego, CA, USA) was included as a standard. DNA libraries were pooled and sequenced on an Illumina MiSeq next-generation sequencing system (Illumina; CA) using a V2 2 × 250 bp paired-end protocol with overlapping reads.

Of note, we included strict quality control processes involving microbial DNA extraction, 16S rRNA gene amplicon amplification, and amplicon sequencing with a set of controls that enabled us to evaluate the potential introduction of contaminants or off-target amplification. Non-template controls (extraction chemistries) were included in the microbial DNA extraction process and the resulting material was subsequently used for PCR amplification. Additionally, at the step of amplification, another set of non-template controls (PCR-mix) was included to evaluate the potential introduction of contamination at this step. Similarly, a positive control comprised of known and previously characterized microbial DNA was included at this step to evaluate the efficiency of the amplification process. Before samples were pooled together, sequencing controls were evaluated, and samples were rejected if the presence of amplicons in any of the non-template controls or the absence of amplicons in the positive control was detected. In the present study, no amplicons were observed in the non-template controls and a negligible number of raw reads were recovered after sequencing.

2.4. Sequencing Data Processing

Sequencing data were processed and analyzed with a QIIME software package v. 2018.2.0 [36]. The sequencing reads with a low quality score (average $Q < 25$) were truncated to 240 bp, followed by filtering using the deblur algorithm with default settings [37]. The remaining high-quality reads were aligned with the reference library using mafft [38]. Next, the aligned reads were masked to remove highly variable positions, and a phylogenetic tree was generated from the masked alignment using the FastTree method [39]. Taxonomy assignment was performed using the feature-classifier method and naïve Bayes classifier trained on the Greengenes 13_8 99% operational taxonomic units (OTUs) (Table S2).

2.5. Biostatistical Analysis

2.5.1. Variables Measured

The main variables are the compositions of oral and gut microbiome, and quantities of microbes on genus and phylum level within each sample (OTUs). Other variables include patients' demographic information, baseline medical conditions, lifestyle factors and clinical indices.

2.5.2. Alpha and Beta Diversity

Alpha diversity was calculated on the basis of the gene profile for each sample based on the Shannon index, Faith's index, and Simpson's evenness index [40–42]. Beta diversity was calculated on the unweighted and weighted UniFrac distances, Jaccard and the Bray–Curtis dissimilarity [43,44]. Alpha and beta-diversity estimates were computed using QIIME2 [36]. Alpha and beta diversity metrics and Principal Component Analysis plots based on the Jaccard distance were generated using default QIIME2 plugins [36,43,45–47].

Kruskal–Wallis tests were used to compare alpha diversity between ASD patients and controls for oral or gut microbiome respectively. A cut off false discovery rate (FDR) of 0.05 based on the Benjamini–Hochberg (BH) method was applied [48]. Comparison of beta diversity indices were calculated by Permutational multivariate analysis of variance (PERMANOVA).

2.5.3. Statistical Analyses of Differentially Enriched Microbiome Taxa

Significant differences in the relative abundance of microbial genera and phyla between individuals with ASD and controls were identified by Kruskal–Wallis tests and BH adjustment for multiple comparisons. In addition, we performed a paired Wilcoxon signed-rank test on the relative abundances with BH adjustment. Furthermore, we explored differential bacteria enrichment on all taxonomy levels using the ANCOM (Analysis of Composition of Microbiomes) method, an algorithm that accounts for compositional constraints to reduce false discoveries in detecting differentially abundant taxa at an ecosystem level, while maintaining high statistical power [49]. An FDR cutoff of 0.2 was applied for taxa-level comparison [50].

2.6. Microbiome Biomarker Discovery

In order to measure whether the relative abundance of gut and oral microbial taxa and the dysbiosis markers could classify ASD and control groups correctly, we created a receiver operator characteristics (ROC) curve using Prism GraphPad (version 7.00 for Mac, GraphPad Software, La Jolla, San Diego, CA, USA, www.graphpad.com). Statistical significance of areas under the curves (AUCs) for dysbiosis markers were performed with the default plugin of Prism GraphPad.

2.7. Microbiome Network Analysis

In order to assess the taxonomic relatedness/association within the gut and oral microbiota as well as between oral and gut microbiota, we performed correlation-based network analysis using the SparCC (Sparse Correlations for Compositional data) method [51,52]. We performed SparCC for microbiome data on phylum and genus level from all subjects, as well as within ASD and control groups, respectively (Correlation coefficient cut-off = 0.3).

2.8. Influence of Clinical and Lifestyle Factors

Kruskal–Wallis tests with BH adjustment for FDR were used to assess differential abundance of dysbiosis markers and bacterial taxa (phylum and genus level) between binary clinical classifiers (i.e., presence or absence of allergy, constipation and abdominal pain) with a FDR cut off of 0.2. Relevant clinical indices were treated as binary even though some data were collected as ordinal (e.g., GSI scores). Analysis was further stratified by ASD and control groups. Genus level analysis was performed with genera that have a relative abundance of at least 0.5%. We compared the dietary habits between ASD patients and neurotypical controls based on numerical scores from baseline survey questions. The responses for each question were recorded on a numerical scale from 0 to 4, where a larger score indicated that the subject exhibited the behavior with greater prominence. We next assessed the correlation between eating habit scores, allergy/autoimmunity scores, GSI total score, and key ASD gut microbiome markers in patients with ASD. We used the Spearman's correlation and an FDR cutoff of 0.05.

2.9. Softwares Used

QIIME software package v. 2018.2.0 [36], RStudio (RStudio Team, 2017), R (R Core Team, 2017) and Prism GraphPad version 7.00 for Mac (GraphPad Software, La Jolla, CA, USA) were used for statistical testing and graph generation. Adobe Illustrator CC was used for figure editing.

3. Results

To characterize the gut and oral microbiota associated with autism, we recruited 20 autistic subjects and 19 controls (Table 1). Of the controls, 8 were neurotypical biological parents and 11 were neurotypical biological siblings. Demographic information is summarized in Table 1. One family had 1 parental control with 2 ASD children. Overall, there were significant inter-subject and inter-pair variabilities in microbiota composition (Figure 1A,B, Figures S1 and S2).

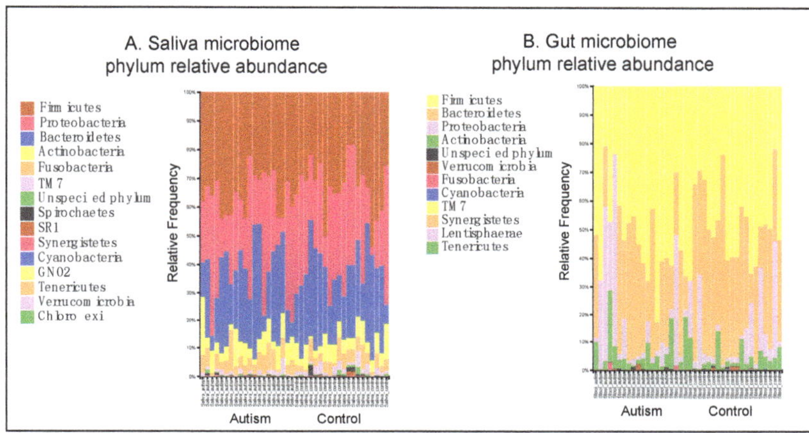

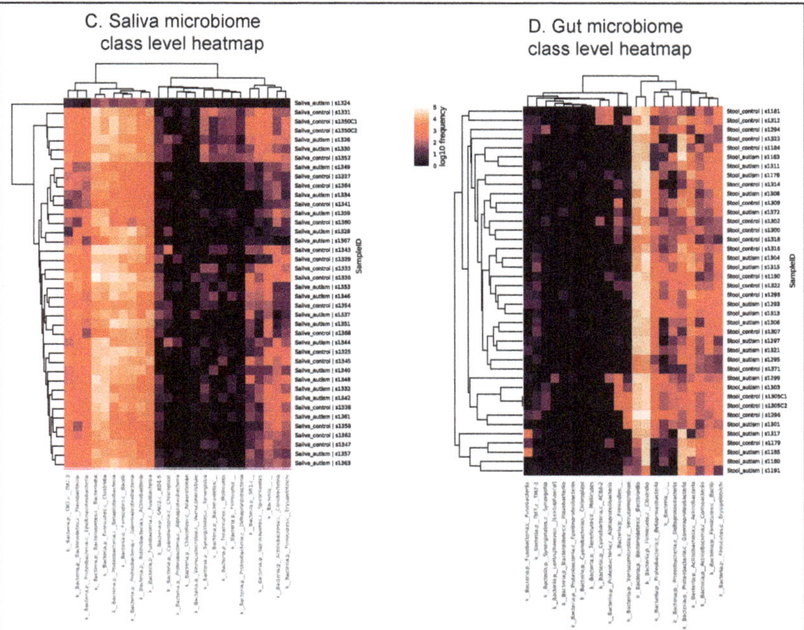

Figure 1. Bar plots of bacterial phylum-level relative abundances of the salivary (**A**) and gut (**B**) microbiomes. Each bar represents one subject. (**C**) Salivary microbiome class-level heatmap expression profile. (**D**) Gut microbiome class-level heatmap expression profile.

3.1. Autistic Subjects Harbor an Altered Oral Microbiota Compared to First Degree-Family Member Controls

Consistent with previous studies, analysis of alpha diversity calculated by the Shannon index revealed no significant differences between autistic and neurotypical subjects' salivary microbiota (Figure S2, Table S3). A heatmap (Figure 1C) visiually demonstrates that the beta diversity calculated on the unweighted, weighted UniFrac distances and the Bray–Curtis dissimilarity revealed no significant difference between the ASD and control groups for oral flora (Figure 2A, Figure S3, Table S4, PERMANOVA). The major phyla that contributed to the oral microbiome in ASD and control groups are summarized in Figure 2C. On the genus level, the ASD and control groups share 9 out of 10 most abundant genera, including *Prevotella, Fusobacterium, Rothia, Haemophilus, Streptococcus, Neisseria, Veillonella*, and an unknown genus in the *Neisseriaceae* family.

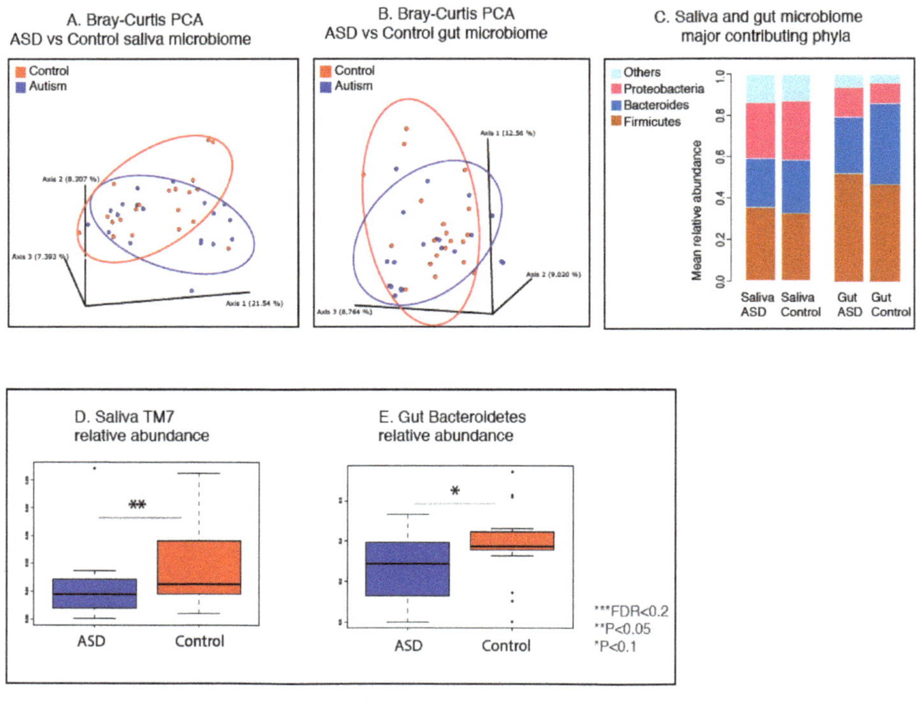

Figure 2. PCA of bacterial beta diversity of saliva (**A**) and gut (**B**) microbiomes based on the Bray–Curtis dissimilarity for ASD and neurotypical subjects. ASD and neurotypical subjects are colored in blue and red, respectively. (**C**) The major contributing phyla of gut and oral microbiome, in ASD and control subjects. The values used to compose the figures represent group mean relative abundances. (**D**,**E**) Box plots depicting relative abundances of the most differentially abundant salivary or gut bacterial phyla between patients with ASD and control subjects. Single asterisk indicates $p < 0.1$ with adjusted FDR > 0.2; double asterisk indicates $p < 0.05$ with adjusted FDR > 0.2, triple asterisk indicates $p < 0.05$ and adjusted FDR < 0.2, Kruskal–Wallis test.

We found differential enrichment of bacterial taxa in the oral microbiota of autistic individuals compared to the controls. On the phylum level, ASD patients showed a trend of lower relative abundance of *TM7* bacteria (Figure 2D, Figure S4). In total, 6 genera showed altered relative abundance between the two groups (Kruskal–Wallis test, $p < 0.05$, Figure 3A, Figure 4B, Table S5). In particular, the relative abundance of an unspecified genus in the class of *Bacilli* was statistically significant after adjusting for the false discovery rate (FDR) (Figure 5B, Table S5).

3.2. Autistic Subjects Harbor an Altered Bacterial Gut Microbiota Compared to First Degree-Family Member Controls

Consistent with previous studies, the analysis of gut alpha and beta diversity as well as principal component analysis (PCA) revealed no significant differences between autistic and neurotypical subjects (Figure 2B, PERMANOVA, Table S4, Figure S3), as visualized by a heatmap (Figure 1D). On the phylum level, *Firmicutes*, *Bacteroidetes* and *Proteobacteria* are the most abundant gut phyla in both ASD patients and control subjects, comprising more than 90% of all operational taxonomic units (OTUs) (Figure 2C). On the genus level, ASD and control groups share 9 out of 10 most abundant genera, including *Bifidobacterium*, *Blautia*, *Prevotella*, *Bacteroides*, *Faecalibacterium*, and unknown genera in *Ruminococcaceae* family, *Lachnospiraceae* family, *Enterobacteriaceae* family and *Clostridiales* order.

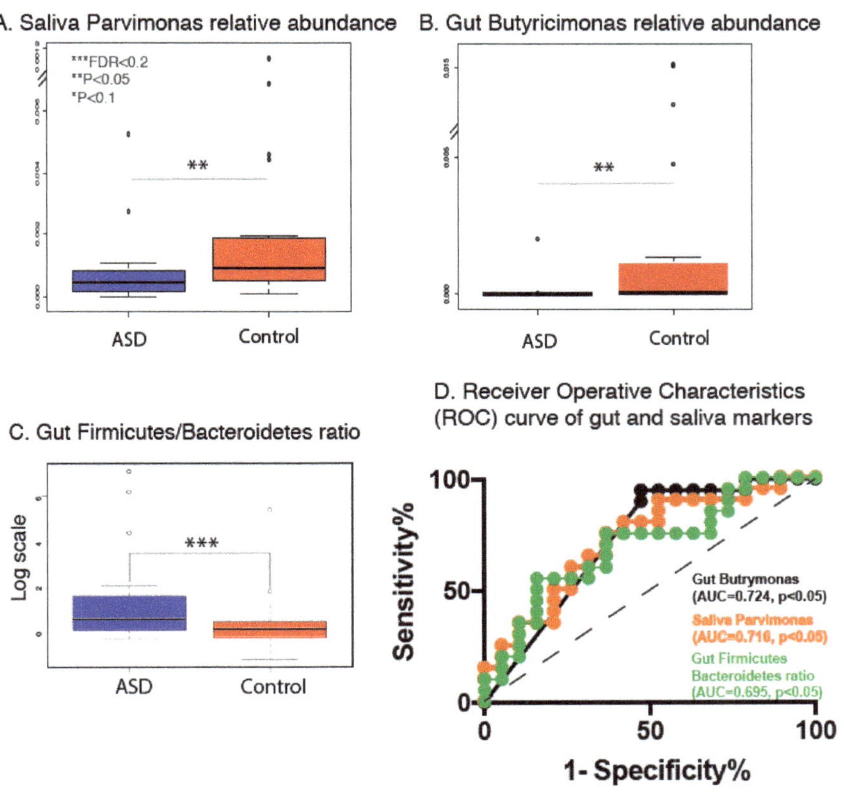

Figure 3. (**A**,**B**) Box plot representations of the relative abundances of differentially abundant salivary or gut bacterial genera in patients with Autism Spectrum Disorder (ASD) and control subjects. (**C**) Box plots representation of gut phylum-level dysbiosis marker *Firmicutes/Bacteroidetes* ratio, in patients with ASD and control subjects. ASD and neurotypical subjects are colored in blue and red, respectively. Single asterisk indicates $p < 0.1$ with adjusted FDR > 0.2; double asterisk indicates $p < 0.05$ with adjusted FDR > 0.2, triple asterisk indicates $p < 0.05$ and adjusted FDR < 0.2, Kruskal–Wallis test. (**D**) receiver operator characteristics (ROC) curve of the 3 differentially abundant gut or oral genera and dysbiosis markers that have the highest area under the curve (AUC), and $p < 0.05$ based on two-sided Z-test for ROC.

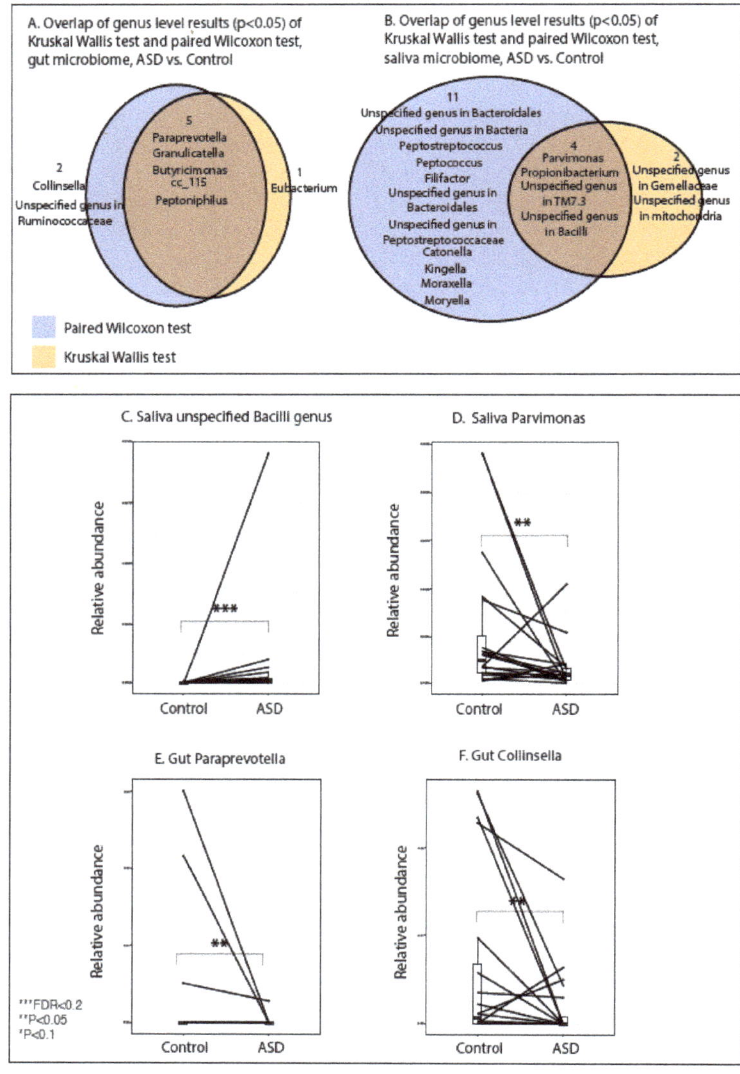

Figure 4. (**A**,**B**) Overlap of differentially abundant gut or salivary genera based on Kruskal–Wallis test and paired Wilcoxon test. Results are for taxa with unadjusted $p < 0.05$. (**C**,**D**) Paired-test representation of the relative abundances of top most differentially abundant salivary bacterial genera between ASD patient–family member control pairs. (**E**,**F**) Paired-test representation of the relative abundances of top most differentially abundant gut bacterial genera between ASD patient–family member control pairs. Single asterisk indicates $p < 0.1$ with adjusted FDR > 0.2; double asterisk indicates $p < 0.05$ with adjusted FDR > 0.2, triple asterisk indicates $p < 0.05$ and adjusted FDR < 0.2, Wilcoxon's paired test.

Further analysis of the dysbiosis markers revealed differences in the gut microbiota of subjects with autism and their family member controls. Several phylum level markers showed statistically significant changes between ASD and control, including *Firmicutes/Bacteroidetes* ratio (Figure 3C) likely driven by *Bacteroidetes* (Figure 2E, Figure S4). The phylum *Proteobacteria* is associated with metabolic syndrome and inflammatory bowel disease (IBD), and normally makes up less than 10% of the gut microbiome in healthy individuals [53]. Among the six subjects with significant *Proteobacteria* overgrowth (with relative abundance values greater than 30%), 4 were ASD patients (Table S1). On the genus level, 6 taxa showed trends of altered abundance between the two groups, including *Paraprevotella, Granulicatella, Butyricimonas, cc_115, Peptoniphilus* and *Eubacterium* (Figure 5A, Table S5).

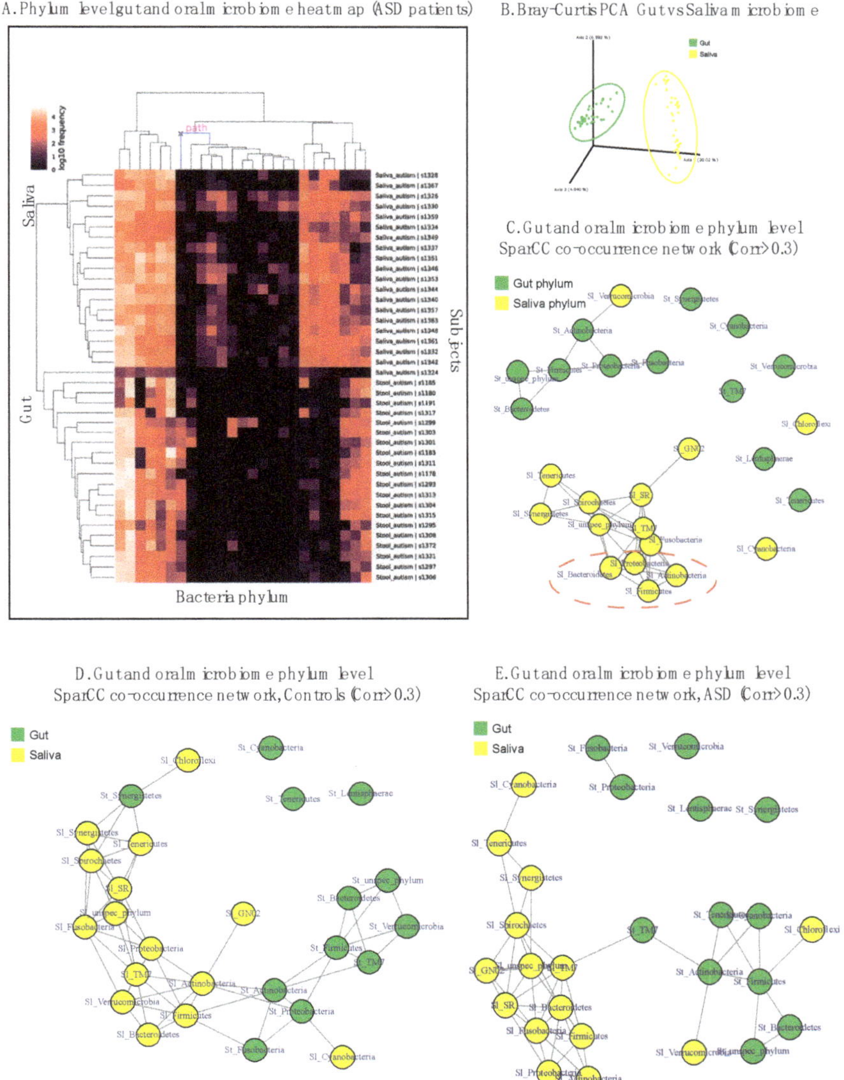

Figure 5. (**A**) Phylum-level heat map expression profiles of gut and oral microbiomes in ASD patients. (**B**) PCA of bacterial beta diversity based on Bray–Curtis dissimilarity for saliva and gut (all subjects are represented). Saliva and gut microbiome are colored in yellow and green, respectively. (**C–E**) Gut and oral microbiome phylum level co-occurrence network using the Sparse Correlations for Compositional data (SparCC) method with a correlation cut-off >0.3 ((**C**) all subjects, (**D**) control only, (**E**) ASD only). Each node represents a saliva (Sl) or stool (St) phylum, and saliva and stool microbiomes are colored in yellow and green, respectively. The dotted red circle highlights a co-occurrence cluster with the greatest inter-nodal correlations.

3.3. Gut and Saliva Biomarkers Can Classify ASD and Control Groups

In order to measure how correctly the relative abundance of gut and oral microbial taxa and the dysbiosis markers could classify two groups of samples, we created a receiver operator characteristics (ROC) curve, which is a common methodology used to evaluate classification performance of potential biomarkers (Figure 3D). The performance of a potential classifier (binary) can be evaluated by measuring the area under the curve (AUC), which represents true versus false positive rates. An AUC value of 0.5 corresponds to random classification and a value of 1.0 corresponds to perfect classification. Taking all gut and saliva genera as well as gut dysbiosis markers that showed statistically significant differential expression (Kruskal–Wallis tests) from previous analyses, two genera (gut *Butyricimonas*, saliva *Parvimonas*, Figure 3A,B) and the well-recognized dysbiosis marker gut *Firmicutes/Bacteroidetes* ratio (Figure 3C), all showed the highest AUC values (up to 0.724) with p value < 0.05 (Figure 3D, Table S8).

3.4. Results of Paired Analysis Overlap Partially with Group Analysis

Due to the nature of paired study design, we also performed paired a Wilcoxon signed-rank test on the relative abundance of the OTUs, in addition to Kruskal–Wallis tests, by subject groups (ASD vs. control). Those with significant Wilcoxon's p values had partial overlap with results from grouped Kruskal–Wallis tests (Figure 4A,B). However, after adjustment for multiple comparison, FDRs from paired analyses were not statistically significant (Table S6). Examples of gut and oral genera that showed the most significant pairwise changes are recorded in Figure 4C–F. Due to high inter-individual variabilities, subsequent analysis consisted of group-wise approaches.

In addition to Kruskal–Wallis tests with FDR adjustment, we explored differential bacteria enrichment on all taxonomy levels using the more conservative ANCOM method [49]. This method did not reveal statistically significant differences in the enrichment patterns detected by the Kruskal–Wallis test (Table S7).

3.5. Exploring the Relationship between Gut–Oral Microbiome and Their Co-Occurrence Network

Since the current project characterized gut and oral microbiota samples from the same subjects, we explored the relationship between gut and oral microbiota within individuals. Consistent with previous publications, we found that the gut and oral microbiome are distinct, based on beta diversity indices and PCA (Figure 5B, PERMANOVA). This can be seen through heatmap clustering (Figure 5A) as well as the OTU level ANCOM analysis (Figure S5, Table S7).

In order to assess the taxonomic association within the gut and oral microbiota as well as between oral and gut microbiota in a non-biased manner, we performed correlation-based network analysis using the Sparse Correlations for Compositional data (SparCC) method [51,52] (Figure 5C). This method is capable of estimating correlation values from compositional data and has been validated as a superior analysis technique than Pearson's correlation methods for compositional data such as 16S rRNA gene amplicon sequencing [51]. The goal of this analysis is to infer any potential synergistic relationships between bacterial taxa within a community and between communities. We also hoped to detect GI dysbiosis purely using salivary microbial markers because, due to high prevalence of constipation in the ASD population, it is much easier to obtain saliva samples than stool samples. The salivary microbiome could then serve as a diagnostic window into the GI environment of the ASD patients. Previously, network correlation analysis has yielded important insights regarding bacterial community structures related to enterotypes [54].

Overall, the oral microbiome exhibits a denser co-occurrence network compared to the gut, both at the phylum and genus level (Figure 5C–E). The same trend holds true when analyzing ASD subjects and control subjects separately (Figure 5D,E). Within the salivary co-occurrence network at the phylum level, the highest correlations are observed in a cluster consisting of *Actinobacteria*, *Proteobacteria*, *Firmicutes* and *Bacteroidetes* (Figure 5C, dotted circle), especially between *Firmicutes* and *Actinobacteria* (Figure S6).

Importantly, some gut and oral phylum show positive inter-community co-occurrence. There is a positive correlation between saliva *Verrucomicrobia* and gut *Actinobacteria* (Figure 5C). In the ASD population but not the controls, gut *Firmicutes*, which is a known dysbiosis marker, showed positive correlation with saliva level of *Chloroflexi* (Figure 5D,E). We then computed the co-occurrence network on the genus level using bacteria genera that make up at least 0.5% of all OTUs. The genus-level co-occurrence density was notably higher compared to phylum level (Figure S6B,C), as many genera demonstrated intra-community co-occurring relationships. In terms of inter-community co-occurrence, several gut genera, including *Bifidobacteria*, *Dialister*, *Escherichia*, *SMB53* and an unspecified genus in *Enterobacteriaceae* all exhibited positive correlation with salivary genera in the control subjects, whereas only *Escherichia* and an unspecified genus of *Clostridiales* showed co-occurrence with saliva genera in ASD patients (Figure S7).

Alpha diversity has been conventionally used as an index for dysbiosis, as low alpha diversity indicates diminished community richness and potentially diminished resilience to disturbances. Alpha diversity shows a positive correlation between the gut and oral microbiota, although it is not statistically significant (Table S3, Figure S8).

3.6. Microbiome Signatures in Clinical Subtypes

Due to the tremendous clinical heterogeneity of the ASD population, we explored the relationship between microbiome and clinical indices as an attempt to extract microbiome signatures assocociated with clinical subtypes. We focused on three major medical comorbidities that have previously reported associations with microbiome, including allergy, GI disturbances and poor diet.

3.7. Allergies

We first investigated whether phylum level dysbiosis markers (including gut *Proteobacteria*, *Firmicutes*, and *Bacteroidetes*, and oral *SR1* and *Synergistetes*) may be associated with disease states. Among all clinical indices assessed, the incidences of allergy were notably higher in the ASD group (7/19 vs. 11/18, Chi-square test, p value < 0.05). The relative abundance of oral *SR1* is significantly lower in ASD patients who also have allergies, in comparison to ASD patients without allergies, but this trend is not present in control subjects (Kruskal–Wallis, Figure 6A). Subjects with allergies also showed increased relative abundance of gut *Proteobacteria*, a phylum previously associated with autoimmune conditions (Kruskal–Wallis, Figure 6B). These differences are detected only in ASD patients and not in controls (Figure 6B). All ASD subjects who had significant gut *Proteobacteria* overgrowth (>30%) also suffered from allergies (4/4), whereas none of the 2 control subjects with *Proteobacteria* overgrowth did (0/2).

We next performed genus level correlation analysis of the oral and gut bacterial relative abundances against allergy status, using bacteria genera that make up at least 0.5% of all OTUs. No salivary or gut genus was significantly and differentially enriched by allergy status, after stratifying by ASD and control group (Table S9).

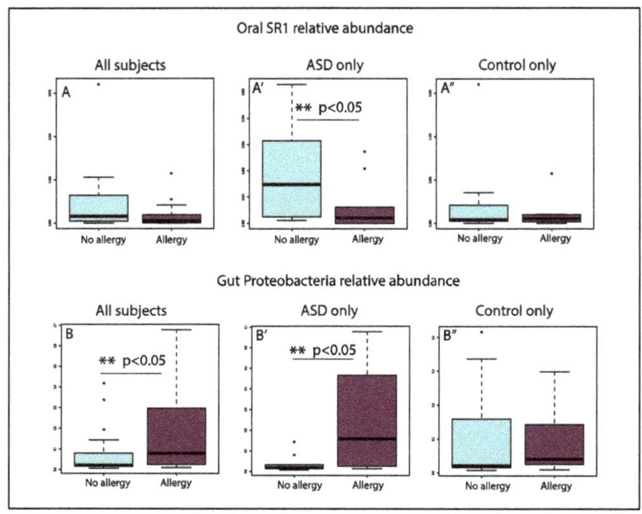

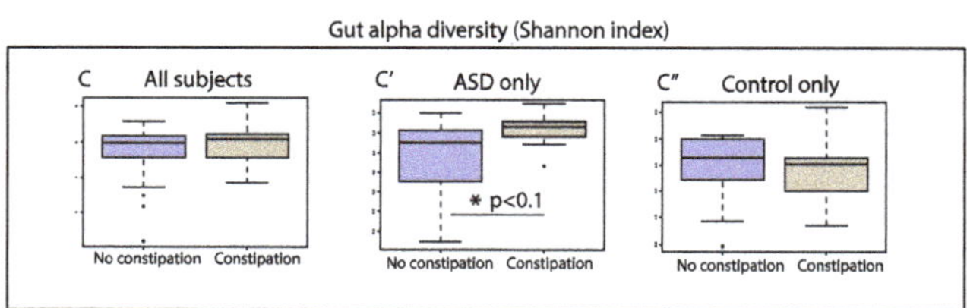

Figure 6. Box plot representation of the relative abundances of oral (**A–A″**) and gut (**B–B″**) bacterial phyla correlating with the allergy status of the subjects enrolled in this study. (**A**) Oral SR1 relative abundance in all subjects with no allergy and those with allergy; (**A′**) Oral SR1 relative abundance in ASD subjects with no allergy and patients with allergy; (**A″**) Oral SR1 relative abundance in neurotypical subjects with no allergy and neurotypical subjects with allergy. (**B**) Gut *Proteobacteria* relative abundance in all subjects with no allergy and those with allergy; (**B′**) Gut *Proteobacteria* relative abundance in ASD subjects with no allergy and patients with allergy; (**B″**) Gut *Proteobacteria* relative abundance in neurotypical subjects with no allergy and neurotypical subjects with allergy. Box plot representation of the gut alpha diversity (Shannon index) that correlated with the allergy status of the subjects enrolled in this study. (**C**) Gut alpha diversity in all subjects with no constipation and those with constipation; (**C′**) Gut alpha diversity in ASD subjects with no constipation and patients with constipation; (**C″**) Gut alpha diversity in neurotypical subjects with no constipation and neurotypical subjects with constipation. Single asterisk indicates $p < 0.1$ with adjusted FDR > 0.2; double asterisk indicates $p < 0.05$ with adjusted FDR > 0.2, triple asterisk indicates $p < 0.05$ and adjusted FDR < 0.2, Kruskal–Wallis test.

3.8. GI Disturbances

Patients with autism suffer from many co-occurring GI conditions [55]. Previous studies found that gut microbiome is associated with and may play important roles in GI symptoms such as constipation and abdominal pain [23]. We performed genus level correlation analysis of the gut bacterial relative abundances by constipation and abdominal pain status, using gut genera that make up at least 0.5% of all OTUs. *Roseburia* and *Bacteroides* were differentially enriched in subjects without abdominal pain (Figure 7A, Kruskal–Wallis, Table S10), and this difference in *Roseburia* remained statistically significant after FDR adjustment (pain 2.7% vs. no pain 5.7%). After stratifying by ASD and control subjects, ASD patients without abdominal pain had significantly higher levels of *Bacteroides*, as compared to ASD patients with abdominal pain, whereas control subjects without abdominal pain had lower levels of *Bacteroides*, as compared to control subjects with abdominal pain (Figure 7A, Kruskal–Wallis, Table S10).

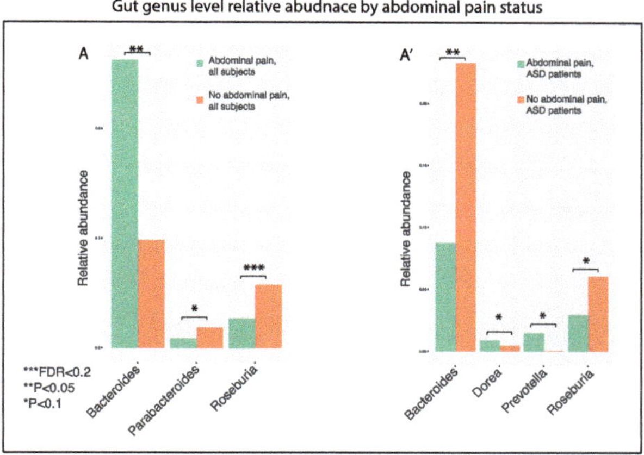

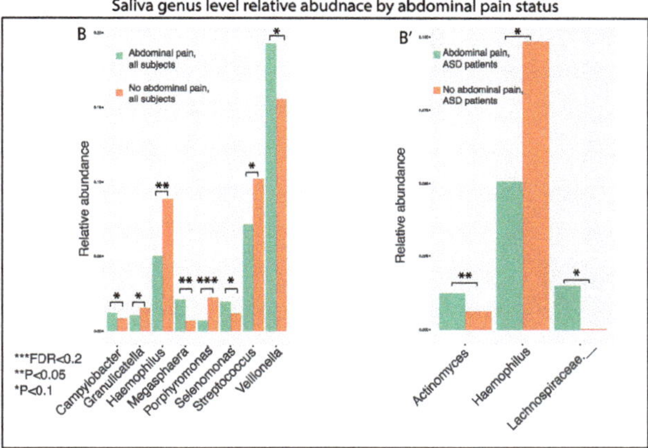

Figure 7. Bar plot representation of the relative abundances of gut (**A**–**A′**) and oral (**B**–**B′**) bacterial genera correlating with the abdominal status of the subjects enrolled in this study. (**A**) The most differentially abundant gut genera in all subjects with no abdominal pain and those with abdominal pain; (**A′**) The most differentially abundant gut genera in ASD patients with no abdominal pain and patients with abdominal pain. (**B**) The most differentially abundant oral genera in all subjects with no abdominal pain and those with abdominal pain; (**B′**) The most differentially abundant oral genera in ASD patients with no abdominal pain and patients with abdominal pain. Single asterisk indicates $p < 0.1$; double asterisk indicates $p < 0.05$, Kruskal–Wallis test.

Given the concordance between the oral microbiome and upper GI microbiome [26], it is possible that the oral microbiome may be associated with upper GI health and contribute to abdominal pain. We explored phylum and genus levels correlation analysis of the oral bacterial relative abundances between subjects with or without abdominal pain. No oral phylum showed differential enrichment, but several oral genera are differentially enriched based on abdominal pain status, including *Porphyromonas*, *Megasphaera*, *Haemophilus* (Figure 7B, Kruskal–Wallis test, Table S10). Remarkably, *Porphyromonas* is significantly less abundant in subjects without abdominal pain after FDR adjustment (pain 0.7% vs. no pain 2.2%). When stratifying based on ASD status, ASD patients with abdominal pain showed a higher trend of *Actinomyces*, as compared to ASD patients without abdominal pain (Figure 7B, Kruskal–Wallis test, Table S10).

The gut alpha diversity showed no difference between the constipated and non-constipated group (Figure 6C). When stratifying patients with ASD from the control group, there was an increased trend of gut alpha diversity in constipated ASD patients but not in constipated controls (Figure 6C′,C″), consistent with a previous study showing increased gut alpha diversity in functional constipation patients [56].

3.9. Dietary Habits and Gut Microbiome Markers

Previous studies indicate dietary challenges in ASD patients, but the association between altered dietary patterns with gut dysbiosis has not been explored in ASD patients. We found that ASD patients exhibit a statistically more restricted diet, while finding it more difficult to accept certain foods and try new foods (Mann–Whitney U test, Figure 8A–C). However, no significant differences were found between groups in respect to the amount, rate, interest, environment, or multitasking habits while eating.

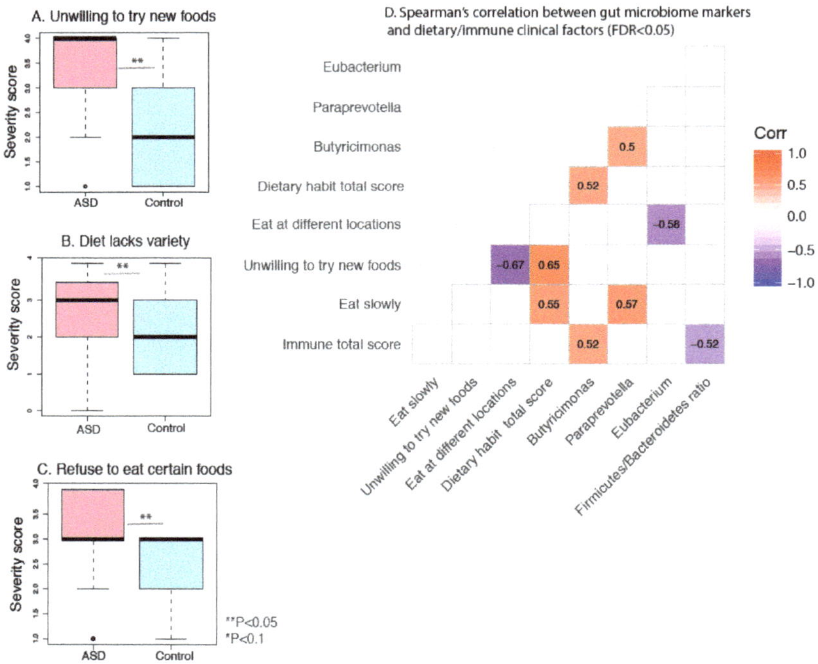

Figure 8. Box plot representation of abnormal dietary habit severity scores in ASD and control subjects. (**A**) Unwilling to try new foods. (**B**) Diet lacks variety. (**C**) Refuse to eat certain foods. Single asterisk indicates $p < 0.1$; double asterisk indicates $p < 0.05$, Mann–Whitney U test. (**D**) Spearman's correlation matrix between habit scores, allergy/autoimmunity scores, gastrointestinal severity indices (GSI) total score, and selected ASD gut microbiome markers in patients with ASD (results with FDR < 0.05 were shown).

We next assessed correlation between eating habit scores, allergy/autoimmunity scores, GSI total score, and key ASD gut microbiome markers in patients with ASD. Examined gut microbiome markers include Shannon alpha diversity index, gut *Firmicutes/Bacteroidetes* ratio and relative abundances of gut butyricimonas, paraprevotella, granulicatella, eubacterium, and cc_115 genera which showed significant difference between ASD and control groups based on previous grouped or paired analysis

(Figure 4). Most notably, we found that ASD individuals uniquely display correlations between gut butyricimonas relative abundance, eating habit total score, and allergy/immune functions (Figure 8D). *Firmicutes/Bacteroidetes* ratio is negatively correlated with allergy/immune function while the same trends are not observed in neurotypical controls. Assessed variables lacking significant correlations with gut microbiome markers are not shown.

4. Discussion

In this cross-sectional study, we conducted a comparative analysis between the gut and oral microbiota of ASD children and that of healthy, first-degree relative co-inhabitant controls. Our study is the first to use a first-degree relative matched subject design combined with high fidelity next generation sequencing technology to investigate the microbiome of ASD individuals. We believe that this study design better controls for variations in genetic background and environmental factors, and therefore has better specificity for detecting ASD-related microbial signatures [23,24]. This paired control scheme has been increasingly used in microbiome studies for diseases that have strong genetic and environmental contributing factors, such as IBD [57].

Our analysis detected differences between ASD and control subjects in both their gut and oral microbiomes. We identified an unspecified oral *Bacilli* genus, the relative abundance of which is significantly different between the ASD and control groups (FDR < 0.05), which has not been described by previous reports [30,31]. Parallel to this observation, amounts of bacteria in the class *Bacilli* were significantly higher in the gut of ASD individuals compared to controls (0.7% vs. 0.4%, Kruskal–Wallis test, $p < 0.05$), consistent with findings of Adams et al. [15]. Previous studies of the gut microbiome have revealed significant increases in facultative anaerobic commensal bacteria belonging to the class *Bacilli* seen in individuals with IBD, supporting a potential connection between *Bacilli* and gut inflammation [58]. It is unknown whether the simultaneous upregulation of *Bacilli* species in the mouth and the gut environment of ASD patients represents any common causal environmental factor (such as diet), or whether overgrowth of *Bacilli* in the mouth could lead to overgrowth of *Bacilli* in the gut. Answers to these questions would help elucidate further the interactions between gut and mouth microbiomes, as well as provide insight into potential ASD pathology.

Consistent with prior reports, ASD patients demonstrated a significantly higher gut *Firmicutes/Bacteroidetes* ratio [59,60], which is a measure associated with inflammatory conditions such as IBD [61,62]. Overgrowth of *Proteobacteria* has been associated with diarrheal diseases, metabolic syndrome and IBD [53], and 4 out of the 6 subjects who exhibited significant *Proteobacteria* overgrowth were ASD patients. *Proteobacteria* overgrowth observed in our study is unlikely due to confounding factors: none of the six subjects were under 5 years-old (age range: 15–45), and none had used antibiotics in the past month. We also explored other putative combined phylum level relative abundance or ratios as dysbiosis markers, which all appear to be abnormal in patients with ASD.

4.1. Microbial Signatures Can Serve as Potential Diagnostic Markers for ASD

Although oral and gut microbiomes are distinct, we showed that analysis of both can be combined to classify ASD subjects from controls. Among the dysbiosis markers and differentially expressed taxa in the present study, three promising candidates stood out from our analysis: gut *Butyricimonas*, saliva *Parvimonas*, and gut *Firmicutes/Bacteroidetes* ratio. In support of our findings is the work done by Kang et al. (2013) which also reported decreased *Butyricimonas* in the gut of ASD patients as compared to controls [24]. *Butyricimonas* is prevalent in healthy individuals and produces butyrate, which has been shown to improve gut health [63]. In addition, recent work on multiple sclerosis suggests that it may play an important role in immune tolerance and prevention against disease pathogenesis and progression [64,65]. *Butyricimonas* had negative correlations with gene expression implicated in cytokine signalling molecules IFN and IL-2, and activation of receptors PPAR and RXR [64]. Given the important association between autoimmune conditions and ASD, it will be important to further explore the role of *Butyricimonas* in the pathogenesis and autoimmune manifestation of ASD patients. Another study reported thedepletion of oral *Parvimonas* in IBD patients, although this has not been reported in ASD patients [66,67].

Currently, ASD diagnosis is guided by criteria in the DSM-5, which are based solely on clinical symptoms without any objective laboratory measures. Utilizing a combination of gut and oral microbiome signatures could improve the diagnosis and screening process of ASD individuals. This could also identify subclinical or clinical subgroups of ASD patients with potential GI involvement, autoimmunity, or inflammation. Future studies should explore whether these microbiome markers can predict a patient's response to treatment. This would be particularly useful to guide treatment with probiotics or drug options during probiotics therapy and anti-inflammatory interventions, as it could individualize treatment and improve outcomes for patients with ASD.

4.2. Gut and Oral Co-Occurrence Network Reveal Possible Connections between Distinct Microbial Communities

Our study is the first to co-analyze stool and oral microbiota in patients with ASD. We explored methodologies to investigate the relationship between the oral and gut microbiomes using unbiased approaches. Our analysis revealed novel co-occurrence networks within and between microbial communities that may hold diagnostic significance for ASD. Given how environmental factors (such as diet) can facilitate competitive and cooperative relationships between microbial groups [68], it is possible that such effects can span across distant communities along the digestive tract. The SparCC co-occurrence network analysis revealed an overall denser correlation network of the saliva microbiome compared to the gut. It is known that inter-individual variability of gut microbiota is higher compared with that of salivary microbiota [26], which may explain this observed difference.

Interestingly, some gut and oral taxa show evidence of co-occurrence despite the distal separation. For example, gut *Firmicutes* and saliva *Chloroflexi* showed strong correlation in the ASD population. From a diagnostic perspective, it would be pertinent to explore whether oral *Chloroflexi* can serve as a read-out for the status of gut *Firmicutes* in patients with ASD, thereby using oral microbiome as a more convenient tool to assess dysbiosis of the gut when stool samples are not readily available. More significant oral–gut co-occurrence clusters were observed at the genus level. The oral microbiome may help predict the levels of *Bifidobacteria*, *Escherichia* and *Clostridiales* genera in the gut, which all showed positive correlations with oral genera and are likely correlated with GI and/or ASD pathophysiology [12,69,70].

4.3. Clinical Correlates of ASD Microbiome

Despite the recognized importance of the gut microbiota in health and disease, our study is the one of the few designed to investigate the relationship between the human microbiota and medical comorbidities of ASD patients. Previously, Plaza-Diaz investigated gut microbiome in ASD patients with or without mental regression and found microbiome signatures associated with different psychiatric subtypes [32]. We analyzed gut alpha diversity, as well oral and gut phylum and genus levels of relative abundance in the context of three common co-occurring medical conditions affecting the ASD individuals: allergies, abdominal pain and poor dietary habits.

We found ASD patients tend to have more unhealthy and restricted dietary habits. This is consistent with previous studies, showing that up to 79% of children with ASD suffer from feeding-related difficulties or nutritional challenges [71] and strong preference for nutrient-poor foods [72]. Given the correlation between severity of poor dietary habits and relative abundances of gut microbiome biomarkers, it is conceivable that the unhealthy dietary habits may be driving gut dysbiosis [73].

Second, we detected a significantly higher prevalence of allergies in ASD patients with than those without. Gut *Proteobacteria* overgrowth is also over-represented in ASD patients and its relative abundance is positively correlated with allergy status. Overgrowth of *Proteobacteria* has been implicated in autoimmune disorders such as IBD [74]. This is opposite to the trend of *Bacteroidetes*, a marker for healthy flora. We also report a negative association between oral *SR1* numbers and allergy status, but this association is only present in ASD patients and not healthy controls.

Little is known about the connection between allergies and autism. In a recent, large population-based, cross-sectional study of data provided by the National Health Interview Survey (NHIS) from 1997–2016, Xu et al. found that children with ASD were more likely to have a food allergy (11.25% versus 4.25%), respiratory allergy (18.73% versus 12.08%), and skin allergy (16.81% versus 9.84%) than neurotypical children. Further, the odds ratio of ASD among children with a food allergy is nearly triple the ratio of ASD among those without a food allergy [75].

The "bi-directional" association between allergies and ASD raises the following questions: (1) whether these dysbiosis markers are simply associated with allergy or whether an abnormal microbiome is involved in the pathogenesis of allergy in ASD, (2) if a pathogenic mechanism could be established, whether ASD patients are more vulnerable to it than neurotypical individuals, and (3) whether there are common underlying mechanisms, potentially involving the dysregulation of the immune system and gut and oral microbiota, that could induce the development of both allergy and ASD. Future studies using animal models, immunology markers, genomics and metabolomics approaches are needed to elucidate the mechanisms of possible causal relationships.

In analyzing the relationship between microbiota and GI pathology, we found significantly higher levels of gut *Roseburia* in subjects without abdominal pain. The genus *Roseburia* consists of obligate Gram-positive anaerobic commensal bacteria that affect one's health in many ways. These bacteria produce short-chain fatty acids such as butyrate, affect colonic motility, maintain the immune response, and contribute anti-inflammatory factors to their environments [76]. Although previous studies have linked *Roseburia* abundance to some disease states such as irritable bowel syndrome and IBD [77], certain species in the genus likely play a positive role in GI health. One recent study found that treatment with the *Roseburia hominis* bacterium provided protection against dextran sulfate sodium (DSS)-induced colitis due to its immunomodulatory properties [78].

Interestingly, the oral genus *Porphyromonas* is significantly more abundant in subjects with abdominal pain. Many members of this genus have been associated with periodontal diseases [79]. The most well-characterized species, *Porphyromonas gingivalis*, has been linked to systemic diseases including upper GI tract inflammation and cancer due to upregulation of systemic cytokine release [80]. Further investigations should consider the mechanistic roles these genera could play in abdominal pain, and whether these gut and oral genera can serve as markers for the diagnosis and treatment monitoring of abdominal symptoms in patients with ASD.

The correlation between abdominal pain status and differential expression of bacterial genera differs between the ASD and control groups. Previously, Strati et al. found that constipation status is correlated with different amounts of bacterial taxa depending on whether an individual has ASD or not [23]. Notably, *Bacteroides* is one genus that shows the most prominent differential patterns: whereas *Bacteroides* appears to be protective against abdominal pain in ASD patients (higher levels are associated with no abdominal pain), the association is the opposite in controls. *Bacteroides* genus harbor species that can have either positive or negative effects on GI health. Some *Bacteroides* species synthesize lipopolysaccharide, an important bacterial virulence factor, and can cause diseases such as GI infection and septicemia in children. Many other *Bacteroides* species can be healthy commensals [81]. A recent meta-analysis concluded that a lower level of *Bacteroides* in the gut microbiota is associated with IBD [82], and functional analysis showed that *Bacteroides* expresses polysaccharide A, which can induce regulatory T-cell growth and cytokine expression to protect against colitis [83]. It is possible that ASD patients may be more prone to positive effects of *Bacteroides* than control subjects, potentially through the action of bacterial metabolites and the gut–brain axis [13]. This is supported by a mouse study which found that administration of *Bacteroides fragilis* corrects gut permeability, alters microbial composition and ameliorates ASD-related defects [13]. The ASD mice also display an altered serum metabolomic profile, and *B. fragilis* modulates levels of several metabolites. Further species-level analysis with higher 16S rRNA gene amplicon sequencing resolution and functional studies could elucidate the roles of different Bacteroides species on abdominal pain in ASD subjects. Future studies should also investigate the relationships between abdominal pain, *Bacteroides* abundance, and the severity of ASD symptoms.

5. Conclusions

In conclusion, our study is the first to use a first degree-relative matched design combined with high fidelity 16S rRNA gene amplicon sequencing technology to characterize the microbiome of patients with ASD compared to neurotypical individuals. To our knowledge, this study is the first to co-analyze the oral and gut microbiomes in patients with ASD, as well as explore the relationship between the two microbial communities and clinical indices. This study identified distinct features of gut and salivary microbiota that differ between individuals with and without an ASD diagnosis. The diagnosis of ASD currently relies on psychological testing with potential high subjectivity and inconsistencies. We suggest improvement of current diagnostic approaches based on gut and oral microbial signatures and co-occurrence networks. Given the emerging role that the human microbiome plays in systemic diseases, we hope that these analyses will provide clues for developing microbial markers for diagnosing ASD and comorbid conditions, and to guide treatment. In particular, ASD patients have disproportional gastrointestinal symptoms compared to neurotypical individuals. Therefore, developing "gut microbiome markers" is particularly important for monitoring GI health or guiding interventions of the gut. For example, these preliminary results can serve as a starting point to test whether changing the microbiome (e.g., with probiotics) would improve co-morbid conditions in patients with ASD and further modify the core and GI symptoms of ASD.

The explorations of causal relationships between microbiomes, ASD status and co-morbidities await future investigations. Further research could explore metabolomics profiles to characterize microbiome-related inflammatory factors and metabolites in the oral and gut cavity such as interleukins and short-chain fatty acids. Other areas of future study should include exploring the role of microbiota in inflammatory conditions such as allergy and autoimmunity, investigating their genetic and/or epigenetic linkage, researching mechanism of the gut–brain axis and relevant neural circuits, and ultimately inquiring more about the pathogenesis of ASD. These indices and studies will improve the algorithm for ASD screening, diagnosis, and treatment monitoring in the future.

Limitations of the current study include: (1) The use of both sibling and parental controls, where age could contribute to the large inter-individual variability. Future studies should focus on only age-matched sibling controls, if possible. (2) The small sample size, which likely contributed to high FDR in the majority of our analyses and the difficulty in distinguishing true differences from noise. Verification of our findings with a larger cohort is required. The current study was not sufficiently powered for detecting clinically relevant biomarkers. However, with the methodologies in hand, we will be able to expand the study to develop clinically biomarkers in the future. That being said, even with the relatively small sample size, we were able to find biomarkers that have withstood rigorous statistical testing and adjustment. (3) Our genus level differential expression patterns showed discrepancies from previous reports that used neurotypical controls [12,23], but this likely reflects the differences in study design [12]. For example, we did not detect changes in *Prevotella*, *Bacteroides*, *Clostridium* cluster I/II, or *Lactobacillus*, which have been reported by some studies to be differentially expressed between ASD and control groups [12], but previous studies using sibling-matched designs also did not detect these differences [84–86].

Supplementary Materials: The following are available online at http://www.mdpi.com/2072-6643/11/9/2128/s1. Figure S1. Saliva and gut microbiome OTU-level relative abundance, all subjects. Figure S2. Box plots of saliva and stool alpha diversity by groups (saliva_ASD, saliva_control, stool_ASD, stool_control), by ASD-control pairs, and by individual subjects. Figure S3. PCA beta diversity plots based on Bray–Curtis, Unweighted Unifrac, and Weighted Unifrac dissimilarity index. Figure S4. Differential abundances of saliva and gut microbiome in ASD and control subjects at phylum level, shown as fold changes in mean relative abundances. Figure S5. ANCOM volcano plots of saliva vs. gut microbiome OTU level differential expression in control and ASD subjects, respectively. Figure S6. Pearson's correlations of salivary Actinobacteria, Bacteroidetes, Firmicutes and Proteobacteria co-occurrence cluster identified from SparCC co-occurrence analysis. Figure S7. Gut and oral microbiome genus level SparCC co-occurrence network (Corr > 0.3 and Corr > 0.7). Figure S8. Correlation between gut and oral alpha diversity (Shannon index). Table S1. Metadata, subject characteristics and clinical indices. Table S2. Salivary and gut phylum and genus level OTU table. (XLS, 195KB). Table S3. Salivary and gut alpha Diversity raw data and statistics (Kruskal–Wallis test). Table S4. Statistics of beta diversity indices (PERMANOVA). Table S5. Stool and saliva phylum and genus level differential relative abundances (group means, p values, FDR adjustment with BH method, Kruskal–Wallis test). Table S6. Saliva and gut genus level paired Wilcoxon test, ASD vs. Control subjects (Wilcoxon's F statistic, p values, FDR adjustment with BH method, Paired Wilcoxon's test). Table S7: Differential abundances by ANCOM (ASD: saliva vs. stool, Control: saliva vs. stool, Saliva: ASD vs. Control, Stool: ASD vs. control). Table S8: ROC curve AUC and test statistics for significant biomarkers that distinguish ASD and control groups. Table S9: Saliva and stool genus level relative abundances based on allergy status using genera with at least 0.5% mean relative abundances (group means, p values, FDR adjustment with BH method, Kruskal–Wallis test). Table S10: Saliva and stool genus level relative abundances based on abdominal pain status using genera with at least 0.5% mean relative abundances (group means, p values, FDR adjustment with BH method, Kruskal–Wallis test).

Author Contributions: Author Contributions: X.K. devised the project and the main conceptual ideas. X.K. and J.L. worked out almost all of the technical details, and X.K., J.L., M.K., H.H., A.A., P.H. and J.Z. performed data collections and measurements. J.L., R.S., and M.C. conceptualized and performed bioinformatics analyses and statistical analyses, with assistance from H.R. and input from X.K. X.K. and J.L. wrote the manuscript, with assistance from M.K.,K.L, and input from A.B.O. J.L. drafted the figures with input from R.S., M.C., and A.B.O., and K.L. X.K., J.L., C.H. and M.K. participated in manuscript editing.

Funding: This study was funded by Massachusetts General Hospital grant number 230361 at the Athinoula A. Martinos Center for Biomedical Imaging, Boston, MA.

Acknowledgments: This work was conducted with support from Harvard Catalyst|The Harvard Clinical and Translational Science Center (National Center for Advancing Translational Sciences, National Institutes of Health Award UL 1TR002541) and financial contributions from Harvard University and its affiliated academic healthcare centers. We thank Georg Gerber for extensive feedback on methodology, Dong Kong and Zuoteng Yu for assistance with manuscript preparation. We thank Bruce Paster and Hattice Hasturk for advice on study design.

Conflicts of Interest: Author Hugh Russell was employed by company Precidiag INC. Xuejun Kong served as a short term consultant for Precidiag more than two years ago. All other authors declare no competing interests. The funding agencies and Precidiag INC have no role in the study design, implementation, the interpretation of the results.

Nomenclature

ANCOM	Analysis of Composition of Microbes
ASD	Autism Spectrum Disorder
AUC	area under the graph
BH	Benjamini–Hochberg
CNS	central nervous system
DSM-5	Diagnostic and Statistical Manual of Mental Disorders
FDR	false discovery rate
GI	gastrointestinal
IBD	inflammatory bowel disease
MGH	Massachusetts General Hospital
OTU	operational taxonomic unit
PCA	principal component analysis
PCR	polymerase chain reaction
PERMANOVA	Permutational multivariate analysis of variance
ROC	receiver operator characteristics
rRNA	ribosomal RNA
SparCC	Sparse Correlations for Compositional data

References

1. Baio, J.; Wiggins, L.; Christensen, D.L.; Maenner, M.J.; Daniels, J.; Warren, Z.; Kurzius-Spencer, M.; Zahorodny, W.; Robinson Rosenberg, C.; White, T.; et al. Prevalence of Autism Spectrum Disorder among Children Aged 8 Years-Autism and Developmental Disabilities Monitoring Network, 11 Sites, United States, 2014. *MMWR Surveill. Summ.* **2018**, *67*, 1–23. [CrossRef]
2. Karimi, P.; Kamali, E.; Mousavi, S.; Karahmadi, M. Environmental factors influencing the risk of autism. *J. Res. Med. Sci.* **2017**, *22*, 27.
3. Wu, M.; Wu, Y.; Yu, L.; Liu, J.; Zhang, M.; Kong, X.; Wu, B. A Survey of Epidemiological Studies and Risk Factors of ASD, with a Focus on China. *N. Am. J. Med. Sci.* **2017**, *10*, 1–9.
4. Siniscalco, D.; Schultz, S.; Brigida, A.; Antonucci, N. Inflammation and Neuro-Immune Dysregulations in Autism Spectrum Disorders. *Pharmaceuticals* **2018**, *11*, 56. [CrossRef]
5. Alam, R.; Abdolmaleky, H.M.; Zhou, J.-R. Microbiome, inflammation, epigenetic alterations, and mental diseases. *Am. J. Med. Genet. B Neuropsychiatr. Genet.* **2017**, *174*, 651–660. [CrossRef]
6. Bengmark, S. Gut microbiota, immune development and function. *Pharmacol. Res.* **2013**, *69*, 87–113. [CrossRef]
7. Bruce-Keller, A.J.; Salbaum, J.M.; Berthoud, H.-R. Harnessing Gut Microbes for Mental Health: Getting from Here to There. *Biol. Psychiatry* **2018**, *83*, 214–223. [CrossRef]
8. Cryan, J.F.; Dinan, T.G. Mind-altering microorganisms: The impact of the gut microbiota on brain and behaviour. *Nat. Rev. Neurosci.* **2012**, *13*, 701–712. [CrossRef]
9. Vuong, H.E.; Hsiao, E.Y. Emerging Roles for the Gut Microbiome in Autism Spectrum Disorder. *Biol. Psychiatry* **2017**, *81*, 411–423. [CrossRef]
10. Petersen, C.; Round, J.L. Defining dysbiosis and its influence on host immunity and disease. *Cell. Microbiol.* **2014**, *16*, 1024–1033. [CrossRef]
11. Rossignol, D.A.; Frye, R.E. Evidence linking oxidative stress, mitochondrial dysfunction, and inflammation in the brain of individuals with autism. *Front. Physiol.* **2014**, *5*, 150. [CrossRef]
12. Ding, H.T. Gut Microbiota and Autism: Key Concepts and Findings. *J. Autism Dev. Disord.* **2017**, *47*, 480–489. [CrossRef]
13. Hsiao, E.Y.; McBride, S.W.; Hsien, S.; Sharon, G.; Hyde, E.R.; McCue, T.; Codelli, J.A.; Chow, J.; Reisman, S.E.; Petrosino, J.F.; et al. Microbiota Modulate Behavioral and Physiological Abnormalities Associated with Neurodevelopmental Disorders. *Cell* **2013**, *155*, 1451–1463. [CrossRef]
14. Buffington, S.A.; Di Prisco, G.V.; Auchtung, T.A.; Ajami, N.J.; Petrosino, J.F.; Costa-Mattioli, M. Microbial Reconstitution Reverses Maternal Diet-Induced Social and Synaptic Deficits in Offspring. *Cell* **2016**, *165*, 1762–1775. [CrossRef]
15. Adams, J.B.; Johansen, L.J.; Powell, L.D.; Quig, D.; Rubin, R.A. Gastrointestinal flora and gastrointestinal status in children with autism–comparisons to typical children and correlation with autism severity. *BMC Gastroenterol.* **2011**, *11*, 22. [CrossRef]
16. Carabotti, M.; Scirocco, A.; Maselli, M.A.; Severi, C. The gut-brain axis: Interactions between enteric microbiota, central and enteric nervous systems. *Ann. Gastroenterol.* **2015**, *28*, 203–209.
17. Van De Sande, M.M.H.; van Buul, V.J.; Brouns, F.J. Autism and nutrition: The role of the gut-brain axis. *Nutr. Res. Rev.* **2014**, *27*, 199–214. [CrossRef]
18. Petra, A.I.; Panagiotidou, S.; Hatziagelaki, E.; Stewart, J.M.; Conti, P.; Theoharides, T.C. Gut-Microbiota-Brain Axis and Its Effect on Neuropsychiatric Disorders With Suspected Immune Dysregulation. *Clin. Ther.* **2015**, *37*, 984–995. [CrossRef]
19. Kang, V.; Wagner, G.C.; Ming, X. Gastrointestinal Dysfunction in Children With Autism Spectrum Disorders. *Autism Res.* **2014**, *7*, 501–506. [CrossRef]
20. Rothschild, D.; Weissbrod, O.; Barkan, E.; Kurilshikov, A.; Korem, T.; Zeevi, D.; Costea, P.I.; Godneva, A.; Kalka, I.N.; Bar, N.; et al. Environment dominates over host genetics in shaping human gut microbiota. *Nature* **2018**, *555*, 210–215. [CrossRef]
21. Yatsunenko, T.; Rey, F.E.; Manary, M.J.; Trehan, I.; Dominguez-Bello, M.G.; Contreras, M.; Magris, M.; Hidalgo, G.; Baldassano, R.N.; Anokhin, A.P.; et al. Human gut microbiome viewed across age and geography. *Nature* **2012**, *486*, 222–227. [CrossRef]

22. Rodríguez, J.M.; Murphy, K.; Stanton, C.; Ross, R.P.; Kober, O.I.; Juge, N.; Avershina, E.; Rudi, K.; Narbad, A.; Jenmalm, M.C.; et al. The composition of the gut microbiota throughout life, with an emphasis on early life. *Microb. Ecol. Health Dis.* **2015**, *26*, 26050. [CrossRef]
23. Strati, F.; Cavalieri, D.; Albanese, D.; De Felice, C.; Donati, C.; Hayek, J.; Jousson, O.; Leoncini, S.; Renzi, D.; Calabrò, A.; et al. New evidences on the altered gut microbiota in autism spectrum disorders. *Microbiome* **2017**, *5*, 24. [CrossRef]
24. Kang, D.-W.; Park, J.G.; Ilhan, Z.E.; Wallstrom, G.; LaBaer, J.; Adams, J.B.; Krajmalnik-Brown, R. Reduced Incidence of Prevotella and Other Fermenters in Intestinal Microflora of Autistic Children. *PLoS ONE* **2013**, *8*, e68322. [CrossRef]
25. Sudhakara, P.; Gupta, A.; Bhardwaj, A.; Wilson, A. Oral Dysbiotic Communities and Their Implications in Systemic Diseases. *Dent. J.* **2018**, *6*, 10. [CrossRef]
26. Tsuda, A.; Suda, W.; Morita, H.; Takanashi, K.; Takagi, A.; Koga, Y.; Hattori, M. Influence of Proton-Pump Inhibitors on the Luminal Microbiota in the Gastrointestinal Tract. *Clin. Transl. Gastroenterol.* **2015**, *6*, e89. [CrossRef]
27. Zhang, D.; Jia, H.; Feng, Q.; Wang, D.; Liang, D.; Wu, X.; Li, J.; Tang, L.; Li, Y.; Lan, Z.; et al. The oral and gut microbiomes are perturbed in rheumatoid arthritis and partly normalized after treatment. *Nat. Med.* **2015**, *21*, 895. [CrossRef]
28. Shoemark, D.K.; Allen, S.J. The microbiome and disease: Reviewing the links between the oral microbiome, aging, and Alzheimer's disease. *J. Alzheimer's Dis.* **2015**, *43*, 725–738. [CrossRef]
29. Lai, B.; Milano, M.; Roberts, M.W.; Hooper, S.R. Unmet Dental Needs and Barriers to Dental Care Among Children with Autism Spectrum Disorders. *J. Autism Dev. Disord.* **2012**, *42*, 1294–1303. [CrossRef]
30. Hicks, S.D.; Uhlig, R.; Afshari, P.; Williams, J.; Chroneos, M.; Tierney-Aves, C.; Wagner, K.; Middleton, F.A. Oral microbiome activity in children with autism spectrum disorder. *Autism Res.* **2018**, *11*, 1286–1299. [CrossRef]
31. Qiao, Y.; Wu, M.; Feng, Y.; Zhou, Z.; Chen, L.; Chen, F. Alterations of oral microbiota distinguish children with autism spectrum disorders from healthy controls. *Sci. Rep.* **2018**, *8*, 1597. [CrossRef]
32. Plaza-Díaz, J.; Gómez-Fernández, A.; Chueca, N.; Torre-Aguilar, M.J.d.L.; Gil, Á.; Perez-Navero, J.L.; Flores-Rojas, K.; Martín-Borreguero, P.; Solis-Urra, P.; Ruiz-Ojeda, F.J.; et al. Autism Spectrum Disorder (ASD) with and without Mental Regression is Associated with Changes in the Fecal Microbiota. *Nutrients* **2019**, *11*, 337.
33. Lee, P.F.; Thomas, R.E.; Lee, P.A. Approach to autism spectrum disorder: Using the new DSM-V diagnostic criteria and the CanMEDS-FM framework. *Can. Fam. Physician* **2015**, *61*, 421–424.
34. Mahapatra, S.; Vyshedsky, D.; Martinez, S.; Kannel, B.; Braverman, J.; Edelson, S.; Vyshedskiy, A. Autism Treatment Evaluation Checklist (ATEC) Norms: A "Growth Chart" for ATEC Score Changes as a Function of Age. *Children* **2018**, *5*, 25. [CrossRef]
35. Koh, H.; Lee, M.J.; Kim, M.J.; Shin, J.I.; Chung, K.S. Simple diagnostic approach to childhood fecal retention using the Leech score and Bristol stool form scale in medical practice. *J. Gastroenterol. Hepatol.* **2010**, *25*, 334–338. [CrossRef]
36. Caporaso, J.G.; Kuczynski, J.; Stombaugh, J.; Bittinger, K.; Bushman, F.D.; Costello, E.K.; Fierer, N.; Peña, A.G.; Goodrich, J.K.; Gordon, J.I.; et al. QIIME allows analysis of high-throughput community sequencing data. *Nat. Methods* **2010**, *7*, 335–336. [CrossRef]
37. Amir, A.; McDonald, D.; Navas-Molina, J.A.; Kopylova, E.; Morton, J.T.; Zech Xu, Z.; Kightley, E.P.; Thompson, L.R.; Hyde, E.R.; Gonzalez, A.; et al. Deblur Rapidly Resolves Single-Nucleotide Community Sequence Patterns. *mSystems* **2017**, *2*, 759. [CrossRef]
38. Katoh, K.; Standley, D.M. MAFFT Multiple Sequence Alignment Software Version 7: Improvements in Performance and Usability. *Mol. Biol. Evol.* **2013**, *30*, 772–780. [CrossRef]
39. Price, M.N.; Dehal, P.S.; Arkin, A.P. FastTree 2-approximately maximum-likelihood trees for large alignments. *PLoS ONE* **2010**, *5*, e9490. [CrossRef]
40. Shannon, C.E. A Mathematical Theory of Communication. *Bell Syst. Tech. J.* **1948**, *27*, 379–423. [CrossRef]
41. Faith, D.P. Phylogenetic Diversity and Conservation Evaluation: Perspectives on Multiple Values, Indices, and Scales of Application. In *Phylogenetic Diversity*; Applications and Challenges in Biodiversity Science; Springer: Cham, Germany, 2018; Volume 302, pp. 1–26.
42. Simpson, E.H. Measurement of Diversity. *Nature* **1949**, *163*, 688. [CrossRef]

43. Lozupone, C.; Knight, R. UniFrac: A New Phylogenetic Method for Comparing Microbial Communities. *Appl. Environ. Microbiol.* **2005**, *71*, 8228–8235. [CrossRef]
44. Chen, J.; Bittinger, K.; Charlson, E.S.; Hoffmann, C.; Lewis, J.; Wu, G.D.; Collman, R.G.; Bushman, F.D.; Li, H. Associating microbiome composition with environmental covariates using generalized UniFrac distances. *Bioinformatics* **2012**, *28*, 2106–2113. [CrossRef]
45. Chang, Q.; Luan, Y.; Sun, F. Variance adjusted weighted UniFrac: A powerful beta diversity measure for comparing communities based on phylogeny. *BMC Bioinform.* **2011**, *12*, 118. [CrossRef]
46. Vázquez-Baeza, Y.; Pirrung, M.; Gonzalez, A.; Knight, R. EMPeror: A tool for visualizing high-throughput microbial community data. *Gigascience* **2013**, *2*, 16. [CrossRef]
47. Vázquez-Baeza, Y.; Gonzalez, A.; Smarr, L.; McDonald, D.; Morton, J.T.; Navas-Molina, J.A.; Knight, R. Bringing the Dynamic Microbiome to Life with Animations. *Cell Host Microbe* **2017**, *21*, 7–10. [CrossRef]
48. Benjamini, Y.; Hochberg, Y. Controlling the False Discovery Rate: A Practical and Powerful Approach to Multiple Testing. *J. R. Stat. Soc. Ser. B (Methodol.)* **1995**, *57*, 289–300. [CrossRef]
49. Mandal, S.; Van Treuren, W.; White, R.A.; Eggesbø, M.; Knight, R.; Peddada, S.D. Analysis of composition of microbiomes: A novel method for studying microbial composition. *Microb. Ecol. Health Dis.* **2015**, *26*, 27663. [CrossRef]
50. Schirmer, M.; Denson, L.; Vlamakis, H.; Franzosa, E.A.; Thomas, S.; Gotman, N.M.; Rufo, P.; Baker, S.S.; Sauer, C.; Markowitz, J.; et al. Compositional and Temporal Changes in the Gut Microbiome of Pediatric Ulcerative Colitis Patients Are Linked to Disease Course. *Cell Host Microbe* **2018**, *24*, 600–610. [CrossRef]
51. Friedman, J.; Alm, E.J. Inferring correlation networks from genomic survey data. *PLoS Comput. Biol.* **2012**, *8*, e1002687. [CrossRef]
52. Fang, H.; Huang, C.; Zhao, H.; Deng, M. CCLasso: Correlation inference for compositional data through Lasso. *Bioinformatics* **2015**, *31*, 3172–3180. [CrossRef]
53. Pollard, K. Proteobacteria explain significant functional variability in the human gut microbiome. *Microbiome* **2017**, *5*, 36.
54. Arumugam, M.; Raes, J.; Pelletier, E.; Le Paslier, D.; Yamada, T.; Mende, D.R.; Fernandes, G.R.; Tap, J.; Bruls, T.; Batto, J.-M.; et al. Enterotypes of the human gut microbiome. *Nature* **2011**, *473*, 174–180. [CrossRef]
55. Muskens, J.B.; Velders, F.P.; Staal, W.G. Medical comorbidities in children and adolescents with autism spectrum disorders and attention deficit hyperactivity disorders: A systematic review. *Eur. Child Adolesc. Psychiatry* **2017**, *26*, 1093–1103. [CrossRef]
56. Mancabelli, L.; Milani, C.; Lugli, G.A.; Turroni, F.; Mangifesta, M.; Viappiani, A.; Ticinesi, A.; Nouvenne, A.; Meschi, T.; Sinderen, D.; et al. Unveiling the gut microbiota composition and functionality associated with constipation through metagenomic analyses. *Sci. Rep.* **2017**, *7*, 9879. [CrossRef]
57. Hedin, C.R.; McCarthy, N.E.; Louis, P.; Farquharson, F.M.; McCartney, S.; Taylor, K.; Prescott, N.J.; Murrells, T.; Stagg, A.J.; Whelan, K.; et al. Altered intestinal microbiota and blood T cell phenotype are shared by patients with Crohn's disease and their unaffected siblings. *Gut* **2014**, *63*, 1578–1586. [CrossRef]
58. Winter, S.E.; Lopez, C.A.; Bäumler, A.J. The dynamics of gut-associated microbial communities during inflammation. *Nat. Publ. Group* **2013**, *14*, 319–327. [CrossRef]
59. Williams, B.L.; Hornig, M.; Buie, T.; Bauman, M.L.; Cho Paik, M.; Wick, I.; Bennett, A.; Jabado, O.; Hirschberg, D.L.; Lipkin, W.I. Impaired Carbohydrate Digestion and Transport and Mucosal Dysbiosis in the Intestines of Children with Autism and Gastrointestinal Disturbances. *PLoS ONE* **2011**, *6*, e24585. [CrossRef]
60. Tomova, A.; Husarova, V.; Lakatosova, S.; Bakos, J.; Vlkova, B.; Babinska, K.; Ostatnikova, D. Gastrointestinal microbiota in children with autism in Slovakia. *Physiol. Behav.* **2015**, *138*, 179–187. [CrossRef]
61. Frank, D.N.; Amand, A.L.S.; Feldman, R.A.; Boedeker, E.C.; Harpaz, N.; Pace, N.R. Molecular-phylogenetic characterization of microbial community imbalances in human inflammatory bowel diseases. *Proc. Natl. Acad. Sci. USA* **2007**, *104*, 13780–13785. [CrossRef]
62. Turnbaugh, P.J.; Hamady, M.; Yatsunenko, T.; Cantarel, B.L.; Duncan, A.; Ley, R.E.; Sogin, M.L.; Jones, W.J.; Roe, B.A.; Affourtit, J.P.; et al. A core gut microbiome in obese and lean twins. *Nature* **2008**, *457*, 480–484. [CrossRef]
63. Anand, S.; Kaur, H.; Mande, S.S. Comparative In silico Analysis of Butyrate Production Pathways in Gut Commensals and Pathogens. *Front. Microbiol.* **2016**, *7*, 1945. [CrossRef]
64. Jangi, S.; Gandhi, R.; Cox, L.M.; Li, N.; von Glehn, F.; Yan, R.; Patel, B.; Mazzola, M.A.; Liu, S.; Glanz, B.L.; et al. Alterations of the human gut microbiome in multiple sclerosis. *Nat. Commun.* **2016**, *7*, 12015. [CrossRef]

65. Kirby, T.; Ochoa-Repáraz, J. The Gut Microbiome in Multiple Sclerosis: A Potential Therapeutic Avenue. *Med. Sci.* **2018**, *6*, 69. [CrossRef]
66. Xun, Z.; Zhang, Q.; Xu, T.; Chen, N.; Chen, F. Dysbiosis and Ecotypes of the Salivary Microbiome Associated With Inflammatory Bowel Diseases and the Assistance in Diagnosis of Diseases Using Oral Bacterial Profiles. *Front. Microbiol.* **2018**, *9*, 1136. [CrossRef]
67. Papageorgiou, S.N.; Hagner, M.; Nogueira, A.V.B.; Franke, A.; Jäger, A.; Deschner, J. Inflammatory bowel disease and oral health: Systematic review and a meta-analysis. *J. Clin. Periodontol.* **2017**, *44*, 382–393. [CrossRef]
68. Flint, H.J.; Duncan, S.H.; Scott, K.P.; Louis, P. Interactions and competition within the microbial community of the human colon: Links between diet and health. *Environ. Microbiol.* **2007**, *9*, 1101–1111. [CrossRef]
69. Luna, R.A.; Oezguen, N.; Balderas, M.; Venkatachalam, A.; Runge, J.K.; Versalovic, J.; Veenstra-VanderWeele, J.; Anderson, G.M.; Savidge, T.; Williams, K.C. Distinct Microbiome-Neuroimmune Signatures Correlate with Functional Abdominal Pain in Children With Autism Spectrum Disorder. *Cell. Mol. Gastroenterol. Hepatol.* **2017**, *3*, 218–230. [CrossRef]
70. Gargari, G.; Taverniti, V.; Gardana, C.; Cremon, C.; Canducci, F.; Pagano, I.; Barbaro, M.R.; Bellacosa, L.; Castellazzi, A.M.; Valsecchi, C.; et al. Fecal Clostridiales distribution and short-chain fatty acids reflect bowel habits in irritable bowel syndrome. *Environ. Microbiol.* **2018**, *20*, 3201–3213. [CrossRef]
71. Malhi, P.; Venkatesh, L.; Bharti, B.; Singhi, P. Feeding Problems and Nutrient Intake in Children with and without Autism: A Comparative Study. *Indian J. Pediatrics* **2017**, *84*, 283–288. [CrossRef]
72. Ahearn, W.H.; Castine, T.; Nault, K.; Green, G. An Assessment of Food Acceptance in Children with Autism or Pervasive Developmental Disorder-Not Otherwise Specified. *J Autism Dev. Disord.* **2001**, *31*, 505–511. [CrossRef]
73. Liu, J.; Zhang, M.; Kong, X. Gut Microbiome and Autism: Recent Advances and Future Perspectives. *N. Am. J. Med. Sci.* **2016**, *9*, 1–12.
74. Rizzatti, G.; Lopetuso, L.R.; Gibiino, G.; Binda, C.; Gasbarrini, A. Proteobacteria: A Common Factor in Human Diseases. *BioMed Res. Int.* **2017**, *2017*, 1–7. [CrossRef]
75. Xu, G.; Snetselaar, L.G.; Jing, J.; Liu, B.; Strathearn, L.; Bao, W. Association of Food Allergy and Other Allergic Conditions With Autism Spectrum Disorder in Children. *JAMA Netw. Open* **2018**, *1*, e180279. [CrossRef]
76. Tamanai-Shacoori, Z.; Smida, I.; Bousarghin, L.; Loreal, O.; Meuric, V.; Fong, S.B.; Bonnaure-Mallet, M.; Jolivet-Gougeon, A. Roseburiaspp.: A marker of health? *Future Microbiol.* **2017**, *12*, 157–170. [CrossRef]
77. Guinane, C.M.; Cotter, P.D. Role of the gut microbiota in health and chronic gastrointestinal disease: Understanding a hidden metabolic organ. *Ther. Adv. Gastroenterol.* **2013**, *6*, 295–308. [CrossRef]
78. Patterson, A.M.; Mulder, I.E.; Travis, A.J.; Lan, A.; Cerf-Bensussan, N.; Gaboriau-Routhiau, V.; Garden, K.; Logan, E.; Delday, M.I.; Coutts, A.G.P.; et al. Human Gut Symbiont Roseburia hominis Promotes and Regulates Innate Immunity. *Front. Immunol.* **2017**, *8*, 1166. [CrossRef]
79. Gibson, F.C.; Genco, C.A. The Genus Porphyromonas. In *The Prokaryotes*; Dworkin, M., Falkow, S., Rosenberg, E., Schleifer, K.-H., Stackebrandt, E., Eds.; Springer: New York, NY, USA, 2006; pp. 428–454.
80. Atanasova, K.R.; Yilmaz, Ö. Looking in the Porphyromonas gingivaliscabinet of curiosities: The microbium, the host and cancer association. *Mol. Oral Microbiol.* **2014**, *29*, 55–66. [CrossRef]
81. Wexler, H.M. Bacteroides: The Good, the Bad, and the Nitty-Gritty. *Clin. Microbiol. Rev.* **2007**, *20*, 593–621. [CrossRef]
82. Zhou, Y.; Zhi, F. Lower Level of Bacteroidesin the Gut Microbiota Is Associated with Inflammatory Bowel Disease: A Meta-Analysis. *BioMed Res. Int.* **2016**, *2016*, 1–9.
83. Houston, S.; Blakely, G.W.; McDowell, A.; Martin, L.; Patrick, S. Binding and degradation of fibrinogen by Bacteroides fragilis and characterization of a 54 kDa fibrinogen-binding protein. *Microbiology* **2010**, *156*, 2516–2526. [CrossRef]
84. De Angelis, M.; Piccolo, M.; Vannini, L.; Siragusa, S.; De Giacomo, A.; Serrazzanetti, D.I.; Cristofori, F.; Guerzoni, M.E.; Gobbetti, M.; Francavilla, R. Fecal Microbiota and Metabolome of Children with Autism and Pervasive Developmental Disorder Not Otherwise Specified. *PLoS ONE* **2013**, *8*, e76993. [CrossRef]

85. Finegold, S.M.; Dowd, S.E.; Gontcharova, V.; Liu, C.; Henley, K.E.; Wolcott, R.D.; Youn, E.; Summanen, P.H.; Granpeesheh, D.; Dixon, D.; et al. Pyrosequencing study of fecal microflora of autistic and control children. *Anaerobe* **2010**, *16*, 444–453. [CrossRef]
86. Son, J.S.; Zheng, L.J.; Rowehl, L.M.; Tian, X.; Zhang, Y.; Zhu, W.; Litcher-Kelly, L.; Gadow, K.D.; Gathungu, G.; Robertson, C.E.; et al. Comparison of Fecal Microbiota in Children with Autism Spectrum Disorders and Neurotypical Siblings in the Simons Simplex Collection. *PLoS ONE* **2015**, *10*, e0137725. [CrossRef]

Article

Effects of Regular Kefir Consumption on Gut Microbiota in Patients with Metabolic Syndrome: A Parallel-Group, Randomized, Controlled Study

Ezgi BELLIKCI-KOYU [1,2], Banu Pınar SARER-YUREKLI [3], Yakut AKYON [4], Fadime AYDIN-KOSE [5], Cem KARAGOZLU [6], Ahmet Gokhan OZGEN [3], Annika BRINKMANN [7], Andreas NITSCHE [7], Koray ERGUNAY [4], Engin YILMAZ [8] and Zehra BUYUKTUNCER [1,*]

1 Department of Nutrition and Dietetics, Faculty of Health Sciences, Hacettepe University, Ankara 06230, Turkey
2 Department of Nutrition and Dietetics, Faculty of Health Sciences, Izmir Katip Celebi University, Izmir 35620, Turkey
3 Department of Endocrinology, Faculty of Medicine, Ege University, Izmir 35040, Turkey
4 Department of Medical Microbiology, Faculty of Medicine, Hacettepe University, Ankara 06230, Turkey
5 Department of Biochemistry, Faculty of Pharmacy, Ege University, Izmir 35040, Turkey
6 Department of Dairy Technology, Faculty of Agriculture, Ege University, Izmir 35040, Turkey
7 Robert Koch Institute; Center for Biological Threats and Special Pathogens 1 (ZBS-1), 13353 Berlin, Germany
8 Department of Medical Biology, Acıbadem Mehmet Ali Aydınlar University, Istanbul 34752, Turkey
* Correspondence: zbtuncer@hacettepe.edu.tr; Tel.: +90-0312-305-1094; Fax: +90-312-309-13-10

Received: 13 August 2019; Accepted: 22 August 2019; Published: 4 September 2019

Abstract: Several health-promoting effects of kefir have been suggested, however, there is limited evidence for its potential effect on gut microbiota in metabolic syndrome This study aimed to investigate the effects of regular kefir consumption on gut microbiota composition, and their relation with the components of metabolic syndrome. In a parallel-group, randomized, controlled clinical trial setting, patients with metabolic syndrome were randomized to receive 180 mL/day kefir ($n = 12$) or unfermented milk ($n = 10$) for 12 weeks. Anthropometrical measurements, blood samples, blood pressure measurements, and fecal samples were taken at the beginning and end of the study. Fasting insulin, HOMA-IR, TNF-α, IFN-γ, and systolic and diastolic blood pressure showed a significant decrease by the intervention of kefir ($p \leq 0.05$, for each). However, no significant difference was obtained between the kefir and unfermented milk groups ($p > 0.05$ for each). Gut microbiota analysis showed that regular kefir consumption resulted in a significant increase only in the relative abundance of *Actinobacteria* ($p = 0.023$). No significant change in the relative abundance of *Bacteroidetes, Proteobacteria or Verrucomicrobia* by kefir consumption was obtained. Furthermore, the changes in the relative abundance of sub-phylum bacterial populations did not differ significantly between the groups ($p > 0.05$, for each). Kefir supplementation had favorable effects on some of the metabolic syndrome parameters, however, further investigation is needed to understand its effect on gut microbiota composition.

Keywords: kefir; gut microbiota; metabolic syndrome

1. Introduction

Metabolic syndrome (MetS) is a pathologic condition that includes abdominal obesity, insulin resistance, dyslipidemia, and arterial hypertension [1]. Each component of MetS is known as a risk factor for the development of type 2 diabetes and cardiovascular diseases. It was found that the risk of type 2 diabetes was five times, the risk of cardiovascular disease was two times, and the risk of death was one-half times higher in individuals with MetS compared to those without the syndrome. Due to its high prevalence and related health problems, the MetS is currently considered as a significant public health problem [1,2].

MetS has a multi-factorial etiology comprising complex interactions between genetic predispositions and environmental factors including diet, physical activity, and other lifestyle factors [3,4]. Since Turnbaugh et al. showed the linked between gut microbiota and obesity, there has been growing evidence that suggests a causal relationship between gut microbiota and the components of MetS [5]. Primary, the low-grade chronic inflammation state in MetS has been explained by the metabolic endotoxemia that was a result of gut dysbiosis [6,7]. Most of the animal and human studies have reported that obesity and insulin resistance are associated with an altered ratio of *Firmicutes* and *Bacteroidetes* [8,9]. In addition to the effects on immune function, the gut microbiota also exerts its role through the influence on host energy metabolism and gut barrier integrity [10,11]. Therefore, the gut microbiota has been suggested as a potential target to modify the risk factors that contribute to conditions of MetS.

The modification of diet using prebiotics and probiotics has been suggested as a useful strategy to improve metabolic health via the modulation of gut microbiota. Although the effects of probiotic and prebiotic supplementation on metabolic health have been examined in previous studies, the results are inconsistent due to the choice of probiotic strain, formulation of the probiotic, outcome of interest, and duration of the intervention [12–14]. Furthermore, ingestion of probiotics through traditional fermented foods has not been widely examined in terms of their efficiency on MetS components. Kefir is a fermented milk product, traditionally produced with kefir grains that have a specific combination of bacteria and yeasts [15,16]. Microbial composition of kefir varies depending upon the type of kefir grains, the type and composition of milk, culture medium, fermentation period and temperature, and also storage conditions [17]. *Lactobacillus*, *Lactococcus*, *Streptococcus Leuconostoc*, and acetic acid bacteria are the most common bacteria; and *Saccharomyces*, *Kluyveromyces*, and *Candida* species are mostly found yeasts in kefir [18]. Animal studies have suggested that kefir has anticarcinogenic, antimicrobial, anti-inflammatory activities, and thus may ameliorate MetS components [19–23]. However, there is still limited clinical evidence for its potential effects on MetS patients. To our knowledge, especially, the effects of kefir on MetS components via the modulation of gut microbiota have not been examined widely in clinical settings. To address the research gap, this study aimed to investigate the effects of daily kefir consumption on gut microbiota composition and their relation with the components of metabolic syndrome in adults with MetS.

2. Materials and Methods

2.1. Subjects

Subjects with MetS, aged 18–65 years, were recruited from the outpatient clinic of the Department of Endocrinology and Metabolism at the Ege University, Izmir, Turkey. MetS was diagnosed using the IDF-2005 guidelines [24]. The eligibility of a subject was confirmed following a physical examination by the research endocrinologist and a nutritional assessment by the research dietitian in the screening period. Adults were excluded if they (1) were using antibiotics in the past 1 month or during the intervention period, (2) were using dietary supplements (probiotic, prebiotic, or symbiotic) during the past three months or during the intervention period, (3) were pregnant or lactating, (4) had severe liver, kidney, heart, or immune deficiency, (5) had chronic gastrointestinal system diseases, type 1 diabetes or cancer, (6) had allergy to the dairy products or lactose intolerance, (7) were currently taking prescribed drugs that can modulate lipid profile or glycaemic control, and (8) did not comply with the consumption of test drinks.

The compliance was assessed by interviewing the participants and reviewing the record of their consumption in each visit. Non-compliance was defined as consuming < 80% of the scheduled serving during the study period.

The study protocol was approved by the Ethics Committee of Clinical Research at Ege University Faculty of Medicine (15-2.1/14) and registered at clinicaltrials.gov (NCT03966846). All procedures were performed in accordance with the Declaration of Helsinki. Written informed consent was obtained from all participants.

2.2. Study Design

A parallel-group, randomized, controlled clinical trial was performed. A total of 40 eligible participants were randomized, and 20 participants in each group were allocated to intervention. Five participants in the kefir group and four participants in the unfermented milk group did not receive the allocated intervention due to medical conditions and not providing the fecal samples, and also three participants in the kefir group and six participants in the unfermented milk group discontinued the intervention due to taking antibiotics and declined consent. Therefore, the study was completed with 22 participants and an allocation ratio of 55%. The recruitment and follow up of participants were conducted between March 2015–July 2017. Participants were randomized into two groups (kefir group and unfermented milk group as control) by the research physicians using a stratified block randomization method. The random allocation sequence was provided by the Department of Biostatistics, Hacettepe University. Participants visited the research center 5 times in total. The first visit included the screening of individuals in terms of inclusion and exclusion criteria. The second visit (Week 0) included recording general characteristics, medical history, and lifestyle behaviors of participants, assessing the nutritional status of participants using 24-h dietary recall and anthropometrical measurements, collecting the initial blood and fecal samples, measuring the blood pressure, and also proving information about the consumption and storage of test drinks. The third (Week 4) and forth (Week 8) visits included the assessment of the compliance in terms of consumption of test drinks and dietary intake. The fifth visit (Week 12) included the assessment of the nutritional status of participants using 24-h dietary recall and anthropometrical measurements, collection of the final blood and fecal samples, and the measurement of blood pressure (Figure 1).

The primary outcome of the study was the change in the relative abundance of microorganisms in gut microbiota by regular kefir consumption. The potential correlations between the changes in dietary intake, anthropometrical measurements, biochemical parameters, or blood pressure and the change in microbiota composition were all secondary outcomes.

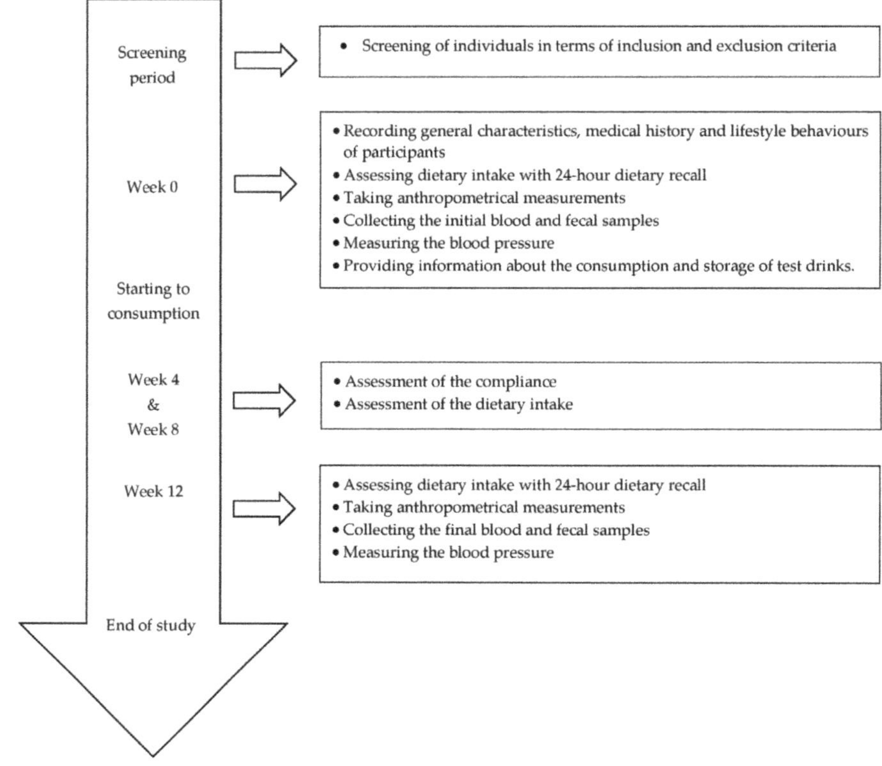

Figure 1. Timeline of the study.

2.3. Intervention

During a 12-week intervention period, kefir group (n = 12) received kefir (180 mL/day) while control group (n = 10) received unfermented milk (180 mL/day) regularly. Participants were asked to maintain their habitual diet and physical activity. Additional products that contain probiotics were not allowed during the intervention period. No dietary supplement use was recorded before or during the study.

2.4. Test Drinks

Two dairy products (kefir and unfermented milk) were tested in parallel groups. Kefir was prepared using the culture of DC1500I (Danisco, Olsztyn, Poland) containing *Lactococcus lactis* subsp. *lactis, Lactococcus lactis* subsp. *cremoris, Lactococcus lactis* subsp. *diacetylactis, Leuconostoc mesenteroides* subsp. *cremoris, Lactobacillus kefir, Kluyveromyces marxianus*, and *Saccharomyces unisporus* at Ege University Faculty of Agriculture, Department of Dairy Technology. Kefir was derived from the full-fat (3.5%) homogenized and pasteurized (at 85 °C) milk that was used as a control drink at the same time. The beverages were distributed and stored at 4 °C. The test drinks were received to participants twice a week, and they consumed the test drinks between 1 and 4 days of post-production.

2.5. Dietary Assessment

Dietary intake was assessed using 24-h dietary recall method by research dietician in each visit. A photographic atlas of food portion sizes was used to clarify the amounts of food items consumed.

Dietary energy, macro- and micronutrient intakes were analyzed using BeBIS software (Ebispro for Windows, Stuttgart, Germany; Turkish Version BeBIS, Nutrition Information System, Version 8).

2.6. Anthropometrical Measurements

Body weight and composition (fat mass and fat-free mass) were measured by Tanita BC418 (USA), and height was measured by a calibrated stadiometer (Nan Tartı, TR). Body Mass Index (BMI) was calculated by dividing body weight (in kilograms) by the square of height (in meter). The waist circumference was measured at the midpoint between the lower ribs and the iliac crest, and hip circumference was measured horizontal at the largest circumference of hip. Waist-to-hip ratio (WHR) was calculated.

2.7. Biochemical Analysis and Blood Pressure

Venous blood samples were drawn after a 10-h overnight fasting excluding only the water at the Visit 1, Visit 2 (Week 0), and Visit 5 (Week 12). Serum glucose, insulin, HbA1c, total cholesterol, high-density lipoprotein (HDL-C), low-density lipoprotein (LDL-C), triglycerides, homocysteine, high-sensitivity C-reactive protein (hs-CRP), alanine aminotransferase (ALT), aspartate aminotransferase (AST), and gamma-glutamyl transferase (GGT) were analyzed at Ege University, Hospital of Medical School, Laboratory of Clinical Biochemistry. All biomarkers were analyzed using routine methods by Roche/Cobas analyzer series. Serum concentrations of tumor necrosis factor-α (TNF-α), interleukin-6 (IL-6), interleukin-10 (IL-10) and interferon-gamma (IFN-γ) were determined by enzyme-linked immunosorbent assay (ELISA) using standard kits, and the analyses were conducted as described by the manufacturer (DIAsource ImmunoAssays, Louvain-la-Neuve, Belgium). Insulin resistance was assessed using Homeostatic Model Assessment (HOMA-IR) model calculated with the equation of "the fasting insulin level (μU/L) × fasting plasma glucose (mg/dL)/405". Systolic (SBP) and diastolic blood pressure (DBP) were measured at the brachial artery of right upper arm after 15 min rest. Both blood pressures were measured twice at 5-min intervals and recorded on average.

2.8. Specimen Processing, 16S rRNA Amplification and Sequencing

Fecal samples from individuals enrolled in the study were collected in sterile containers and kept frozen at −80 °C. A sterile spatula was used to obtain 4–5 pieces of frozen chunks from the surface and internal portions of the specimen. They were combined for a 150–200 mg total weight for each and mixed by vortexing. Following a bead-beated step described by Tomas et al. and Wu et al. [25,26], DNA was extracted using Qiagen Stool Mini Kit (Qiagen, Hilden, Germany) as directed by the manufacturer. DNA amount of 50 ng/μL was prepared for each specimen, using a NanoDrop 2000 spectrophotometer (Thermo Fisher Scientific, Hennigsdorf, Germany).

The 16S rRNA sequences were amplified using previously-described primers targeting the V3-V4 region, frequently used to study bacterial diversity [27], with Illumina adapter overhang sequences added, as directed by the manufacturer. Attachment of sequencing adapters to PCR products, amplification and library preparation were performed using the Nextera XT Index and Nextera DNA Library Prep kits (Illumina, San Diego, CA, United States), as suggested by the manufacturer. Product clean-up, library quantification, and optimization were carried out using the Agencourt AMPure XP reagent (Beckman Coulter Biosciences, Krefeld, Germany) standard protocol and a Qubit 2.0 Fluorometer (Thermo Fisher Scientific, Rochester, NY, USA). The sequencing runs were performed in an Illumina MiSeq sequencer (Illumina Inc., New York, USA).

2.9. Data Handling, Phylogenetic and Statistical Analyses

The raw sequencing data were de-multiplexed and extracted in fastq format. Sequence data handling and taxonomic assignment were carried out using Geneious v11.1 (Biomatters Ltd., Auckland, New Zealand), MALT V0.3.8 and MEGAN v6.11 [28]. Trimming for read quality and length and adaptor sequence removal were performed using Trimmomatic v0.35 [29,30]. Trimmed reads were mapped to the NCBI-NT RefSeq 16S database via MALT V0.3.8, with hits down to 95% identity. For the operational taxonomic unit (OTU) identification and taxonomic binning, LCA-assignment algorithm (with 95% minimum identity) and 16s percent identity filter (species assignment at 99% identity) were employed. Relative bacterial abundance on the genus and species levels were calculated using the reads numbers of the corresponding OTUs.

Various alpha and beta diversity metrics were calculated for bacterial diversity and composition analyses. For this purpose, raw data were imported into QIIME2 [31], filtered and controlled for quality and chimeric sequences using DADA2, q2-demux, and dblur scripts [32,33]. The trimmed reads were subsequently mapped to the GreenGenes [34] and SILVA [35] databases for OTU identification and taxonomic binning. Faith phylogenetic diversity (PD), Pielou, Shannon, and Jaccard indices and Bray–Curtis and UniFrac distances were computed and evaluated using Kruskal–Wallis, Spearman, or permutational multivariate analysis of variance (PERMANOVA) tests as appropriate.

Data were analyzed using the Statistical Package for the Social Sciences (SPSS), version 22.0. (SPSS Inc., Chicago, IL, USA). Data normality was tested by Shapiro–Wilk test prior to further analyses. Kruskal–Wallis and Mann–Whitney U tests were employed for comparisons among groups where appropriate. Spearman's rank-order correlation was used to analyze the correlation analysis between microbial taxa and biochemical and blood pressure measurements. A value of $p < 0.05$ was considered as significant.

3. Results

Baseline characteristics of the participants were summarized in Table 1. There were no differences in terms of age, dietary intake, anthropometrical measurements, and biochemical parameters except serum insulin levels between groups.

Table 1. Baseline characteristics of kefir and unfermented milk groups.

Characteristics	Kefir Group	Unfermented Milk Group	
	Baseline	**Baseline**	p
Sex (Female/Male)	10/2	6/4	0.348
Age (year)	52.00 (47.50–60.50)	53.00 (45.00–60.00)	0.821
Dietary intake			
Energy (kcal/day)	1694.16 (1590.92–1936.72)	1655.35 (1423.52–2026.12)	0.821
Carbohydrate (g)	182.87 (166.79–205.74)	155.79 (141.61–224.90)	0.283
Protein (g)	73.12 (59.29–83.91)	65.23 (47.55–75.36)	0.254
Fat (g)	73.89 (68.16–97.67)	85.21 (68.06–105.53)	0.418
Fibre (g)	26.11 (18.42–36.90)	23.28 (17.11–26.23)	0.418
Anthropometrical measurements			
Weight (kg)	84.05 (69.23–88.78)	87.65 (75.60–100.60)	0.180
Body mass index (kg/m^2)	30.67 (26.94–34.66)	32.38 (29.18–34.59)	0.381
Body fat mass (%)	37.05 (31.33–44.05)	37.45 (27.05–41.45)	0.582
Waist circumference (cm)	100.50 (90.75–110.00)	106.75 (102.25–119.00)	0.228
Hip circumference (cm)	111.50 (106.00–116.50)	112.00 (106.00–119.25)	0.771
Waist–to–hip ratio	0.92 (0.86–0.99)	0.97 (0.92–1.00)	0.203
Lipid profile			
Total cholesterol (mg/dL)	243.50 (217.25–265.25)	220.00 (199.75–249.00)	0.228
HDL cholesterol (mg/dL)	45.00 (39.00–55.75)	42.50 (34.50–56.25)	0.456
LDL cholesterol (mg/dL)	154.50 (135.75–177.00)	141.00 (114.50–177.50)	0.283
Triglycerides (mg/dL)	185.00 (114.50–216.75)	164.50 (126.25–220.75)	0.923
Homocysteine (μmoL/L)	10.01 (8.64–12.40)	13.10 (10.73–15.25)	0.050

Table 1. *Cont.*

Characteristics	Kefir Group	Unfermented Milk Group	
	Baseline	Baseline	p
Glycaemic status			
Glucose (mg/dL)	105.00 (93.75–109.75)	101.50 (97.00–107.25)	>0.99
Insulin (mU/L)	15.94 (11.75–17.64)	19.04 (18.09–25.49)	0.011 *
HbA1c (%)	5.60 (5.25–5.88)	5.65 (5.20–6.03)	0.872
HOMA–IR	4.18 (2.86–4.59)	4.52 (4.29–6.65)	0.180
Inflammation–related indicators			
hs–CRP (mg/dL)	0.22 (0.69–0.80)	0.27 (0.21–0.41)	0.722
TNF–α (pg/mL)	12.01 (0.76–43.05)	8.51 (0.49–25.85)	0.418
IL–6 (pg/mL)	15.82 (11.52–29.75)	19.73 (13.85–28.71)	0.418
IL–10 (pg/mL)	4.38 (1.13–32.90)	1.45 (1.13–9.34)	0.456
IFN–γ (IU/mL)	1.23 (0.12–2.19)	0.56 (0.02–3.04)	>0.99
ALT (U/L)	18.50 (16.50–24.00)	25.00 (20.75–31.25)	0.140
AST (U/L)	19.00 (18.00–20.00)	19.00 (18.00–20.25)	0.923
GGT (U/L)	15.00 (10.75–23.00)	19.00 (16.00–40.50)	0.169
Blood pressure			
Systolic blood pressure (mmHg)	134.50 (115.25–140.50)	132.50 (123.75–144.00)	0.722
Diastolic blood pressure (mmHg)	85.00 (77.50–92.00)	89.00 (81.00–92.00)	0.497

Data are given as median (25th percentile–75th percentile). Mann–Whitney U test was used to compare differences between groups. Fisher's exact test was used to compare gender between groups. HDL: High Density Lipoprotein, LDL: Low Density Lipoprotein, HbA1c: Glycosylated Hemoglobin, HOMA-IR: Homeostasis Model of Assessment Insulin Resistance, hs-CRP: High-sensitivity C-reactive Protein, TNF: Tumor Necrosis Factor, IL: Interleukin, IFN: Interferon, ALT: Alanine Aminotransferase, AST: Aspartate Aminotransferase, GGT: Gamma-glutamyl Transferase. * $p < 0.05$.

The changes in dietary intake, anthropometrical measurements, biochemical parameters, and blood pressure during the intervention period were given in Table 2. Intakes of energy and macronutrients did not change significantly during the intervention period in either kefir or unfermented milk groups ($p > 0.05$, for each). In terms of anthropometrical measurements, body weight and fat mass showed slight reductions after the 12-weeks intervention of kefir compared to unfermented milk, however, the changes in any of anthropometrical measurement from baseline to after the intervention did not differ significantly between the groups ($p > 0.05$, for each). Among the biochemical biomarkers, almost all parameters of lipid profile and glycaemic status showed amelioration by the intervention kefir group, however, only the difference in fasting insulin and thereby HOMA-IR from baseline to after intervention was significant ($p = 0.050$). Furthermore, TNF-α and IFN-γ showed a significant decrease after the intervention of kefir ($p = 0.015$ and $p = 0.013$, respectively), whereas IL-6 showed a slightly larger decrease in the unfermented milk group ($p = 0.047$). Both systolic blood pressure and diastolic blood pressure decreased significantly after the intervention in the kefir group (respectively $p = 0.041$ and $p = 0.019$), while only systolic blood pressure showed a modest decrease in the unfermented milk group ($p = 0.047$).

Table 2. Dietary intake, anthropometrical measurements, biochemical parameters, and blood pressure in kefir and unfermented milk groups.

Characteristics	Kefir Group			Unfermented Milk Group			
	Baseline	Week 12	p^1	Baseline	Week 12	p^2	p^3
Dietary intake							
Energy (kcal/day)	1694.16 (1590.92–1936.72)	1995.73 (1567.23–2351.80)	0.347	1655.35 (1423.52–2026.12)	1979.09 (1606.15–2123.62)	0.575	0.821
Carbohydrate (%)	44.00 (38.00–45.75)	42.50 (37.75–48.75)	0.964	41.00 (32.00–45.25)	46.00 (38.50–52.50)	0.214	0.283
Protein (%)	16.50 (15.25–19.00)	13.50 (12.00–18.50)	0.066	15.50 (12.75–17.25)	15.00 (12.00–16.50)	0.717	0.254
Fat (%)	39.50 (37.00–44.75)	41.00 (38.00–46.00)	0.666	43.50 (41.75–48.00)	40.00 (32.75–44.25)	0.167	0.228
Fibre (g)	26.11 (18.42–36.90)	26.81 (21.58–32.65)	0.814	23.28 (17.11–26.23)	25.17 (16.67–33.61)	0.646	0.722
Anthropometrical measurements							
Weight (kg)	84.05 (69.23–88.78)	83.50 (66.90–88.75)	0.695	87.65 (75.60–100.60)	88.55 (74.33–96.65)	0.207	0.418
Body mass index (kg/m^2)	30.67 (26.94–34.66)	30.58 (26.24–34.31)	0.754	32.38 (29.18–34.59)	31.90 (29.05–33.71)	0.241	0.418
Body fat mass (%)	37.05 (31.33–44.05)	35.85 (30.58–44.23)	0.248	37.45 (27.05–41.45)	38.30 (29.63–43.98)	0.241	0.069
Waist circumference (cm)	100.50 (90.75–110.00)	102.25 (90.00–109.00)	0.407	106.75 (102.25–119.00)	106.75 (100.50–118.50)	0.952	0.722
Hip circumference (cm)	111.50 (106.00–116.50)	110.00 (106.25–118.63)	0.813	112.00 (106.00–119.25)	111.75 (105.13–116.25)	0.483	0.228
Waist–to–hip ratio	0.92 (0.86–0.99)	0.92 (0.86–0.95)	0.929	0.97 (0.92–1.00)	0.99 (0.92–1.03)	0.386	0.497
Lipid profile							
Total cholesterol (mg/dL)	243.50 (217.25–265.25)	222.00 (201.25–275.00)	0.209	220.00 (199.75–249.00)	226.50 (198.75–240.25)	0.953	0.539
HDL cholesterol (mg/dL)	45.00 (39.00–55.75)	46.00 (41.00–63.00)	0.271	42.50 (34.50–56.25)	43.50 (36.00–58.00)	0.412	0.346
LDL cholesterol (mg/dL)	154.50 (135.75–177.00)	144.00 (116.50–188.75)	0.170	141.00 (114.50–177.50)	147.50 (115.75–167.50)	0.959	0.314
Triglycerides (mg/dL)	185.00 (114.50–216.75)	152.50 (116.50–191.25)	0.530	164.50 (126.25–220.75)	161.50 (117.00–236.75)	0.878	1.000
Homocysteine (μmol/L)	10.01 (8.64–12.40)	9.31 (7.45–12.70)	0.182	13.10 (10.73–15.25)	12.00 (11.05–14.35)	0.213	0.710
Glycaemic status							
Glucose (mg/dL)	105.00 (93.75–109.75)	100.50 (96.50–103.00)	0.157	101.50 (97.00–107.25)	98.50 (97.50–116.25)	0.918	0.159
Insulin (mU/L)	15.94 (11.75–17.64)	13.64 (7.33–16.36)	0.050 *	19.04 (18.09–25.49)	22.08 (15.05–28.54)	0.386	0.123
HbA1c (%)	5.60 (5.25–5.88)	5.65 (5.50–5.98)	0.157	5.65 (5.20–6.03)	5.70 (5.10–5.90)	0.918	0.123
HOMA–IR	4.18 (2.86–4.59)	3.42 (1.93–4.22)	0.050 *	4.52 (4.29–6.65)	5.52 (3.38–8.49)	0.445	0.159

Table 2. *Cont.*

Characteristics	Kefir Group			Unfermented Milk Group			
	Baseline	Week 12	p^1	Baseline	Week 12	p^2	p^3
Inflammation–related indicators							
hs–CRP (mg/dL)	0.22 (0.69–0.80)	0.16 (0.10–0.46)	0.533	0.27 (0.21–0.41)	0.24 (0.13–0.50)	0.917	0.733
TNF–α (pg/mL)	12.01 (0.76–43.05)	1.13 (0.49–8.33)	0.015 *	8.51 (0.49–25.85)	4.12 (0.49–13.03)	0.401	0.123
IL–6 (pg/mL)	15.82 (11.52–29.75)	13.47 (5.65–21.39)	0.099	19.73 (13.85–28.71)	10.03 (6.16–16.45)	0.047 *	0.872
IL–10 (pg/mL)	4.38 (1.13–32.90)	1.91 (1.13–14.77)	0.386	1.45 (1.13–9.34)	1.13 (1.13–15.95)	0.735	0.539
IFN–γ (IU/mL)	1.23 (0.12–2.19)	0.38 (0.04–0.85)	0.013 *	0.56 (0.02–3.04)	0.49 (0.18–1.19)	0.086	0.628
ALT (U/L)	18.50 (16.50–24.00)	22.00 (19.50–24.00)	0.288	25.00 (20.75–31.25)	24.50 (18.75–29.00)	0.215	0.180
AST (U/L)	19.00 (18.00–20.00)	19.00 (17.00–22.50)	0.887	19.00 (18.00–20.25)	17.50 (17.00–19.50)	0.136	0.203
GGT (U/L)	15.00 (10.75–23.00)	14.50 (12.00–23.75)	0.371	19.00 (16.00–40.50)	19.00 (14.25–29.25)	0.065	0.169
Blood pressure							
Systolic blood pressure (mmHg)	134.50 (115.25–140.50)	118.00 (103.25–137.75)	0.041 *	132.50 (123.75–144.00)	118.00 (105.75–137.00)	0.047 *	0.974
Diastolic blood pressure (mmHg)	85.00 (77.50–92.00)	78.50 (69.00–80.00)	0.019 *	89.00 (81.00–92.00)	78.50 (66.75–89.50)	0.059	1.000

Data are given as median (25th percentile–75th percentile). HDL: High Density Lipoprotein, LDL: Low Density Lipoprotein, HbA1c: Glycosylated Hemoglobin, HOMA-IR: Homeostasis Model of Assessment Insulin Resistance, hs-CRP: High-sensitivity C-reactive Protein, TNF: Tumor Necrosis Factor, IL: Interleukin, IFN-γ: Interferon-γ, ALT: Alanine Aminotransferase, AST: Aspartate Aminotransferase, GGT: Gamma-glutamyl Transferase. p^1 gives the differences between baseline and after intervention in the kefir group. p^2 gives the differences between baseline and after intervention in the unfermented group. p^3 gives the comparison of the changes from baseline to week 12th between groups. p^1 and p^2 were analysed by the Wilcoxon test while p^3 was analyzed by the Mann–Whitney U test. * $p \leq 0.05$.

In regard to the analysis of gut microbiota composition, the mean number of total reads per sample was 90627 (Standart Deviation (SD): 44912; range: 31198–183068) at baseline and 118025 (SD:38831; range 39248–171915) after the intervention in kefir group, while it was recorded as 138775 (SD: 29961, range: 100922–195456) at the baseline and 95058 (SD: 23740, range: 55853–124889) after the intervention in the unfermented milk group. The gut microbiome of the participants was composed of phyla *Bacteroidetes* (51%), *Firmicutes* (30%), *Proteobacteria* (11%), *Verrucomicrobia* (0.02%) and *Actinobacteria* (0.003%) at the baseline in kefir group; the relative abundance of these phyla were detected as 39%, 39%, 10%, 0.03%, and 0.04% respectively after the intervention of kefir. Only the increase in the relative abundance of *Actinobacteria* was found to be statistically significant ($p = 0.023$). In the unfermented milk group, the phyla *Bacteroidetes* (66%), *Firmicutes* (27%), *Verrucomicrobia* (0.01%), *Actinobacteria* (0.01%), and *Proteobacteria* (0.03%) were detected at the baseline, however, the relative abundance of these phyla were changed to 33%, 56%, 0.03%, 0.04%, and 0.02% respectively after the intervention (Figure 2). When the changes in the relative abundance of each phyla distribution from after the intervention to baseline were compared between the groups, no significant difference was obtained ($p > 0.05$, for each). In the kefir group, the median *of Firmicutes/Bacteroidetes* ratio was 0.62 (range:0.06–10.01) at the baseline, and 1.77 (range:0.14–46.16) after the intervention ($p = 0.388$). This ratio was 0.30 (range:0.03–25.27) and 2.22 (range:0.34–12.04) respectively at baseline and end of the intervention in the unfermented milk group ($p = 0.333$). No significant difference was obtained between the changes in *Firmicutes* to *Bacteroidetes* ratio of groups ($p > 0.05$).

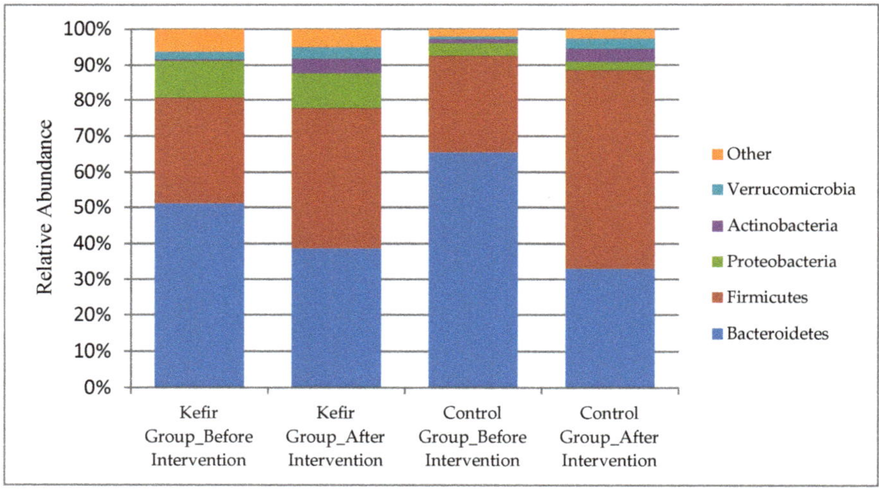

Figure 2. Gut microbiota composition before and after the intervention in each group.

The changes in the relative abundance of *Bacteriodetes* and *Firmucutes* by the consumption of test drinks were given in Figure 3. The phyla *Bacteroidetes* was composed of five dominant genera; *Bacteroides*, *Odoribacteraceae*, *Porphyromonadaceae*, *Prevotellaceae*, and *Alistipes* in the gut microbiome of participants. *Bacteroides* (54% in kefir group and 44% in unfermented milk group), *Prevotellacea* (26% in kefir group and 40% in unfermented milk group) and *Alistipes* (11% in kefir group and 7% in unfermented milk group) were respectively most abundant families among *Bacteroidetes* at baseline. The relative abundance of *Bacteroides* changed to 59% in kefir group and 50% in unfermented milk group; *Prevotellacea* changed to 25% in kefir group and 29% in unfermented milk group; *Alistipes* changed to 9% in kefir group and 10% in unfermented milk group after the intervention. Despite the modest changes in the relative abundance of some genera, no significant difference was obtained when the changes in the relative abundances were compared between groups ($p > 0.05$, for each). Among *Firmicutes*, *Clostridia*, *Erysipelotrichaceae*, *Veillonellaceae*, and *Lactobacillales* were obtained in the fecal samples of participants.

Although an increase in the relative abundance of *Clostridia* (from 73% to 85%) and *Lactobacillales* (2% to 5%), and also a decrease in the relative abundance of *Veillonellaceae* (from 9% to 6%) were obtained from baseline to after the intervention in kefir group, none of these changes were statistically significant ($p > 0.05$, for each). In the unfermented milk group, the relative abundance of *Clostridia* was increased from 75% to 89%, whereas *Lactobacillales* (5% to 2%) and *Veillonellaceae* (from 5% to 4%) were decreased from baseline to after the intervention. Similar to the phyla *Bacteriodetes*, no significant difference was obtained in the changes of the relative abundance of *Firmicutes* at genus level between the kefir and unfermented milk groups ($p > 0.05$, for each) (Figure 3). Among the phyla *Actinobacteria*, the relative abundance of *Bifidobacterium* was increased from 31% to 39% by the intervention of kefir, and from 23% to 32% by the intervention of unfermented milk. However, these changes were not found significant when compared between the groups ($p > 0.05$). Furthermore, *Bifidobacterium* species were detectable in only 50% of participants' the fecal samples at the baseline, whereas they could be detected in 91.7% after the intervention of kefir (data not shown). On the contrary, *Verrucomicrobia* was obtained less frequently from baseline (75% of participants) to after the intervention (58.3% of participants) in the kefir group.

The correlation between the change in gut microbiota composition and the change in physiological characteristics, including anthropometrical measurements, biochemical markers, or blood pressure were conducted to examine the potential associations. Correlations between the changes in anthropometrical measurements and fecal microbiota composition at phylum and subphylum level were summarized in Table 3. The body weight and BMI were positively correlated with the relative abundance of *Firmicutes* and *Proteobacteria*. However, they were negatively correlated with the relative abundance of *Clostridia* ($p < 0.05$, for each). The body fat mass was negatively correlated with the relative abundance of *Bacteroidetes* ($p < 0.01$), and positively correlated with the relative abundance of *Porphyromonadaceae*, *Firmicutes*, and *Actinobacteria* ($p < 0.05$, for each). The waist circumference was negatively correlated with the relative abundance of *Clostridia* and positively correlated with the relative abundance *Veillonellaceae* ($p < 0.05$, for each).

Table 4 summarizes the correlation between the change in gut microbiota composition and biochemical markers. The changes in the relative abundance of *Bacteroides* was negatively correlated with both total and LDL cholesterol while the changes in the relative abundance of *Veillonellaceae* was negatively correlated with only LDL cholesterol ($p < 0.05$, for each). The change in the relative abundance of *Bacteroidetes* was negatively, and the change in the relative abundance of *Odoribacteraceae* and *Alistipes* groups was positively correlated with the change in serum glucose. The change in the relative abundance of *Verrucomicrobia* was positively correlated with changes in serum homocysteine and insulin ($p < 0.05$, for each). When we analyzed the correlation between the change in gut microbiota composition and blood pressure in the phylum level, the change in the relative abundance of *Actinobacteria* and *Proteobacteria* was positive, whereas *Bacteroidetes* was negatively correlated with the change in blood pressure. In the sub–phylum level, the change in the relative abundance *Lactobacillales* was positively correlated with the change in systolic and diastolic blood pressure ($p < 0.05$).

We further assessed several alpha and beta diversity metrics to assess bacterial biodiversity in specimens collected in week 0 and 12 from kefir and unfermented milk groups. No statistically significant differences were observed in OTU counts (Figure 4). Alpha diversity indices, indicating species richness and evenness with/without phylogenetic relations; namely, Shannon, Jaccard, and Faith PD indices, were similar between the study groups (Figure 5) (Jaccard plots not provided). Biodiversity between study cohorts, assessed by Bray–Curtis and weighted/unweighted UniFrac distances revealed no significant variation among study groups or in different time points. The PCoA plots of the unweighted UniFrac distances were given in Figure 5. No differences in OTU counts, alpha or beta diversity measures were observed when specimens from week 0 and 12 were assessed, regardless of the study group.

No side effect was reported by the participants during or after the intervention period that included the consumption of kefir.

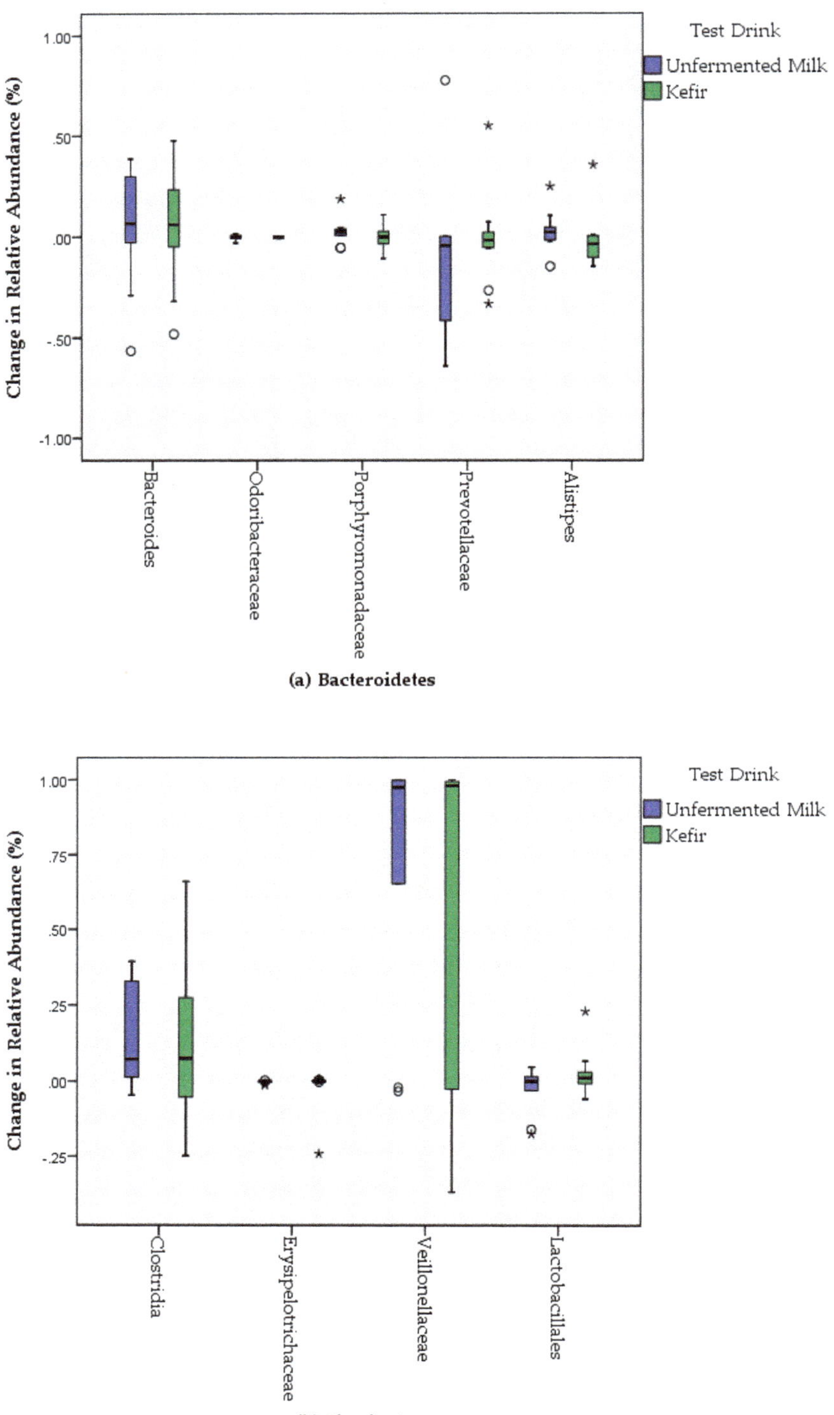

Figure 3. Bacterial changes in the relative abundance of Bacteriodetes (**a**) and Firmucutes (**b**) by the consumption of test drinks.

Table 3. Correlations between changes in anthropometrical measurements and fecal microbiota composition at phylum and subphylum level ≠.

	Bacteroidetes	**Bacteroides**	**Odoribacteraceae**	**Porphyromonadaceae**	**Prevotellaceae**	**Alistipes**	**Firmicutes**	**Clostridia**	**Erysipelotrichaceae**	**Veillonellaceae**	**Lactobacillales**	**Verrucomicrobia**	**Actinobacteria**	**Bifidobacterium**	**Proteobacteria**
Anthropometric measurements															
Weight (kg)	−0.384	0.070	0.176	0.160	−0.131	0.133	0.433 *	−0.432 *	−0.338	0.295	0.352	0.132	0.400	0.030	0.456 *
Body mass index (kg/m^2)	−0.383	0.103	0.189	0.172	−0.168	0.139	0.434 *	−0.455 *	−0.341	0.291	0.347	0.095	0.382	0.047	0.461 *
Body fat mass (%)	−0.563 **	0.223	0.285	0.468 *	−0.388	0.402	0.599 **	0.017	−0.070	0.210	0.038	0.420	0.536 *	−0.018	0.371
Waist circumference (cm)	−0.151	0.267	−0.321	−0.009	−0.217	−0.047	0.242	−0.505 *	−0.287	0.432 *	0.135	0.247	0.128	0.044	0.332
Hip circumference (cm)	0.082	0.029	0.326	0.164	−0.149	−0.078	0.070	−0.096	−0.264	0.272	0.215	−0.115	0.127	−0.367	0.081
Waist–to–hip ratio	−0.239	0.112	−0.526 *	−0.086	−0.038	0.014	0.168	−0.251	0.046	0.066	−0.043	0.376	0.090	0.223	0.208

≠ Values indicate Spearman correlation coefficients (n = 22). * Correlation is significant at the 0.05 level (2–tailed). ** Correlation is significant at the 0.01 level (2–tailed).

Table 4. Correlations between changes in biochemical markers and fecal microbiota composition at phylum and sub–phylum level ≠.

	Bacteroidetes	Bacteroides	Odoribacteraceae	Porphyromonadaceae	Prevotellaceae	Alistipes	Firmicutes	Clostridia	Erysipelotrichaceae	Veillonellaceae	Lactobacillales	Verrucomicrobia	Actinobacteria	Bifidobacterium	Proteobacteria
Lipid profile															
Total cholesterol (mg/dL)	−0.151	−0.444 *	−0.025	−0.026	0.334	−0.007	−0.101	0.389	0.205	−0.408	0.009	0.059	0.190	−0.164	−0.096
HDL cholesterol (mg/dL)	−0.003	−0.418	0.130	−0.066	0.414	−0.004	−0.248	−0.224	0.161	0.154	0.259	0.175	−0.016	−0.012	0.130
LDL cholesterol (mg/dL)	−0.016	−0.535 *	−0.021	−0.023	0.305	0.075	−0.298	0.322	0.330	−0.431 *	0.005	0.152	0.042	−0.034	−0.094
Triglycerides (mg/dL)	−0.080	−0.007	−0.118	−0.079	0.197	−0.263	0.251	0.334	−0.211	−0.242	−0.082	−0.176	0.199	−0.292	−0.210
Homocysteine (µmoL/L)	−0.162	0.077	−0.186	0.141	−0.135	−0.023	−0.057	−0.044	0.232	0.224	0.149	0.477 *	0.110	0.127	0.359
Glycaemic status															
Glucose (mg/dL)	−0.590 **	0.092	0.423 *	0.370	−0.374	0.629 **	0.387	0.317	0.056	−0.037	−0.033	0.328	0.365	0.090	0.187
Insulin (mU/L)	−0.331	0.069	0.344	0.244	−0.265	0.361	0.357	−0.098	−0.159	0.312	0.010	0.428 *	0.284	−0.218	0.123
HbA1c (%)	0.189	0.062	−0.168	−0.266	0.004	−0.123	0.021	0.126	−0.372	0.211	−0.282	−0.003	−0.280	0.068	−0.277
HOMA–IR	−0.356	0.030	0.406	0.207	−0.255	0.458 *	0.346	−0.119	−0.148	0.269	0.067	0.395	0.294	−0.209	0.123
Inflammation Related Indicators															
hs–CRP (mg/dL)	−0.145	0.110	−0.294	0.238	−0.255	−0.032	0.096	0.145	0.358	0.078	−0.015	0.380	0.321	−0.289	0.078
TNF–α (pg/mL)	−0.297	0.087	−0.305	−0.034	−0.224	0.051	0.010	0.317	0.093	0.023	−0.062	0.392	0.281	−0.323	0.322
IL–6 (pg/mL)	−0.016	0.381	0.001	0.013	−0.287	−0.142	0.313	−0.320	−0.333	0.399	−0.170	0.039	−0.043	0.091	0.274
IL–10 (pg/mL)	0.086	−0.130	−0.414	−0.314	0.165	−0.216	−0.054	−0.289	−0.040	0.015	0.273	0.151	−0.006	0.008	−0.054
IFN–γ (IU/mL)	0.066	−0.241	−0.089	−0.074	0.111	0.291	−0.182	−0.024	0.074	−0.162	0.084	0.093	−0.029	0.086	−0.076
AST	−0.343	0.292	0.016	0.058	−0.159	−0.178	0.283	−0.160	−0.031	−0.006	0.203	−0.328	0.190	−0.187	0.169
ALT	−0.251	0.235	−0.089	0.067	−0.081	−0.423 *	0.391	0.138	−0.120	0.253	0.098	0.158	0.252	−0.402	0.175
GGT	0.255	−0.264	0.112	−0.099	0.493 *	−0.385	−0.093	0.025	−0.285	−0.089	−0.093	−0.457 *	−0.016	−0.169	−0.208
Blood Pressure															
Systolic blood pressure (mmHg)	−0.531 *	−0.148	−0.080	0.137	0.092	0.076	0.243	−0.172	0.144	−0.198	0.536 *	0.205	0.710 **	−0.168	0.379
Diastolic blood pressure (mmHg)	−0.491 *	0.169	0.183	0.271	−0.348	0.171	0.244	−0.205	0.239	−0.128	0.561 **	0.096	0.452 *	0.013	0.469 *

≠ Values indicate Spearman correlation coefficients ($n = 22$). * Correlation is significant at the 0.05 level (2–tailed). ** Correlation is significant at the 0.01 level (2–tailed). HDL: High Density Lipoprotein, LDL: Low Density Lipoprotein, HbA1c: Glycosylated Hemoglobin, HOMA-IR: Homeostasis Model of Assessment Insulin Resistance, hs-CRP: High-sensitivity C-reactive Protein, TNF: Tumor Necrosis Factor, IL: Interleukin, IFN: Interferon, ALT: Alanine Aminotransferase, AST: Aspartate Aminotransferase, GGT: Gamma-glutamyl Transferase.

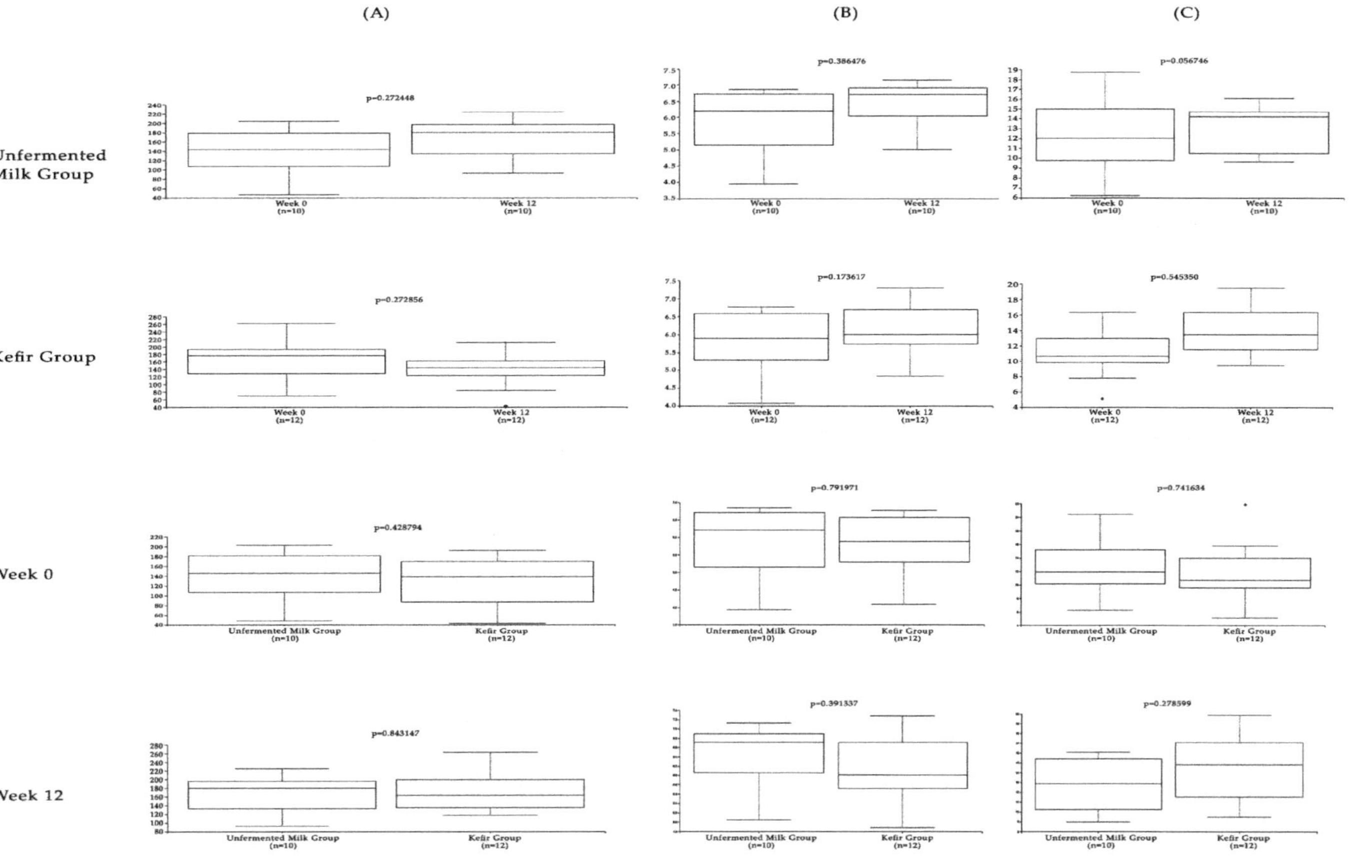

Figure 4. Box–and–whisker plots of the alpha diversity metrics calculated for the study groups. The box represents the median and the interquartile range, whereas whiskers indicate the 90th and 10th percentiles (**A**: Observed OTUs, **B**: Shannon index, **C**: Faith phylogenetic diversity index).

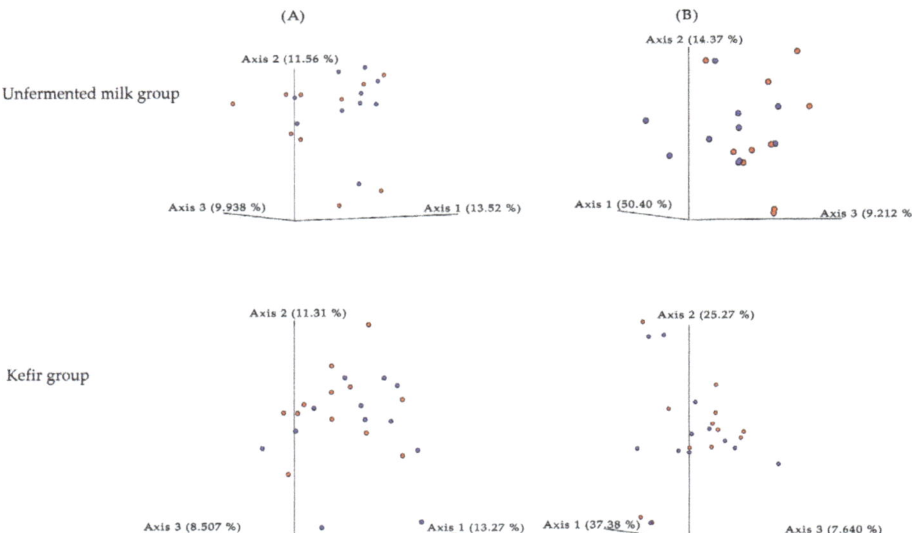

Figure 5. Principal coordinate analysis (PCoA) plot of the unweighted (**A**) and weighted (**B**) UniFrac distance matrices in the study groups. The plots were generated using EMPeror [36]. Axis titles indicate the percentage variations. The colors indicate sampling time (red: week 0, blue: week 12).

4. Discussion

In this parallel–group randomized controlled study, regular kefir consumption during 12 weeks provided some improvements in anthropometrical measurements, lipid profile, glycaemic status, and inflammation in participants with MetS. In particular, insulin and HOMA–IR levels were significantly decreased, and also pro–inflammatory cytokines (TNF–α and IFN–γ) and blood pressure were ameliorated by kefir consumption. However, the magnitude of the improvements stayed insignificant when compared to unfermented milk. The effects of kefir on metabolic status were previously investigated in both animal models and human studies [37–39]. Some animal models suggested that kefir might have a potential to benefit the management of MetS by reducing body weight, fasting blood glucose, insulin, total, and LDL cholesterol, triacylglycerol, and pro–inflammatory cytokines, including IL–1β and IL–6 [38]. However, the evidence from human studies has been controversial. For instance, Ostadrahimi et al. reported that consumption of 600 mL/d kefir containing *Lactobacillus casei*, *Lactobacillus acidophilus*, and *Bifidobacteria* species had beneficial effects on fasting blood glucose and HbA1c compared to the control drink in patients with type 2 diabetes [37]. On the other hand, St–Onge et al. showed that 500 mL/day of kefir consumption for four weeks had no effect on lipid profile [40]. Furthermore, Fathi et al. showed that two servings of kefir in a day during eight weeks led a similar improvement both in lipid profile and weight management compared with milk [41,42]. The variation in response to the kefir consumption could be mainly explained by the variation of kefir composition, and the characteristics of study samples in different studies. Many different bacteria and yeast might be used for kefir production, and this might lead to distinct effects on metabolism and gut microbiota. Kefir used in this study contained *Lactococcus lactis* subsp. *lactis, Lactococcus lactis* subsp. *cremoris, Lactococcus lactis* subsp. *diacetylactis, Leuconostoc mesenteroides* subsp. *cremoris, Lactobacillus kefir, Kluyveromyces marxianus*, and *Saccharomyces unisporus*, and differed from the kefir samples used in other studies [37,38]. Furthermore, the initial metabolic profile of the participants was suggested an essential factor for the efficacy of probiotic interventions. Fuentes et al. showed that probiotics are more effective in patients with high baseline total cholesterol levels (251–300 mg/dL) compared to the patients with low baseline total cholesterol levels (200–250 mg/dL) [43]. Similarly, Nikbakht found out that probiotic supplementation was only effective in patients with baseline fasting blood glucose level above 126 mg/dL [44]. In our study, the median of total cholesterol levels was 243.50 mg/dL and 220.00 mg/dL, and the baseline glucose levels were 105.00 mg/dL and 101.50 mg/dL for kefir group and unfermented milk group, respectively. This may partly explain the lack of efficacy of kefir on metabolic status in our study.

Alterations in gut microbiota diversity, composition, and function were suggested to play a significant role in the development of MetS [12]. Ameliorating the intestinal dysbiosis with prebiotics and probiotics have gained considerable attention in recent years for the management of MetS [45]. However, the studies have yielded inconsistent results regarding the influence of probiotics on fecal microbial diversity and composition [46–52]. Furthermore, the effects of kefir as a probiotic on gut microbiota have been examined very limited and mainly with animal studies. Kim et al. revealed that three-week oral administration of kefir provided a decrease in the number of *Firmicutes, Proteobacteria,* and *Enterobacteriaceae*, and an increase in the number of *Bacteroidetes, Lactobacillus, Lactococcus,* and total yeast compared to milk group in mice [53]. However, in their follow–up study, no significant difference apart from the increase in *Lactobacillus/Lactococcus* populations was observed in the kefir group compared to control [39]. Similarly, an increase in *Lactobacillus* and *Bifidobacterium* populations and a reduction in *Clostridium* populations by consumption of kefir have been reported in mice previously [54–56]. The present study is one of the first reports showing the impact of kefir on human microbiota composition in patients with MetS. In this study, regular kefir and milk consumption for 12 weeks resulted in some alterations in the gut microbiota composition. For instance, *Lactobacillus* and *Bifidobacterium spp.* were increased by kefir consumption. However, apart from the increase in the relative abundance of *Actinobacteria*, no significant change by kefir consumption was recorded. Furthermore, the changes in the relative abundance of bacterial populations did not differ significantly between the groups. In some studies, following the probiotic supplementation, increases in the supplemented genera without an additional impact on the main microbial groups were observed [57,58]. Previously, Yılmaz et al. showed that 400 mL/day kefir consumption for four weeks in patients with inflammatory bowel diseases resulted in the significant increase of Lactobacillus bacterial load in feces [59]. Our study also showed an increase from 2% to 5% in the relative abundance of *Lactobacillales* by kefir consumption, albeit lacking statistical significance. It has suggested that the change of the microbiota composition may be related to several factors such as age, gender, initial microbiota composition, dietary intake, lifestyle factors, menopausal status, and medical therapy of the individuals [60,61]. Moreover, the microbiota composition of the product (kefir) that was tested [17,18] and the consumption pattern, including the period and frequency of consumption and amount of the product should be considered as the factors that have potential to influence the magnitude of the changes in gut microbiota [62]. Therefore, evaluating the effect of probiotics or fermented foods such as kefir on an individual basis may be set as a goal for future studies.

In this study, the correlations between changes in microbiota and anthropometric measurements or biochemical status were demonstrated. Our results pointed out a negative correlation between body fat mass and abundance of *Bacteroidetes*, whereas a positive correlation with the abundance of *Firmicutes* and *Actinobacteria* were observed. Although the data regarding the abundance of *Bacteriodetes* and *Firmicutes* phyla in obese and lean individuals is inconsistent, an overall analysis of results indicates an increase in *Firmicutes* with obesity [63]. Our results supported the previous reports, which revealed increased *Firmicutes* and decreased the abundance of *Bacteroidetes* are associated with obesity [64]. Turnbaugh et al. revealed a higher proportion of *Actinobacteria* in obese individuals compared to lean individuals [65]. In this study, we observed a positive correlation between body fat mass and abundance of *Actinobacteria,* which is in line with Turnbaugh et al.'s work. Members of the phylum of *Proteobacteria* are gram–negative bacteria and include several common human pathogens. An association between the increased relative abundance of *Proteobacteria* and increased risk of cardio–metabolic disorders was suggested previously [66]. In parallel with these findings, our study showed a positive correlation between an increased relative abundance of Proteobacteria and both body weight and diastolic blood pressure.

In terms of glycaemic status, Larsen et al. showed a lower abundance of *Firmicutes* and *Clostridia*, and a higher abundance of *Bacteroidetes* and *Betaproteobacteria* in diabetic patients compared to the non–diabetics [67]. Accordingly, we observed a lower abundance of *Firmicutes* and a higher abundance of *Bacteroidetes* in both groups at the beginning of the study. However, only the change in the relative abundance of *Bacteroidetes* by the dietary intervention was negatively correlated with the change in fasting blood glucose. This was parallel to results of the study conducted by Egshatyan et al., which found that microbiota of glucose–intolerant subjects were represented by *Firmicutes* phylum and to a lesser degree by *Bacteroidetes* phylum [68]. In the subphylum level, studies mostly indicate higher levels of *Bacteroides* and *Prevotella* and lower levels of butyrate producing–bacteria in type 2 diabetic patients [68,69]. In this study, we have observed a positive correlation between the relative abundance of *Odoribacteraceae* and *Alistipes*, and fasting plasma glucose. The correlation between *Alistipes* and blood glucose was also observed in a previous study [70].

Many researchers demonstrated a link between dysbiosis of gut microbiota and blood pressure. Yang et al. reported an increase in *Firmicutes/Bacteroidetes* ratio in hypertensive rats and humans. They also recorded a lower abundance of *Actinobacteria* as well as acetate– and butyrate–producing bacteria [71]. Yan et al. indicated higher levels of *Proteobacteria* but lower levels of *Actinobacteria* in hypertensive subjects [72]. In this study, the change in the relative abundance of *Bacteroidetes* was negatively correlated with the change in systolic and diastolic blood pressure as reported previously. Surprisingly, we observed a strong positive correlation with systolic and a weak positive correlation with diastolic blood pressure, respectively, with *Actinobacteria* abundance. The phylum *Actinobacteria* includes *Bifidobacterium* genera, possessing probiotic features [73]. Studies that report reductions of *Actinobacteria* in hypertensive patients explained this relationship mostly with *Bifidobacterium* levels [71,74]. In our study, when we analyzed the association between *Bifidobacterium* and blood pressure, no significant correlation was noted. This may be due to the lack of significant changes in *Bifidobacterium* abundance after the intervention. The species other than *Bifidobacterium* within the *Actinobacteria* phylum might be further investigated in terms of their contributions to hypertension.

Apart from fermentation, unfermented dairy products may also affect the gut microbiota [75,76]. In our study, regular milk consumption that was used as control also led to some changes in microbiota composition compared to the baseline. *Firmucutes* and *Verrucomicrobiota* were increased with milk consumption. However, *Bacteroidetes* group was decreased compared to the baseline. In accordance with our results, Ntemiri et al. also found out that whole milk consumption was associated with an increase in taxons belonging to *Firmicutes* and a higher *Firmicutes/Bacteroidetes* ratio [76]. In another randomized cross–over study, consumption of probiotic yogurt and milk acidified with D-(+)-glucono-δ-lactone showed some distinct effects on microbiota composition. In both groups, the abundance of *Bilophila wadsworthia* was reduced. However, only the abundance of *Bifidobacterium* species was increased with acidified milk intake, and it was suggested that gluconic acid in milk might possess prebiotic activity [75]. In a like manner, exopolysaccharides such as kefiran derived from kefir also suggested as bioactive compounds due to their potential prebiotic effects and relation to alteration of intestinal microbiota [77]. These results support a strong interaction between diet and microbiota even without probiotic intervention.

Studies showed that not only the composition of microbiota but also its functionality plays a role in the metabolic status [78,79]. This study focused only on the composition of gut microbiota; any consideration of its functionality was not taken into account. Using the metabolites of microbiota, such as postbiotics, as markers of the efficiency might have provided a better understanding. This should be noted as the main limitation of the study. The small number of participants in each arm could also be considered as the other limitation of the study.

5. Conclusions

In conclusion, to our knowledge, this was the first report exploring the effect of kefir on microbiota composition in patients with metabolic syndrome. This study indicated that kefir consumption could provide some potential improvements, especially in glycaemic status, inflammation–related indicators, and blood pressure, however, none of these improvements might stay significant when compared the changes led by unfermented milk consumption. Regarding to microbiota composition, the relative abundance of *Actinobacteria* phylum were increased in the kefir group compared to the baseline, even though a similar change by unfermented milk was also reported. Furthermore, this study underlined the potential alterations in gut microbiota composition that can be correlated with some indicators of the metabolic status led by both kefir and milk consumption, even if the magnitude of the efficiency remained limited. Further studies, especially randomized controlled trials, are needed to clarify the efficiency of kefir on gut microbiota and its link to metabolic status.

Author Contributions: Conceptualization, E.B.-K. and Z.B.; Data curation, E.B.-K. and Z.B.; Formal analysis, E.B.-K., K.E. and Z.B.; Funding acquisition, Y.A., E.Y. and Z.B.; Investigation, E.B.-K., B.P.S.-Y., C.K. and A.G.O.; Methodology, E.B.-K., B.P.S.-Y., Y.A., F.A.-K., C.K., A.B., A.N., K.E. and E.Y.; Writing—original draft, E.B.-K., K.E. and Z.B.; Writing—review & editing, C.K., K.E. and Z.B.

Funding: This research was funded by the Turkish Council of Higher Education.

Acknowledgments: We thank all study participants for their cooperation.

Conflicts of Interest: The authors declare no conflict of interest.

References

1. Alberti, K.G.; Eckel, R.H.; Grundy, S.M.; Zimmet, P.Z.; Cleeman, J.I.; Donato, K.A.; Fruchart, J.C.; James, W.P.; Loria, C.M.; Smith, S.C., Jr.; et al. Harmonizing the metabolic syndrome: a joint interim statement of the International Diabetes Federation Task Force on Epidemiology and Prevention; National Heart, Lung, and Blood Institute; American Heart Association; World Heart Federation; International Atherosclerosis Society; and International Association for the Study of Obesity. *Circulation* **2009**, *120*, 1640–1645. [CrossRef] [PubMed]
2. Mottillo, S.; Filion, K.B.; Genest, J.; Joseph, L.; Pilote, L.; Poirier, P.; Rinfret, S.; Schiffrin, E.L.; Eisenberg, M.J. The metabolic syndrome and cardiovascular risk a systematic review and meta–analysis. *J. Am. Coll. Cardiol.* **2010**, *56*, 1113–1132. [CrossRef] [PubMed]
3. Elder, S.J.; Lichtenstein, A.H.; Pittas, A.G.; Roberts, S.B.; Fuss, P.J.; Greenberg, A.S.; McCrory, M.A.; Bouchard, T.J., Jr.; Saltzman, E.; Neale, M.C. Genetic and environmental influences on factors associated with cardiovascular disease and the metabolic syndrome. *J. Lipid. Res.* **2009**, *50*, 1917–1926. [CrossRef] [PubMed]
4. Stancakova, A.; Laakso, M. Genetics of metabolic syndrome. *Rev. Endocr. Metab. Disord.* **2014**, *15*, 243–252. [CrossRef] [PubMed]
5. Turnbaugh, P.J.; Ley, R.E.; Mahowald, M.A.; Magrini, V.; Mardis, E.R.; Gordon, J.I. An obesity–associated gut microbiome with increased capacity for energy harvest. *Nature* **2006**, *444*, 1027–1031. [CrossRef] [PubMed]
6. Burcelin, R.; Luche, E.; Serino, M.; Amar, J. The gut microbiota ecology: a new opportunity for the treatment of metabolic diseases? *Front Biosci* **2009**, *14*, 5107–5117. [CrossRef]
7. Cani, P.D.; Amar, J.; Iglesias, M.A.; Poggi, M.; Knauf, C.; Bastelica, D.; Neyrinck, A.M.; Fava, F.; Tuohy, K.M.; Chabo, C.; et al. Metabolic endotoxemia initiates obesity and insulin resistance. *Diabetes* **2007**, *56*, 1761–1772. [CrossRef]
8. Ley, R.E.; Backhed, F.; Turnbaugh, P.; Lozupone, C.A.; Knight, R.D.; Gordon, J.I. Obesity alters gut microbial ecology. *Proc. Natl. Acad. Sci. USA* **2005**, *102*, 11070–11075. [CrossRef]
9. Backhed, F.; Ley, R.E.; Sonnenburg, J.L.; Peterson, D.A.; Gordon, J.I. Host–bacterial mutualism in the human intestine. *Science* **2005**, *307*, 1915–1920. [CrossRef]
10. Shen, J.; Obin, M.S.; Zhao, L. The gut microbiota, obesity and insulin resistance. *Mol. Asp. Med.* **2013**, *34*, 39–58. [CrossRef]

11. Esteve, E.; Ricart, W.; Fernandez-Real, J.M. Gut microbiota interactions with obesity, insulin resistance and type 2 diabetes: did gut microbiote co–evolve with insulin resistance? *Curr. Opin. Clin. Nutr. Metab. Care* **2011**, *14*, 483–490. [CrossRef]
12. Festi, D.; Schiumerini, R.; Eusebi, L.H.; Marasco, G.; Taddia, M.; Colecchia, A. Gut microbiota and metabolic syndrome. *World J. Gastroenterol.* **2014**, *20*, 16079–16094. [CrossRef] [PubMed]
13. Le Barz, M.; Anhe, F.F.; Varin, T.V.; Desjardins, Y.; Levy, E.; Roy, D.; Urdaci, M.C.; Marette, A. Probiotics as Complementary Treatment for Metabolic Disorders. *Diabetes Metab. J.* **2015**, *39*, 291–303. [CrossRef] [PubMed]
14. Cani, P.D.; Van Hul, M. Novel opportunities for next–generation probiotics targeting metabolic syndrome. *Curr. Opin. Biotechnol.* **2015**, *32*, 21–27. [CrossRef] [PubMed]
15. Ahmed, Z.; Wang, Y.; Ahmad, A.; Khan, S.T.; Nisa, M.; Ahmad, H.; Afreen, A. Kefir and health: A contemporary perspective. *Crit. Rev. Food Sci. Nutr.* **2013**, *53*, 422–434. [CrossRef] [PubMed]
16. Koyu, E.B.; Buyuktuncer Demirel, Z. A functional food: Kefir. *J. Nutr. Diet.* **2018**, *46*, 166–175. [CrossRef]
17. Rosa, D.D.; Dias, M.M.S.; Grzeskowiak, L.M.; Reis, S.A.; Conceicao, L.L.; Peluzio, M. Milk kefir: nutritional, microbiological and health benefits. *Nutr. Res. Rev.* **2017**, *30*, 82–96. [CrossRef]
18. Bourrie, B.C.; Willing, B.P.; Cotter, P.D. The microbiota and health promoting characteristics of the fermented beverage kefir. *Front. Microbiol.* **2016**, *7*, 647. [CrossRef] [PubMed]
19. Liu, J.R.; Chen, M.J.; Lin, C.W. Antimutagenic and antioxidant properties of milk–kefir and soymilk–kefir. *J. Agric. Food Chem.* **2005**, *53*, 2467–2474. [CrossRef] [PubMed]
20. Lee, M.Y.; Ahn, K.S.; Kwon, O.K.; Kim, M.J.; Kim, M.K.; Lee, I.Y.; Oh, S.R.; Lee, H.K. Anti–inflammatory and anti–allergic effects of kefir in a mouse asthma model. *Immunobiology* **2007**, *212*, 647–654. [CrossRef]
21. Turan, I.; Dedeli, O.; Bor, S.; Ilter, T. Effects of a kefir supplement on symptoms, colonic transit, and bowel satisfaction score in patients with chronic constipation: a pilot study. *Turk. J. Gastroenterol.* **2014**, *25*, 650–656. [CrossRef] [PubMed]
22. Friques, A.G.F.; Arpini, C.M.; Kalil, I.C.; Gava, A.L.; Leal, M.A.; Porto, M.L.; Nogueira, B.V.; Dias, A.T.; Andrade, T.U.; Pereira, T.M.C.; et al. Chronic administration of the probiotic kefir improves the endothelial function in spontaneously hypertensive rats. *J. Transl. Med.* **2015**, *13*, 390. [CrossRef] [PubMed]
23. Hadisaputro, S.; Djokomoeljanto, R.R.; Judiono; Soesatyo, M.H. The effects of oral plain kefir supplementation on proinflammatory cytokine properties of the hyperglycemia Wistar rats induced by streptozotocin. *Acta Med. Indones.* **2012**, *44*, 100–104. [PubMed]
24. Zimmet, P.; KG, M.M.A.; Serrano Rios, M. A new international diabetes federation worldwide definition of the metabolic syndrome: the rationale and the results. *Rev. Esp. Cardiol.* **2005**, *58*, 1371–1376. [CrossRef]
25. Thomas, V.; Clark, J.; Dore, J. Fecal microbiota analysis: an overview of sample collection methods and sequencing strategies. *Future Microbiol.* **2015**, *10*, 1485–1504. [CrossRef]
26. Wu, W.K.; Chen, C.C.; Panyod, S.; Chen, R.A.; Wu, M.S.; Sheen, L.Y.; Chang, S.C. Optimization of fecal sample processing for microbiome study—The journey from bathroom to bench. *J. Formos. Med. Assoc.* **2019**, *118*, 545–555. [CrossRef]
27. Klindworth, A.; Pruesse, E.; Schweer, T.; Peplies, J.; Quast, C.; Horn, M.; Glockner, F.O. Evaluation of general 16S ribosomal RNA gene PCR primers for classical and next–generation sequencing–based diversity studies. *Nucleic Acids Res.* **2013**, *41*, e1. [CrossRef]
28. Huson, D.H.; Beier, S.; Flade, I.; Górska, A.; El-Hadidi, M.; Mitra, S.; Ruscheweyh, H.J.; Tappu, R. MEGAN Community Edition– Interactive exploration and analysis of large–scale microbiome sequencing data. *PLoS Comput. Biol.* **2016**, *12*, e1004957. [CrossRef]
29. Bolger, A.M.; Lohse, M.; Usadel, B. Trimmomatic: A flexible trimmer for Illumina sequence data. *Bioinformatics* **2014**, *30*, 2114–2120. [CrossRef]
30. Langmead, B.; Salzberg, S.L. Fast gapped–read alignment with Bowtie 2. *Nat. Methods* **2012**, *9*, 357–359. [CrossRef]
31. Bolyen, E.; Rideout, J.R.; Dillon, M.R.; Bokulich, N.A.; Abnet, C.; Al-Ghalith, G.A.; Alexander, H.; Alm, E.J.; Arumugam, M.; Asnicar, F.; et al. QIIME 2: Reproducible, interactive, scalable, and extensible microbiome data science. *PeerJ Prepr.* **2018**, *6*, e27295v27292. [CrossRef] [PubMed]
32. Callahan, B.J.; McMurdie, P.J.; Rosen, M.J.; Han, A.W.; Johnson, A.J.; Holmes, S.P. DADA2: High–resolution sample inference from Illumina amplicon data. *Nat. Methods* **2016**, *13*, 581–583. [CrossRef] [PubMed]

33. Amir, A.; McDonald, D.; Navas-Molina, J.A.; Kopylova, E.; Morton, J.T.; Zech Xu, Z.; Kightley, E.P.; Thompson, L.R.; Hyde, E.R.; Gonzalez, A.; et al. Deblur rapidly resolves single–nucleotide community sequence patterns. *MSystems* **2017**, *2*, e00191-16. [CrossRef] [PubMed]
34. McDonald, D.; Price, M.N.; Goodrich, J.; Nawrocki, E.P.; DeSantis, T.Z.; Probst, A.; Andersen, G.L.; Knight, R.; Hugenholtz, P. An improved Greengenes taxonomy with explicit ranks for ecological and evolutionary analyses of bacteria and archaea. *ISME J.* **2012**, *6*, 610–618. [CrossRef] [PubMed]
35. Yilmaz, P.; Parfrey, L.W.; Yarza, P.; Gerken, J.; Pruesse, E.; Quast, C.; Schweer, T.; Peplies, J.; Ludwig, W.; Glockner, F.O. The SILVA and "All–species Living Tree Project (LTP)" taxonomic frameworks. *Nucleic Acids Res.* **2014**, *42*, D643–D648. [CrossRef] [PubMed]
36. Vazquez-Baeza, Y.; Pirrung, M.; Gonzalez, A.; Knight, R. EMPeror: a tool for visualizing high–throughput microbial community data. *GigaScience* **2013**, *2*, 16. [CrossRef]
37. Ostadrahimi, A.; Taghizadeh, A.; Mobasseri, M.; Farrin, N.; Payahoo, L.; Beyramalipoor Gheshlaghi, Z.; Vahedjabbari, M. Effect of probiotic fermented milk (kefir) on glycemic control and lipid profile in type 2 diabetic patients: a randomized double–blind placebo–controlled clinical trial. *Iran. J. Public Health* **2015**, *44*, 228–237.
38. Rosa, D.D.; Grzeskowiak, L.M.; Ferreira, C.L.; Fonseca, A.C.; Reis, S.A.; Dias, M.M.; Siqueira, N.P.; Silva, L.L.; Neves, C.A.; Oliveira, L.L.; et al. Kefir reduces insulin resistance and inflammatory cytokine expression in an animal model of metabolic syndrome. *Food Funct.* **2016**, *7*, 3390–3401. [CrossRef]
39. Kim, D.H.; Kim, H.; Jeong, D.; Kang, I.B.; Chon, J.W.; Kim, H.S.; Song, K.Y.; Seo, K.H. Kefir alleviates obesity and hepatic steatosis in high–fat diet–fed mice by modulation of gut microbiota and mycobiota: targeted and untargeted community analysis with correlation of biomarkers. *J. Nutr. Biochem.* **2017**, *44*, 35–43. [CrossRef]
40. St-Onge, M.P.; Farnworth, E.R.; Savard, T.; Chabot, D.; Mafu, A.; Jones, P.J. Kefir consumption does not alter plasma lipid levels or cholesterol fractional synthesis rates relative to milk in hyperlipidemic men: A randomized controlled trial [ISRCTN10820810]. *BMC Complement. Altern. Med.* **2002**, *2*, 1. [CrossRef]
41. Fathi, Y.; Ghodrati, N.; Zibaeenezhad, M.J.; Faghih, S. Kefir drink causes a significant yet similar improvement in serum lipid profile, compared with low–fat milk, in a dairy–rich diet in overweight or obese premenopausal women: A randomized controlled trial. *J. Clin. Lipidol.* **2017**, *11*, 136–146. [CrossRef] [PubMed]
42. Fathi, Y.; Faghih, S.; Zibaeenezhad, M.J.; Tabatabaei, S.H. Kefir drink leads to a similar weight loss, compared with milk, in a dairy–rich non–energy–restricted diet in overweight or obese premenopausal women: A randomized controlled trial. *Eur. J. Nutr.* **2016**, *55*, 295–304. [CrossRef] [PubMed]
43. Fuentes, M.C.; Lajo, T.; Carrion, J.M.; Cune, J. Cholesterol–lowering efficacy of Lactobacillus plantarum CECT 7527, 7528 and 7529 in hypercholesterolaemic adults. *Br. J. Nutr.* **2013**, *109*, 1866–1872. [CrossRef] [PubMed]
44. Nikbakht, E.; Khalesi, S.; Singh, I.; Williams, L.T.; West, N.P.; Colson, N. Effect of probiotics and synbiotics on blood glucose: a systematic review and meta–analysis of controlled trials. *Eur. J. Nutr.* **2018**, *57*, 95–106. [CrossRef] [PubMed]
45. Mallappa, R.H.; Rokana, N.; Duary, R.K.; Panwar, H.; Batish, V.K.; Grover, S. Management of metabolic syndrome through probiotic and prebiotic interventions. *Indian J. Endocrinol. Metab.* **2012**, *16*, 20–27. [CrossRef] [PubMed]
46. Zhang, H.; Sun, J.; Liu, X.; Hong, C.; Zhu, Y.; Liu, A.; Li, S.; Guo, H.; Ren, F. Lactobacillus paracasei subsp. paracasei LC01 positively modulates intestinal microflora in healthy young adults. *J. Microbiol.* **2013**, *51*, 777–782. [CrossRef] [PubMed]
47. Toscano, M.; De Grandi, R.; Stronati, L.; De Vecchi, E.; Drago, L. Effect of Lactobacillus rhamnosus HN001 and Bifidobacterium longum BB536 on the healthy gut microbiota composition at phyla and species level: A preliminary study. *World J. Gastroenterol.* **2017**, *23*, 2696–2704. [CrossRef] [PubMed]
48. Larsen, N.; Vogensen, F.K.; Gobel, R.J.; Michaelsen, K.F.; Forssten, S.D.; Lahtinen, S.J.; Jakobsen, M. Effect of Lactobacillus salivarius Ls-33 on fecal microbiota in obese adolescents. *Clin. Nutr.* **2013**, *32*, 935–940. [CrossRef] [PubMed]
49. Stadlbauer, V.; Leber, B.; Lemesch, S.; Trajanoski, S.; Bashir, M.; Horvath, A.; Tawdrous, M.; Stojakovic, T.; Fauler, G.; Fickert, P.; et al. Lactobacillus casei Shirota Supplementation Does Not Restore Gut Microbiota Composition and Gut Barrier in Metabolic Syndrome: A Randomized Pilot Study. *PLoS ONE* **2015**, *10*, e0141399. [CrossRef]

50. Nova, E.; Pérez de Heredia, F.; Gómez-Martínez, S.; Marcos, A. The Role of Probiotics on the Microbiota: Effect on Obesity. *Nutr. Clin. Pract.* **2016**, *31*, 387–400. [CrossRef]
51. Costabile, A.; Buttarazzi, I.; Kolida, S.; Quercia, S.; Baldini, J.; Swann, J.R.; Brigidi, P.; Gibson, G.R. An in vivo assessment of the cholesterol–lowering efficacy of Lactobacillus plantarum ECGC 13110402 in normal to mildly hypercholesterolaemic adults. *PLoS ONE* **2017**, *12*, e0187964. [CrossRef]
52. Kristensen, N.B.; Bryrup, T.; Allin, K.H.; Nielsen, T.; Hansen, T.H.; Pedersen, O. Alterations in fecal microbiota composition by probiotic supplementation in healthy adults: a systematic review of randomized controlled trials. *Genome Med.* **2016**, *8*, 52. [CrossRef] [PubMed]
53. Kim, D.-H.; Chon, J.-W.; Kim, H.; Seo, K.-H. Modulation of intestinal microbiota in mice by kefir administration. *Food Sci. Biotechnol.* **2015**, *24*, 1397–1403. [CrossRef]
54. Liu, J.-R.; Wang, S.-Y.; Chen, M.-J.; Yueh, P.-Y.; Lin, C.-W. The anti–allergenic properties of milk kefir and soymilk kefir and their beneficial effects on the intestinal microflora. *J. Sci. Food Agric.* **2006**, *86*, 2527–2533. [CrossRef]
55. Carasi, P.; Racedo, S.M.; Jacquot, C.; Romanin, D.E.; Serradell, M.A.; Urdaci, M.C. Impact of kefir derived Lactobacillus kefiri on the mucosal immune response and gut microbiota. *J. Immunol. Res.* **2015**, *2015*, 361604. [CrossRef] [PubMed]
56. Hamet, M.F.; Medrano, M.; Perez, P.F.; Abraham, A.G. Oral administration of kefiran exerts a bifidogenic effect on BALB/c mice intestinal microbiota. *Benef. Microbes* **2016**, *7*, 237–246. [CrossRef] [PubMed]
57. Lahtinen, S.J.; Forssten, S.; Aakko, J.; Granlund, L.; Rautonen, N.; Salminen, S.; Viitanen, M.; Ouwehand, A.C. Probiotic cheese containing Lactobacillus rhamnosus HN001 and Lactobacillus acidophilus NCFM(R) modifies subpopulations of fecal lactobacilli and Clostridium difficile in the elderly. *Age* **2012**, *34*, 133–143. [CrossRef] [PubMed]
58. Bogovic Matijasic, B.; Obermajer, T.; Lipoglavsek, L.; Sernel, T.; Locatelli, I.; Kos, M.; Smid, A.; Rogelj, I. Effects of synbiotic fermented milk containing Lactobacillus acidophilus La–5 and Bifidobacterium animalis ssp. lactis BB–12 on the fecal microbiota of adults with irritable bowel syndrome: A randomized double–blind, placebo–controlled trial. *J. Dairy Sci.* **2016**, *99*, 5008–5021. [CrossRef]
59. Yilmaz, I.; Dolar, M.E.; Ozpinar, H. Effect of administering kefir on the changes in fecal microbiota and symptoms of inflammatory bowel disease: A randomized controlled trial. *Turk. J. Gastroenterol.* **2019**. [CrossRef]
60. Uyeno, Y.; Sekiguchi, Y.; Kamagata, Y. Impact of consumption of probiotic lactobacilli–containing yogurt on microbial composition in human feces. *Int. J. Food Microbiol.* **2008**, *122*, 16–22. [CrossRef]
61. Santos-Marcos, J.A.; Rangel-Zuniga, O.A.; Jimenez-Lucena, R.; Quintana-Navarro, G.M.; Garcia-Carpintero, S.; Malagon, M.M.; Landa, B.B.; Tena-Sempere, M.; Perez-Martinez, P.; Lopez-Miranda, J.; et al. Influence of gender and menopausal status on gut microbiota. *Maturitas* **2018**, *116*, 43–53. [CrossRef] [PubMed]
62. Gerritsen, J.; Smidt, H.; Rijkers, G.T.; de Vos, W.M. Intestinal microbiota in human health and disease: The impact of probiotics. *Genes Nutr.* **2011**, *6*, 209–240. [CrossRef] [PubMed]
63. Chakraborti, C.K. New–found link between microbiota and obesity. *World J. Gastrointest. Pathophysiol.* **2015**, *6*, 110–119. [CrossRef] [PubMed]
64. Ley, R.E.; Turnbaugh, P.J.; Klein, S.; Gordon, J.I. Human gut microbes associated with obesity. *Nature* **2006**, *444*, 1022–1023. [CrossRef] [PubMed]
65. Turnbaugh, P.J.; Hamady, M.; Yatsunenko, T.; Cantarel, B.L.; Duncan, A.; Ley, R.E.; Sogin, M.L.; Jones, W.J.; Roe, B.A.; Affourtit, J.P.; et al. A core gut microbiome in obese and lean twins. *Nature* **2008**, *457*, 480. [CrossRef]
66. Rizzatti, G.; Lopetuso, L.; Gibiino, G.; Binda, C.; Gasbarrini, A.J.B.r.i. Proteobacteria: A common factor in human diseases. *Biomed. Res. Int.* **2017**, *2017*. [CrossRef] [PubMed]
67. Larsen, N.; Vogensen, F.K.; van den Berg, F.W.; Nielsen, D.S.; Andreasen, A.S.; Pedersen, B.K.; Al-Soud, W.A.; Sorensen, S.J.; Hansen, L.H.; Jakobsen, M. Gut microbiota in human adults with type 2 diabetes differs from non–diabetic adults. *PLoS ONE* **2010**, *5*, e9085. [CrossRef]
68. Egshatyan, L.; Kashtanova, D.; Popenko, A.; Tkacheva, O.; Tyakht, A.; Alexeev, D.; Karamnova, N.; Kostryukova, E.; Babenko, V.; Vakhitova, M.; et al. Gut microbiota and diet in patients with different glucose tolerance. *Endocr. Connect.* **2016**, *5*, 1–9. [CrossRef]
69. Qin, J.; Li, Y.; Cai, Z.; Li, S.; Zhu, J.; Zhang, F.; Liang, S.; Zhang, W.; Guan, Y.; Shen, D.; et al. A metagenome–wide association study of gut microbiota in type 2 diabetes. *Nature* **2012**, *490*, 55. [CrossRef]

70. Huang, G.; Xu, J.; Lefever, D.E.; Glenn, T.C.; Nagy, T.; Guo, T.L. Genistein prevention of hyperglycemia and improvement of glucose tolerance in adult non–obese diabetic mice are associated with alterations of gut microbiome and immune homeostasis. *Toxicol. Appl. Pharmacol.* **2017**, *332*, 138–148. [CrossRef]
71. Yang, T.; Santisteban, M.M.; Rodriguez, V.; Li, E.; Ahmari, N.; Carvajal, J.M.; Zadeh, M.; Gong, M.; Qi, Y.; Zubcevic, J.; et al. Gut dysbiosis is linked to hypertension. *Hypertension* **2015**, *65*, 1331–1340. [CrossRef] [PubMed]
72. Yan, Q.; Gu, Y.; Li, X.; Yang, W.; Jia, L.; Chen, C.; Han, X.; Huang, Y.; Zhao, L.; Li, P.; et al. Alterations of the Gut Microbiome in Hypertension. *Front. Cell. Infect. Microbiol.* **2017**, *7*, 381. [CrossRef] [PubMed]
73. Ventura, M.; Canchaya, C.; Tauch, A.; Chandra, G.; Fitzgerald, G.F.; Chater, K.F.; van Sinderen, D. Genomics of Actinobacteria: tracing the evolutionary history of an ancient phylum. *Microbiol. Mol. Biol. Rev.* **2007**, *71*, 495–548. [CrossRef] [PubMed]
74. Santisteban, M.M.; Qi, Y.; Zubcevic, J.; Kim, S.; Yang, T.; Shenoy, V.; Cole-Jeffrey, C.T.; Lobaton, G.O.; Stewart, D.C.; Rubiano, A.; et al. Hypertension–Linked Pathophysiological Alterations in the Gut. *Circulation Res.* **2017**, *120*, 312–323. [CrossRef] [PubMed]
75. Burton, K.J.; Rosikiewicz, M.; Pimentel, G.; Butikofer, U.; von Ah, U.; Voirol, M.J.; Croxatto, A.; Aeby, S.; Drai, J.; McTernan, P.G.; et al. Probiotic yogurt and acidified milk similarly reduce postprandial inflammation and both alter the gut microbiota of healthy, young men. *Br. J. Nutr.* **2017**, *117*, 1312–1322. [CrossRef]
76. Ntemiri, A.; Ribiere, C.; Stanton, C.; Ross, R.P.; O'Connor, E.; O'Toole, P.W. Retention of microbiota diversity by lactose–free milk in a mouse model of elderly gut microbiota. *J. Agric. Food Chem.* **2019**. [CrossRef] [PubMed]
77. Lim, J.; Kale, M.; Kim, D.H.; Kim, H.S.; Chon, J.W.; Seo, K.H.; Lee, H.G.; Yokoyama, W.; Kim, H. Antiobesity Effect of Exopolysaccharides Isolated from Kefir Grains. *J. Agric. Food Chem.* **2017**, *65*, 10011–10019. [CrossRef]
78. Hemarajata, P.; Versalovic, J. Effects of probiotics on gut microbiota: mechanisms of intestinal immunomodulation and neuromodulation. *Therap. Adv. Gastroenterol.* **2013**, *6*, 39–51. [CrossRef]
79. Sanchez, B.; Delgado, S.; Blanco-Miguez, A.; Lourenco, A.; Gueimonde, M.; Margolles, A. Probiotics, gut microbiota, and their influence on host health and disease. *Mol. Nutr. Food Res.* **2017**, *61*. [CrossRef]

Article

The Interplay between Maternal and Post-Weaning High-Fat Diet and Gut Microbiota in the Developmental Programming of Hypertension

Chien-Ning Hsu [1,2], Chih-Yao Hou [3], Chien-Te Lee [4], Julie Y.H. Chan [5] and You-Lin Tain [5,6,*]

1 Department of Pharmacy, Kaohsiung Chang Gung Memorial Hospital, Kaohsiung 833, Taiwan
2 School of Pharmacy, Kaohsiung Medical University, Kaohsiung 807, Taiwan
3 Department of Seafood Science, National Kaohsiung University of Science and Technology, Kaohsiung 811, Taiwan
4 Division of Nephrology, Kaohsiung Chang Gung Memorial Hospital, Kaohsiung 833, Taiwan
5 Institute for Translational Research in Biomedicine, Kaohsiung Chang Gung Memorial Hospital and Chang Gung University College of Medicine, Kaohsiung 833, Taiwan
6 Department of Pediatrics, Kaohsiung Chang Gung Memorial Hospital and Chang Gung University College of Medicine, Kaohsiung 833, Taiwan
* Correspondence: tainyl@cgmh.org.tw; Tel.: +886-975-056-995; Fax: +886-7733-8009

Received: 26 July 2019; Accepted: 21 August 2019; Published: 22 August 2019

Abstract: Excessive intake of saturated fat has been linked to hypertension. Gut microbiota and their metabolites, short-chain fatty acids (SCFAs), are known to be involved in the development of hypertension. We examined whether maternal and post-weaning high-fat (HF) diet-induced hypertension in adult male offspring is related to alterations of gut microbiota, mediation of SCFAs and their receptors, and downregulation of nutrient-sensing signals. Female Sprague–Dawley rats received either a normal diet (ND) or HF diet (D12331, Research Diets) during pregnancy and lactation. Male offspring were put on either the ND or HF diet from weaning to 16 weeks of age, and designated to four groups (maternal diet/post-weaning diet; n = 8/group): ND/ND, HF/ND, ND/HF, and HF/HF. Rats were sacrificed at 16 weeks of age. Combined HF/HF diets induced elevated blood pressure (BP) and increased body weight and kidney damage in male adult offspring. The rise in BP is related to a downregulated AMP-activated protein kinase (AMPK)–peroxisome proliferator-activated receptor co-activator 1α (PGC-1α) pathway. Additionally, HF/HF diets decreased fecal concentrations of propionate and butyrate and decreased G protein-coupled receptor 41 (GPR41), but increased olfactory receptor 78 (Oflr78) expression. Maternal HF diet has differential programming effects on the offspring's microbiota at 3 and 16 weeks of age. Combined HF/HF diet induced BP elevation was associated with an increased *Firmicutes* to *Bacteroidetes* ratio, increased abundance of genus *Akkermansia* and phylum *Verrucomicrobia*, and reduced abundance in genus *Lactobacillus*. Maternal gut microbiota-targeted dietary interventions might be reprogramming strategies to protect against programmed hypertension in children and their mothers on consumption of a fat-rich diet.

Keywords: AMP-activated protein kinase; butyrate; developmental origins of health and disease (DOHaD); gut microbiota; high fat diet; hypertension; nutrient-sensing signals; propionate; short chain fatty acids

---

1. Introduction

Non-communicable diseases (NCDs) are increasingly becoming the leading causes of global morbidity and mortality [1]. Among NCDs, hypertension-related diseases are the most common causes of deaths. Despite substantial advances in therapy, the global epidemic rise of NCDs remains a significant challenge. Early-life exposure can program the onset of chronic NCDs [2], now framed as the "developmental origins of health and disease" (DOHaD) [3].

Perinatal nutrition affects fetal development and long-term health of the offspring. Imbalanced maternal diet may induce fetal programming that permanently alters the morphology and function of fetal organs and systems, leading to various NCDs, including hypertension [4]. The high-fat (HF) diet model has been used to study obesity-related disorders like hypertension [5,6]. Along these lines, using a rat model of maternal plus post-weaning HF diets, we have demonstrated that adult male offspring exposed to HF intake develop hypertension [7].

Among the proposed mechanisms linking maternal nutritional insults to offspring adverse outcomes, changes of gut microbiota and their metabolites have recently received more attention [4,8]. Diet is an instrumental factor in shaping the gut microbiota. Increasing evidence links gut microbiota dysbiosis to the development of a variety of diseases [9]. During pregnancy, the diet–gut microbiota interactions can mediate epigenetic regulation of gene expression not only in mother but also in the fetus via the contact with their metabolites [10]. The offspring gut microbiota is highly sensitive to the early-life environmental stimuli. Accordingly, maternal diet can influence the gut microbiota of mothers and their offspring, consequently driving developmental programming of chronic diseases in adult offspring [11,12]. Although several microbial markers have been reported related to HF consumption, like increased abundance of phylum *Firmicutes* and decreased *Bacteroidetes* [9], whether a similar pattern of results can be obtained from offspring born to mothers fed with HF diet is largely unknown.

The gut microbiota produces a variety of metabolites like short-chain fatty acids detectable in host circulation [13]. Short-chain fatty acids (SCFAs, e.g., acetate, butyrate, and propionate) and their receptors are reported to be involved in the regulation of blood pressure (BP) [14]. In line with this, a recent study from our laboratory reported that prebiotic or probiotic therapy can alter gut microbiota, regulate SCFAs and their receptors, and mediate nutrient-sensing signals to protect adult male offspring against hypertension programmed by high-fructose diet [15].

Nutrient-sensing signals are regarded as key players in the developmental programming of hypertension, such as 5′-adenosine monophosphate-activated protein kinase (AMPK), peroxisome proliferator-activated receptor (PPAR), and PPARγ co-activator 1α (PGC-1α) [4,16]. Activation of AMPK by resveratrol can affect PGC-1α activity to regulate the downstream expression of PPAR target genes [17]. Our recently published study demonstrated that HF diet-induced hypertension is correlated to inhibitory AMPK/PGC-1α pathway and altered gut microbiota [18].

Our objective in this study was to examine whether maternal and post-weaning HF diet cause differential effects on BP, gut microbiota, SCFAs and their receptors, and nutrient-sensing signals in adult offspring.

2. Materials and Methods

2.1. Animal Model

This study was followed the Guide for the Care and Use of Laboratory Animals of the National Institutes of Health. The protocol was approved by the Institutional Animal Care and Use Committee of the Kaohsiung Chang Gung Memorial Hospital (IACUC permit number: 201721408). Virgin female Sprague–Dawley (SD) rats (n = 12) were obtained from BioLASCO Taiwan Co., Ltd. (Taipei, Taiwan) and maintained in an Association for Assessment and Accreditation of Laboratory Animal Care International (AAALAC)-approved animal facility in our hospital. The rats were housed in a in a controlled environment with 12:12 light-dark cycle and humidity of 55%, throughout the study. Male SD rats were caged with female rats until mating. The presence of the plug confirmed mating. Female rats were weight-matched and assigned to receive either a normal diet with regular rat chow (ND; Fwusow Taiwan Co., Ltd., Taichung, Taiwan; 52% carbohydrates, 23.5% protein, 4.5% fat, 10% ash, and 8% fiber) or a 58% high-fat diet (D12331, Research Diets, Inc., New Brunswick, NJ, USA; 58% fat (hydrogenated coconut oil), 25.5% carbohydrate, 16.4% protein, and 0% fiber) during pregnancy and lactation. After birth, litters were culled to eight from each mother to standardize the received quantity of milk and maternal pup care. Since men are much more likely to be hypertensive than women at a younger age [19], only male offspring were used. Male offspring were weaned at 3 weeks of age, and onto either the normal diet (ND) or HF diet ad libitum from weaning to 16 weeks of age. Rats were assigned to four experimental groups (maternal diet/postweaning diet; n = 8/group): ND/ND, HF/ND, ND/HF, and HF/HF.

We used BP-2000 tail-cuff system (BP-2000, Visitech Systems, Inc., Apex, NC, USA) to measure BP in conscious rats at 3, 4, 8, 12, and 16 weeks of age [7]. Rats were allowed to adapt to restraint and tail-cuff inflation for 1 week prior to the experiment. Rats were placed on the specimen platform. Their tails were passed through a cuff and immobilized by adhesive tape. Following a 10-min warm-up period, 10 preliminary cycles were performed to allow the rats to adjust to the inflating cuff. For each rat, three stable measures were taken and averaged. Fresh feces samples were collected at 3 and 16 weeks of age, frozen, and stored at −80 °C until use. At 16 weeks of age, rats were anesthetized by intraperitoneally injecting ketamine (50 mg/kg body weight) and xylazine (10 mg/kg body weight) and were euthanized by intraperitoneally injecting an overdose of pentobarbital for sacrifice. Blood samples were collected. Kidneys were harvested and stored at −80 °C in a freezer for further analysis.

2.2. Gas Chromatography-Flame Ionization Detector (GC-FID)

We used gas chromatography-mass spectrometry (GCMS-QP2010; Shimadzu, Kyoto, Japan) with a flame ionization detector (FID) to measure levels of acetate, butyrate, and propionate in the plasma and feces [15]. We used internal standards in analytical standard grades for acetate and propionate (from Sigma-Aldrich, St. Louis, MO, USA), and for butyrate (from Chem Service, West Chester, PA, USA). The working solutions of used as internal and external standards were prepared at the concentration of 10 mM. These solutions were kept at −20 °C in a freezer. Dry air, nitrogen, and hydrogen were supplied to the FID at 300, 20 and 30 mL/min, respectively. A 2-μL aliquot of sample was injected into the column. The inlet and FID temperature were set at 200 and 240 °C, respectively. The total running time was 17.5 min.

2.3. Analysis of Gut-Microbiota Composition

Metagenomic DNA was extracted from frozen fecal samples after centrifugation. All polymerase chain-reaction amplicons were mixed together and sent to the Genomic and Proteomic Core Laboratory, Kaohsiung Chang Gung Memorial Hospital (Kaohsiung, Taiwan) for sequencing using an Illumina Miseq platform (Illumina, San Diego, CA, USA) [15]. Amplicons were prepared according to the 16S Metagenomics Sequencing Library Preparation protocol (Illumina, San Diego, CA, USA), and sequenced with the Illumina MiSeq platform (Illumina, San Diego, CA, USA). Sequences (Illumina, San Diego, CA, USA) with a distance-based similarity of 97% or greater were grouped into operational taxonomic units (OTUs) using the USEARCH algorithm. To determine the significantly differential taxa, we applied linear discriminant analysis effect size (LEfSe) to compare samples between groups. The LEfSe uses linear discriminant analysis (LDA) to estimate the effect size of each differentially abundant feature. The threshold of the linear discriminant was set to two.

2.4. Western Blot

Western blot analysis was performed using the methods published previously [7]. Protein samples (200-μg kidney cortex) were boiled with gel-loading buffer for 5 min, subjected to 10–15% SDS-PAGE, and then transferred to a nitrocellulose membrane (GE Healthcare Bio-Sciences Corp., Piscataway, NJ, USA). To verify equal loading, the membranes were incubated with Ponceau S red (PonS) stain solution (Sigma-Aldrich, St. Louis, MO, USA) for 10 min on the rocker. Two nutrient-sensing signals, AMPKα2 and PGC-1α, were analyzed. Additionally, we determined the protein abundance of three SCFA receptors, including G protein-coupled receptor 41 (GPR41), GPR43, and olfactory receptor 78 (Olfr78). We used the following primary antibodies: a rabbit polyclonal anti-rat phosphorylated

AMPKα1/2 antibody (1:1000, overnight incubation; Santa Cruz Biotechnology), a rabbit polyclonal anti-PGC-1α antibody (1:1000, overnight incubation; Abcam, Cambridge, MA, USA), a rabbit polyclonal anti-GPR41 antibody (1:500, overnight incubation; USBiological, Salem, MA, USA), a rabbit polyclonal anti-GPR43 antibody (1:500, overnight incubation; Millipore, Burlington, MA, USA), and a rabbit polyclonal anti-Olfr78 antibody (1:500, overnight incubation; Assay Biotech, Fremont, CA, USA). Next, the membrane was washed five times with 0.1% T-TBS, incubated for 1h with a peroxidase-labeled secondary antibody diluted 1:1000 in T-TBS, and then developed using Chemi Doc (Bio-rad Image Lab 5.0). Bands were quantified by densitometry as integrated optical density (IOD). IOD was then normalized to total protein PonS staining. The protein abundance was represented as IOD/PonS.

2.5. Immunohistochemistry Staining

Paraffin-embedded tissues sectioned at 3-μm thickness were deparaffinized in xylene and rehydrated in a graded ethanol series to phosphate-buffered saline. Following blocking with immunoblock (BIOTnA Biotech., Kaohsiung, Taiwan), the sections were incubated for 2 h at room temperature with an anti-phosphorylated AMPKα2 antibody (1:400, Cell Signaling, Danvers, MA, USA) or an anti-PGC-1α antibody (1:200, Abcam, Cambridge, MA, USA). Immunoreactivity was revealed using the polymer-horseradish peroxidase (HRP) labeling kit (BIOTnA Biotech). For the substrate–chromogen reaction, 3,30-diaminobenzidine (DAB) was used. An identical staining protocol omitting incubation with primary antibody was employed to prepare samples that were used as negative controls. Renal cells positive for immunostaining were examined in 10 randomly selected ×400 microscopic fields per section. The number of immunostained cells was expressed as we described previously [18].

2.6. Statistical Analysis

Data are reported as the mean ± standard error of mean (SEM). A value of $p < 0.05$ was considered statistically significant. Statistical analysis was conducted with one-way analysis of variance (ANOVA) with a Tukey post hoc test for multiple comparisons. Analyses were performed using the SPSS software 14.0 (SPSS Inc., Chicago, IL, USA).

3. Results

3.1. The Effects of Maternal and Post-Weaning HF Diet on Morphological Values and BPs

Post-weaning consumption of HF diet caused a greater body weight (BW) compared with controls and the HF/ND group, with the greatest BW in the HF/HF group (Table 1). The kidney weights and the ratios of kidney weight-to-body weight were lower in the ND/HF and HF/HF groups compared to controls and the HF/ND groups. At 16 weeks of age, maternal and post-weaning HF diet increased systolic BP by 5 and 11 mmHg compared to controls, respectively. There is a synergistic effect of maternal and post-weaning HF diet on systolic BP, resulting in an increase of ~26 mmHg in the HF/HF group versus control. Similarly, diastolic BP and mean arterial pressure were higher in the HF/ND and ND/HF group compared with those in the control group, with the highest in the HF/HF group. Figure 1 shows the systolic BPs of ND/HF and HF/HF group were significantly higher than those in the control group from 8 to 16 weeks. By 12 weeks of age, the systolic BP had significantly increased in the HF/HF group relative to the other three groups. The plasma creatinine level was higher in HF/HF group compared to the controls. These findings indicate that maternal or post-weaning HF diet more or less caused a rise in BW and BPs, which was enhanced to a greater extent in the combined HF/HF diets. However, only combined HF/HF diet resulted in kidney damage, represented by elevated creatinine levels.

Table 1. Measures of morphological values, blood pressure, and renal function in 16-week-old male offspring exposed to high-fat diet (HF).

Groups	ND/ND	HF/ND	ND/HF	HF/HF
Body weight (BW) (g)	580 ± 8	561 ± 10	680 ± 22 [a,b]	715 ± 26 [a,b,c]
Left kidney weight (g)	2.44 ± 0.06	2.17 ± 0.07 [a]	2.14 ± 0.09 [a]	2.14 ± 0.09 [a]
Left kidney weight/100 g BW	0.42 ± 0.01	0.39 ± 0.01	0.32 ± 0.01 [a,b]	0.30 ± 0.01 [a,b]
Systolic blood pressure (mm Hg)	142 ± 0	147 ± 1 [a]	153 ± 1 [a,b]	168 ± 1 [a,b,c]
Diastolic blood pressure (mm Hg)	65 ± 2	70 ± 3	73 ± 2 [a]	76 ± 2 [a]
Mean arterial pressure (mm Hg)	91 ± 1	96 ± 2 [a]	99 ± 2 [a]	107 ± 2 [a,b,c]
Creatinine (μM)	14.5 ± 0.9	16.2 ± 1.1	17.2 ± 1.1	20 ± 1.8 [a]

ND/ND, maternal plus post-weaning normal diet; HF/ND, maternal high-fat diet plus post-weaning normal diet; ND/HF, maternal normal diet plus post-weaning high-fat diet; HF/HF, maternal plus post-weaning high-fat diet. BW, body weight; $n = 8$/group; [a] $p < 0.05$ vs. ND/ND; [b] $p < 0.05$ vs. HF/ND; [c] $p < 0.05$ vs. ND/HF.

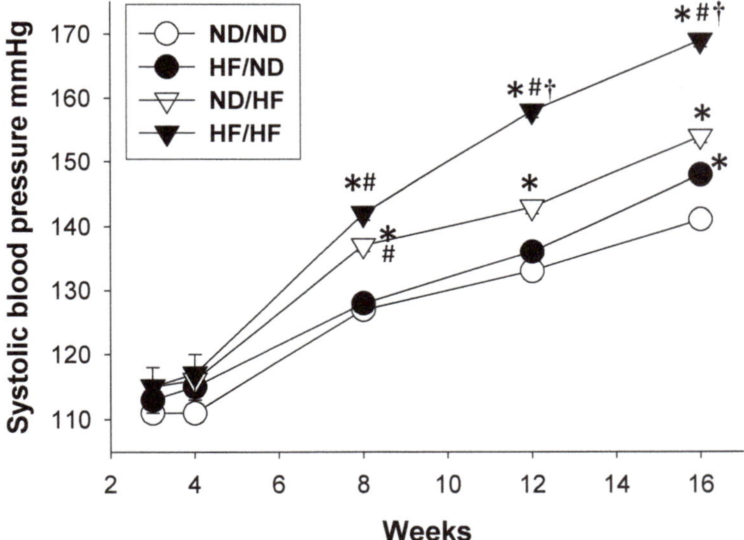

Figure 1. Effects of maternal and postnatal high-fat (HF) diet on systolic blood pressure in male offspring from 3 to 16 weeks. ND/ND, maternal plus post-weaning normal diet; HF/ND, maternal high-fat diet plus post-weaning normal diet; ND/HF, maternal normal diet plus post-weaning high-fat diet; HF/HF, maternal plus post-weaning high-fat diet. * $p < 0.05$ vs. ND/ND; # $p < 0.05$ vs. HF/ND; † $p < 0.05$ vs. ND/HF.

3.2. The Effects of Maternal and Post-Weaning HF Diet on Nutrient-Sensing Signals

We evaluated key elements in the nutrient-sensing pathway, including phosphorylated AMPKα2 and PGC-1α. As shown in Figure 2, the renal protein level of phosphorylated AMPKα2 (Figure 2B) was lower in the HF/ND, ND/HF, and HF/HF group compared with that in the ND/ND group. Additionally, the HF/HF diet caused a significant reduction of PGC-1α versus the controls in offspring kidneys (Figure 2C). We next evaluated phosphorylated AMPKα2 (Figure 3) and PGC-1α (Figure 4) in the offspring kidneys by immunohistochemistry.

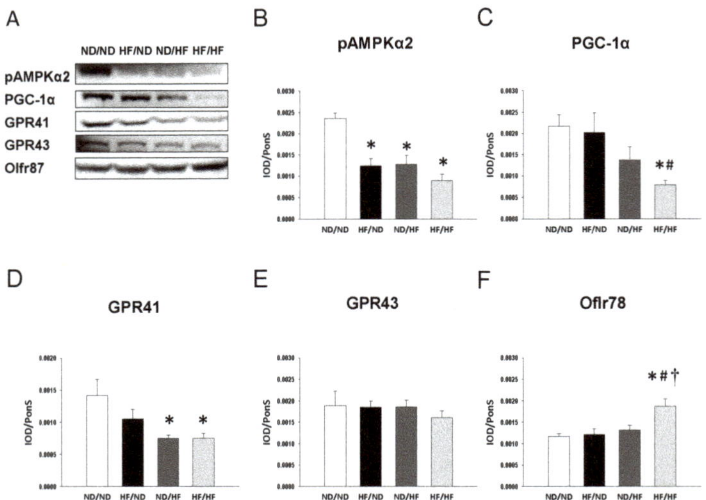

Figure 2. (**A**) Representative western blots showing phosphorylated AMP-activated protein kinase (AMPKα2, ~63kDa), peroxisome proliferator-activated receptor co-activator 1α (PGC-1α, ~90kDa), G protein-coupled receptor 41 (GPR41, ~38kDa), GPR43 (~47kDa), and olfactory receptor 78 (Oflr78) (~35kDa) bands in offspring kidneys at 16 weeks of age. Relative abundance of renal cortical (**B**) phosphorylated AMPKα2, (**C**) PGC-1α, (**D**) GPR41, (**E**) GPR43, and (**F**) Oflr78 were quantified. ND/ND, maternal plus post-weaning normal diet; HF/ND, maternal high-fat diet plus post-weaning normal diet; ND/HF, maternal normal diet plus post-weaning high-fat diet; HF/HF, maternal plus post-weaning high-fat diet. n = 8/group. * $p < 0.05$ vs. ND/ND; # $p < 0.05$ vs. HF/ND; † $p < 0.05$ vs. ND/HF.

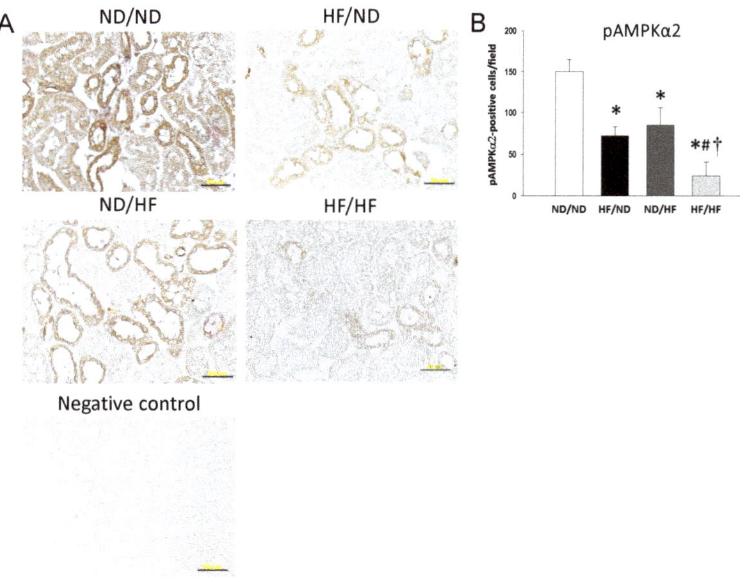

Figure 3. (**A**) Light microscopic findings of phosphorylated AMPKα2 immunostaining in the kidney cortex in 16-week-old male offspring. Bar = 50 μm; (**B**) Quantitative analysis of phosphorylated AMPKα2-positive cells per microscopic field (400×); * $p < 0.05$ vs. ND/ND; # $p < 0.05$ vs. HF/ND; † $p < 0.05$ vs. ND/HF.

Immunostaining of phosphorylated AMPKα2 in the glomeruli and renal tubules indicated intense staining in the ND/ND group (150 ± 15 positive cells), an intermediate level of staining in the HF/ND group (72 ± 11 positive cells) and ND/HF group (85 ± 21 positive cells), and little staining in the HF/HF group (24 ± 17 positive cells) (Figure 3B). Similar to phosphorylated AMPKα2, maternal or post-weaning HF diet significantly decreased PGC-1α expression in the HF/ND group (112 ± 14 positive cells) and the ND/HF group (121 ± 21 positive cells) vs. the ND/ND group (220 ± 29 positive cells) (Figure 4A). Combined maternal and post-weaning HF diets caused the reduction of PGC-1α expression to a greater extent (36 ± 14 positive cells) (Figure 4B). Taken together, these findings indicated that HF/HF diet synergistically downregulated AMPK–PGC-1α pathway.

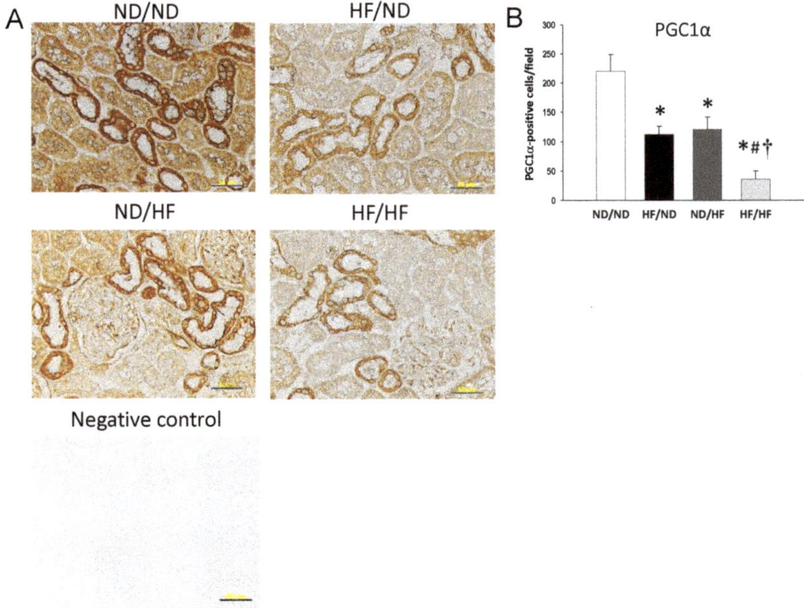

Figure 4. (**A**) Light microscopic findings of PGC-1α immunostaining in the kidney cortex in 16-week-old male offspring. Bar = 50 μm; (**B**) Quantitative analysis of PGC-1α-positive cells per microscopic field (400×); * $p < 0.05$ vs. ND/ND; # $p < 0.05$ vs. HF/ND; † $p < 0.05$ vs. ND/HF.

3.3. *The Effects of Maternal and Post-Weaning HF Diet on SCFAs and Their Receptors*

It was reported previously that SCFAs are involved in the development of hypertension [14]. We investigated whether HF diet causes a rise in BP is related to alterations of SCFAs production and the expression of SCFA receptors. Our results demonstrated that post-weaning HF diet decreased fecal concentrations of acetate compared to the ND/ND and HF/ND group (Table 2). Fecal propionate and butyrate levels were lower in the ND/HF group than those in the ND/ND and HF/ND group. Similarly, combined maternal and post-weaning HF reduced fecal concentrations of propionate and butyrate compared to controls. We next evaluated the protein levels of SCFA receptors. Renal GPR41 expression was lower in the ND/HF and HF/HF group compared to that in the ND/ND group (Figure 2D). GPR43 protein level in offspring kidney was not different among the four groups (Figure 2E). However, combined HF/HF diets resulted in a significant increase of renal Olfr78 expression compared to the other three groups (Figure 2F).

Table 2. Fecal levels of acetate, propionate, and butyrate in in male offspring exposed to high-fat diet (HF) at 16 weeks of age.

Group	ND/ND	HF/ND	ND/HF	HF/HF
Acetate, mM/g feces	3.68 ± 0.13	3.62 ± 0.22	1.34 ± 0.13 ^{a,b}	2.46 ± 0.57
Propionate, mM/g feces	0.84 ± 0.06	0.75 ± 0.05	0.25 ± 0.05 ^{a,b}	0.47 ± 0.14 ^{a}
Butyrate, mM/g feces	1.68 ± 0.21	1.51 ± 0.29	0.22 ± 0.02 ^{a,b}	0.27 ± 0.05 ^{a,b}

ND/ND, maternal plus post-weaning normal diet; HF/ND, maternal high-fat diet plus post-weaning normal diet; ND/HF, maternal normal diet plus post-weaning high-fat diet; HF/HF, maternal plus post-weaning high-fat diet. BW, body weight; n = 8/group; ^{a} $p < 0.05$ vs. ND/ND; ^{b} $p < 0.05$ vs. HF/ND.

3.4. The Effects of Maternal and Post-Weaning HF Diet on Gut Microbiota

We further analyzed bacterial populations in the gut at the phylum and genus levels at 3 weeks (Figure 5) and 16 weeks of age (Figure 6). At 3 weeks, the age of weaning, the main phyla in the offspring born of dams fed with regular chow (ND) or HF diet were *Firmicutes*, *Bacteroidetes*, *Verrucomicrobia*, *Proteobacteria*, and *Actinobacteria*. Maternal HF intake caused a remarkable increase in the phylum *Firmicutes* (72.3 ± 4% vs. 52.9 ± 3.7%; $p = 0.002$), but a decrease in the *Verrucomicrobia* (10.3 ± 2.6% vs. 25.9 ± 4.5%; $p = 0.007$) and *Proteobacteria* (2.4 ± 0.2% vs. 4.8 ± 0.4%; $p < 0.001$) (Figure 5A). The *Firmicutes* to *Bacteroidetes* ratio has been considered a signature for hypertension [20,21]. In the current study, the *Firmicutes* to *Bacteroidetes* ratio was higher in the HF group (8.3 ± 1.6) compared to that in the control group (4.2 ± 0.6, $p = 0.03$) (Figure 5B). Additionally, the main bacterial genera were *Akkermansia*, *Blautia*, *Clostridium*, *Parabacteroides*, *Lactobacillus*, *Alkaliphilus*, *Ruminococcus*, *Sarcina*, *Natronincola*, and *Flavobacterium* (Figure 5C). Among them, maternal HF diet decreased abundance of genus *Akkermansia* (9.7 ± 2.4% vs. 25 ± 4.3%; $p = 0.006$) (Figure 5D). Conversely, abundance of genus *Clostridium* was induced in the HF group (19.9 ± 2.6%) compared with that in control (10.4 ± 1.2%; $p = 0.03$).

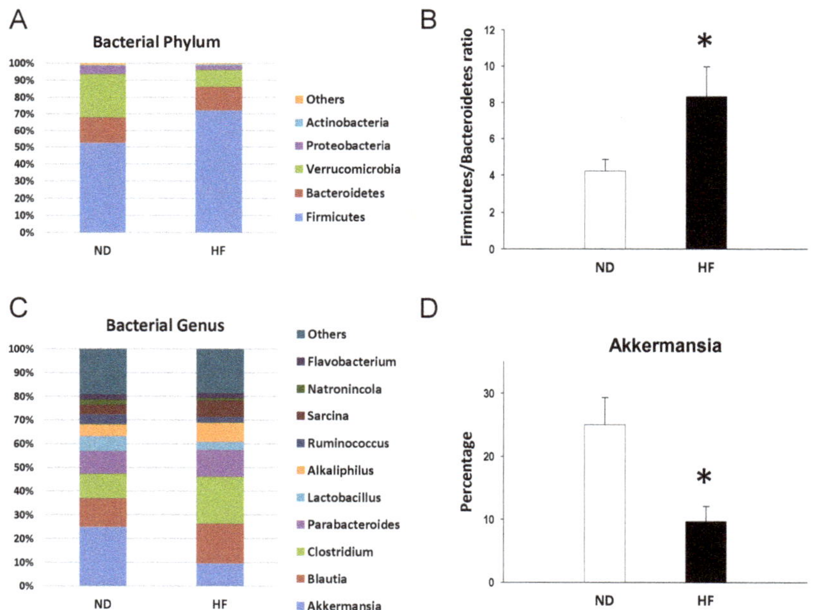

Figure 5. Effect of maternal high-fructose (HF) diet on offspring gut microbiota at 3 weeks of age. (**A**) Relative abundances of the top five phyla. (**B**) The *Firmicutes* to *Bacteroidetes* ratio. (**C**) Relative abundances of the top 10 genera. (**D**) Relative abundances of the genus *Akkermansia*. n = 16/group. * $p < 0.05$ vs. ND.

As shown in Figure 6A, the main phyla in the offspring gut microbiota at 16 weeks were identical to those at 3 weeks of age. Combined HF/HF diet significantly reduced the abundance of the phylum *Bacteroidetes* (15.1 ± 1.7% vs. 30.2 ± 0.7%; $p < 0.001$), while augmenting the abundance of the *Verrucomicrobia* (15.6 ± 2.2% vs. 0.5 ± 0.2%; $p < 0.001$). Additionally, the *Firmicutes* to *Bacteroidetes* ratio was the highest in the HF/HF group compared to that in the other three groups (All $p < 0.05$) (Figure 6B).

Maternal HF intake decreased the abundance of genera *Lactobacillus* (HF/ND vs. ND/ND = 4.3 ± 0.8% vs. 13.7 ± 2.5%, $p = 0.005$) and *Turicibacter* (HF/ND vs. ND/ND = 0.9 ± 0.3% vs. 2.1 ± 0.4%, $p = 0.035$). The post-weaning HF diet caused an increase of genus *Akkermansia* (ND/HF vs. ND/ND = 9.4 ± 3.8% vs. 0.4 ± 0.2%, $p = 0.002$), and decreased the abundance of genera *Lactobacillus* (3 ± 0.5%, $p = 0.002$) and *Turicibacter* (0.6 ± 0.1%, $p = 0.01$). Combined HF/HF diet caused increases of several bacterial genera, including *Akkermansia*, *Clostridium*, and *Alkaliphilus* (Figure 6C; all $p < 0.05$). Conversely, the abundance of genera *Parabacteroides*, *Lactobacillus*, and *Ruminococcus* was reduced by HF/HF exposure (Figure 6C; all $p < 0.05$). Of note is that maternal (HF/ND: 4.3 ± 0.8%) and post-weaning HF diet (ND/HF: 3 ± 0.5%) both resulted in the reduced abundance in genus *Lactobacillus* compared to the ND/ND group (13.7 ± 2.5%; both $p < 0.05$). The combined HF/HF diet caused the reduction of genus *Lactobacillus* abundance to a greater extent (0.8 ± 0.3%, all $p < 0.05$) (Figure 6D).

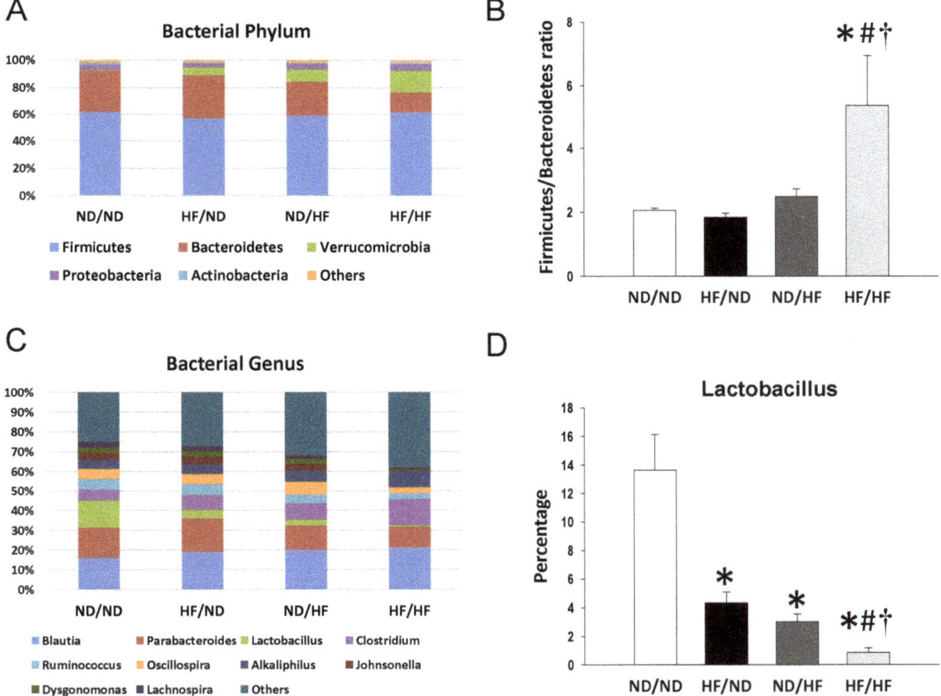

Figure 6. Effect of maternal and post-weaning high-fructose (HF) diet on offspring gut microbiota at 16 weeks of age. (**A**) Relative abundances of the top five phyla. (**B**) The *Firmicutes* to *Bacteroidetes* ratio. (**C**) Relative abundances of the top 10 genera. (**D**) Relative abundances of the genus *Lactobaccilus*. n = 8/group. * $p < 0.05$ vs. ND/ND; # $p < 0.05$ vs. HF/ND; † $p < 0.05$ vs. ND/HF.

The main bacterial species modified by the maternal HF diet were *Leptolyngbya laminosa* (LDA score = −3.1), *Enterococcus avium* (LDA score = −2.3), and *Enterococcus casseliflavus* (LDA score = −2.2) (Figure 7A). The post-weaning HF diet showed an increase in species *Lactococcus lactis* (LDA score = 2.6) and *Streptococcus dentirousetti* (LDA score = 2), and caused a decrease in the species *Leptolyngbya laminosa*

(LDA score = −2.5) and *Enterococcus casseliflavus* (LDA score = −2.1) as compared to the ND/ND group (Figure 7B). Of note, there was a remarkable decrease in several species of *Lactobaccilus* in the HF/HF group vs. the ND/ND group (Figure 7C). Conversely, HF/HF diet caused an increase of in species *Akkermansia muciniphila* (LDA score = 2.1).

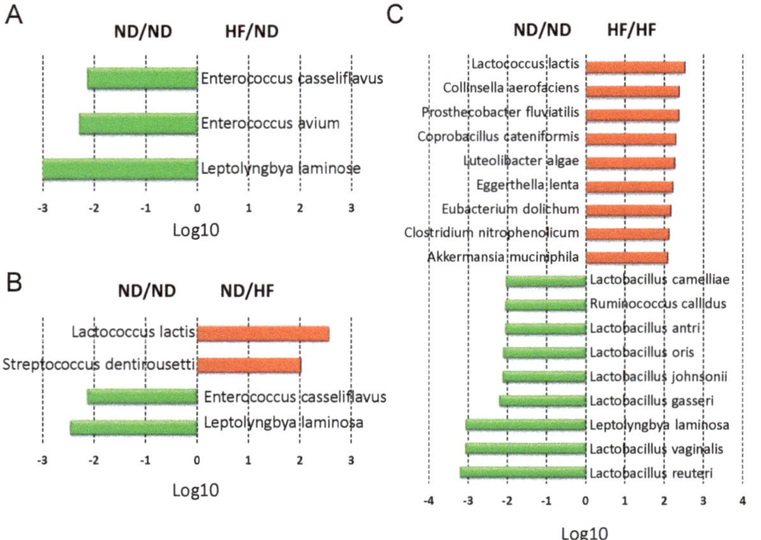

Figure 7. Effect of maternal and post-weaning high-fructose (HF) diet on 16-week-old offspring gut microbiota at the species level. Linear discriminant analysis (LDA), along with effect size measurements, was applied to identify enriched bacterial species. Most enriched and depleted species (LDA score (log10) > 2.0) in the (**A**) HF/ND (red) vs. ND/ND (green), (**B**) ND/HF (red) vs. ND/ND (green), and (**C**) HF/HF (red) vs. ND/ND (green). n = 8/group.

4. Discussion

This study provides a novel insight into the mechanisms responsible for the development of hypertension programmed by maternal and post-weaning HF diet with particular emphasis on gut microbiota-derived metabolites SCFAs and nutrient-sensing signals. The main findings of this study are as follows: (1) combined maternal plus postweaning HF diets induced elevated BP and increased BW and kidney damage in male adult offspring; (2) The combined HF/HF diets caused a rise in BP, which is related to a downregulated AMPK–PGC-1α pathway; (3) The offspring exposed to HF/HF diets had decreased fecal concentrations of propionate and butyrate, decreased renal GPR41 protein levels, and increased renal Oflr78 expression; (4) At 3 weeks of age, the maternal HF diet increased the *Firmicutes* to *Bacteroidetes* ratio and abundance of genus *Clostridium*, and decreased the abundance of genus *Akkermansia* in the gut microbiota in offspring; and (5) The HF/HF diet caused the rise of BP at 16 weeks of age, which was associated with the increased *Firmicutes* to *Bacteroidetes* ratio and reduced abundance in genus *Lactobacillus*.

In line with previous studies showing that maternal HF intake induces elevated BP in offspring [22,23], our results demonstrated that systolic BP was approximately 5 mmHg higher in the HF/ND group than that in the ND/ND group. Maternal HF diet-induced programmed hypertension may be related to a downregulated AMPK–PGC-1α pathway, an increased *Firmicutes* to *Bacteroidetes* ratio, and a decreased abundance of the genera *Akkermansia* and *Lactobacillus*. Additionally, we found that there is a synergistic effect between maternal and post-weaning HF diet causing a rise in BP and body weight, in support of our previous study showing that effect of maternal nutritional insults on the fetus are not set in stone and can be amplified by changes in the postnatal environment [24,25].

The observed effect of maternal HF diet on BP increase may be related to the inhibition of AMPK–PGC-1α pathway. The interplay between AMPK and other nutrient-sensing signals, driven by maternal nutritional insults, is known to regulate PPARs and their target genes, thus leading to programming of hypertension [17]. AMPKα2 knockout mice expressed activation of the renin-angiotensin system (RAS) to favor the development of hypertension [26]. Also, uni-nephrectomized rats developed hypertension, which was associated with decreased AMPK expression and activation of the RAS [27]. On the contrary, the AMPK activation has been shown to regulate the RAS, resulting in protection from hypertension in different models of programming [28,29]. Recently, AMPK activation has emerged as a reprogramming strategy, via regulating other nutrient-sensing signals like PGC-1α, to protect against hypertension and kidney disease with developmental origins [30]. In the current study, maternal or post-weaning HF diet reduced phosphorylated AMPKα2 and PGC-1α expression. Remarkably, combined maternal and post-weaning HF diets caused the reduction of phosphorylated AMPKα2 and PGC-1α expression to a greater extent in the HF/HF group. These results reconfirmed our previous study showing that combined HF/HF diet-induced hypertension is associated with reduced phosphorylated AMPKα2 and PGC-1α expression. These changes were restored by AMPK activation through resveratrol treatment [7]. These observations suggest that pharmacological therapies aimed at AMPKα2 as a reprogramming intervention to prevent hypertension programmed by maternal HF intake deserve further evaluation.

The results of this study showed that changes of SCFAs and their receptors are another mechanism contributing to HF/HF-induced hypertension. Although maternal HF diet had a neglectable effect on fecal SCFA levels and their receptors, post-weaning HF diet significantly reduced fecal propionate and butyrate concentrations. Propionate and butyrate have been reported to induce vasodilatation via mediating GPR41 and GPR43 receptor [14]. Conversely, acetate is a ligand for Olfr78 to raise BP [14]. Our report showed that combined HF/HF diet decreased fecal propionate and butyrate levels, decreased GPR41 expression, and increased Oflr78 expression in adult offspring kidneys, all of which may favor the development of hypertension. AMPK can be activated by SCFAs, like propionate and butyrate [31,32]. SCFAs have been report to protect against ethanol-induced gut leakiness via AMPK activation [33]. On the other hand, AMPK activation altered microbial populations, which promotes SCFA production [34]. In line with increasing evidence of a link between gut microbiota, SCFAs, and AMPK [35], our study demonstrated that HF/HF-induced hypertension is associated with inactivation of AMPK signaling and the reduction of SCFA production. Additional studies warranted to clarify whether microbiota-derived SCFAs regulate AMPK signaling contributing to hypertension programmed by HF diet.

Additionally, we observed the major acetate-producing bacteria could be either decreased (e.g., *Lactobacillus*) or increased (e.g., *Clostridiums* and *Akkermansia*) in the HF/HF group with hypertension. Unlike a previous study demonstrating that hypertension-associated dysbiosis is characterized by increases in lactate-producing bacteria [21], results of this study showed that the abundance of genera *Lactobacillus* and *Turicibacter*, which are lactate-producing bacteria, were decreased in the ND/HF and HF/ND-induced hypertension groups. Thus, additional studies are required to clarify whether the imbalance of gut acetate-, butyrate-, and propionate-producing bacterial populations directly contribute to BP control in a variety of programming hypertension models.

The detrimental effects of HF diet may also relate to alterations in gut microbiota composition. Emerging evidence shows that the development of hypertension is related to gut microbiota dysbiosis in animal models of hypertension [20,21]. Microbiota dysbiosis in early life has deleterious effects and may have long-term consequences leading to many diseases in later life [31]. Our results go beyond previous studies, demonstrating that altered gut microbiota links early-life HF intake to the developmental programming of hypertension. Although the interactions between dietary fat with the gut microbiota have been well explored in human and experimental studies [36], little is known about the impact of maternal fat intake on the offspring gut microbiota. Previous studies showed that maternal HF consumption can alter the offspring microbiome in various animal species [7,37,38]. In line with this, our study demonstrated that maternal HF diet resulted in a considerable impact on the infant microbiota (i.e., 3 weeks of age), as reflected in a higher *Firmicutes* to *Bacteroidetes* ratio, higher abundance of genus *Clostridium*, and lower abundance of genus *Akkermansia*. However, these changes in microbiota compositions seem not persistent until adulthood (i.e., 16 weeks of age). An increased *Firmicutes* to *Bacteroidetes* ratio has been related to obesity in animals fed with saturated fat [36]. Our results go beyond previous reports, showing that mother rats exposed to HF intake caused an increase of the *Firmicutes* to *Bacteroidetes* ratio in their offspring's microbiota. We found it notable that a certain change in this ratio was persistent until adulthood in the HF/HF group, which had significant BP increases. Given previously published studies using this ratio as a microbial marker for hypertension [20,21], we speculate that this ratio might be a marker to predict hypertension of developmental origins.

Akkermansia muciniphila is the main genus classified in the *Verrucomicrobia* phylum, and recent studies revealed its beneficial effects against obesity and cardiometabolic disease [39]. According to our data, maternal HF diet reduced abundance of genus *Akkermansia* in 3-week-old offspring microbiota. Conflicting with previous reports showing that *Akkermansia muciniphila* abundance inversely correlated with obesity and hypertension [40,41], our results demonstrated that the combined HF/HF diet caused a more than a 100-fold increase of *Akkermansia muciniphila* abundance. Also, HF/HF diet increased abundance of genus *Akkermansia* and phylum *Verrucomicrobia* in offspring microbiota at 16 weeks of age. One possible reason was because we were experimenting in the model of developmental programming, which is more complex than the established disease models. Thus, further studies are needed to elucidate whether *Akkermansia muciniphila* may serve as a microbial marker for hypertension in other developmental programming models. Additionally, we observed that combined HF/HF diets caused a remarkable decrease in abundance of *Lactobacillus*, which is generally considered as a beneficial microbe [42]. Certain probiotic strains like *Lactobacillus* have shown hypotensive effects [43]. As we observed, several *Lactobacillus* spp. were depleted in the HF/HF group, and our previous study demonstrated that maternal *Lactobacillus casei* treatment protected adult offspring against programmed hypertension [15], there is a need to further explore whether early probiotic supplementation may serve as a reprogramming strategy to prevent hypertension programmed by HF/HF diets as well as in other programming models.

Our study has a few limitations. First, we did not examine serial changes in the composition of offspring microbiota. The alterations in gut microbiota we observed in adult offspring may reflect postnatal plasticity rather than programmed processes. Second, we did not analyze other organs controlling BP. The hypertensive effect of HF diet might be attributed to other organs, such as the heart, brain, and vasculature. Third, we employed 16S rRNA gene amplicon analysis to determine proportional changes among bacterial taxonomies. Further studies addressing gene functions contributed by the gut microbiome rather than abundance of taxa to hypertension of developmental origin are required. With the exception of hypertension, maternal HF diet has been used to model other DOHaD-related NCDs [5,44]. It remains to be determined whether changes in microbial composition and their metabolite SCFAs observed in the current study are involved in the pathogenesis of other NCDs. Last, only male offspring were studied in the present study. Given that sex differences appear in gut microbiota and hypertension [19,45], additional studies are required to clarify whether sex-specific interactions between gut microbiota and hypertension exist in mechanisms underlying hypertension programmed by HF diet.

5. Conclusions

In conclusion, several important mechanisms are involved in the development of hypertension programmed by maternal and post-weaning HF diet, including alterations of gut microbiota, SCFAs and their receptors, and nutrient-sensing signals. Targeting AMPK signaling, gut microbiota, and SCFAs might be a reprogramming strategy to reverse the development of hypertension programmed by high fat consumption. Although reprogramming strategies from animal models still await further clinical translation, our findings highlight that pregnant women and children's caretakers must pay attention to avoid excessive foods that have high fat content.

Author Contributions: C.-N.H.: contributed to concept generation, data interpretation, drafting of the manuscript, critical revision of the manuscript and approval of the article; C.-Y.H.: contributed to data interpretation, critical revision of the manuscript and approval of the article; C.-T.L.: contributed to data interpretation, critical revision of the manuscript and approval of the article; J.Y.H.C.: contributed to methodology and approval of the article; Y.-L.T.: contributed to concept generation, data interpretation, drafting of the manuscript, critical revision of the manuscript, and approval of the article.

Funding: This work was supported by grant MOST 107-2314-B-182-045-MY3 from the Ministry of Science and Technology, Taiwan.

Acknowledgments: We would like to thank the Genomic & Proteomic Core Laboratory, Department of Medical Research and Development, Kaohsiung Chang Gung Memorial Hospital, Kaohsiung, Taiwan, for gut microbiota profiling.

Conflicts of Interest: The authors declare no conflict of interest.

References

1. Zarocostas, J. Need to increase focus on non-communicable diseases in global health, says WHO. *Br. Med. J.* **2010**, *341*, c7065. [CrossRef] [PubMed]
2. Hanson, M.; Gluckman, P. Developmental origins of noncommunicable disease: Population and public health implications. *Am. J. Clin. Nutr.* **2011**, *94*, 1754S–1758S. [CrossRef] [PubMed]
3. Hanson, M. The birth and future health of DOHaD. *J. Dev. Orig. Health Dis.* **2015**, *6*, 434–437. [CrossRef] [PubMed]
4. Hsu, C.N.; Tain, Y.L. The Double-Edged Sword Effects of Maternal Nutrition in the Developmental Programming of Hypertension. *Nutrients* **2018**, *10*, 1917. [CrossRef] [PubMed]
5. Williams, L.; Seki, Y.; Vuguin, P.M.; Charron, M.J. Animal models of in utero exposure to a high fat diet: A review. *Biochim. Biophys. Acta* **2014**, *1842*, 507–519. [CrossRef] [PubMed]
6. Khan, I.Y.; Taylor, P.D.; Dekou, V.; Seed, P.T.; Lakasing, L.; Graham, D.; Dominiczak, A.F.; Hanson, M.A.; Poston, L. Gender-linked hypertension in offspring of lard-fed pregnant rats. *Hypertension* **2003**, *41*, 168–175. [CrossRef] [PubMed]
7. Tain, Y.L.; Lin, Y.J.; Sheen, J.M.; Lin, I.C.; Yu, H.R.; Huang, L.T.; Hsu, C.N. Resveratrol prevents the combined maternal plus postweaning high-fat-diets-induced hypertension in male offspring. *J. Nutr. Biochem.* **2017**, *48*, 120–127. [CrossRef] [PubMed]
8. Chu, D.M.; Meyer, K.M.; Prince, A.L.; Aagaard, K.M. Impact of maternal nutrition in pregnancy and lactation on offspring gut microbial composition and function. *Gut Microbes* **2016**, *7*, 459–470. [CrossRef]
9. Portune, K.J.; Benítez-Páez, A.; Del Pulgar, E.M.; Cerrudo, V.; Sanz, Y. Gut microbiota, diet, and obesity-related disorders-The good, the bad, and the future challenges. *Mol. Nutr. Food Res.* **2017**, *61*, 1600252. [CrossRef]
10. Mulligan, C.M.; Friedman, J.E. Maternal modifiers of the infant gut microbiota: Metabolic consequences. *J. Endocrinol.* **2017**, *235*, R1–R12. [CrossRef]
11. Tamburini, S.; Shen, N.; Wu, H.C.; Clemente, J.C. The microbiome in early life: Implications for health outcomes. *Nat. Med.* **2016**, *22*, 713–722. [CrossRef] [PubMed]
12. Stiemsma, L.T.; Michels, K.B. The role of the microbiome in the developmental origins of health and disease. *Pediatrics* **2018**, *141*, e20172437. [CrossRef] [PubMed]
13. Nicholson, J.K.; Holmes, E.; Kinross, J.; Burcelin, R.; Gibson, G.; Jia, W.; Pettersson, S. Host-gut microbiota metabolic interactions. *Science* **2012**, *336*, 1262–1267. [CrossRef] [PubMed]
14. Pluznick, J.L. Microbial Short-Chain Fatty Acids and Blood Pressure Regulation. *Curr. Hypertens. Rep.* **2017**, *19*, 25. [CrossRef] [PubMed]
15. Hsu, C.N.; Lin, Y.J.; Hou, C.Y.; Tain, Y.L. Maternal Administration of Probiotic or Prebiotic Prevents Male Adult Rat Offspring against Developmental Programming of Hypertension Induced by High Fructose Consumption in Pregnancy and Lactation. *Nutrients* **2018**, *10*, 1229. [CrossRef] [PubMed]
16. Efeyan, A.; Comb, W.C.; Sabatini, D.M. Nutrient-sensing mechanisms and pathways. *Nature* **2015**, *517*, 302–310. [CrossRef] [PubMed]

17. Tain, Y.L.; Hsu, C.N.; Chan, J.Y. PPARs Link Early Life Nutritional Insults to Later Programmed Hypertension and Metabolic Syndrome. *Int. J. Mol. Sci.* **2015**, *17*, 20. [CrossRef] [PubMed]
18. Chen, H.E.; Lin, Y.J.; Lin, I.C.; Yu, H.R.; Sheen, J.M.; Tsai, C.C.; Huang, L.T.; Tain, Y.L. Resveratrol prevents combined prenatal NG-Nitro-L-arginine-methyl ester (L-NAME) treatment plus postnatal high-fat diet induced programmed hypertension in adult rat offspring: Interplay between nutrient-sensing signals, oxidative stress and gut microbiota. *J. Nutr. Biochem.* **2019**, *70*, 28–37. [CrossRef] [PubMed]
19. Reckelhoff, J.F. Gender differences in the regulation of blood pressure. *Hypertension* **2001**, *37*, 1199–1208. [CrossRef] [PubMed]
20. Al Khodor, S.; Reichert, B.; Shatat, I.F. The Microbiome and Blood Pressure: Can Microbes Regulate Our Blood Pressure? *Front. Pediatr.* **2017**, *5*, 138. [CrossRef] [PubMed]
21. Yang, T.; Santisteban, M.M.; Rodriguez, V.; Li, E.; Ahmari, N.; Carvajal, J.M.; Zadeh, M.; Gong, M.; Qi, Y.; Zubcevic, J.; et al. Gut dysbiosis is linked to hypertension. *Hypertension* **2015**, *65*, 1331–1340. [CrossRef] [PubMed]
22. Resende, A.C.; Emiliano, A.F.; Cordeiro, V.S.; de Bem, G.F.; de Cavalho, L.C.; de Oliveira, P.R.; Neto, M.L.; Costa, C.A.; Boaventura, G.T.; de Moura, R.S. Grape skin extract protects against programmed changes in the adult rat offspring caused by maternal high-fat diet during lactation. *J. Nutr. Biochem.* **2013**, *24*, 2119–2126. [CrossRef] [PubMed]
23. Torrens, C.; Ethirajan, P.; Bruce, K.D.; Cagampang, F.R.; Siow, R.C.; Hanson, M.A.; Byrne, C.D.; Mann, G.E.; Clough, G.F. Interaction between maternal and offspring diet to impair vascular function and oxidative balance in high fat fed male mice. *PLoS ONE* **2012**, *7*, e50671. [CrossRef] [PubMed]
24. Tain, Y.L.; Lee, W.C.; Leu, S.; Wu, K.; Chan, J. High salt exacerbates programmed hypertension in maternal fructose-fed male offspring. *Nutr. Metab. Cardiovasc. Dis.* **2015**, *25*, 1146–1151. [CrossRef] [PubMed]
25. Tain, Y.L.; Lee, W.C.; Wu, K.; Leu, S.; Chan, J. Maternal high fructose intake increases the vulnerability to post-weaning high-fat diet-induced programmed hypertension in male offspring. *Nutrients* **2018**, *10*, 56. [CrossRef]
26. Kohlstedt, K.; Trouvain, C.; Boettger, T.; Shi, L.; Fisslthaler, B.; Fleming, I. AMP-activated protein kinase regulates endothelial cell angiotensin-converting enzyme expression via p53 and the post-transcriptional regulation of microRNA-143/145. *Circ. Res.* **2013**, *112*, 1150–1158. [CrossRef]
27. Yang, K.K.; Sui, Y.; Zhou, H.R.; Shen, J.; Tan, N.; Huang, Y.M.; Li, S.S.; Pan, Y.H.; Zhang, X.X.; Zhao, H.L. Cross-talk between AMP-activated protein kinase and renin-angiotensin system in uninephrectomised rats. *J. Renin Angiotensin Aldosterone Syst.* **2016**, *17*, 1470320316673231. [CrossRef]
28. Tain, Y.L.; Wu, K.L.H.; Lee, W.C.; Leu, S.; Chan, J.Y.H. Prenatal Metformin Therapy Attenuates Hypertension of Developmental Origin in Male Adult Offspring Exposed to Maternal High-Fructose and Post-Weaning High-Fat Diets. *Int. J. Mol. Sci.* **2018**, *19*, 1066. [CrossRef]
29. Kim, E.N.; Kim, M.Y.; Lim, J.H.; Kim, Y.; Shin, S.J.; Park, C.W.; Kim, Y.S.; Chang, Y.S.; Yoon, H.E.; Choi, B.S. The protective effect of resveratrol on vascular aging by modulation of the renin-angiotensin system. *Atherosclerosis* **2018**, *270*, 123–131. [CrossRef]
30. Tain, Y.L.; Hsu, C.N. AMP-Activated Protein Kinase as a Reprogramming Strategy for Hypertension and Kidney Disease of Developmental Origin. *Int. J. Mol. Sci.* **2018**, *19*, 1744. [CrossRef]
31. Si, X.; Shang, W.; Zhou, Z.; Strappe, P.; Wang, B.; Bird, A.; Blanchard, C. Gut Microbiome-Induced Shift of Acetate to Butyrate Positively Manages Dysbiosis in High Fat Diet. *Mol. Nutr. Food Res.* **2018**, *62*. [CrossRef] [PubMed]
32. Yoshida, H.; Ishii, M.; Akagawa, M. Propionate suppresses hepatic gluconeogenesis via GPR43/AMPK signaling pathway. *Arch. Biochem. Biophys.* **2019**, *672*, 108057. [CrossRef] [PubMed]
33. Elamin, E.E.; Masclee, A.A.; Dekker, J.; Pieters, H.J.; Jonkers, D.M. Short-chain fatty acids activate AMP-activated protein kinase and ameliorate ethanol-induced intestinal barrier dysfunction in Caco-2 cell monolayers. *J. Nutr.* **2013**, *143*, 1872–1881. [CrossRef] [PubMed]
34. Zhang, X.; Zhao, Y.; Xu, J.; Xue, Z.; Zhang, M.; Pang, X.; Zhang, X.; Zhao, L. Modulation of gut microbiota by berberine and metformin during the treatment of high-fat diet-induced obesity in rats. *Sci. Rep.* **2015**, *5*, 14405. [CrossRef] [PubMed]
35. Sun, X.; Zhu, M.J. AMP-activated protein kinase: A therapeutic target in intestinal diseases. *Open Biol.* **2017**, *7*, 170104. [CrossRef]

36. Mokkala, K.; Houttu, N.; Cansev, T.; Laitinen, K. Interactions of dietary fat with the gut microbiota: Evaluation of mechanisms and metabolic consequences. *Clin. Nutr.* **2019**. [CrossRef]
37. Ma, J.; Prince, A.L.; Bader, D.; Hu, M.; Ganu, R.; Baquero, K.; Blundell, P.; Harris, R.A.; Frias, A.E.; Grove, K.L.; et al. High-fat maternal diet during pregnancy persistently alters the offspring microbiome in a primate model. *Nat. Commun.* **2014**, *5*, 3889. [CrossRef]
38. Xie, R.; Sun, Y.; Wu, J.; Huang, S.; Jin, G.; Guo, Z.; Zhang, Y.; Liu, T.; Liu, X.; Cao, X.; et al. Maternal High Fat Diet Alters Gut Microbiota of Offspring and Exacerbates DSS-Induced Colitis in Adulthood. *Front. Immunol.* **2018**, *9*, 2608. [CrossRef]
39. Cani, P.D.; de Vos, W.M. Next-Generation Beneficial Microbes: The Case of Akkermansia muciniphila. *Front. Microbiol.* **2017**, *8*, 1765. [CrossRef]
40. Li, J.; Zhao, F.; Wang, Y.; Chen, J.; Tao, J.; Tian, G.; Wu, S.; Liu, W.; Cui, Q.; Geng, B.; et al. Gut microbiota dysbiosis contributes to the development of hypertension. *Microbiome* **2017**, *5*, 14. [CrossRef]
41. Gomes, A.C.; Hoffmann, C.; Mota, J.F. The human gut microbiota: Metabolism and perspective in obesity. *Gut Microbes* **2018**, *9*, 308–325. [CrossRef] [PubMed]
42. DiRienzo, D.B. Effect of probiotics on biomarkers of cardiovascular disease: Implications for heart-healthy diets. *Nutr. Rev.* **2014**, *72*, 18–29. [CrossRef] [PubMed]
43. Kang, Y.; Cai, Y. Gut microbiota and hypertension: From pathogenesis to new therapeutic strategies. *Clin. Res. Hepatol. Gastroenterol.* **2018**, *42*, 110–117. [CrossRef] [PubMed]
44. Hsu, C.N.; Tain, Y.L. The Good, the Bad, and the Ugly of Pregnancy Nutrients and Developmental Programming of Adult Disease. *Nutrients* **2019**, *11*, 894. [CrossRef] [PubMed]
45. Razavi, A.C.; Potts, K.S.; Kelly, T.N.; Bazzano, L.A. Sex, gut microbiome, and cardiovascular disease risk. *Biol. Sex Differ.* **2019**, *10*, 29. [CrossRef] [PubMed]

Article

Lactobacillus reuteri V3401 Reduces Inflammatory Biomarkers and Modifies the Gastrointestinal Microbiome in Adults with Metabolic Syndrome: The PROSIR Study

Carmen Tenorio-Jiménez [1], María José Martínez-Ramírez [2,3], Isabel Del Castillo-Codes [4], Carmen Arraiza-Irigoyen [2], Mercedes Tercero-Lozano [4], José Camacho [5], Natalia Chueca [6,7], Federico García [6,7], Josune Olza [8], Julio Plaza-Díaz [7,8,9,10], Luis Fontana [7,8,9], Mónica Olivares [11], Ángel Gil [7,8,9,10] and Carolina Gómez-Llorente [7,8,9,10,*]

1 Endocrinology and Nutrition Clinical Management Unit, University Hospital Virgen de las Nieves, 18014 Granada, Spain
2 Endocrinology and Nutrition Clinical Management Unit, University Hospital of Jaén, 23007 Jaén, Spain
3 Department of Health Sciences, School of Health Sciences, University of Jaén, 23071 Jaén, Spain
4 Digestive Diseases Clinical Management Unit, University Hospital of Jaén, 23007 Jaén, Spain
5 Department of Signal Theory, Networking, and Communications, University of Granada, 18071 Granada, Spain
6 Department of Microbiology, University Hospital Campus de la Salud, 18016 Granada, Spain
7 Instituto de Investigación Biosanitaria ibs. GRANADA, 18012 Granada, Spain
8 Department of Biochemistry and Molecular Biology II, School of Pharmacy, University of Granada, 18071 Granada, Spain
9 Institute of Nutrition and Food Technology "José Mataix", Center of Biomedical Research, University of Granada, 18016 Granada, Spain
10 CIBEROBN (CIBER Physiopathology of Obesity and Nutrition), Instituto de Salud Carlos III, 28029 Madrid, Spain
11 Biosearch Life, 18004 Granada, Spain
* Correspondence: gomezll@ugr.es; Tel.: +34-958-241-000 (ext. 40092)

Received: 27 June 2019; Accepted: 29 July 2019; Published: 31 July 2019

Abstract: Previous studies have reported that probiotics may improve clinical and inflammatory parameters in patients with obesity and metabolic syndrome (MetS). *Lactobacillus (L.) reuteri* V3401 has shown promising results on the components of MetS in animal studies. We aimed to evaluate the effects of *L. reuteri* V3401 together with healthy lifestyle recommendations on adult patients with MetS. Methods: We carried out a randomized, crossover, placebo-controlled, single-center trial in which we included 53 adult patients newly diagnosed with MetS. Patients were block randomly allocated by body mass index (BMI) and sex to receive a capsule containing either the probiotic *L. reuteri* V3401 (5×10^9 colony-forming units) or a placebo once daily for 12 weeks. Anthropometric variables, biochemical and inflammatory biomarkers, as well as the gastrointestinal microbiome composition were determined. Results: There were no differences between groups in the clinical characteristics of MetS. However, we found that interleukin-6 (IL-6) and soluble vascular cell adhesion molecule 1 (sVCAM-1) diminished by effect of the treatment with *L. reuteri* V3401. Analysis of the gastrointestinal microbiome revealed a rise in the proportion of *Verrucomicrobia*. Conclusions: Consumption of *L. reuteri* V3401 improved selected inflammatory parameters and modified the gastrointestinal microbiome. Further studies are needed to ascertain additional beneficial effects of other probiotic strains in MetS as well as the mechanisms by which such effects are exerted.

Keywords: metabolic syndrome; gastrointestinal microbiome; *Lactobacillus reuteri* V3401; probiotics; obesity

1. Introduction

Obesity is a chronic disease, affecting developed and developing countries, that has multiple comorbidities and deteriorates quality of life. It is characterized by an increase of fat mass, which can consequently produce hypertrophy of the adipocytes, leading to an altered adipose tissue functionality. Individuals who are obese can develop an insulin resistance syndrome, also called metabolic syndrome (MetS). MetS is defined by insulin resistance, dyslipidemia, hypertension, and increased abdominal circumference, and it is associated with the development of type 2 diabetes (DM2), cardiovascular disease (CVD), and nonalcoholic fatty liver disease (NAFLD). This condition is associated with a two-fold increase in the risk of coronary heart disease, cerebrovascular disease, and a 1.5-fold increase in the risk of all-cause mortality [1], constituting a major public health challenge worldwide.

Nowadays, there is sound evidence linking the metabolic dysfunction seen in MetS to a proinflammatory state. Adipose tissue is, in part, responsible of this low-grade inflammatory state through the increasing release of proinflammatory molecules, such as leptin and tumor necrosis factor α (TNF-α), and the inhibition of adiponectin secretion, an anti-inflammatory adipokine [2]. In recent years, it has become evident that alteration of the gastrointestinal microbiome, also called gastrointestinal dysbiosis, may also contribute to the development of insulin resistance associated with obesity [3–5]. Furthermore, different studies have linked gastrointestinal dysbiosis with the development of obesity and other hallmarks of MetS [6,7]. In this sense, a decreased ratio of *Bacteroidetes/Firmicutes* has been described in individuals who are obese compared to normal-weight individuals [7]. Likewise, individuals with a low bacterial richness have more dyslipidemia, insulin resistance, inflammatory phenotype, and overall adiposity than individuals with high bacterial richness [8]. In addition, an aberrant gastrointestinal microbiome can promote subacute systemic inflammation, insulin resistance, and increased risk of CVD by mechanisms that include exposure to bacterial products, such as lipopolysaccharide (LPS), which is responsible for the metabolic endotoxemia related to MetS [9].

In the last years, treatment of the hallmarks of MetS with probiotics has emerged as a promising therapy. Probiotics are living microorganisms that confer health benefits to the host when administered in adequate amounts [10]. *Bifidobacterium* and *Lactobacillus* are the most frequently used genera of probiotics used in humans. Some of the beneficial effects of probiotics are mediated by their capacity to normalize the gastrointestinal microbiome, reinforce the gut barrier function composition [11,12], and their immunomodulatory actions [12,13]. Therefore, the addition of probiotics to a healthy diet could represent an interesting tool to fight obesity, MetS, and associated inflammation when used alongside dietary management and lifestyle modifications (e.g., increased physical activity). In this regard, some studies have found an improvement of anthropometric parameters and a decrease in inflammatory biomarkers in this disease after probiotic administration [14]. However, the beneficial effects of probiotics on MetS components are contradictory [15], probably because of the different probiotic strains, doses, and clinical study designs.

Lactobacillus (L.) reuteri V3401 strain, deposited in the Spanish Type Culture Collection (CECT) with accession number CECT 8695, was isolated from cow's raw milk on Mark, Rogosa and Sharper (MRS) agar medium, and 16S gene sequence analysis was carried out for its identification. In addition, its carbohydrate fermentation ability was characterized by the Analytical Profile Index (API) CH50 test [16]. This strain has been shown to reduce the absorption of fluoresterol, a fluorescent cholesterol analogue, by HT-29 human enterocytes [16]. Furthermore, Wistar rats fed a hypercholesterolemic diet supplemented with the probiotic strain for 57 days showed HDL levels similar to those of a healthy control group fed a standard diet [16]. Regarding glycemic levels, hypercholesterolemic animals supplemented with the probiotic strain showed similar values to those of normocholesterolemic mice, whereas animals under a hypercholesterolemic diet without the probiotic strain exhibited higher levels than normocholesterolemic mice [16]. Higher glucose levels are related to insulin resistance, which is normally associated with hypercholesterolemia and low HDL levels, both of them components of MetS. In this setting, supplementation with *L. reuteri* V3401 might offer an additional metabolic advantage together with healthy diet and exercise recommendations in patients with MetS.

All things considered, the present study aimed to evaluate whether the consumption of the probiotic strain *L. reuteri* V3401, together with healthy lifestyle (hypocaloric diet and physical activity) recommendations, was capable of improving MetS components. For this purpose, we designed a double-blind, crossover, placebo-controlled, single-center, randomized clinical trial (RCT).

2. Materials and Methods

2.1. Ethical Statement

All research and procedures performed during the study complied with the Declaration of Helsinki and the Guidelines of Good Clinical Practice. After receiving a complete verbal description of the study, patients signed a written informed consent. The study protocol was approved by the local Ethics Committee of both Granada and Jaén (references CEI-Jaén 25022016 and CEI-Granada 28022016, respectively).

2.2. Subjects and Experimental Design

We performed a randomized, double-blind, crossover, placebo-controlled, single-center trial in patients with a new diagnosis of MetS, according to the criteria of the International Diabetes Federation (IDF). The complete study design including sample size, randomization, and the trial protocol have been previously published [17] and registered at www.clinicaltrials.gov as NCT02972567. The study was conducted in agreement with the Standard Protocol Items: Recommendations for Interventional Trials (SPIRIT) guidelines.

Sample size was calculated based on the range and median value of lipopolysaccharide (LPS) and assuming a power of 80% and a significant level of 5% [17]. In brief, a total of 53 out of 60 adult patients were recruited at the Endocrinology and Nutrition Clinical Management Unit, University Hospital of Jaén (Jaén, Spain) by qualified personnel. Patients were block randomly allocated, by BMI and sex, in a 1:1 ratio to receive a capsule containing either the probiotic *L. reuteri* V3401 (5×10^9 colony-forming units) or the placebo (maltodextrin) once daily for 12 weeks.

Both capsules, probiotic and placebo, were provided by Biosearch Life (Granada, Spain). In addition, participants received an intensive lifestyle intervention program that included nutritional and physical counseling to achieve and maintain a 7% loss of initial body weight and increase moderate-intensity physical activity for at least 150 min/week. In Figure 1 we summarized the experimental design of the study.

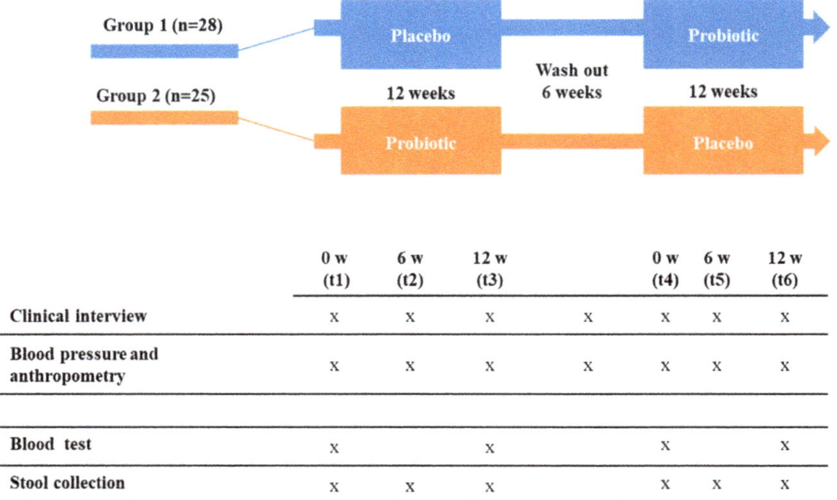

	0 w (t1)	6 w (t2)	12 w (t3)		0 w (t4)	6 w (t5)	12 w (t6)
Clinical interview	x	x	x	x	x	x	x
Blood pressure and anthropometry	x	x	x	x	x	x	x
Blood test	x		x		x		x
Stool collection	x	x	x		x	x	x

Figure 1. Illustration of the experimental design: A randomized, double-blind, crossover, placebo-controlled, single-center trial comparing the effect of consumption of *Lactobacillus reuteri* V3401 for 12 weeks on various clinical, biochemical, and inflammatory biomarkers and gastrointestinal microbiota. w: weeks; t: time; x: sample collection.

2.3. Anthropometric, Biochemical, Inflammatory, and Cardiovascular Data

We performed a systematic symptom evaluation on each visit, with special emphasis on gastrointestinal symptoms, and a physical examination. Body weight (kg), height (cm), and waist circumference (cm) were measured by the same person using standardized procedures. Blood pressure was taken 3 times by the same person, and the mean of the three values was included. The biochemical analyses, including lipid and glucose metabolism, were performed at the University Hospital of Jaén following internationally accepted quality control protocols. Homeostasis assessment model for insulin resistance (HOMA-IR) was calculated using fasting plasma glucose and insulin values.

Blood samples were collected from each patient and after 12 h of fasting, at the beginning and the end of each intervention period. Serum and plasma samples were collected by centrifugation of blood samples and kept at −80 °C until analysis.

Plasma adipokines as well as cardiovascular and inflammatory biomarkers—adiponectin, leptin, resistin, IL-6, IL-8, TNF-α, total plasminogen activator inhibitor-1 (PAI-1), hepatocyte growth factor (HGF), monocyte chemoattractant protein 1 (MCP-1), soluble intracellular adhesion molecule 1 (sICAM-1), soluble vascular cell adhesion molecule 1 (sVCAM-1), and myeloperoxidase (MPO)—were analyzed on a Luminex 200 system (Luminex Corporation, Austin, TX, USA) with human monoclonal antibodies (EMD Millipore Corp, Billerica, MA, USA) using MILLIplex™ kits (HADK1MAG-16K, HSTCMAG-28SK, HAD2MAG-61K, HCVD2MAG-67K) according to the manufacturer's recommendations.

LPS and LPS-binding protein (LBP) were determined in serum samples using CEB526GE and SEB406 HU ELISA kits (Cloud-Clone Corp, TX, USA), respectively, following the manufacturer's instructions.

2.4. Fecal Samples, DNA Extraction, and Next-Generation Sequencing

Fecal samples were collected from each patient at each time (t1, t2, t2, t4, t5, and t6). Fecal samples were placed inside of a sterile plastic bottle and kept at −80 °C until analysis. DNA was extracted using a QIAamp DNA stool Mini Kit (QIAGEN, Barcelona, Spain) according to the manufacturer's instructions, with the exception that samples were incubated with the lysis buffer at 95 °C instead of

70 °C to guarantee the lysis of both Gram-positive and Gram-negative bacteria. Extracted DNA samples were sequenced at facilities of the Department of Microbiology, University Hospital Campus de la Salud (Granada, Spain). A 16S metagenomics sequencing was performed following the Illumina protocol.

In summary, the V3-V4 region of the bacterial 16S rRNA gene was amplified using the primers described by Klindworth et al., 2013 [18]. The PCR mixture was composed of 5 μL for each forward and reverse primers (1 μM, Macrogen, Seoul, Korea), 2.5 μL of DNA template samples, and 12.5 μL of 1x Hot Master Mix (KAPA HiFi HS RM, Roche, Basilea, Switzerland) to a final volume of 25 μL. Five microliters of elution solution was used for the negative control. The PCR conditions were: initial denaturation at 95 °C for 3 min, followed by 25 cycles of denaturation at 95 °C for 30 s, primer annealing at 55 °C for 30 s, extension at 72 °C for 30 s, and a final elongation at 72 °C for 5 min. The PCR products were demonstrated by electrophoresis on a 2% agarose gel. No amplification product was observed in the negative control. The amplifications were subjected to purification using Ampure beads (Agencourt Bioscience, La Jolla, CA, USA), the eluted DNA product was quantified using the assays of the Qubit kit (Invitrogen, Life Technologies, Waltham, Massachisetts, USA), and then all samples were pooled in equal concentrations for sequencing. Bioanalyzer 2100 was used with the DNA 1000 Chip kit (Agilent, Palo Alto, CA, USA) to evaluate the quality of the final products for each sample individually. Sequencing was carried out using Illumina MiSeq paired-end sequencing in an Illumina MiSeq device (Illumina Inc., San Diego, CA, USA) with 600 cycles (300 cycles for each paired reading and 12 cycles for the sequence of bar codes) according to the manufacturer's instructions. Sequence analysis was performed using the metagenomic workflow based on 16S of MiSeq Reporter v2.3 (Illumina Inc., San Diego, CA, USA).

2.5. Taxonomic Analysis

The "Quantitative Insights Into Microbial Ecology 2" (QUIIME 2) package was used to analyze sequence data [19]. Denoising quality, chimera check, and clustering were performed using the DADA2 plugins implemented in QUIIME 2. Amplicon sequence variants (ASVs) with a relative proportion lower than 0.1% were eliminated; as a result, the total numbers of ASV were reduced to 2015 but with a very low impact on the total data. The GreenGenes database (version 13.8), together with the naïve Bayes algorithm, was used as the reference 16S database.

2.6. Statistical Analysis

For the anthropometric, biochemical, and inflammatory biomarkers, results are presented as the mean values ± standard deviation (SD), unless otherwise indicated. For those variables not following a normal distribution, we applied the logarithmic transformation (insulin, HOMA index, glycated hemoglobin, total cholesterol, triacylglycerols, alanine aminotransferase (GPT), gamma glutamiltransferase (γGT), C reactive protein (CRP), IL-6, IL-8, adiponectin, resistin, HGF, sICAM, sVCAM and LBP) or the inverse transformation (high-density lipoprotein (HDL), aspartate aminotransferase (GOT).

Only patients with less than 5 missing data were considered, resulting in a final number of 34 patients. Missing data in these patients were imputed using principal component analysis (PCA) and trimmed score regression (TSR) [20]. The treatment effect in anthropometric, biochemical, and inflammatory biomarkers was evaluated according to the approach described by Wellek et al. [21]. Two tests were carried out: (i) a pretest for significance of carryover effects, and (ii) a test for significance of treatment effects. The treatment effects were considered significant for those biomarkers for which the null hypothesis of the pretest was not rejected and the null hypothesis of the test was rejected ($p < 0.05$), confirming that the biomarker presented statistically significant differences only due to treatment effects. p-value computations were confirmed with different state-of-the-art multivariate approaches, including multivariate analysis of variance (MANOVA) [22], partial least-squares discriminant analysis (PLS-DA) [23], and ANOVA simultaneous component analysis (ASCA) [24]. TheMEDA toolbox (https://github/josecamachop/MEDA-Toolbox) and the MANCOVAN toolbox (http://www.mathworks.

com/matlabcentral/fileexchange/27014-mancovan) in Matlab (Mathworks) were used to perform the statistical analysis.

For the gastrointestinal microbiota analysis, the generated sequences, ASV, were normalized by means of the rarefaction method (Figure S1). The alpha diversity was measured by means of the Shannon index, whereas the unique fraction metric (Unifrac), both weighted and unweighted, was used to determine the beta diversity. When comparing the incremental of relative bacteria proportions before and after treatment (delta), a pairwise Wilcoxon signed-rank test was used. p-values were adjusted by False discovery rate-FDR (q-values).

3. Results

3.1. Anthropometric, Biochemical, and Inflammatory Data

Anthropometric and biochemical characteristics of the subjects are described in Table 1, whereas in Table 2 the inflammatory biomarkers determined in blood samples are described.

In the case of BMI, diastolic blood pressure, GOT, and LBP, we found that the washout period was not long enough to avoid the carryover effects. We found significant differences for Il-6, sVCAM (Figure 2), and insulin levels (Table 1); however, we did not find any significant results for HOMA index (Table 1).

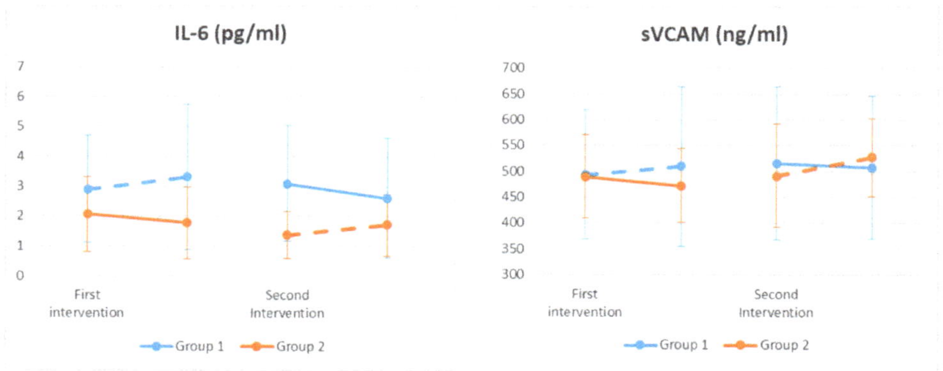

Figure 2. Inflammatory biomarkers throughout the study. The levels of interleukin 6 (IL-6) and soluble vascular cell adhesion molecule 1 (sVCAM) were modified by the probiotic consumption ($p < 0.05$). Continuous line: probiotic group. Discontinuous line: placebo group.

3.2. Gastrointestinal Microbiome Composition

We characterized the gastrointestinal microbiome composition of the participants at the beginning, middle, and end of each intervention period (Table S1). However, we were unable to determine the specific presence of the *L. reuteri* V3401 strain in fecal samples due to the lack of specific primers for this strain. As shown in Figure 3, at the beginning of the intervention, the most abundant phyla were *Firmicutes* and *Bacteroidetes* followed by *Proteobacteria*, *Actinobacteria*, *Verrucomicrobia*, and *Cyanobacteria*.

Table 1. Anthropometric and biochemical characteristics of the patients.

	Group 1				Group 2			
	Placebo		Probiotic		Probiotic		Placebo	
	t1	t3	t4	t6	t1	t3	t4	t6
Weight (kg)	109.02 ± 26.7	105.70 ± 26.2	101.50 ± 24.5	101.08 ± 24.2	103.49 ± 15.2	96.56 ± 16.2	93.91 ± 16.9	92.0 ± 17.3
BMI (kg/m^2)	38.76 ± 7.2	37.57 ± 7.1	36.77 ± 6.8	36.56 ± 6.6	38.30 ± 7.3	35.69 ± 7.1	34.57 ± 6.9	33.80 ± 6.6
SBP (mm Hg)	137.68 ± 16.9	133.28 ± 15.4	133.11 ± 20.4	132.21 ± 14.6	139 ± 22.6	129.95 ± 16.0	131.30 ± 20.0	131.40 ± 18.6
DBP (mm Hg)	84.28 ± 9.6	81.96 ± 7.7	81.68 ± 11.0	82.11 ± 10.5	87.95 ± 14.3	78.18 ±10.4	78.85 ± 12.4	81.60 ± 11.2
Glucose (mg/dL)	103.29 ± 11.0	108.08 ± 11.5	106.74 ± 8.9	105.53 ± 10.5	101.0 ± 13.9	103.68 ± 13.4	101.22 ± 11.8	103.78 ± 16.5
Insulin (mU/mL)	17.50 ± 10.6	16.18 ± 11.3	22.44 ± 10.3	21.74 ± 11.7	14.24 ± 8.5	12.42 ± 10.9	14.04 ± 6.1	17.47 ± 7.8
HOMA index	4.48 ± 2.8	4.41 ± 3.3	5.91 ± 2.8	5.66 ± 3.5	3.71 ± 2.7	3.52 ± 3.6	3.64 ± 1.7	4.46 ± 2.2
Glycated Hemoglobin (%)	5.59 ± 0.4	6.04 ± 2.2	5.44 ± 0.3	5.49 ± 0.3	5.68 ± 0.4	5.90 ± 1.9	5.46 ± 0.3	5.44 ± 0.3
Total cholesterol (mg/dL)	232.42 ± 43.0	207.08 ± 36.0	202.79 ± 45.8	224.16 ± 45.5	233.41 ± 46.5	203.64 ± 37.9	209.56 ± 58.0	220.89 ± 53.8
Triacylglycerols (mg/dL)	119.25 ± 47.6	122.46 ± 59.9	109.00 ± 47.3	118.89 ± 52.2	130.55 ± 47.5	128.23 ± 57.6	112.89 ± 42.7	100.56 ± 62.6
LDL (mg/dL)	156.79 ± 35.7	131.71 ± 30.0	128.11 ± 32.5	144.42 ± 39.0	161.0 ± 41.6	132.91 ± 32.1	136.22 ± 47.7	145.72 ± 44.5
HDL (mg/dL)	50.54 ± 14.6	50.46 ± 12.2	52.47 ± 13.3	54.11 ± 10.2	45.95 ± 9.5	44.68 ± 7.9	50.44 ± 10.3	54.61 ± 11.5
GOT (U/L)	25.75 ± 7.7	22.83 ± 6.3	22.63 ± 6.8	22.32 ± 5.7	23.55 ± 12.4	25.50 ± 14.9	20.56 ± 6.1	21.33 ± 6.2
GPT (U/L)	34.79 ± 17.4	28.17 ± 14.3	22.42 ± 11.9	27.89 ± 12.7	24.38 ± 8.9	22.24 ± 8.4	18.00 ± 7.5	22.50 ± 11.0
γ-GT (U/L)	36.29 ± 13.6	37.08 ± 16.7	38.42 ± 18.9	36.74 ± 21.2	26.05 ± 12.7	24.91 ± 14.2	24.89 ± 14.3	24.00 ± 15.4

Values are expressed as means ± SD. BMI: body mass index; SBP: systolic blood pressure; DBP: diastolic blood Pressure; HOMA-IR: Homeostasis assessment model for insulin resistance; LDL: low-density lipoprotein; HDL: high-density lipoprotein; GOT: aspartate aminotransferase; GPT: alanine aminotransferase; γ-GT: gamma glutamyltransferase.

Table 2. Inflammatory biomarkers.

	Group 1				Group 2			
	Placebo		Probiotic		Probiotic		Placebo	
	t1	t2	t3	t5	t1	t2	t3	t5
CRP (mg/dL)	5.13 ± 3.7	5.78 ± 4.8	6.97 ± 7.4	5.52 ± 4.4	5.88 ± 4.4	6.12 ± 6.3	3.66 ± 2.7	4.24 ± 3.8
IL-6 (pg/mL)	2.91 ± 1.8	3.33 ± 2.4	3.12 ± 1.9	2.62 ± 2.0	2.07 ± 1.2	1.79 ± 1.2	1.37 ± 0.8	1.72 ± 1.0
IL-8 (pg/mL)	2.86 ± 1.7	2.80 ± 1.2	4.11 ± 7.1	4.23 ± 9.4	2.66 ± 1.1	2.73 ± 1.2	2.27 ± 1.0	2.28 ± 1.1
TNF-α (pg/mL)	4.70 ± 2.5	4.91 ± 2.6	4.59 ± 2.1	3.51 ± 1.7	4.05 ± 1.9	4.15 ± 2.1	3.05 ± 1.2	3.28 ± 2.0
Adiponectin (mg/L)	6.55 ± 5.1	5.95 ± 4.7	5.82 ± 3.8	6.56 ± 3.4	5.69 ± 3.7	6.91 ± 6.5	7.20 ± 4.4	8.26 ± 6.1
tPAI1 (μg/L)	9.24 ± 4.9	9.55 ± 4.3	10.55 ± 4.8	11.56 ± 7.1	9.31 ± 3.2	8.36 ± 2.6	9.08 ± 3.2	9.51 ± 3.6
P-selectin (ng/mL)	46.78 ± 19.7	46.70 ± 21.2	49.26 ± 22.8	63.47 ± 38.3	48.06 ± 16.7	40.51 ± 11.4	60.73 ± 35.4	58.09 ± 21.6
Resistin (μg/L)	17.71 ± 8.1	17.70 ± 13.7	16.99 ± 5.2	17.45 ± 10.4	15.33 ± 7.2	11.89 ± 5.2	11.08 ± 4.0	12.60 ± 5.2
HGF (pg/mL)	161.12 ± 97.9	155.73 ± 93.6	131.45 ± 65.9	162.45 ± 88.3	175.06 ± 75.1	170.17 ± 72.7	160.47 ± 68.8	157.95 ± 57.5
Leptin (μg/L)	28.56 ± 14.8	24.07 ± 12.3	23.71 ± 13.8	18.42 ± 10.5	21.97 ± 11.8	17.68 ± 10.9	13.67 ± 9.3	14.24 ± 10.4
MCP-1 (pg/mL)	107.53 ± 39.3	106.57 ± 31.7	120.07 ± 60.1	118.20 ± 46.8	108.86 ± 39.3	114.61 ± 51.5	116.31 ± 43.6	112.51 ± 41.2
sICAM (ng/mL)	73.65 ± 37.2	73.71 ± 41.1	67.86 ± 35.2	65.47 ± 32.2	74.80 ± 26.9	71.50 ± 33.1	75.17 ± 40.2	73.0 ± 41.2
MPO (ng/mL)	17.69 ± 5.9	19.96 ± 10.7	20.70 ± 13.2	30.53 ± 21.2	15.56 ± 7.8	18.14 ± 13.0	17.31 ± 12.4	19.46 ± 10.2
sVCAM (ng/mL)	494.22 ± 125.1	511.04 ± 154.8	516.47 ± 149.1	507.61 ± 138.7	489.68 ± 80.9	472.72 ± 71.4	491.62 ± 99.6	527.43 ± 74.6
LPS (ng/mL)	285.81 ± 107.2	277.90 ± 116.3	312.22 ± 126.0	326.19 ± 166.0	321.82 ± 105.4	316.83 ± 124.0	309.91 ± 136.7	308.70 ± 131.4
LBP (ng/mL)	731.26 ± 512.2	782.55 ± 323.0	635.96 ± 294.2	747.39 ± 272.6	837.47 ± 423.5	742.81 ± 349.2	833.63 ± 560.5	855.78 ± 663.1

Values are expressed as means ± SD. CRP: C reactive protein; IL: interleukin; TNF-α: tumor necrosis factor alpha; tPAI1: plasminogen activator inhibitor-1; HGF: hepatocyte growth factor; MCP-1: monocyte chemoattractant protein 1; sICAM: soluble intracellular adhesion molecules 1; sVCAM: soluble vascular cell adhesion molecule 1; MPO: myeloperoxidase; LPS: lipopolysaccharide; LBP: lipopolysaccharide-binding protein.

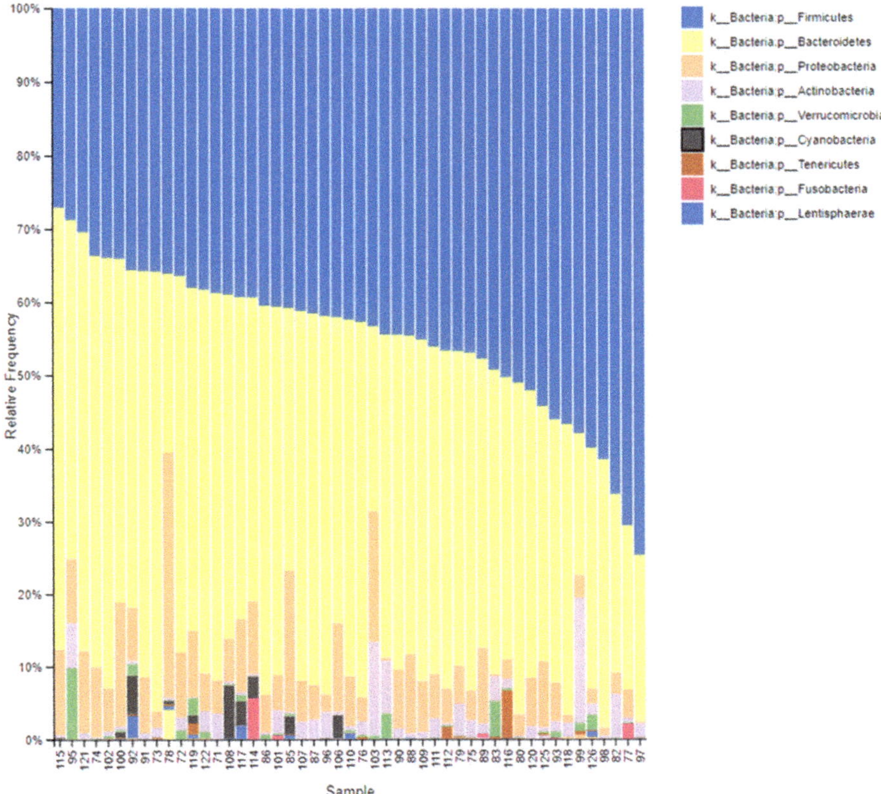

Figure 3. Baseline gastrointestinal microbiome composition. Taxonomic composition of the gastrointestinal communities at the beginning of the intervention. The figure shows bar charts of the relative abundance of bacteria at the phylum level. Each column represents a participant.

Regarding the bacterial diversity, we did not find significant differences in the alpha diversity throughout the study, measured as the Shannon index (H) (Figure S2), or in the beta diversity (Figure S3). Therefore, our next analysis was to determine the evolution of the relative proportion of specific taxa, namely *Firmicutes*, *Bacteroidetes*, *Verrucomicrobia*, *Actinobacteria*, *Proteobacteria*, *Fusobacteria*, *Cyanobacteria*, *Elusimicrobia*, *Tenericutes*, and *Lentisphaerae*. It is worth mentioning there was an increase in the relative proportion of the *Verrucomicromia* phylum in the participants that consumed the probiotic strain (Figure 4A). The same results were found in the *Akkermansia* genus (Figure 4B).

Based on the results described above, we decided to determine whether there were significant differences in the relative abundance of these taxa due to the treatment (probiotic versus placebo). We, therefore, performed a pairwise comparison [25]. During the first intervention (t1, t2, and t3) we observed a significant increase in the delta values (t3–t1) in the *Verrucomicrobia* phylum due to the treatment (probiotic versus placebo). However, during the crossover intervention, the differences (t6–t4) were not statistically significant, although we found a significant trend (FDR $p = 0.07$) (Figure 5).

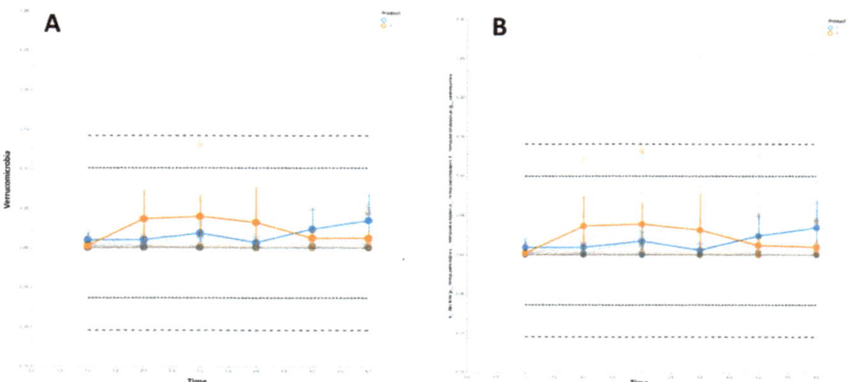

Figure 4. Trend of the relative proportion of *Verrucomicrobia* phylum and *Akkermansia* genus. The thick lines with error bars represent the means of both groups (Group 1: blue lines; group 2: orange lines). Dashed lines represent the means ± 2 and 3 standard deviations. Group 1 started the intervention receiving the placebo (t1, t2, and t3) and then was switched to receive the probiotic strain (t4, t5, and t6). Group 2 started the intervention receiving the probiotic strain (t1, t2, and t3) and then was switched to receive the placebo (t4, t5, and t6). (**A**) Temporal trend of *Verrucomicrobia* phylum; (**B**) Temporal trend of *Akkermansia* genus.

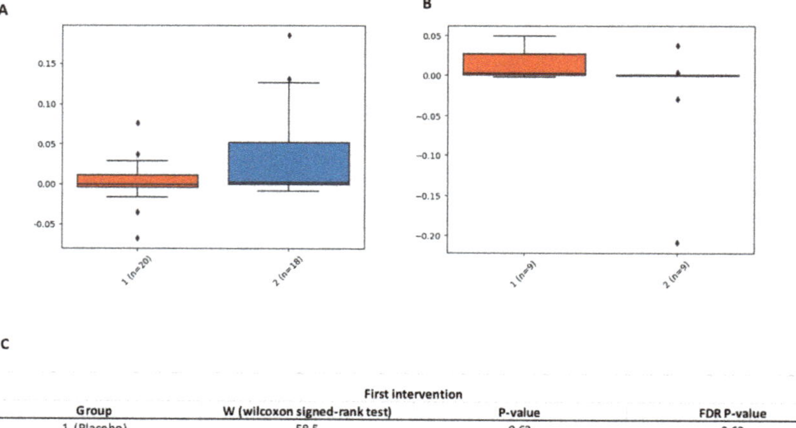

Group	W (wilcoxon signed-rank test)	P-value	FDR P-value
	First intervention		
1 (Placebo)	58.5	0.62	0.62
2 (*L. reuteri* V3401)	17.5	0.02	0.03
	Second intervention		
Group	**W (wilcoxon signed-rank test)**	**P-value**	**FDR P-value**
1 (*L. reuteri* V3401)	3	0.04	0.07
2 (Placebo)	9.5	0.83	0.83

Figure 5. Relative abundance of the delta values of *Verrucomicrobia* phylum. The box plots indicate the relative abundance of *Verrucomicrobia* phylum between the final and the beginning points of each intervention—t3-t1 (**A**) and t6-t4 (**B**)—due to the treatment and the number of patients (n) of each group and treatment. (**A**) shows data for the first part of this intervention study (First intervention, 1: placebo; 2: *L. reuteri* V3401), whereas (**B**) shows data from the intervention after the crossover (Second intervention, 1: *L. reuteri* V3401; 2: placebo). (**C**) describes the Wilcoxon signed-rank test values and the significant levels by means of *p* and FDR *p*-values. FDR: False discovery rate; ♦: outliers data values.

4. Discussion

To the best of our knowledge, the PROSIR study is the first randomized, crossover clinical trial in humans that evaluates whether the strain *L. reuteri* V3401 is capable of improving the components of MetS in humans when added to a healthy lifestyle. We did not find any differences in the clinical features of the syndrome between groups. This may be due to the fact that all subjects included in the study lost weight and improved their metabolic status as a result of the counseling to follow a healthy lifestyle that included diet and physical activity. However, we did find a decrease in IL-6 and sVCAM levels in patients who consumed the probiotic strain, together with a modification of the gastrointestinal microbiome, in particular, an increase in the *Verrucomicrobia* phylum.

Other studies have shown that consumption of *Lactobacillus casei* Shirota reduces sVCAM-1 levels in individuals who suffer from MetS, although in this study no effects on insulin sensitivity, endothelial function, or the inflammatory biomarkers were observed [26]. Bernini et al. [14] showed in another work that consumption of fermented milk enriched with *Bifidobacterium lactis* HN019 resulted in a reduction in BMI, an improvement in the lipid profile, and a significant decrease in proinflammatory cytokines (TNF-α and IL-6).

Systemic low-grade inflammation has an important role in the development of MetS. In this sense, IL-6 is a cytokine that has been associated with insulin resistance. Specifically, IL-6 is able to induce insulin resistance in both liver and adipocytes through reduction of phosphorylation of the insulin receptor substrate (IRS), or by transcription inhibition of the IRS [27,28]. In addition, adhesion molecules, such as sVCAM-1, are necessary for normal development and function of the heart and blood vessels, and they have been related to the development of CVD [29]. In the adult Spaniard population, impaired glucose metabolism has been related to increased levels of sVCAM-1 [30].

Results regarding the utility of probiotics in the treatment of MetS have been contradictory. This may be due to various facts: (i) the particular probiotic strain used in each trial; (ii) the experimental design—most of the studies have been parallel-group randomized trials, whereas a crossover study is a more appropriated approach to determine health benefits of clinical interventions; in crossover studies each participant serves as their own control, but in addition, this clinical design demands a lower sample size than parallel-group studies [21]; and (iii) the duration of the treatment. In our study, 12 weeks might not have been a long enough treatment to reverse, or at least improve, a chronic proinflammatory state as the one observed in MetS.

In recent years, it has become clear that the gut microbiota plays a role in the development of MetS. Specific bacterial groups have been described to be involved in obesity and related metabolic diseases. Among these bacteria, *Akkermansia muciniphila* has been proposed as a contributor to the maintenance of gut health and glucose homeostasis [31]. Administration of *A. muciniphila* to diet-induced obese animals improve their metabolic endotoxemia adipose tissue inflammation and insulin resistance [32]. It is worth mentioning that *Akkermansia* was the only genus of the *Verrucomicrobia* phylum present in the gastrointestinal samples [33]. In humans, *A. muciniphila* has been found to be decreased in prediabetic patients compared to normal glucose tolerance subjects [34]. Conversely, other studies have shown an increase of *A. muciniphila* in type 2 diabetes [35]. More recently, *A. muciniphila* has been reported to be associated with a healthy metabolic status in overweight and obese individuals, in agreement with previous results in murine studies [31]. Additionally, higher *A. muciniphila* abundance has been described in subjects with high bacterial gene richness, which is associated with a healthier metabolic status, in French and Danish population [8,31]. Although we found an increase in the proportion of the *Verrucomicrobia* phylum in the group that received the probiotic, we did not find any significant correlation between the delta proportions of *Verrucomicrobia* and any inflammatory biomarker.

It is important to highlight the age of the subjects that participated in our study. Our patients were younger compared to other previously published trials [14,15,36], and still we observed a decrease of inflammatory biomarkers and an increase in the abundance of *Verrucomicrobia* phylum in this pretreatment phase of the disease. This is consistent in both intervention periods, although in the case of gastrointestinal microbiome, the results only showed a trend in the second intervention, probably because of the dropout number we had, which is usual in large clinical intervention studies.

5. Conclusions

In conclusion, our data point to a beneficial effect of supplementation with *L. reuteri* V3401 in subjects with MetS when added to a hypocaloric diet and regular physical activity. In particular, these effects may be mediated by an improvement of dysbiosis and a decreased proinflammatory state, both features of this condition. However, further studies with longer periods of intervention are needed, in animals and clinical studies, to confirm these results and to elucidate the underlying mechanisms of action.

Supplementary Materials: The following are available online at http://www.mdpi.com/2072-6643/11/8/1761/s1: Figure S1, Rarefaction curves; Figure S2, Alpha diversity measured by means of the Shannon index (H); Figure S3, Bacterial beta diversity; Table S1, Gastrointestinal microbiome normalized data.

Author Contributions: M.J.M.-R., M.O., Á.G. and C.G.-L. conceived and designed the study. C.T.-J., M.J.M.-R., I.D.C.-C., M.T.-L. and C.A.-I. were responsible for the clinical intervention and the follow-up of the patients, as part of the clinical team. J.P.-D., J.O., N.C. and F.G. performed the analysis. J.O., J.C., Á.G., C.T.-J., L.F. and C.G.-L. reviewed and did the statistics. C.T.-J. and C.G.-L. wrote the manuscript. L.F. revised the English and helped in writing the manuscript. All the authors read and approved the final manuscript.

Funding: This study is part of the grant entitled "Guía para la sustanciación de declaraciones de salud en alimentos: funciones inmune, cognitiva y síndrome metabólico", funded by the company Biosearch life (Granada, Spain), reference 3006, managed by Fundación General Universidad de Granada, Spain.

Acknowledgments: This paper will be part of Carmen Tenorio Jimenez's doctorate, which is being completed as part of Programa de Doctorado en Seguridad de los Alimentos at the University of Jaén, Spain. We also wish to thank all the participants who generously contributed to the study.

Conflicts of Interest: M.O. is Head of the Department of Research and Development at Biosearch Life. The funders had no role in the recruitment, biological sample analysis, statistical analysis, data interpretation, writing of the manuscript, nor in the decision to publish the results.

References

1. Engin, A. The Definition and Prevalence of Obesity and Metabolic Syndrome. *Results Probl. Cell Differ.* **2017**, *960*, 1–17.
2. Fasshauer, M.; Blüher, M. Adipokines in health and disease. *Trends Pharmacol. Sci.* **2015**, *36*, 461–470. [CrossRef] [PubMed]
3. Bäckhed, F.; Ding, H.; Wang, T.; Hooper, L.V.; Koh, G.Y.; Nagy, A.; Semenkovich, C.F.; Gordon, J.I. The gut microbiota as an environmental factor that regulates fat storage. *Proc. Natl. Acad. Sci. USA* **2004**, *101*, 15718–15723. [CrossRef] [PubMed]
4. Bäckhed, F.; Manchester, J.K.; Semenkovich, C.F.; Gordon, J.I. Mechanisms underlying the resistance to diet-induced obesity in germ-free mice. *Proc. Natl. Acad. Sci. USA* **2007**, *104*, 979–984. [CrossRef] [PubMed]
5. Turnbaugh, P.J.; Ley, R.E.; Mahowald, M.A.; Magrini, V.; Mardis, E.R.; Gordon, J.I. An obesity-associated gut microbiome with increased capacity for energy harvest. *Nature* **2006**, *444*, 1027–1031. [CrossRef] [PubMed]
6. Zupancic, M.L.; Cantarel, B.L.; Liu, Z.; Drabek, E.F.; Ryan, K.A.; Cirimotich, S.; Jones, C.; Knight, R.; Walters, W.A.; Knights, D.; et al. Analysis of the gut microbiota in the older order Amish and its relation to the metabolic syndrome. *PLoS ONE* **2012**, *7*, e43052. [CrossRef] [PubMed]
7. Turnbaugh, P.J.; Hamady, M.; Yatsunenko, T.; Cantarel, B.L.; Duncan, A.; Ley, R.E.; Sogin, M.L.; Jones, W.J.; Roe, B.A.; Affourtit, J.P. A core gut microbiome in obese and lean twins. *Nature* **2009**, *457*, 480–484. [CrossRef] [PubMed]
8. Le Chatelier, E.; MetaHIT Consortium; Nielsen, T.; Qin, J.; Prifti, E.; Hildebrand, F.; Falony, G.; Almeida, M.; Arumugam, M.; Batto, J.M.; et al. Richness of human gut microbiome correlates with metabolic markers. *Nature* **2013**, *500*, 541–546. [CrossRef]
9. Cani, P.D.; Bibiloni, R.; Knauf, C.; Waget, A.; Neyrinck, A.M.; Delzenne, N.M.; Burcelin, R. Changes in gut microbiota control metabolic endotoxemia-induced inflammation inhigh-fat diet induced obesity and diabetes in mice. *Diabetes* **2008**, *57*, 1470–1481. [CrossRef]
10. Sanders, M.E. Probiotics: Definition, sources, selection, and uses. *Clin. Infact. Dis.* **2008**, *46*, S58–S61. [CrossRef]

11. Karczewski, J.; Troost, F.J.; Konings, I.; Dekker, J.; Kleerebezem, M.; Brummer, R.-J.M.; Wells, J.M. Regulation of human epithelial tight junction proteins by Lactobacillus plantarum in vivo and protective effects on the epithelial barrier. *Am. J. Physiol. Liver Physiol.* **2010**, *298*, G851–G859. [CrossRef] [PubMed]
12. Miglioranza Scavuzzi, B.; Miglioranza, L.H.; Henrique, F.C.; Pitelli Paroschi, T.; Lozovoy, M.A.; Simao, A.N.; Dichi, I. The role of probiotics on each component of the metabolic syndrome and other cardiovascular risks. *Expert Opin. Ther. Targets* **2015**, *19*, 1127–1138. [CrossRef] [PubMed]
13. Ren, C.; Zhang, Q.; De Haan, B.J.; Zhang, H.; Faas, M.M.; De Vos, P. Identification of TLR2/TLR6 signalling lactic acid bacteria for supporting immune regulation. *Sci. Rep.* **2016**, *6*, 34561. [CrossRef] [PubMed]
14. Bernini, L.J.; Simão, A.N.; Alfieri, D.F.; Lozovoy, M.A.; Mari, N.L.; de Souza, C.H.; Dichi, I.; Costa, G.N. Beneficial effects of Bifidobacterium lactis on lipid profile and cytokines in patients with metabolic syndrome: A randomized trial. Effects of probiotics on metabolic syndrome. *Nutrition* **2016**, *32*, 716–719. [CrossRef] [PubMed]
15. Barreto, F.M.; Simão, A.N.; Morimoto, H.K.; Lozovoy, M.A.; Dichi, I.; Miglioranza, H.D.S. Beneficial effects of Lactobacillus plantarum on glycemia and homocysteine levels in postmenopausal women with metabolic syndrome. *Nutrition* **2014**, *30*, 939–942. [CrossRef] [PubMed]
16. Sañudo Otero, A.I.; Criado García, R.; Rodríguez Nogales, A.; Garach Domech, A.; Olivares Martín, M.; Gálvez Peralta, J.J.; De La Escalera Hueso, S.; Duarte Pérez, J.M.; Zarzuelo Zurita, A.; Bañuelos Hortigüela, O. Probiotic Strains Having Cholesterol Absorbing Capacity, Methods and Uses Thereof. Google Patents No. EP3031930A1, 15 June 2016.
17. Tenorio-Jiménez, C.; Martínez-Ramírez, M.J.; Tercero-Lozano, M.; Arraiza-Irigoyen, C.; Del Castillo-Codes, I.; Olza, J.; Plaza-Díaz, J.; Fontana, L.; Migueles, J.H.; Olivares, M.; et al. Evaluation of the effect of Lactobacillus reuteri V3401 on biomarkers of inflammation, cardiovascular risk and liver steatosis in obese adults with metabolic syndrome: A randomized clinical trial (PROSIR). *BMC Complement. Altern. Med.* **2018**, *18*, 306. [CrossRef]
18. Klindworth, A.; Pruesse, E.; Schweer, T.; Peplies, J.; Quast, C.; Horn, M.; Glöckner, F.O. Evaluation of general 16S ribosomal RNA gene PCR primers for classical and next-generation sequencing-based diversity studies. *Nucleic Acids Res.* **2013**, *7*, e1. [CrossRef]
19. Bolyen, E.; Rideout, J.R.; Dillon, M.R.; Bokulich, N.A.; Abnet, C.; Al-Ghalith, G.A.; Alexander, H.; Alm, E.J.; Arumugam, M.; Asnicar, F.; et al. QUIIME 2: Reproducible, interactive, scalable, and extensible microbiome data science. *Peer J. Prepr.* **2018**, *6*, e27295v2.
20. Arteaga, F.; Ferrer, F. Dealing with missing data in MSPC: Several methods, different interpretations, some examples. *J. Chemom.* **2002**, *16*, 408–418. [CrossRef]
21. Wellek, S.; Blettner, M. On the proper use of the crossover design in clinical trials: Part 18 of a series on evaluation of scientific publications. *Dtsch. Arztebl. Int.* **2012**, *109*, 276–281. [CrossRef]
22. Bibby, J.; Kent, J.; Mardia, K. *Multivariate Analysis*; Academic Press: London, UK, 1979.
23. Barker, M.; Rayens, W. Partial least squares for discrimination. *J. Chemom.* **2003**, *17*, 166–173. [CrossRef]
24. Smilde, A.K.; Jansen, J.J.; Hoefsloot, H.C.; Lamers, R.J.; Van Der Greef, J.; Timmerman, M.E. ANOVA-simultaneous component analysis (ASCA): A new tool for analyzing designed metabolomics data. *Bioinformatics* **2005**, *21*, 3043–3048. [CrossRef] [PubMed]
25. Bokulich, N.A.; Dillon, M.R.; Zhang, Y.; Rideout, J.R.; Bolyen, E.; Li, H.; Albert, P.S.; Caporaso, J.G. q2-longitudinal: Longitudinal and Paired-Sample Analyses of Microbiome Data. *mSystems* **2018**, *3*, e00219-18. [CrossRef] [PubMed]
26. Tripolt, N.; Leber, B.; Blattl, D.; Eder, M.; Wonisch, W.; Scharnagl, H.; Stojakovic, T.; Obermayer-Pietsch, B.; Wascher, T.; Pieber, T.; et al. Short communication: Effect of supplementation with Lactobacillus casei Shirota on insulin sensitivity, β-cell function, and markers of endothelial function and inflammation in subjects with metabolic syndrome—A pilot study. *J. Dairy Sci.* **2013**, *96*, 89–95. [CrossRef] [PubMed]
27. Jaganathan, R.; Ravindran, R.; Dhanasekaran, S. Emerging Role of Adipocytokines in Type 2 Diabetes as Mediators of Insulin Resistance and Cardiovascular Disease. *Can. J. Diabetes* **2018**, *42*, 446–456.e1. [CrossRef] [PubMed]
28. Weigert, C.; Hennige, A.M.; Lehmann, R.; Brodbeck, K.; Baumgartner, F.; Schaüble, M.; Häring, H.U.; Schleicher, E.D. Direct Cross-talk of Interleukin-6 and Insulin Signal Transduction via Insulin Receptor Substrate-1 in Skeletal Muscle Cells. *J. Biol. Chem.* **2006**, *281*, 7060–7067. [CrossRef]

29. Glowinska, B.; Urban, M.; Peczynska, J.; Florys, B.; Głowińska-Olszewska, B. Soluble adhesion molecules (sICAM-1, sVCAM-1) and selectins (sE selectin, sP selectin, sL selectin) levels in children and adolescents with obesity, hypertension, and diabetes. *Metabolism* **2005**, *54*, 1020–1026. [CrossRef]
30. Guzmán-Guzmán, I.P.; Zaragoza-García, O.; Vences-Valázquez, A.; Castro-Alarcón, N.; Muñoz-Valle, J.F.; Parra-Rojas, I. Circulating levels of MCP-1, VEGF-A, sICAM-1, sVCAM-1, sE-selectin and sVE-cadherin: Relationship with components of metabolic síndrome in young population. *Med. Clin. (Bar)* **2016**, *147*, 427–434. [CrossRef]
31. Dao, M.C.; Everard, A.; Aron-Wisnewsky, J.; Sokolovska, N.; Prifti, E.; Verger, E.O.; Kayser, B.D.; Levenez, F.; Chilloux, J.; Hoyles, L.; et al. Akkermansia muciniphila and improved metabolic health during a dietary intervention in obesity: Relationship with gut microbiome richness and ecology. *Gut* **2016**, *65*, 426–436. [CrossRef]
32. Everard, A.; Belzer, C.; Geurts, L.; Ouwerkerk, J.P.; Druart, C.; Bindels, L.B.; Guiot, Y.; Derrien, M.; Muccioli, G.G.; Delzenne, N.M.; et al. Cross-talk between Akkermansia muciniphila and intestinal epithelium controls diet-induced obesity. *Proc. Natl. Acad. Sci. USA* **2013**, *110*, 9066–9071. [CrossRef]
33. de Vos, W.M. Microbe profile: Akkermansia muciniphila: A conserved intestinal symbiont that acts as the gatekeeper of our mucosa. *Microbiology* **2017**, *163*, 646–648. [CrossRef] [PubMed]
34. Zhang, X.; Shen, D.; Fang, Z.; Jie, Z.; Qiu, X.; Zhang, C.; Chen, Y.; Ji, L. Human gut microbiota changes reveal the progression of glucose intolerance. *PLoS ONE* **2013**, *27*, e71108. [CrossRef] [PubMed]
35. Qin, J.; Li, Y.; Cai, Z.; Li, S.; Zhu, J.; Zhang, F.; Liang, S.; Zhang, W.; Guan, Y.; Shen, D.; et al. A metagenome-wide association study of gut microbiota in type 2 diabetes. *Nature* **2012**, *490*, 55–60. [CrossRef] [PubMed]
36. Stadlbauer, V.; Leber, B.; Lemesch, S.; Trajanoski, S.; Bashir, M.; Horvath, A.; Tawdrous, M.; Stojakovic, T.; Fauler, G.; Fickert, P.; et al. Lactobacillus casei Shirota supplementation does not restore gut microbiota composition and gut barrier in metabolic syndrome: A randomized pilot study. *PLoS ONE* **2015**, *10*, e0141399. [CrossRef] [PubMed]

Article

Probiotic Ingestion, Obesity, and Metabolic-Related Disorders: Results from NHANES, 1999–2014

Eva Lau [1,2,*], João Sérgio Neves [1,3], Manuel Ferreira-Magalhães [2,4], Davide Carvalho [1,5] and Paula Freitas [1,5]

1 Department of Endocrinology, Diabetes and Metabolism, Centro Hospitalar Univesrsitário São João, 4200 Porto, Portugal
2 CINTESIS—Centre for Health Technologies and Information Systems Research, Faculty of Medicine, University of Porto, 4200 Porto, Portugal
3 Department of Surgery and Physiology, Cardiovascular Research Centre, Faculty of Medicine, University of Porto, 4200 Porto, Portugal
4 Health Information and Decision Sciences Department, Faculty of Medicine, Porto University, 4200 Porto, Portugal
5 I3S—Instituto de Investigação e Inovação em Saúde, Faculty of Medicine, University of Porto, 4200 Porto, Portugal
* Correspondence: evalau.med@gmail.com; Tel.: +351-919-624-956

Received: 24 May 2019; Accepted: 26 June 2019; Published: 28 June 2019

Abstract: Gut microbiota dysbiosis has been recognized as having key importance in obesity- and metabolic-related diseases. Although there is increasing evidence of the potential benefits induced by probiotics in metabolic disturbances, there is a lack of large cross-sectional studies to assess population-based prevalence of probiotic intake and metabolic diseases. Our aim was to evaluate the association of probiotic ingestion with obesity, type 2 diabetes, hypertension, and dyslipidemia. A cross-sectional study was designed using data from the National Health and Nutrition Examination Survey (NHANES), 1999–2014. Probiotic ingestion was considered when a subject reported consumption of yogurt or a probiotic supplement during the 24-h dietary recall or during the Dietary Supplement Use 30-Day questionnaire. We included 38,802 adults and 13.1% reported probiotic ingestion. The prevalence of obesity and hypertension was lower in the probiotic group (obesity-adjusted Odds Ratio (OR): 0.84, 95% CI 0.76–0.92, $p < 0.001$; hypertension-adjusted OR: 0.79, 95% CI 0.71–0.88, $p < 0.001$). Accordingly, even after analytic adjustments, body mass index (BMI) was significantly lower in the probiotic group, as were systolic and diastolic blood pressure and triglycerides; high-density lipoprotein (HDL) was significantly higher in the probiotic group for the adjusted model. In this large-scale study, ingestion of probiotic supplements or yogurt was associated with a lower prevalence of obesity and hypertension.

Keywords: intestinal microbiota; probiotics; nutrients

1. Introduction

Obesity is a pro-inflammatory state that plays a central role in the progression of several diseases, such as type 2 diabetes, hypertension, and dyslipidemia [1]. The pathophysiology of obesity and metabolic-related diseases is complex, resulting from the imbalance between environmental and genetic factors. The human gastrointestinal tract is populated by a complex ecosystem—the gut microbiota—which is responsible for the regulation of essential functions for the maintenance of health, including protective, structural, and histological functions [2]. New insights emphasize the role of gut microbiota in energy homeostasis, giving rise to the "The Metagenome Hypothesis" as a key player in the comprehension of metabolic diseases [3–5]. Accordingly, recent studies have shown the relationship between intestinal dysbiosis, which is defined as a change in the composition of gut microbiota and glucose and lipid metabolism deregulation in obesity and type 2 diabetes [6,7]. Obesity has been linked to an increase in Firmicutes and a decrease in Bacteroidetes [5,8]. Likewise, two large metagenome-wide association studies reported that type 2 diabetes had a lower proportion of butyrate-producing *Clostridiales* (*Roseburia* and *Faecalibacterium prausnitzii*) and greater proportions of *Clostridiales* that do not produce butyrate [7,9]. On one hand, distinct differences in gut microbiota result in a greater increase in harvesting energy from the diet by fermentation and the absorption of dietary polyssacharides, promoting hepatic lipogenesis. On the other hand, gut microbiota regulate intestinal permeability and an increase in the translocation of lipopolysaccharide-containing gut microbiota increases the inflammatory state, which is named metabolic endotoxemia, accompanied by weight gain and insulin resistance [10]. Understanding this interplay between gut microbiota and the host has created interest in shaping microbiota to prevent, treat, or delay obesity, type 2 diabetes, and metabolic-associated complications.

Probiotics are food components or supplements with living microorganisms that confer health advantages to the host [11]; specific strains have been increasingly studied as a potential therapeutic approach to shape gut microbiota composition, with possible benefits for weight control and diabetes management [12–15]. The administration of probiotics may restore the crosstalk between human host and gut microbiota, controlling homeostatic functions during obesity and metabolic-related disorders. Although there is increasing evidence of the potential benefits of probiotics in metabolic diseases, there is a lack of large cross-sectional studies to evaluate the population-based prevalence of probiotic intake and metabolic differences in those exposed to probiotics compared to those who are not. Large population surveys can create powerful information about population health status and trends. To date, there are no published studies about the large-scale use of probiotic supplements and yogurts and possible associations with metabolic diseases. This type of analysis can produce high-quality data for the real-life use of these types of food and supplements. Our aim was to assess the association of probiotic ingestion, through yogurt or supplements, with the prevalence of obesity and associated metabolic disturbances, namely type 2 diabetes, hypertension, and dyslipidemia.

2. Materials and Methods

2.1. Study Design and Settings

We designed a cross-sectional analysis, using data from the National Health and Nutrition Examination Survey (NHANES). NHANES is a national research survey designed to collect demographic, socio-economic, health, and nutritional statuses from a representative sample of the non-institutionalized civilian resident population of the United States of America. NHANES is a major program of the National Centre for Health Statistics (NCHS), which is part of the Centers for Disease Control and Prevention (CDC), and the detailed methodology is described in the literature [16]. NHANES was approved by the NHANES Institutional Review Board (IRB) and the NCHS Research Ethics Review Board (ERB) (after 2003).

2.2. Participants and Data Collection

We included adults aged 18 years or older, who had been included in NHANES between 1999 and 2014. Pregnant women were excluded. NHANES participants without physical examination or laboratory data and with no dietary data or implausible dietary data (24-h dietary recall) were also excluded. Figure S1 (supplementary material) shows the flowchart of the study population. NHANES data collection was performed through an in-home interview for demographic and basic health information data collection, together with a health examination in a Mobile Examination Centre (MEC), where participants were examined and surveyed. NHANES MEC examinations included anthropometric measurements, blood pressure assessment, and blood workup. Data were collected by a trained interviewer who had completed an intensive training course administered by the US Department of Agriculture and the US Department of Health and Human Services.

2.3. Assessment and Definition of Probiotic Exposure

In all NHANES cycles, from 1999 to 2014, a 24-h dietary recall was collected. Using an automated multiple-pass method, a detailed dietary intake (quality and quantity) for the 24-h period before the interview was recorded. For participants in the 1999–2002 NHANES, only one in-person 24-h dietary recall was performed. From 2003 onward, an additional telephone dietary recall interview was also performed 3 to 10 days following the in-person dietary interview. For the participants of the 2003–2014 NHANES, we used the mean of the nutritional information from both recalls (in-person recall and telephone recall). To assess probiotic supplementation exposure, we also used the Dietary Supplement Use 30-Day (DSQ), which assesses food supplement use during the preceding 30 days. Table S1 (supplementary materialsupplementary material) lists the probiotic supplements included.

Probiotic ingestion was considered when a subject reported consumption of a probiotic supplement or yogurt (as a dietary source of probiotics) during the 24-h dietary recall or of a probiotic supplement during the DSQ. Non-yogurt foods containing probiotics were classified as probiotic supplements.

2.4. Definition of Metabolic Comorbidities, Smoking, and Physical Activity

Obesity was defined as a body mass index (BMI) ≥30 Kg/m^2. Type 2 diabetes was defined as glycated hemoglobin (HbA1c) ≥6.5%, fasting plasma glucose level ≥126 mg/dL, or current glucose-lowering drug use. Dyslipidemia was assumed if participants had low-density cholesterol (LDL) ≥160 mg/dL, high-density cholesterol (HDL) <40 mg/dL, triglycerides ≥200mg/dL, total cholesterol ≥240 mg/dL, or if they were being treated with lipid-lowering drugs. Systolic and diastolic blood pressures (BP) were determined by the mean of 3 or 4 consecutive blood pressure readings. Hypertension was defined as systolic blood pressure ≥140 mmHg, or diastolic blood pressure ≥90 mmHg, or current medication for hypertension.

Smoking status was classified as former and current smokers. Current smokers were those who reported smoking at least 100 cigarettes during their lifetime and were currently smoking every day, or some days. Former smokers were those who reported smoking at least 100 cigarettes during their lifetime, but do not currently smoke.

Physical activity was measured differently along NHANES cycles. We classified participants using variables that allowed categorization of physical activity level into three categories (low, intermediate, and high). From 1999 to 2006 the physical activity level was assessed with the question "compare activity with others of the same age" (participants answering "less active" were classified into category "low", "about the same" into category "intermediate", and "more active" into category "high"). From 2007 to 2014, the weekly metabolic equivalents (MET) minutes of physical activity (accounting for vigorous work-related activity, moderate work-related activity, walking or bicycling for transportation, vigorous leisure-time physical activity, and moderate leisure-time physical activity) was divided into tertiles (participants were classified as "low" if included in the lower MET-minute tertile, as "intermediate" if in the middle MET-minute tertile, and as "high" if in the higher MET-minute tertile).

2.5. Statistical Analysis

Statistical analysis took into account the complex survey design of the NHANES dataset and was performed according to the CDC analytic recommendations [17].

Continuous variables were described as mean ± standard deviation (SD) and categorical variables were described as absolute and relative frequencies. To assess the association between probiotic exposure and metabolic comorbidities (obesity, diabetes, hypertension, and dyslipidemia), we performed unadjusted and adjusted logistic regression models. To evaluate the association between probiotic exposure and cardiomatebolic parameters (BMI, HbA1c, fasting plasma glucose, systolic BP, diastolic BP, LDL, HDL, and triglycerides), we performed unadjusted and adjusted multivariate linear regression models. We excluded those participants who were being treated with anti-hypertensive drugs from systolic and diastolic BP analysis, participants treated with anti-dyslipidemic drugs from analysis concerning lipid profiles, and patients treated with antidiabetic drugs from HbA1c and fasting plasma glucose analysis.

In the adjusted analyses, we used the following models (Table S2): Model 1, including age, sex, ethnicity (Mexican American, other Hispanic, non-Hispanic white), annual family income (<$25,000, $25,000 to $75,000, >$75,000), and education (<9th grade, ≥9th grade); model 2, including all model 1 covariates plus alcohol intake, smoking status (never a smoker, current smoker, or former smoker), physical activity (low, intermediate, high), ingested kcal per day, ingested carbohydrates/kcal per day, ingested protein/kcal per day, ingested fiber/kcal per day, and ingested polyunsaturated/saturated fatty acids ratio. In model 2, we also included BMI in all analyses except in the obesity analysis and sodium intake per day only in the hypertension and blood pressure analyses. As a supplementary analysis, we performed an additional model (model 3) that classified the diet pattern using the Dietary Approaches to Stop Hypertension (DASH) score. In model 3 we included all model 1 covariates plus alcohol intake, smoking status, physical activity, and the DASH dietary pattern score. Model 3 also included BMI in all analyses except the obesity analysis and sodium intake per day only in the hypertension and blood pressure analyses. The DASH score is based on 9 target nutrients (sodium, total fat, saturated fat, protein, fiber, cholesterol, calcium, magnesium, and potassium), as previously described [18]. Individuals meeting the DASH goal were given a score of 1.0 for that nutrient; if they attained an intermediate goal, they were given a score of 0.5 for that nutrient. The DASH score is the sum of the score for each individual nutrient. Logistic regression results were expressed as an odds ratio (OR) and a 95% confidence interval (95% CI). A two-sided p-value of <0.05 was considered statistically significant. Analyses were performed with Stata (version 14.2).

3. Results

3.1. Baseline Characteristics According to Probiotic Consumption

We included 38,802 adults, of whom 13.1% had exposure to probiotic supplements or yogurt. Baseline population characteristics according to probiotic consumption are described in Table 1. Participants in the group exposed to probiotic supplements or yogurt were more likely to be female, older, non-Hispanic white and to have a higher income and education level. Ingestion of kcal/day was similar between groups.

Table 1. Baseline population characteristics according to probiotic consumption (n = 38,802).

	No Exposure to Probiotics	Exposure to Probiotics	p Value
Participants, n (%)	33,719 (86.9%)	5083 (13.1%)	n.a.
Socio-economic characteristics			
Male gender, %	50.2 %	35.0 %	<0.001 *
Age, years ± SD	46.0 ± 15.3	48.9 ± 13.2	<0.001 *
Annual family income <$25000, %	30.7%	19.6%	<0.001 *
Education level less than 9th grade, %	6.4%	2.7%	<0.001 *
Ethnicity			<0.001 *
Non-Hispanic White, %	68.4%	79.5%	
Non-Hispanic Black, %	11.9%	5.4%	
Mexican American, %	8.5%	5.0%	
Other Hispanic, %	5.2%	4.1%	
Other ethnicities, %	6.0%	5.9%	

Table 1. *Cont.*

	No Exposure to Probiotics	Exposure to Probiotics	*p* Value
Risk factors			
Current smokers, %	20.2%	7.6%	<0.001 *
Former smokers, %	28.3%	31.7%	0.002 *
Alcohol consumption >20 g/day, %	15.7%	13.9%	0.028 *
Physical activity level #			0.028 *
Low	28.8%	26.9%	
Intermediate	36.6%	39.3%	
High	34.6%	33.8%	
Nutritional characteristics			
Kcal/day, kcal ± SD	2060.7 ± 648.9	2042.4 ± 529.3	0.163
Carbohydrates/day, g/100 kcal ± SD	12.3 ± 2.3	12.6 ± 1.8	<0.001 *
Protein/day, g/100 kcal ± SD	3.9 ± 0.9	4.2 ±0.8	<0.001 *
Fiber/day, g/100 kcal ± SD	0.8 ± 0.3	0.9 ± 0.3	<0.001 *
Polyunsaturated/saturated fatty acids ratio ± SD	0.7 ± 0.3	0.8 ± 0.3	0.001 *
Sodium per day, mg	3389.2 ± 1272.3	3282.3 ± 1024.4	<0.001 *
DASH score (0–9)	2.66 ± 1.15	3.33 ± 1.12	<0.001 *
Cardiometabolic parameters			
BMI, kg/m^2	28.5 ± 5.8	27.8 ± 5.1	<0.001 *
HbA1c, %	5.6 ± 0.8	5.5 ± 0.6	<0.001 *
Glucose, mg/dL	97.9 ± 28.4	96.0 ± 22.4	0.001 *
Systolic BP, mmHg	122.8 ± 15.7	120.5 ± 13.9	<0.001 *
Diastolic BP, mmHg	71.2 ± 10.1	70.2 ± 8.4	<0.001 *
HDL, mg/dL	51.9 ± 13.6	56.7 ± 12.6	<0.001 *
LDL, mg/dL	116.5 ± 29.5	115.5 ± 25.5	0.398
Triglycerides, mg/dL	139.1 ± 100.1	121.1 ± 61.5	<0.001 *

BMI: body mass index; BP: blood pressure; HDL: high-density lipoprotein; LDL: low-density lipoprotein; n.a.: Not applicable; * statistically significant. # See methods regarding physical activity level definition.

3.2. *Prevalence of Metabolic Comorbidities According to Probiotic Consumption*

The prevalence of metabolic comorbidities according to probiotic supplement or yogurt exposure is represented in Figure 1. All four studied comorbidities were lower in the exposed group. The comorbidities that showed higher differences in prevalence between groups were obesity and hypertension; 5.4% and 2.9% lower, respectively. Diabetes prevalence difference was only 1.6%, but this was still significant in our analysis.

3.3. *Modulation of Metabolic Comorbidities According to Probiotic Consumption*

Table 2 summarizes the results of metabolic disturbances by using unadjusted and adjusted models of the prevalence of comorbidities according to probiotic supplement or yogurt exposure. For unadjusted analysis, participants exposed to probiotics manifested a 22% reduction in the odds of having obesity (OR: 0.78, 95% CI 0.71–0.86; $p < 001$), a 16% reduction in the odds of having diabetes (OR: 0.84, 95% CI 0.73–0.96; $p = 0.020$), and a 12% reduction in the odds of having hypertension (OR: 0.88, 95% CI 0.81–0.96; $p = 0.004$). No significant differences were found in dyslipidemia prevalence.

After adjusting for potential confounders, obesity and hypertension prevalence remained significantly lower in the probiotic exposed group (Table 2). In turn, diabetes prevalence became similar between groups.

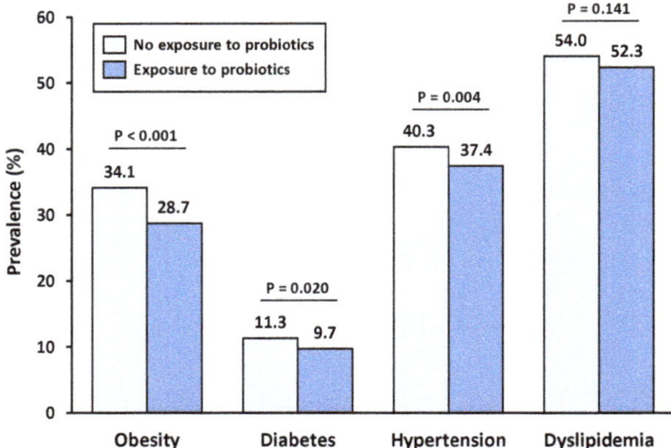

Figure 1. Prevalence of obesity, diabetes, hypertension, and dyslipidemia, according to probiotic exposure.

Table 2. Odds ratios of disease in subjects exposed to probiotics compared to non-exposed subjects.

	Unadjusted, OR (95% CI)	*p* Value	Model 1, OR (95% CI)	*p* Value	Model 2, OR (95% CI)	*p* Value
Obesity	0.78 (0.71–0.86)	<0.001	0.82 (0.75–0.90)	<0.001	0.83 (0.76–0.92)	<0.001 *
Diabetes	0.84 (0.73–0.97)	0.020	0.96 (0.82–1.13)	0.650	0.97 (0.81–1.17)	0.783
Hypertension	0.88 (0.81–0.96)	0.004	0.76 (0.68–0.84)	<0.001	0.79 (0.71–0.89)	<0.001 *
Dyslipidemia	0.93 (0.85–1.02)	0.141	0.95 (0.86–1.05)	0.356	1.01 (0.90–1.13)	0.863

Model 1: Age, sex, race, income and education; model 2: Model 1 + alcohol intake, smoking status, physical activity, kcal per day, carbohydrates/kcal per day, protein/kcal per day, fiber/kcal per day, and polyunsaturated/saturated fatty acids ratio. Model 2 also includes BMI in all analyses except in the obesity analysis and includes sodium intake per day only in the hypertension analysis. * Statistically significant.

Table 3 summarizes the results of cardiometabolic parameters according to probiotic supplement or yogurt exposure. In the unadjusted analysis, all the studied markers were lower in the probiotic-exposed group, with the exception of HDL (which was higher in the exposed group) and LDL (no differences were seen). After adjusting for potential confounders, BMI, and systolic BP, diastolic BP and triglycerides remained significantly lower in the probiotic-exposed group and HDL remained significantly higher (Table 3). Tables S3 sand S4 summarize the odds ratios of disease and the variation of cardiometabolic parameters, respectively, according to probiotic supplement or yogurt exposure, after accounting for the DASH dietary pattern score. The associations of probiotic supplement or yogurt ingestion with cardiometabolic parameters were not different after adjusting for DASH diet adherence.

Table 3. Variation of cardiometabolic parameters in participants exposed to probiotics compared to non-exposed participants.

	Unadjusted	*p* Value	Model 1	*p* Value	Model 2	*p* Value
BMI, kg/m^2	−0.74 (−1.01 to −0.46)	<0.001 *	−0.47 (−0.75 to −0.20)	0.001 *	−0.41 (−0.67 to −0.15)	0.002 *
HbA1c [a], %	−0.03 (−0.05 to −0.01)	0.003 *	−0.01 (−0.03 to 0.01)	0.382	0.01 (−0.01 to 0.03)	0.590
Glucose [a], mg/dL	−0.94 (−1.71 to −0.18)	0.016 *	−0.55 (−1.29 to 0.19)	0.146	−0.17 (−0.90 to −0.55)	0.641
Systolic BP [b], mmHg	−2.43 (−3.33 to −1.53)	<0.001 *	−1.99 (−2.83 to −1.16)	<0.001 *	−1.48 (−2.31 to −0.66)	<0.001 *
Diastolic BP [b], mmHg	−0.92 (−1.46 to −0.38)	0.001 *	−1.13 (−1.71 to −0.55)	<0.001 *	−0.86 (−1.45 to −0.27)	0.005 *
LDL [c], mg/dL	0.03 (−2.53 to 2.59)	0.980	−0.93 (−3.50 to 1.63)	0.472	−0.02 (−2.59 to 2.55)	0.988
HDL [c], mg/dL	4.93 (4.10 to 5.75)	<0.001 *	1.89 (1.12 to 2.66)	<0.001 *	1.43 (0.69 to 2.17)	<0.001 *
Triglycerides [c], mg/dL	−16.82 (−22.64 to −10.99)	<0.001 *	−11.74 (−18.14 to −5.33)	<0.001 *	−8.52 (−15.18 to −1.86)	0.013 *

Model 1: Age, sex, race, income and education; model 2: Model 1 + alcohol intake, smoking status, physical activity, kcal per day, carbohydrates/kcal per day, protein/kcal per day, fiber/kcal per day, and polyunsaturated/saturated fatty acids ratio. Model 2 also includes BMI in all analyses except in the BMI analysis and includes sodium intake per day only in the BP analyses. [a] Excluding participants treated with anti-hypertensive drugs. [b] Excluding participants treated with anti-dyslipidemic drugs. [c] Excluding participants treated with antidiabetic drugs. * Statistically significant.

The odds of metabolic comorbidities according to the origin of probiotics is presented in Figure 2. Among those participants exposed to probiotics, 95.7% were exposed to yogurt and 5.4% were exposed to probiotic supplements. The odds of obesity, diabetes, hypertension, and dyslipidemia for participants exposed to yogurt or to probiotic supplements alone were similar to the odds for participants exposed to any type of probiotic. Although the confidence intervals were wider in the analysis of probiotic supplements (due to the smaller number of exposed individuals), the point estimates for the association with comorbidities was similar to the association with exposure to any probiotic or exposure to yogurt, with the exception of diabetes. Although the association in both analyses was not significant, the odds ratio for diabetes was 0.92 (95% CI 0.77–1.10) among participants exposed to yogurt and 1.28 (95% CI 0.71–2.23) among those exposed to probiotic supplements.

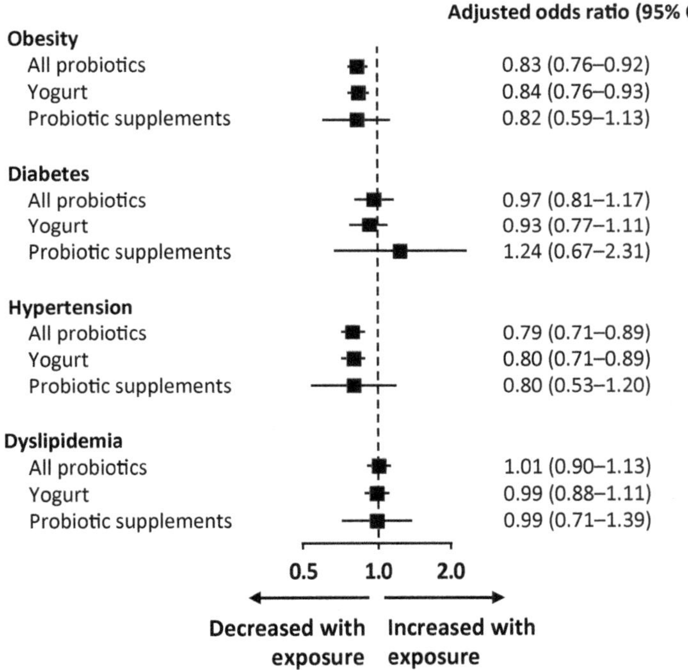

Figure 2. Odds ratio of disease in participants exposed to probiotics compared to those not exposed, according to the origin of probiotics (all probiotics, yogurt, or probiotic supplements). Logarithmic regression models adjusted for age, sex, race, income, education, alcohol intake, smoking status, physical activity, carbohydrates/kcal per day, protein/kcal per day, fiber/kcal per day, and polyunsaturated/saturated fatty acids ratio (model 2). Model 2 also includes BMI in all analyses except in the obesity analysis and includes sodium intake per day only in the hypertension analysis.

4. Discussion

We conducted a cross-sectional analysis on a large and representative US population, for a total of 38,802 adults, and found that 13.1% reported the use of probiotic supplements or yogurt ingestion. Although there are several studies addressing the possible beneficial associations of probiotic ingestion and several metabolic outcomes, there is a lack of large cross-sectional studies to objectively assess population-based prevalence of probiotic intake and metabolic differences in those exposed and not exposed to probiotics. To our knowledge, this was the first large cross-sectional analysis aiming to assess the association of probiotic ingestion, either by probiotic supplements or yogurt, with metabolic disturbances. Probiotic ingestion was associated with a 17% lower prevalence of obesity and a

21% lower prevalence of hypertension. Furthermore, HDL cholesterol was significantly higher and triglyceride levels were significantly lower in the probiotic group.

Probiotics modulate gut microbial communities, exerting beneficial metabolic effects through the regulation of multitudinous physiological metabolic pathways. Among the molecular mechanisms, the regulation of adipogenesis, stimulation of insulin signaling, improvement of gut barrier function, reduction of metabolic endotoxemia, and down-regulation of cholesterol levels are some of the suggested key players in the crosstalk between probiotics and metabolic disorders [19].

4.1. Obesity

We found that probiotic ingestion, via supplements or yogurt, was associated with a lower prevalence of obesity (17% reduction), before and after adjusting for demographics and potential confounders. Although subjects that consumed probiotics had a higher consumption of carbohydrates, fiber, and protein, the effects of probiotics on BMI were significant, even after adjustment for confounders (−0.41 Kg/m^2 between groups in model 2). Furthermore, there were no differences in total energy intake per day between groups and, also, there was no linear association of physical activity level (low, intermediate, and high) and probiotic ingestion. The only thing observed was a higher proportion of intermediate physical activity level, but lower proportions of high and low physical activity levels in the probiotic-exposed group. Putting this all together, these results support our hypothesis of the beneficial impact of probiotic ingestion per se on metabolic health, namely its effects on the regulation of body weight. In agreement with our study, a recent meta-analysis of randomized clinical trials with 957 subjects, with a mean BMI of 27.6 kg/m^2, showed that probiotic administration significantly reduced body weight by −0.60 kg and BMI by −0.27 kg/m^2 [20]. NHANES's data between 1999–2004 also showed that yogurt consumption was associated with a lower likelihood of having obesity (OR: 0.57, 95% CI 0.40–0.82; $p < 0.05$) [21]. A prospective study including 120,877 US individuals evaluated lifestyle factors and weight change at four-year intervals, with multivariable adjustments. The four-year weight change was negatively associated with yogurt ingestion [22], which further supports our results.

4.2. Diabetes

In the unadjusted analysis, probiotic supplement and yogurt consumers had lower odds of having diabetes and, accordingly, lower glycemia and HbA1c levels. However, when adjusted for individual characteristics and confounders, the difference was no longer significant. One possible explanation for this is that diabetes is largely determined by individual demographic characteristics, such as age, ethnicity and, mainly, BMI [23]. Therefore, there are no longer differences after adjusting for individual factors. Furthermore, in another meta-analysis of randomized controlled trials, Ruan I et al. [14]. concluded that probiotics had a greater effect on fasting blood glucose for people with diabetes. On the contrary, those without diabetes only show a trend of a glucose-lowering effect, which shows that probiotic supplementation may have a greater benefit for individuals with higher fasting glucose levels [14]. In three prospective cohorts in the US, yogurt intake was consistently and inversely associated with type 2 diabetes risk [24]. Probiotic supplementation seems to be more effective in reducing HbA1c in diabetic patients with higher baseline BMIs and, furthermore, probiotic supplements with greater bacterial species may be more effective [25], which we could not evaluate in our study. Yao K et al. [26] conducted a meta-analysis in patients with type 2 diabetes to investigate the effects of probiotics on glucose metabolism, and, similar to our results, they did not find a significant effect on fasting blood glucose levels, although HbA1c was improved with probiotic supplementation. In the analysis according to the origin of probiotics, we found a non-significant trend to higher odds of diabetes in participants exposed to probiotic supplements, which was not seen in participants exposed to yogurt. This may be explained by a reverse-causality relationship. Participants with diabetes may be more prone to consuming probiotic supplements due to their known potential to improve glucose control in diabetes.

4.3. Dyslipidemia

We did not find differences in the odds of dyslipidemia according to probiotic ingestion, however, we did observe some beneficial aspects in the lipid profile of the probiotic-exposed group. HDL was significantly increased and triglycerides were significantly decreased in the probiotic group, even after adjustment. Our results are in line with a study by Fu et al. [27], with 893 subjects, which showed that gut microbiota was associated with a 4.5% variance in BMI, a 6% variance in blood levels of triglycerides, and a 4% variance in HDL, but had little effect on LDL or total cholesterol. In contrast, a meta-analysis including 1624 participants (828 in the probiotic and 796 in the placebo group) demonstrated that probiotics reduced total cholesterol and LDL cholesterol by 7.8 mg/dL and 7.3 mg/dL, respectively but had no significant effects on HDL cholesterol or triglycerides [28]. Human clinical studies have yielded different results on the association of lipid profiles with probiotic supplementation. Differences in the type of probiotics and in the experimental designs, including the clinical heterogeneity of participants, namely their baseline levels of blood lipids, may affect the role of probiotics in lipid metabolism and explain the different results. Given the clinical correlation between obesity and related metabolic disorders, it is possible that the observed associations between gut bacterial composition and lipid levels can be mediated, in part, through the effects on BMI.

4.4. Hypertension

Ingestion of probiotic supplements or yogurt resulted in a 21% reduction in the odds of hypertension after adjusting for potential confounders. Both systolic BP and diastolic BP were significantly lower. A meta-analysis of randomized clinical trials supports our results, showing that probiotic consumption significantly reduced systolic BP by −3.56 mmHg and diastolic blood pressure by −2.38 mmHg, compared with control groups [29]. In our study, for the adjusted model, probiotic ingestion was associated with a lower systolic BP by 1.48 mmHg and a lower diastolic BP by 0.86 mmHg. The modulation of BP by probiotics may be linked to several mechanisms, including their capacity to improve lipid profiles, to reduce BMI, and to produce bioactive peptides with angiotensin-converting inhibitory activity [30–32].

4.5. Strengths and Limitations

Our study has a number of strengths and limitations, which need to be highlighted. We conducted an analysis of a large cross-sectional survey, which was representative of the US population. To the best of our knowledge, this is the first cross-sectional study aiming to assess the association of probiotic ingestion, through supplements or yogurt, and metabolic disturbances. A previous study on the NHANES population was carried out, aiming to evaluate the association between dairy products (which included yogurt ingestion) with obesity and other disturbances of metabolic syndrome; however it did not evaluate probiotic supplementation [21]. We also developed an analytic strategy that included adjusted logistic regression models to obviate confounders such as physical activity, alcohol consumption, and smoking status. One of the limitations of this type of analysis is that we cannot deduce causation; however, we were able to show strong associations between probiotic ingestion and the prevalence of some metabolic disturbances. The participants in the probiotic-exposed group were more likely to be non-Hispanic white. It has been previously stated that gut microbiome varies by geographic ancestry [33], which may limit the extrapolation of these results to other ethnic groups. Furthermore, inter-individual differences in the composition of gut microbiota were previously associated with different responses to probiotics [34], including non-responders to gut microbiota modulation. The absence of individual gut microbiota analyses in our study may have hampered the analysis of these type of responders. The assessment of probiotic exposure was based on self-reported information, however, NHANES only provides dietary information that is considered to be reliable [35]. In our study, we defined probiotic ingestion as being either yogurt or probiotic supplement consumption. Assuming that there could be differences between the types of ingestion, we performed a sensitivity analysis based on the origin of probiotics. There were no significant differences in associations according to the source of exposure to probiotics. Probiotic supplements and yogurts vary in the amounts of bacteria per serving they are composed of. In our study, the population was classified according to whether or not they were exposed to probiotic supplements or yogurt. The duration and quantity of exposure were not taken into account, which may have diluted the magnitude of the association between probiotic consumption and metabolic disturbance.

In summary, our study supports the beneficial association of probiotic supplement or yogurt ingestion with metabolic health, specifically obesity and hypertension. Furthermore, probiotic ingestion was significantly associated with higher HDL cholesterol and lower triglyceride levels. Our study supports the possibility of gut microbiota modulation by the use of probiotics as an attractive therapeutic target to prevent and treat obesity and related cardiometabolic disorders. Future research should focus on understanding the gut microbiota ecosystem and on identifying individuals who benefit the most from selective modulation of microbiota.

Supplementary Materials: The following are available online at http://www.mdpi.com/2072-6643/11/7/1482/s1. Figure S1: Flowchart of the study population; Table S1: Probiotic supplements included; Table S2: Variables included in the adjusted models; Table S3: Odds ratio of disease in subjects exposed to probiotics compared to non-exposed (model 3); Table S4: Variation of cardiometabolic parameters in participants exposed to probiotics compared to non-exposed (model 3).

Author Contributions: Conceptualization, E.L. and J.S.N.; methodology, E.L., J.S.N., and M.F.-M.; validation, P.F. and D.C.; formal analysis, J.S.N. and M.F.-M.; writing—original draft preparation, E.L.; writing—review and editing, P.F. and D.C.

Funding: This research received no external funding.

Acknowledgments: All data sets used in the present study were provided by the U.S. Centers for Disease Control and Prevention as part of the ongoing National Health and Nutrition Examination Survey (NHANES).

Conflicts of Interest: The authors declare no conflict of interest.

References

1. Prospective Studies Collaboration; Whitlock, G.; Lewington, S.; Sherliker, P.; Clarke, R.; Emberson, J.; Halsey, J.; Qizilbash, N.; Collins, R.; Peto, R. Body-mass index and cause-specific mortality in 900 000 adults: Collaborative analyses of 57 prospective studies. *Lancet* **2009**, *373*, 1083–1096.
2. Prakash, S.; Rodes, L.; Coussa-Charley, M.; Tomaro-Duchesneau, C. Gut microbiota: Next frontier in understanding human health and development of biotherapeutics. *Biologics* **2011**, *5*, 71–86. [CrossRef] [PubMed]
3. Cani, P.D.; Delzenne, N.M. The Role of the Gut Microbiota in Energy Metabolism and Metabolic Disease. *Curr. Pharm. Des.* **2009**, *15*, 1546–1558. [CrossRef] [PubMed]
4. Lau, E.; Marques, C.; Pestana, D.; Santoalha, M.; Carvalho, D.; Freitas, P.; Calhau, C. The role of I-FABP as a biomarker of intestinal barrier dysfunction driven by gut microbiota changes in obesity. *Nutr. Metab (Lond)* **2016**, *13*, 31. [CrossRef] [PubMed]
5. Ley, R.E.; Turnbaugh, P.J.; Klein, S.; Gordon, J.I. Microbial ecology: Human gut microbes associated with obesity. *Nature* **2006**, *444*, 1022–1023. [CrossRef] [PubMed]
6. Ley, R.E.; Bäckhed, F.; Turnbaugh, P.; Lozupone, C.A.; Knight, R.D.; Gordon, J.I. Obesity alters gut microbial ecology. *Proc. Natl. Acad. Sci. USA* **2005**, *102*, 11070–11075. [CrossRef] [PubMed]
7. Qin, J.; Li, Y.; Cai, Z.; Li, S.; Zhu, J.; Zhang, F.; Liang, S.; Zhang, W.; Guna, Y.; Shen, D.; et al. A metagenome-wide association study of gut microbiota in type 2 diabetes. *Nature* **2012**, *490*, 55–60. [CrossRef] [PubMed]
8. Turnbaugh, P.J.; Ley, R.E.; Mahowald, M.A.; Magrini, V.; Mardis, E.R.; Gordon, J.I. An obesity-associated gut microbiome with increased capacity for energy harvest. *Nature* **2006**, *444*, 1027–1031. [CrossRef] [PubMed]
9. Karlsson, F.H.; Tremaroli, V.; Nookaew, I.; Bergström, G.; Behre, C.J.; Fagerberg, B.; Nielsen, J.; Bäckhed, F. Gut metagenome in European women with normal, impaired and diabetic glucose control. *Nature* **2013**, *498*, 99–103. [CrossRef] [PubMed]

10. Meijnikman, A.S.; Gerdes, V.E.; Nieuwdorp, M.; Herrema, H. Evaluating Causality of Gut Microbiota in Obesity and Diabetes in Humans. *Endocr. Rev.* **2018**, *39*, 133–153. [CrossRef]
11. Hill, C.; Guarner, F.; Reid, G.; Gibson, G.R.; Merenstein, D.J.; Pot, B.; Morelli, L.; Canani, R.B.; Flint, H.J.; Salminen, S.J. The International Scientific Association for Probiotics and Prebiotics consensus statement on the scope and appropriate use of the term probiotic. *Nat. Rev. Gastroenterol. Hepatol.* **2014**, *11*, 506–514. [CrossRef] [PubMed]
12. Crovesy, L.; Ostrowski, M.; Ferreira, D.M.T.P.; Rosado, E.L.; Soares-Mota, M. Effect of Lactobacillus on body weight and body fat in overweight subjects: A systematic review of randomized controlled clinical trials. *Int. J. Obes.* **2017**, *41*, 1607–1614. [CrossRef] [PubMed]
13. Gomes, A.C.; Bueno, A.A.; de Souza, R.G.M.; Mota, J.F. Gut microbiota, probiotics and diabetes. *Nutr. J.* **2014**, *13*, 60. [CrossRef] [PubMed]
14. Ruan, Y.; Sun, J.; He, J.; Chen, F.; Chen, R.; Chen, H. Effect of Probiotics on Glycemic Control: A Systematic Review and Meta-Analysis of Randomized, Controlled Trials. *PLoS ONE* **2015**, *10*, e0132121. [CrossRef] [PubMed]
15. Everard, A.; Belzer, C.; Geurts, L.; Ouwerkerk, J.P.; Druart, C.; Bindels, L.B.; Guiot, Y.; Derrien, M.; Muccioli, G.G.; Delzenne, N.M.; et al. Cross-talk between Akkermansia muciniphila and intestinal epithelium controls diet-induced obesity. *Proc. Natl. Acad. Sci. USA* **2013**, *110*, 9066–9071. [CrossRef] [PubMed]
16. NHANES—National Health and Nutrition Examination Survey Homepage. Available online: https://www.cdc.gov/nchs/nhanes/index.htm (accessed on 7 June 2018).
17. NHANES—Survey Methods and Analytic Guidelines. Available online: https://wwwn.cdc.gov/nchs/nhanes/AnalyticGuidelines.aspx. (accessed on 10 June 2018).
18. Kim, H.; Andrade, F.C.D. Diagnostic status of hypertension on the adherence to the Dietary Approaches to Stop Hypertension (DASH) diet. *Prev. Med. Rep.* **2016**, *4*, 525–531. [CrossRef] [PubMed]
19. He, M.; Shi, B. Gut microbiota as a potential target of metabolic syndrome: The role of probiotics and prebiotics. *Cell Biosci.* **2017**, *7*, 54. [CrossRef]
20. Borgeraas, H.; Johnson, L.K.; Skattebu, J.; Hertel, J.K.; Hjelmesaeth, J. Effects of probiotics on body weight, body mass index, fat mass and fat percentage in subjects with overweight or obesity: A systematic review and meta-analysis of randomized controlled trials. *Obes. Rev.* **2018**, *19*, 219–232. [CrossRef]
21. Beydoun, M.A.; Gary, T.L.; Caballero, B.H.; Lawrence, R.S.; Cheskin, L.J.; Wang, Y. Ethnic differences in dairy and related nutrient consumption among US adults and their association with obesity, central obesity, and the metabolic syndrome. *Am. J. Clin. Nutr.* **2008**, *87*, 1914–1925. [CrossRef]
22. Mozaffarian, D.; Hao, T.; Rimm, E.B.; Willett, W.C.; Hu, F.B. Changes in diet and lifestyle and long-term weight gain in women and men. *N. Engl. J. Med.* **2011**, *364*, 2392–2404. [CrossRef]
23. Menke, A.; Rust, K.F.; Fradkin, J.; Cheng, Y.J.; Cowie, C.C. Associations Between Trends in Race/Ethnicity, Aging, and Body Mass Index With Diabetes Prevalence in the United States. *Ann. Intern. Med.* **2014**, *161*, 328. [CrossRef]
24. Chen, M.; Sun, Q.; Giovannucci, E.; Mozaffarian, D.; Manson, J.E.; Willett, W.C.; Hu, F.B. Dairy consumption and risk of type 2 diabetes: 3 cohorts of US adults and an updated meta-analysis. *BMC Med.* **2014**, *12*, 215. [CrossRef] [PubMed]
25. Akbari, V.; Hendijani, F. Effects of probiotic supplementation in patients with type 2 diabetes: Systematic review and meta-analysis. *Nutr. Rev.* **2016**, *74*, 774–784. [CrossRef]
26. Yao, K.; Zeng, L.; He, Q.; Wang, W.; Lei, J.; Zou, X. Effect of Probiotics on Glucose and Lipid Metabolism in Type 2 Diabetes Mellitus: A Meta-Analysis of 12 Randomized Controlled Trials. *Med. Sci. Monit.* **2017**, *23*, 3044–3053. [CrossRef]
27. Fu, J.; Bonder, M.J.; Cenit, M.C.; Tigchelaar, E.F.; Maatman, A.; Dekens, J.A.; Brandsma, E.; Marczynska, J.; Imhann, F.; Weersma, R.K.; et al. The Gut Microbiome Contributes to a Substantial Proportion of the Variation in Blood Lipids. *Circ. Res.* **2015**, *117*, 817–824. [CrossRef]
28. Cho, Y.A.; Kim, J. Effect of Probiotics on Blood Lipid Concentrations. *Medicine (Baltimore)* **2015**, *94*, e1714. [CrossRef] [PubMed]
29. Khalesi, S.; Sun, J.; Buys, N.; Jayasinghe, R. Effect of probiotics on blood pressure: A systematic review and meta-analysis of randomized, controlled trials. *Hypertens (Dallas Tex. 1979)* **2014**, *64*, 897–903. [CrossRef]
30. Upadrasta, A.; Madempudi, R.S. Probiotics and blood pressure: Current insights. *Integr. Blood Press. Control* **2016**, *9*, 33–42.

31. Mohanty, D.P.; Mohapatra, S.; Misra, S.; Sahu, P.S. Milk derived bioactive peptides and their impact on human health—A review. *Saudi J. Biol. Sci.* **2016**, *23*, 577–583. [CrossRef]
32. Donkor, O.N.; Henriksson, A.; Singh, T.K.; Vasiljevic, T.; Shah, N.P. ACE-inhibitory activity of probiotic yoghurt. *Int. Dairy J.* **2007**, *17*, 1321–1331. [CrossRef]
33. Gupta, V.K.; Paul, S.; Dutta, C. Geography, Ethnicity or Subsistence-Specific Variations in Human Microbiome Composition and Diversity. *Front. Microbiol.* **2017**, *8*, 1162. [CrossRef] [PubMed]
34. Lampe, J.W.; Navarro, S.L.; Hullar, M.A.J.; Shojaie, A. Inter-individual differences in response to dietary intervention: Integrating omics platforms towards personalised dietary recommendations. *Proc. Nutr. Soc.* **2013**, *72*, 207–218. [CrossRef] [PubMed]
35. Raper, N.; Perloff, B.; Ingwersen, L.; Steinfeldt, L.; Anand, J. An overview of USDA's Dietary Intake Data System. *J. Food Compos. Anal.* **2004**, *17*, 545–555. [CrossRef]

Article

Aerobic Exercise Training with Brisk Walking Increases Intestinal Bacteroides in Healthy Elderly Women

Emiko Morita [1,2], Hisayo Yokoyama [1,3,*], Daiki Imai [1,3], Ryosuke Takeda [3], Akemi Ota [4], Eriko Kawai [1], Takayoshi Hisada [5], Masanori Emoto [6], Yuta Suzuki [1,3] and Kazunobu Okazaki [1,3]

1 Department of Environmental Physiology for Exercise, Osaka City University Graduate School of Medicine, 3-3-138 Sugimoto, Sumiyoshi-ku, Osaka-shi, Osaka 558-8585, Japan; e-morita@pt-u.aino.ac.jp (E.M.); dimai@sports.osaka-cu.ac.jp (D.I.); kawai@respiratorycontrol.com (E.K.); suzuki@sports.osaka-cu.ac.jp (Y.S.); okazaki@sports.osaka-cu.ac.jp (K.O.)

2 Department of Physical Therapy Faculty of Health Science, Aino University, 4-5-4 Higashiohda, Ibaraki-shi, Osaka 567-0012, Japan

3 Research Center for Urban Health and Sports, Osaka City University, 3-3-138 Sugimoto, Sumiyoshi-ku, Osaka-shi, Osaka 558-8585, Japan; wind-05@med.osaka-cu.ac.jp

4 Department of Health and Sports Science, Osaka Electro-communication University, 1130-70 Kiyotaki, Shijonawate-shi, Osaka 575-0063, Japan; ota@osakac.ac.jp

5 TechnoSuruga Laboratory Company, Ltd., 330 Nagasaki, Shimizu-ku, Shizuoka-shi, Shizuoka 424-0065, Japan; tsl-contact@tecsrg.co.jp

6 Metabolism, Endocrinology and Molecular Medicine, Osaka City University Graduate School of Medicine, 1-4-3, Asahi-Machi, Abeno-ku, Osaka-shi, Osaka 545-8586, Japan; memoto@med.osaka-cu.ac.jp

* Correspondence: yokoyama@sports.osaka-cu.ac.jp; Tel.: +81-06-6605-2947

Received: 19 March 2019; Accepted: 16 April 2019; Published: 17 April 2019

Abstract: This study examined the effect of an exercise intervention on the composition of the intestinal microbiota in healthy elderly women. Thirty-two sedentary women that were aged 65 years and older participated in a 12-week, non-randomized comparative trial. The subjects were allocated to two groups receiving different exercise interventions, trunk muscle training (TM), or aerobic exercise training (AE). AE included brisk walking, i.e., at an intensity of ≥ 3 metabolic equivalents (METs). The composition of the intestinal microbiota in fecal samples was determined before and after the training period. We also assessed the daily physical activity using an accelerometer, trunk muscle strength by the modified Kraus–Weber (K-W) test, and cardiorespiratory fitness by a 6-min. walk test (6MWT). K-W test scores and distance achieved during the 6MWT (6MWD) improved in both groups. The relative abundance of intestinal *Bacteroides* only significantly increased in the AE group, particularly in subjects showing increases in the time spent in brisk walking. Overall, the increases in intestinal *Bacteroides* following the exercise intervention were associated with increases in 6MWD. In conclusion, aerobic exercise training that targets an increase of the time spent in brisk walking may increase intestinal *Bacteroides* in association with improved cardiorespiratory fitness in healthy elderly women.

Keywords: intestinal microbiota; intestinal *Bacteroides*; cardiorespiratory fitness; trunk muscle training; aerobic exercise training; brisk walking

1. Introduction

"All disease begins in the gut"., a quotation from the ancient Greek physician Hippocrates, highlights the potential roles of intestinal microbiota in various disease risks, which have recently attracted considerable attention from researchers. The presence of an imbalanced, low-diversity, intestinal microbiota is known as dysbiosis and it is associated with a variety of pathologies, including constipation [1], obesity [2], diabetes [3], colon cancer [4], coronary artery disease [5], inflammatory bowel disease [6], and depression [7]. Aging also strongly affects the composition of the intestinal microbiota. In general, the intestinal microbiota of the elderly show reduced species diversity [8]. In addition, intestinal *Bifidobacterium* and *Bacteroides*, which are known to be related to obesity, are also reduced [8], which potentially contributes to the high prevalence of obesity in the elderly population. Overall, the intestinal microbiota could be regarded as an indicator of host health.

Multiple factors, including host genetics [9], method of childbirth (i.e., by vaginal delivery or caesarian section) [10], age [8], nutrition [11], and antibiotic intake [8], have been suggested to affect the composition of the intestinal microbiota. Recent studies demonstrated the association between exercise training, i.e., a low-cost health strategy, and lower risks of colon cancer [12,13], a disease that is known to at least partly arise from imbalanced intestinal microbiota [4]. Therefore, exercise may also have potential for modifying the composition of the intestinal microbiota, although these studies did not directly examine the effect of exercise on intestinal microbiota.

In fact, animal studies have demonstrated the changes in the composition of the intestinal microbiota by exercise training [14–16]. A number of cross-sectional human studies have confirmed the associations between physical activity or cardiorespiratory fitness and the composition of the intestinal microbiota [17–19]. For example, rugby players were found to have a greater diversity of intestinal microbiota and an enlarged abundance of *Akkermansia*—which is known to prevent diabetes—when compared to sedentary adults [17]. Other studies showed that cardiorespiratory fitness or physical activity level is associated with greater microbial diversity in healthy humans [18–20]. Furthermore, trained elite race walkers show increased relative abundance of *Bacteroides,* in combination with high fat diet [21]. However, these studies did not examine the effect of exercise alone on intestinal microbiota independent of the dietary habits that may have the greater impact on intestinal microbiota than exercise. Therefore, the potential impact of exercise interventions on human intestinal microbiota has not been fully clarified.

In the present study, we examined the effects of exercise interventions on intestinal microbiotic composition in healthy elderly women. We hypothesized that an improvement of cardiorespiratory fitness would be crucial to exercise-induced changes in the intestinal microbiota. We compared the effects of two exercise modalities on the intestinal microbiota: aerobic exercise, which specifically enhances cardiorespiratory fitness, and trunk muscle training as a control condition to verify this hypothesis.

2. Materials and Methods

2.1. Subjects

Thirty-two healthy sedentary women that were aged 65 years and over were recruited from the residents of Osaka City, Japan, by an advertisement in a local magazine. The selected 32 subjects voluntarily opted for enrollment in either of the two exercise programs, aerobic exercise training (AE) or trunk muscle training (control condition; TM). Prior to the study, none of the subjects engaged in a regular exercise for more than 1 h per week. Health status and the use of medication were assessed by structured interview. Applicants presenting a history of ischemic heart disease, chronic heart failure, stroke, severe hypertension, diabetes, or neuropsychiatric disorder were excluded from the study. Applicants who were judged by a physician to be unable or ill-equipped to participate in the exercise program were also excluded. Consequently, none of the 32 subjects was excluded. The Institutional Review Board of Osaka City University Graduate School of Medicine approved the study protocol (approval no. 3501, approved on August 30, 2016). The authors also confirm that all of the ongoing and related trials for this intervention are registered in the University Hospital Medical Information Network Clinical Trials Registry (UMIN 000023930). Written informed consent was obtained from all of the participants after explanation of the study purpose. The study protocol also conformed to the ethical guidelines of the 1975 Declaration of Helsinki.

2.2. Study Design

The study design involved a 12-week non-randomized, comparative trial, in which the allocation of the participants to either of the two exercise groups, AE and TM, was based on their own preference. This study was conducted between the first recruitment of the participants on 12 September 2016 and the final follow-up of the participants on 24 January 2018. Before study enrollment, all of the applicants visited our research center at Osaka City University for baseline measurements, e.g., body composition, motor ability, and clinical laboratory analyses, as well as an assessment of daily physical activity levels, nutrient intake, and bowel habits. In addition, fecal samples were collected. All of the baseline assessments were conducted at least 1 week before the first training session. Finally, 18 and 14 applicants who met the inclusion criteria were enrolled in the AE group and the TM group, respectively, after which they were started on the selected 12-week exercise programs. The measurements during the baseline session were repeated at least one week after the final session of the exercise program.

2.3. Exercise Intervention

The subjects in the TM group received a 1-h group training weekly for 12 weeks, which aimed at strengthening the trunk muscles. All of the sessions were held at Sumiyoshi Sports Center, a gymnasium located in Osaka City, and supervised by a trained instructor. A training session comprised 5–10 min. of warm-up, followed by 45 min. of targeted resistance training of the trunk muscles and 5–10 min. of cool down exercises. Figure 1 shows examples of the trunk muscle training. The training was composed of several kinds of exercises, including arching–swaying, plank, pelvic rotation in the supine position, and diagonal lifting while standing on all fours. The contraction duration was set at 3 to 5 s, and each exercise was performed in two sets of 10 repetitions. The subjects were also instructed to work out at home daily. Adherence to group sessions, as well as to the home exercises, was recorded weekly by the instructor throughout the 12-week intervention period.

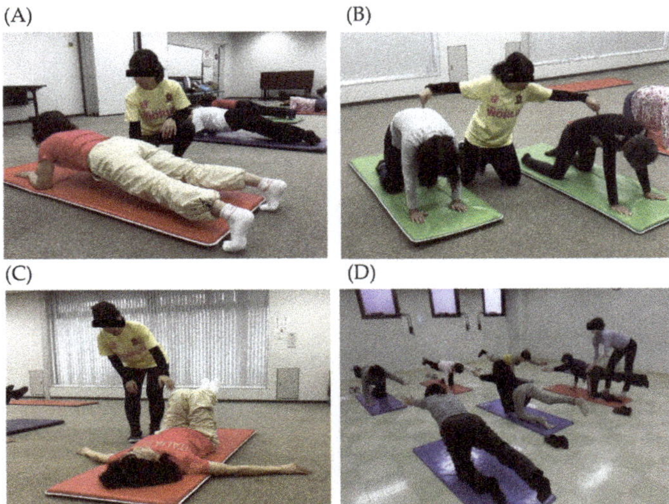

Figure 1. Exercises during trunk muscle training. (**A**) Arching–swaying while standing on all fours, (**B**) plank, (**C**) lying pelvic rotation, and (**D**) diagonal lifting while standing on all fours.

The subjects in the AE group were instructed to perform 60 min. of brisk walking at an intensity of ≥ 3 metabolic equivalents (METs) daily for 12 weeks. They wore a three-axis accelerometer (Mediwalk® MT-KT02DZ, TERUMO, Tokyo, Japan [22,23]) throughout the intervention period, except while sleeping and bathing, to record their daily number of steps and time that is spent in brisk walking.

The instructor shared the accelerometer data with the participants once a week and was encouraged them to increase the intensity and duration of their brisk walking regimen gradually as much as possible. The subjects were also instructed to keep good posture while walking.

2.4. Analysis of Intestinal Microbiota

The fecal samples were collected in a container with guanidine thiocyanate as a preservative solution (TechnoSuruga Laboratory, Shizuoka, Japan) and refrigerated at 4 °C until transfer to the laboratory within seven days. We conformed to the protocol [24] for the representative extraction of DNA from bacterial populations in feces. Terminal restriction fragment length polymorphism (T-RFLP) analyses to determine the relative abundance of intestinal microbiota phylogenetic groups from each fecal sample were performed at the TechnoSuruga Laboratory (Shizuoka, Japan) [25,26]. T-RFLP analysis is one of the most well-established and reliable 16S ribosomal RNA-based methods, especially when considering its high throughput and reproducibility. Briefly, the fecal samples (approximately 4 mg each) were suspended in a 1200 µL solution containing 100 mM Tris-HCl (pH 9.0), 40 mM ethylenediaminetetraacetic acid, 4 M guanidine thiocyanate, and 0.001% bromothymol blue. A FastPrep 24 device homogenized the Fecal solids in the suspension (MP Biomedicals, Irvine, CA, USA) with zirconia beads being set at 5 m/s for 2 min. DNA was then extracted from a 200 µL suspension using magLEAD 12gC (Precision System Science; Chiba, Japan). MagDEA® Dx SV (Precision System Science) was used as the reagent in automatic nucleic acid extraction. PCR was performed with a Takara Thermal Cycler Dice TP650 (Takara Bio, Shiga, Japan) in 20 µL of a reaction mixture containing 1× PCR buffer, with each deoxynucleotide triphosphate at a concentration of 200 µM, 1.5 mM $MgCl_2$, each primer at a concentration of 0.2 µM, 10 ng of fecal DNA, and 0.2 U of HotStarTaq DNA polymerase (Qiagen, Hilden, Germany). 5′ FAM-labeled 516f (5′-TGC-CAGCAGCCGCGGTA-3′; *Escherichia coli* positions 516–532) and 1510r (5′-GGTTACCTTGTTACGA-CTT-3′; *E. coli* positions 1510–1492) were the primers used. The amplification program used was as follows: preheating at 95 °C for 15 min, 35 cycles of denaturation at 95 °C for 30 s, annealing at 50 °C for 30 s, extension at 72 °C for 90 s, and finally, terminal extension at 72 °C for 10 min. Electrophoresis and purified using a MultiScreen PCR µ96 Filter Plate verified amplified DNA (Millipore, Billerica, MA, USA). The purified 16S rDNA amplicons were treated with 10 U of FastDigest BseLI (Thermo Fisher Scientific, Waltham, MA, USA) for 10 min. An ABI PRISM 3130xl genetic analyzer (Thermo Fisher Scientific) was used to analyze the resultant DNA fragments, i.e., fluorescent-labeled terminal restriction fragments (T-RFs). GeneMapper software (Thermo Fisher Scientific) was used to determine the T-RF length and the peak area for each sample. T-RFs were divided into 29 operational taxonomic units (OTUs). The individual OTUs were quantified as the percentage of all OTUs combined based on the area under the curve (% AUC). The reference database, Human Fecal Microbiota T-RFLP profiling (http://www.tecsrg-lab.jp/t_rflp_hito_OTU.html), was used to putatively match the bacteria in each classification unit to the corresponding OTU. T-RFLP analyses enabled the classification of the sampled intestinal microbiota into the following 10 groups: *Bifidobacterium*, *Lactobacillales*, *Bacteroides*, *Prevotella*, *Clostridium* cluster IV, *Clostridium* subcluster XIVa, *Clostridium* cluster IX, *Clostridium* cluster XI, *Clostridium* cluster XVIII, and others.

2.5. Anthropometrical Measurements

The body mass index (BMI) was calculated as body weight/(height)^2, as expressed in kg/m^2. Bioelectrical impedance analysis using a body composition analyzer estiated the percentages of fat and muscle mass of the trunk and lower extremities (Nippon Shooter Ltd., Physion MD, Tokyo, Japan).

2.6. Physiological Performance

Quadriceps muscle strength was assessed using a strain gage dynamometer (ST-200S, MUL-TECH, Tokyo, Japan). Each subject performed two attempts on each leg and the maximum value of these four trials was marked for later analysis. The modified Kraus–Weber (K-W) test was used to assess trunk muscle strength [27]. This simple exercise test was based on the K-W Minimum test that was developed by Drs. Hans Kraus and Sonja Weber in the 1950s [28] to assess the strength and endurance of the trunk muscles. The trunk muscle strength of each subject was rated based on the total scores (full marks = 40) of the test (Supplementary Figure S1).

Four physical performance tests were conducted to evaluate motor ability and fitness: maximal step length (MSL), Timed Up and Go (TUG) test, single-leg standing, and the 6-min. walk test (6MWT). MSL was determined as the maximum possible stride per step of a subject. In the TUG test, we measured the time that is required for a subject to stand up from a chair, walk 3 m, turn, walk back to the chair, and sit down. In single-leg standing, we measured the maximum time that a subject could stand on one leg. In case a subject continued single-leg standing for over 120 s, the test was discontinued. All of the functional tests were conducted twice and the best scores were marked for analysis. Cardiorespiratory fitness was evaluated by the 6MWT according to the guidelines of the American Thoracic Society [29]. In short, he subjects were instructed to walk back and forth on a 25-m course as fast as possible for 6 min under the supervision of a medical doctor. They were permitted to stop and rest in case of fatigue. The investigator encouraged the subjects with routine phrases (e.g., "you are doing well" and "keep up the good work") once per minute during the test. The total distance (in meters) walked after 6 min. (6MWD) was recorded and used as an indicator of cardiorespiratory fitness, since performance on the 6MWD strongly correlates with peak oxygen uptake [30,31].

2.7. Daily Physical Activity Level

The parameters reflecting the daily physical activity level of the participants included the number of steps and the time spent in brisk walking—i.e., at an intensity of three METs or more—was estimated using the same three-axis accelerometer as that used during training in the AE group. This device also automatically calculates ethe nergy expenditure (EE) from METs based on a widely-accepted formula (EE (kcal) = 1.05 × METs × time (h) × body weight (kg) [32]). All of the subjects were instructed to wear the accelerometer throughout the one-week measurement period, except while sleeping and bathing, and to continue with daily activities as usual. The assessments were conducted before and after the 12-week intervention. The data, which were automatically stored on the device, were subsequently transferred to a computer while using specialized software (HR Joint® Smile Data Vision, TERUMO, Tokyo, Japan). The mean daily values of all parameters recorded during the one-week monitoring period were used for further analysis.

2.8. Laboratory Measurements

The blood samples were collected at 9 AM under standardized 12-h fasting conditions. Serum samples were stored at −80 °C until further analysis. The hexokinase UV method measured the plasma glucose levels, whereas serum insulin levels were determined by chemiluminescent enzyme immunoassay. Serum triglycerides, low-density lipoprotein cholesterol (LDL-C), and high-density lipoprotein cholesterol (HDL-C) were determined by enzymatic methods. The homeostasis model assessment of insulin resistance (HOMA-IR), which is an established surrogate index of insulin resistance [33], was also determined. The HOMA-IR was obtained from fasting plasma glucose (FPG) and serum insulin (FIRI) levels according to the original method by Matthews et al. [34] while using the following formula:

$$\text{HOMA-IR} = \text{FPG (mmol/L)} \times \text{FIRI } (\mu\text{U/mL})/22.5 \quad (1)$$

A higher HOMA-IR value represents higher insulin resistance.

2.9. Nutrient Intake

Nutrient intake was estimated using a food frequency questionnaire (FFQ), which the Japan Public Health Center-based Prospective Study developed and previously validated [35]. The FFQ consists of 138 food and beverage items and measures nine intake frequency categories: never or seldom, 1–3 times/month, 1–2 times/week, 3–4 times/week, 5–6 times/week, once/day, 2–3 times/day, 4–6 times/day, and more than seven times/day. All of the subjects were asked to complete the questionnaire before and after the 12-week intervention. FFQ data were analyzed with the help of Education Software Co., Ltd. (Tokyo, Japan), and then converted to quantitative estimates of the daily consumed amounts of energy, protein, lipid, carbohydrates, saturated fat, and dietary fiber.

2.10. Defecation Assessment

Defecation patterns were assessed using the Japanese version of the Constipation Assessment Scale (CAS-J), which was modified from the original scale that was developed by McMillan et al. [36] to assess constipation in Japanese populations [37]. The CAS-J comprises eight questions, i.e., "The abdomen appears distended or swollen", "The amount of flatus", "The frequency of defecation" "The rectum appears to be filled with feces", "Pain of the anus during defecation", "The amount of feces", "Ease of defecation", and "Diarrhea or watery stools". Each item includes a three-point rating scale: 0 ("no problem"), 1 ("some problem"), and 2 ("severe problem"). Thus, the maximum possible CAS-J score is 16, with higher scores indicating more severe cases of constipation.

2.11. Statistical Analyses

The data are presented as median and interquartile ranges. Changes in clinical parameters and relative abundances of specific classes of intestinal microbiota following intervention in each group were examined by the Wilcoxon Signed-rank test. The Spearman's rank correlation coefficient test examined the relationships between the parameters and changes in the relative abundance of specific types of intestinal microbiota. Stepwise regression analysis was also performed to identify the factors that determined the change in the relative abundance of specific microbiota. Finally, the Mann–Whitney U-test was used to compare the changes in the relative abundance of specific types of intestinal microbiota between the exercise groups according to the increase in time spent in brisk walking. All of the statistical procedures were performed using SPSS statistical software (version 24.0, IBM, New York, NY, USA). P values less than 0.05 were considered to be statistically significant.

3. Results

3.1. Clinical Characteristics of the Subjects

Figure 2 shows the procedural flowchart of the enrollment, measurement, intervention, and data analysis of this study. Two participants in the TM group and one in the AE group dropped out during the intervention period. A total of 12 participants in the TM group and 17 participants in the AE group completed the study. We could confirm that all of the subjects in the TM group participated in 90% or more of the sessions and that the mean adherence to the home exercise was 96.0%. The mean percentage of attendance at weekly meetings with the instructor was 97.1% in the AE group. Table 1 summarizes the clinical characteristics of both groups. The median age was 70 (65–77) years in the TM group and 70 (66–75) years in the AE group.

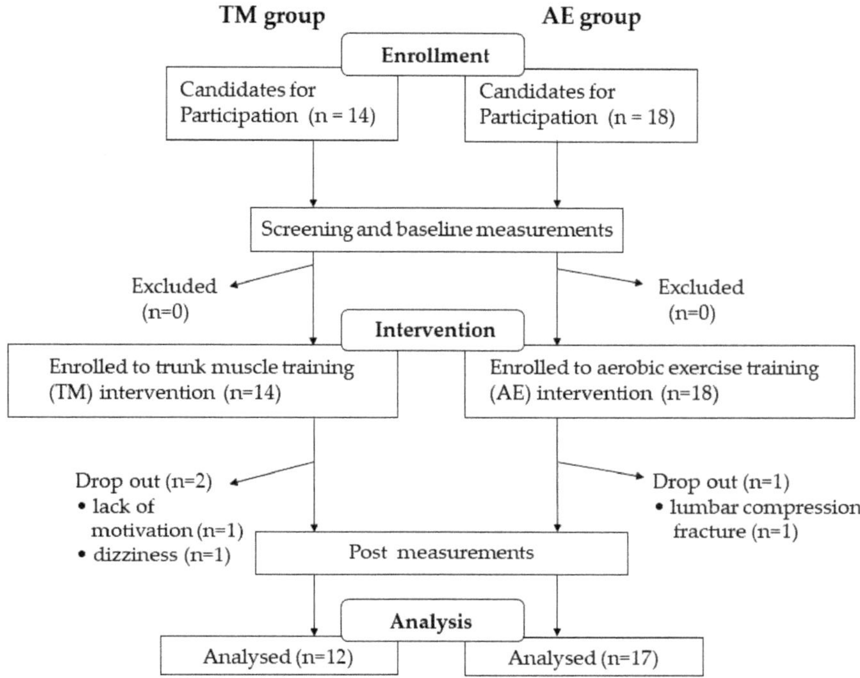

Figure 2. Flowchart of the screening, enrollment, intervention, and data analysis of the study. Abbreviations: TM, trunk muscle training; AE, aerobic exercise training.

Table 1. Clinical characteristics of the subjects.

		Total	TM Group	AE Group
n		29	12	17
Age	(years)	70 (66–75)	70 (66–77)	70 (66–75)
BW	(kg)	51.8 (47.8–56.5)	49.8 (48.3–56.8)	52.0 (46.9–56.0)
BMI	(kg/m^2)	21.4 (18.8–23.1)	20.6 (18.7–24.0)	21.7 (18.9–23.1)
Body fat	(%)	29.0 (23.6–32.7)	26.6 (22.9–32.2)	30.6 (25.1–33.0)
SBP	(mmHg)	141 (120–152)	129 (114–151)	142 (124–154)
DBP	(mmHg)	82 (74–92)	81 (74–86)	85 (74–93)
Present illness	*n* (%)			
No		17 (58.6)	9 (75.0)	8 (47.1)
Yes		12 (41.4)	3 (25.0)	9 (52.9)
Past history	*n* (%)			
No		15 (51.7)	7 (58.3)	8 (47.1)
Yes		14 (48.3)	5 (41.7)	9 (52.9)
Medication	*n* (%)			
No		19 (65.5)	10 (83.3)	9 (52.9)
Yes		10 (34.5)	2 (16.7)	8 (47.1)

Data are presented as median (interquartile range) for age, BW, body fat, SBP, and DBP, and as *n* (%) for present illness, past history, and medication. Abbreviations: TM, trunk muscle training; AE, aerobic exercise training; BW, body weight; BMI, body mass index; SBP, systolic blood pressure; DBP, diastolic blood pressure.

3.2. Changes in Body Composition, Muscle Strength, Physical Performance, and Daily Physical Activity Following the Intervention

Table 2 shows the changes in body composition, muscle strength, physical performance, and daily physical activity following the intervention in both groups.

Table 2. Changes in the parameters following the intervention.

		TM Group (n = 12)		AE Group (n = 17)	
		Baseline	**Post**	**Baseline**	**Post**
BMI	(kg/m^2)	20.6 (18.7–24.0)	20.8 (18.8–23.8)	21.7 (18.9–23.1)	21.3 (18.8–23.5)
Body fat	(%)	26.6 (22.9–32.2)	27.4 (23.7–31.9)	30.6 (25.1–33.0)	28.6 (25.1–33.75)
Leg muscle mass	(kg)	8.08 (7.06–8.29)	7.82 (6.80–8.16)	7.29 (7.03–8.08)	7.44 (7.12–8.25)
K-W test score	(/40)	15.5 (8.5–24.8)	27.5 (22.0–31.8) *	13.0 (9.0–16.5)	21.0 (15.5–29.0) *
Quad. muscle strength	(kg)	22.7 (20.1–29.2)	23.5 (22.1–30.8)	26.2 (19.9–32.5)	24.8 (20.6–29.2)
MSL	(cm)	111.6 (107.6–123.2)	111.5 (107.0–125.5)	112.9 (108.9–120.0)	113.1 (104.3–119.5)
TUG	(sec)	6.19 (5.60–6.77)	5.80 (5.40–6.50)	6.14 (5.50–6.80)	5.87 (5.59–6.42)
Single-leg standing	(sec)	28.6 (12.3–120.0)	70.9 (32.3–120.0)	98.5 (39.9–120.0)	120.0 (79.0–120.0)
6MWD	(m)	540.8 (521.0–570.0)	567.5 (538.0–627.6) *	550.0 (510.9–579.7)	582.7 (541.0–618.7) *
Number of steps	(steps/day)	6348 (5256–7267)	6438 (4443–8073)	7869 (6456–10246)	10297 (7396–14117) *
Time spent in brisk walking	(min/day)	10 (2–15)	9 (2–17)	16 (8–30)	45 (16–52) *
Total EE	(kcal/day)	1561.0 (1418.3–1672.8)	1561.5 (1406.3–1613.3) *	1598.0 (1478.0–1724.0)	1633.0 (1469.5–1844.0) *
Exercise-induced EE	(kcal/day)	125.5 (99.5–140.0)	125.5 (85.5–154.0)	161.0 (118.5–211.5)	228.0 (153.5–318.0) *
FPG	(mmol/L)	5.9 (5.5–7.0)	5.7 (5.3–6.8)	5.8 (5.2–6.1)	5.3 (5.1–6.3)
TG	(mmol/L)	1.08 (0.87–1.27)	1.07 (0.91–1.54)	0.89 (0.75–1.17)	1.06 (0.91–1.53)
LDL-C	(mmol/L)	3.45 (3.23–3.77)	3.40 (2.95–4.25)	3.72 (3.25–4.19)	3.72 (3.21–4.24)
HDL-C	(mmol/L)	1.60 (1.27–2.26)	1.66 (1.29–2.43)	1.73 (1.42–2.03)	1.68 (1.44–2.06)
Insulin	(pmol/L)	29.8 (21.7–33.7)	32.3 (25.8–60.4)	38.0 (26.2–54.5)	40.2 (25.1–59.6)
HOMA-IR		1.10 (0.74–1.45)	1.14 (0.86–2.55)	1.36 (0.84–2.05)	1.31 (0.80–2.32)

All values are presented as median (interquartile range). Changes in clinical parameters following intervention in each group were examined by the Wilcoxon Signed-rank test. *: $p < 0.05$ compared with baseline. Abbreviations: TM, trunk muscle training; AE, aerobic exercise training; BMI, body mass index; K-W test score, Kraus–Weber test score; Quad. muscle strength, Quadriceps muscle strength; MSL, maximal step length; TUG, Timed Up & Go; 6MWD, distance in the 6-min. walk test; EE, energy expenditure; FPG, fasting plasma glucose; TG, triglyceride; LDL-C, low-density lipoprotein cholesterol; HDL-C, high-density lipoprotein cholesterol; HOMA-IR, homeostatic model assessment of insulin resistance.

The number of steps (p = 0.004) and the time spent in brisk walking (p = 0.003), as well as the exercise-induced EE (p = 0.003), were significantly increased following the intervention in the AE group only. Total EE was significantly increased in the AE group (p = 0.012), while it was decreased in the TM group (p = 0.049) following the intervention. The K-W test scores (TM group: p = 0.008; AE group: $p < 0.001$) and 6MWD (TM group: p = 0.028; AE group: p = 0.001) were equally improved following the intervention in both groups. No further significant changes were observed in other parameters of motor ability following the interventions in either group.

3.3. Changes in Laboratory Measurements Following the Intervention

Table 2 shows the changes in the laboratory measurements following the intervention in both groups. FPG and blood levels of triglycerides, LDL-C, HDL-C, and insulin as well as HOMA-IR remained unchanged after the intervention in both groups.

3.4. Changes in Nutrient Intake and Defecation Pattern Following the Intervention

Table 3 shows the changes in nutrient intake and the CAS-J scores following the intervention. Significant differences in nutrient intake patterns as well as total energy intake were found neither at baseline nor after the interventions. Regarding the patterns of defecation, the CAS-J scores were significantly decreased in the AE group only ($p = 0.036$) following the interventions. For individual components of the CAS-J, the score on "Ease of defecation" was significantly decreased following the intervention in the AE group. On the other hand, the score in "The rectum appears to be filled with feces" was significantly improved following the intervention in the TM group only.

Table 3. Changes in nutrient intake and defecation pattern following the intervention.

		TM Group ($n = 12$)		AE Group ($n = 17$)	
		Baseline	**Post**	**Baseline**	**Post**
Nutrient intake					
Total energy	(kcal/day)	1863 (1827–1908)	1878 (1839–1942)	1874 (1795–1956)	1828 (1796–1942)
Carbohydrates	(g/day)	244.8 (237.6–252.7)	248.0 (243.0–255.3)	246.7 (240.8–258.1)	243.4 (238.9–255.2)
Protein	(g/day)	76.5 (74.2–83.1)	76.8 (74.4–84.2)	75.6 (71.6–82.9)	75.3 (71.8–82.1)
Lipid	(g/day)	59.2 (57.8–60.5)	59.9 (59.1–64.5)	58.9 (56.2–64.0)	58.5 (55.8–63.9)
Saturated fat	(g/day)	17.1 (16.7–17.7)	17.7 (16.7–20.0)	17.7 (16.1–19.1)	16.9 (16.0–19.1)
Fiber	(g/day)	17.6 (17.1–17.9)	18.2 (17.3–18.8)	17.6 (17.2–18.5)	17.7 (17.0–18.2)
Defecation pattern					
CAS-J	(/16)	3.50 (2.25–5.75)	3.50 (2.00–5.75)	2.00 (1.00–4.50)	2.00 (0.00–3.00) *
Abdomen appears distended or swollen	(/2)	0.0 (0.0–1.0)	0.0 (0.0–1.0)	0.0 (0.0–1.0)	0.0 (0.0–0.0)
Amount of flatus	(/2)	1.0 (0.0–1.0)	1.0 (0.0–2.0)	0.0 (0.0–1.0)	0.0 (0.0–1.0)
Frequency of defecation	(/2)	0.0 (0.0–1.0)	0.0 (0.0–1.0)	0.0 (0.0–1.0)	0.0 (0.0–0.5)
Rectum appears to be filled with feces	(/2)	1.0 (0.0–1.0)	0.0 (0.0–0.8) *	0.0 (0.0–1.0)	0.0 (0.0–0.0)
Pain of the anus during defecation	(/2)	0.0 (0.0–1.0)	0.0 (0.0–0.8)	0.0 (0.0–0.5)	0.0 (0.0–0.0)
Amount of feces	(/2)	0.0 (0.0–0.8)	0.0 (0.0–1.0)	0.0 (0.0–0.0)	0.0 (0.0–0.0)
Ease of defecation	(/2)	0.5 (0.0–1.0)	0.0 (0.0–1.0)	0.0 (0.0–1.0)	0.0 (0.0–1.0) *
Diarrhea or watery stools	(/2)	0.0 (0.0–1.0)	0.0 (0.0–0.0)	0.0 (0.0–1.0)	0.0 (0.0–0.0)

All values are presented as median (interquartile range). Changes in clinical parameters following intervention in each group were examined by the Wilcoxon Signed-rank test. *: $p < 0.05$ compared with baseline. Abbreviations: TM, trunk muscle training; AE, aerobic exercise training; CAS-J, Japanese version of the Constipation Assessment Scale.

3.5. Composition of Intestinal Microbiota

Figure 3 shows the composition of the intestinal microbiota in both groups. Following the interventions, the relative abundance of *Bacteroides* was significantly increased, and that of *Clostridium* subcluster XIVa was only decreased in the AE group. The relative abundance of *Clostridium* cluster IX was only significantly increased in the TM group. After the interventions, the relative abundance of other microbiota groups remained unchanged in both of the groups.

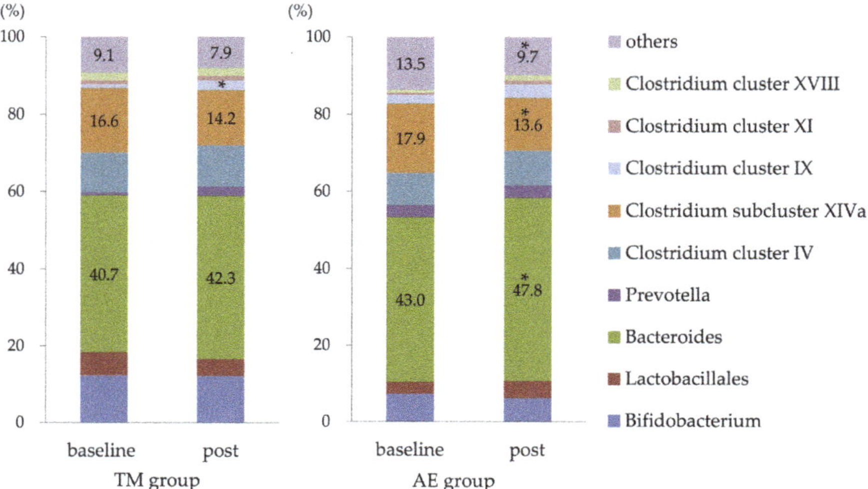

Figure 3. Changes in the composition of the intestinal microbiota following the intervention. The relative abundance of intestinal *Bacteroides* was significantly increased, and that of the *Clostridium* subcluster XIVa was decreased only in the AE group (by the Wilcoxon Signed-rank test). The relative abundance of *Clostridium* cluster IX was significantly increased only in the TM group. *: $p < 0.05$ compared with baseline. Abbreviations: TM, trunk muscle training; AE, aerobic exercise training.

3.6. Relationship between Changes in the Parameters and Change in the Relative Abundance of Intestinal Bacteroides after the Intervention

We examined the relationships between age, the relative abundance of intestinal *Bacteroides* before the intervention (pre-*Bacteroides*), or the changes in the parameters that were modulated by the exercise intervention and the change in the relative intestinal abundance of *Bacteroides* (Δ%*Bacteroides*). Pre-*Bacteroides* was negatively correlated with Δ%*Bacteroides* ($r = -0.519$, $p = 0.004$) when analyzing all of the subjects combined. A significant positive correlation was also found between the change in 6MWD (Δ6MWD; $r = 0.431$, $p = 0.020$) or that in time spent in brisk walking (ΔTime spent in brisk walking; $r = 0.371$, $p = 0.047$) and the Δ%*Bacteroides* following the intervention in all subjects (Figure 4). There were no significant correlations between changes in other parameters and Δ%*Bacteroides* among all the subjects combined (Table 4).

To identify the factors that contribute to Δ%*Bacteroides*, we performed stepwise multiple regression analysis, in which Δ%*Bacteroides* was included as the dependent variable and age, pre-*Bacteroides*, Δ6MWD, and ΔTime spent in brisk walking were included as the possible independent variables. In this analysis, Δ6MWD ($\beta = 0.370$, $p = 0.034$) and pre-*Bacteroides* ($\beta = -0.356$, $p = 0.041$) were found to be independent contributors ($R^2 = 0.317$).

Table 4. Correlation coefficients in simple regression analysis between clinical factors and the changes in the relative abundance of intestinal *Bacteroides* following the exercise intervention in all subjects.

Related Factors	Correlation Coefficient	*p* Value
Age	−0.343	0.068
Pre-*Bacteroides*	−0.519	0.004 *
ΔK-W test score	0.327	0.083
Δ6MWD	0.431	0.020 *
ΔNumber of steps	0.210	0.275
ΔTime spent in brisk walking	0.371	0.047 *
ΔTotal EE	0.216	0.261
ΔExercise-induced EE	0.250	0.191
ΔCAS-J	0.071	0.715

The relationships between the parameters and changes in the relative abundance of specific types of intestinal microbiota were examined by Spearman's rank correlation coefficient test. *: $p < 0.05$. Abbreviations: Pre-*Bacteroides*, the relative abundance of the intestinal *Bacteroides* before the intervention; K-W test score, Kraus–Weber test score; 6MWD, distance in the 6-min. walk test; EE, energy expenditure; CAS-J, Japanese version of the Constipation Assessment Scale.

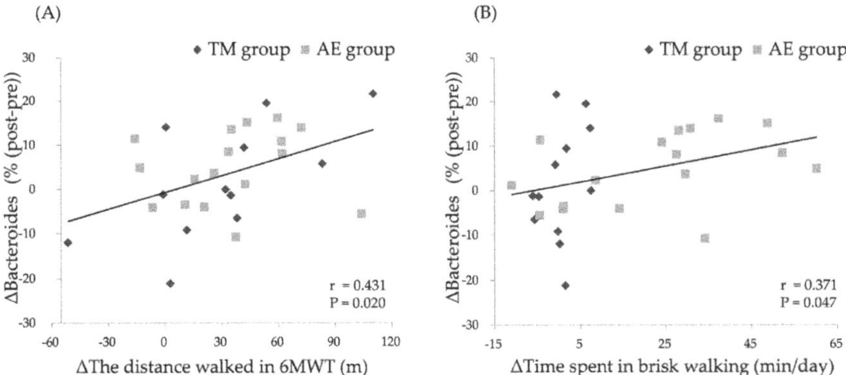

Figure 4. Relationship between changes in the distance during the 6MWT (6MWD) (**A**), changes in the time spent in brisk walking (**B**), and changes in the relative abundance of intestinal *Bacteroides* by the intervention. Improvements in 6MWD and time spent in brisk walking were positively correlated with increases in the relative abundance of intestinal *Bacteroides* in all subjects. Abbreviations: TM, trunk muscle training; AE, aerobic exercise training; 6MWT, 6-min. walk test; 6MWD, distance in the 6MWT.

3.7. *Effect of Increased Daily Physical Activity on Changes in the Relative Abundance of Intestinal Bacteroides Following the Intervention in the AE Group*

Although the improvement of 6MWD was observed in each group, a significant increase in the relative abundance of intestinal *Bacteroides* was only found in the AE group. Therefore, we focused on the effect of the increased time spent in brisk walking on Δ%*Bacteroides* following the intervention in the AE group. The subjects in the AE group were divided into two groups according to whether they had increased their time spent in brisk walking by more or less than 20 min. following the intervention. As shown in Figure 5, Δ%*Bacteroides* in subjects who added > 20 min. of time spent in brisk walking was greater than that in the subjects who added ≤ 20 min. (9.7% (4.7%–14.2%), $n = 10$ vs. −3.5% (−4.2% − 2.4%), $n = 7$; $p = 0.025$).

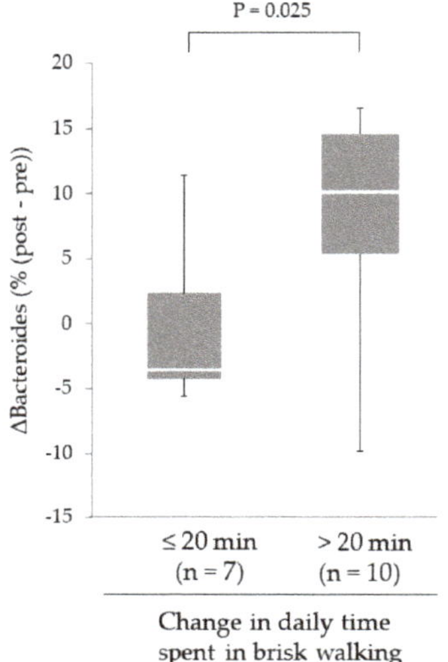

Figure 5. Effect of increased daily physical activity on changes in the relative abundance of intestinal *Bacteroides* following the intervention in the AE group. Increases in intestinal *Bacteroides* in subjects who increased the daily time spent in brisk walking for 20 min. or more were greater than in those who did not (by the Mann–Whitney *U*-test). Horizontal bars indicate the minimum values, the 25th, 50th, 75th percentile levels, and the maximum values. Abbreviations: AE, aerobic exercise training.

4. Discussion

The aim of the present study was to investigate whether exercise intervention modifies the composition of intestinal microbiota in healthy elderly women. Our main findings were that a 12-week aerobic exercise program that consists of daily episodes of brisk walking increased the relative abundance of intestinal *Bacteroides,* while improving cardiorespiratory fitness without any changes to nutrient intake. Moreover, the increase relative abundance of intestinal *Bacteroides* was especially marked in subjects who increased the time spent in brisk walking by more than 20 min. We also found that aerobic exercise improved the pattern of defecation independently of Δ%*Bacteroides.* Meanwhile, the elderly subjects who engaged in the trunk muscle training showed neither a significant Δ%*Bacteroides* nor changes in the defecation pattern.

To date, the primary findings of animal studies suggested that the level of exercise may modulate the composition of the intestinal microbiota. In rodents, six days of wheel running exercise increased *Bifidobacterium* and *Lactobacillus,* which are widely recognized as health-promoting intestinal bacteria [14]. It was also reported that the exercise-induced changes of the intestinal microbiota in mice depend on the exercise modalities (voluntary wheel running or forced treadmill running) and that voluntary wheel running reduced *Turicibacter* spp., which are associated with immune dysfunction and bowel diseases [15]. Another study demonstrated that a six-week schedule of interval treadmill running in mice enhanced the diversity of intestinal microbiota, with marked increases in the relative abundance of *Bacteroidetes* [16]. On the other hand, few data in humans have been published regarding the effect of exercise interventions on the intestinal microbiota. In a recent report by Allen et al. a six-week aerobic exercise training altered the intestinal microbiota differently, depending on body weight status [38]. The present study could further elaborate these results by demonstrating that a 12-week aerobic exercise program that consists of brisk walking—in contrast to training of trunk muscles—increased the relative abundance of intestinal *Bacteroides*. This suggests that aerobic exercise may beneficially modify the intestinal microbiota in healthy elderly women. In previous studies, the maintenance of an optimal intestinal environment has been shown to contribute to the prevention of various types of diseases [1–7]. The results of our study suggest a practical approach, i.e., aerobic exercise, as a strategy to attain the optimization of the intestinal microbiota in humans.

The 12-week aerobic exercise training increased the relative abundance of intestinal *Bacteroides*, in parallel with an improvement in cardiorespiratory fitness. Interestingly, a cross-sectional study has shown that cardiorespiratory fitness is associated with a larger proportion of *Bacteroides* in the intestinal microbiota of premenopausal women [18]. The results from our interventional study are consistent with—and augment—the significance of this observation. We also demonstrated that increases in the relative abundance of *Bacteroides* in the large intestine were greater in the subjects who improved the daily time spent in brisk walking at an intensity of ≥ 3 METs by more than 20 min. On the basis of the findings of the Nakanojo Study, brisk walking at an intensity of > 3 METs for 20 min. or more on most days are recommended for the elderly to reduce the risk of lifestyle-related diseases [39]. In particular, such exercise levels reduce the incidence of osteoporosis [40], metabolic syndrome, hypertension, and hyperglycemia [41]. Based on these considerations, we initially set the target volume of brisk walking for the AE group in our study at 20 min. per day. Our results demonstrate that this regimen also effectively modifies and optimizes the composition of the intestinal microbiota.

By contrast, trunk muscle training did not change the composition of the intestinal microbiota in subjects within our TM group, although it did improve the cardiorespiratory fitness. An improvement in cardiorespiratory fitness in the TM group may have resulted from the strengthening of the respiratory muscles by the trunk muscle training [42]. It may have also occurred because the subjects commuted to the sports center once weekly during the study period. However, the improvement in the cardiorespiratory fitness in the TM group did not coincide with a changed composition of the intestinal microbiota. Furthermore, in the present study, the increase in the time spent in brisk walking was positively correlated with the increase in the relative abundance of intestinal *Bacteroides*. To put these result into perspective, cardiorespiratory fitness may need to be improved by aerobic exercise, such as brisk walking, when the goal is to modify the intestinal microbiota.

There are some candidate mechanisms by which aerobic exercise might increase intestinal *Bacteroides*. Changes in the colonic transit time result in changes in pH within the colonic lumen that may be key in affecting the composition of the intestinal microbiota. Prolonged colonic transit time is known to limit the diversity of intestinal microbiota [43], and this coincides with a greater rise in pH during transit from the proximal to the distal colon [44]. Aerobic exercises, such as jogging and cycling at a moderate intensity, decrease intestinal transit time in healthy people [45] as well as middle-aged patients with chronic constipation [46], probably via increases in the visceral blood flow, increased release of gastrointestinal hormones, mechanical stimulation, and strengthening of the abdominal muscles [46]. Furthermore, aerobic exercise increases the fecal concentrations of the short-chain fatty acids (SCFA) [47], which slightly lowers the colonic-luminal pH [48]. *Bacteroides* species prefer mildly acidic conditions (pH 6.7) for their survival in the colonic lumen, whereas they grow poorly at pH 5.5 [49]. This may explain why aerobic exercise increases intestinal *Bacteroides*, although, of course, a more detailed analysis of the underlying factors remains necessary.

Bacteroides species are opportunistic bacteria: whether they positively or negatively affect host health depends on the characteristics of their intestinal environment. *Bacteroides* spp. play a role in protecting against inflammatory bowel disease [50,51], whereas they may increase infants' susceptibility to chronic allergic disease, such as early-onset atopic eczema [52]. Thus, future studies will need to detail the clinical consequences of aerobic-exercise-induced increases in *Bacteroides*. Nonetheless, it is widely accepted that lower levels of *Bacteroides* are associated with the higher prevalence of obesity and metabolic syndrome and that *Bacteroides* species may help in suppressing metabolic dysfunction [53,54], although we unfortunately could not evaluate waist circumference as a surrogate index of visceral fat accumulation in the present study. However, in the present study, the increase in intestinal *Bacteroides* in the AE group did not decrease insulin resistance, as assessed by HOMA-IR. This may be because the sedentary but healthy subjects that were included in the present study presented with good insulin sensitivity at baseline, making further improvements in insulin sensitivity following exercise difficult to attain. Further studies should clarify whether aerobic exercise might improve insulin sensitivity through the increase of intestinal *Bacteroides* in obese and/or insulin-resistant subjects.

It is widely accepted that *Bifidobacteria* and *Lactobacillales* contribute to intestinal health, preventing diarrhea and various infectious, allergic, and inflammatory conditions [55]. The relative abundances of these bacteria are decreased in elderly people [8], which may at least partly result in intestinal barrier dysfunction in the population [56]. In addition to some factors, such as probiotics and dietary fiber [56,57], vigorous exercise also has the potential for increasing these bacteria based on rodent studies [14]. However, in the present study, the relative abundance of *Bifidobacterium*, the only *Bifidobacteria* that can be identified by our T-RFLP analysis, as well as *Lactobacillales* remained unchanged in both groups. This may be because the quantity (time and intensity) of our brisk walking was not enough to increase these bacteria. Exploring an exercise prescription that can increase these bacteria will benefit intestinal health in elderly people.

A few limitations of the present study should be noted. First, our non-randomized study design with a relatively small subject sample size may have been insufficiently powered to detect differences in efficacy between the two exercise programs to affect the clinical outcomes, such as trunk muscle strength, cardiorespiratory fitness, defecation pattern, and the composition of the intestinal microbiota. Second, we confirmed that the participants had no substantial exercise habits before exercise intervention. However, exercise-induced energy expenditure at the baseline was greater in the AE than in the TM group. The reason for this may have been that the subjects who opted for the AE training were more aware of the health benefits of walking, which may have resulted in a superior effect of brisk walking to trunk muscle training on increasing intestinal *Bacteroides*. Finally, we classified fecal intestinal microbiota into only 10 major groups, which were present in the fecal samples of all the subjects. Therefore, it was impossivly to evaluate the effects of the exercise intervention on the diversity of the intestinal microbiota. A greater diversity of the intestinal microbiota is generally considered to provide various health benefits. It might have been possible to detect an increase in the diversity of the intestinal microbiota following the exercise intervention if the adopted microbiotic classification scheme had included several hundred subdivisions.

5. Conclusions

Aerobic exercise training targeting an increase of the time spent in brisk walking may have a potential for increasing intestinal *Bacteroides,* while also improving cardiorespiratory fitness in healthy elderly women. Exercise intervention may provide a practical means of acquiring a more optimal composition of intestinal microbiota. Further studies are needed to clarify the mechanism by which exercise exerts it effect on the composition of the intestinal microbiota.

Supplementary Materials: The following are available online at http://www.mdpi.com/2072-6643/11/4/868/s1, Figure S1: Modified Kraus–Weber test.

Author Contributions: Conceptualization, H.Y.; Data curation, E.M., H.Y. and T.H.; Formal analysis, E.M. and H.Y.; Funding acquisition, H.Y.; Investigation, E.M., H.Y., D.I., R.T., A.O., E.K. and T.H.; Methodology, H.Y. and T.H.; Project administration, H.Y.; Supervision, H.Y. and K.O.; Writing—original draft, E.M., H.Y. and T.H.; Writing—review & editing, H.Y., D.I., M.E., Y.S. and K.O.

Funding: The present study was funded by Nakatomi Foundation and by TIPNESS Inc. in Tokyo, Japan. The present research was also supported by the Osaka City University (OCU) Strategic Research Grant 2016 for basic researches, and in part by the OCU "Think globally, act locally" Research Grant for Female Researchers 2017 and 2018 through the hometown donation fund of Osaka City.

Acknowledgments: We are grateful to TIPNESS Inc. for skillfully instructing the study subjects during the exercise programs. We would also like to thank TERUMO Co., Ltd., for providing us with Mediwalk® devices to monitor the physical activity levels of the participants in the present study. We also acknowledge Nooshin Naghavi for her skillful technical assistance.

Conflicts of Interest: The authors declare no conflict of interest.

References

1. Zhu, L.; Liu, W.; Alkhouri, R.; Baker, R.D.; Bard, J.E.; Quigley, E.M.; Baker, S.S. Structural changes in the gut microbiome of constipated patients. *Physiol. Genom.* **2014**, *46*, 679–686. [CrossRef]
2. Turnbaugh, P.J.; Ley, R.E.; Mahowald, M.A.; Magrini, V.; Mardis, E.R.; Gordon, J.I. An obesity-associated gut microbiome with increased capacity for energy harvest. *Nature* **2006**, *444*, 1027–1031. [CrossRef] [PubMed]
3. Larsen, N.; Vogensen, F.K.; van den Berg, F.W.; Nielsen, D.S.; Andreasen, A.S.; Pedersen, B.K.; Al-Soud, W.A.; Sorensen, S.J.; Hansen, L.H.; Jakobsen, M. Gut microbiota in human adults with type 2 diabetes differs from non-diabetic adults. *PLoS ONE* **2010**, *5*, e9085. [CrossRef] [PubMed]
4. Sobhani, I.; Tap, J.; Roudot-Thoraval, F.; Roperch, J.P.; Letulle, S.; Langella, P.; Corthier, G.; Tran Van Nhieu, J.; Furet, J.P. Microbial dysbiosis in colorectal cancer (CRC) patients. *PLoS ONE* **2011**, *6*, e16393. [CrossRef]
5. Emoto, T.; Yamashita, T.; Sasaki, N.; Hirota, Y.; Hayashi, T.; So, A.; Kasahara, K.; Yodoi, K.; Matsumoto, T.; Mizoguchi, T.; et al. Analysis of Gut Microbiota in Coronary Artery Disease Patients: A Possible Link between Gut Microbiota and Coronary Artery Disease. *J. Atheroscler. Thromb.* **2016**, *23*, 908–921. [CrossRef]
6. Tamboli, C.P.; Neut, C.; Desreumaux, P.; Colombel, J.F. Dysbiosis as a prerequisite for IBD. *Gut* **2004**, *53*, 1057. [PubMed]
7. Cenit, M.C.; Sanz, Y.; Codoner-Franch, P. Influence of gut microbiota on neuropsychiatric disorders. *World J. Gastroenterol.* **2017**, *23*, 5486–5498. [CrossRef]
8. Woodmansey, E.J.; McMurdo, M.E.; Macfarlane, G.T.; Macfarlane, S. Comparison of compositions and metabolic activities of fecal microbiotas in young adults and in antibiotic-treated and non-antibiotic-treated elderly subjects. *Appl. Environ. Microbiol.* **2004**, *70*, 6113–6122. [CrossRef]
9. Dicksved, J.; Halfvarson, J.; Rosenquist, M.; Järnerot, G.; Tysk, C.; Apajalahti, J.; Engstrand, L.; Jansson, J.K. Molecular analysis of the gut microbiota of identical twins with Crohn's disease. *ISME J.* **2008**, *2*, 716–727. [CrossRef]
10. Salminen, S.; Gibson, G.R.; McCartney, A.L.; Isolauri, E. Influence of mode of delivery on gut microbiota composition in seven year old children. *Gut* **2004**, *53*, 1388–1389. [CrossRef] [PubMed]
11. David, L.A.; Maurice, C.F.; Carmody, R.N.; Gootenberg, D.B.; Button, J.E.; Wolfe, B.E.; Ling, A.V.; Devlin, A.S.; Varma, Y.; Fischbach, M.A.; et al. Diet rapidly and reproducibly alters the human gut microbiome. *Nature* **2014**, *505*, 559–563. [CrossRef]
12. Takahashi, H.; Kuriyama, S.; Tsubono, Y.; Nakaya, N.; Fujita, K.; Nishino, Y.; Shibuya, D.; Tsuji, I. Time spent walking and risk of colorectal cancer in Japan: The Miyagi Cohort study. *Eur. J. Cancer Prev.* **2007**, *16*, 403–408. [CrossRef]
13. Brown, J.C.; Rhim, A.D.; Manning, S.L.; Brennan, L.; Mansour, A.I.; Rustgi, A.K.; Damjanov, N.; Troxel, A.B.; Rickels, M.R.; Ky, B.; et al. Effects of exercise on circulating tumor cells among patients with resected stage I-III colon cancer. *PLoS ONE* **2018**, *13*, e0204875. [CrossRef]
14. Queipo-Ortuno, M.I.; Seoane, L.M.; Murri, M.; Pardo, M.; Gomez-Zumaquero, J.M.; Cardona, F.; Casanueva, F.; Tinahones, F.J. Gut microbiota composition in male rat models under different nutritional status and physical activity and its association with serum leptin and ghrelin levels. *PLoS ONE* **2013**, *8*, e65465. [CrossRef]
15. Allen, J.M.; Berg Miller, M.E.; Pence, B.D.; Whitlock, K.; Nehra, V.; Gaskins, H.R.; White, B.A.; Fryer, J.D.; Woods, J.A. Voluntary and forced exercise differently alters the gut microbiome in C57BL/6J mice. *J. Appl. Physiol.* **2015**, *118*, 1059–1066. [CrossRef]
16. Denou, E.; Marcinko, K.; Surette, M.G.; Steinberg, G.R.; Schertzer, J.D. High-intensity exercise training increases the diversity and metabolic capacity of the mouse distal gut microbiota during diet-induced obesity. *Am. J. Physiol. Endocrinol. Metab.* **2016**, *310*, E982–E993. [CrossRef]
17. Clarke, S.F.; Murphy, E.F.; O'Sullivan, O.; Lucey, A.J.; Humphreys, M.; Hogan, A.; Hayes, P.; O'Reilly, M.; Jeffery, I.B.; Wood-Martin, R.; et al. Exercise and associated dietary extremes impact on gut microbial diversity. *Gut* **2014**, *63*, 1913–1920. [CrossRef]
18. Yang, Y.; Shi, Y.; Wiklund, P.; Tan, X.; Wu, N.; Zhang, X.; Tikkanen, O.; Zhang, C.; Munukka, E.; Cheng, S. The Association between Cardiorespiratory Fitness and Gut Microbiota Composition in Premenopausal Women. *Nutrients* **2017**, *9*, 792. [CrossRef]
19. Estaki, M.; Pither, J.; Baumeister, P.; Little, J.P.; Gill, S.K.; Ghosh, S.; Ahmadi-Vand, Z.; Marsden, K.R.; Gibson, D.L. Cardiorespiratory fitness as a predictor of intestinal microbial diversity and distinct metagenomic functions. *Microbiome* **2016**, *4*, 42. [CrossRef]

20. Mahnic, A.; Rupnik, M. Different host factors are associated with patterns in bacterial and fungal gut microbiota in Slovenian healthy cohort. *PLoS ONE* **2018**, *13*, e0209209. [CrossRef]
21. Murtaza, N.; Burke, L.M.; Vlahovich, N.; Charlesson, B.; O'Neill, H.; Ross, M.L.; Campbell, K.L.; Krause, L.; Morrison, M. The Effects of Dietary Pattern during Intensified Training on Stool Microbiota of Elite Race Walkers. *Nutrients* **2019**, *11*, 261. [CrossRef]
22. Sakaki, S.; Takahashi, T.; Matsumoto, J.; Kubo, K.; Matsumoto, T.; Hishinuma, R.; Terabe, Y.; Ando, H. Characteristics of physical activity in patients with critical limb ischemia. *J. Phys. Ther. Sci.* **2016**, *28*, 3454–3457. [CrossRef]
23. Moriyama, N.; Urabe, Y.; Onoda, S.; Maeda, N.; Oikawa, T. Effect of Two month Intervention to Improve Physical Activity of Evacuees in Temporary Housing after the Great East Japan Earthquake: Pilot Study. *Int. J. Community Med. Health. Educ.* **2017**, *7*, 528.
24. Hosomi, K.; Ohno, H.; Murakami, H.; Natsume-Kitatani, Y.; Tanisawa, K.; Hirata, S.; Suzuki, H.; Nagatake, T.; Nishino, T.; Mizuguchi, K.; et al. Method for preparing DNA from feces in guanidine thiocyanate solution affects 16S rRNA-based profiling of human microbiota diversity. *Sci. Rep.* **2017**, *7*, 4339. [CrossRef]
25. Nagashima, K.; Hisada, T.; Sato, M.; Mochizuki, J. Application of new primer-enzyme combinations to terminal restriction fragment length polymorphism profiling of bacterial populations in human feces. *Appl. Environ. Microbiol.* **2003**, *69*, 1251–1262. [CrossRef]
26. Nagashima, K.; Mochizuki, J.; Hisada, T.; Suzuki, S.; Shimoura, K. Phylogenetic analysis of 16S ribosomal RNA gene sequences from Human fecal microbiota and improved utility of terminal restriction fragment length polymorphism profiling. *Biosci. Microflora* **2006**, *25*, 99–107. [CrossRef]
27. Koyama, Y.; Ishikawa, S.; Sukigara, S. Trunk Fitness of Female University Student evaluated by modified Kraus-Weber Test. *J. Ibaraki Christian Univ. II Soc. Nat. Sci.* **2007**, *41*, 211–220. (In Japanese)
28. Kraus, H.; Hirschland, R. Minimum muscular fitness of the school children. *Res. Quart.* **1954**, *25*, 178–188. [CrossRef]
29. ATS Committee on Proficiency Standards for Clinical Pulmonary Function Laboratories. ATS statement: Guidelines for the six-minute walk test. *Am. J. Respir. Crit. Care Med.* **2002**, *166*, 111–117. [CrossRef]
30. Cahalin, L.P.; Mathier, M.A.; Semigran, M.J.; Dec, G.W.; DiSalvo, T.G. The six-minute walk test predicts peak oxygen uptake and survival in patients with advanced heart failure. *Chest* **1996**, *110*, 325–332. [CrossRef]
31. Ross, R.M.; Murthy, J.N.; Wollak, I.D.; Jackson, A.S. The six minute walk test accurately estimates mean peak oxygen uptake. *BMC. Pulm. Med.* **2010**, *10*, 31. [CrossRef] [PubMed]
32. Garber, C.E.; Blissmer, B.; Deschenes, M.R.; Franklin, B.A.; Lamonte, M.J.; Lee, I.M.; Nieman, D.C.; Swain, D.P. American College of Sports Medicine. American College of Sports Medicine position stand. Quantity and quality of exercise for developing and maintaining cardiorespiratory, musculoskeletal, and neuromotor fitness in apparently healthy adults: Guidance for prescribing exercise. *Med. Sci. Sports Exerc.* **2011**, *43*, 1334–1359. [PubMed]
33. Yokoyama, H.; Emoto, M.; Fujiwara, S.; Motoyama, K.; Morioka, T.; Komatsu, M.; Tahara, H.; Shoji, T.; Okuno, Y.; Nishizawa, Y. Quantitative insulin sensitivity check index and the reciprocal index of homeostasis model assessment in normal range weight and moderately obese type 2 diabetic patients. *Diabetes Care* **2003**, *26*, 2426–2432. [CrossRef] [PubMed]
34. Matthews, D.R.; Hosker, J.P.; Rudenski, A.S.; Naylor, B.A.; Treacher, D.F.; Turner, R.C. Homeostasis model assessment: Insulin resistance and beta-cell function from fasting plasma glucose and insulin concentrations in man. *Diabetologia* **1985**, *28*, 412–419. [CrossRef]
35. Tsubono, Y.; Takamori, S.; Kobayashi, M.; Takahashi, T.; Iwase, Y.; Iitoi, Y.; Akabane, M.; Yamaguchi, M.; Tsugane, S. A data-based approach for designing a semiquantitative food frequency questionnaire for a population-based prospective study in Japan. *J. Epidemiol.* **1996**, *6*, 45–53. [CrossRef] [PubMed]
36. McMillan, S.C.; Williams, F.A. Validity and reliability of the Constipation Assessment Scale. *Cancer Nurs.* **1989**, *12*, 183–188. [CrossRef] [PubMed]
37. Fukai, K.; Sugita, A.; Tanaka, M. A developmental study of the Japanese version of the constipation assessment scale. *Jpn. J. Nurs. Res.* **1995**, *28*, 201–208.
38. Allen, J.M.; Mailing, L.J.; Niemiro, G.M.; Moore, R.; Cook, M.D.; White, B.A.; Holscher, H.D.; Woods, J.A. Exercise Alters Gut Microbiota Composition and Function in Lean and Obese Humans. *Med. Sci. Sports. Exerc.* **2018**, *50*, 747–757. [CrossRef]

39. Aoyagi, Y.; Shephard, R.J. Habitual physical activity and health in the elderly: The Nakanojo Study. Geriatr. *Gerontol. Int.* **2010**, *10*, S236–S243.
40. Shephard, R.J.; Park, H.; Park, S.; Aoyagi, Y. Objective Longitudinal Measures of Physical Activity and Bone Health in Older Japanese: The Nakanojo Study. *J. Am. Geriatr. Soc.* **2017**, *65*, 800–807. [CrossRef]
41. Park, S.; Park, H.; Togo, F.; Watanabe, E.; Yasunaga, A.; Yoshiuchi, K.; Shephard, R.J.; Aoyagi, Y. Year-long physical activity and metabolic syndrome in older Japanese adults: Cross-sectional data from the Nakanojo Study. *J. Gerontol. A. Biol. Sci. Med. Sci.* **2008**, *63*, 1119–1123. [CrossRef] [PubMed]
42. Volianitis, S.; McConnell, A.K.; Koutedakis, Y.; McNaughton, L.; Backx, K.; Jones, D.A. Inspiratory muscle training improves rowing performance. *Med. Sci. Sports. Exerc.* **2001**, *33*, 803–809. [CrossRef] [PubMed]
43. Tottey, W.; Feria-Gervasio, D.; Gaci, N.; Laillet, B.; Pujos, E.; Martin, J.F.; Sebedio, J.L.; Sion, B.; Jarrige, J.F.; Alric, M.; et al. Colonic Transit Time Is a Driven Force of the Gut Microbiota Composition and Metabolism: In Vitro Evidence. *J. Neurogastroenterol. Motil.* **2017**, *23*, 124–134. [CrossRef] [PubMed]
44. Abbas, A.; Wilding, G.E.; Sitrin, M.D. Does Colonic Transit Time Affect Colonic pH? *J. Gastroenterol. Hepatol. Res.* **2014**, *3*, 1103–1107. [CrossRef]
45. Oettle, G.J. Effect of moderate exercise on bowel habit. *Gut* **1991**, *32*, 8, 941–944. [CrossRef]
46. De Schryver, A.M.; Keulemans, Y.C.; Peters, H.P.; Akkermans, L.M.; Smout, A.J.; De Vries, W.R.; van Berge-Henegouwen, G.P. Effects of regular physical activity on defecation pattern in middle-aged patients complaining of chronic constipation. *Scand. J. Gastroenterol.* **2005**, *40*, 422–429. [CrossRef]
47. Matsumoto, M.; Inoue, R.; Tsukahara, T.; Ushida, K.; Chiji, H.; Matsubara, N.; Hara, H. Voluntary running exercise alters microbiota composition and increases n-butyrate concentration in the rat cecum. *Biosci. Biotechnol. Biochem.* **2008**, *72*, 572–576. [CrossRef]
48. Walker, A.W.; Duncan, S.H.; McWilliam Leitch, E.C.; Child, M.W.; Flint, H.J. pH and peptide supply can radically alter bacterial populations and short-chain fatty acid ratios within microbial communities from the human colon. *Appl. Environ. Microbiol.* **2005**, *71*, 3692–3700. [CrossRef] [PubMed]
49. Duncan, S.H.; Louis, P.; Thomson, J.M.; Flint, H.J. The role of pH in determining the species composition of the human colonic microbiota. *Environ. Microbiol.* **2009**, *11*, 2112–2122. [CrossRef]
50. Mazmanian, S.K.; Round, J.L.; Kasper, D.L. A microbial symbiosis factor prevents intestinal inflammatory disease. *Nature* **2008**, *453*, 620–625. [CrossRef]
51. Shukla, R.; Ghoshal, U.; Dhole, T.N.; Ghoshal, U.C. Fecal Microbiota in Patients with Irritable Bowel Syndrome Compared with Healthy Controls Using Real-Time Polymerase Chain Reaction: An Evidence of Dysbiosis. *Dig. Dis. Sci.* **2015**, *60*, 2953–2962. [CrossRef]
52. Kirjavainen, P.V.; Arvola, T.; Salminen, S.J.; Isolauri, E. Aberrant composition of gut microbiota of allergic infants: A target of bifidobacterial therapy at weaning? *Gut* **2002**, *51*, 51–55. [CrossRef]
53. Ley, R.E.; Turnbaugh, P.J.; Klein, S.; Gordon, J.I. Microbial ecology: Human gut microbes associated with obesity. *Nature* **2006**, *444*, 1022–1023. [CrossRef]
54. Santacruz, A.; Collado, M.C.; Garcia-Valdes, L.; Segura, M.T.; Martin-Lagos, J.A.; Anjos, T.; Marti-Romero, M.; Lopez, R.M.; Florido, J.; Campoy, C.; et al. Gut microbiota composition is associated with body weight, weight gain and biochemical parameters in pregnant women. *Br. J. Nutr.* **2010**, *104*, 83–92. [CrossRef]
55. Vlasova, A.N.; Kandasamy, S.; Chattha, K.S.; Rajashekara, G.; Saif, L.J. Comparison of probiotic lactobacilli and bifidobacteria effects, immune responses and rotavirus vaccines and infection in different host species. *Vet. Immunol. Immunopathol.* **2016**, *172*, 72–84. [CrossRef]
56. Lahtinen, S.J.; Tammela, L.; Korpela, J.; Parhiala, R.; Ahokoski, H.; Mykkänen, H.; Salminen, S.J. Probiotics modulate the Bifidobacterium microbiota of elderly nursing home residents. *Age (Dordr.)* **2009**, *31*, 59–66. [CrossRef]
57. Bressa, C.; Bailen-Andrino, M.; Perez-Santiago, J.; González-Soltero, R.; Pérez, M.; Montalvo-Lominchar, M.G.; Maté-Muñoz, J.L.; Domínguez, R.; Moreno, D.; Larrosa, M. Differences in gut microbiota profile between women with active lifestyle and sedentary women. *PLoS ONE* **2017**, *12*, e0171352. [CrossRef]

Article

Diet Supplemented with Antioxidant and Anti-Inflammatory Probiotics Improves Sperm Quality after Only One Spermatogenic Cycle in Zebrafish Model

David G. Valcarce [1], Marta F. Riesco [1], Juan M. Martínez-Vázquez [1] and Vanesa Robles [1,2,*]

[1] IEO, Spanish Institute of Oceanography, Planta de Cultivos El Bocal, Barrio Corbanera, 39012 Monte, Santander, Spain; dgvalcarce@gmail.com (D.G.V.); riesco.mf@gmail.com (M.F.R.); juanma.martinez@ieo.es (J.M.M.-V.)

[2] MODCELL GROUP, Department of Molecular Biology, Universidad de León, 24071 León, Spain

* Correspondence: vanesa.robles@gmail.com; Tel.: +34-9423-48397

Received: 20 March 2019; Accepted: 11 April 2019; Published: 13 April 2019

Abstract: Infertility is a medical concern worldwide and could also have economic consequences in farmed animals. Developing an efficient diet supplement with immediate effects on sperm quality is a promising tool for human reproduction and for domesticated animal species. This study aims at elucidating the effect of a short-time probiotic supplementation consisting of a mixture of two probiotic bacteria with proven antioxidant and anti-inflammatory activities on zebrafish sperm quality and male behavior. For this purpose, three homogeneous groups of males in terms of motility (<60%) were established. The control group was fed with a normal standard diet. The other received supplements: One group (vehicle control) was fed with maltodextrin and the other received a probiotic preparation based on a mixture (1:1) of *Lactobacillus rhamnosus* CECT8361 and *Bifidobacterium longum* CECT7347. The feeding regime was 21 days corresponding with a single spermatogenesis in zebrafish. The preparation did not modify animal weight, positively affected the number of fluent males, increased sperm concentration, total motility, progressive motility, and fast spermatozoa subpopulations. Moreover, the animals fed with the supplement showed different behavior patterns compared to control groups. Our results suggest a diet-related modulation on the exploration activity indicating a lower stress-like conduct. The studied formulation described here should be considered as advantageous in male reproductive biotechnology.

Keywords: sperm quality; probiotics; zebrafish; motility; behavior

1. Introduction

Infertility is a highly ubiquitous global health problem. It has been recognized as a public health issue worldwide by the World Health Organization (WHO) and is predicted to affect 9% of the world population on average [1]. Altered production of functional and motile spermatozoa is a causal factor in up to 70% of infertility cases [2]. Sperm concentration has reduced by half in Western countries in the last four decades, without evidence of improvement [3]. Moreover, a high proportion of young men have sperm counts below the fertile threshold [4]. A variety of environmental factors may be contributing to reduced semen quality [4]. One of them is diet [5]. For example, there exists vast evidence regarding the adverse consequences of high-fat diets on male reproductive success [6]. Describing the dietary factors that can influence male fertility potential is of high interest. Nowadays, there is strong and consistent evidence about antioxidants as essential factors for sperm defense [7]. Selenium, vitamin E, vitamin C, folate, carotenoids, zinc, or carnitine are antioxidants naturally found in semen samples. The main function of these molecules is helping to overcome reactive oxygen species

(ROS) production from free radicals [8]. Since subfertile men have been identified as having lower levels of these scavengers in their semen [9], they have become an object of study for reproductive biologists [10].

Probiotics are living microorganisms that improve animal health status when integrated in the diet [11]. These microorganisms act by balancing the gut microbiota, regulating the intestinal transit, modulating intestinal villi, and protecting nutrient digestion and absorption. The intestinal microbiome is a complex ecosystem, which provides numerous crucial functions to the host organism [12]. During the last decades, gut microbiota (16S rRNA surveys are used to taxonomically identify the microorganisms in the environment [13]) has emerged as a key factor which regulates host metabolism and different gut microbiome phenotypes (the genes and genomes of the microbiota, as well as the products of the microbiota and the host environment [13]) have been associated with diseases [14]. Therefore, being able to regulate intestinal microbiota is of huge interest for scientists due to the potential implications in several fields of knowledge. To date, a myriad of species from *Bacillus*, *Enterococcus*, *Lactococcus*, *Streptococcus*, *Bifidobacterium*, or *Lactobacillus* have been used as probiotics [15], with the last two genera being the most used for this purpose [16].

Nowadays, zebrafish (*Danio rerio*) is accepted by the scientific community as a vertebrate model for the study of genetics, development, and diseases among others [17,18]. This teleost is also a good model for probiotic-related experiments [19] since zebrafish microbiota is comparable to that of human as well as gut colonization [20,21]. The aim of this study was to provide, by in vivo experimentation, new insights into the potential positive effects of probiotics on male reproductive biology. The finding of new ways to increase sperm quality would be useful for clinical protocols and the possible applications derived from the potential beneficial effects of probiotics on reproductive biology will be of interest not only in human reproduction, but also in animal production where nutrition is a key element.

Our hypothesis is that probiotics during a single spermatogenesis cycle can improve sperm quality as well as animal welfare. In order to validate this hypothesis, in this work we use zebrafish since it is an optimal model for reproductive biology because of its easy reproduction, low-cost maintenance, and fast cystic spermatogenesis (21 days) [22]. To verify whether a short-term exposure to probiotics has effects on sperm quality and animal behaviour, we exposed zebrafish adult males with different initial sperm quality to a multistrain probiotic combination containing two previously described bacteria with antioxidant and anti-inflammatory activities: *Lactobacillus rhamnosus* CECT8361 and *Bifidobacterium longum* CECT7347 [23,24].

2. Materials and Methods

2.1. Ethics Statement

The institutional Animal Care and Use Committee at the Marine Culture Plant El Bocal of the Spanish Institute of Oceanography in Santander (Spain) approved the experimental design and all protocols and procedures including animals (PI-10-16). All animals were manipulated in accordance with the Guidelines of the European Union Council (86/609/EU, modified by 2010/62/EU), following Spanish regulations (RD/1201/2005, abrogated by RD/2013) for the use of laboratory animals.

2.2. Animals

Wild-type zebrafish (Ab strain) were housed in the Marine Culture Plant El Bocal zebrafish platform of the Spanish Institute of Oceanography in Santander (Spain). Fish were bred and maintained according to standard protocols. In all trials, males were anesthetized in 110 mg/L buffered tricaine methane sulfonate (MS222). All efforts were made to reduce suffering and a humane endpoint was applied with a lethal dose of anesthetic if fish reached a moribund state.

2.3. Visible Implant Elastomer Tags (VIE) Tagging

VIE tags were prepared following manufacturer's indications adapted for the minimal volume (Northwest Marine Technology, Shaw Island, WA, USA). Green and red fluorescent elastomers (viscoelastic polymers) were used in the experiment. A code combining colors, number of tags, and positions (taking as reference the anteroposterior and dorsoventral axis) was generated and individually associated to a specific male in the zebrafish colony (Figure 1A). An expert hand injected small amounts of elastomers (dot shaped) in each anesthetized male. After injection, tag retention and injury evaluation (not registered) was evaluated in the recovery tanks. Health status was checked daily.

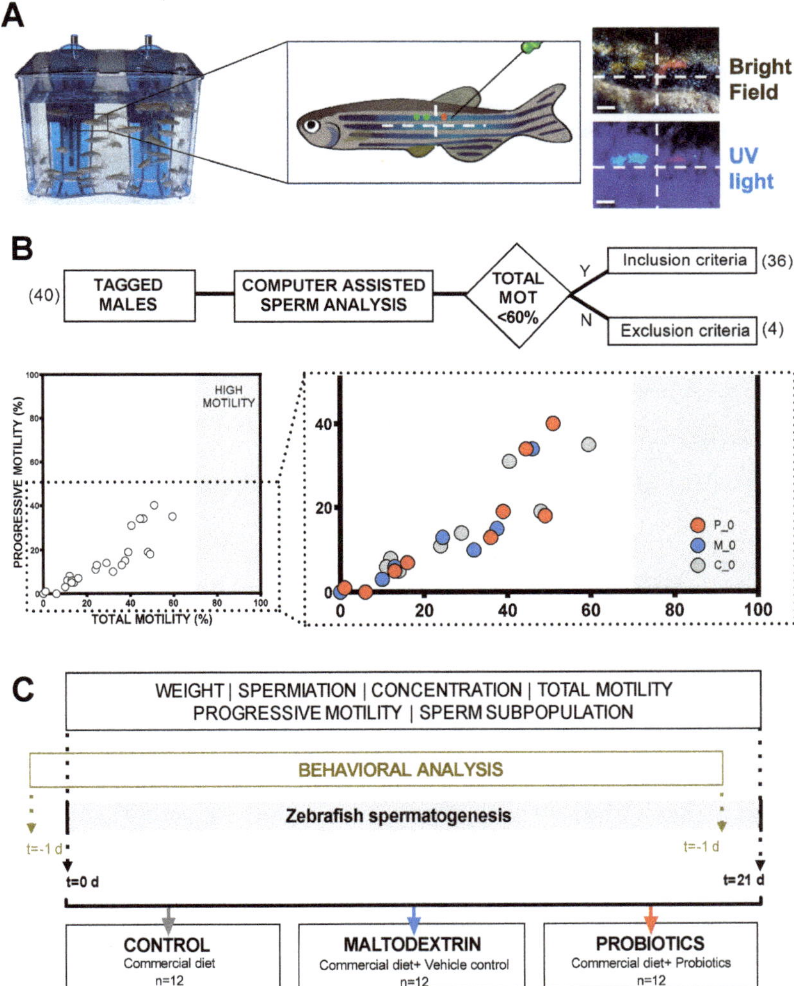

Figure 1. Study design summary. (**A**) Animals participating in the experiment ($n = 40$) were tracked with fluorescent visible implant elastomers. Each male carried a unique code visible under white and UV light. (**B**) Only males that reached the inclusion criteria described in the flowchart ($n = 36$) were selected for creating homogeneous experimental groups in terms of total motility. (**C**) Each group ($n = 12$) received a different diet during 21 days corresponding to a *Danio rerio* spermatogenesis cycle. "C", "M", and "P" refer to the experimental groups: Control, maltodextrin, and probiotics, respectively.

2.4. Inclusion Criteria, Experimental Group Definition, Study Design, and Feeding Regimes

Adult zebrafish males were anesthetized and sperm samples were collected and evaluated with a CASA (computed assisted sperm analysis) system (see below for procedure). Only males showing a total motility under 60% were selected for the experiment (Figure 1B). The inclusion criterion was established this way to analyze the effect of the probiotic strains mixture on diverse quality sperm samples. The cutoff value of 60% was chosen with the aim that there would be an improvement margin in the samples after the treatment. Semen samples over 60% can be considered acceptable samples in terms of motility. Males reaching the inclusion criteria were used to generate three homogeneous groups (n = 12) in terms of motility (Figure 1B). Each group had a different feeding regime: (1) The control group "CTRL" ingested only a commercial diet; (2) the vehicle control group "MALTO" received the commercial diet and two doses of 0.11 g of maltodextrin; and (3) the experimental group "PROBIO" received the commercial diet and a probiotic treatment consisting of a daily 10^9 Colony Forming Units (CFU) mixture (1:1) of lyophilized *L. rhamnosus* CECT8361 and *B. longum* CECT7347 strains carried in 0.22 g of maltodextrin. Strains were kindly provided by the company Biopolis S.L. (Valencia, Spain) and the commercial diet provided to all animals within the experiment twice a day (55% min. crude protein, 15% min. crude fat, 1.5% max. crude fiber, and 12% max. moisture) was purchased from Aquatic Animals (Apopka, FL, USA). In order to guarantee the ingestion, supplements (vehicle or probiotic mixture) were provided to experimental males in rearing water 30 min before each routine feeding. All experimental groups were held under the same conditions during all experiments. The experiment was replicated three times, including four males per experimental group each time (final population for each experimental group, n = 12). Sperm analysis was performed at t = 0 days and t = 21 days of each experimental replicate. Feeding regimes were maintained during the 21 days according to a spermatogenesis cycle in the species (Figure 1C).

2.5. BiometricAnalysis

At day 0 and day 21, fish weight was determined using a microbalance (Mettler MT5, Mettler Toledo, Spain).

2.6. Sperm Sampling

At 0 days and at 21 days, semen was collected approximately 1h after the lights of the zebrafish facility were turned on. Each fish was identified by checking its VIE tag code and after that, they were anesthetized one by one. Once the absence of reflexes was corroborated, the animals were gently located on a sponge, the surrounding area of the urogenital pore was dried and sperm collection was performed by abdominal massage using glass flat forceps as tools to smoothly press both sides according to routine protocols. Ejaculates were collected with a micropipette and diluted in 10 µL of buffered Hank's solution (0.137 M NaCl; 5.4 mM KCl; 0.25 mM Na_2HPO_4; 0.44 mM KH_2PO_4; 1.3 mM $CaCl_2$; 1.0 mM $MgSO_4$; 4.2 mM $NaHCO_3$). The diluted samples were stored at 22°C until analysis (5 min).

2.7. CASA Sperm Analysis

The activation of motility was performed by diluting 1 µL of sperm with 9 µL of system water (~300 mOsm/L) at 28 °C. Sperm motility, kinetics, and concentration were analyzed using a CASA system with ISAS software (ISAS, PROiSERR+D, S.L. Spain). Activated sperm was loaded into a Makler counting chamber (10 µm depth; Sefi Medical Instruments, Haifa, Israel). The CASA system consisted of a tri-ocular optical phase-contrast Nikon Eclipse Ts2R microscope (Nikon, Tokyo, Japan) using a 10× objective equipped with Basler A312fc digital camera (Basler Vision Technologies, Ahrensburg, Germany). The ISAS software was used with specific settings for fish spermatozoa (1 μm^2 < particle area < 20 μm^2; cell description according to VCL (curvilinear velocity): 10 µm/s < slow < 45 µm/s < medium < 100 µm/s < fast); and it rendered the following parameters: (1) Concentration; (2) percentage of motile spermatozoa (MOT,%); (3) percentage of progressive spermatozoa (P-MOT,%) defined as the percentage of spermatozoa which swim forward in 80% of a straight line; (4) curvilinear velocity (VCL, μm/s) defined as the time per average velocity of a sperm head along its actual curvilinear trajectory; (5) average path velocity (VAP, μm/s) defined as the time per average velocity of a sperm head along its spatial average trajectory; (6) straight-line velocity (VSL, μm/s) defined as the time per average velocity of a sperm head along the straight line between its first-detected position and its last position; (7) linearity of the curvilinear path (LIN, %), expressed as VSL/VCL; (8) straightness (STR,%) defined as VSL/VAP; (9) wobble (WOB,%) expressed as VAP/VCL; (10) amplitude of the lateral head displacement (ALH, μm); and (11) beat cross frequency (BCF, Hz) based on VCL crossing VAP per second. Motility parameters were evaluated at 15 s after activation to avoid drifting and to corroborate that all samples were measured at an exact equal time post activation. At least, 200 spermatozoa were analyzed for each sample. Three fields per sample were evaluated. If samples reported very low concentrations, more than three fields were captured.

2.8. Behavior Analysis

To test the exploratory behavior of the animals, we used a novel tank test (NTT), which evokes motivational conflict between the "protective" diving behavior and subsequent vertical examination following established procedures. Briefly, each animal was individually placed in the evaluation arena (20 cm (x) × 18 cm (y) × 8 cm (z); swimming volume: 3.5 L). Males were let to acclimate to the new environment for 3 min and right after animal behaviour was filmed (1920 × 1080 px) for 3 min. Individual male swimming activity was monitored using the free digital video tracking software Tracker (physlets.org/tracker/). The actual position of the animal was manually located every 20 frames to avoid possible inaccuracies of the automatic option of the software. Then, each resulting track was evaluated using a virtual grid pattern with upper and lower subareas in order to allow quantification and comparison between experimental groups. For each animal we quantified two estimates of exploratory behavior: Number of crossings between the upper-half subarea and the lower one and the percentage of time spent in the upper half of the arena.

2.9. Data Analysis

Results are expressed as the mean ± standard error. Statistical differences between mean values of each variable at 0 and 21 days were determined using a t-Student test for correlated variables for normally distributed variables or a Wilcoxon test for paired samples for non-parametric variables. A principal component analysis was performed for the set of observed variables for CASA parameters. All statistical analysis were performed using Prism 8 (GraphPad Software, San Diego, CA, USA) and SPSS V. 22 (SPSS Inc., Chicago, IL, USA). P-values < 0.0500 were considered statistically significant.

3. Results

3.1. Effects of Probiotic Mixture Supplementation on Total Body Weight and Spermiation Capacity

In order to investigate the effects of probiotic supplementation in male zebrafish on growth parameters, we weighed the animals included in the experiment at t = 0 days and t = 1 days (Figure 1). As expected, taking into account the short temporal frame of our experiment, our analysis revealed no statistical differences ($p > 0.0500$) at day 21 in any of the experimental groups: Control (C; CTRL); maltodextrin, the vehicle control (M; MALTO); or probiotic-fed (P; PROBIO) (Figure 2A; Supplementary File Table S1). These data provide confidence about the suitability of maltodextrin as a carrier in our probiotic-fed group.

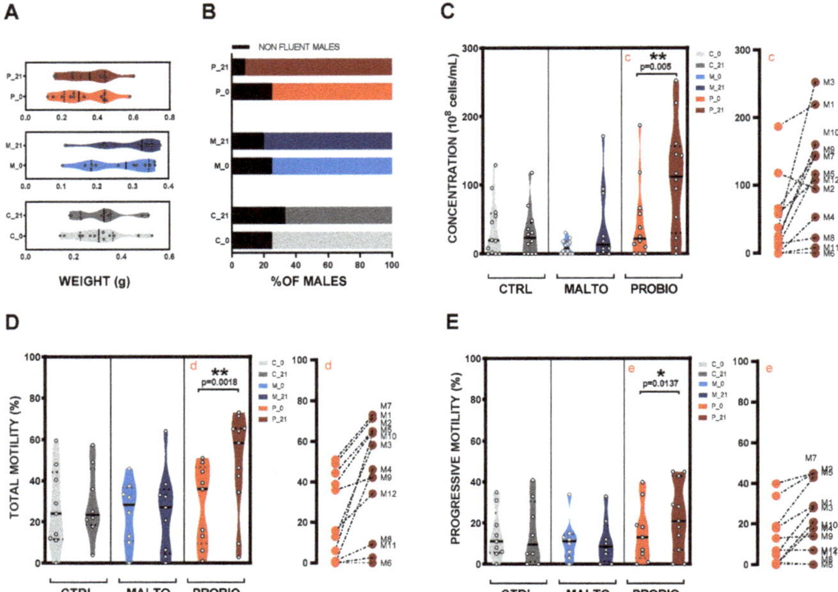

Figure 2. Probiotic mixture supplementation effects on zebrafish males and sperm quality after a cycle of spermatogenesis exposure. (**A**) None of the studied diets modified male total body weight. (**B**) Spermiation ability of studied males before and after the experiment. (**C**) Concentration, (**D**) total motility, and (**E**) progressive motility at 0 and 21 days obtained for each experimental group represented with violin graphs. "C", "M", and "P" refer to the experimental groups: Control, maltodextrin, and probiotics respectively. Furthermore, "c", "d", and "e" are before–after graphs for the PROBIO group where "M#" indicates the number of the male. Asterisks show statistically significant differences: * ($p < 0.0500$), ** ($p < 0.0100$).

As a first general parameter regarding spermatogenesis, we focused our attention on spermiation ability. As a result of a correct spermatogenesis, mature spermatozoa are released from cysts into the lumen of the tubules and therefore ejaculated. At day 0, we arranged population homogeneously with three (25%; CTRL), four (33%; MALTO), and three (25%; PROBIO) non-spermiating males in each group. After a single cycle of spermatogenesis (21 days) the non-fluent male percentage changed as follows: four (33%; CTRL), two (20%; MALTO), and one (8.33%; PROBIO). Interestingly, a spermiation modulation was suggested with this data in males supplemented with bacteria strains (PROBIO). In this group, only one male did not provide sperm at day 21 sampling (Figure 2B).

3.2. *Effects of the Probiotic Mixture on Concentration, Total Sperm Motility, and Progressive Motility*

To study the effects of the ingested probiotic strains on a single cycle of spermatogenesis, we studied individually, using VIE tagging for male tracking, the sperm samples in terms of concentration, total motility, and progressive motility. At day 0, all males included in this study presented an initial total motility below 60%. Groups were created including a wide range of sperm motility values from 0% motility to the 60% threshold (Figure 1B). Results regarding sperm concentration, total motility, and progressive motility are presented in Figure 2.

Bacteria ingestion strongly modified (p = 0.0050) sperm count (10^8 cells/mL; mean ± s.e.) in the PROBIO group after 21 days of supplementation (Figure 2C; Table S2). The mean value for this variable increased from 44.58 ± 16.40 to 110.10 ± 23.13. Controls reported lower concentrations at day 21: 30.19 ± 10.15 (CTRL) and 41.57 ± 18.16 (MALTO), respectively.

Regarding total motility (%; mean ± s.e.), controls showed similar values ($p > 0.0500$) before and after the experiment. Mean values were: 26.44 ± 6.528 (day 0) vs. 28.97 ±6.194 (day 21) for the CTRL group and 24.56 ± 5.53 (day 0) vs. 24.97 ±6.77 (day 21) for the MALTO group (Figure 2D; Table S2). In contrast, the animals with a feeding regime supplemented with probiotics (PROBIO) revealed a substantial rise ($p = 0.0018$) in total motility from 28.39 ± 6.46 (day 0) to 48.36 ± 7.32 (day 21). When a detailed individual evaluation of data was performed, results showed strong increments (>40%) in 11 of 12 studied males. Four fish increased their sperm motility more than 100%: M3, M8, M10, and M12 (Figure 2D.d). These results evidence a very strong positive effect of probiotic ingestion on zebrafish sperm quality. Please find individual before–after graphs for CTRL and MALTO groups in Figure S1.

Moreover, and concomitantly with total motility, progressive motility (P-MOT) was also significantly raised ($p = 0.0137$) after 21 days in the PROBIO group from 15.22% ± 4.71% to 22.73% ± 5.09% (Figure 2E; Table S2). P-MOT is another key parameter in sperm quality since it influences fertilization success, and it is a focus of attention in zebrafish research [25,26]. In our experiment, we considered progressive those cell in which swimming track was forwards in 80% of a straight line.

3.3. *Effects of the Probiotic Mixture on Sperm Kinematic Parameters*

We studied sperm kinetics in depth since it has been reported that external-fertilizing fish have the highest known intensity of sperm competition of any external fertilizing vertebrates. Thus, the presence of fast subpopulations within the motile cells seems to be an advantage and, therefore, it can be considered a parameter of sperm quality. In this experiment subpopulations were clustered in terms of VCL. Four groups were established and set up in the CASA software: Static, slow, medium, and fast according to the following thresholds: 10 μm/s < slow < 45 μm/s< medium < 100 μm/s < fast. Interestingly, only the PROBIO group showed statistical differences at the end of the experiment (Figure 3A) in the four subpopulations: Static ($p = 0.006$), slow ($p = 0.0208$), medium ($p = 0.0270$), and fast ($p = 0.0323$) cells.

Concerning kinematic parameters, there was no overall difference ($p > 0.0500$) in sperm velocities (VCL, VSL, VAP), linearity (LIN), straightness (STR), wobble (WOB), amplitude of the lateral head displacement (ALH), or beat cross frequency (BCF). Figure 3B shows these results for the PROBIO experimental group. These results suggest that the effect of probiotic bacteria do not fine tune zebrafish sperm kinetics. PCA results can be found in Figure S2.

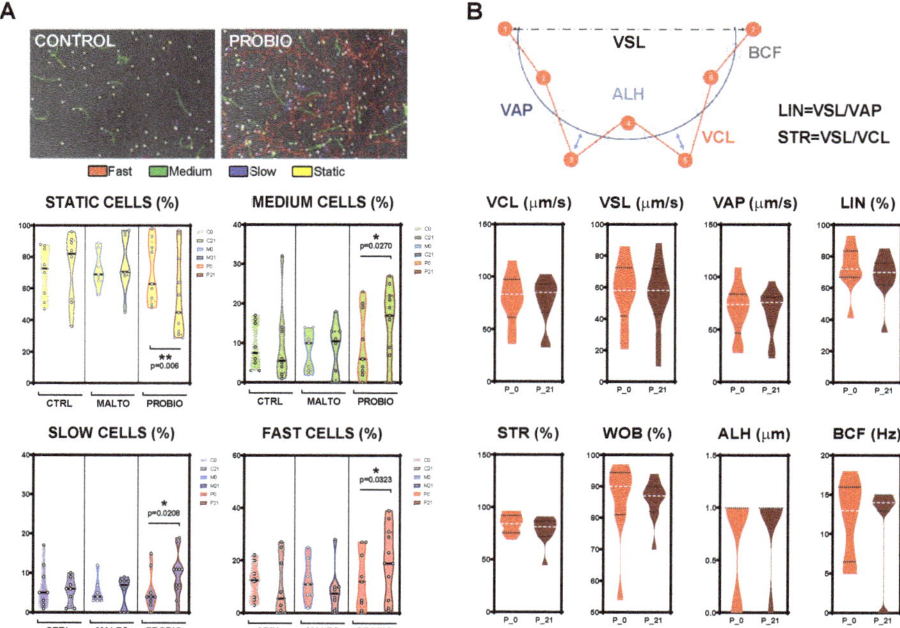

Figure 3. Probiotic mixture supplementation effects on zebrafish sperm kinematics after a cycle of spermatogenesis exposure. (**A**) Sperm subpopulations within the motile population according to speed parameters before and after the experiment for each group. (**B**) Detailed sperm kinetics rendered by the CASA system for the PROBIO group. Asterisks show statistically significant differences; * ($p < 0.0500$). Abbreviations: VCL—curvilinear velocity, VSL—straight line velocity, VAP—average path velocity, LIN—linearity of the curvilinear path, STR—straightness, WOB—wobble, ALH—amplitude of the lateral head displacement, BCF—beat cross frequency.

3.4. *Effects of the Probiotic Mixture on Male Behavior*

After three minutes of adaptation time to a new environment, tracking analysis were performed to evaluate the anxiety status of the fish. Quantification of the novel tank test (NTT) was carried out attending to two variables: (1) The percentage of positions scored in each of the two virtual zones (upper and lower) of the novel tank and (2) the number of crossings from one to another (Figure 4A). The novel tank test (NTT) is the conceptual equivalent of the rodent open field (OF) paradigm; NTT induces motivational conflict between the "defensive" diving behavior and following vertical exploration [27,28]. Each male was analyzed at the beginning and at the end of the experiment a day before sperm squeezing ($t = -1$ day and $t = 20$ days). Summaries of each animal behavior were created for easier evaluation (Figure 4B). As can be checked in Figure 4C,E, there was a non-statistical ($p > 0.0500$) trend toward PROBIO fish spending more time in the top of the tank, close to the 50% in mean values, at day 21 (47.92% ± 8.37%) compared to day 0 (39.25% ± 4.71%). The number-of-crossings evaluation revealed again significant differences ($p = 0.0373$) only in the PROBIO population doubling the mean values of the variable (Figure 4D; Table S3). Moreover, a moderate correlation ($p = 0.0018$; $r = 0.5595$) between fish behavior and total motility of squeezed ejaculates was observed (Figure 4F).

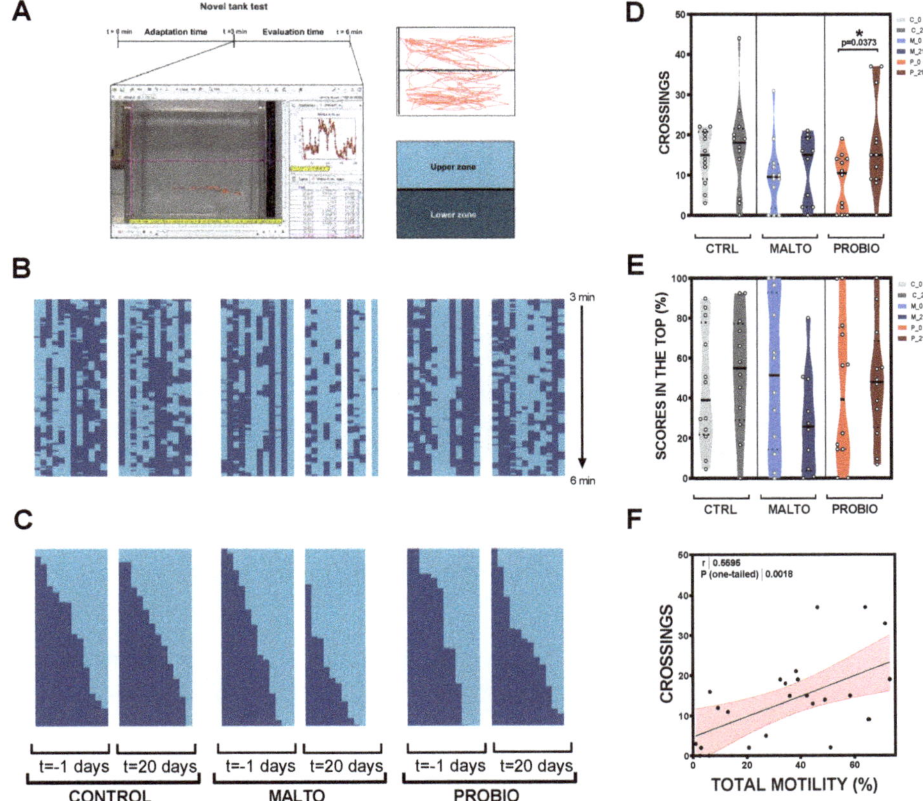

Figure 4. Probiotic bacteria ingestion modulates male behavior in zebrafish. (**A**) Schematic representation of the novel tank test (NTT). (**B**) Individual summaries of the track for each animal in the experiment at 0 and 21 days. (**C**) Organized summaries from higher to lower scores in the lower subarea. (**D**) Comparison of the scores registered in the upper zone before and after treatment. (**E**) Comparison of the number of crossings between the two subareas of the arena at 0 and 21 days. (**F**) Correlation between "number of crossings" and "total motility" variables integrating data from the three experimental groups after probiotic treatment. Asterisk shows statistically significant differences * ($p < 0.0500$).

4. Discussion

Nutrition could have a positive or negative impact on reproduction. Nowadays, decrease in sperm quality could be considered a global health problem. Indeed, asthenozoospermia is one of the male subfertility pathologies described by the WHO (2010) as a condition in which the percentage of progressively motile sperm is abnormally low [29]. Since the development and optimization during the last decades of the artificial reproductive technologies (ARTs): Intracytoplasmic sperm injection (ICSI), ovarian stimulation, intrauterine insemination (IUI), or in vitro fertilization (IVF) many infertile couples have found a solution to conceive. Although they have become a major worldwide focus of attention, these techniques are expensive and invasive. Possible clinical approaches may include antioxidant ingestion as a preliminary or concomitant treatment to reproductive techniques to improve fertility outcomes.

The definition of oxidative stress (OS) is the overabundance of reactive oxygen species (ROS) or a deficiency of antioxidants [30]. The imbalance produced by ROS causes cell damage. The deleterious effects of this damage on spermatozoa have been known since the 80 s [31]. There exists evidence regarding the need of certain amounts of ROS for normal sperm functions of both in mammals [32] or teleost [33] mainly produced by the mitochondria. However, excessive quantities become pathophysiological and lead to DNA damage and even apoptosis [7]. Endogenous or exogenous factors may be the cause of high levels of ROS. The most common exogenous causes of OS are obesity, smoking, environmental contaminants, alcohol intake, and malnutrition [9]. Natural antioxidants can scavenge ROS, inactivate them, and repair the cellular damage [34]. Spermatozoa, due to their high specialization, do not present cytoplasm after spermatogenesis and they depend on seminal plasma, which is rich in antioxidants [35]. Moreover, in spermatozoa, polyunsaturated fatty acids (a highly oxidizable substrate) enrich the cell membrane, provoking a high vulnerability to lipid peroxidation from ROS both in mammals [5] and teleost [36]. Oxidative damage affects the sperm flexibility and therefore motility, which is the excellence parameter to assess sperm quality. The spermatozoal heightened vulnerability to OS has caused enormous interest in the role of diet antioxidants in the management of infertile men [7].

Nowadays, the probiotic market is increasing globally as a cheap and well accepted (by the consumers) supplement source all around the world. There exists an increase in the demand of these kinds of products to improve health or prevent human illness. The developing observation that the gut microbiota plays a central role in regulating the host's physiology has supported the significance of the probiotic concept. The modulation of the intestinal microbiota composition has been proposed as one of the main mechanism of probiotic activity [37]. In a previous study, our group reported that the effects of a commercial probiotic diet supplement (Bactocell®, Lallemand Animal Nutrition S.A., Blagnac, France), containing a lactic acid bacteria strain (*Pediococcus acidilactici*) improved molecular sperm quality markers in zebrafish testicular cells after a short period (10 days) [38], providing initial data regarding the potential use of probiotic supplementation on zebrafish male reproductive performance. In the present study, the ingestion of a supplement containing probiotic strains on a single cycle spermatogenesis evidenced a positive effect of the host's sperm quality after a single cycle of spermatogenesis in zebrafish model. Specifically, the present study was undertaken to evaluate the combined effects of two strains: *L. rhamnosus* CECT8361 and *B. longum* CECT7347. These strains were selected because they belong to the most-used genera as probiotics nowadays [16] and they have been previously described as strains with antioxidant activity [23]. Additionally, *B. longum* CECT7347 has been assigned with anti-inflammatory activity. This strain has been described to reduce the inflammatory effects of the dysbiotic intestinal microbiota of individuals with coeliac disease on peripheral blood mononuclear cells partially via the induction of IL-10 production [24,39]. *B. longum* CECT7347 has also been demonstrated to decrease the cytotoxic and inflammatory effects of gliadin peptides on epithelial cell in vitro degradation [40,41]. Furthermore, in the gliadin-induced enteropathy animal model, this strain has been shown to reduce the peripheral CD4$^+$ T cells, rise IL-10, and shrink TNF-α production [42]. The other strain used in this experiment belongs to the *Lactobacillus* genera, which has been repeatedly shown as the predominant bacteria in the semen, accompanied by a flexible composition of other taxa [43–45]. *L. rhamnosus* species has been described as a highly adhesive bacteria in zebrafish [21]. The specific mechanism by which these bacterial strains are modifying fish behaviour and sperm quality in the present study is unclear. It is known that the ingestion of antioxidants can improve sperm motility [7], but the anti-inflammatory properties of *B. longum* CECT7347 could not be ignored and cannot be separately evaluated in our study. In fact, in humans, it has been reported that the ingestion of a combination of the two strains *L. rhamnosus* and *B. longum* modulated the gut microbiota composition, leading to a significant reduction of potentially harmful bacteria and an increase of beneficial ones [46]. Indeed, the combination of specific bacterial strains belonging to these two species can act in optimal synergy for restoring the intestinal balance [47] even better than individually [48].

In the present study, it was demonstrated that in the zebrafish model, males fed with the probiotic formulation increased sperm quality. In particular, in terms of sperm counts, 11 of the 12 males within the PROBIO group showed an improvement in concentration (Figure 2C), independent of initial values after 21 days of ingestion. The individual track of animals was available thanks to the use of VIE tagging. This technique is starting to spread among facilities since this inert, non-immunogenic polymer is useful for many purposes. The results achieved by this in vivo study clearly showed that the ingestion of the combination of *L. rhamnosus* CECT8361 and *B. longum* CECT7347 increased the percentage of motile cells after a single cycle of spermatogenesis (Figure 2D). After 21 days of exposure, a clear induction of total motility was found in all males within the PROBIO group, contrary to control cohorts (Figure 2D). These results are in line with our previous observations in asthenozoospermic human samples [23] on the ability of the same couple of probiotics reporting an increase of total motility after treatment. In the current study, total motility improved with a 1.7-fold change. Concomitantly to the increment in total motility values, progressive motile cells were also improved (Figure 2D), although the fold-change before and after the probiotics ingestion was lower. In our results, the increment of total motility was also accompanied by a modulation of sperm subpopulations within the motile population (Figure 3). All slow, medium, and fast motile cell populations were increased after 21 days in the probiotic-fed animals contrary to diet-controlled and vehicle-fed ones (Figure 3). Interestingly, the spermatozoa kinetics did not show differences after the experimental time indicating that probiotic bacteria are not able to alter these parameters in the zebrafish model. Altogether, the capability of *L. rhamnosus* CECT8361 and *B. longum* CECT7347 to modulate sperm quality was remarkably corroborated. Our results regarding sperm quality improvement are further supported by some studies [49,50] involving the use of probiotics in other animal models, which reported a potentially positive effect of probiotics in terms of sperm quality parameters. In particular, Dardmeh and colleagues [50] demonstrated that *L. rhamnosus* PB01 (DSM 14870) may have an effect on weight after eight weeks of treatment as well as a modulation of sperm kinetics and hormone levels in mice with diet-induced obesity. It has also been suggested that the use of *Bacillus amyloliquefaciens* TOA5001 as a probiotic has potential positive effect on broiler breeders, since the strain was able to increase sperm count and sperm viability after six weeks of treatment [49].

The scientific community is starting to elucidate the mechanisms provoking these beneficial effects of probiotic ingestion on sperm quality. Recently, Kelton Tremellen has published a novel theory [51], the GELDING theory (Gut Endotoxin Leading to a Decline IN Gonadal function) in which it is postulated that "the trans-mucosal passage of bacterial lipopolysaccharide from the gut lumen into the circulation is a key inflammatory trigger underlying male hypogonadism". The author has also linked the theory to a described positive effect of probiotics on human sperm samples from infertile patients [52]. This new and interesting theory is remarkable after analysing our results. The synergy between antioxidant and anti-inflammatory properties of the two bacteria used in the present study may explain the registered improvement in zebrafish sperm quality. According to Tremellen's theory, this assumption may be accepted.

Noticeably, in the present study, we offer evidence that in adult male zebrafish, short ingestion of probiotics modulates behavior pattern (Figure 4). Zebrafish is an interesting model organism to investigate behavior. Founded on geotaxis—an innate escape "diving" behavior of fish in novel environments—the novel tank test (NTT) has long been used to analyze adult zebrafish behaviors [53] and drug responses [54]. Adult zebrafish initially spend more time at the lower part of the tank when they are exposed to a novel environment. Concomitantly, they reduce "top" swimming and reveal more unpredictable movements and show freezing/immobility events [55]. Subsequent, because of habituation to the NTT novelty, animals gradually explore the top area (theoretically less safe for zebrafish in their wild habitats due to predator risk) [28]. Although our results did not report statistically significant differences in the number of scores in the upper subarea of the novel arena before and after probiotic administration, the number of crossing between the bottom and the upper area revealed differences (Figure 4). These results suggest that *L. rhamnosus* CECT8361 and *B. longum* CECT7347 modulate the exploration activity of the males after only 21 days of exposure showing a lower stress-like conduct. The microbiota signals to the central nervous system (CNS) via several potential pathways [56]. Probable mechanisms of communication embrace production of various metabolites that cross the intestinal barrier into the circulatory system, and/or microbe-derived metabolites that can signal through the immune system [57]. Moreover afferent pathways of the vagus nerve from the enteric nervous system (ENS) to the CNS have been associated as a key route of communication concerning the microbiota and CNS [58]. Our results are in accordance with a number of recent findings reporting that the use of various *Lactobacillus* and/or *Bifidobacterium* strains can lighten anxiety-and depressive-like behavior and alleviate stress responses in animal models [58–61].

5. Conclusions

In conclusion, our study showed that 21 days of treatment (a spermatogenesis cycle) with a probiotic mixture with described antioxidant and anti-inflammatory activities significantly improved zebrafish sperm quality and increased the number of fluent males. These data highlight the promising use of this probiotic mixture to improve reproductive performance in different quality sperm samples by increasing sperm total motility, progressive motility, concentration, and fast sperm populations. Furthermore, behavior analysis revealed a modulation in probiotics-fed males suggesting a lower anxiety-like pattern, which may be correlated with the improvement of sperm quality parameters in this model. Considering the simplicity and economical effectiveness of the studied multistrain product, the results presented here strengthen the potential use of this preparation in male reproductive biotechnology, which may be useful in the aquaculture industry and reproductive biology fields.

Supplementary Materials: The following are available online at http://www.mdpi.com/2072-6643/11/4/843/s1, Figure S1: Before–after graphs for (**A**) concentration, (**B**) total motility, and (**C**) progressive motility at 0 and 21 days obtained for each experimental group. "M" and "P" refers to the experimental groups: Control and maltodextrin respectively. "M#" indicates the number of the male; Figure S2: (**A**) PCA analysis for Computer Assisted Sperm Analysis (CASA) variables for probiotic-fed group. (**B**) Representation of the experimental group in a principal component plane; Table S1: Weight values (mean ± s.e.) for the groups before and after treatment; Table S2: Sperm quality values (mean ± s.e.) for the groups before and after treatment; Table S3: novel tank values for behavior estimators (mean ± s.e.) for the groups before and after treatment.

Author Contributions: Conceptualization, V.R., D.G.V., and M.F.R.; data curation, D.G.V., M.F.R., and J.M.M.-V.; formal analysis, D.G.V. and M.F.R.; funding acquisition, V.R.; investigation, V.R., D.G.V., and M.F.R.; methodology, V.R., D.G.V., M.F.R., and J.M.M.-V.; project administration, V.R.; resources, V.R.; software, D.G.V. and J.M.M.-V.; supervision, V.R.; validation, D.G.V.; visualization, D.G.V.; writing—original draft, V.R., D.G.V., and M.F.R.; writing—review and editing, V.R. and D.G.V.

Funding: This work was supported by project AGL2015 68330-C2-1-R (MINECO/FEDER).

Acknowledgments: The authors acknowledge contract PTA2016-11987-I (MINECO/FEDER), Biopolis S.L. (Valencia, Spain), AQUA-CIBUS International Net 318RT0549 (CYTED), and the staff from Planta de Cultivos El Bocal (IEO).

Conflicts of Interest: The authors declare that they have no competing interests.

References

1. Boivin, J.; Bunting, L.; Collins, J.A.; Nygren, K.G. International estimates of infertility prevalence and treatment-seeking: Potential need and demand for infertility medical care. *Hum. Reprod.* **2007**, *22*, 1506–1512. [CrossRef] [PubMed]
2. Agarwal, A.; Mulgund, A.; Hamada, A.; Chyatte, M.R. A unique view on male infertility around the globe. *Reprod. Biol. Endocrinol.* **2015**, *13*, 37. [CrossRef] [PubMed]
3. Levine, H.; Jørgensen, N.; Martino-Andrade, A.; Mendiola, J.; Weksler-Derri, D.; Mindlis, I.; Pinotti, R.; Swan, S.H. Temporal trends in sperm count: A systematic review and meta-regression analysis. *Hum. Reprod. Update* **2017**, *23*, 646–659. [CrossRef] [PubMed]
4. Virtanen, H.E.; Jørgensen, N.; Toppari, J. Semen quality in the 21st century. *Nat. Rev. Urol.* **2017**, *14*, 120–130. [CrossRef] [PubMed]
5. Nassan, F.L.; Chavarro, J.E.; Tanrikut, C. Diet and men's fertility: Does diet affect sperm quality? *Fertil. Steril.* **2018**, *110*, 570–577. [CrossRef] [PubMed]
6. Crean, A.J.; Senior, A.M. High-fat diets reduce male reproductive success in animal models: A systematic review and meta-analysis. *Obes. Rev.* **2019**. [CrossRef] [PubMed]

7. Smits, R.M.; Mackenzie-Proctor, R.; Fleischer, K.; Showell, M.G. Antioxidants in fertility: Impact on male and female reproductive outcomes. *Fertil. Steril.* **2018**, *110*, 578–580. [CrossRef]
8. Talevi, R.; Barbato, V.; Fiorentino, I.; Braun, S.; Longobardi, S.; Gualtieri, R. Protective effects of in vitro treatment with zinc, d-aspartate and coenzyme q10 on human sperm motility, lipid peroxidation and DNA fragmentation. *Reprod. Biol. Endocrinol.* **2013**, *11*, 81. [CrossRef] [PubMed]
9. Tremellen, K. Oxidative stress and male infertility—A clinical perspective. *Hum. Reprod. Update* **2008**, *14*, 243–258. [CrossRef] [PubMed]
10. Cardoso, J.P.; Cocuzza, M.; Elterman, D. Optimizing male fertility: Oxidative stress and the use of antioxidants. *World J. Urol.* **2019**. [CrossRef]
11. Fuller, R. Probiotics in man and animals. *J. Appl. Bacteriol.* **1989**, *66*, 365–378. [PubMed]
12. Zmora, N.; Suez, J.; Elinav, E. You are what you eat: Diet, health and the gut microbiota. *Nat. Rev. Gastroenterol. Hepatol.* **2019**, *16*, 35–56. [CrossRef] [PubMed]
13. Whiteside, S.A.; Razvi, H.; Dave, S.; Reid, G.; Burton, J.P. The microbiome of the urinary tract—A role beyond infection. *Nat. Rev. Urol.* **2015**, *12*, 81–90. [CrossRef] [PubMed]
14. Hall, A.B.; Tolonen, A.C.; Xavier, R.J. Human genetic variation and the gut microbiome in disease. *Nat. Rev. Genet.* **2017**, *18*, 690–699. [CrossRef] [PubMed]
15. Hill, C.; Guarner, F.; Reid, G.; Gibson, G.R.; Merenstein, D.J.; Pot, B.; Morelli, L.; Canani, R.B.; Flint, H.J.; Salminen, S.; et al. The International Scientific Association for Probiotics and Prebiotics consensus statement on the scope and appropriate use of the term probiotic. *Nat. Rev. Gastroenterol. Hepatol.* **2014**, *11*, 506–514. [CrossRef]
16. Fijan, S. Microorganisms with claimed probiotic properties: An overview of recent literature. *Int. J. Environ. Res. Public Health.* **2014**, *11*, 4745–4767. [CrossRef] [PubMed]
17. Grunwald, D.J.; Eisen, J.S. Headwaters of the zebrafish—Emergence of a new model vertebrate. *Nat. Rev. Genet.* **2002**, *3*, 717–724. [CrossRef] [PubMed]
18. Keller, E.T.; Murtha, J.M. The use of mature zebrafish (*Danio rerio*) as a model for human aging and disease. *Comp. Biochem. Physiol. C Toxicol. Pharmacol.* **2004**, *138*, 335–341. [CrossRef] [PubMed]
19. Carnevali, O.; Avella, M.A.; Gioacchini, G. Effects of probiotic administration on zebrafish development and reproduction. *Gen. Comp. Endocrinol.* **2013**, *188*, 297–302. [CrossRef] [PubMed]
20. Gioacchini, G.; Giorgini, E.; Olivotto, I.; Maradonna, F.; Merrifield, D.L.; Carnevali, O. The influence of probiotics on zebrafish *Danio rerio* innate immunity and hepatic stress. *Zebrafish* **2014**, *11*, 98–106. [CrossRef]
21. Zhou, Z.; Wang, W.; Liu, W.; Gatlin, D.M.; Zhang, Y.; Yao, B.; Ringø, E. Identification of highly-adhesive gut *Lactobacillus* strains in zebrafish (*Danio rerio*) by partial rpoB gene sequence analysis. *Aquaculture* **2012**, *370–371*, 150–157. [CrossRef]
22. Schulz, R.W.; de França, L.R.; LeGac, F.; Chiarini-Garcia, H.; Nobrega, R.H.; Miura, T. Spermatogenesis in fish. *Gen. Comp. Endocrinol.* **2010**, *165*, 390–411. [CrossRef] [PubMed]
23. Valcarce, D.G.; Genovés, S.; Riesco, M.F.; Martorell, P.; Herráez, M.P.; Ramón, D.; Robles, V. Probiotic administration improves sperm quality in asthenozoospermic human donors. *Benef. Microbes* **2017**, *8*, 193–206. [CrossRef] [PubMed]
24. Medina, M.; De Palma, G.; Ribes-Koninckx, C.; Calabuig, M.; Sanz, Y. *Bifidobacterium* strains suppress in vitro the pro-inflammatory milieu triggered by the large intestinal microbiota of coeliac patients. *J. Inflamm.* **2008**, *5*, 19. [CrossRef]
25. Guo, W.; Xie, B.; Xiong, S.; Liang, X.; Gui, J.-F.; Mei, J. miR-34a Regulates Sperm Motility in Zebrafish. *Int. J. Mol. Sci.* **2017**, *18*, 2676. [CrossRef]
26. Sadeghi, S.; Pertusa, J.; Yaniz, J.L.; Nuñez, J.; Soler, C.; Silvestre, M.A. Effect of different oxidative stress degrees generated by hydrogen peroxide on motility and DNA fragmentation of zebrafish (*Danio rerio*) spermatozoa. *Reprod. Domest. Anim.* **2018**, *53*, 1498–1505. [CrossRef]
27. Kalueff, A.V.; Stewart, A.; Kadri, F.; Dileo, J.; Chung, K.M.; Cachat, J.; Goodspeed, J.; Suciu, C.; Roy, S.; Gaikwad, S.; et al. The Developing Utility of Zebrafish in Modeling Neurobehavioral Disorders. *Int. J. Comp. Psychol.* **2010**, *23*, 104–120.
28. Kysil, E.V.; Meshalkina, D.A.; Frick, E.E.; Echevarria, D.J.; Rosemberg, D.B.; Maximino, C.; Lima, M.G.; Abreu, M.S.; Giacomini, A.C.; Barcellos, L.J.G.; et al. Comparative Analyses of Zebrafish Anxiety-Like Behavior Using Conflict-Based Novelty Tests. *Zebrafish* **2017**, *14*, 197–208. [CrossRef]

29. Cooper, T.G.; Noonan, E.; Von Eckardstein, S.; Auger, J.; Gordon Baker, H.W.; Behre, H.M.; Haugen, T.B.; Kruger, T.; Wang, C.; Mbizvo, M.T.; et al. World Health Organization reference values for human semen characteristics. *Hum. Reprod. Update* **2010**, *16*, 231–245. [CrossRef]
30. Agarwal, A.; Rana, M.; Qiu, E.; AlBunni, H.; Bui, A.D.; Henkel, R. Role of oxidative stress, infection and inflammation in male infertility. *Andrologia* **2018**, *50*, e13126. [CrossRef]
31. Aitken, R.J.; Clarkson, J.S. Cellular basis of defective sperm function and its association with the genesis of reactive oxygen species by human spermatozoa. *J. Reprod. Fertil.* **1987**, *81*, 459–469. [CrossRef] [PubMed]
32. Aitken, J.; Fisher, H. Reactive oxygen species generation and human spermatozoa: The balance of benefit and risk. *Bioessays* **1994**, *16*, 259–267. [CrossRef]
33. Valcarce, D.G.; Robles, V. Effect of captivity and cryopreservation on ROS production in *Solea senegalensis* spermatozoa. *Reproduction* **2016**, *152*, 439–446. [CrossRef]
34. Mirończuk-Chodakowska, I.; Witkowska, A.M.; Zujko, M.E. Endogenous non-enzymatic antioxidants in the human body. *Adv. Med. Sci.* **2018**, *63*, 68–78. [CrossRef] [PubMed]
35. Zini, A.; de Lamirande, E.; Gagnon, C. Reactive oxygen species in semen of infertile patients: Levels of superoxide dismutase- and catalase-like activities in seminal plasma and spermatozoa. *Int. J. Androl.* **1993**, *16*, 183–188. [CrossRef] [PubMed]
36. Cabrita, E.; Martínez-Páramo, S.; Gavaia, P.J.; Riesco, M.F.; Valcarce, D.G.; Sarasquete, C.; Herráez, M.P.; Robles, V. Factors enhancing fish sperm quality and emerging tools for sperm analysis. *Aquaculture* **2014**, *432*, 389–401. [CrossRef]
37. Hemarajata, P.; Versalovic, J. Effects of probiotics on gut microbiota: Mechanisms of intestinal immunomodulation and neuromodulation. *Ther. Adv. Gastroenterol.* **2013**, *6*, 39–51. [CrossRef]
38. Valcarce, D.G.; Pardo, M.Á.; Riesco, M.F.; Cruz, Z.; Robles, V. Effect of diet supplementation with a commercial probiotic containing *Pediococcus acidilactici* (Lindner, 1887) on the expression of five quality markers in zebrafish (*Danio rerio* (Hamilton, 1822)) testis. *J. Appl. Ichthyol.* **2015**, *31*, 18–21. [CrossRef]
39. De Palma, G.; Kamanova, J.; Cinova, J.; Olivares, M.; Drasarova, H.; Tuckova, L.; Sanz, Y. Modulation of phenotypic and functional maturation of dendritic cells by intestinal bacteria and gliadin: Relevance for celiac disease. *J. Leukoc. Biol.* **2012**, *92*, 1043–1054. [CrossRef] [PubMed]
40. Laparra, J.M.; Sanz, Y. Bifidobacteria inhibit the inflammatory response induced by gliadins in intestinal epithelial cells via modifications of toxic peptide generation during digestion. *J. Cell. Biochem.* **2010**, *109*, 801–807. [CrossRef]
41. Olivares, M.; Laparra, M.; Sanz, Y. Influence of *Bifidobacterium longum* CECT 7347 and gliadin peptides on intestinal epithelial cell proteome. *J. Agric. Food Chem.* **2011**, *59*, 7666–7671. [CrossRef]
42. Laparra, J.M.; Olivares, M.; Gallina, O.; Sanz, Y. Bifidobacterium longum CECT 7347 modulates immune responses in a gliadin-induced enteropathy animal model. *PLoS ONE* **2012**, *7*, e30744. [CrossRef]
43. Hou, D.; Zhou, X.; Zhong, X.; Settles, M.L.; Herring, J.; Wang, L.; Abdo, Z.; Forney, L.J.; Xu, C. Microbiota of the seminal fluid from healthy and infertile men. *Fertil. Steril.* **2013**, *100*, 1261-9. [CrossRef]
44. Weng, S.-L.; Chiu, C.-M.; Lin, F.-M.; Huang, W.-C.; Liang, C.; Yang, T.; Yang, T.-L.; Liu, C.-Y.; Wu, W.-Y.; Chang, Y.-A.; et al. Bacterial communities in semen from men of infertile couples: Metagenomic sequencing reveals relationships of seminal microbiota to semen quality. *PLoS ONE* **2014**, *9*, e110152. [CrossRef]
45. Mändar, R.; Punab, M.; Korrovits, P.; Türk, S.; Ausmees, K.; Lapp, E.; Preem, J.-K.; Oopkaup, K.; Salumets, A.; Truu, J. Seminal microbiome in men with and without prostatitis. *Int. J. Urol.* **2017**, *24*, 211–216. [CrossRef] [PubMed]
46. Toscano, M.; De Grandi, R.; Stronati, L.; De Vecchi, E.; Drago, L. Effect of *Lactobacillus rhamnosus* HN001 and *Bifidobacterium longum* BB536 on the healthy gut microbiota composition at phyla and species level: A preliminary study. *World J. Gastroenterol.* **2017**, *23*, 2696–2704. [CrossRef] [PubMed]
47. Khailova, L.; Petrie, B.; Baird, C.H.; Dominguez Rieg, J.A.; Wischmeyer, P.E. *Lactobacillus rhamnosus* GG and *Bifidobacterium longum* attenuate lung injury and inflammatory response in experimental sepsis. *PLoS ONE* **2014**, *9*, e97861. [CrossRef] [PubMed]
48. Inturri, R.; Stivala, A.; Furneri, P.M.; Blandino, G. Growth and adhesion to HT-29 cells inhibition of Gram-negatives by *Bifidobacterium longum* BB536 e *Lactobacillus rhamnosus* HN001 alone and in combination. *Eur. Rev. Med. Pharmacol. Sci.* **2016**, *20*, 4943–4949. [PubMed]
49. Inatomi, T.; Otomaru, K. Effect of dietary probiotics on the semen traits and antioxidative activity of male broiler breeders. *Sci. Rep.* **2018**, *8*, 5874. [CrossRef]

50. Dardmeh, F.; Alipour, H.; Gazerani, P.; van der Horst, G.; Brandsborg, E.; Nielsen, H.I. *Lactobacillus rhamnosus* PB01 (DSM 14870) supplementation affects markers of sperm kinematic parameters in a diet-induced obesity mice model. *PLoS ONE* **2017**, *12*, e0185964. [CrossRef]
51. Tremellen, K. Gut Endotoxin Leading to a Decline IN Gonadal function (GELDING)—A novel theory for the development of late onset hypogonadism in obese men. *Basic Clin. Androl.* **2016**, *26*, 7. [CrossRef] [PubMed]
52. Tremellen, K.; Pearce, K. Probiotics to improve testicular function (Andrology 5:439-444, 2017)—A comment on mechanism of action and therapeutic potential of probiotics beyond reproduction. *Andrology* **2017**, *5*, 1052–1053. [CrossRef] [PubMed]
53. Levin, E.D.; Bencan, Z.; Cerutti, D.T. Anxiolytic effects of nicotine in zebrafish. *Physiol. Behav.* **2007**, *90*, 54–58. [CrossRef] [PubMed]
54. Stewart, A.M.; Braubach, O.; Spitsbergen, J.; Gerlai, R.; Kalueff, A.V. Zebrafish models for translational neuroscience research: From tank to bedside. *Trends Neurosci.* **2014**, *37*, 264–278. [CrossRef] [PubMed]
55. Cachat, J.; Stewart, A.; Grossman, L.; Gaikwad, S.; Kadri, F.; Chung, K.M.; Wu, N.; Wong, K.; Roy, S.; Suciu, C.; et al. Measuring behavioral and endocrine responses to novelty stress in adult zebrafish. *Nat. Protoc.* **2010**, *5*, 1786–1799. [CrossRef] [PubMed]
56. Davis, D.J.; Bryda, E.C.; Gillespie, C.H.; Ericsson, A.C. Microbial modulation of behavior and stress responses in zebrafish larvae. *Behav. Brain Res.* **2016**, *311*, 219–227. [CrossRef]
57. Sampson, T.R.; Mazmanian, S.K. Control of Brain Development, Function, and Behavior by the Microbiome. *Cell Host Microbe* **2015**, *17*, 565–576. [CrossRef] [PubMed]
58. Bravo, J.A.; Forsythe, P.; Chew, M.V.; Escaravage, E.; Savignac, H.M.; Dinan, T.G.; Bienenstock, J.; Cryan, J.F. Ingestion of *Lactobacillus* strain regulates emotional behavior and central GABA receptor expression in a mouse via the vagus nerve. *Proc. Natl. Acad. Sci. USA* **2011**, *108*, 16050–16055. [CrossRef] [PubMed]
59. Messaoudi, M.; Lalonde, R.; Violle, N.; Javelot, H.; Desor, D.; Nejdi, A.; Bisson, J.-F.; Rougeot, C.; Pichelin, M.; Cazaubiel, M.; et al. Assessment of psychotropic-like properties of a probiotic formulation (*Lactobacillus helveticus* R0052 and *Bifidobacterium longum* R0175) in rats and human subjects. *Br. J. Nutr.* **2011**, *105*, 755–764. [CrossRef]
60. Savignac, H.M.; Kiely, B.; Dinan, T.G.; Cryan, J.F. Bifidobacteria exert strain-specific effects on stress-related behavior and physiology in BALB/c mice. *Neurogastroenterol. Motil.* **2014**, *26*, 1615–1627. [CrossRef] [PubMed]
61. Savignac, H.M.; Tramullas, M.; Kiely, B.; Dinan, T.G.; Cryan, J.F. Bifidobacteria modulate cognitive processes in an anxious mouse strain. *Behav. Brain Res.* **2015**, *287*, 59–72. [CrossRef] [PubMed]

Review

Plausible Biological Interactions of Low- and Non-Calorie Sweeteners with the Intestinal Microbiota: An Update of Recent Studies

Julio Plaza-Diaz [1,2,3], Belén Pastor-Villaescusa [1,4,5], Ascensión Rueda-Robles [1,2], Francisco Abadia-Molina [2,6] and Francisco Javier Ruiz-Ojeda [1,3,7,*]

1. Department of Biochemistry and Molecular Biology II, School of Pharmacy, University of Granada, 18071 Granada, Spain
2. Institute of Nutrition and Food Technology "José Mataix", Center of Biomedical Research, University of Granada, Avda. del Conocimiento s/n., 18016 Armilla, Granada, Spain
3. Instituto de Investigación Biosanitaria IBS.GRANADA, Complejo Hospitalario Universitario de Granada, 18014 Granada, Spain
4. LMU–Ludwig-Maximilians-University of Munich, Division of Metabolic and Nutritional Medicine, von Hauner Children's Hospital, University of Munich Medical Center, 80337 Munich, Germany
5. Institute of Epidemiology, Helmholtz Zentrum München–German Research Centre for Environmental Health, 85764 Neuherberg, Germany
6. Department of Cell Biology, School of Sciences, University of Granada, 18071 Granada, Spain
7. RG Adipocytes and metabolism, Institute for Diabetes and Obesity, Helmholtz Diabetes Center at Helmholtz Center Munich, 85764 Neuherberg, Munich, Germany

* Correspondence: francisco.ruiz@helmholtz-muenchen.de; Tel.: +49-89 3187-2036

Received: 8 April 2020; Accepted: 15 April 2020; Published: 21 April 2020

Abstract: Sweeteners that are a hundred thousand times sweeter than sucrose are being consumed as sugar substitutes. The effects of sweeteners on gut microbiota composition have not been completely elucidated yet, and numerous gaps related to the effects of nonnutritive sweeteners (NNS) on health still remain. The NNS aspartame and acesulfame-K do not interact with the colonic microbiota, and, as a result, potentially expected shifts in the gut microbiota are relatively limited, although acesulfame-K intake increases Firmicutes and depletes *Akkermansia muciniphila* populations. On the other hand, saccharin and sucralose provoke changes in the gut microbiota populations, while no health effects, either positive or negative, have been described; hence, further studies are needed to clarify these observations. Steviol glycosides might directly interact with the intestinal microbiota and need bacteria for their metabolization, thus they could potentially alter the bacterial population. Finally, the effects of polyols, which are sugar alcohols that can reach the colonic microbiota, are not completely understood; polyols have some prebiotics properties, with laxative effects, especially in patients with inflammatory bowel syndrome. In this review, we aimed to update the current evidence about sweeteners' effects on and their plausible biological interactions with the gut microbiota.

Keywords: nonnutritive sweeteners; sweetening agents; gut microbiota

---

1. Introduction

Excessive sugar consumption has become an important public health concern due to its adverse effects on health and metabolic consequences such as obesity, insulin resistance, metabolic syndrome, cardiovascular diseases, and type 2 diabetes. One century ago, sweetening agents or sweeteners—sugar substitutes that mimic the sweet taste—emerged as an alternative to sucrose and glucose–fructose syrups consumption to reduce energy intake [1,2]. However, the impact of sugar consumption on health continues to be a controversial topic in relation to its effects on metabolic disease [3]. Some contradictory results were published in 2015 about sweeteners and gut microbiota. Suez et al. concluded that some sweeteners may affect the human microbiome, and consequently more studies are needed [4]. In contrast, Frankelfeld et al. [5] reported no differences in median bacterial abundance across consumers and non-consumers of sweeteners.

Sweeteners are between several hundred to thousands of times sweeter than sucrose and they do not contain too many calories. They include nonnutritive sweeteners (NNS), which have a higher sweetening intensity than other sweeteners, such as acesulfame K (ace-K), advantame, aspartame, aspartame–acesulfame salt, cyclamate, neohesperidin dihydrochalcone, neotame, saccharin, steviol glycosides (including 10 different glycosides), sucralose, and thaumatin, low-calorie sweeteners (LCS), such as polyols or sugar alcohols and other new sugars which are low-digestible carbohydrates derived from the hydrogenation of their sugar or syrup sources. Polyols are around 25%–100% as sweet as sugar and include erythritol, hydrogenated starch hydrolysates (sometimes listed as maltitol syrup, hydrogenated glucose syrup, polyglycitol, polyglucitol, or simply HSH), isomalt, lactitol, maltitol, mannitol, sorbitol, and xylitol. All of them are considered safe for human consumption as long as they are consumed within the acceptable daily intake [6]. This safety was claimed by the European Food Safety Authority (EFSA) except for cyclamate, which is not approved by the US Food and Drug Administration [1,7,8].

In 2019, we reviewed the effects of sweeteners on the gut microbiota, considering both experimental studies and clinical trials, and we reported that, among NNS, only saccharin and sucralose shift the populations of the gut microbiota, although more human studies are needed to clarify those observations. Within nutritive sweeteners (NS), only stevia extracts seem to affect gut microbiota composition, while some polyols, such as isomalt and maltitol which can reach the colon, increase *Bifidobacterium* in healthy subjects and might act as prebiotics. Besides, lactitol can decrease *Bacteroides*, *Clostridium*, coliforms, and *Eubacterium*, increasing butyrate and IgA secretion in humans [1]. Thus, we concluded that still more studies are needed; however, as the plausible biological interaction between sweeteners and intestinal microbiota has not been reported elsewhere, we aimed to review and update the current knowledge about sweeteners and gut microbiota interactions in humans.

A comprehensive literature search was conducted in PubMed, Embase®, and Scopus using different combinations of the following keywords: aspartame, acesulfame-K, cyclamate, sucralose, saccharin, steviol glycosides, erythritol, isomalt, lactitol, maltitol, sorbitol, mannitol, xylitol, and microbiota, with special attention and interest to what was published from February 2018 to March 2020.

2. Biological Plausibility: Which Low- and Non-Calorie Sweetener (LNCS) Could Potentially Affect the Colonic Microbiota?

Biological plausibility is one component of a method of reasoning that can establish a cause-and-effect relationship between a biological factor and a particular disease or adverse effect based on assessing the strength of evidence, since the work of Bradford Hill [9]. Here, we will assess biological plausibility between different sweeteners and gut microbiota composition. Although we usually refer to the different low- and non-calorie sweeteners (LNCS) as if they were a single molecule, it is well known that they do not share their absorption, distribution, metabolism, and excretion (ADME) profiles. Therefore, the extrapolation of the effect of a particular LNCS on the intestinal microbiota to all LNCS is unappropriated. These differences are crucial to understanding if each LNCS has the potential to alter the intestinal microbiota, directly or indirectly. For more detailed information on the metabolic fate of each LNCS beyond its relationship with the colonic microbiota, the excellent review by Magnuson et al. (2016) [10] can be consulted.

2.1. *Effects of Non-Nutritive Low-Calorie Sweeteners on the Gut Microbiota*

2.1.1. Aspartame

Aspartame is a methyl ester of a dipeptide composed of L-phenylalanine and aspartic acid. When ingested, this dipeptide undergoes enzymatic hydrolysis in the gastrointestinal lumen and in the cells of the internal intestinal mucosa (by peptidases and intestinal esterases), so that virtually no aspartame enters the general circulation [11,12]. Hence, aspartame as an intact molecule cannot interact directly with the colonic microbiota. The three digestion products (aspartic acid, L-phenylalanine, and methanol) are rapidly absorbed in the duodenum and jejunum [12], reaching the systemic circulation without passing through the colon [10]. These degradation products are presented in the same way as when they are absorbed from vegetables, fruits, dairy, or meat, and at much lower concentrations than when they are derived from such foods [11]. These products follow their usual metabolic pathways.

Methanol enters the portal circulation into the liver and, by the enzymatic action of alcohol dehydrogenase, is metabolized to formaldehyde, which in turn, by the action of formaldehyde dehydrogenase, is oxidized to formic acid. Formic acid can be eliminated by the respiratory tract as carbon dioxide or excreted into the urine [10,11]. Aspartate undergoes a transamination reaction in the enterocytes, becoming oxalacetate. Oxalacetate and aspartate are interconverted in the body and can participate in the urea cycle and gluconeogenesis in the liver. Excess aspartate is eliminated in the urine [10]. Phenylalanine is absorbed in the gastrointestinal tract mucosa. It enters the liver through portal circulation, where, by the action of phenylalanine hydroxylase, can be converted into tyrosine. Phenylalanine that reaches the systemic circulation can be distributed throughout the body [11]. Its excess is excreted in the urine [13].

Based on the abovementioned information, the finding of a positive association between intake of aspartame and alteration of the colonic microbiota of rodents could be in fact due to the effect of what the animals stop eating rather than to the effect of aspartame intake itself. This last concept applies to all LNCS. However, a recent study carried out in female Sprague Dawley rats subjected to a high-fat/sucrose (HFSD), a HFSD + aspartame (5–7 mg kg^{-1} day^{-1}), or a HFSD + stevia (2–3 mg kg^{-1} day^{-1}) diet showed an increase of body fat in the offspring at weaning following maternal consumption of aspartame and stevia in the HSFD. In addition, glucose tolerance was altered, particularly with aspartame. *Akkermansia muciniphila* and *Enterobacteriaceae* concentrations were higher in mothers compared with their offspring. Regarding the cecal microbiota, a reduced abundance of *Enterococcaceae*, *Enterococcus*, and *Parasutterella* and an increased abundance of *Clostridium* cluster IV were found in the aspartame group. Moreover, fecal transplantation from offspring to germ-free mice produced an altered gut microbiota, causing impaired adiposity and glucose tolerance. In addition, increased concentrations of *Porphyromonadaceae* in males and females obese–aspartame and obese–stevia offspring were found [14]. In contrast, in the study by Suez et al., food intake in mice assigned to a water group with LNCS (aspartame, sucralose, and saccharin) was reduced by up to 50%. It is known that dietary factors are key determinants of the composition of the intestinal microbiota; indeed, differences in both total caloric intake and the type of food consumed can lead to a different microbial composition [15–17].

Thus, the intestinal microbiota might have been altered by a reduced consumption of fiber, protein, fat, and carbohydrates; therefore, it seems uncertain that the reported change in the intestinal microbiota was caused by the LNCS, and the changes that diet per se may provoke in the intestine should be considered. Nonetheless, there are studies that reveal possible modifications of the microbiota due to the use of aspartame. The study by Mahmud et al. analyzes the combined and individual effects of the administration of low concentrations of aspartame and Ace-K. Induction of *Escherichia coli* growth and expression of some important genes which may be related to its colonization in the gut were observed [18]. In another study with human fecal samples, aspartame administration significantly increased *Bifidobacterium* and *Blautia coccoides* growth and decreased the *Bacteroides*/*Prevotella* ratio; nevertheless, the aspartame-based sweetener used in this study was rich in maltodextrin, thus, the authors did not study the effect of aspartame alone [19].

2.1.2. Potassium Acesulfame (Ace-K)

After its intake, Ace-K is absorbed almost completely in the small intestine as an intact molecule and distributed by the blood to different tissues. Without undergoing any metabolization, more than 99% of Ace-K is excreted in the urinary tract within the first 24 h, with less than 1% being eliminated in the feces [10,20]. The minimum amount of Ace-K ingested, its rapid absorption, and its urinary excretion causes the Ace-K concentration that reaches the fecal or colonic bacteria to be negligible [10,21]. Therefore, it is extremely unlikely that this LNCS could have a direct effect on the colonic microbiota [16]. However, some studies have reported small shifts in the gut microbiota composition following Ace-K intake.

A cross-sectional study was conducted in humans and showed no modifications in the intestinal microbiota nor significant differences by sex, contrary to the study conducted in rats by Bian X et al. [22]. Other studies also indicate that Ace-K causes changes in the microbiota and their metabolites, such as butyrate and pyruvate [22,23]. The study carried out by Uebanso et al. suggests that the daily intake of maximum adequate diary intake (ADI) levels of Ace-K does not affect the relative amount of the *Clostridium* cluster XIVa in the fecal microbiome [24]. In contrast, a study in mice that received 150 mg kg^{-1} of Ace-K by free drinking during 8 weeks, showed that lymphocyte recruitment was increased, with augmented expression of inflammatory cytokines and adhesion molecules [25]. Recent studies in rats indicate that administration of a mixture of sucralose and Ace-K at concentrations near the upper limit of ADI for human consumption during mice pregnancy has consequences on the progeny, causing metabolic and microbiome alterations. The authors observed an increase in *Firmicutes* and a depletion of *A. muciniphila*, which is a beneficial bacterium inversely correlated with fat mass gain, type 1 diabetes, and inflammatory bowel syndrome (IBS) [26]. The researchers also indicated an increase in the variety of species in the microbiota; however, *A. muciniphila* was significantly depleted, suggesting that the divergence between mothers' and pups' microbiomes was due to increasing NNS concentrations [27].

Regarding the bacteriostatic effect of Ace-K, this sweetener shows a strong inhibitory effect on the growth of *E. coli* HB101 and *E. coli* K-12 [28]. In contrast, using a concentration of Ace-K of 2.5 mg/mL, the result was an induction in *E. coli* growth, whereas the growth stimulation decreased gradually when higher concentrations of sweetener were used [18].

2.1.3. Cyclamate

Cyclamate is the sodium or calcium salt of cyclamic acid (cyclohexanesulfamic acid), which itself is prepared by the sulfonation of cyclohexylamine and is eliminated in the feces [29]. In a study carried out by Vamanu et al. [30], the authors determined the effect of sweeteners on the microbiota pattern using an in vitro model. In this study, the total quantity of synthesized short-chain fatty acids (SCFA) and the number of microorganisms were decreased, and a negative influence on the fermentative profile was observed, although with an increase of *Bifidobacterium*. The ratio of butyric/propionic acids was also affected, indicating that those SCFA could affect the gut microbiota composition. Cyclamate also exerts a positive effect, producing an inhibitory anaerobic fermentation of glucose in a rat model of intestinal gut microbiota [28].

Cyclamate and sucralose can alter the ratio between butyric and propionic acids [30]. SCFA have multiple effects on human health. Butyric acid has anti-obesogenic effects, reduces insulin resistance, and improves dyslipidemia [31]. Lower concentrations of propionic and butyric acids have been positively correlated with the four subtypes of IBS and can be harmful to people with that disease [32]. Overall, it seems that cyclamate has some effects on gut microbiota composition, but more studies on its possible effect on human health are needed.

2.1.4. Sucralose

Sucralose has a very low level of absorption (less than 15%) and it is practically not metabolized. Therefore, after intake, more than 85% of sucralose reaches the colon unchanged [10]. The small proportion of sucralose that is absorbed is eliminated in the urine mainly unchanged, though two glucuronides of sucralose were also detected in a small proportion (approximately 2%) [33].

Although more than 85% of the ingested sucralose contacts the colonic microbiota, between 94% and 99% of this LNCS is recovered in the feces without any structural change, thus indicating little or no metabolism by the gut microbiota [10]. Thus, sucralose does not appear to be a substrate for the colonic microbiota [16]. Nevertheless, considering the practically null microbial metabolism of sucralose, we must be cautious when interpreting the results of studies that indicate an alteration of the intestinal microbiota after sucralose consumption [17]. In those cases, it will be worth investigating whether pure sucralose or a commercial formulation was used in the research, since these formulations usually contain around 1% of sucralose and 99% of the carriers maltodextrins [16].

On the other hand, it has been shown that sucralose promotes inflammation in a mouse model of human Crohn's disease-like ileitis as well as dysbiosis of the gut microbiota [34]. Furthermore, sucralose causes a decrease in the number of Firmicutes species [35]. This result is the opposite to that reporter by Olivier-Van Stichelen et al., who found that Firmicutes doubled, including the Clostridiales families Lachnospiraceae and Ruminococcaceae (e.g., Oscillospira), in mice's pups [27]. Wang et al. observed an increase of Firmicutes and a tendency to decrease for Bacteroidetes [28]. These authors did not observe changes in Actinobacteria and Proteobacteria phyla in mice fed with a chow diet, but they reported a synergistic effect when sucralose was provided in the context of a high-fat diet. On the other hand, a chow diet might cause a significant increase in *Bifidobacterium* [28]. A study carried out in humans examined the short-term effect of sucralose consumption on glucose homeostasis and gut microbiome in healthy male volunteers. The authors concluded that no changes occurred in the gut microbiome due to sucralose intake [36]. In contrast, another study shows an increase in the abundance of pro-inflammatory bacteria like *Turicibacter*, which was associated with hepatic inflammation, after sucralose administration [37].

Splenda administration in mice was associated with a high presence of Bacteroidetes, an enhanced overgrowth of *E. coli*, and the expansion of Proteobacteria [38]. The effect of sucralose was analyzed in fecal samples from 13 healthy volunteers. The authors found increased abundances of *Escherichia*, *Shigella*, and *Bilophila*. With regard to SCFA, increased production of valeric acid was observed [19].

A recent publication evaluated the short-term effect of sucralose consumption on glycemic control and its interaction with the intestinal microbiota (comparison before/after the intervention by 16S rRNA sequencing) in healthy subjects. This study concluded that consumption of high doses of sucralose (75% of the ADI) for 7 days did not alter glycemic control, insulin resistance, or intestinal microbiome at the phylum level [36].

Although previous human studies showed similar results concerning glycemic control (glycosylated hemoglobin, fasting blood glucose, C-peptide), both in diabetic [39] and in non-diabetic populations [40], this is the first time that a randomized, controlled, double-blind study concomitantly evaluated the composition of the intestinal microbiome in healthy subjects, thus providing a better level of evidence in comparison to other earlier published trials.

2.1.5. Saccharin

After intake, more than 85% of saccharin is absorbed as an intact molecule, since it does not undergo gastrointestinal metabolism. Once absorbed, it binds to plasma proteins and is distributed throughout the body. Finally, it is eliminated by urine through active tubular transport [10,41,42]. The small percentage of non-absorbed saccharin is excreted into the feces, indicating that high concentrations of this LNCS could lead to changes in the composition of the intestinal microbial population [16]. It is important to highlight that one of the main studies that reported an alteration of the intestinal microbiota with the consumption of saccharin [17] was carried out by administering the full ADI of saccharin, which does not correspond to what happens with habitual human consumption.

In an in vitro model study, saccharin produced an increase in *Bifidobacterium*. Not only saccharin but also sucralose caused a decrease in the number of Firmicutes species, directly correlated with the SCFA level [30].

Some herbicides, which are considered nowadays safe, can change the gut microbiota of animals in the early stages of embryonic development. Indeed, exposure to glyphosate and glyphosate in combination with saccharin contributes to the broader reproduction of pathogenic bacteria such as *Klebsiella*, *Citrobacter*, *Enterobacter*, and *Pseudomonas* [43]. On the other hand, studies show that saccharin administration can also disrupt monolayer integrity and alter paracellular permeability in a Caco-2 cell monolayer model [44].

Overall, saccharin administration also promotes Bacteroidetes, Turicibacter, and Clostridiales and reduces Firmicutes abundances. The Turicibacter bacteria increases have been related to a pro-inflammatory effect of saccharin [37].

The effect of a mixture of fiber–prebiotics and saccharin–eugenol has been evaluated in dogs. Four diets were prepared: control diet, containing 5% of cellulose; diet containing a 5% fiber and prebiotic blend; diet containing 0.02% of saccharin (sweetener SUCRAM) and eugenol and 5% of a fiber and prebiotic blend plus 0.02% of saccharin and eugenol. The use of saccharine did not affect species richness measured by alpha-diversity or alter the proportions of bacterial phyla. No changes were observed in fecal microbial communities [45]. More studies are needed to confirm these saccharin effects using different concentrations and animal models.

2.1.6. Steviol Glycosides

Steviol glycosides can be extracted from the leaves of *Stevia rebaudiana*. They all have a central steviol structure, conjugated with different sugar residues, such as stevioside and rebaudioside A, which all are steviol glycosides. Steviol glycosides are hydrolyzed neither by enzymes nor by the acid present in the upper gastrointestinal tract [46]. Therefore, they pass through the upper portion of the gastrointestinal tract without being absorbed and enter the colon as intact molecules [47]. In the colon, bacteria of the Bacteroidacea family eliminate the sugar residues that are conjugated to steviol [47,48]. While these sugar residues may represent a source of energy for the microbiota [49], it is worth noting that the energy contribution is negligible, given the low total daily intake of steviol glycosides [50]. The resulting steviol is not a substrate for the intestinal microbiota, since it is resistant to bacterial degradation [48]. Hence, steviol is completely absorbed and reaches the liver where it is conjugated with glucuronic acid. Steviol glucuronide is mainly excreted in the urine in humans [51,52].

While steviol glycosides interact with the colonic microbiota, there are no reports indicating that these compounds could affect bacteria negatively [30]. A recent study showed that steviol incubation in the GIS1–phase 2 system, an in vitro system that simulates the human intestinal microbial ecosystem, reduced the ammonium level and *Bifidobacterium* and exerted a negative influence on the fermentative profile, resulting in higher pH and SCFA ratio [30].

S. rebaudiana is another natural steviol glycoside 250 times sweeter than sucrose [53]. In Europe, only the purified steviol glycosides are approved for use in food, and the ADI of 4 mg kg^{-1} of body weight per day is safe (EU Regulation (EU) 1129/2011) [1,7,54]. Another study recently reported that a low dose of stevia rebaudioside A alters gut microbiota composition and reduces nucleus accumbens tyrosine hydroxylase and dopamine transporter mRNA levels in rebaudioside A-supplemented rats. Nonetheless, the oligofructose-enriched inulin prebiotic, in the presence or absence of rebaudioside A, reduced fat mass, food intake, gut permeability, and cecal SCFA concentration. However, only stevia rebaudioside A increased SCFAs acetate and valerate, which are positively correlated with fat mass and total weight. Hence, stevia rebaudioside A seems to decrease the "healthy" status of the gut microbiota [55].

Chronic stevia consumption has effects on gut microbiota and immunity in the small intestine of young mice. In 21-day-old mice treated with sucrose, Splenda, and stevia, mice preferred the consumption of Splenda and stevia. Besides, those mice showed an increase in $CD3^+$ lymphocytes in Peyer's patches, but only stevia induced an increase in the lamina propria. Both Splenda and stevia elevated leptin, C-peptide, IL-6, and IL-17 and decreased resistin. Stevia modified the predominantly genera *Bacillus* such as *Bacillus aerius*, *Bacillus circulans*, *Bacillus licheniformis*, and *Bacillus safensis*, although the authors observed effects on *Streptococcus saliviloxodontae*, *Oceanobacillus sojae*, and *Staphylococcus lugdunensis*. Even though the results of this study are significant, they have some limitations. The modifications observed in the immune system of the mucous membranes and in the microbiota of the small intestine in young mice after weaning depend on age and diet. This study used culture media and not metagenomic approaches, and some results might be related to some carriers present in the evaluated products, such as maltodextrins [56].

Recently, by testing stevia glycosides and erythritol, which are often combined in food preparation to minimize changes in the organoleptic profile, in an in vivo *Cebus apella* model, changes in bacteria growth and gut microbial structure and diversity have been observed [57]. Overall, stevia seems to modify the gut microbiota; however, further studies are needed to clarify its specific effects.

Although different changes in the intestinal microbiota have been described in relation to the influence of sweeteners on the immune system, the wide use of aspartame, ace-K, cyclamate, sucralose, saccharin and steviol glycosides makes it necessary to carry out other analyses to complete the picture of the influence that these sweeteners have on the intestinal microbiota.

2.2. Effects of Nutritive Low-Calorie Sweeteners on the Gut Microbiota

2.2.1. Polyols

Polyols are a group of compounds used in an increasingly wide variety of commercial foods as additives. They are quite stable at high temperatures and various pH and do not interfere in Maillard reactions, conferring organoleptic characteristics to the foods. Polyols are naturally present in fruits, vegetables, and mushrooms and are used to produce food without added sugar, reducing the sugar content in recipes. In addition, polyols are non-cariogenic, do not induce salivation, and do not interfere with insulin and glucose levels in the blood. Nevertheless, the excessive consumption of polyols causes gastrointestinal symptoms and laxative effects, which can be even worse in patients with IBS. As we described previously, the FDA, the Codex Alimentarius, and the EFSA have approved eight different polyols, i.e., erythritol, hydrogenated starch hydrolysates, isomalt, lactitol, maltitol, mannitol, sorbitol, and xylitol, for use as bulk sweeteners in human foods [1,58]. Indeed, in September 2019, EFSA launched an open consultation on the "Protocol for the assessment of hazard identification and characterization of the sweeteners", which will be used for the evaluation of the safety of sweeteners under the re-evaluation program of food additives. The evaluation should be completed by the end of 2020 [59].

2.2.2. Erythritol

Erythritol (E-968) is a four-carbon sugar alcohol that has a fast absorption through the small intestine with a very low metabolization and it is over 90% excreted unchanged in the urine [58]. Furthermore, an unabsorbed part (~10%) is fermented in the large intestine by the colonic microbiota, which consequently rarely leads to gas production [60]. Hence, the limited amount of erythritol that reaches the colon could be the explanation of the lack of evidence of effects of erythritol on the gut microbiota in humans, based on clinical trials as we previously reported [1]. Nevertheless, a recent in vitro study demonstrated that low doses of erythritol (25 $\mu g\ mL^{-1}$, 50 $\mu g\ mL^{-1}$, and 100 $\mu g\ mL^{-1}$) did not exert any effect on the growth of *Escherichia*, *Enterococcus*, *Lactobacillus*, *Ruminococcus* and Bacteroides in the human gut microbiota. Moreover, erythritol doses did not disrupt alpha and beta diversities or the composition of the human gut microbial community [57]. In contrast, butyric and pentanoic acids were increased significantly after erythritol consumption, indicating that this polyol may be able to affect the function of the human gut microbiota. Indeed, the authors reported that this change in SCFAs production was due to the 10% of erythritol that reaches the human colon [57].

2.2.3. Isomalt

Hydrogenated isomalt, isomaltitol (E-953), is not absorbed by the small intestine and is easily fermented in the colon by the microbiota [61]. This fermented fraction of ingested isomalt is approximately 90% [62]. Therefore, it is expected that isomalt is capable of altering the bacterial population. Isomalt has been proposed as a prebiotic carbohydrate that might contribute to a healthy luminal colonic mucosal environment, with bifidogenic properties and high butyrate production [63]. Accordingly, besides evidence reported in Ruiz-Ojeda et al., 2019 [1], a recent study based on the administration of buckwheat honey to human gut microbes cultures reported that the principal constituents of buckwheat honey are oligosaccharides with a low degree of polymerization, including isomalt and isomaltotriose, which may serve as food to promote the growth of indigenous intestinal probiotics such as *Bifidobacterium* [64]. In addition, an increase in the abundance of *Escherichia/Shigella* and *Streptococcus* was also reported, while the alpha diversity, as well as the abundance of *Prevotella*, *Faecalibacterium*, and *Lachnospiraceae incertae sedis*, were decreased, thus fostering a reduction of pathogenic bacteria in the gut tract [64]. However, this might be also explained by the polyphenols composition of the buckwheat honey studied, since polyphenols also markedly affect the gut microbiota [64]. Indeed, the authors concluded that phenolic compounds and oligosaccharides in buckwheat honey appear to synergistically impact human intestinal microbes to enhance the growth of probiotics. More efforts, especially in vivo, are required to elucidate the possible specific impact of isomalt on the gut microbiota.

2.2.4. Lactitol

Lactitol (E-966) is a disaccharide normally not absorbed in the small intestine [65] that therefore reaches the lower gut where it is fermented, producing both gases and SCFA [66]. Lactitol mitigates pathogenic translocation in the small intestine by the reduction of permeability and stimulates the growth of bifidobacteria and lactobacilli [67]. Thus, similarly to isomalt, lactitol could act as a prebiotic, enhancing the composition of the intestinal microbiota, even when consumed at low doses as a sweetener, normally 10 grams [68]. Nevertheless, it is important to highlight that lactitol, due to its limited sweetening power, is usually used in combination with other intense sweeteners [69] or a set of prebiotics [70], and this could disturb the results concerning its effect on the intestinal microbiota. Furthermore, it has also been studied as a synbiotic product along with *Lactobacillus acidophilus* NCFM and jointly promoted beneficial changes since it led to a decrease in the abundance of the *Blautia coccoides–Eubacterium rectale* bacterial group and *Clostridium* cluster XIVab counts in the elderly population [71]. Since 2018, two studies were identified regarding lactitol and the gut microbiota. One trial was based on the administration of probiotics, synbiotics, probiotics together with lactitol, or only lactitol to mice with acute colitis. The authors found that the lactitol group showed higher levels of *Akkermansia* compared with the control, probiotic (*Bifidobacterium* and *Lactobacillus*), and synbiotic (probiotics and inulin) groups. It is worth highlighting this work, since *Akkermansia* seems to ameliorate the inflammatory response and insulin resistance in obese and diabetic patients [72], protecting the intestinal epithelial cells and enhancing the mucosal barrier function [73]. As the genome of *Akkermansia* was proved to be able to encode a wide variety of secretory proteins such as glycohydrolyzases [74], the authors speculated that *Akkermansia* might be able to decompose lactitol and promote its own proliferation [75]. Furthermore, the supplementation of probiotics and prebiotics, including lactitol, induced an increment of the proportion of helpful bacteria and regulated the balance of the intestinal microbiota [75]. For instance, the abundance of *Bifidobacterium* was increased in all the experimental groups in comparison with the control. However, the observed effect might not be exerted by lactitol itself [75]. Another study was performed in Korean adults to evaluate the efficacy of supplementation with the prebiotic UG1601 (based on inulin (61.5%), lactitol (34.6%), and an aloe vera gel (3.9%)) for 4 weeks to alleviate the symptoms of constipation associated with the gut microbiota [70].

Here, the clinical trial showed that the prebiotic UG1601 in patients with mild constipation resulted in decreased serum concentrations of the bacterial endotoxin lipopolysaccharide and its receptor CD 14. Additionally, it increased the abundance of *Roseburia hominis*, a major butyrate producer, which could be related to the observed reduction of the levels of these endotoxemia markers [70]. In summary, lactitol along with other compounds, may induce changes in the gut microbiota, but further studies are needed to demonstrate whether lactitol itself triggers an effect on the gut microbiota.

2.2.5. Maltitol

Maltitol (E-965) is obtained through the hydrolysis, reduction, and hydrogenation of starch. This polyol has a very slow absorption rate, being fermented in the colon. Thus, as we previously mentioned, it is expected that maltitol is susceptible fermentation by the gut microbiota [1]. To date, only one clinical trial has been reported which studied the effect of maltitol present in experimental chocolate on the gut microbiota. The authors concluded that both maltitol and polydextrose, as well as maltitol alone, increased the amount of fecal bifidobacteria, lactobacilli, and SCFA compared with the control chocolate [26]. Besides evidence reported by Ruiz-Ojeda et al., 2019 [1], there are no additional studies. Although maltitol could be a good alternative with high sweetening capacity (~90%), safe, and non-cariogenic, data to determine the specific effects of maltitol on the gut microbiota are not still sufficient.

2.2.6. Sorbitol

Sorbitol or D-glucitol (E-420) is partially absorbed in the upper gastrointestinal tract, where it undergoes digestion, while the non-absorbed portion is extensively fermented to SCFA and gases by the colonic microbiota [62]. Consumers can suffer slight gastrointestinal symptoms, such as flatulence or bloating, or more severe symptoms when it is ingested at high doses as 20 g d^{-1} [76]. Overall, studies on this isomeric polyol and its effect on the gastrointestinal tract are mostly focused on the symptomatology induced by sorbitol than on its possible capacity to alter the gut microbiota. Since the 1930s, it is known that sorbitol can be fermented by bacteria like *E. coli*, *Lactobacillus* spp., and *Streptococcus* spp. [77] which are present in our intestinal microbiota. However, so far, there has been no thorough study and there is not enough evidence to define the specific effects of sorbitol on the gut microbiota.

2.2.7. Mannitol

Mannitol (E-421) is an isomer of sorbitol, and both are listed as hydrogenated monosaccharides. Approximately, 75% of ingested mannitol reaches the large intestine [78]. The intestinal bacteria metabolize D-mannitol to butyrate and propionate in animal models. Indeed, D-mannitol has been suggested as a prebiotic, due to its stimulation of colonic butyrate and propionate production [79]. Although no data are available so far about the effects of mannitol on the gut microbiota, its role as a substrate reflects an interaction between this polyol and the intestinal microbiota that should be studied more deeply.

2.2.8. Xylitol

Xylitol (E-967) is a five-carbon polyol obtained from the hydrogenation of D-xylose, called wood sugar or birch sugar. Xylitol can be directly metabolized mainly in the liver, remaining unchanged in the gastrointestinal tract [80]. Furthermore, only a certain proportion of the ingested xylitol is absorbed slowly from the intestinal lumen and fermented by the intestinal microbiota. Besides minor amounts of gases such as H_2, CH_4, and CO_2, the end products of the bacterial metabolism of xylitol are mainly SCFA, (i.e., acetate, propionate, and butyrate). Xylitol might cause osmotic diarrhea when the amounts consumed are too high [81]. Hence, it is expected that this polyol is capable of altering the intestinal microbiota. Interesting results were reported, as previously mentioned, in our recent review [1], but further studies were not reported since then.

In summary, according to the new findings reported from February 2018, erythritol, lactitol, and maltitol have shown to exert beneficial effects on the gut microbiota by themselves. Nevertheless, because of the promising effect of lactitol to enhance *Akkermansia* proliferation in mice with acute colitis, we encourage corroborating this finding by further studies in humans. Overall, the latest evidence is not still enough to establish firm conclusions in relation to how polyols influence the gut microbiota. In addition, it is necessary to highlight that some polyols could induce laxative effects, and it would be more reliable to evaluate their effects separately. Figure 1 summarizes the effects of different sweeteners on intestinal microbiota.

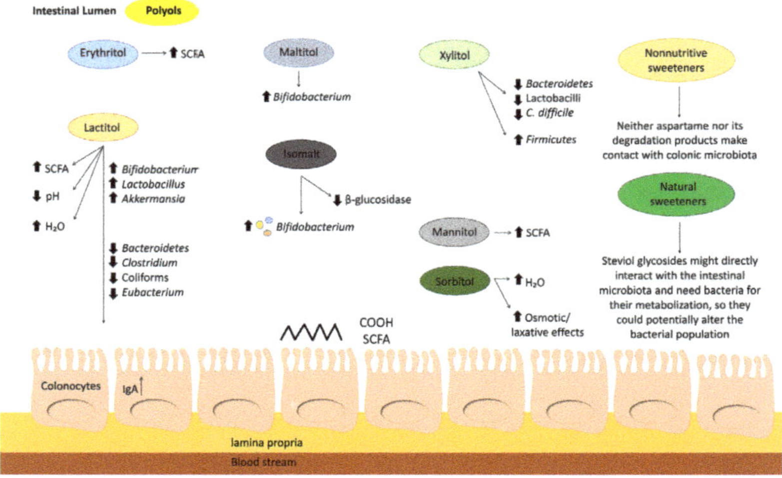

Figure 1. Schematic representation of sweeteners' effects on the gut microbiota. Abbreviations. IgA, Immunoglobulin A; N/A, not available information; SCFA, short-chain fatty acids.

3. Conclusions and Future Perspectives

The effects of sweeteners on gut microbiota composition are still in discussion. Even though there are some gaps in the evidence related to the health effects of NNS in both healthy and non-healthy populations, authorities such as FDA, EFSA, and Codex Alimentarius consider them safe and well-tolerated, as long as the appropriate ADI is not exceeded. Regarding NNS, neither aspartame nor its degradation products make contact with the colonic microbiota. In contrast, though Ace-K is absorbed and eliminated by urine and almost does not contact the colonic microbiota, surprisingly, it increases Firmicutes and depletes *A. muciniphila*. However, further research is required in order to firmly establish an effect in humans. We previously reported that saccharin and sucralose seem to change the composition of the gut microbiota. However, it is necessary to take account that only 15% of the consumed saccharin contacts the colonic microbiota, so only high doses could alter the intestinal microbiota composition. On the contrary, more than 85% of the consumed sucralose reaches the colon; therefore, sucralose could potentially either alter or change the gut microbiota composition, but it is not practically metabolized by intestinal bacteria. On the other hand, steviol glycosides directly interact with the intestinal microbiota and need bacteria for their metabolization, so they could potentially alter the bacterial population.

In summary, in the absence of biological plausibility, results indicating a possible alteration of the intestinal bacteria population after the consumption of LNCS should be explained by alternative mechanisms, such as alterations in the dietary pattern, administration of exaggerated LNCS doses, and co-administration of carriers.

Author Contributions: J.P.-D. and F.J.R.-O. design the work, participated in the bibliographic search, discussion, and writing of the manuscript. B.P.-V. and A.R.-R. participated in the bibliographic search, discussion, and writing of the manuscript. F.A.-M. revised the manuscript. All authors have read and agreed to the published version of the manuscript.

Funding: This research received no external funding.

Acknowledgments: Julio Plaza-Díaz is part of the "UGR Plan Propio de Investigación 2016" and the "Excellence actions: Unit of Excellence on Exercise and Health (UCEES), University of Granada". Francisco J. Ruiz-Ojeda and Belén Pastor-Villaescusa are supported by a grant to postdoctoral researchers at foreign universities and research centers from the "Fundación Alfonso Martín-Escudero", Madrid, Spain. We are grateful to Belen Vazquez-Gonzalez for assistance with the illustration service.

Conflicts of Interest: The authors declare no conflict of interest.

References

1. Ruiz-Ojeda, F.J.; Plaza-Diaz, J.; Saez-Lara, M.J.; Gil, A. Effects of Sweeteners on the Gut Microbiota: A Review of Experimental Studies and Clinical Trials. *Adv. Nutr.* **2019**, *10*, S31–S48. [CrossRef] [PubMed]
2. Lohner, S.; Toews, I.; Meerpohl, J.J. Health outcomes of non-nutritive sweeteners: Analysis of the research landscape. *Nutr. J.* **2017**, *16*, 55. [CrossRef] [PubMed]
3. Stanhope, K.L. Sugar consumption, metabolic disease and obesity: The state of the controversy. *Crit. Rev. Clin. Lab. Sci.* **2016**, *53*, 52–67. [CrossRef] [PubMed]
4. Suez, J.; Korem, T.; Zilberman-Schapira, G.; Segal, E.; Elinav, E. Non-caloric artificial sweeteners and the microbiome: Findings and challenges. *Gut Microbes* **2015**, *6*, 149–155. [CrossRef]
5. Frankenfeld, C.L.; Sikaroodi, M.; Lamb, E.; Shoemaker, S.; Gillevet, P.M. High-intensity sweetener consumption and gut microbiome content and predicted gene function in a cross-sectional study of adults in the United States. *Ann. Epidemiol.* **2015**, *25*, 736–742. [CrossRef] [PubMed]
6. Fitriakusumah, Y.; Lesmana, C.R.A.; Bastian, W.P.; Jasirwan, C.O.M.; Hasan, I.; Simadibrata, M.; Kurniawan, J.; Sulaiman, A.S.; Gani, R.A. The role of Small Intestinal Bacterial Overgrowth (SIBO) in Non-alcoholic Fatty Liver Disease (NAFLD) patients evaluated using Controlled Attenuation Parameter (CAP) Transient Elastography (TE): A tertiary referral center experience. *BMC Gastroenterol.* **2019**, *19*, 43. [CrossRef]
7. Food Standards Agency, U.K. Food Additives. Different Food Additives and Advice on Regulations and the Safety of Additives in Food. Available online: https://www.food.gov.uk/safety-hygiene/food-additives (accessed on 24 December 2019).
8. FDA. High-Intensity Sweeteners. Available online: https://www.fda.gov/food/food-additives-petitions/high-intensity-sweeteners (accessed on 24 December 2019).
9. Hill, A.B. The Environment and Disease: Association or Causation? *Proc. R. Soc. Med.* **1965**, *58*, 295–300. [PubMed]
10. Magnuson, B.A.; Carakostas, M.C.; Moore, N.H.; Poulos, S.P.; Renwick, A.G. Biological fate of low-calorie sweeteners. *Nutr. Rev.* **2016**, *74*, 670–689. [CrossRef]
11. Butchko, H.H.; Stargel, W.W.; Comer, C.P.; Mayhew, D.A.; Benninger, C.; Blackburn, G.L.; de Sonneville, L.M.; Geha, R.S.; Hertelendy, Z.; Koestner, A.; et al. Aspartame: Review of safety. *Regul. Toxicol. Pharmacol.* **2002**, *35*, S1–S93.
12. Stegink, L.D.; Wolf-Novak, L.C.; Filer, L.J., Jr.; Bell, E.F.; Ziegler, E.E.; Krause, W.L.; Brummel, M.C. Aspartame-sweetened beverage: Effect on plasma amino acid concentrations in normal adults and adults heterozygous for phenylketonuria. *J. Nutr.* **1987**, *117*, 1989–1995. [CrossRef]
13. Sanchez de Medina Contreras, F.; Suárez Ortega, M.D. *Tratado de Nutrición*; Metabolismo de los aminoácidos; Editorial Médica Panamericana: Madrid, Spain, 2017; Chapter 8, pp. 173–195.
14. Nettleton, J.E.; Cho, N.A.; Klancic, T.; Nicolucci, A.C.; Shearer, J.; Borgland, S.L.; Johnston, L.A.; Ramay, H.R.; Noye Tuplin, E.; Chleilat, F.; et al. Maternal low-dose aspartame and stevia consumption with an obesogenic diet alters metabolism, gut microbiota and mesolimbic reward system in rat dams and their offspring. *Gut* **2020**. [CrossRef]
15. David, L.A.; Maurice, C.F.; Carmody, R.N.; Gootenberg, D.B.; Button, J.E.; Wolfe, B.E.; Ling, A.V.; Devlin, A.S.; Varma, Y.; Fischbach, M.A.; et al. Diet rapidly and reproducibly alters the human gut microbiome. *Nature* **2014**, *505*, 559–563. [CrossRef] [PubMed]

16. Lobach, A.R.; Roberts, A.; Rowland, I.R. Assessing the in vivo data on low/no-calorie sweeteners and the gut microbiota. *Food Chem. Toxicol.* **2019**, *124*, 385–399. [CrossRef]
17. Suez, J.; Korem, T.; Zeevi, D.; Zilberman-Schapira, G.; Thaiss, C.A.; Maza, O.; Israeli, D.; Zmora, N.; Gilad, S.; Weinberger, A.; et al. Artificial sweeteners induce glucose intolerance by altering the gut microbiota. *Nature* **2014**, *514*, 181–186. [CrossRef] [PubMed]
18. Mahmud, R.; Shehreen, S.; Shahriar, S.; Rahman, M.S.; Akhteruzzaman, S.; Sajib, A.A. Non-Caloric Artificial Sweeteners Modulate the Expression of Key Metabolic Genes in the Omnipresent Gut Microbe Escherichia coli. *J. Mol. Microbiol. Biotechnol.* **2019**, 1–14. [CrossRef] [PubMed]
19. Gerasimidis, K.; Bryden, K.; Chen, X.; Papachristou, E.; Verney, A.; Roig, M.; Hansen, R.; Nichols, B.; Papadopoulou, R.; Parrett, A. The impact of food additives, artificial sweeteners and domestic hygiene products on the human gut microbiome and its fibre fermentation capacity. *Eur. J. Nutr.* **2019**. [CrossRef] [PubMed]
20. Renwick, A.G. The metabolism of intense sweeteners. *Xenobiotica Fate Foreign Compd. Biol. Syst.* **1986**, *16*, 1057–1071. [CrossRef] [PubMed]
21. Pfeffer, M.; Ziesenitz, S.C.; Siebert, G. Acesulfame K, cyclamate and saccharin inhibit the anaerobic fermentation of glucose by intestinal bacteria. *Z. Ernahr.* **1985**, *24*, 231–235. [CrossRef]
22. Bian, X.; Chi, L.; Gao, B.; Tu, P.; Ru, H.; Lu, K. The artificial sweetener acesulfame potassium affects the gut microbiome and body weight gain in CD-1 mice. *PLoS ONE* **2017**, *12*, e0178426. [CrossRef]
23. Bueno-Hernández, N.; Vázquez-Frías, R.; Abreu, A.; Almeda-Valdés, P.; Barajas-Nava, L.; Carmona-Sánchez, R.; Chávez-Sáenz, J.; Consuelo-Sánchez, A.; Espinosa-Flores, A.; Hernández-Rosiles, V. Revisión de la evidencia científica y opinión técnica sobre el consumo de edulcorantes no calóricos en enfermedades gastrointestinales. *Rev. Gastroenterol. Méx.* **2019**, *84*, 492–510.
24. Uebanso, T.; Ohnishi, A.; Kitayama, R.; Yoshimoto, A.; Nakahashi, M.; Shimohata, T.; Mawatari, K.; Takahashi, A. Effects of Low-Dose Non-Caloric Sweetener Consumption on Gut Microbiota in Mice. *Nutrients* **2017**, *9*, 560. [CrossRef]
25. Hanawa, Y.; Higashiyama, M.; Sugihara, N.; Wada, A.; Inaba, K.; Horiuchi, K.; Furuhashi, H.; Kurihara, C.; Okada, Y.; Shibuya, N. Su1775–Artificial Sweetener Acesulfame Potassium Enhanced Lymphocyte Migration to Intestinal Microvessels by Enhancing Expression of Adhesion Molecules Through Dysbiosis. *Gastroenterology* **2019**, *156*, S-606. [CrossRef]
26. Beards, E.; Tuohy, K.; Gibson, G. A human volunteer study to assess the impact of confectionery sweeteners on the gut microbiota composition. *Br. J. Nutr.* **2010**, *104*, 701–708. [CrossRef]
27. Stichelen, O.-V.; Rother, K.I.; Hanover, J.A. Maternal exposure to non-nutritive sweeteners impacts progeny's metabolism and microbiome. *Front. Microbiol.* **2019**, *10*, 1360. [CrossRef] [PubMed]
28. Wang, Q.P.; Browman, D.; Herzog, H.; Neely, G.G. Non-nutritive sweeteners possess a bacteriostatic effect and alter gut microbiota in mice. *PLoS ONE* **2018**, *13*, e0199080. [CrossRef] [PubMed]
29. Price, J.M.; Biava, C.G.; Oser, B.L.; Vogin, E.E.; Steinfeld, J.; Ley, H.L. Bladder tumors in rats fed cyclohexylamine or high doses of a mixture of cyclamate and saccharin. *Science* **1970**, *167*, 1131–1132. [CrossRef] [PubMed]
30. Vamanu, E.; Pelinescu, D.; Gatea, F.; Sarbu, I. Altered in Vitro Metabolomic Response of the Human Microbiota to Sweeteners. *Genes* **2019**, *10*, 535. [CrossRef]
31. Farup, P.G.; Lydersen, S.; Valeur, J. Are Nonnutritive Sweeteners Obesogenic? Associations between Diet, Faecal Microbiota, and Short-Chain Fatty Acids in Morbidly Obese Subjects. *J. Obes.* **2019**, *2019*, 4608315. [CrossRef]
32. Hills, R.D., Jr.; Pontefract, B.A.; Mishcon, H.R.; Black, C.A.; Sutton, S.C.; Theberge, C.R. Gut Microbiome: Profound Implications for Diet and Disease. *Nutrients* **2019**, *11*, 1613. [CrossRef]
33. Roberts, A.; Renwick, A.G.; Sims, J.; Snodin, D.J. Sucralose metabolism and pharmacokinetics in man. *Food Chem. Toxicol.* **2000**, *38* (Suppl. 2), S31–S41. [CrossRef]
34. Wang, X.; Guo, J.; Liu, Y.; Yu, H.; Qin, X. Sucralose Increased Susceptibility to Colitis in Rats. *Inflamm. Bowel Dis.* **2019**, *25*, e3–e4. [CrossRef]
35. Bian, X.; Chi, L.; Gao, B.; Tu, P.; Ru, H.; Lu, K. Gut Microbiome Response to Sucralose and Its Potential Role in Inducing Liver Inflammation in Mice. *Front. Physiol.* **2017**, *8*, 487. [CrossRef]

36. Thomson, P.; Santibanez, R.; Aguirre, C.; Galgani, J.E.; Garrido, D. Short-term impact of sucralose consumption on the metabolic response and gut microbiome of healthy adults. *Br. J. Nutr.* **2019**, *122*, 856–862. [CrossRef] [PubMed]
37. Golonka, R.M.; San Yeoh, B.; Vijay-Kumar, M. Dietary Additives and Supplements Revisited: The Fewer, the Safer for Gut and Liver Health. *Curr. Pharmacol. Rep.* **2019**, *5*, 303–316. [CrossRef]
38. Rodriguez-Palacios, A.; Harding, A.; Menghini, P.; Himmelman, C.; Retuerto, M.; Nickerson, K.P.; Lam, M.; Croniger, C.M.; McLean, M.H.; Durum, S.K.; et al. The Artificial Sweetener Splenda Promotes Gut Proteobacteria, Dysbiosis, and Myeloperoxidase Reactivity in Crohn's Disease-Like Ileitis. *Inflamm. Bowel Dis.* **2018**, *24*, 1005–1020. [CrossRef]
39. Grotz, V.L.; Henry, R.R.; McGill, J.B.; Prince, M.J.; Shamoon, H.; Trout, J.R.; Pi-Sunyer, F.X. Lack of effect of sucralose on glucose homeostasis in subjects with type 2 diabetes. *J. Am. Diet. Assoc.* **2003**, *103*, 1607–1612. [CrossRef] [PubMed]
40. Grotz, V.L.; Pi-Sunyer, X.; Porte, D., Jr.; Roberts, A.; Trout, J.R. A 12-week randomized clinical trial investigating the potential for sucralose to affect glucose homeostasis. *Regul. Toxicol. Pharmacol.* **2017**, *88*, 22–33. [CrossRef] [PubMed]
41. Byard, J.; McChesney, E.; Golberg, L.; Coulston, F. Excretion and metabolism of saccharin in man. II. Studies with 14C-labelled and unlabelled saccharin. *Food Cosmet. Toxicol.* **1974**, *12*, 175–184. [CrossRef]
42. Renwick, A.G. The disposition of saccharin in animals and man—A review. *Food Chem. Toxicol.* **1985**, *23*, 429–435. [CrossRef]
43. Bilan, M.; Lieshchova, M.; Tishkina, N.; Brygadyrenko, V. Combined effect of glyphosate, saccharin and sodium benzoate on the gut microbiota of rats. *Regul. Mech. Biosyst.* **2019**, *10*, 228–232. [CrossRef]
44. Santos, P.; Caria, C.; Gotardo, E.; Ribeiro, M.; Pedrazzoli, J.; Gambero, A. Artificial sweetener saccharin disrupts intestinal epithelial cells' barrier function in vitro. *Food Funct.* **2018**, *9*, 3815–3822. [CrossRef]
45. Nogueira, J.P.S.; He, F.; Mangian, H.F.; Oba, P.M.; De Godoy, M.R.C. Dietary supplementation of a fiber-prebiotic and saccharin-eugenol blend in extruded diets fed to dogs. *J. Anim. Sci* **2019**, *97*, 4519–4531. [CrossRef] [PubMed]
46. Hutapea, A.M.; Toskulkao, C.; Buddhasukh, D.; Wilairat, P.; Glinsukon, T. Digestion of stevioside, a natural sweetener, by various digestive enzymes. *J. Clin. Biochem. Nutr.* **1997**, *23*, 177–186. [CrossRef]
47. Koyama, E.; Kitazawa, K.; Ohori, Y.; Izawa, O.; Kakegawa, K.; Fujino, A.; Ui, M. In Vitro metabolism of the glycosidic sweeteners, stevia mixture and enzymatically modified stevia in human intestinal microflora. *Food Chem. Toxicol.* **2003**, *41*, 359–374. [CrossRef]
48. Gardana, C.; Simonetti, P.; Canzi, E.; Zanchi, R.; Pietta, P. Metabolism of stevioside and rebaudioside A from Stevia rebaudiana extracts by human microflora. *J. Agric. Food Chem.* **2003**, *51*, 6618–6622. [CrossRef]
49. Carakostas, M.C.; Curry, L.L.; Boileau, A.C.; Brusick, D.J. Overview: The history, technical function and safety of rebaudioside A, a naturally occurring steviol glycoside, for use in food and beverages. *Food Chem. Toxicol.* **2008**, *46* (Suppl. 7), S1–S10. [CrossRef]
50. Renwick, A.G.; Tarka, S.M. Microbial hydrolysis of steviol glycosides. *Food Chem. Toxicol.* **2008**, *46* (Suppl. 7), S70–S74. [CrossRef]
51. Geuns, J.M.; Buyse, J.; Vankeirsbilck, A.; Temme, E.H.; Compernolle, F.; Toppet, S. Identification of steviol glucuronide in human urine. *J. Agric. Food Chem.* **2006**, *54*, 2794–2798. [CrossRef]
52. Wheeler, A.; Boileau, A.C.; Winkler, P.C.; Compton, J.C.; Prakash, I.; Jiang, X.; Mandarino, D.A. Pharmacokinetics of rebaudioside A and stevioside after single oral doses in healthy men. *Food Chem. Toxicol.* **2008**, *46* (Suppl. 7), S54–S60. [CrossRef]
53. Bian, X.; Tu, P.; Chi, L.; Gao, B.; Ru, H.; Lu, K. Saccharin induced liver inflammation in mice by altering the gut microbiota and its metabolic functions. *Food Chem. Toxicol.* **2017**, *107*, 530–539. [CrossRef]
54. Additives, E.P.O.F.; Food, N.S.A.T.; Younes, M.; Aggett, P.; Aguilar, F.; Crebelli, R.; Dusemund, B.; Filipič, M.; Frutos, M.J.; Galtier, P.; et al. Safety in use of glucosylated steviol glycosides as a food additive in different food categories. *EFSA J.* **2018**, *16*, e05181.
55. Nettleton, J.E.; Klancic, T.; Schick, A.; Choo, A.C.; Shearer, J.; Borgland, S.L.; Chleilat, F.; Mayengbam, S.; Reimer, R.A. Low-Dose Stevia (Rebaudioside A) Consumption Perturbs Gut Microbiota and the Mesolimbic Dopamine Reward System. *Nutrients* **2019**, *11*, 1248. [CrossRef] [PubMed]

56. Martinez-Carrillo, B.E.; Rosales-Gomez, C.A.; Ramirez-Duran, N.; Resendiz-Albor, A.A.; Escoto-Herrera, J.A.; Mondragon-Velasquez, T., Valdes-Ramos, R.; Castillo-Cardiel, A. Effect of Chronic Consumption of Sweeteners on Microbiota and Immunity in the Small Intestine of Young Mice. *Int. J. Food Sci.* **2019**, *2019*, 9619020. [CrossRef] [PubMed]
57. Mahalak, K.K.; Firrman, J.; Tomasula, P.M.; Nunez, A.; Lee, J.J.; Bittinger, K.; Rinaldi, W.; Liu, L.S. Impact of Steviol Glycosides and Erythritol on the Human and Cebus apella Gut Microbiome. *J. Agric. Food Chem.* **2020**. [CrossRef]
58. Grembecka, M. Sugar alcohols—Their role in the modern world of sweeteners: A review. *Eur. Food Res. Technol.* **2015**, *241*, 1–14. [CrossRef]
59. EFSA. Public Consultation on the Draft Protocol for the Assessment of Hazard Identification and Characterisation of the Sweeteners. Available online: http://www.efsa.europa.eu/en/consultations/call/public-consultation-draft-protocol-assessment (accessed on 24 December 2019).
60. Oku, T.; Okazaki, M. Laxative threshold of sugar alcohol erythritol in human subjects. *Nutr. Res.* **1996**, *16*, 577–589. [CrossRef]
61. Caballero, B.; Finglas, P.; Toldrá, F. *Encyclopedia of Food and Health*; Academic Press: Cambridge, MA, USA, 2015.
62. Livesey, G. Health potential of polyols as sugar replacers, with emphasis on low glycaemic properties. *Nutr. Res. Rev.* **2003**, *16*, 163–191. [CrossRef]
63. Gostner, A.; Blaut, M.; Schäffer, V.; Kozianowski, G.; Theis, S.; Klingeberg, M.; Dombrowski, Y.; Martin, D.; Ehrhardt, S.; Taras, D. Effect of isomalt consumption on faecal microflora and colonic metabolism in healthy volunteers. *Br. J. Nutr.* **2006**, *95*, 40–50. [CrossRef]
64. Jiang, L.; Xie, M.; Chen, G.; Qiao, J.; Zhang, H.; Zeng, X. Phenolics and Carbohydrates in Buckwheat Honey Regulate the Human Intestinal Microbiota. *Evid. Based Complement. Altern. Med.* **2020**, *2020*, 6432942. [CrossRef]
65. Schauber, J.; Weiler, F.; Gostner, A.; Melcher, R.; Kudlich, T.; Lührs, H.; Scheppach, W. Human rectal mucosal gene expression after consumption of digestible and non-digestible carbohydrates. *Mol. Nutr. Food Res.* **2006**, *50*, 1006–1012. [CrossRef]
66. Ballongue, J.; Schumann, C.; Quignon, P. Effects of lactulose and lactitol on colonic microflora and enzymatic activity. *Scand. J. Gastroenterol.* **1997**, *32*, 41–44. [CrossRef]
67. Nath, A.; Haktanirlar, G.; Varga, A.; Molnar, M.A.; Albert, K.; Galambos, I.; Koris, A.; Vatai, G. Biological Activities of Lactose-Derived Prebiotics and Symbiotic with Probiotics on Gastrointestinal System. *Medicina* **2018**, *54*, 18. [CrossRef]
68. Finney, M.; Smullen, J.; Foster, H.A.; Brokx, S.; Storey, D.M. Effects of low doses of lactitol on faecal microflora, pH, short chain fatty acids and gastrointestinal symptomology. *Eur. J. Nutr.* **2007**, *46*, 307. [CrossRef] [PubMed]
69. Carocho, M.; Morales, P.; Ferreira, I.C. Sweeteners as food additives in the XXI century: A review of what is known, and what is to come. *Food Chem. Toxicol.* **2017**, *107*, 302–317. [CrossRef] [PubMed]
70. Chu, J.R.; Kang, S.Y.; Kim, S.E.; Lee, S.J.; Lee, Y.C.; Sung, M.K. Prebiotic UG1601 mitigates constipation-related events in association with gut microbiota: A randomized placebo-controlled intervention study. *World J. Gastroenterol.* **2019**, *25*, 6129–6144. [CrossRef] [PubMed]
71. Björklund, M.; Ouwehand, A.C.; Forssten, S.D.; Nikkilä, J.; Tiihonen, K.; Rautonen, N.; Lahtinen, S.J. Gut microbiota of healthy elderly NSAID users is selectively modified with the administration of Lactobacillus acidophilus NCFM and lactitol. *Age* **2012**, *34*, 987–999. [CrossRef] [PubMed]
72. Hansen, C.H.; Krych, L.; Nielsen, D.S.; Vogensen, F.K.; Hansen, L.H.; Sorensen, S.J.; Buschard, K.; Hansen, A.K. Early life treatment with vancomycin propagates Akkermansia muciniphila and reduces diabetes incidence in the NOD mouse. *Diabetologia* **2012**, *55*, 2285–2294. [CrossRef] [PubMed]
73. Reunanen, J.; Kainulainen, V.; Huuskonen, L.; Ottman, N.; Belzer, C.; Huhtinen, H.; de Vos, W.M.; Satokari, R. Akkermansia muciniphila Adheres to Enterocytes and Strengthens the Integrity of the Epithelial Cell Layer. *Appl. Environ. Microbiol.* **2015**, *81*, 3655–3662. [CrossRef]
74. Singla, V.; Chakkaravarthi, S. Applications of prebiotics in food industry: A review. *Food Sci. Technol. Int.* **2017**, *23*, 649–667. [CrossRef]
75. Wang, Y.N.; Meng, X.C.; Dong, Y.F.; Zhao, X.H.; Qian, J.M.; Wang, H.Y.; Li, J.N. Effects of probiotics and prebiotics on intestinal microbiota in mice with acute colitis based on 16S rRNA gene sequencing. *Chin. Med. J.* **2019**, *132*, 1833–1842. [CrossRef]

76. Hyams, J.S. Sorbitol intolerance: An unappreciated cause of functional gastrointestinal complaints. *Gastroenterology* **1983**, *84*, 30–33. [CrossRef]
77. Robbins, G.B.; Lewis, K.H. Fermentation of sugar acids by bacteria. *J. Bacteriol.* **1940**, *39*, 399. [CrossRef] [PubMed]
78. Bernier, J.; Pascal, G. Valeur énergetique des polyols (sucres-alcools). *Méd. Nutr.* **1990**, *26*, 221–238.
79. Maekawa, M.; Maekawa, M.; Ushida, K.; Hoshi, S.; Kashima, N.; Ajisaka, K.; Maekawa, M.; Ushida, K.; Hoshi, S.; Kashima, N. Butyrate and propionate production from D-mannitol in the large intestine of pig and rat. *Microb. Ecol. Health Dis.* **2005**, *17*, 169–176. [CrossRef]
80. Mäkinen, K.K. History, Safety, and Dental Properties of Xylitol. 2004. Available online: http://www.naturallysweet.com.au/uploads/50072/ufiles/download_info/History_Safety_and_Dental_Properties_of_Xylitol.pdf (accessed on 31 March 2020).
81. Dubach, U.; Feiner, E.; Forgó, I. Oral tolerance of Xylit in subjects with normal metabolism. *Schweiz. Med. Wochenschr.* **1969**, *99*, 190–194.

Review

The Gut Microbiota and Its Implication in the Development of Atherosclerosis and Related Cardiovascular Diseases

Estefania Sanchez-Rodriguez [1,2,*], Alejandro Egea-Zorrilla [2], Julio Plaza-Díaz [1,2,3,*], Jerónimo Aragón-Vela [4], Sergio Muñoz-Quezada [5,6], Luis Tercedor-Sánchez [7] and Francisco Abadia-Molina [2,8,*]

1 Department of Biochemistry and Molecular Biology II, School of Pharmacy, University of Granada, 18071 Granada, Spain
2 Institute of Nutrition and Food Technology "José Mataix", Center of Biomedical Research, University of Granada, Avda. del Conocimiento s/n., 18016 Armilla, Granada, Spain; alejandroegezor@gmail.com
3 Instituto de Investigación Biosanitaria IBS.GRANADA, Complejo Hospitalario Universitario de Granada, 18014 Granada, Spain
4 Department of Nutrition, Exercise and Sports (NEXS), Section of Integrative Physiology, University of Copenhagen, Nørre Allé 51, DK-2200 Copenhagen, Denmark; jeroav@ugr.es
5 Departamento de Farmacia, Facultad de Química, Pontificia Universidad Católica de Chile, Santiago 6094411, Chile; chechomu@hotmail.com
6 National Agency for Medicines (ANAMED), Public Health Institute, Santiago 7780050, Chile
7 Cardiology Unit, Hospital Universitario Virgen de la Nieves, 18014 Granada, Spain; luis.tercedor.sspa@juntadeandalucia.es
8 Department of Cell Biology, School of Sciences, University of Granada, 18071 Granada, Spain
* Correspondence: estefaniasr@outlook.com (E.S.-R.); jrplaza@ugr.es (J.P.-D.); fmolina@ugr.es (F.A.-M.); Tel.: +34-9-5824-1000 (ext. 41599) (J.P.-D.)

Received: 25 January 2020; Accepted: 21 February 2020; Published: 26 February 2020

Abstract: The importance of gut microbiota in health and disease is being highlighted by numerous research groups worldwide. Atherosclerosis, the leading cause of heart disease and stroke, is responsible for about 50% of all cardiovascular deaths. Recently, gut dysbiosis has been identified as a remarkable factor to be considered in the pathogenesis of cardiovascular diseases (CVDs). In this review, we briefly discuss how external factors such as dietary and physical activity habits influence host-microbiota and atherogenesis, the potential mechanisms of the influence of gut microbiota in host blood pressure and the alterations in the prevalence of those bacterial genera affecting vascular tone and the development of hypertension. We will also be examining the microbiota as a therapeutic target in the prevention of CVDs and the beneficial mechanisms of probiotic administration related to cardiovascular risks. All these new insights might lead to novel analysis and CVD therapeutics based on the microbiota.

Keywords: cardiovascular diseases; atherosclerosis; gut microbiota; microbiome

1. Introduction

Cardiovascular diseases (CVDs) are a group of disorders of heart and blood vessels, including hypertension (high blood pressure), coronary heart disease (disorder of the blood vessels supplying the cardiac muscle), cerebrovascular disease (disorder of the blood vessels supplying the brain), peripheral vascular disease, heart failure, rheumatic heart disease, congenital heart disease and cardiomyopathies [1]. Globally, CVD is the major cause of morbidity and mortality [2]; an estimated 17.9 million people died from CVDs in 2016, representing 31% of all global deaths. Of these deaths,

85% were due to ischemic heart disease and stroke [3]. Atherosclerosis, the precursor of myocardial infarction, or coronary artery disease, happens over periods and is related to long-term and accumulative contact to causal changeable risk factors. Different processes such as endothelial dysfunction, chronic inflammation, hyperglycemia and oxidative stress cause atherosclerosis, a complex process present in CVDs in which an inflammation response to injury is caused [4]. In atherosclerosis, the early start leading to the onset is characterized by the increase of lipids and fibrous tissue to the internal lining of arterial walls. Increased intimal thickening may eventually lead to reduced or complete occlusion of blood flow to vital organs such as the heart and brain, resulting in myocardial infarction or stroke, respectively [5].

The atherosclerosis development is defined as the formation and accumulation of foam cells within the lipid-rich subendothelial space of the affected artery. Monocytes attracted to the area will differentiate into tissue macrophages. Due to lipid metabolic pathways dysregulation, lipid-dense macrophages called foam cells are accumulated inside the arterial lining and a characteristic 'fatty streak' with atherogenic functions, including the release of extracellular-matrix-degrading enzymes, leading to a greater likelihood of plaque rupture and consequently blood vessel occlusion [5].

Low- and middle-income countries are affected by CVDs, out of the 17 million premature deaths (under the age of 70) due to noncommunicable diseases in 2015, 82% were in low- and middle-income countries, of which 37% were caused by CVDs [1] and happened nearly equally in men and women. Although preventive measures such as reductions in smoking, blood pressure and atherogenic lipids and advances in treatments have led to a major reduction in age-standardized death rates for CVD in high-income regions, its prevalence is rising in developing countries [1]. The factors that influence to the progress of CVD are genetic sources and epigenetic factors, environmental sources, or a combination of both [6]. On the one hand, a lesser amount of one-fifth of attributable CVD risk has been accounted for genetic determinants [7,8]. On the other hand, among environmental CVD risk factors, are contaminants (e.g., atmospheric pollution and noise), tobacco smoking, physical activity, sedentariness and what we eat, the diet. If atherosclerosis remains, it is also frequently accompanied by body weight increase, blood pressure changes, lipidemia, serum glucose, endothelial dysfunction, inflammation and thrombosis [9].

Studies in human populations and model organisms have shown that intestinal microbiota changes might be associated with CVD [10,11]. In addition, some obesity-associated comorbidities, namely type 2 diabetes (T2D) and nonalcoholic fatty liver disease (NAFLD), also exhibit perturbation of the intestinal microbiota [12,13]. Microbiota communication generates complex pathways via intestinal microbiota-generated metabolites and has been shown to disturb relevant phenotypes to CVD, covering from inflammation, insulin resistance and obesity to more direct processes similar to atherosclerosis and thrombosis susceptibility [10,14–20].

This review discusses the role of the human intestinal microbiota in the development of CVDs with special emphasis on atherosclerosis, some nutritional aspects, microbiota targeted therapeutics and prevention of CVD.

2. Relationship between Microbiota and Cardiovascular Diseases

When we talk about microbiota we refer to the ecological community of commensal, symbiotic and pathogenic microorganisms that coexist on and within an organism [21]. This comprises bacteria, archaea, fungi, protozoa and even viruses [22]; bacteria are in the spotlight due to the lack of efficient methods to study the other organisms. However, the resolution is increasing in all omics-based profiling, and the cost is decreasing as as well, which is facilitating the characterization of these other organisms [23].

The research on the human microbiota, especially gut microbiota, has come to be one of the most innovative areas when it comes to the study of different pathologies [16]. It has been demonstrated that specific microbial communities may be related with the development of several diseases like obesity [24,25], cancer [26–29], inflammatory bowel disease [30,31] and rheumatic disease [32,33]; some experiments have shown a direct connection between changes in gut microbiome and cardiovascular health and disease [15,34–38]. The presence of microbes in our intestine endows us with a protective milieu by inhabiting biological places that may otherwise be colonized by potentially pathogenic microorganisms [39]. Also, it is known that the microbial community exerts an effect on the host immune response and that this is an important aspect to take into consideration in the study of autoimmune diseases [40]. Besides, the microbial community has the potential for providing microbiota-derived specific molecules, such as short-chain fatty acids (SCFAs) which directly feed colonocytes and thus prevent inflammation and gut leakage [41–43], increase nutrient harvest [44] and alter appetite signaling [45]. The quality and quantity of each SCFA depend not only on the diet's indigestible fraction [46] but also on a cross-feeding mechanism established in the bacterial community [47,48]. The most abundant SCFAs are acetic, propionic and butyric acids, which together represent nearly 90–95% of the SCFA present in the colon [49]. Acetate is a net product of carbohydrate fermentation of most anaerobic bacteria, while propionic and butyric acid are generated from carbohydrate or protein fermentation by a distinct subset of bacteria [50,51].

Each person can present a wide variety of microorganisms in the gut depending on several things, like their lifestyle [52–55]. It is known that the microbiota varies widely during the first year of life, then it stabilizes as a consortium that resembles that of adults [56]. The major taxa present in gut microbiota are Firmicutes and Bacteroidetes, whose magnitudes seem to remain remarkably constant over time [57,58].

In connection to the vast diversity of microbes among individuals, the nutritional status has a strong impact in gut microbiota modeling [59], to such an extent that specific diets such as those high in fats or sugars might lead to variations in the microbial population that, eventually, might facilitate the development of diseases [60]. Furthermore, exercise training is also considered a physical activity that modifies the gut microbiota composition and functional capacity [61]. Another important factor to take into account is the mental status of the individual since the presence of disorders such as anxiety [62] or depression are related with fluctuations in the gut microbiome [63]. The focus on the bidirectional association between the brain and gut microbiota, also known as gut-microbiota–brain axis, in neuropsychiatric disorders is a current field in the research of human microbiota [64,65].

2.1. Diet, Gut Microbiota and Cardiovascular Diseases

The gut microbiota might influence multiple metabolic and physiological processes and the modifications in these microbial structures are related with the progress of metabolic disorders such as obesity [16,66–70], insulin resistance [16,66–70] and atherosclerosis.

Foods abundant in fats (saturated, polyunsaturated and monounsaturated) are frequently copious in dietary nutrients possessing trimethylamine (TMA) moiety, such as phosphatidylcholine (PC) (lecithin), choline and L-carnitine. Mammals do not have TMA lyases, and the use of these enzymes by gut microbes, which are able to leave the C-N bond of the aforementioned nutrients, release the TMA moiety as a remaining product, so that gut microbiota are able to use these nutrients as a carbon fuel source. Portal circulation transport carries the TMA to a cluster of hepatic enzymes, the flavin-monooxygenase-3-dependent FMOs (particularly FMO3), that efficiently oxidize TMA, thus forming TMA-N-oxide (TMAO) [71–74].

Direct ingestion of PC, principally found in meat, poultry, fish, dairy foods, pasta, rice and egg-based dishes and the main nutritional source of choline in omnivores [75–77], was shown to be accompanied with increases in choline, betaine and TMAO levels [78]. Studies have shown that TMAO plasma levels are related with CVD risk [78]. Nevertheless, in other human studies, these elevated TMAO plasma levels have been independently associated with the prevalence of CVD and incident risks for myocardial infarction, stroke, death and revascularization, so more research is needed to understand the current mechanism [78–83].

Other studies have shown that L-carnitine, an another TMA-containing nutrient found almost completely in red meat, works as a nutritional precursor to gut microbial production of TMA and

TMAO in mice and humans [79]. Foods abundant in cholesterol and fats, such as red meat, liver and egg yolk have the highest levels of choline and L-carnitine, and, despite the fact that many large-scale epidemiologic studies have related red meat consumption with intensified mortality and CVD risks, the association between egg ingestion and CVD risks [84,85] has shown contradictory results [84–93]. A recent study has investigated the relationship between acute consumption of egg yolk and increased plasma and urine TMAO concentrations [94]. Whereas plasma levels of choline, betaine and TMAO were related with increased CVD risk in 1876 subjects with cardiac risk evaluation [78], further analyses in cohorts exposed that the predictive significance was mostly limited to the TMAO formation, especially from choline and L-carnitine [79,95].

In a prospective clinical study employing more than 4000 subjects undertaking elective coronary angiography, high TMAO levels projected major adverse cardiac events such as death, myocardial infarction and stroke over 3 years. The major differences were observed in the patients in the upper quartile for TMAO levels with a 2.5-fold increased risk of suffering a cardiac event compared with the lowest quartile [80].

Some evidence shows a clear positive correlation of *Atopobium* to different anthropometric variables like waist circumference, weight and body mass index and also to fat and protein intake reported in a 24 h dietary recall study [96]. Additionally, metagenomic studies have demonstrated a positive correlation between *Clostridium* and TMAO formation [79] and a positive correlation between *Clostridium histolyticum/perfringens* and waist circumference, weight, body mass index and fat mass [96]; these studies suggest that the *Clostridium* species mentioned above and *Atopobium* may be considered as markers of inflammation and CVD risk.

Gut microbiota metabolites might contribute to both hypertension and inflammation [97]. Blood pressure and plasminogen activator inhibitor-1 (PAI-1) levels have been associated to the gut microbiota composition in overweight and obese pregnant women. The butyrate-producing genus *Odoribacter* abundance has been oppositely correlated with systolic blood pressure. Butyrate production capacity is lower and PAI-1 concentrations higher in obese pregnant women. In addition, PAI-1 levels have been conversely correlated with the butyrate kinase expression and *Odoribacter* abundance [97]. A recent meta-analysis from prospective studies has concluded that elevated TMAO concentrations and its precursors were related with increased risks of major adverse cardiovascular events and all-cause mortality independently of traditional risk causes [98]. After the administration of wine with polyphenols, the authors reported a significant increase of 4-hydroxyphenylacetate in a healthy cluster. Other results in humans have shown gut microbiota responsive phenotypes to wine polyphenols intervention [99].

Preliminary results of the Prevention with Mediterranean Diet (PREDIMED) study have shown that baseline plasma concentrations of choline and hydroxyproline were associated with higher CVD risk independent of traditional risk factors, while no significant association between plasma concentrations of TMAO and CVD was found. The plasma concentrations of choline and hydroxyproline were associated with a 2.13-fold higher risk of CVD across extreme quartiles and a 1.99-fold higher risk of stroke, and baseline betaine/choline ratio was inversely associated with CVD. Compared to participants with a score below the median and randomized to the Mediterranean diet, the hazard ratio of developing CVD was 2.56 for participants with a gut microbiota score above the median and randomized to the control group [100]. TMAO levels have also been correlated with brain-type natriuretic peptide and associated with both heart failure severity and heart failure mortality [101].

Three recent meta-analyses have established that elevated TMAO blood levels are related with increased CVD risks and all-cause mortality [98,102,103], nevertheless, some criticism exists about the TMAO and CVD relationship because fish could contain high concentrations of TMAO and TMA [104]. However, fish consumption is related with heart health [105–107]. Also, there is a study without association with measures of atherosclerosis and TMAO [108]. More randomized clinical trials and larger studies are needed to clarify if TMAO is a marker or mediator in CVD. The contribution of gut microbiota in our health, immune function and disease development, continue to be generally unknown areas.

Other compounds such as intestinal-derived endogenous endotoxins, e.g., lipopolysaccharides [109], indoxyl sulfate [110] and para-cresyl sulfate [111], have been suggested to play important metabolic roles in conditions ranging from atherosclerosis to cardio-renal dysfunction [112–114].

A recent study in mice showed that the products generated by the intestinal microbiota such as SCFAs, secondary bile acids, endotoxins and tryptophan metabolites, are often altered in diets rich in fat (coconut oil and soybean oil) and low in fiber and would then impact L-cell glucagon-like peptide (GLP)-1 secretion [115].

2.2. Microbiota and Cardiovascular Diseases

It has been investigated whether metabolites derived from microbiota could influence the composition of fluids within the human body. such as blood and urine, and whether they may regulate fat absorption and bile acid/cholesterol metabolism among other physiological functions [116]. The gut microbiota can possibly affect host blood pressure through multiple mechanisms. Bacteria belonging to *Bifidobacterium*, *Lactobacillus*, *Streptococcus* and *Escherichia* genera can produce neurotransmitters within the autonomic nervous system [117]. Modifications in the prevalence of these bacteria might change the vascular tone and contribute to the hypertension development or other CVD [97,118–120]. Metabolomics approach have shown that dietary lipid phosphatidylcholine and its metabolites betaine, TMAO and choline are risk factors for CVD [78]. One study that comprised three different groups of men distributed according to the European Society of Hypertension criteria based on 24-h ambulatory blood pressure measurements supports this since the results obtained indicated a positive correlation between blood pressure and SCFA levels [121]. Another study on patients with prehypertension and stage 1 hypertension found that hippurate, phenylacetylglutamine and 4-cresyl sulfate found in urine were related with blood pressure [122].

A systematic review of human studies has reported that the *Faecalibacterium*, *Bifidobacterium*, *Ruminococcus* and *Prevotella* abundances are conversely linked to different low-grade inflammation markers such as high sensitivity C-reactive protein and interleukin (IL)-6. The existing relationships between the gut microbiota and low-grade inflammation markers in humans and the benefit of a therapeutic strategy to prevent and treat atherosclerotic CVD considering the gut microbiota and its relation with the innate and adaptive immune system [123] underline the importance of the investigation into the human gut microbiota as a potential diagnostic tool.

It has been observed how the bacteria located in the oral cavity might be related to CVD [124]. A study in patients admitted for acute coronary syndrome showed a higher subgingival bacterial load when compared to controls; the species that were mostly increased in this study were *Streptococcus intermedius*, *S. sanguis*, *S. anginosus*, *Tannerella forsythensis*, *T. denticola* and *Porphyromonas gingivalis*. Hence, these species could be risk issues for the development of acute coronary syndrome [125]. Furthermore, a possible association between *Actinobacillus actinomycetemcomitans* present in the oral cavity and both coronary heart disease and stroke has been described after several sero-epidemiologic studies [126,127].

2.3. Microbiota and Atherosclerosis

In mice, a choline-rich diet increased TMAO levels and atherosclerosis, depending on gut microbiota activity, as shown by broad-spectrum antibiotics treatment [128]. On the other hand, gut microbiota influences the host inflammatory response, altering endothelial function, which can influence host blood pressure. SCFAs production by the gut microbiota is associated with hypertension, as a result of the influence of SCFA on vascular tone [43,109,129].

Another study on men with atherosclerotic plaque on the carotid wall who consumed a drink with high numbers of *Lactobacillus plantarum* (DSM9843), showed an increased bacterial diversity compared to the placebo group as well as a decrease in the concentration of some SCFA [130], suggesting that the consumption of this strain might be a strategy to favor the intestinal diversity in patients with atherosclerotic plaque on the carotid wall.

Recent studies have directly related high levels of TMAO with an increase in cardiovascular risk and its severity [131,132]. Accordingly, TMAO levels have been correlated both with atherosclerotic plaque size and cardiovascular events [78]. Other research studies have observed that atherosclerotic plaques contain bacterial DNA, and the bacterial taxa observed were also present in the gut and oral microbiota of the same individuals [133,134]. Several epidemiological studies have associated periodontal disease and CVD [135–137]; an oral microbiota role in the CVD pathophysiology has also been studied [133,137–139].

In addition, metagenomic analyses have shown that microbial composition is altered in patients with unstable compared with stable plaques; unstable plaques are related with reduced *Roseburia* fecal levels and both increased theoretical capacity of the microbiome to produce proinflammatory peptidoglycans and reduced production of anti-inflammatory carotenes [11].

Other examples of the relationships between microbiota and atherosclerosis are the administration of metformin and whole grains; metformin is a biguanide antidiabetic drug widely used in adults that have shown to exert positive effects to fight against CVD risk and that might be used safely in patients with heart failure and even reduce its occurrence or mortality, not only by direct effects [140], but also because of the possible effects produced through gut microbiota remodeling [141]. Also, the diet seems to be a potential therapy to diminish the risk of CVD since a study on a specific population of Danish adults showed that a diet abundant in whole grain compared to refined grain reduces body mass and systemic inflammation [142], which are risk factors to a bad prognosis of CVD.

2.4. *Other Microbiota Aspects Related to Cardiovascular Diseases*

One of the most studied pathogen-associated molecular patterns (PAMPs) concerning cardiovascular function and the increase in CVD risk is lipopolysaccharide (LPS), a Gram-negative bacterial cell wall component [143,144]. Circulating LPS is raised in at-risk individuals and predicts future CVD [145–147]; accordingly, administration of low-doses of LPS induces vascular inflammation and atherosclerosis in experimental animals [148,149]. Another significant PAMP is the peptidoglycan that can trigger the nucleotide-binding oligomerization domain (NOD) receptors. NOD receptors can identify bacterial determinants once they are phagocytosed by macrophages and dendritic cells. NOD2-deficient mice fed with a high-fat diet have shown increased bacterial translocation and insulin resistance [150]. Additionally, human genetic and mouse knockout studies have investigated the role of NOD2 in atherosclerosis [151,152].

2.4.1. Microbiota, Choline and Homocysteine Cycle

Recently, a study reported that microorganisms in the gut microbiota hydrolyze PC to obtain choline for downstream metabolism [153,154]. A previous study reported that gut microorganisms can anaerobically convert choline to TMA, which is further metabolized by the host to TMAO [78]. Analysis in gnotobiotic mice has revealed that specific bacteria might increase the TMAO formation [155].

Choline is an essential nutrient that is usually grouped within the vitamin B complex. Choline and its metabolite betaine are methyl donors along with folate, and are metabolically linked to transmethylation pathways including synthesis of the CVD risk factor homocysteine [78]. Deficiency in both choline and betaine has been suggested to produce epigenetic changes in genes linked to atherosclerosis [156,157], and acute choline and methionine deficiency in rodent models causes lipid accumulation in liver (steatohepatitis), heart and arterial tissues [158]. Homocysteine, a sulfhydryl-containing amino acid produced via demethylation of methionine and essential for intravascular metabolism [159], has been supposed as a reasonable risk issue for the atherosclerotic vascular disease leading to CVD and stroke [160]. Highly elevated homocysteine levels in genetic hyperhomocysteinemia are pathogenic to the vascular system, and homocysteine, at comparably high levels, also exerts proinflammatory effects on vascular cells in vitro [161].

Higher dietary choline intake was associated with a lower risk of incident ischemic stroke in African-American participants; also, higher dietary betaine intake was associated with a nonlinear higher risk of incident coronary heart disease [162].

2.4.2. Vitamin B-Complex and Microbiota

Commensal bacteria are suppliers and consumers of B vitamins and vitamin K. While dietary B vitamins are generally absorbed through the small intestine, bacterial B vitamins are produced and absorbed mainly through the colon [163,164], showing that dietary and gut microbiota-derived B vitamins are probably controlled differently by the human body.

In a prospective study with Korean men, the authors found that higher dietary intake levels of vitamin B6 were associated with a reduced CVD risk [165]. Vitamin B3 might increase all-cause mortality, which was probably associated with its adverse effects on glycemic response [166,167].

Vitamin B9 and B-vitamin complex reduced risk for stroke, and vitamin B9 reduced risk for total CVD events. There was no evidence of a reduction of CVD risk with any other vitamins or supplements, and no supplements reduced mortality [168]. A recent meta-analysis found that vitamin B9 supplementation significantly reduced the risk of stroke in patients with CVD [169].

2.4.3. Low-Grade Inflammation

Evidence exists that inflammation and oxidative stress are influenced by the diet, and it may, therefore, be possible to reduce or delay the effects of age-related changes in these parameters through appropriate dietary intervention and/or use of nutraceutical dietary supplements [170].

Some studies have investigated the relationship between gut microbiota and markers of chronic low-grade inflammation in humans. An opposite association among *Prevotella* and inflammatory markers and an increased abundance of certain *Prevotella* species were associated with low-grade inflammation in systemic diseases, such as rheumatoid arthritis [171]. In addition, *Prevotella* abundance was inversely associated with LPS and high sensitivity C-reactive protein. Furthermore, individuals with obesity have a lower abundance of *Prevotella* species in their gut [171].

The RISTOMED project is an open-label study that investigated the diet as a means to improve health-related quality of life for older people and to prevent aging-related diseases and also the concomitant administration of VSL#3, a mixture of probiotic strains, in the possible reduction of high-sensitivity C-reactive protein plasma concentration and microbiota changes due to the fact that this protein is defined as a cardiovascular risk in this population by the American Heart Association [170]. Changes in the aforementioned outcomes were observed in a subgroup analysis in participants with low-grade inflammation. The RISTOMED diet plus VSL#3 administration has shown a reduction in high-sensitivity C-reactive protein and also an increase in *Bifidobacterium* species [170,172]. Further analyses with more participants in the study have shown similar results in high-sensitivity C-reactive protein and microbiota [173]. Similar studies, involving the administration of probiotics in elderly human trials have shown no effects on inflammatory outcomes [174–178], augmented levels of fecal prostaglandin E_2 [179] and diminished plasma endotoxin, the soluble cluster of differentiation 14 and LPS binding protein levels [180].

Recently, Gil-Cruz et al. reported that mimic peptides from commensal bacteria can promote inflammatory cardiomyopathy in genetically susceptible individuals [181].

In brief, several studies have established the relationship between homocysteine, PAMP, low-grade inflammation, microbiota and CVD. In this regard, further studies are needed to determine the specific factors and the underlying mechanism in the progression and prevention of CVD.

3. Microbiota-Targeted Therapeutics

3.1. Physical Activity, Microbiota and Cardiovascular Diseases

In the last 10 years, it has been observed that there is a possible relationship between the intestinal microbiota and the cardiovascular system [11,144,182,183]. Human cardiometabolic health has been related with variations in the gut microbiota composition (dysbiosis) [184]. Kelly et al. reported that subjects with a high lifetime burden of CVD risk factors had less microbial wealth compared to those with a low lifetime burden, identifying a high number of Bacteroidetes and Firmicutes [185].

The benefits of regular physical activity against cardiovascular problems are widely known. Recent studies have revealed how physical exercise affects gut microbiota [186–190]. Increased levels of Bacteroidetes and decreased of Firmicutes were observed in obese adults who had moderate to severe aerobic exercise for 10 weeks [191]. Although no direct evidence supports the idea that physical exercise prevents atherosclerotic CVD through changing the gut microbiota and by improving systematic inflammation, many studies have supported this hypothesis [186]. Zuheng Liu et al. reported that the changes in the gut microbial organization that are produced by physical exercise are associated with cardiac function in myocardial infarction mice [192].

It is known that voluntary running exercise modifies the microbiota composition of the cecum and increases the n-butyrate concentration in the cecal content [189]. Butyrate is one of the three most important SCFAs, and several studies have shown that it may have effects on cardiovascular function [97,144,183]. Nevertheless, more studies are needed to explore the principal physiological mechanisms that relate regular exercise to SCFA levels and its effect on blood pressure and inflammation.

3.2. Probiotic Administration, Microbiota, Bile Acids and Cardiovascular Diseases

It has been demonstrated that probiotics can affect the structure of gut microbiota and the interaction with the microbial community and the host health through different mechanisms [12,193–195]. These effects are mediated by the direct or indirect action of probiotics and can involve the modulation of the immune system or that of remote organs like the brain and liver due to the production of metabolites finally localized in these organs [193,196–199].

Obesity is one of the primary risk factors for the development of CVD and presents a major risk for T2D, hypertension and hyperlipidemia and predisposes to coronary heart disease [200,201]. Hypercholesterolemia is directly associated with the prevalence of ischemic heart disease in both men and women [202]. Dietary modifications are the first line of treatment and offer an effective means of reducing blood cholesterol levels. However, the low rate of patient dietary compliance means that drug administration is one of the most effective treatments to control plasma cholesterol, triacylglycerols and blood sugar levels.

There is evidence supporting that probiotics can improve some parameters of the risk factors of CVD, like obesity. A recent systematic review reported that specific strains from *Lactobacillus* and *Bifidobacterium* have been generally used as probiotic treatment in well-established animal models of obesity [203] and in blood lipid index, T2D and hypertension [200,201,204–206].

The potential probiotic mechanisms related to the hypocholesterolemic effect could involve active bile salt hydrolase (BSH), cholesterol co-precipitation with deconjugated bile salts, bacterial cell membrane assimilation and incorporation of cholesterol, conversion of cholesterol to coprostanol through the cholesterol reductase enzyme and SCFA production [207]. The BSH increased fecal excretion of free bile acids, preventing their reabsorption and compensatory increased use of cholesterol to produce bile acids, which could lead to a reduction in the cholesterol present in serum. SCFAs can inhibit the hepatic activity of the 3-hydroxymethyl-3-glutaryl-CoA reductase, the hepatic enzyme in the process of hepatic cholesterol synthesis, while the propionate can stimulate the bile salts hepatic synthesis through increasing the activity of 7α-hydroxylase [208].

The antihypertensive effects of probiotics have been related to their metabolites; some studies have reported specific bioactive tripeptides. These compounds have an angiotensin-converting enzyme

(ACE)-inhibitory properties [209]. However, other studies have related these bioactive peptides with up to 12 peptides in length of fermented milk containing probiotics with similar antihypertensive effects [210].

Probiotics improve T2D symptoms, glucose biomarkers and insulin resistance by restoring homeostasis of gut microbiota. Furthermore, a meta-analysis suggests that the supplementation of probiotics has a modest effect on the serum level of fasting blood sugar as well as oxidative stress biomarkers [211]. Mechanisms that have been proposed are as follows: improved intestinal integrity, decreased systemic lipopolysaccharide levels, decreased endoplasmic reticulum stress and improved peripheral insulin sensitivity [204,212]. Data from clinical studies and animal models have shown a reduction in lipopolysaccharide translocation, endotoxemia and inflammation, reducing stimulation of the proinflammatory genes like tumor necrosis factor alpha (TNF-α), IL-6 and IL-1β [204,213].

In contrast, negative results have been found in other studies and meta-analyses regarding the effectiveness of probiotics in diarrhea prevention in children [214], adults [215] and the elderly [216,217]. A consistent and sufficient probiotic consumption might produce a number of health benefits including reducing CVD risk factors [201], nevertheless, further studies in different models are necessary for a better understanding of the beneficial mechanisms of probiotic administration in CVD risk alone or accompanied by foods; certainly, probiotics act in a strain-specific manner and are often used as coadjuvant therapy. Likewise, recent individual studies and meta-analyses should be, mainly due to the different probiotic strains used, carefully interpreted.

3.3. Fecal Microbiota Transplantation

Fecal microbiota transplantation (FMT) has become popular in recent years. FMT is the transplantation of functional bacteria from feces of healthy donors into the gastrointestinal tract of patients to repair the balance of the intestinal microbiota [218]. The process involves the collection of filtered stools collected from either a healthy donor or from the recipient himself (autologous FMT) at a time point before initiation of disease and associated dysbiosis and its installation into the intestinal tract of a patient suffering from a certain medical condition [219]. FMT is effective in the treatment *Clostridium difficile* infection (CDI) in humans [220,221]. The first report of FMT application in the treatment of CDI dated from 1983 [222]; in 2010, the United States Infectious Diseases Society of America and Society for Healthcare Epidemiology of America recommended FMT as a treatment plan for CDI in their clinical guidelines [223]. Recently, some studies have shown that there is a very strong potential application for FMT in the field of cardiometabolic disorders [224,225], such as atherosclerosis, metabolic syndrome and T2D [226]. However, FMT is currently restricted due to its associated risks, including the possible transfer of endotoxins or infectious agents that could cause new gastrointestinal complications [227,228].

Further studies are needed to examine whether FMT might be extended to other facets of cardiometabolic disorders. Instead of fecal contents, the transplantation of only a defined group of bacteria may be a rational alternative to FMT. Also, further research is needed to better define the optimal fecal microbial preparation, dosing and method of delivery.

3.4. Personalized Nutrition

Evidence shows that variations occasioned by dietary interventions in host metabolism are person-specific [229], and, because not all individuals respond to diet in the same way (e.g., weight gain, postprandial glucose, etc.), personalized nutrition is a new therapeutic possibility for prevention and control of disease [230]. Recent studies of cohorts have revealed great differences in post-meal glucose levels between individuals eating the same mealtimes [229,231].

Healthy participants who exhibited enhanced glucose metabolism following barley kernel-based bread (BKB) consumption were related with a greater *Prevotella* abundance [232]. Another study in humans has reported that the whole grains ingestion induced anti-inflammatory responses and blood glucose level changes of different magnitudes; participants with greater blood IL-6 improvements had higher *Dialister* levels and lower Coriobacteriaceae species in their stools [233]. Furthermore, on a calorie-restricted diet, overweight and obese adults with higher levels of baseline *Akkermansia muciniphila* presented a greater improvement in insulin sensitivity and lipid metabolism, as well as a greater reduction in body fat [234]. Another cohort including 800 overweight or obese nondiabetic individuals showed high interpersonal variability in the postprandial glycemic response to identical foods, which was predicted accurately by different factors, i.e., the gut microbiome, dietary habits, blood parameters and anthropometrics, using a machine learning approach [229]. These results concluded that microbiota-based nutrition can be used to expect variable clinical phenotypes in metabolic syndrome as well as gastrointestinal disorders; individuals can then be classified into responders and nonresponders based on different outcomes such as dietary components, age, serum parameters and the microbiome, all contributing to personalized predictions [230].

Interindividual variability regarding the efficacy of certain nutrients in optimizing an individual's health and the identification of factors that give to an individual's response to diet, as well as developing methods of personalizing dietary references, are shown to be critical [235]. Figure 1 summarizes the relationship between microbiota and related metabolites and CVD.

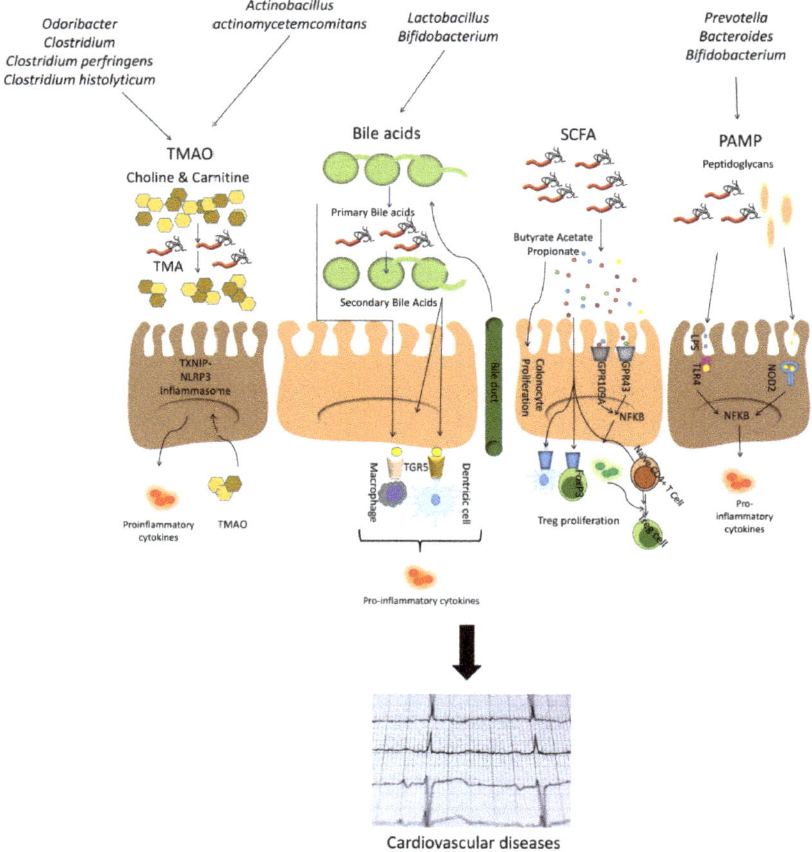

Figure 1. Schematic representation of the relationships between microbiota and cardiovascular disease (CVD). Abbreviations. LPS, lipopolysaccharide; PAMP, pathogen-associated molecular patterns; NFκB, nuclear factor kappa-B; SCFA, short-chain fatty acids; TLR, toll-like receptor; TMAO, trimethylamine N-oxide.

4. Prevention of Cardiovascular Diseases

The balance between pathogenic and nonpathogenic microorganisms in the gut is critical to maintaining the lifelong health of humans. As previously mentioned, the diet is an external factor that influences the gut microbiota composition. Different studies have analyzed diets and their implications with the gut microbiota and the prevention of CVD. The Mediterranean diet, based on the regular ingestion of plant foods, the moderate consumption of fish, seafood and dairy, a low-to-moderate alcohol (mostly red wine) intake, balanced by a comparatively limited use of red meat and other meat products, with olive oil being the main source of fat consumed in this diet, is nowadays universally recognized as beneficial to health by medical professionals and could be an emerging medical prescription [236] based on the reduction incidences of insulin resistance, hypertension, CVD, T2D and metabolic syndrome [236]. Other important diets for the prevention of CVD are plant-based diets, which are characterized by high consumption of seeds, cereals, fruit, berries, nuts and vegetables. Both diets are important sources of fibers and bioactive compounds, which are metabolized by microbes to produce different metabolites [237] such as acetate, propionate and butyrate, which are involved in suppressing inflammatory responses. The mechanisms by which these diets exert their beneficial effects remain to be elucidated, but their bioactive food components such as unsaturated fatty acids [238], complex carbohydrates and fibers [237] and polyphenols [239] are very implicated.

Unsaturated fatty acids, in particular n-3 polyunsaturated fatty acids, are generally considered cardiovascular-protective. Fish oil is the main source of animal oil whereas flaxseed oil is obtained from plants [238]. Both fish oil and flaxseed oil could modulate gut microbiota and enhance the microbial production of SCFAs, with fish oil being more effective than flaxseed oil in promoting the growth of SCFA-producing bacteria and lowering microbial generation of LPS; both oils are implicated in the reduction of TMAO, with fish oil being the most effective in exacerbating atherogenesis [238]. Beta-glucan, a natural polysaccharide from the plant cell walls, belongs to one of the dietary fiber fractions considered to be a prebiotic which stimulates the growth of beneficial intestinal bacteria [240], produces SCFA [237] and reduces cholesterol and glucose concentrations in the blood, all of which reduces the risk of CVD and diabetes.

Polyphenols, mainly founded in the Mediterranean and plant-based diets, are a group of phytochemicals abundant in the human diet and considered to be very important in the prevention of diseases by their ability to modulate the microbiota. In animal models, polyphenols increase bacteria that cause SCFA production and decrease bacteria that produce LPS. The most important polyphenols groups, mainly founded in fruits, are flavonoids, flavones and flavonols [239]. Accordingly, it has been shown that the intake of whole fruits is a good strategy for the prevention of diseases by increasing the growth of beneficial bacteria (i.e., *Bifidobacterium* and *Lactobacillus*) [241], which is in agreement with previous research with pomegranate polyphenol extracts [242] and in animal studies [239].

In conclusion, the consumption of a healthy diet based on unsaturated fatty acids, fruits and vegetables is the best strategy in the prevention and treatment of diseases that are modulated by gut microbiota.

5. Further Directions and Perspectives

The gut microbiota influences drug responses altering both pharmacodynamics and pharmacokinetics. Activity from the gut microbiota can thus result in altered drug pharmacokinetics, activation of prodrugs and the unwanted formation of toxic metabolites or inactivation of drugs [243]. Each patient displays significant variations in response to treatment and drug-associated injurious effects, which results in considerable variations in morbidity and mortality [244–246].

Personalized nutritional approaches can be established to change an individual's microbiome and further develop the response to a specific diet. The future of personalized nutrition will allow for the rational design of diets. A prior step would include the individual analysis of the microbiome, the prediction of particular responders and nonresponders and the identification of beneficial foods for the different microbiome types and desired outcomes. In relation to CVD and atherosclerosis, the personalized diet recommendation would depend on the patient microbiota, the TMAO blood levels and the family history [230].

A greater understanding of the interactions between the patient microbiome and the response to treatments will be fundamental for the improvement of CVD therapies and the development of novel approaches targeting the microbiota in CVDs.

Author Contributions: All authors participated in the bibliographic search, discussion and writing of the manuscript. All authors have read and agreed to the published version of the manuscript.

Funding: This research received no external funding.

Acknowledgments: Julio Plaza-Díaz is part of the "UGR Plan Propio de Investigación 2016" and the "Excellence actions: Unit of Excellence on Exercise and Health (UCEES), University of Granada". We are grateful to Belen Vazquez-Gonzalez for assistance with the illustration service.

Conflicts of Interest: The authors declare no conflict of interest.

References

1. World Health Organization. Cardiovascular Disease. Available online: https://www.who.int/cardiovascular_diseases/about_cvd/en/ (accessed on 13 November 2019).
2. Benjamin, E.J.; Virani, S.S.; Callaway, C.W.; Chamberlain, A.M.; Chang, A.R.; Cheng, S.; Chiuve, S.E.; Cushman, M.; Delling, F.N., Deo, R.; et al. Heart disease and stroke statistics-2018 update: A report from the American heart association. *Circulation* **2018**, *137*, e67–e492. [CrossRef]
3. The, L. GBD 2017: A fragile world. *Lancet (Lond. Engl.)* **2018**, *392*, 1683.
4. Ross, R. Atherosclerosis—An inflammatory disease. *N. Engl. J. Med.* **1999**, *340*, 115–126. [CrossRef] [PubMed]
5. Maguire, E.M.; Pearce, S.W.A.; Xiao, Q. Foam cell formation: A new target for fighting atherosclerosis and cardiovascular disease. *Vasc. Pharmacol.* **2019**, *112*, 54–71. [CrossRef] [PubMed]
6. Nabel, E.G. Cardiovascular disease. *N. Engl. J. Med.* **2003**, *349*, 60–72. [CrossRef]
7. Ardissino, D.; Berzuini, C.; Merlini, P.A.; Mannuccio Mannucci, P.; Surti, A.; Burtt, N.; Voight, B.; Tubaro, M.; Peyvandi, F.; Spreafico, M.; et al. Influence of 9p21.3 genetic variants on clinical and angiographic outcomes in early-onset myocardial infarction. *J. Am. Coll. Cardiol.* **2011**, *58*, 426–434. [CrossRef]
8. Ripatti, S.; Tikkanen, E.; Orho-Melander, M.; Havulinna, A.S.; Silander, K.; Sharma, A.; Guiducci, C.; Perola, M.; Jula, A.; Sinisalo, J.; et al. A multilocus genetic risk score for coronary heart disease: Case-control and prospective cohort analyses. *Lancet* **2010**, *376*, 1393–1400. [CrossRef]
9. Labarthe, D.R. *Epidemiology and Prevention of Cardiovascular Diseases: A Global Challenge*; Jones & Bartlett Publishers: Sudbury, MA, USA, 2010.
10. Jie, Z.; Xia, H.; Zhong, S.L.; Feng, Q.; Li, S.; Liang, S.; Zhong, H.; Liu, Z.; Gao, Y.; Zhao, H.; et al. The gut microbiome in atherosclerotic cardiovascular disease. *Nat. Commun.* **2017**, *8*, 845. [CrossRef]
11. Karlsson, F.H.; Fåk, F.; Nookaew, I.; Tremaroli, V.; Fagerberg, B.; Petranovic, D.; Bäckhed, F.; Nielsen, J. Symptomatic atherosclerosis is associated with an altered gut metagenome. *Nat. Commun.* **2012**, *3*, 1245. [CrossRef]
12. Plaza-Diaz, J.; Ruiz-Ojeda, F.J.; Gil-Campos, M.; Gil, A. Mechanisms of action of probiotics. *Adv. Nutr.* **2019**, *10*, S49–S66. [CrossRef]
13. Mouzaki, M.; Comelli, E.M.; Arendt, B.M.; Bonengel, J.; Fung, S.K.; Fischer, S.E.; McGilvray, I.D.; Allard, J.P. Intestinal microbiota in patients with nonalcoholic fatty liver disease. *Hepatology* **2013**, *58*, 120–127. [CrossRef] [PubMed]
14. Tang, W.H.W.; Backhed, F.; Landmesser, U.; Hazen, S.L. Intestinal microbiota in cardiovascular health and disease: JACC state-of-the-art review. *J. Am. Coll. Cardiol.* **2019**, *73*, 2089–2105. [CrossRef] [PubMed]
15. Tang, W.W.; Kitai, T.; Hazen, S.L. Gut microbiota in cardiovascular health and disease. *Circ. Res.* **2017**, *120*, 1183–1196. [CrossRef]
16. Turnbaugh, P.J.; Ley, R.E.; Mahowald, M.A.; Magrini, V.; Mardis, E.R.; Gordon, J.I. An obesity-associated gut microbiome with increased capacity for energy harvest. *Nature* **2006**, *444*, 1027–1031. [CrossRef] [PubMed]
17. Qin, J.; Li, Y.; Cai, Z.; Li, S.; Zhu, J.; Zhang, F.; Liang, S.; Zhang, W.; Guan, Y.; Shen, D.; et al. A metagenome-wide association study of gut microbiota in type 2 diabetes. *Nature* **2012**, *490*, 55–60. [CrossRef]

18. Reinhardt, C. The gut microbiota as an influencing factor of arterial thrombosis. *Hamostaseologie* **2019**, *39*, 173–179. [CrossRef]
19. Jackel, S.; Kiouptsi, K.; Lillich, M.; Hendrikx, T.; Khandagale, A.; Kollar, B.; Hormann, N.; Reiss, C.; Subramaniam, S.; Wilms, E.; et al. Gut microbiota regulate hepatic von Willebrand factor synthesis and arterial thrombus formation via Toll-like receptor-2. *Blood* **2017**, *130*, 542–553. [CrossRef]
20. Álvarez-Mercado, A.I.; Navarro-Oliveros, M.; Robles-Sánchez, C.; Plaza-Díaz, J.; Sáez-Lara, M.J.; Muñoz-Quezada, S.; Fontana, L.; Abadía-Molina, F. Microbial population changes and their relationship with human health and disease. *Microorganisms* **2019**, *7*, 68. [CrossRef]
21. Group, N.H.W.; Peterson, J.; Garges, S.; Giovanni, M.; McInnes, P.; Wang, L.; Schloss, J.A.; Bonazzi, V.; McEwen, J.E.; Wetterstrand, K.A.; et al. The NIH human microbiome project. *Genome Res.* **2009**, *19*, 2317–2323. [CrossRef]
22. Roy, S.; Trinchieri, G. Microbiota: A key orchestrator of cancer therapy. *Nat. Rev. Cancer* **2017**, *17*, 271–285. [CrossRef]
23. Micheel, C.M.; Nass, S.J.; Omenn, G.S. Omics-based clinical discovery: Science, technology, and applications. In *Evolution of Translational Omics: Lessons Learned and the Path Forward*; National Academies Press: Washington, DC, USA, 2012.
24. Maruvada, P.; Leone, V.; Kaplan, L.M.; Chang, E.B. The human microbiome and obesity: Moving beyond associations. *Cell Host Microbe* **2017**, *22*, 589–599. [CrossRef] [PubMed]
25. Miyamoto, J.; Igarashi, M.; Watanabe, K.; Karaki, S.I.; Mukouyama, H.; Kishino, S.; Li, X.; Ichimura, A.; Irie, J.; Sugimoto, Y.; et al. Gut microbiota confers host resistance to obesity by metabolizing dietary polyunsaturated fatty acids. *Nat. Commun.* **2019**, *10*, 4007. [CrossRef] [PubMed]
26. Rajagopala, S.V.; Vashee, S.; Oldfield, L.M.; Suzuki, Y.; Venter, J.C.; Telenti, A.; Nelson, K.E. The human microbiome and cancer. *Cancer Prev. Res.* **2017**, *10*, 226–234. [CrossRef] [PubMed]
27. Liang, W.; Yang, Y.; Wang, H.; Wang, H.; Yu, X.; Lu, Y.; Shen, S.; Teng, L. Gut microbiota shifts in patients with gastric cancer in perioperative period. *Medicine (Baltimore)* **2019**, *98*, e16626. [CrossRef]
28. Kostic, A.D.; Gevers, D.; Pedamallu, C.S.; Michaud, M.; Duke, F.; Earl, A.M.; Ojesina, A.I.; Jung, J.; Bass, A.J.; Tabernero, J.; et al. Genomic analysis identifies association of Fusobacterium with colorectal carcinoma. *Genome Res.* **2012**, *22*, 292–298. [CrossRef] [PubMed]
29. McCoy, A.N.; Araujo-Perez, F.; Azcarate-Peril, A.; Yeh, J.J.; Sandler, R.S.; Keku, T.O. Fusobacterium is associated with colorectal adenomas. *PLoS ONE* **2013**, *8*, e53653. [CrossRef] [PubMed]
30. McIlroy, J.; Ianiro, G.; Mukhopadhya, I.; Hansen, R.; Hold, G.L. Review article: The gut microbiome in inflammatory bowel disease-avenues for microbial management. *Aliment. Pharmacol. Ther.* **2018**, *47*, 26–42. [CrossRef]
31. Sitkin, S.; Demyanova, E.; Vakhitov, T.; Pokrotnieks, J. Altered sphingolipid metabolism and its interaction with the intestinal microbiome is another key to the pathogenesis of inflammatory bowel disease. *Inflamm. Bowel Dis.* **2019**, *25*, e157–e158. [CrossRef]
32. Yeoh, N.; Burton, J.P.; Suppiah, P.; Reid, G.; Stebbings, S. The role of the microbiome in rheumatic diseases. *Curr. Rheumatol. Rep.* **2013**, *15*, 314. [CrossRef]
33. van Dijkhuizen, E.H.P.; Del Chierico, F.; Malattia, C.; Russo, A.; Pires Marafon, D.; Ter Haar, N.M.; Magni-Manzoni, S.; Vastert, S.J.; Dallapiccola, B.; Prakken, B.; et al. Microbiome analytics of the gut microbiota in patients with juvenile idiopathic arthritis: A longitudinal observational cohort study. *Arthritis Rheumatol.* **2019**, *71*, 1000–1010. [CrossRef]
34. Ascher, S.; Reinhardt, C. The gut microbiota: An emerging risk factor for cardiovascular and cerebrovascular disease. *Eur. J. Immunol.* **2018**, *48*, 564–575. [CrossRef] [PubMed]
35. Miele, L.; Giorgio, V.; Alberelli, M.A.; De Candia, E.; Gasbarrini, A.; Grieco, A. Impact of gut microbiota on obesity, diabetes, and cardiovascular disease risk. *Curr. Cardiol. Rep.* **2015**, *17*, 120. [CrossRef] [PubMed]
36. Koeth, R.A.; Wang, Z.; Levison, B.S.; Buffa, J.A.; Org, E.; Sheehy, B.T.; Britt, E.B.; Fu, X.; Wu, Y.; Li, L.; et al. Intestinal microbiota metabolism of L-carnitine, a nutrient in red meat, promotes atherosclerosis. *Nat. Med.* **2013**, *19*, 576–585. [CrossRef] [PubMed]
37. Zhu, W.; Gregory, J.C.; Org, E.; Buffa, J.A.; Gupta, N.; Wang, Z.; Li, L.; Fu, X.; Wu, Y.; Mehrabian, M.; et al. Gut microbial metabolite TMAO enhances platelet hyperreactivity and thrombosis risk. *Cell* **2016**, *165*, 111–124. [CrossRef] [PubMed]

38. Benakis, C.; Brea, D.; Caballero, S.; Faraco, G.; Moore, J.; Murphy, M.; Sita, G.; Racchumi, G.; Ling, L.; Pamer, E.G.; et al. Commensal microbiota affects ischemic stroke outcome by regulating intestinal gammadelta T cells. *Nat. Med.* **2016**, *22*, 516–523. [CrossRef]
39. Lin, L.; Zhang, J. Role of intestinal microbiota and metabolites on gut homeostasis and human diseases. *BMC Immunol.* **2017**, *18*, 2. [CrossRef]
40. Kohling, H.L.; Plummer, S.F.; Marchesi, J.R.; Davidge, K.S.; Ludgate, M. The microbiota and autoimmunity: Their role in thyroid autoimmune diseases. *Clin. Immunol.* **2017**, *183*, 63–74. [CrossRef]
41. Chambers, E.S.; Preston, T.; Frost, G.; Morrison, D.J. Role of gut microbiota-generated short-chain fatty acids in metabolic and cardiovascular health. *Curr. Nutr. Rep.* **2018**, *7*, 198–206. [CrossRef]
42. Pluznick, J.L.; Protzko, R.J.; Gevorgyan, H.; Peterlin, Z.; Sipos, A.; Han, J.; Brunet, I.; Wan, L.X.; Rey, F.; Wang, T.; et al. Olfactory receptor responding to gut microbiota-derived signals plays a role in renin secretion and blood pressure regulation. *Proc. Natl. Acad. Sci. USA* **2013**, *110*, 4410–4415. [CrossRef]
43. Natarajan, N.; Hori, D.; Flavahan, S.; Steppan, J.; Flavahan, N.A.; Berkowitz, D.E.; Pluznick, J.L. Microbial short chain fatty acid metabolites lower blood pressure via endothelial G protein-coupled receptor 41. *Physiol. Genom.* **2016**, *48*, 826–834. [CrossRef]
44. Wang, B.; Jiang, X.; Cao, M.; Ge, J.; Bao, Q.; Tang, L.; Chen, Y.; Li, L. Altered fecal microbiota correlates with liver biochemistry in nonobese patients with non-alcoholic fatty liver disease. *Sci. Rep.* **2016**, *6*, 32002. [CrossRef]
45. Perry, R.J.; Peng, L.; Barry, N.A.; Cline, G.W.; Zhang, D.; Cardone, R.L.; Petersen, K.F.; Kibbey, R.G.; Goodman, A.L.; Shulman, G.I. Acetate mediates a microbiome-brain-beta-cell axis to promote metabolic syndrome. *Nature* **2016**, *534*, 213–217. [CrossRef] [PubMed]
46. Koh, A.; De Vadder, F.; Kovatcheva-Datchary, P.; Backhed, F. From dietary fiber to host physiology: Short-Chain fatty acids as key bacterial metabolites. *Cell* **2016**, *165*, 1332–1345. [CrossRef] [PubMed]
47. Rios-Covian, D.; Gueimonde, M.; Duncan, S.H.; Flint, H.J.; de los Reyes-Gavilan, C.G. Enhanced butyrate formation by cross-feeding between Faecalibacterium prausnitzii and Bifidobacterium adolescentis. *FEMS Microbiol. Lett.* **2015**, *362*. [CrossRef]
48. Hoek, M.; Merks, R.M.H. Emergence of microbial diversity due to cross-feeding interactions in a spatial model of gut microbial metabolism. *BMC Syst. Biol.* **2017**, *11*, 56. [CrossRef] [PubMed]
49. Cummings, J.H.; Pomare, E.W.; Branch, W.J.; Naylor, C.P.; Macfarlane, G.T. Short chain fatty acids in human large intestine, portal, hepatic and venous blood. *Gut* **1987**, *28*, 1221–1227. [CrossRef]
50. Louis, P.; Flint, H.J. Formation of propionate and butyrate by the human colonic microbiota. *Environ. Microbiol.* **2017**, *19*, 29–41. [CrossRef]
51. Granado-Serrano, A.B.; Martin-Gari, M.; Sanchez, V.; Riart Solans, M.; Berdun, R.; Ludwig, I.A.; Rubio, L.; Vilaprinyo, E.; Portero-Otin, M.; Serrano, J.C.E. Faecal bacterial and short-chain fatty acids signature in hypercholesterolemia. *Sci. Rep.* **2019**, *9*, 1772. [CrossRef]
52. Yamashita, M.; Okubo, H.; Kobuke, K.; Ohno, H.; Oki, K.; Yoneda, M.; Tanaka, J.; Hattori, N. Alteration of gut microbiota by a Westernized lifestyle and its correlation with insulin resistance in non-diabetic Japanese men. *J. Diabetes Investig.* **2019**, *10*, 1463–1470. [CrossRef]
53. Zhong, H.; Penders, J.; Shi, Z.; Ren, H.; Cai, K.; Fang, C.; Ding, Q.; Thijs, C.; Blaak, E.E.; Stehouwer, C.D.A.; et al. Impact of early events and lifestyle on the gut microbiota and metabolic phenotypes in young school-age children. *Microbiome* **2019**, *7*, 2. [CrossRef]
54. Ruggles, K.V.; Wang, J.; Volkova, A.; Contreras, M.; Noya-Alarcon, O.; Lander, O.; Caballero, H.; Dominguez-Bello, M.G. Changes in the gut microbiota of urban subjects during an immersion in the traditional diet and lifestyle of a rainforest village. *mSphere* **2018**, *3*. [CrossRef] [PubMed]
55. Dubois, G.; Girard, C.; Lapointe, F.J.; Shapiro, B.J. The Inuit gut microbiome is dynamic over time and shaped by traditional foods. *Microbiome* **2017**, *5*, 151. [CrossRef] [PubMed]
56. Palmer, C.; Bik, E.M.; DiGiulio, D.B.; Relman, D.A.; Brown, P.O. Development of the human infant intestinal microbiota. *PLoS Biol.* **2007**, *5*, e177. [CrossRef] [PubMed]
57. Human Microbiome Project, C. Structure, function and diversity of the healthy human microbiome. *Nature* **2012**, *486*, 207–214. [CrossRef] [PubMed]
58. Faith, J.J.; Guruge, J.L.; Charbonneau, M.; Subramanian, S.; Seedorf, H.; Goodman, A.L.; Clemente, J.C.; Knight, R.; Heath, A.C.; Leibel, R.L.; et al. The long-term stability of the human gut microbiota. *Science* **2013**, *341*, 1237439. [CrossRef] [PubMed]

59. Wu, Y.; Wan, J.; Choe, U.; Pham, Q.; Schoene, N.W.; He, Q.; Li, B.; Yu, L.; Wang, T.T.Y. Interactions between food and gut microbiota: Impact on human health. *Annu. Rev. Food Sci. Technol.* **2019**, *10*, 389–408. [CrossRef] [PubMed]
60. Proctor, C.; Thiennimitr, P.; Chattipakorn, N.; Chattipakorn, S.C. Diet, gut microbiota and cognition. *Metab. Brain Dis.* **2017**, *32*, 1–17. [CrossRef]
61. Mailing, L.J.; Allen, J.M.; Buford, T.W.; Fields, C.J.; Woods, J.A. Exercise and the gut microbiome: A review of the evidence, potential mechanisms, and implications for human health. *Exerc. Sport Sci. Rev.* **2019**, *47*, 75–85. [CrossRef]
62. Neufeld, K.A.; Kang, N.; Bienenstock, J.; Foster, J.A. Effects of intestinal microbiota on anxiety-like behavior. *Commun. Integr. Biol.* **2011**, *4*, 492–494. [CrossRef]
63. Lach, G.; Schellekens, H.; Dinan, T.G.; Cryan, J.F. Anxiety, depression, and the microbiome: A role for gut peptides. *Neurother. J. Am. Soc. Exp. Neurother.* **2018**, *15*, 36–59. [CrossRef]
64. Petra, A.I.; Panagiotidou, S.; Hatziagelaki, E.; Stewart, J.M.; Conti, P.; Theoharides, T.C. Gut-microbiota-brain axis and its effect on neuropsychiatric disorders with suspected immune dysregulation. *Clin. Ther.* **2015**, *37*, 984–995. [CrossRef] [PubMed]
65. Kim, N.; Yun, M.; Oh, Y.J.; Choi, H.J. Mind-altering with the gut: Modulation of the gut-brain axis with probiotics. *J. Microbiol.* **2018**, *56*, 172–182. [CrossRef] [PubMed]
66. Dumas, M.E.; Kinross, J.; Nicholson, J.K. Metabolic phenotyping and systems biology approaches to understanding metabolic syndrome and fatty liver disease. *Gastroenterology* **2014**, *146*, 46–62. [CrossRef] [PubMed]
67. Dumas, M.-E.; Barton, R.H.; Toye, A.; Cloarec, O.; Blancher, C.; Rothwell, A.; Fearnside, J.; Tatoud, R.; Blanc, V.; Lindon, J.C. Metabolic profiling reveals a contribution of gut microbiota to fatty liver phenotype in insulin-resistant mice. *Proc. Natl. Acad. Sci. USA* **2006**, *103*, 12511–12516. [CrossRef]
68. Ridaura, V.K.; Faith, J.J.; Rey, F.E.; Cheng, J.; Duncan, A.E.; Kau, A.L.; Griffin, N.W.; Lombard, V.; Henrissat, B.; Bain, J.R. Gut microbiota from twins discordant for obesity modulate metabolism in mice. *Science* **2013**, *341*, 1241214. [CrossRef] [PubMed]
69. Turnbaugh, P.J.; Gordon, J.I. The core gut microbiome, energy balance and obesity. *J. Physiol.* **2009**, *587*, 4153–4158. [CrossRef]
70. Turnbaugh, P.J.; Hamady, M.; Yatsunenko, T.; Cantarel, B.L.; Duncan, A.; Ley, R.E.; Sogin, M.L.; Jones, W.J.; Roe, B.A.; Affourtit, J.P. A core gut microbiome in obese and lean twins. *Nature* **2009**, *457*, 480. [CrossRef]
71. Barrett, E.; Kwan, H. Bacterial reduction of trimethylamine oxide. *Annu. Rev. Microbiol.* **1985**, *39*, 131–149. [CrossRef]
72. Smith, J.L.; Wishnok, J.S.; Deen, W.M. Metabolism and excretion of methylamines in rats. *Toxicol. Appl. Pharmacol.* **1994**, *125*, 296–308. [CrossRef]
73. Zhang, A.; Mitchell, S.; Smith, R. Dietary precursors of trimethylamine in man: A pilot study. *Food Chem. Toxicol.* **1999**, *37*, 515–520. [CrossRef]
74. Bain, M.A.; Fornasini, G.; Evans, A.M. Trimethylamine: Metabolic, pharmacokinetic and safety aspects. *Curr. Drug Metab.* **2005**, *6*, 227–240. [CrossRef] [PubMed]
75. Li, Z.; Vance, D.E. Thematic review series: Glycerolipids. Phosphatidylcholine and choline homeostasis. *J. Lipid Res.* **2008**, *49*, 1187–1194. [CrossRef] [PubMed]
76. Geiger, O.; López-Lara, I.M.; Sohlenkamp, C. Phosphatidylcholine biosynthesis and function in bacteria. *Biochim. Et Biophys. Acta (BBA) Mol. Cell Biol. Lipids* **2013**, *1831*, 503–513. [CrossRef] [PubMed]
77. Tang, W.H.; Hazen, S.L. The contributory role of gut microbiota in cardiovascular disease. *J. Clin. Investig.* **2014**, *124*, 4204–4211. [CrossRef] [PubMed]
78. Wang, Z.; Klipfell, E.; Bennett, B.J.; Koeth, R.; Levison, B.S.; DuGar, B.; Feldstein, A.E.; Britt, E.B.; Fu, X.; Chung, Y.-M. Gut flora metabolism of phosphatidylcholine promotes cardiovascular disease. *Nature* **2011**, *472*, 57. [CrossRef] [PubMed]
79. Koeth, R.A.; Lam-Galvez, B.R.; Kirsop, J.; Wang, Z.; Levison, B.S.; Gu, X.; Copeland, M.F.; Bartlett, D.; Cody, D.B.; Dai, H.J.; et al. L-Carnitine in omnivorous diets induces an atherogenic gut microbial pathway in humans. *J. Clin. Invest.* **2019**, *129*. [CrossRef]
80. Tang, W.W.; Wang, Z.; Levison, B.S.; Koeth, R.A.; Britt, E.B.; Fu, X.; Wu, Y.; Hazen, S.L. Intestinal microbial metabolism of phosphatidylcholine and cardiovascular risk. *N. Engl. J. Med.* **2013**, *368*, 1575–1584. [CrossRef]

81. Lever, M.; George, P.M.; Slow, S.; Bellamy, D.; Young, J.M.; Ho, M.; McEntyre, C.J.; Elmslie, J.L.; Atkinson, W.; Molyneux, S.L. Betaine and trimethylamine-N-oxide as predictors of cardiovascular outcomes show different patterns in diabetes mellitus: An observational study. *PLoS ONE* **2014**, *9*, e114969. [CrossRef]
82. Mente, A.; Chalcraft, K.; Ak, H.; Davis, A.D.; Lonn, E.; Miller, R.; Potter, M.A.; Yusuf, S.; Anand, S.S.; McQueen, M.J. The relationship between trimethylamine-N-oxide and prevalent cardiovascular disease in a multiethnic population living in Canada. *Can. J. Cardiol.* **2015**, *31*, 1189–1194. [CrossRef]
83. Trøseid, M.; Ueland, T.; Hov, J.R.; Svardal, A.; Gregersen, I.; Dahl, C.P.; Aakhus, S.; Gude, E.; Bjørndal, B.; Halvorsen, B.; et al. Microbiota-dependent metabolite trimethylamine-N-oxide is associated with disease severity and survival of patients with chronic heart failure. *J. Intern. Med.* **2015**, *277*, 717–726. [CrossRef]
84. Kaluza, J.; Åkesson, A.; Wolk, A. Processed and unprocessed red meat consumption and risk of heart failure: Prospective study of men. *Circ. Heart Fail.* **2014**, *7*, 552–557. [CrossRef] [PubMed]
85. Pan, A.; Sun, Q.; Bernstein, A.M.; Schulze, M.B.; Manson, J.E.; Stampfer, M.J.; Willett, W.C.; Hu, F.B. Red meat consumption and mortality: Results from 2 prospective cohort studies. *Arch. Intern. Med.* **2012**, *172*, 555–563. [PubMed]
86. Spence, J.D.; Jenkins, D.J.; Davignon, J. Egg yolk consumption and carotid plaque. *Atherosclerosis* **2012**, *224*, 469–473. [CrossRef] [PubMed]
87. Kaluza, J.; Wolk, A.; Larsson, S.C. Red meat consumption and risk of stroke: A meta-analysis of prospective studies. *Stroke* **2012**, *43*, 2556–2560. [CrossRef]
88. Larsson, S.C.; Orsini, N. Red meat and processed meat consumption and all-cause mortality: A meta-analysis. *Am. J. Epidemiol.* **2013**, *179*, 282–289. [CrossRef]
89. Djoussé, L.; Gaziano, J.M. Egg consumption in relation to cardiovascular disease and mortality: The Physicians' Health Study. *Am. J. Clin. Nutr.* **2008**, *87*, 964–969. [CrossRef]
90. Djoussé, L.; Gaziano, J.M. Egg consumption and risk of heart failure in the Physicians' Health Study. *Circulation* **2008**, *117*, 512. [CrossRef]
91. Qureshi, A.I.; Suri, M.F.K.; Ahmed, S.; Nasar, A.; Divani, A.A.; Kirmani, J.F. Regular egg consumption does not increase the risk of stroke and cardiovascular diseases. *Med. Sci. Monit.* **2007**, *13*, 8.
92. Scrafford, C.G.; Tran, N.L.; Barraj, L.M.; Mink, P.J. Egg consumption and CHD and stroke mortality: A prospective study of US adults. *Public Health Nutr.* **2011**, *14*, 261–270. [CrossRef]
93. Tran, N.L.; Barraj, L.M.; Heilman, J.M.; Scrafford, C.G. Egg consumption and cardiovascular disease among diabetic individuals: A systematic review of the literature. *Diabetes Metab. Syndr. Obes. Targ. Ther.* **2014**, *7*, 121. [CrossRef]
94. Miller, C.A.; Corbin, K.D.; da Costa, K.-A.; Zhang, S.; Zhao, X.; Galanko, J.A.; Blevins, T.; Bennett, B.J.; O'Connor, A.; Zeisel, S.H. Effect of egg ingestion on trimethylamine-N-oxide production in humans: A randomized, controlled, dose-response study. *Am. J. Clin. Nutr.* **2014**, *100*, 778–786. [CrossRef] [PubMed]
95. Wang, Z.; Tang, W.W.; Buffa, J.A.; Fu, X.; Britt, E.B.; Koeth, R.A.; Levison, B.S.; Fan, Y.; Wu, Y.; Hazen, S.L. Prognostic value of choline and betaine depends on intestinal microbiota-generated metabolite trimethylamine-N-oxide. *Eur. Heart J.* **2014**, *35*, 904–910. [CrossRef] [PubMed]
96. Klinder, A.; Shen, Q.; Heppel, S.; Lovegrove, J.A.; Rowland, I.; Tuohy, K.M. Impact of increasing fruit and vegetables and flavonoid intake on the human gut microbiota. *Food Funct.* **2016**, *7*, 1788–1796. [CrossRef] [PubMed]
97. Gomez-Arango, L.F.; Barrett, H.L.; McIntyre, H.D.; Callaway, L.K.; Morrison, M.; Dekker Nitert, M. Increased systolic and diastolic blood pressure is associated with altered gut microbiota composition and butyrate production in early pregnancy. *Hypertension* **2016**, *68*, 974–981. [CrossRef] [PubMed]
98. Heianza, Y.; Ma, W.; Manson, J.E.; Rexrode, K.M.; Qi, L. Gut microbiota metabolites and risk of major adverse cardiovascular disease events and death: A systematic review and meta-analysis of prospective studies. *J. Am. Heart Assoc.* **2017**, *6*. [CrossRef] [PubMed]
99. Vázquez-Fresno, R.; Llorach, R.; Perera, A.; Mandal, R.; Feliz, M.; Tinahones, F.J.; Wishart, D.S.; Andres-Lacueva, C. Clinical phenotype clustering in cardiovascular risk patients for the identification of responsive metabotypes after red wine polyphenol intake. *J. Nutr. Biochem.* **2016**, *28*, 114–120. [CrossRef]
100. Guasch-Ferre, M.; Hu, F.B.; Ruiz-Canela, M.; Bullo, M.; Yu, E.; Zheng, Y.; Toledo, E.; Wang, D.D.; Hruby, A.; Corella, D.; et al. Gut microbiota related plasma metabolites and risk of cardiovascular disease in the PREDIMED study. *Circulation* **2017**, *135*, A23.

101. Kerley, C.P. Dietary patterns and components to prevent and treat heart failure: A comprehensive review of human studies. *Nutr. Res. Rev.* **2018**. [CrossRef]
102. Qi, J.; You, T.; Li, J.; Pan, T.; Xiang, L.; Han, Y.; Zhu, L. Circulating trimethylamine N-oxide and the risk of cardiovascular diseases: A systematic review and meta-analysis of 11 prospective cohort studies. *J. Cell. Mol. Med.* **2018**, *22*, 185–194. [CrossRef] [PubMed]
103. Yao, M.E.; Liao, P.D.; Zhao, X.J.; Wang, L. Trimethylamine-N-oxide has prognostic value in coronary heart disease: A meta-analysis and dose-response analysis. *BMC Cardiovasc. Disord.* **2020**, *20*, 7. [CrossRef]
104. Abbasi, J. TMAO and heart disease: The new red meat risk? *JAMA* **2019**, *321*, 2149–2151. [CrossRef] [PubMed]
105. McCarty, M.F. L-carnitine consumption, its metabolism by intestinal microbiota, and cardiovascular health. *Mayo. Clin. Proc.* **2013**, *88*, 786–789. [CrossRef] [PubMed]
106. Landfald, B.; Valeur, J.; Berstad, A.; Raa, J. Microbial trimethylamine-N-oxide as a disease marker: Something fishy? *Microb. Ecol. Health Dis.* **2017**, *28*, 1327309. [CrossRef] [PubMed]
107. Ussher, J.R.; Lopaschuk, G.D.; Arduini, A. Gut microbiota metabolism of L-carnitine and cardiovascular risk. *Atherosclerosis* **2013**, *231*, 456–461. [CrossRef]
108. Meyer, K.A.; Benton, T.Z.; Bennett, B.J.; Jacobs, D.R., Jr.; Lloyd-Jones, D.M.; Gross, M.D.; Carr, J.J.; Gordon-Larsen, P.; Zeisel, S.H. Microbiota-dependent metabolite trimethylamine N-oxide and coronary artery calcium in the Coronary Artery Risk Development in Young Adults Study (CARDIA). *J. Am. Heart Assoc.* **2016**, *5*. [CrossRef]
109. Cani, P.D.; Amar, J.; Iglesias, M.A.; Poggi, M.; Knauf, C.; Bastelica, D.; Neyrinck, A.M.; Fava, F.; Tuohy, K.M.; Chabo, C.; et al. Metabolic endotoxemia initiates obesity and insulin resistance. *Diabetes* **2007**, *56*, 1761–1772. [CrossRef]
110. Taki, K.; Tsuruta, Y.; Niwa, T. Indoxyl sulfate and atherosclerotic risk factors in hemodialysis patients. *Am. J. Nephrol.* **2007**, *27*, 30–35. [CrossRef]
111. Meijers, B.K.; Verbeke, K.; Dehaen, W.; Vanrenterghem, Y.; Hoylaerts, M.F.; Evenepoel, P. The uremic retention solute p-cresyl sulfate and markers of endothelial damage. *Am. J. Kidney Dis.* **2009**, *54*, 891–901. [CrossRef]
112. Buffie, C.G.; Pamer, E.G. Microbiota-mediated colonization resistance against intestinal pathogens. *Nat. Rev. Immunol.* **2013**, *13*, 790–801. [CrossRef]
113. Kamada, N.; Seo, S.-U.; Chen, G.Y.; Núñez, G. Role of the gut microbiota in immunity and inflammatory disease. *Nat. Rev. Immunol.* **2013**, *13*, 321. [CrossRef]
114. Collins, S.M. A role for the gut microbiota in IBS. *Nat. Rev. Gastroenterol. Hepatol.* **2014**, *11*, 497. [CrossRef] [PubMed]
115. Tsai, S.; Winer, S.; Winer, D.A. Gut T cells feast on GLP-1 to modulate cardiometabolic disease. *Cell Metab.* **2019**, *29*, 787–789. [CrossRef] [PubMed]
116. Alphonse, P.A.S.; Jones, P.J.H. Revisiting human cholesterol synthesis and absorption: The reciprocity paradigm and its key regulators. *Lipids* **2016**, *51*, 519–536. [CrossRef] [PubMed]
117. Lyte, M. Probiotics function mechanistically as delivery vehicles for neuroactive compounds: Microbial endocrinology in the design and use of probiotics. *Bioessays* **2011**, *33*, 574–581. [CrossRef]
118. Lan, T.H.; Huang, X.Q.; Tan, H.M. Vascular fibrosis in atherosclerosis. *Cardiovasc. Pathol. Off. J. Soc. Cardiovasc. Pathol.* **2013**, *22*, 401–407. [CrossRef]
119. Karbach, S.H.; Schönfelder, T.; Brandão, I.; Wilms, E.; Hörmann, N.; Jäckel, S.; Schüler, R.; Finger, S.; Knorr, M.; Lagrange, J. Gut microbiota promote angiotensin II–induced arterial hypertension and vascular dysfunction. *J. Am. Heart Assoc.* **2016**, *5*, e003698. [CrossRef]
120. Geurts, L.; Lazarevic, V.; Derrien, M.; Everard, A.; Van Roye, M.; Knauf, C.; Valet, P.; Girard, M.; Muccioli, G.G.; François, P. Altered gut microbiota and endocannabinoid system tone in obese and diabetic leptin-resistant mice: Impact on apelin regulation in adipose tissue. *Front. Microbiol.* **2011**, *2*, 149. [CrossRef]
121. Huart, J.; Leenders, J.; Taminiau, B.; Descy, J.; Saint-Remy, A.; Daube, G.; Krzesinski, J.M.; Melin, P.; de Tullio, P.; Jouret, F. Gut microbiota and fecal levels of short-chain fatty acids differ upon 24-hour blood pressure levels in men. *Hypertension* **2019**, *74*, 1005–1013. [CrossRef]
122. Loo, R.L.; Zou, X.; Appel, L.J.; Nicholson, J.K.; Holmes, E. Characterization of metabolic responses to healthy diets and association with blood pressure: Application to the Optimal Macronutrient Intake Trial for Heart Health (OmniHeart), a randomized controlled study. *Am. J. Clin. Nutr.* **2018**, *107*, 323–334. [CrossRef]

123. van den Munckhof, I.C.L.; Kurilshikov, A.; ter Horst, R.; Riksen, N.P.; Joosten, L.A.B.; Zhernakova, A.; Fu, J.; Keating, S.T.; Netea, M.G.; de Graaf, J.; et al. Role of gut microbiota in chronic low-grade inflammation as potential driver for atherosclerotic cardiovascular disease: A systematic review of human studies. *Obes. Rev.* **2018**, *19*, 1719–1734. [CrossRef]
124. Lam, O.L.T.; Zhang, W.; Samaranayake, L.P.; Li, L.S.W.; McGrath, C. A systematic review of the effectiveness of oral health promotion activities among patients with cardiovascular disease. *Int. J. Cardiol.* **2011**, *151*, 261–267. [CrossRef] [PubMed]
125. Renvert, S.; Pettersson, T.; Ohlsson, O.; Persson, G.R. Bacterial profile and burden of periodontal infection in subjects with a diagnosis of acute coronary syndrome. *J. Periodontol.* **2006**, *77*, 1110–1119. [CrossRef] [PubMed]
126. Pussinen, P.J.; Alfthan, G.; Rissanen, H.; Reunanen, A.; Asikainen, S.; Knekt, P. Antibodies to periodontal pathogens and stroke risk. *Stroke* **2004**, *35*, 2020–2023. [CrossRef]
127. Beck, J.D.; Eke, P.; Heiss, G.; Madianos, P.; Couper, D.; Lin, D.; Moss, K.; Elter, J.; Offenbacher, S. Periodontal disease and coronary heart disease: A reappraisal of the exposure. *Circulation* **2005**, *112*, 19–24. [CrossRef]
128. Yang, S.; Li, X.; Yang, F.; Zhao, R.; Pan, X.; Liang, J.; Tian, L.; Li, X.; Liu, L.; Xing, Y.; et al. Gut Microbiota-Dependent Marker TMAO in Promoting Cardiovascular Disease: Inflammation Mechanism, Clinical Prognostic, and Potential as a Therapeutic Target. *Front Pharmacol.* **2019**, *10*, 1360. [CrossRef] [PubMed]
129. Mell, B.; Jala, V.R.; Mathew, A.V.; Byun, J.; Waghulde, H.; Zhang, Y.; Haribabu, B.; Vijay-Kumar, M.; Pennathur, S.; Joe, B. Evidence for a link between gut microbiota and hypertension in the Dahl rat. *Physiol. Genom.* **2015**, *47*, 187–197. [CrossRef] [PubMed]
130. Karlsson, C.; Ahrné, S.; Molin, G.; Berggren, A.; Palmquist, I.; Fredrikson, G.N.; Jeppsson, B. Probiotic therapy to men with incipient arteriosclerosis initiates increased bacterial diversity in colon: A randomized controlled trial. *Atherosclerosis* **2010**, *208*, 228–233. [CrossRef] [PubMed]
131. Fernández-Ruiz, I. Acute coronary syndromes: Microbial-dependent TMAO as a prognostic marker in ACS. *Nat. Rev. Cardiol.* **2017**, *14*, 128. [CrossRef]
132. Organ, C.L.; Otsuka, H.; Bhushan, S.; Wang, Z.; Bradley, J.; Trivedi, R.; Polhemus, D.J.; Tang, W.W.; Wu, Y.; Hazen, S.L. Choline diet and its gut microbe–derived metabolite, trimethylamine N-oxide, exacerbate pressure overload–induced heart failure. *Circ. Heart Fail.* **2016**, *9*, e002314. [CrossRef]
133. Koren, O.; Spor, A.; Felin, J.; Fåk, F.; Stombaugh, J.; Tremaroli, V.; Behre, C.J.; Knight, R.; Fagerberg, B.; Ley, R.E. Human oral, gut, and plaque microbiota in patients with atherosclerosis. *Proc. Natl. Acad. Sci. USA* **2011**, *108*, 4592–4598. [CrossRef]
134. Ott, S.J.; El Mokhtari, N.E.; Musfeldt, M.; Hellmig, S.; Freitag, S.; Rehman, A.; Kuhbacher, T.; Nikolaus, S.; Namsolleck, P.; Blaut, M. Detection of diverse bacterial signatures in atherosclerotic lesions of patients with coronary heart disease. *Circulation* **2006**, *113*, 929–937. [CrossRef] [PubMed]
135. Mattila, K.J.; Nieminen, M.S.; Valtonen, V.V.; Rasi, V.P.; Kesäniemi, Y.A.; Syrjälä, S.L.; Jungell, P.S.; Isoluoma, M.; Hietaniemi, K.; Jokinen, M.J. Association between dental health and acute myocardial infarction. *BMJ* **1989**, *298*, 779–781. [CrossRef] [PubMed]
136. Hyvärinen, K.; Mäntylä, P.; Buhlin, K.; Paju, S.; Nieminen, M.S.; Sinisalo, J.; Pussinen, P.J. A common periodontal pathogen has an adverse association with both acute and stable coronary artery disease. *Atherosclerosis* **2012**, *223*, 478–484. [CrossRef] [PubMed]
137. Fåk, F.; Tremaroli, V.; Bergström, G.; Bäckhed, F. Oral microbiota in patients with atherosclerosis. *Atherosclerosis* **2015**, *243*, 573–578. [CrossRef]
138. e Silva Filho, W.S.; Casarin, R.C.; Junior, E.L.N.; Passos, H.M.; Sallum, A.W.; Gonçalves, R.B. Microbial diversity similarities in periodontal pockets and atheromatous plaques of cardiovascular disease patients. *PLoS ONE* **2014**, *9*, e109761. [CrossRef]
139. Jonsson, A.L.; Backhed, F. Role of gut microbiota in atherosclerosis. *Nat. Rev. Cardiol.* **2017**, *14*, 79–87. [CrossRef]
140. Dziubak, A.; Wojcicka, G.; Wojtak, A.; Beltowski, J. Metabolic effects of metformin in the failing heart. *Int. J. Mol. Sci.* **2018**, 19. [CrossRef]
141. McCreight, L.J.; Bailey, C.J.; Pearson, E.R. Metformin and the gastrointestinal tract. *Diabetologia* **2016**, *59*, 426–435. [CrossRef]

142. Munch Roager, H.; Vogt, J.K.; Kristensen, M.; Hansen, L.B.S.; Ibrügger, S.; Maerkedahl, R.B.; Bahl, M.I.; Lind, M.V.; Nielsen, R.L.; Frøkiaer, H.; et al. Whole grain-rich diet reduces body weight and systemic low-grade inflammation without inducing major changes of the gut microbiome: A randomised cross-over trial. *Gut* **2019**, *68*, 83–93. [CrossRef]
143. Plaza-Diaz, J.; Gomez-Llorente, C.; Abadia-Molina, F.; Saez-Lara, M.J.; Campana-Martin, L.; Munoz-Quezada, S.; Romero, F.; Gil, A.; Fontana, L. Effects of Lactobacillus paracasei CNCM I-4034, Bifidobacterium breve CNCM I-4035 and Lactobacillus rhamnosus CNCM I-4036 on hepatic steatosis in Zucker rats. *PLoS ONE* **2014**, *9*, e98401. [CrossRef]
144. Battson, M.L.; Lee, D.M.; Weir, T.L.; Gentile, C.L. The gut microbiota as a novel regulator of cardiovascular function and disease. *J. Nutr. Biochem.* **2018**, *56*, 1–15. [CrossRef] [PubMed]
145. Kiechl, S.; Egger, G.; Mayr, M.; Wiedermann, C.J.; Bonora, E.; Oberhollenzer, F.; Muggeo, M.; Xu, Q.; Wick, G.; Poewe, W.; et al. Chronic infections and the risk of carotid atherosclerosis: Prospective results from a large population study. *Circulation* **2001**, *103*, 1064–1070. [CrossRef] [PubMed]
146. Niebauer, J.; Volk, H.D.; Kemp, M.; Dominguez, M.; Schumann, R.R.; Rauchhaus, M.; Poole-Wilson, P.A.; Coats, A.J.; Anker, S.D. Endotoxin and immune activation in chronic heart failure: A prospective cohort study. *Lancet* **1999**, *353*, 1838–1842. [CrossRef]
147. Miller, M.A.; McTernan, P.G.; Harte, A.L.; Silva, N.F.; Strazzullo, P.; Alberti, K.G.; Kumar, S.; Cappuccio, F.P. Ethnic and sex differences in circulating endotoxin levels: A novel marker of atherosclerotic and cardiovascular risk in a British multi-ethnic population. *Atherosclerosis* **2009**, *203*, 494–502. [CrossRef]
148. Li, J.; Lin, S.; Vanhoutte, P.M.; Woo, C.W.; Xu, A. Akkermansia muciniphila protects against atherosclerosis by preventing metabolic endotoxemia-induced inflammation in apoe-/- mice. *Circulation* **2016**, *133*, 2434–2446. [CrossRef]
149. Lehr, H.A.; Sagban, T.A.; Ihling, C.; Zahringer, U.; Hungerer, K.D.; Blumrich, M.; Reifenberg, K.; Bhakdi, S. Immunopathogenesis of atherosclerosis: Endotoxin accelerates atherosclerosis in rabbits on hypercholesterolemic diet. *Circulation* **2001**, *104*, 914–920. [CrossRef]
150. Denou, E.; Lolmede, K.; Garidou, L.; Pomie, C.; Chabo, C.; Lau, T.C.; Fullerton, M.D.; Nigro, G.; Zakaroff-Girard, A.; Luche, E.; et al. Defective NOD2 peptidoglycan sensing promotes diet-induced inflammation, dysbiosis, and insulin resistance. *EMBO Mol. Med.* **2015**, *7*, 259–274. [CrossRef]
151. Galluzzo, S.; Patti, G.; Dicuonzo, G.; Di Sciascio, G.; Tonini, G.; Ferraro, E.; Spoto, C.; Campanale, R.; Zoccoli, A.; Angeletti, S. Association between NOD2/CARD15 polymorphisms and coronary artery disease: A case-control study. *Hum. Immunol.* **2011**, *72*, 636–640. [CrossRef]
152. Yuan, H.; Zelkha, S.; Burkatovskaya, M.; Gupte, R.; Leeman, S.E.; Amar, S. Pivotal role of NOD2 in inflammatory processes affecting atherosclerosis and periodontal bone loss. *Proc. Natl. Acad. Sci. USA* **2013**, *110*, E5059–E5068. [CrossRef]
153. Patel, P.; Mendall, M.A.; Carrington, D.; Strachan, D.P.; Leatham, E.; Molineaux, N.; Levy, J.; Blakeston, C.; Seymour, C.A.; Camm, A.J.; et al. Association of Helicobacter pylori and Chlamydia pneumoniae infections with coronary heart disease and cardiovascular risk factors. *BMJ* **1995**, *311*, 711–714. [CrossRef]
154. Wright, A.T. Gut commensals make choline too. *Nat. Microbiol.* **2019**, *4*, 4–5. [CrossRef] [PubMed]
155. Romano, K.A.; Vivas, E.I.; Amador-Noguez, D.; Rey, F.E. Intestinal microbiota composition modulates choline bioavailability from diet and accumulation of the proatherogenic metabolite trimethylamine-N-oxide. *mBio* **2015**, *6*, e02481. [CrossRef] [PubMed]
156. Dong, C.; Yoon, W.; Goldschmidt-Clermont, P.J. DNA methylation and atherosclerosis. *J. Nutr.* **2002**, *132*, 2406S–2409S. [CrossRef] [PubMed]
157. Zaina, S.; Lindholm, M.W.; Lund, G. Nutrition and aberrant DNA methylation patterns in atherosclerosis: More than just hyperhomocysteinemia? *J. Nutr.* **2005**, *135*, 5–8. [CrossRef] [PubMed]
158. Salmon, W.D.; Newberne, P.M. Cardiovascular disease in choline-deficient rats. Effects of choline deficiency, nature and level of dietary lipids and proteins, and duration of feeding on plasma and liver lipid values and cardiovascular lesions. *Arch. Pathol.* **1962**, *73*, 190–209. [PubMed]
159. Mudd, S.H.; Finkelstein, J.D.; Refsum, H.; Ueland, P.M.; Malinow, M.R.; Lentz, S.R.; Jacobsen, D.W.; Brattstrom, L.; Wilcken, B.; Wilcken, D.E.; et al. Homocysteine and its disulfide derivatives: A suggested consensus terminology. *Arterioscler. Thromb. Vasc. Biol.* **2000**, *20*, 1704–1706. [CrossRef]
160. McCully, K.S.; Ragsdale, B.D. Production of arteriosclerosis by homocysteinemia. *Am. J. Pathol.* **1970**, *61*, 1–11.

161. Marti-Carvajal, A.J.; Sola, I.; Lathyris, D.; Dayer, M. Homocysteine-lowering interventions for preventing cardiovascular events. *Cochrane Database Syst. Rev.* **2017**, *8*, CD006612. [CrossRef]
162. Millard, H.R.; Musani, S.K.; Dibaba, D.T.; Talegawkar, S.A.; Taylor, H.A.; Tucker, K.L.; Bidulescu, A. Dietary choline and betaine; associations with subclinical markers of cardiovascular disease risk and incidence of CVD, coronary heart disease and stroke: The Jackson Heart Study. *Eur. J. Nutr.* **2018**, *57*, 51–60. [CrossRef]
163. Magnúsdóttir, S.; Ravcheev, D.; de Crécy-Lagard, V.; Thiele, I. Systematic genome assessment of B-vitamin biosynthesis suggests co-operation among gut microbes. *Front. Genet.* **2015**, *6*, 148. [CrossRef]
164. Said, H.M. Intestinal absorption of water-soluble vitamins in health and disease. *Biochem. J.* **2011**, *437*, 357–372. [CrossRef] [PubMed]
165. Jeon, J.; Park, K. Dietary vitamin B6 intake associated with a decreased risk of cardiovascular disease: A prospective cohort study. *Nutrients* **2019**, *11*, 1484. [CrossRef] [PubMed]
166. Group, H.T.C.; Landray, M.J.; Haynes, R.; Hopewell, J.C.; Parish, S.; Aung, T.; Tomson, J.; Wallendszus, K.; Craig, M.; Jiang, L.; et al. Effects of extended-release niacin with laropiprant in high-risk patients. *N. Engl. J. Med.* **2014**, *371*, 203–212. [CrossRef]
167. Jenkins, D.J.A.; Spence, J.D.; Giovannucci, E.L.; Kim, Y.I.; Josse, R.; Vieth, R.; Blanco Mejia, S.; Viguiliouk, E.; Nishi, S.; Sahye-Pudaruth, S.; et al. Supplemental vitamins and minerals for CVD prevention and treatment. *J. Am. Coll. Cardiol.* **2018**, *71*, 2570–2584. [CrossRef] [PubMed]
168. Aung, K.; Htay, T. Review: Folic acid may reduce risk for CVD and stroke, and B-vitamin complex may reduce risk for stroke. *Ann. Intern. Med.* **2018**, *169*, JC44. [CrossRef] [PubMed]
169. Wang, Y.; Jin, Y.; Wang, Y.; Li, L.; Liao, Y.; Zhang, Y.; Yu, D. The effect of folic acid in patients with cardiovascular disease: A systematic review and meta-analysis. *Medicine (Baltimore)* **2019**, *98*, e17095. [CrossRef] [PubMed]
170. Valentini, L.; Pinto, A.; Bourdel-Marchasson, I.; Ostan, R.; Brigidi, P.; Turroni, S.; Hrelia, S.; Hrelia, P.; Bereswill, S.; Fischer, A.; et al. Impact of personalized diet and probiotic supplementation on inflammation, nutritional parameters and intestinal microbiota—The "RISTOMED project": Randomized controlled trial in healthy older people. *Clin. Nutr.* **2015**, *34*, 593–602. [CrossRef]
171. Larsen, J.M. The immune response to Prevotella bacteria in chronic inflammatory disease. *Immunology* **2017**, *151*, 363–374. [CrossRef]
172. Pearson, T.A.; Mensah, G.A.; Alexander, R.W.; Anderson, J.L.; Cannon III, R.O.; Criqui, M.; Fadl, Y.Y.; Fortmann, S.P.; Hong, Y.; Myers, G.L. Markers of inflammation and cardiovascular disease: Application to clinical and public health practice: A statement for healthcare professionals from the Centers for Disease Control and Prevention and the American Heart Association. *Circulation* **2003**, *107*, 499–511. [CrossRef]
173. Ostan, R.; Bene, M.C.; Spazzafumo, L.; Pinto, A.; Donini, L.M.; Pryen, F.; Charrouf, Z.; Valentini, L.; Lochs, H.; Bourdel-Marchasson, I.; et al. Impact of diet and nutraceutical supplementation on inflammation in elderly people. Results from the RISTOMED study, an open-label randomized control trial. *Clin. Nutr.* **2016**, *35*, 812–818. [CrossRef]
174. Simone, C.D.; Ciardi, A.; Grassi, A.; Gardini, S.L.; Tzantzoglou, S.; Trinchieri, V.; Moretti, S.; Jirillo, E. Effect of Bifidobacterium bifidum and Lactobacillus acidophilus on gut mucosa and peripheral blood B lymphocytes. *Immunopharmacol. Immunotoxicol.* **1992**, *14*, 331–340. [CrossRef] [PubMed]
175. Guillemard, E.; Tondu, F.; Lacoin, F.; Schrezenmeir, J. Consumption of a fermented dairy product containing the probiotic Lactobacillus casei DN-114001 reduces the duration of respiratory infections in the elderly in a randomised controlled trial. *Br. J. Nutr.* **2010**, *103*, 58–68. [CrossRef] [PubMed]
176. Mane, J.; Pedrosa, E.; Loren, V.; Gassull, M.A.; Espadaler, J.; Cune, J.; Audivert, S.; Bonachera, M.A.; Cabre, E. A mixture of Lactobacillus plantarum CECT 7315 and CECT 7316 enhances systemic immunity in elderly subjects. A dose-response, double-blind, placebo-controlled, randomized pilot trial. *Nutr. Hosp.* **2011**, *26*, 228–235. [PubMed]
177. Dong, H.; Rowland, I.; Thomas, L.V.; Yaqoob, P. Immunomodulatory effects of a probiotic drink containing Lactobacillus casei Shirota in healthy older volunteers. *Eur. J. Nutr.* **2013**, *52*, 1853–1863. [CrossRef]
178. Moro-Garcia, M.A.; Alonso-Arias, R.; Baltadjieva, M.; Fernandez Benitez, C.; Fernandez Barrial, M.A.; Diaz Ruisanchez, E.; Alonso Santos, R.; Alvarez Sanchez, M.; Saavedra Mijan, J.; Lopez-Larrea, C. Oral supplementation with Lactobacillus delbrueckii subsp. bulgaricus 8481 enhances systemic immunity in elderly subjects. *Age* **2013**, *35*, 1311–1326. [CrossRef]

179. Ouwehand, A.C.; Tiihonen, K.; Saarinen, M.; Putaala, H.; Rautonen, N. Influence of a combination of Lactobacillus acidophilus NCFM and lactitol on healthy elderly: Intestinal and immune parameters. *Br. J. Nutr.* **2009**, *101*, 367–375. [CrossRef]
180. Schiffrin, E.J.; Parlesak, A.; Bode, C.; Bode, J.C.; van't Hof, M.A.; Grathwohl, D.; Guigoz, Y. Probiotic yogurt in the elderly with intestinal bacterial overgrowth: Endotoxaemia and innate immune functions. *Br. J. Nutr.* **2008**, *101*, 961–966. [CrossRef]
181. Gil-Cruz, C.; Perez-Shibayama, C.; De Martin, A.; Ronchi, F.; van der Borght, K.; Niederer, R.; Onder, L.; Lutge, M.; Novkovic, M.; Nindl, V.; et al. Microbiota-derived peptide mimics drive lethal inflammatory cardiomyopathy. *Science* **2019**, *366*, 881–886. [CrossRef]
182. Adnan, S.; Nelson, J.W.; Ajami, N.J.; Venna, V.R.; Petrosino, J.F.; Bryan Jr, R.M.; Durgan, D.J. Alterations in the gut microbiota can elicit hypertension in rats. *Physiol. Genom.* **2016**, *49*, 96–104. [CrossRef]
183. Yang, T.; Santisteban, M.M.; Rodriguez, V.; Li, E.; Ahmari, N.; Carvajal, J.M.; Zadeh, M.; Gong, M.; Qi, Y.; Zubcevic, J. Gut dysbiosis is linked to hypertension. *Hypertension* **2015**, *65*, 1331–1340. [CrossRef]
184. Guzmán-Castañeda, S.J.; Ortega-Vega, E.L.; de la Cuesta-Zuluaga, J.; Velásquez-Mejía, E.P.; Rojas, W.; Bedoya, G.; Escobar, J.S. Gut microbiota composition explains more variance in the host cardiometabolic risk than genetic ancestry. *Gut Microbes* **2019**, 1–14. [CrossRef] [PubMed]
185. Kelly, T.N.; Bazzano, L.A.; Ajami, N.J.; He, H.; Zhao, J.; Petrosino, J.F.; Correa, A.; He, J. Gut microbiome associates with lifetime cardiovascular disease risk profile among bogalusa heart study participants. *Circ. Res.* **2016**, *119*, 956–964. [CrossRef] [PubMed]
186. Chen, J.; Guo, Y.; Gui, Y.; Xu, D. Physical exercise, gut, gut microbiota, and atherosclerotic cardiovascular diseases. *Lipids Health Dis.* **2018**, *17*, 17. [CrossRef] [PubMed]
187. Denou, E.; Marcinko, K.; Surette, M.G.; Steinberg, G.R.; Schertzer, J.D. High-intensity exercise training increases the diversity and metabolic capacity of the mouse distal gut microbiota during diet-induced obesity. *Am. J. Physiol. Endocrinol. Metab.* **2016**, *310*, E982–E993. [CrossRef]
188. Lambert, J.E.; Myslicki, J.P.; Bomhof, M.R.; Belke, D.D.; Shearer, J.; Reimer, R.A. Exercise training modifies gut microbiota in normal and diabetic mice. *Appl. Physiol. Nutr. Metab.* **2015**, *40*, 749–752. [CrossRef]
189. Matsumoto, M.; Inoue, R.; Tsukahara, T.; Ushida, K.; Chiji, H.; Matsubara, N.; Hara, H. Voluntary running exercise alters microbiota composition and increases n-butyrate concentration in the rat cecum. *Biosci. Biotechnol. Biochem.* **2008**, *72*, 572–576. [CrossRef]
190. Queipo-Ortuno, M.I.; Seoane, L.M.; Murri, M.; Pardo, M.; Gomez-Zumaquero, J.M.; Cardona, F.; Casanueva, F.; Tinahones, F.J. Gut microbiota composition in male rat models under different nutritional status and physical activity and its association with serum leptin and ghrelin levels. *PLoS ONE* **2013**, *8*, e65465. [CrossRef]
191. Clarke, S.F.; Murphy, E.F.; O'Sullivan, O.; Lucey, A.J.; Humphreys, M.; Hogan, A.; Hayes, P.; O'Reilly, M.; Jeffery, I.B.; Wood-Martin, R.; et al. Exercise and associated dietary extremes impact on gut microbial diversity. *Gut* **2014**, *63*, 1913–1920. [CrossRef]
192. Liu, Z.; Liu, H.-Y.; Zhou, H.; Zhan, Q.; Lai, W.; Zeng, Q.; Ren, H.; Xu, D. Moderate-intensity exercise affects gut microbiome composition and influences cardiac function in myocardial infarction mice. *Front. Microbiol.* **2017**, *8*, 1687. [CrossRef]
193. Bermudez-Brito, M.; Plaza-Díaz, J.; Munoz-Quezada, S.; Gomez-Llorente, C.; Gil, A. Probiotic mechanisms of action. *Ann. Nutr. Metab.* **2012**, *61*, 160–174. [CrossRef]
194. Plaza-Díaz, J.; Fernández-Caballero, J.Á.; Chueca, N.; García, F.; Gómez-Llorente, C.; Sáez-Lara, M.J.; Fontana, L.; Gil, Á. Pyrosequencing analysis reveals changes in intestinal microbiota of healthy adults who received a daily dose of immunomodulatory probiotic strains. *Nutrients* **2015**, *7*, 3999–4015. [CrossRef] [PubMed]
195. Chen, D.; Jin, D.; Huang, S.; Wu, J.; Xu, M.; Liu, T.; Dong, W.; Liu, X.; Wang, S.; Zhong, W. Clostridium butyricum, a butyrate-producing probiotic, inhibits intestinal tumor development through modulating Wnt signaling and gut microbiota. *Cancer Lett.* **2020**, *469*, 456–467. [CrossRef] [PubMed]
196. Plaza-Díaz, J.; Ruiz-Ojeda, F.J.; Vilchez-Padial, L.M.; Gil, A. Evidence of the anti-inflammatory effects of probiotics and synbiotics in intestinal chronic diseases. *Nutrients* **2017**, *9*, 555. [CrossRef]
197. Plaza-Díaz, J.; Robles-Sánchez, C.; Abadía-Molina, F.; Morón-Calvente, V.; Sáez-Lara, M.J.; Ruiz-Bravo, A.; Jiménez-Valera, M.; Gil, Á.; Gómez-Llorente, C.; Fontana, L. Adamdec1, Ednrb and Ptgs1/Cox1, inflammation genes upregulated in the intestinal mucosa of obese rats, are downregulated by three probiotic strains. *Sci. Rep.* **2017**, *7*, 1939. [CrossRef] [PubMed]

198. Munoz-Quezada, S.; Bermudez-Brito, M.; Chenoll, E.; Genovés, S.; Gomez-Llorente, C.; Plaza-Diaz, J.; Matencio, E.; Bernal, M.J.; Romero, F.; Ramón, D. Competitive inhibition of three novel bacteria isolated from faeces of breast milk-fed infants against selected enteropathogens. *Br. J. Nutr.* **2013**, *109*, S63–S69. [CrossRef]
199. Plaza-Diaz, J.; Gomez-Llorente, C.; Campaña-Martin, L.; Matencio, E.; Ortuño, I.; Martínez-Silla, R.; Gomez-Gallego, C.; Periago, M.J.; Ros, G.; Chenoll, E. Safety and immunomodulatory effects of three probiotic strains isolated from the feces of breast-fed infants in healthy adults: SETOPROB study. *PLoS ONE* **2013**, *8*, e78111. [CrossRef]
200. Ebel, B.; Lemetais, G.; Beney, L.; Cachon, R.; Sokol, H.; Langella, P.; Gervais, P. Impact of probiotics on risk factors for cardiovascular diseases. A review. *Crit. Rev. Food Sci. Nutr.* **2014**, *54*, 175–189. [CrossRef]
201. Thushara, R.M.; Gangadaran, S.; Solati, Z.; Moghadasian, M.H. Cardiovascular benefits of probiotics: A review of experimental and clinical studies. *Food Funct.* **2016**, *7*, 632–642. [CrossRef]
202. Hassan, A.; Din, A.U.; Zhu, Y.; Zhang, K.; Li, T.; Wang, Y.; Luo, Y.; Wang, G. Updates in understanding the hypocholesterolemia effect of probiotics on atherosclerosis. *Appl. Microbiol. Biotechnol.* **2019**, *103*, 5993–6006. [CrossRef]
203. Ejtahed, H.-S.; Angoorani, P.; Soroush, A.-R.; Atlasi, R.; Hasani-Ranjbar, S.; Mortazavian, A.M.; Larijani, B. Probiotics supplementation for the obesity management; A systematic review of animal studies and clinical trials. *J. Funct. Foods* **2019**, *52*, 228–242. [CrossRef]
204. Salgaço, M.K.; Oliveira, L.G.S.; Costa, G.N.; Bianchi, F.; Sivieri, K. Relationship between gut microbiota, probiotics, and type 2 diabetes mellitus. *Appl. Microbiol. Biotechnol.* **2019**, *103*, 9229–9238. [CrossRef] [PubMed]
205. DiRienzo, D.B. Effect of probiotics on biomarkers of cardiovascular disease: Implications for heart-healthy diets. *Nutr. Rev.* **2014**, *72*, 18–29. [CrossRef] [PubMed]
206. Sáez-Lara, M.J.; Robles-Sanchez, C.; Ruiz-Ojeda, F.J.; Plaza-Diaz, J.; Gil, A. Effects of probiotics and synbiotics on obesity, insulin resistance syndrome, type 2 diabetes and non-alcoholic fatty liver disease: A review of human clinical trials. *Int. J. Mol. Sci.* **2016**, *17*, 928. [CrossRef] [PubMed]
207. Reis, S.; Conceição, L.; Rosa, D.; Siqueira, N.; Peluzio, M. Mechanisms responsible for the hypocholesterolaemic effect of regular consumption of probiotics. *Nutr. Res. Rev.* **2017**, *30*, 36–49. [CrossRef]
208. Arora, T.; Sharma, R.; Frost, G. Propionate. Anti-obesity and satiety enhancing factor? *Appetite* **2011**, *56*, 511–515. [CrossRef]
209. Korhonen, H. Milk-derived bioactive peptides: From science to applications. *J. Funct. Foods* **2009**, *1*, 177–187. [CrossRef]
210. Ahtesh, F.B.; Stojanovska, L.; Apostolopoulos, V. Anti-hypertensive peptides released from milk proteins by probiotics. *Maturitas* **2018**, *115*, 103–109. [CrossRef]
211. Ardeshirlarijani, E.; Tabatabaei-Malazy, O.; Mohseni, S.; Qorbani, M.; Larijani, B.; Jalili, R.B. Effect of probiotics supplementation on glucose and oxidative stress in type 2 diabetes mellitus: A meta-analysis of randomized trials. *Daru J. Pharm. Sci.* **2019**, *27*, 827–837. [CrossRef]
212. Balakumar, M.; Prabhu, D.; Sathishkumar, C.; Prabu, P.; Rokana, N.; Kumar, R.; Raghavan, S.; Soundarajan, A.; Grover, S.; Batish, V.K. Improvement in glucose tolerance and insulin sensitivity by probiotic strains of Indian gut origin in high-fat diet-fed C57BL/6J mice. *Eur. J. Nutr.* **2018**, *57*, 279–295. [CrossRef]
213. Amar, J.; Chabo, C.; Waget, A.; Klopp, P.; Vachoux, C.; Bermúdez-Humarán, L.G.; Smirnova, N.; Bergé, M.; Sulpice, T.; Lahtinen, S. Intestinal mucosal adherence and translocation of commensal bacteria at the early onset of type 2 diabetes: Molecular mechanisms and probiotic treatment. *EMBO Mol. Med.* **2011**, *3*, 559–572. [CrossRef]
214. Olek, A.; Woynarowski, M.; Ahren, I.L.; Kierkus, J.; Socha, P.; Larsson, N.; Onning, G. Efficacy and safety of lactobacillus plantarum DSM 9843 (LP299V) in the prevention of antibiotic-associated gastrointestinal symptoms in children-randomized, double-blind, placebo-controlled study. *J. Pediatr.* **2017**, *186*, 82–86. [CrossRef] [PubMed]
215. Pereg, D.; Kimhi, O.; Tirosh, A.; Orr, N.; Kayouf, R.; Lishner, M. The effect of fermented yogurt on the prevention of diarrhea in a healthy adult population. *Am. J. Infect. Control* **2005**, *33*, 122–125. [CrossRef] [PubMed]

216. Jafarnejad, S.; Shab-Bidar, S.; Speakman, J.R.; Parastui, K.; Daneshi-Maskooni, M.; Djafarian, K. Probiotics Reduce the Risk of Antibiotic-Associated Diarrhea in Adults (18–64 Years) but Not the Elderly (>65 Years): A Meta-Analysis. *Nutr. Clin. Pract. Off. Publ. Am. Soc. Parenter. Enter. Nutr.* **2016**, *31*, 502–513. [CrossRef]
217. Allen, S.J.; Wareham, K.; Wang, D.; Bradley, C.; Hutchings, H.; Harris, W.; Dhar, A.; Brown, H.; Foden, A.; Gravenor, M.B.; et al. Lactobacilli and bifidobacteria in the prevention of antibiotic-associated diarrhoea and *Clostridium difficile* diarrhoea in older inpatients (PLACIDE): A randomised, double-blind, placebo-controlled, multicentre trial. *Lancet* **2013**, *382*, 1249–1257. [CrossRef]
218. Zeng, W.; Shen, J.; Bo, T.; Peng, L.; Xu, H.; Nasser, M.I.; Zhuang, Q.; Zhao, M. Cutting edge: Probiotics and fecal microbiota transplantation in immunomodulation. *J. Immunol. Res.* **2019**, *2019*, 1603758. [CrossRef] [PubMed]
219. Leshem, A.; Horesh, N.; Elinav, E. Fecal microbial transplantation and its potential application in cardiometabolic syndrome. *Front. Immunol.* **2019**, *10*, 1341. [CrossRef] [PubMed]
220. Brandt, L.J.; Aroniadis, O.C.; Mellow, M.; Kanatzar, A.; Kelly, C.; Park, T.; Stollman, N.; Rohlke, F.; Surawicz, C. Long-term follow-up of colonoscopic fecal microbiota transplant for recurrent *Clostridium difficile* infection. *Am. J. Gastroenterol.* **2012**, *107*, 1079–1087. [CrossRef]
221. van Nood, E.; Vrieze, A.; Nieuwdorp, M.; Fuentes, S.; Zoetendal, E.G.; de Vos, W.M.; Visser, C.E.; Kuijper, E.J.; Bartelsman, J.F.; Tijssen, J.G.; et al. Duodenal infusion of donor feces for recurrent *Clostridium difficile*. *N. Engl. J. Med.* **2013**, *368*, 407–415. [CrossRef]
222. Schwan, A.; Sjolin, S.; Trottestam, U.; Aronsson, B. Relapsing *Clostridium difficile* enterocolitis cured by rectal infusion of homologous faeces. *Lancet* **1983**, *2*, 845. [CrossRef]
223. Cohen, S.H.; Gerding, D.N.; Johnson, S.M.; Kelly, C.P.; Loo, V.G.; McDonald, L.C. Clinical practice guidelines for *Clostridium difficile* infection in adults: 2010 update by the society for healthcare epidemiology of America (SHEA) and the infectious diseases society of America (IDSA). *Infect. Control Hosp. Epidemiol.* **2010**, *31*, 431–455. [CrossRef]
224. Gregory, J.C.; Buffa, J.A.; Org, E.; Wang, Z.; Levison, B.S.; Zhu, W.; Wagner, M.A.; Bennett, B.J.; Li, L.; DiDonato, J.A.; et al. Transmission of atherosclerosis susceptibility with gut microbial transplantation. *J. Biol. Chem.* **2015**, *290*, 5647–5660. [CrossRef] [PubMed]
225. Vrieze, A.; Van Nood, E.; Holleman, F.; Salojarvi, J.; Kootte, R.S.; Bartelsman, J.F.; Dallinga-Thie, G.M.; Ackermans, M.T.; Serlie, M.J.; Oozeer, R.; et al. Transfer of intestinal microbiota from lean donors increases insulin sensitivity in individuals with metabolic syndrome. *Gastroenterology* **2012**, *143*, 913–916.e917. [CrossRef] [PubMed]
226. Choi, H.H.; Cho, Y.S. Fecal microbiota transplantation: Current applications, effectiveness, and future perspectives. *Clin. Endosc.* **2016**, *49*, 257–265. [CrossRef] [PubMed]
227. De Leon, L.M.; Watson, J.B.; Kelly, C.R. Transient flare of ulcerative colitis after fecal microbiota transplantation for recurrent *Clostridium difficile* infection. *Clin. Gastroenterol. Hepatol. Off. Clin. Pract. J. Am. Gastroenterol. Assoc.* **2013**, *11*, 1036–1038. [CrossRef] [PubMed]
228. Brandt, L.J. FMT: First step in a long journey. *Am. J. Gastroenterol.* **2013**, *108*, 1367–1368. [CrossRef]
229. Zeevi, D.; Korem, T.; Zmora, N.; Israeli, D.; Rothschild, D.; Weinberger, A.; Ben-Yacov, O.; Lador, D.; Avnit-Sagi, T.; Lotan-Pompan, M.; et al. Personalized nutrition by prediction of glycemic responses. *Cell* **2015**, *163*, 1079–1094. [CrossRef]
230. Kolodziejczyk, A.A.; Zheng, D.; Elinav, E. Diet-microbiota interactions and personalized nutrition. *Nat. Rev. Microbiol.* **2019**, *17*, 742–753. [CrossRef]
231. Mendes-Soares, H.; Raveh-Sadka, T.; Azulay, S.; Ben-Shlomo, Y.; Cohen, Y.; Ofek, T.; Stevens, J.; Bachrach, D.; Kashyap, P.; Segal, L.; et al. Model of personalized postprandial glycemic response to food developed for an Israeli cohort predicts responses in Midwestern American individuals. *Am. J. Clin. Nutr.* **2019**, *110*, 63–75. [CrossRef]
232. Kovatcheva-Datchary, P.; Nilsson, A.; Akrami, R.; Lee, Y.S.; De Vadder, F.; Arora, T.; Hallen, A.; Martens, E.; Bjorck, I.; Backhed, F. Dietary fiber-induced improvement in glucose metabolism is associated with increased abundance of prevotella. *Cell Metab.* **2015**, *22*, 971–982. [CrossRef]
233. Martinez, I.; Lattimer, J.M.; Hubach, K.L.; Case, J.A.; Yang, J.; Weber, C.G.; Louk, J.A.; Rose, D.J.; Kyureghian, G.; Peterson, D.A.; et al. Gut microbiome composition is linked to whole grain-induced immunological improvements. *ISME J.* **2013**, *7*, 269–280. [CrossRef]

234. Dao, M.C.; Everard, A.; Aron-Wisnewsky, J.; Sokolovska, N.; Prifti, E.; Verger, E.O.; Kayser, B.D.; Levenez, F.; Chilloux, J.; Hoyles, L.; et al. Akkermansia muciniphila and improved metabolic health during a dietary intervention in obesity: Relationship with gut microbiome richness and ecology. *Gut* **2016**, *65*, 426–436. [CrossRef] [PubMed]
235. Hughes, R.L.; Marco, M.L.; Hughes, J.P.; Keim, N.L.; Kable, M.E. The role of the gut microbiome in predicting response to diet and the development of precision nutrition models-part I: Overview of current methods. *Adv. Nutr.* **2019**, *10*, 953–978. [CrossRef]
236. Lacatusu, C.M.; Grigorescu, E.D.; Floria, M.; Onofriescu, A.; Mihai, B.M. The mediterranean diet: From an environment-driven food culture to an emerging medical prescription. *Int. J. Environ. Res. Public Health* **2019**, *16*, 942. [CrossRef]
237. Ciecierska, A.; Drywien, M.E.; Hamulka, J.; Sadkowski, T. Nutraceutical functions of beta-glucans in human nutrition. *Rocz. Panstw. Zakl. Hig.* **2019**, *70*, 315–324. [CrossRef]
238. He, Z.; Hao, W.; Kwek, E.; Lei, L.; Liu, J.; Zhu, H.; Ma, K.Y.; Zhao, Y.; Ho, H.M.; He, W.S.; et al. Fish oil is more potent than flaxseed oil in modulating gut microbiota and reducing trimethylamine-N-oxide-exacerbated atherogenesis. *J. Agric. Food Chem.* **2019**, *67*, 13635–13647. [CrossRef] [PubMed]
239. Liu, J.; He, Z.; Ma, N.; Chen, Z.Y. Beneficial effects of dietary polyphenols on high-fat diet-induced obesity linking with modulation of gut microbiota. *J. Agric. Food Chem.* **2020**, *68*, 33–47. [CrossRef]
240. Hamaker, B.R.; Tuncil, Y.E. A perspective on the complexity of dietary fiber structures and their potential effect on the gut microbiota. *J. Mol. Biol.* **2014**, *426*, 3838–3850. [CrossRef] [PubMed]
241. Han, Y.; Song, M.; Gu, M.; Ren, D.; Zhu, X.; Cao, X.; Li, F.; Wang, W.; Cai, X.; Yuan, B.; et al. Dietary intake of whole strawberry inhibited colonic inflammation in dextran-sulfate-sodium-treated mice via restoring immune homeostasis and alleviating gut microbiota dysbiosis. *J. Agric. Food Chem.* **2019**, *67*, 9168–9177. [CrossRef] [PubMed]
242. Larrosa, M.; Gonzalez-Sarrias, A.; Yanez-Gascon, M.J.; Selma, M.V.; Azorin-Ortuno, M.; Toti, S.; Tomas-Barberan, F.; Dolara, P.; Espin, J.C. Anti-inflammatory properties of a pomegranate extract and its metabolite urolithin-A in a colitis rat model and the effect of colon inflammation on phenolic metabolism. *J. Nutr. Biochem.* **2010**, *21*, 717–725. [CrossRef] [PubMed]
243. Sousa, T.; Paterson, R.; Moore, V.; Carlsson, A.; Abrahamsson, B.; Basit, A.W. The gastrointestinal microbiota as a site for the biotransformation of drugs. *Int. J. Pharm.* **2008**, *363*, 1–25. [CrossRef]
244. Tuteja, S.; Ferguson, J.F. Gut microbiome and response to cardiovascular drugs. *Circ. Genom. Precis. Med.* **2019**, *12*, 421–429. [CrossRef] [PubMed]
245. Spanogiannopoulos, P.; Bess, E.N.; Carmody, R.N.; Turnbaugh, P.J. The microbial pharmacists within us: A metagenomic view of xenobiotic metabolism. *Nat. Rev. Microbiol.* **2016**, *14*, 273–287. [CrossRef] [PubMed]
246. Haiser, H.J.; Turnbaugh, P.J. Developing a metagenomic view of xenobiotic metabolism. *Pharmacol. Res.* **2013**, *69*, 21–31. [CrossRef] [PubMed]

Review

Reviewing the Composition of Vaginal Microbiota: Inclusion of Nutrition and Probiotic Factors in the Maintenance of Eubiosis

Antonio Barrientos-Durán [1], Ana Fuentes-López [1], Adolfo de Salazar [1], Julio Plaza-Díaz [2,3,4,†] and Federico García [1,*,†]

1 Hospital Clínico Universitario San Cecilio, Servicio de Microbiología, Instituto de Investigación ibs. GRANADA, Avenida de la Ilustración S/N, 18016 Granada, Spain; antoniobarrientosduran@gmail.com (A.B.-D.); anafuenteslop@gmail.com (A.F.-L.); adolsalazar@gmail.com (A.d.S.)

2 Department of Biochemistry and Molecular Biology II, School of Pharmacy, University of Granada, 18071 Granada, Spain; jrplaza@ugr.es

3 Institute of Nutrition and Food Technology "José Mataix", Biomedical Research Center, University of Granada, Armilla, 18016 Granada, Spain

4 Instituto de Investigación Biosanitaria ibs GRANADA, Complejo Hospitalario Universitario de Granada, 18014 Granada, Spain

* Correspondence: fegarcia@ugr.es

† These authors have equally contributed to this work.

Received: 27 November 2019; Accepted: 4 February 2020; Published: 6 February 2020

Abstract: The vaginal microbiota has importance in preserving vaginal health and defending the host against disease. The advent of new molecular techniques and computer science has allowed researchers to discover microbial composition in depth and associate the structure of vaginal microbial communities. There is a consensus that vaginal flora is grouped into a restricted number of communities, although the structure of the community is constantly changing. Certain Community-State Types (CSTs) are more associated with poor reproductive outcomes and sexually transmitted diseases (STDs) meanwhile, CSTs dominated by *Lactobacillus* species—particularly *Lactobacillus crispatus*—are more related to vaginal health. In this work, we have reviewed how modifiable and non-modifiable factors may affect normal vaginal microbiota homeostasis—including sexual behavior, race or ethnicity, and hygiene. Special interest has been given to how the use of probiotics, diet intake, and use of hormone replacement therapies (HRTs) can potentially impact vaginal microbiota composition.

Keywords: vaginal microbiome; bacterial communities; vaginal dysbiosis; bacterial vaginosis; risk factors; nutrition; probiotics; hormone replacement therapy

1. Introduction

The human body accommodates ecological communities of commensal, symbiotic and pathogenic organisms—known as the microbiota—that reside on surfaces and cavities exposed or not to the exterior environment [1]. The kinds of organisms present include bacteria, archaea, protists, fungi and viruses, and these may differ greatly between body sites and between individuals [2]. The impact that microbiota communities have on the host human body was revealed by studies led by the National Institute of Health in 2008 with the development of the Human Microbiome Project (HMP). Results from this project focused on two main facts: (i) the healthy human body is habited by a large diverse microbiota with more genetic material—a presence that exceeds ours in a 10:1 ratio [3,4]—than the host itself; and (ii) the use of new molecular techniques and statistical methods that use high-performance DNA and RNA sequencing technology instead of culture-dependent techniques make possible the identification of complex microbial communities of microorganisms, demonstrating the great impact of microbiota on the host at different levels: metabolic homeostasis, nutrients acquisition, programmed acquisition of immunity and protection against pathogens among others [1,5–7]. In the context of genomics, the term microbiome denotes either the collective genetic material of microbiota microorganisms that reside in an environmental niche or the microorganisms themselves. This term, microbiome, has generated some controversy in the scientific community since its definition. Recently reviewed in [1], it was proposed that this term should refer to an entire habitat that includes the belonging microorganisms, their genomes (i.e., genes), and the surrounding environmental conditions in contrast to the definition that simply considers a mere collection of genes and genomes of the members of a microbiota. The new concept is based on that of "biome", the biotic and abiotic factors of given environments. It is argued that this is the definition of the metagenome, which, combined with the environment, constitutes the microbiome.

2. Materials and Methods

For the review, a search of the scientific literature was conducted using PubMed/Medline with the following search keywords: "vaginal microbiota (or microbiome)", "vaginal dysbiosis", "bacterial vaginosis", "bacterial vaginosis and age", "bacterial vaginosis and ethnicity (or race)", "bacterial vaginosis and stress", "bacterial vaginosis and pelvic inflammatory disease", "bacterial vaginosis and preterm birth (or pregnancy)", "probiotics and vaginal microbiota", and "hormone replacement therapy and vaginal microbiota". Pertinent original articles and reviews that were peer reviewed, indexed in PubMed/Medline and written in English were included. The publication dates were not limited in order to fully review the literature available until September 2019.

3. Results

3.1. *Human Vaginal Microbiota*

3.1.1. Human Vaginal Microbiota: Role as a Natural Barrier

At the histological level, the vagina is a fibromuscular structure that has three main layers or tunics known as mucosa, muscle, and adventitia. The mucous layer forms numerous transverse folds called "wrinkles" or vaginal folds that, in turn, have two layers: stratified squamous epithelium and lamina propria, an unattached connective tissue that joins the epithelium with the muscle layers. Fundamentally, it is in this squamous epithelium where microorganism communities, formerly called the vaginal microbiota, reside. This vaginal microbiota might play a crucial role in gynecologic wellness and in healthy women, consists classically of a diversity of anaerobic and aerobic microorganisms, with the lactobacilli species being the most predominant microorganisms with a determinant function in preventing urogenital diseases such as bacterial vaginosis (BV), yeast infections, STDs, urinary tract infections, and Human Immunodeficiency Virus (HIV) infections [8–20].

The use of new generation molecular sequencing techniques has revealed that vaginal bacterial communities are grouped in three to nine discrete groups—the majority of which are led by lactobacilli [21–23]. Ravel et al. [24] analyzed the vaginal microbiota in a cohort of 396 non-pregnant, fertile and asymptomatic North American women from four ethnic groups. Vaginal bacterial communities found in these women were grouped into five main types of CSTs (Table 1). Four of these types of CSTs, found in 73% of women, were dominated by different species of *Lactobacillus* (*L. crispatus*, CST I; *L. gasseri*, CST II; *L. iners*, CST III; and *L. jensenii*, CST V). The last 27% of the communities (CST IV) were varied and formed by a great proportion of obligate anaerobic bacteria, including *Atopobium*, *Gardnerella*, *Prevotella* spp. and other bacterial species (Table 1). These communities are frequently found in asymptomatic healthy women—mainly of the Black and Latin races—but they are also commonly related with high Nugent score [25], a Gram stain commonly conducted in the diagnosis of BV. High Nugent score or changes in the vaginal microbiota have been related with a high risk of STDs, HIV infections, preterm birth (PTB), adverse pregnancy outcomes such as post-abortion sepsis, early, late and recurrent abortions, adverse perinatal outcomes due to PTB and/or histological chorioamnionitis and postpartum endometritis [26–31]. Subsequent studies have tuned CTS IV into subgroups IV-A and IV-B (Table 1); both varied in composition, although CST IV-B containing fewer lactobacilli and more anaerobic bacterial taxonomic groups (here *Gardnerella*, *Atopobium*, *Leptotrichia*, *Sneathia* spp. and other organisms related with BV have been included). Many studies have also reported that the important finding that around 20% to 30% of women at any given time have a diverse microbiome deficient in *Lactobacillus*, which historically has not been considered healthy [24,32–34].

Table 1. Organization of the vaginal flora into community types (CSTs) in asymptomatic non-pregnant, sexually active women and their contribution to homeostasis (wellness) or vaginal dysbiosis (disease).

CST	Vaginal pH	Ethnic Group	Type of Bacteria	Microorganism's Contribution to Homeostasis or Dysbiosis	References
I	4.0 ± 0.3	White	*L. crispatus*	[1] Lactobacillus spp. beneficial impact reside on: Preventing urogenital diseases. Adhesion to epithelial cells. Production compounds with antimicrobial properties (i.e., H_2O_2) Stimulated Lactic Dehydrogenase→Decrease pH→Protective Environment. Production of Bacteriocins.	[8–20,22,25,35–41]
II	5.0		*L. gasseri*		
III	4.4	Asian	*L. iners*		
IV A			*G. vaginalis* *A. vaginae* *Prevotella spp.*	[2] Common properties linked to dysbiotic statement: Production of Biofilms → Adhesion to Epithelial Cells→ Antibiotic Tolerance, Resistance to Host Immune Defence. Production of Cytolysins. Production of Amines → pH Alkalinisation. Activates NF-κB cascade. Secretion of Collagenase and Fibrinolysins→Enhance Mucosal Surface Degradation →Detachment of Epithelial Cells.	[42–46]
	5.3 ± 0.6	Black Hispanic		[3] Common properties linked to dysbiotic statement:	
IV B			*A. vaginae* *Leptotrichia spp.* *Mobiluncus spp.*	Secretion of Collagenase and Fibrinolysins→Enhance Mucosal Surface Degradation →Detachment of Epithelial Cells. Adhesins that contribute to Epithelial Colonization. Hemolysin→Cytotoxic Activity. [3] plus: Malic Acid and Trimethylamine Production→vaginal irritation.	[42–46] [46]
V	4.4		*L. jensenii*	[1]	[8–20,22,25,35–41]

From left to right, columns depict: organization into CSTs, associated pH, Human Races attributable to, predominant microorganisms of these CSTs, properties of them that impact to homeostatic or dysbiotic statement and, given references from literature. Five CSTs are generally accepted, being four of them dominated by different species of *Lactobacillus* (*L. crispatus, L. gasseri, L. inners and L. jensenii*), which are associated to a healthy statement of the vagina (referred as homeostasis). CST IV is split into CST IV-A and IV-B communities where *Lactobacillus* spp. are not predominant and by contrast, a diversity of facultative anaerobes have been identified. Main members of these subgroups are species of: *Gardnerella, Atopobium, Mobiluncus, Prevotella and Leptotrichia.* Diversity of these two sub-CSTs is linked to a dysbiotic statement being BV the most single common manifestation of disease (mentioned in the body text). Common properties of *Lactobacillus* spp. contributing to vaginal wellness as well as features of anaerobe microorganisms that contribute to vaginal colonization and, hence, to dysbiosis are given. This table is an updated and adapted version of that published by Ravel J et al. [24].

Lactobacillus spp. are Gram-positive anaerobic bacteria capable of colonizing the vaginal mucosa, preventing the establishment or excessive development of other microorganisms that may become potentially pathogenic for the host. This protection is performed through two mechanisms: (i) by the specific adhesion to epithelial cells and, (ii) by the production of compounds with antimicrobial properties. In the first place, the ability of the lactobacilli to self-aggregate and adhere to the vaginal epithelium through glycoproteins present on the surface of the epithelial cells (i.e., fibronectin) in a binding that is favored by an acidic pH environment has been described [47]. Although further studies are necessary, it is thought that, in addition to the cellular epithelium of the host, proteins, carbohydrates, glycoproteins, lipoteic acids and divalent cations from microbiota species also play an important role [35].

The presence of lactic acid is key to a healthy homeostasis of the vagina and its production comes from two different sources: by the vaginal epithelium (mainly L-lactate representing 20% of the total lactic acid) and by the microbiota, responsible for metabolizing approximately 80% of glycogen producing the two isoforms of lactic acid with a predominance of D-lactic acid [48,49]. When the squamous epithelium requires energy in the form of ATP, the glycogen from the vaginal epithelial cells is converted to glucose, then to pyruvate, and from this to lactic acid, which is released into the vaginal lumen as the epithelium undergoes desquamation [50,51]. This production of lactic acid is performed under the control of the estrogen levels present in the blood, as these promote maturation and deposition in the vaginal epithelial cells. Therefore, due to the known change in estrogens production throughout the woman's life cycle, the vaginal ecosystem can be subjected to modifications [50]. The second and main mechanism for producing lactic acid comes from the glycogen found in the vaginal lumen, which is catabolized by alpha amylases to produce maltose, maltotriose and alpha dextrins, which are subsequently converted into lactic acid, due to the action of the Lactobacillus-stimulated lactic dehydrogenase [50,52]. The presence of lactic acid in the vaginal lumen has the consequence that the vaginal pH remains acidic, at levels of approximately 3.5–4.5, generating a protective environment in the mucosa that, partially or totally, inhibits the growth of pathogenic microorganisms [36,53]. Other compounds produced by lactobacilli that play a secondary control in the vaginal flora are hydrogen peroxide (H_2O_2) and bacteriocins (reported in Table 1) [36,37]. It has been described that some strains of vaginal lactobacilli can produce H_2O_2 protecting the mucosa against alterations caused by opportunistic microorganisms, including those that cause sexually transmitted infections (STIs). On the other hand, bacteriocins are polypeptides synthesized at the ribosomal level whose antimicrobial activity has only been proven in vitro [36,54].

3.1.2. Composition of Vaginal Microbiota Is Defined but Highly Dynamic

Vaginal microbiota composition can be highly dynamic in some women. In short periods, it can go from being dominated by communities led by *Lactobacillus* species to other communities lacking such abundant numbers in these species, while in other women this does not occur, being relatively stable [27]. In both scenarios, there is a certain consensus in the scientific community that vaginal microbial composition has important compositional fluctuations during the woman's life cycle: birth, puberty, menopause, and transition stages, where steroid sex hormones play a key role in the maintenance of the composition and stability of this microbiota [55–58]. Among the changes that the vaginal microbiome may undergo, some reports have focused on that these changes may be preferred from one CST towards a specific community condition [34,38]. Also, there is evidence that the CST I community tends to be the most stable in promoting the stability of the vaginal community [22,33,34,38], while CST IV seems to have frequent transitions to many other conditions [38]. On the other hand, it has been reported that a microbiome controlled by *Lactobacillus* species, different from *L. iners* is optimal for vaginal wellness [22,39]. In this sense, it has been shown that the existence of lactobacilli, especially *L. crispatus*, is strongly related with the lack of BV [25,38–40]. Very interestingly, it has been observed that the production of lactic acid is an indicative marker in all healthy vaginal communities [59]. Lactic acid has inhibitory properties over pathogenic bacteria [39,56], altering bacterial cell membranes and also improving the host immunity when bacterial lipopolysaccharide is present [60]. In a more precise manner, it has been described that the L-lactic acid isomeric form—either produced by lactobacilli and by epithelial vaginal cells of the host—activates a certain type of immune cells and may encourage that epithelial cells release pro-inflammatory cytokines [61].

Importantly, results obtained on vaginal microbiota composition during pregnancy are of special interest. To date, results are still scarce, and few authors have analyzed the vaginal microbiota composition of pregnant women with methods independent of culture [62,63]. Initially, Verstraelen et al., using a methodology based on Gram stain, culture and terminal restriction of polymorphism fragments, revealed that *L. crispatus* and *L. gasseri* species were important in the maintenance of stable vaginal microbiota in a female population collected once in each trimester [62], being this the agreement accepted and extracted from similar methodologies on the vaginal microbiota during a normal gestation. Based on 16S rRNA gene sequencing methods, several papers are considered of reference. On the one hand, the studies carried out by Romero and collaborators showed that: (i) the vaginal microbiota of healthy pregnant women is different from that of non-pregnant women in composition and stability and, (ii) the microbiota is similar in pregnant women whose pregnancy ended at term or prematurely [64]. Contrary to these latest results, DiGiulio et al. observed variations in vaginal microbiota composition in women who finally had a preterminal termination. Here, an imbalance in the *Lactobacillus* species normally found in the vaginal microbiota was observed, a proliferation of other non-native organisms other than the members of the genus *Lactobacillus*, as well as it was demonstrated that a period of less than 1 year between gestations constituted a high risk of preterm pregnancy due to infection of the amniotic cavity [65].

3.2. *Vaginal Dysbiosis: An Imbalance in Vaginal Microbiota Composition*

Sometimes the concentrations of lactobacilli within the vaginal community are modified, producing an imbalance statement or dysbiosis of the microbiota, which is generally defined as a polymicrobial condition characterized by a low prevalence of *Lactobacillus* spp. and by an increase in anaerobic microorganisms. The most common form of dysbiosis is bacterial vaginosis (BV). This condition is described for three main changes in the environment in vagina [66,67]: (i) a change in vaginal microbiota composition from *Lactobacillus* spp. to facultative anaerobes; (ii) the production of amino compounds by the new bacterial microbiota and; (iii) an increase in vaginal pH to more than 4.5. These are the conditions that mainly favor the development of opportunistic microorganisms that behavior like pathogens, whether they are usually found in the vagina or if they come exogenously [36]. Therefore, diversity in the vaginal microbiota, so-called unhealthy microbiome, is less resistance to alteration and more susceptibility to diseases, including the acquisition of STDs and reproductive and obstetric outcomes [20,33,38,68].

3.2.1. Risk Factors Associated with Vaginal Dysbiosis

Here, in this section, the main risk factors associated with vaginal dysbiosis (VD) are reviewed. Like in other diseases, risk factors can be categorized as those inherent to the human condition (known as non-modifiable factors) and those related to social conduct or habitats, so-called modifiable factors (Figure 1).

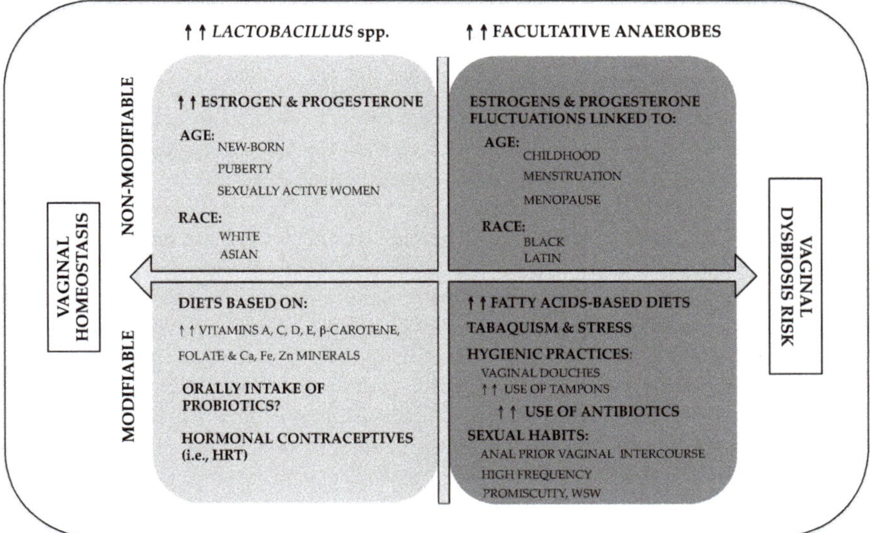

Figure 1. Modifiable and non-modifiable risk factors associated with vaginal homeostasis and dysbiosis. Inherent human conditions linked to vaginal homeostasis and associated with vaginal dysbiosis risk are depicted in the top part of the panel, while modifiable factors are shown in the bottom part. Top and bottom left sections -defined by a double pointed arrow- report those factors that contribute positively to homeostasis. Conversely, right top & bottom sections report those factors associated to vaginal dysbiosis risk. Both of them are associated with a microbiota rich in diverse facultative anaerobes microorganism opposite to those rich in *Lactobacillus* spp. (left sections).

Age and Hormone Physiology

Vaginal microbiota composition changes over time. It is well established that vaginal physiology is modified not only due to estrogen production and concentration—which in turn favors the existence of glycogen—but also to vaginal microbiota composition. During pregnancy, it has been thought for years that the fetus develops sterile (now under consideration), and the first microbial colonization occurs at the time of delivery, which comes from the vagina or skin, depending on the route of birth. In newborns, the vulva and the vagina of the infant are influenced by the presence of transplacental estrogenic residues and these favor glycogen supply, which is metabolized by endogenous bacteria, lowering the vaginal pH. As these estrogens are metabolized, a loss of the vaginal glycogen content occurs and thus, the pH is neutralized or alkalized [69]. With respect to childhood, it has been determined that vaginal pH remains neutral or alkaline, with a diphteroid's colonization (*Corynebacterium* spp. 78%), *Staphylococcus epidermidis* (73%) as well as by *Mycoplasma* spp. [70]. During puberty, maturation of adrenal glands and gonads provoke a rising in the levels of estrogens increasing as well the intracellular production. Two predominant colonies have been determined at this stage of life: *Lactobacillus* spp. and *Atopobium* and *Streptococcus* spp. [70].

In women at reproductive stage, it has been reported that meanwhile menstruation and sexual activity have undesirable consequences on the vaginal microbiota stability—estrogen levels decreased, pH is closed to neutrality which it difficult the growth of lactobacilli (reported in Figure 1)—the secretory phase of the menstrual cycle (described for higher estrogen and progesterone concentrations) is more stable in terms of microbiota composition which correlates high levels of steroid sex hormones [34,71,72]. In the same group of women, hormonal contraceptive administration has been associated with a decrease in the risk of presenting BV, because it generates greater estrogenic stability [36,70,73,74]. Subsequently, as estrogens decrease until menopause, the dominance of *Lactobacillus* decreases and it is stabilized [52]. In postmenopausal women, the decrease in estrogen causes again an increase in pH, which facilitates the presence of enteric bacteria (Figure 1) [75].

Ethnicity

It is a fact that the prevalence of suffering from BV, the main consequence of a dysbiotic statement, varies according to the ethnic group. The reasons for the aforementioned differences are not fully understood, although it is speculated that genetic differences determined by the host could govern the composition of species in the vaginal communities [36,52,74,76]. The acquisition of BV has long been associated with black race in the United States (US) (Figure 1) [52,74–76] and that this association persists even after using adjustments with variables associated with sexual practices and other confounding factors [52,74,76–78]. In other locations such as the United Kingdom and Canada, the prevalence of BV was also higher among Afro-Caribbean and Aboriginal populations, respectively, while in studies performed in countries such as Spain and China, BV prevalence was found higher in Gipsy and Tibetan ethnic groups, respectively [74]. Supporting these results, it has also been reported in other studies about the composition of species in the vaginal microbiome of black and white women born in the US, a significant difference between these two groups, in which black women have a greater microbial diversity and a lower probability of lactobacilli colonization than white women [24,79]. In other studies, conducted in sub-Saharan African countries, a smaller proportion of *L. crispatus* in the vaginal communities compared to women of European or Asian descent has been found [24,79,80]. Here, African communities were dominated by *L. iners* and by a variable mixture of facultative anaerobic bacteria [24,68,81]. Similarly, in a Dutch study about the composition of vaginal microbiome it was significantly associated with ethnic groups where women from African descents had the main occurrence of clusters determined by *Gardnerella vaginalis* or *dysbiosis* [82].

There is even reported evidence that the genetic variation of the host—which can sometimes associate with race or ethnic groups—may be able to affect the microbiome composition. At this point, a large study using metagenomic data from the HMP revealed several associations with key genes of the host related with immune function and abundance of specific microbial taxonomic groups at four distinct locations in the body, although the association with the vagina was not included [83]. Finally, in a retrospective cross-sectional study performed with a small cohort of black South African women, the Black women had a different cervical microbiota without *Lactobacillus* predominance; nevertheless, additional studies are needed to examine whether this microbiota represents abnormal, intermediate or variant states of health [84].

Tobacco

Smoking cigarettes has been related with the increased BV prevalence in several epidemiological studies and occasionally in a dose-dependent manner [70,85]. Certainly, a number of compounds resulting from smoking have been identified in the cervical mucus of smokers [70]. Data analysis from sequences have shown an association between smoking and VD even after adjusting for confounding factors (reported within modifiable factors in Figure 1) [79]. In this sense, a 2014 study found two shreds of evidence: (i) that it was significantly more likely that smokers' vaginal microbiota had a low *Lactobacillus* prevalence and; (ii) metabolites produced during smoking were increased in higher Nugent scores [70]. Recently, the vaginal metabolome of smokers and non-smokers was compared in a cross-sectional study. Smoking was related with differences in vaginal metabolites. Among women categorized to the CST-IV community, biogenic amines were higher in smokers; these amines can affect the virulence of infective pathogens and contribute to vaginal malodor [86].

Stress

Stress is defined as any physical or psychological challenge that threatens or has the potential of threatening the balance, homeostasis, of an organism's internal background [87–89]. These challenges can be lifetime events, emotions, and relations that unfavorably affect the individual's comfort or generate perceived detrimental responses. Very recently, the role of stress over the female lower genital tract has been reviewed [90]. For instance, working with animal models, it has been reported that the persistent exposure to psychosocial stress can lead to an encouragement of the hypothalamic–pituitary–adrenal and sympathetic–adrenal–medullary axes. This, in turn, drives a cortisol-induced inhibition of glycogen deposition in vagina, which is translated into an interruption in epithelial maturation that is crucial to keep vaginal homeostasis (included as negative modifiable factor in Figure 1). This phenomenon is especially relevant during pregnancy, where local production of high levels of corticotropin-releasing hormone occurs in the decidua, fetal membranes and placenta [90].

3.2.2. Other Factors That Influence Vaginal Dysbiosis

Sexual Activity

The number of reports published in recent years that try to find associations between human sexual behavior and BV, as the main form of VD, is growing and diverse. Related to the number of vaginal coitus, it has been found that a higher frequency is related with a major risk of suffering from BV [85]. Related to the fact of having multiple, new, or numerous male partners, there is a direct association with BV [81,91]. The maintenance of unprotected sex has been associated with a risk—greater than double—of suffering BV and recurrent BV [91,92], which is adversely related with the quantity and presence of healthy *Lactobacillus* species [81].

Regarding sexual contacts with people of the same gender, a significant relationship between BV and female sexual mates has been found [78], because women in homosexual relationship seem to be at greater risk (Figure 1) compared to women who have heterosexual sex [78,93]. Other reports have studied the impact of certain sexual practices on BV. Although they are moderately limited, strong associations have been found. For example, the association is direct with BV when vaginal intercourse is performed immediately after receptive anal intercourse [85]. On the other hand, there is some controversy about the relationship between receptive oral sex and BV [93]. The increase in the recognition and copy number of *G. vaginalis* genes in the oral cavity among women who have homosexual sex with BV enhances approximately biological plausibleness to a direct association [94]. However, other studies have been unsuccessful to demonstrate such a relationship with receptive oral sex [93]. The controversy persists in studies that seek to find a relationship between BV and receptive oral or anal sex [93]. Finally, digital receptive sex (either vaginal or anal) does not appear to be related with BV [93].

Lifestyle and Daily Practices

There are certain types of daily practices that can influence the levels of vaginal acidity, which may significantly predispose the excessive proliferation of opportunistic pathogens [75,95]. These practices can be classified into local and systemic. Within local practices, the use of feminine hygiene products including the use of tampons could alter the vaginal immune barrier having an impact on cellular integrity. Others, like vaginal showers have long been associated with the acquisition of BV (Figure 1). In this regard, longitudinal studies suggest that women who attend these practices have an increased risk of BV incident [96]. The consequences of other intravaginal procedures are not entirely well understood, being some of them more associated with risks of suffering from BV than others [97,98]. Given the heterogeneity of the type of intravaginal practices, the variety of products for this purpose and their wide dissemination among cultures/races, additional research is desirable to clarify the effects on resident microbial communities of vaginal flora [79,81,83]. In addition, it has also been reported that the alkalinity of menstruation or semen neutralizes vaginal pH temporarily and could impact the vaginal microbiota [36,76,90]. Relative to systemic practices, the improper or prolonged use of antibiotics can permeate vaginal exudate, causing alteration in the ecosystem of the vagina. Within this category the aforementioned smoking cigarettes impact and nutrition habits (see below) are also included.

3.3. Pathogenesis Associated with Bacterial Vaginosis

As mentioned above, BV is a VD that is defined by a lack of lactic acid—producing lactobacilli and proliferation of facultative and strict anaerobes [99]. BV is the most frequent cause of vaginal discharge [78] and is related with distinct adverse consequences, including an increased risk of PTB, pelvic inflammatory disease (PID), as well as the acquirement of human immunodeficiency virus (HIV) and other sexually transmitted pathogens [20,33,38,68]. Here, in this section, the pathogenesis of BV is reviewed.

3.3.1. Bacterial Vaginosis and Sexually Transmitted Diseases (STDs)

STDs are produced by a wide variety of microorganisms comprising bacteria, protozoa, viruses and fungi. Among bacteria, epidemiological studies have related BV with an increased risk of infection by gonorrhea and chlamydia [20]. In vitro, for instance, it has been proven how vaginal lactobacilli inhibit the growth of *Neisseria gonorrhoeae* [100,101] and other bacterial pathogens [102]. Patients with Nugent scores higher than three were related to a four-fold gain in the gonorrhea infection risk and a triple increase in chlamydial infection risk in a cross-sectional study [15]. In this sense, several longitudinal studies have also established this relationship, being the main study one that shows an augmented risk of chlamydia and gonorrhea incident in women with scores of Nugent greater than 3 [103]. Furthermore, the treatment of asymptomatic BV with intravaginal metronidazole was considerably related with a decrease, by more than triple, in incidental chlamydia in a randomized study [104]; however, recent data from a randomized prospective study showed that detection at home and treatment for BV did not decrease the incidence of chlamydia or gonorrhea [105], results that, taking into account previous research, question the design of this study.

Trichomonas vaginalis infection has also been closely associated with BV [103]. In a National Health and Medical Examination Survey performed in 2001–2004, concurrence happened in, around 50% of the women infected with *T. vaginalis* [106]. Trichomoniasis has been linked to low levels of healthy vaginal characterized for the presence of lactobacilli and has been positively associated with an increase in Nugent score [107]. An in vitro evidence shows that the presence of *T. vaginalis* decreases the lactobacilli linked with the epithelium but not the species related with BV [108]. In longitudinal analyses, it has been proven that a Nugent score higher than three was related to a higher risk of *T. vaginalis* infection [109]. To date, few studies use sequencing techniques focusing on the presence of *T. vaginalis* and vaginal microbiome's composition. In one of these few studies, it was found that the CST-IV community type was considerably related with the detection of *T. vaginalis* [110]. In addition, *T. vaginalis* and BV were independently associated with an increase in the spread of HIV-1 in the vagina, and their concurrence was greatly related with increased probabilities of vaginal spread [111].

BV and herpes simplex virus (HSV) have been epidemiologically connected in multiple cross-sectional and prospective studies. Initial research by Cherpes and colleagues—in a study with 670 women during a year—found that the BV diagnosis was related with a double risk of HSV-2 seroconversion [11]. Subsequently to this, a meta-analysis reported that this relationship could be bidirectional: HSV-2 infection was related in a dependent manner with episodes of BV in sex workers and demonstrated a relative risk of 1.55 for BV incident in women infected with HSV-2 [112]. At the population level, Nugent scores of four or higher have been related with a 32% increase in concurrent HSV-2 and, an 8% increase in HSV-1 [113]. In addition, a meta-analysis has reported that the prevalence of BV was 60% higher among HSV-2 women compared to negative HSV-2 [114]. A recent study revealed that antibiotic-induced VD in mice resulted in a fall of antiviral protection against HSV-2 infection [115]. Furthermore, the association between BV and HSV-2 has also been confirmed in a recent study in South Africa with a large (n = 2750) cohort of patients [116]. In this study, women who had an HSV-2 infection at enrolment were shown to be at increased risk for incident BV infections and, certain risk factors like young age, unmarried and having a partner that has other partners, were significantly related with subsequent BV.

The relationship between BV and Human Papilloma Virus (HPV) is also consistent and well reported in the literature as it is reviewed now. Early longitudinal studies showed a greater relationship of prevalent and incident of HPV in women with both intermediate microbiota and BV [14]. A small, but significant increase in the risk of prevalent HPV, an increase in the chances of incident HPV and late HPV disappearance in women with a Nugent score of seven or higher was reported [117]. In two molecular-based analysis, researchers found that women with HPV positive had a minor fraction of lactobacilli than HPV negative diagnosed women [118,119]. In addition, women with vaginal microbiota dominated by *L. gasseri* appeared to have augmented HPV disappearance rates [119]. Furthermore, other studies have shown that intraepithelial dysplasia severity was significantly related with an increased in microbial diversity in vagina, regardless of HPV condition and showed that the type of community condition had a significant relationship with predominant HPV and that the CST IV-B was linked with HPV positivity [120]. In a retrospective study between 2012 and 2017 with 7081 HPV available cases, authors found that there is a significant association between BV, positive HPV infection, and great-score of squamous intraepithelial lesions [121]. In this study, BV patients with negative HPV infection showed more squamous abnormalities than BV-negative HPV-negative patients [121]. Prevalence of HPV genotypes (HPV59, HPV73, HPV52, and HPV58) increases in women presenting cervical cytological abnormalities has also recently described [122].

There is substantial information that correlates VD with a gained risk of HIV-1 acquisition and transmission. A meta-analysis showed that BV was related with a 60% increment in the risk of contract HIV-1; this comprised four longitudinal studies that inspected HIV-1 incident infection [123]. A model of vaginal mucosa has shown that lactobacilli, predominantly *L. crispatus*, repressed HIV-1 replication [41]. The cervicovaginal mucus with augmented levels of D-lactic acid and a microbiome dominated by *L. crispatus* efficiently stuck HIV-1 in a better way than mucus dominated by distinct microorganisms [41], in addition to the fact that lactic acid at the concentrations obtained in the vagina can incapacitate HIV much effectively in vitro than other acids [124]. Notably, a study in Rwandan sex workers showed that those with a microbiota dominated by *L. crispatus* had a lower incidence of HIV and STIs and that dysbiosis augmented the risk of contracting HIV and STDs in a dose-response manner; in addition, significantly less HIV positive women with microbiota dominated by *Lactobacillus* spp. had demonstrable cervicovaginal levels of HIV-1 [125]. Very recent research has focused on the identification of specific bacterial taxa in the vaginal niche and an increased HIV risk [126]. This analysis demonstrates associations between individual bacterial taxa and pro-inflammatory cytokines (tumor necrosis factor alpha (TNF-α) and interleukin (IL)-1β), suggesting that individual bacterial taxa might show an important role in determining the inflammatory state of the vagina and hence, an increased HIV risk [126].

3.3.2. Bacterial Vaginosis and Pelvic Inflammatory Disease (PID)

PID, infection and inflammation of the uterine lining (endometritis) and fallopian tubes (salpingitis), is a common condition between young women that regularly have the following consequences: tubal factor infertility, chronic pelvic pain and recurrent PID disease [127]. Although PID is a recognized complication of *Chlamydia trachomatis* and *Neisseria gonorrhoeae* infections [128,129], the etiology of up to 70% of cases may be diverse: other cervical, enteric, BV-associated, and respiratory pathogens, including *Mycobacterium tuberculosis* [130,131], may be involved. Truly, early studies revealed that PID frequently occurs in the lack of understood STDs and its etiology [132–134]. For instance, in a large longitudinal cohort study it was reported that vaginal transport of organisms associated with BV double increased the risk of incident PID [135]. The application of 16S rRNA bacterial gene sequencing has revealed the presence of specific novel bacterial species in BV [23], some of them are Gram-negative anaerobes such as *Sneathia (Leptotrichia) sanguinegens/amnionii*, have been related in case reports of postpartum fever [42], endometritis [42], tubo ovarian abscesses [136], amnionitis and preterm labor [137] and; Gram-positive anaerobes such as *Atopobium vaginae* has been related with tubo ovarian abscess, tubal factor infertility [43] endometritis [44] and fetal death [45]. Hebb et al.

identified bacterial 16S sequences in the fallopian tubes of the 24% of women with salpingitis but in none of the controls [138] including phylotypes closely related to *Leptotrichia* and *A. vaginae*.

In last years, a prospective study has demonstrated that *S. sanguinegens*, *S. amnionii*, BV-associated bacterium 1 (BVAB1) and *A. vaginae* were related with PID, disappointment of the Centers for Disease Control and Prevention-recommended treatment to eliminate short term endometritis, recurrent PID and infertility, suggesting that optimal antibiotic regimens for PID might need the treatment of new BV-associated microbes [139]. Very recently, in a cross-sectional analysis nested within the PID Evaluation and Clinical Health study has been evaluated if Toll-Like Receptor (TLR) genetic variants are or not related with particular BV-associated microbes that are connected with infertility following pelvic PID. TLRs are part of the native immune system and cooperate in the elimination of pathogens through nuclear factor kappa-b (NF-kB) signaling. Results from this study suggested a modest association of host gene variants in TLR2 signaling pathways with cervical *A. vaginae*—through excessive inflammatory responses—in women with clinical PID [140].

3.3.3. Bacterial Vaginosis and Pregnancy

PTB is the leading cause of neonatal morbidity and mortality and constitutes an important cost cargo on medical management [141]. The PTB etiology is multifactorial although the culmination-shared pathway is always the same. Infection and/or inflammation are major causes [141–143] representing up to 40% of the cases [142,144] and probably this is much greater in initial gestations where mortality and morbidity are more frequent [145,146]. BV has been related with adverse repercussions in childbirth. However, the mechanism by which dysbiosis might alter pregnancy continues unclear [147]. Importantly, it has also pointed out that some organisms could affect pregnancy outcomes in a different manner than others and, even that they could impact pregnancy at different gestational ages [148].

Some initial studies have reported that VD cases due to BV in pregnant women have a 5-fold increased risk of PTB before 34 completed gestation weeks [149] and a 7-fold increased risk if BV is detected before 16 weeks [150]. In these cases, clindamycin administration before 22 completed weeks of gestation was related with an 80% reduction in the rate of miscarriage and a significant 40% decrease in PTB [151,152]. Molecular-based techniques have brought a shed of light about how vaginal microbiota impact wellness and illness [153]. The more recent the studies have been conducted, the greater is the association originated between microbiota composition and PTB [38,154–158]. Nevertheless, to date, evidence is limited, and results are, in certain cases, contradictive. Truly, a certain association between preterm labor with diverse vaginal communities has been revealed [38,155]. In this sense, a longitudinal study reported that no woman with delivery at term had CST IV-B [64]. Meanwhile, in a study with a large cohort of pregnant women with intermediate vaginal microbiota, the lack of lactobacilli was related with preterm delivery [159].

In the design of molecular-based studies, the importance of the type of species belonging to *Lactobacillus* genera that resides abundantly on vaginal communities was recently shown. In this study, *L. iners* was significantly overrepresented in vaginal microbiota from women who delivered early preterm (67%), compared to those who delivered late preterm (31%), or at term (29%). On the other hand, *L. crispatus* was related with subsequent term birth compared with early PTB, and a comparatively longer duration of pregnancy than that associated with *L. iners* [160]. Plausible explanation to this association can be because *L. crispatus* takes benefits over *L. iners* with respect to the chirality ratio between the productions of the D- and L-isomer of lactic acid, having these major functional implications [161]. Another hypothesis could be that occasionally *L. iners* might be potentially more pathogenic than a vaginal symbiont [162,163].

3.4. Impact of Nutrition in Maintaining Vaginal Homeostasis

3.4.1. Dietary Intake Consequences on Vaginal Homeostasis

Genital tract infections are common in women, with BV being the single most public inferior reproductive tract infection in a population of childbearing age. As already seen here, risk factors for BV include some socio demographic factors, including race and, lifestyle/behavioral factors such as smoking, contraceptive use, douching, sexual behavior and stress. In recent years, researchers have begun to theorize that nutrition is another recognized factor for BV. While little is known about how nutrition may impact vaginal homeostasis, nevertheless, in other body locations such as gut microbiome studies have revealed the surprising effect of the diet on the composition and function of the bacterial community which appears to have a deep impact on human well-being and related diseases including: metabolic disorders, obesity, inflammatory bowel disease and cancer [164]. Furthermore, pro-inflammatory effects of altered intestinal microbiota on distal systems of the body are increasingly recognized [164]. In addition, it is known that the intestine can work as an extravaginal reservoir for lactobacilli and bacteria associated with BV [94].

The investigations published by Neggers et al. and Tohill et al. have constituted the first critical evidence about the role of suboptimal nutrition in BV and other gynecological infections in women of childbearing age. In the first, Neggers et al. have described that subclinical iron and vitamin D deficiencies during pregnancy are related with an increased BV risk [75]. This was also suggested in the studies of Verstraelen et al. [165]. Parallel to the studies, Tohill et al. demonstrated that lesser serum concentrations of vitamins A, C, and E, and β-carotene were associated with BV, and lower iron status was related with increased prevalence of *Candida* colonization in a large cross-sectional study of women with or at risk of HIV. In this work, higher serum zinc concentrations were related with a minor risk of HPV [166].

Subsequently to these studies, Bodnar et al. revealed the contrary relationship between vitamin D and the BV risk during the first trimester of pregnancy [167]. Despite evidence, Klebanoff and Turner [168], in a large longitudinal study, did not find a relationship between vitamin D and BV using a statistical seasonal variable. However, recent results by Akoh et al. again have suggested that minor vitamin D maternal status can increase the infection risk across gestation [169]. Being more precise, authors have observed significant inverse associations between vitamin D and IL-6 and TNF-α in the mother at delivery and between vitamin D and IL-6 and hepcidin in the neonate at birth. Furthermore, authors have revealed that the existence of BV influenced the relationship between IL-6 and vitamin D at delivery suggesting that vitamin D could influence changes in pro inflammatory cytokine production during pregnancy and infections might moderate these relationships. Deeping in the association between nutrients and vaginal health, on certain subsets of women, it has been found an association between the increased of fat in the diet, a higher glycemic load and lower nutritional density [75,170] with BV, and in addition, an contrary relationship between BV and the bigger folate, vitamin E and calcium intake [75]. Besides this, in the latter, the glycemic load was related with the progress and perseverance of BV [170].

BV has also been epidemiologically connected with obesity [78]. In fact, it has been proposed that the increase in saturated fat consumption increases the incidence of BV, and on the other hand, the folate, vitamin E and calcium consumption decreases the BV risk. In the case of pregnant women, iron and vitamin D deficiencies have been related with an increased BV risk [170].

3.4.2. Probiotics Influence on Vaginal Microbiota

It is thought that the vaginal microbiota is mostly formed by the rise of microbes from the rectum. In the vagina, the quantities and categories of residing microbes fluctuate according to certain factors such as hormone levels, sexual contact, douching practices, diet, among others [171]. Vaginal microbiota is a critical actor in gynecologic health, in which bacteria are able to change to a dysbiotic state causing a pathogenic process [172]. BV, the main cause of VD, is the most common genital tract infection in women throughout their reproductive life and it has been related with serious adverse reproductive and obstetric health outcomes, such as PTB and acquisition or transmission of several sexually transmitted agents [170]. Being polymicrobial in nature, BV is considered by a decrease in positive lactobacilli and a significant increase in number of anaerobic bacteria, including *G. vaginalis*, *A. vaginae*, *Mobiluncus* spp., *Bacteroides* spp. and *Prevotella* spp. BV includes the existence of a thick vaginal multi-species biofilm, where *G. vaginalis* is the predominant specie [173]. The standard of-care for BV, an antibiotic therapy based on metronidazole or clindamycin, is incapable to completely eradicate vaginal biofilms, which may explain the existence of high recurrence rates of BV [174]. In addition, prolonged antibiotic therapy can also harm the healthy vaginal microbiota [173]. These issues generated the interesting emerging different therapeutic strategies such as the use of prebiotics and/or probiotics [175]. Probiotics are extensively used to progress gastrointestinal health, but they might also be beneficial to prevent or treat gynecological disorders. In obstetrics and gynecology, probiotics are living microorganisms—mostly formed by *Lactobacillus* spp.—mainly used to restore the physiologic vaginal microbiota in order to treat, besides BV, vulvovaginal candidiasis (VVC) and PTB [176]. Despite this, considerable heterogeneity in probiotic's effectiveness has been detected during clinical trials [174], which are reviewed in this section.

Probiotics in Non-Pregnant Women

The recognized favorable effect of probiotic administration for the BV and VVC treatments has been evaluated in numerous meta-analysis [177–180] and recently reviewed in [176]. Relative to VVC, it is estimated that approximately seven women out of ten women will live at least one experience of VVC in their lives [181], where recurrence is quite often. This fact has made probiotics a real option to be considered together with current antifungal therapies. In a Cochrane systematic review [180], the efficiency of probiotic treatment for VVC in non-pregnant women was recently under evaluation. The conclusions from 10 randomized controlled trial (RCTs) (1656 participants) studying the influence of probiotics used by oral and vaginal routes, as a coadjuvant therapy to antifungal drugs, were that probiotics slightly enhanced the temporary clinical and mycological cure rate and reduced the 1 month relapse rate. Nevertheless, no influence of probiotic administration was observed on continuing clinical or mycological cure rate (3 month post-treatment evaluation). In addition to this, one of the main and unsolved topics related to the VVC treatment is the extraordinary proportion of reappearances even after the use antifungal (azoles) treatment [182,183], a fact that might be due to augmented presence of azole-drug resistance [184]. For these cases of azole-resistance, it has been proven the effectiveness-protecting role of specific *Lactobacillus* species, an example of this is *L. plantarum* P17630 [184].

Relative to the treatment of BV with probiotics, in 2013, a systematic review by Huang et al. [179] already reinforced the possible favorable effect of probiotics for the treatment of BV. The analysis included 12 RCTs where probiotics were implemented either orally or vaginally with continuation periods from 4 weeks to 6 months. The results revealed that probiotic administration was capable to increase the cure rate in adult BV patients, although some subgroup of analysis failed to prove a positive effect of probiotic administration in long-term treatment (>1 month) [177]. In further analysis, authors investigated the effect of metronidazole administration alone or in combination with probiotics. Five RCTs containing a total of 1186 participants were chosen, and the benefit of combined therapy was proven over metronidazole alone on BV.

Probiotics in Pregnant Women

The recognized role of probiotics administered orally on the vaginal niche in the prevention of PTB has been suggested in several studies [185,186]. The rates of PTB differ through different countries, ranging from 5% to 9% in Europe to 13% in US [142]. Although PTB has a multifactorial etiology, it has been expected that approximately one-third of cases are due to intrauterine inflammation [142] triggered by migrant ascending vaginal infections. Remarkably, pre-existing BV give the impression to be intensely related with PTB [187]. Due to this, it has been hypothesized that probiotics could display the possible capacity to transfer and kill resident pathogens in a dysbiotic vagina. Mechanisms in which probiotics might be involved comprise the progress of anti-inflammatory cytokines and the decline of the vaginal pH favoring a vaginal environment that becomes suitable for the growth of healthy bacteria [186,188]. In addition, it has also suggested that, during pregnancy, probiotics might recover maternal glucosidic metabolism over the variation of gut microbial composition and function, as well as an insulin sensitivity improvement [189].

However, the latest published studies do not agree that probiotics have a significantly beneficial role during pregnancy. Some of them are chronologically summarized now. Gille et al. [190] examined the recognized character of oral probiotics on vaginal micro-environment in 320 pregnant women in a triple-blind RCT with oral probiotic supplementation or placebo. After eight weeks of treatment, oral probiotics did not rise the quantity of normal vaginal microbiota compared to placebo.

Subsequently to this work, Jarde et al. [185] have achieved a systematic review and meta-analysis about PTB risk and others unpleasant pregnancy outcomes in pregnant women receiving probiotics. Five studies (1017 women) examined the risk of preterm birth before 34 weeks of gestation, whereas in eleven studies (2484 women) the risk < 37 weeks. Conclusions from these highlighted that the use of probiotics during pregnancy neither decreased nor increased the PTB risk before 34 or before 37 weeks. In addition, it was not seen a protecting effect of probiotic administration over gestational diabetes, preterm premature rupture of membrane (PPROM), and small and large for gestational age infants. Conversely with these results, Daskalakis and Karambelas have previously shown some positive effects in women with PPROM after probiotic administration [191]. In their study, patients were distributed to receive vaginal probiotic in with antibiotic prophylaxis or standard antibiotic treatment alone for 10-days. Women that received the double regimen have higher mean gestational age at birth (35.49 vs. 32.53 weeks) and latency period (5.60 vs. 2.48 weeks) in comparison to control group, although the size sample in this study is questionable ($n = 59$ and $n = 57$, respectively).

Very recently, in a prospective study, Nordqvist et al. [192] evaluated the possible relationship among the probiotic milk consumption and the appearance of PTB and preeclampsia incidences. Maternal inflammatory response is a common background of these two pathologic conditions, and the potential anti-inflammatory effect of probiotics represents the criterion for their selection [193,194]. The study revealed that consumption of probiotic milk in late pregnancy was related with a preeclampsia-reduced risk. Regarding PTB, the probiotic milk ingestion of during early pregnancy was related with a decrease in the PTB risk. In both cases, no dose-response manner was found. Despite these promising results, in both cases, no relationship has been found between the dose applied and the obtained respond. Finally, the results from the studies of Haahr et al. [195] and Olsen et al. [196] do not support the probiotic treatment of BV-positive pregnant women with the objective of (i) diminishing the spontaneous PTB risk and, (ii) reducing the colonization rate of Group B Streptococcal (GBS) on the vagina.

In summary, from the aforementioned latest studies it appears that the use of probiotics during pregnancy neither decreased nor increased the risk of PTB before 34 or before 37 weeks. In a similar manner, no clear profits from the probiotic administration have emerged for PPROM, and for the gestational age of infants.

Other Results Obtained with Probiotics

In the success or failure of a probiotic therapy, a good selection of *Lactobacillus* species seems to be crucial. For instance, the putative beneficial effect as probiotic of *L. rhamnosus* BPL005 was recently proven in an in vitro model of bacterial colonization of primary endometrial epithelial cells with the presence of anaerobe microbes such as *A. vaginae*, *G. vaginalis*, *P. acnes*, and *S. agalactiae* [1]. When co-cultured with these pathogens, the *L. rhamnosus* BPL005 was capable at low pH and produced organic acids, producing a significant decrease in *P. acnes* and *S. agalactiae* levels, in contrast, *A. vaginae* and *G. vaginalis* strains were not affected for lactobacilli strain. Furthermore, it has been proven that the *L. rhamnosus* BPL005 colonization in the culture diminished IL-6, IL-8, MCP-1—increased in the existence of pathogens- and raised IL-1RA and IL-1β abundance [172].

3.5. Restoration of Vaginal Microbiota through Hormone Replacement Therapy (HRT)

Sex hormones, in particular estrogens, appear to have a significant importance in vaginal health, stimulating the growing of lactobacilli by encouraging glycogen accumulation in the vaginal mucosa [58,197]. In healthy pregnant women, high levels of estrogens contribute to the stability of the microbiota increasing the prevalence of *Lactobacillus* spp. [198]. On the other hand, during menstruation it has been reported a significant microbiota alteration, although this may depend on the type of community [34,40]. Following menopause, the deterioration in estrogen excretion might harmfully affect the vaginal mucosa, leading to vaginal atrophy and reduced glycogen levels that result in low abundance of vaginal lactobacilli. Thus, it has been shown that postmenopausal women who are not under hormonal treatment have significantly inferior free glycogen levels and lower levels and diversity of *Lactobacillus* spp., compared with those using hormonal treatment with higher levels of *Lactobacillus* spp. [199].

In one meta-analysis it was demonstrated that all routes of estrogen administration are effective for relief of menopausal symptoms, especially hot flashes [200]. Focusing in oral administration, one study examined the composition of microbiota of 19 postmenopausal women who were already taking oral estrogen therapy (Premarin-conjugated equine estrogen; CEE). After three months, results from the analysis of vaginal swabs revealed that all the patients were populated by *Lactobacillus* species, especially for *L. iners* and *L. crispatus* [201]. Supporting this, additional studies have found a minor presence of anaerobic bacteria in women under hormonal treatment compared to results from women without a replacement therapy and, equally to first evidence, all women on therapy had *Lactobacillus* existing species in their vagina [202,203]. Focusing in the treatment on symptoms like vaginal dryness and concurrent irritation, a study of women treated with CEE reported improvement subsequently with a treatment of three months, (placebo vs CEE treatment) [204]. Similarly, it has been demonstrated that women who use vaginal estrogen for symptoms of dyspareunia and vulvovaginal atrophy (VVA) score much higher on scales measuring quality of life and sexual health than those women who do not use a hormone replacement-based therapy [205].

3.6. Impact of Contraceptives on Vaginal Microbiota

Contraception methods may include the use of estrogen hormones (i.e., estradiol or ethynyl estradiol) or not by progestins, such as medroxyprogesterone acetate (MPA). Routes of administration can be oral, injectable (depot medroxyprogesterone acetace DMPA, or ethinyl estradiol (Net-EN) implants (levonorgestrel or etonogestrel) and intrauterine devices IUDs (such as cupper intrauterine devices).

Relative to BV treatment there is a stable relationship between the use of oral contraceptives and a reduction in BV prevalent [78,206,207]. Together with this latter, a recent meta-analysis has demonstrated a robust undesirable relationship between any hormonal contraception, regardless of type (excluding intrauterine devices), and prevalent, incident, or BV recurrent [208]. However, it has also been reported that certain kinds of hormonal contraceptives may alter vaginal microbiota in a negative manner. For instance, some studies have shown a reduction in prevalent BV in women who use injectable or implanted depot MPA [206]. However, it has also been observed that this contraceptive decreases vaginal Lactobacillus [78,209] and is associated in some studies with an augmented risk of acquisition and transmission of HIV possibly partly intermediated by effects of the microbiota on cervicovaginal inflammation [210]. Comparing the effects on vaginal microbiota from the use of oral contraceptives versus the use of intrauterine systems (IUS) Brooks et al. reported that women using oral contraceptives had a microbiota less colonized by BV-associated microorganisms, meanwhile in patients using levonorgestrel (LNG)-releasing intrauterine systems (IUS) microbiota was colonized by BV-associated microorganisms [211]. Conversely to this latter, Bassis et al. did not find changes in the microbiome consistent with BV in women using the LNG-IUS [212]. Finally, Achilles et al. [213] have recently reported that the use of hormonal contraceptives did not change vaginal microbiota in a period of 6 months, while the use of copper-IUD was related with an increase in the risk of BV and its associated microbiota, including *G. vaginalis* and *A. vaginae* bacteria.

Since contraceptive methods are used extensively by women worldwide, the development of refined research that better elucidates the impact on vaginal microbiota and risk of suffering from BV should be desired. Future research should be focused on precise factors such as the nature of the contraceptives alone or combined with, including a range of applied doses, improvement in routes of administration and extension in the duration of their application—all of them in well-designed controlled population groups of study to achieve more consistent applied results.

4. Conclusions

Microbial populations are essential for vaginal wellness. The advance in the characterization of the communities of microorganisms that inhabit the vagina has been extremely fast in recent years although important research gaps still remain unclear. For instance, it is significant to achieve a better understanding of the metabolic interactions between microbiota members and between them and the host. In this regard, multiple studies have begun to clarify the functionality of the microbiome [214] although up to now further evaluation about protein transcription of both microorganisms and the host is needed. This fact will contribute to filling gaps of information over the pathogenesis of interactions between dysbiosis, microorganisms, and the host that lead to adverse clinical consequences, plus to the evaluation of interventions that attempt to maintain or repair a healthy vaginal environment.

The impact of the diet on the composition of vaginal microbiota has also been considered. Being non-inherent in nature, the female population need to start thinking that lactobacilli-based microbiota is favored following healthy practices of alimentation. Summarized here, it has been reported that diets enriched in nutrients such as vitamins (A, C, D, E), B-carotene and minerals (such Ca and Zn) have been positively related with vaginal wellness, including a reduction in the prevalence of BV and HPV. Meanwhile, diets deficient in these nutrients and hence enriched in sugars (glycemic load) or fats (fatty acids) have negative consequences on homeostasis as well as being related with BV [75,165–170].

BV is the most frequent single infection of the lower reproductive tract. Since BV current cure rates range between 50% and 80% after treatment with metronidazole, recurrence being very common [215], more effective treatments are needed. The consequences of the alteration of the biological films—mainly colonized by anaerobic *G. vaginalis*—[216] and the benefits of the administration of probiotics [217] should be studied in more detail to achieve a better cure and prevention of recurrent infections, respectively. The primary aim of probiotics in obstetrics and gynecology is the restoration of a functional vaginal microbiome. However, given the inconclusive results for the use of probiotics, some international guidelines, such as the Centers for Disease Control and Prevention [218], do not support the use of any available lactobacilli-based formulations as probiotics as coadjuvant therapy in women with VVC and BV. Very surprisingly, the guideline for probiotics differs between countries, without a universal background [219]. Indeed, if probiotics are prescribed in the treatment with specific disorders, they should be regulated as drugs rather than foods or supplements. Under this formula, adverse consequences connected to the use of probiotics should be shared and registered by health authorities [220]. Nowadays, probiotic effects seem to be strain specific and dose dependent, and the lack of standardized manufacturing procedures affect multiple factors such as microbial survival, their growth, and their viability [220]. At the research level, active work in the field is needed and well-designed studies in the future should also focus on other aspects such as: (i) the efficacy and search of distinct mixtures of strains of probiotic species in the restoration of vaginal microbiota, (ii) a consensus in the duration of the treatment with probiotics and colony-forming units employed for restoration in launched studies, and (iii) a better understanding of the combination of antibiotics and probiotics when both are provided together [39].

Relative to risk factors associated with BV and other pathogenesis linked with dysbiosis, from now and in the immediate future, the performance of studies that focus on the impact of social sexual networks in the confcrmation and transmission of the vaginal microbiota and the prevalence of BV is significant. Given the importance of the structure of current social sexual networks for the transmission and prevalence of STDs [221], it is possible that these factors are similarly significant in the composition of the vaginal microbiota. For instance, it would be necessary to conduct further comprehensive longitudinal studies based in the consequences and effect of overlapping couples and how the duration of concurrent couples may have on the vaginal microbiota in distinct populations and cultures [46,74]. These studies could greatly contribute to explaining that racial differences are seen consistently in vaginal microbiota. In addition, other studies that focus on sexual habits, such as order of sexual acts and coital frequency might contribute to explaining variances in the composition of the vaginal microbiota and, in parallel, might facilitate relevant information to reduce risks of dysbiosis for women. The treatment of sexual companions of women with recurrent BV has not diminished recurrence in several RCTs, although this could be due to limitations of the study design and ineffective treatments [221] so profound research would be needed on the efficiency of the management of sexual partners. In addition, the mode of birth effect on the creation and maintenance of a healthy vaginal microbiome may be important an important research area since it has been shown that cesarean sections significantly affect the composition of the intestinal microbiome [222,223], and thus its possible influence on vaginal health.

Hormone replacement therapy-based studies outlined herein reported women having a vaginal microbiota dominated by *Lactobacillus* species, which corroborates that levels of estrogens have a profound effect on vaginal community and structural bacteria. Indeed, estrogens not only improve vaginal symptoms such as dryness and VVA but permit re-colonization of the postmenopausal vagina with lactobacilli and, hence, reduce the risk of BV and VVC among others. Hormone replacement therapy has also been correlated to improve sexual quality of life of postmenopausal women, perhaps linked to the aforementioned lactobacilli presence; however, there is a lack in holistic studies that correlate fluctuations in the vaginal microbiota directly to improved sexual wellness and quality of life [224]. Special care with the hormone replacement therapy (HRT) should be taken, cause not all the formulations works properly, as seen here in the case of medroxyprogesterone acetate [206] treatment with negative consequences for vaginal microbiota.

Contraception is a widely used practice in women worldwide and thus knowing how it impacts on microbiota is of great importance. To date, hormonal contraception seems to have more beneficial results over vaginal homeostasis and hence diminishing the risk of suffering from BV, by mean of favoring a lactobacilli-based microbiota as reported in [211]. Conversely, research about the use of IUs (i.e., levonorgestrel (LNG)-releasing intrauterine systems or Cooper intrauterine device, Cu-IUD) has revealed that even at mid-term (i.e., 180 days) abundance of anaerobic bacteria associated with BV increase and scoring higher in Nugent Gram stain [212,213]. However, frequently found in the literature are weak points of research related to the presence of contraceptives (if alone or combined with), the dose applied, questioning the routes of administration and the difficulties for the selection of controlled groups of population to perform very consistent results.

Author Contributions: Writing—original draft preparation, A.B.-D., A.F.-L., A.d.S.; writing—review, and editing, A.B.-D., F.G. and J.P.-D.; funding acquisition, F.G. All authors have read and agreed to the published version of the manuscript.

Funding: This research received no external funding.

Acknowledgments: Julio Plaza-Díaz is part of the "UGR Plan Propio de Investigación 2016" and the "Excellence actions: Unit of Excellence on Exercise and Health (UCEES), University of Granada".

Conflicts of Interest: The authors declare no conflict of interest.

References

1. Marchesi, J.R.; Ravel, J. The vocabulary of microbiome research: A proposal. *Microbiome* **2015**, *3*, 31. [CrossRef] [PubMed]
2. Human Microbiome Project Consortium. Structure, function and diversity of the healthy human microbiome. *Nature* **2012**, *486*, 207–214. [CrossRef] [PubMed]
3. Turnbaugh, P.J.; Ley, R.E.; Hamady, M.; Fraser-Liggett, C.M.; Knight, R.; Gordon, J.I. The human microbiome project. *Nature* **2007**, *449*, 804–810. [CrossRef] [PubMed]
4. *NIH Human Microbiome Project*; National Institutes of Health: Bethesda, MD, USA. Available online: https://commonfund.nih.gov/hmp (accessed on 12 July 2019).
5. Relman, D.A. The human microbiome and the future practice of medicine. *JAMA* **2015**, *314*, 1127–1128. [CrossRef]
6. Fox, C.; Eichelberger, K. Maternal microbiome and pregnancy outcomes. *Fertil. Steril.* **2015**, *104*, 1358–1363. [CrossRef]
7. Dethlefsen, L.; McFall-Ngai, M.; Relman, D.A. An ecological and evolutionary perspective on human-microbe mutualism and disease. *Nature* **2007**, *449*, 811–818. [CrossRef]
8. Donders, G.G.; Bosmans, E.; Dekeersmaecker, A.; Vereecken, A.; Van Bulck, B.; Spitz, B. Pathogenesis of abnormal vaginal bacterial flora. *Am. J. Obstet. Gynecol.* **2000**, *182*, 872–878. [CrossRef]
9. Gupta, K.; Stapleton, A.E.; Hooton, T.M.; Roberts, P.L.; Fennell, C.L.; Stamm, W.E. Inverse association of H2O2-producing lactobacilli and vaginal Escherichia coli colonization in women with recurrent urinary tract infections. *J. Infect. Dis.* **1998**, *178*, 446–450. [CrossRef]
10. Pybus, V.; Onderdonk, A.B. Microbial interactions in the vaginal ecosystem, with emphasis on the pathogenesis of bacterial vaginosis. *Microbes Infect.* **1999**, *1*, 285–292. [CrossRef]
11. Cherpes, T.L.; Meyn, L.A.; Krohn, M.A.; Lurie, J.G.; Hillier, S.L. Association between acquisition of herpes simplex virus type 2 in women and bacterial vaginosis. *Clin. Infect. Dis.* **2003**, *37*, 319–325. [CrossRef]
12. Martin, H.L.; Richardson, B.A.; Nyange, P.M.; Lavreys, L.; Hillier, S.L.; Chohan, B.; Mandaliya, K.; Ndinya-Achola, J.O.; Bwayo, J.; Kreiss, J. Vaginal lactobacilli, microbial flora, and risk of human immunodeficiency virus type 1 and sexually transmitted disease acquisition. *J. Infect. Dis.* **1999**, *180*, 1863–1868. [CrossRef] [PubMed]
13. Sobel, J.D. Is there a protective role for vaginal flora? *Curr. Infect. Dis. Rep.* **1999**, *1*, 379–383. [CrossRef] [PubMed]
14. Watts, D.H.; Fazzari, M.; Minkoff, H.; Hillier, S.L.; Sha, B.; Glesby, M.; Levine, A.M.; Burk, R.; Palefsky, J.M.; Moxley, M.; et al. Effects of bacterial vaginosis and other genital infections on the natural history of human papillomavirus infection in HIV-1-infected and high-risk HIV-1-uninfected women. *J. Infect. Dis.* **2005**, *191*, 1129–1139. [CrossRef] [PubMed]
15. Wiesenfeld, H.C.; Hillier, S.L.; Krohn, M.A.; Landers, D.V.; Sweet, R.L. Bacterial vaginosis is a strong predictor of Neisseria gonorrhoeae and Chlamydia trachomatis infection. *Clin. Infect. Dis.* **2003**, *36*, 663–668. [CrossRef] [PubMed]
16. Lai, S.K.; Hida, K.; Shukair, S.; Wang, Y.Y.; Figueiredo, A.; Cone, R.; Hope, T.J.; Hanes, J. Human immunodeficiency virus type 1 is trapped by acidic but not by neutralized human cervicovaginal mucus. *J. Virol.* **2009**, *83*, 11196–11200. [CrossRef]
17. Taha, T.E.; Hoover, D.R.; Dallabetta, G.A.; Kumwenda, N.I.; Mtimavalye, L.A.; Yang, L.P.; Liomba, G.N.; Broadhead, R.L.; Chiphangwi, J.D.; Miotti, P.G. Bacterial vaginosis and disturbances of vaginal flora: Association with increased acquisition of HIV. *AIDS* **1998**, *12*, 1699–1706. [CrossRef]
18. Ryckman, K.K.; Simhan, H.N.; Krohn, M.A.; Williams, S.M. Predicting risk of bacterial vaginosis: The role of race, smoking and corticotropin-releasing hormone-related genes. *Mol. Hum. Reprod.* **2009**, *15*, 131–137. [CrossRef]
19. Sobel, J.D. Bacterial vaginosis. *Annu. Rev. Med.* **2000**, *51*, 349–356. [CrossRef]
20. Brotman, R.M. Vaginal microbiome and sexually transmitted infections: An epidemiologic perspective. *J. Clin. Investig.* **2011**, *121*, 4610–4617. [CrossRef]
21. Zhou, X.; Brown, C.J.; Abdo, Z.; Davis, C.C.; Hansmann, M.A.; Joyce, P.; Foster, J.A.; Forney, L.J. Differences in the composition of vaginal microbial communities found in healthy Caucasian and black women. *ISME J.* **2007**, *1*, 121–133. [CrossRef]

22. Van de Wijgert, J.H.; Borgdorff, H.; Verhelst, R.; Crucitti, T.; Francis, S.; Verstraelen, H.; Jespers, V. The vaginal microbiota: What have we learned after a decade of molecular characterization? *PLoS ONE* **2014**, *9*, e105998. [CrossRef] [PubMed]
23. Fredricks, D.N.; Fiedler, T.L.; Marrazzo, J.M. Molecular identification of bacteria associated with bacterial vaginosis. *N. Engl. J. Med.* **2005**, *353*, 1899–1911. [CrossRef] [PubMed]
24. Ravel, J.; Gajer, P.; Abdo, Z.; Schneider, G.M.; Koenig, S.S.; McCulle, S.L.; Karlebach, S.; Gorle, R.; Russell, J.; Tacket, C.O.; et al. Vaginal microbiome of reproductive-age women. *Proc. Natl. Acad. Sci. USA* **2011**, *108* (Suppl. 1), 4680–4687. [CrossRef]
25. Jespers, V.; Menten, J.; Smet, H.; Poradosu, S.; Abdellati, S.; Verhelst, R.; Hardy, L.; Buve, A.; Crucitti, T. Quantification of bacterial species of the vaginal microbiome in different groups of women, using nucleic acid amplification tests. *BMC Microbiol.* **2012**, *12*, 83. [CrossRef] [PubMed]
26. Allsworth, J.E.; Peipert, J.F. Severity of bacterial vaginosis and the risk of sexually transmitted infection. *Am. J. Obstet. Gynecol.* **2011**, *205*, 113.e111–e116. [CrossRef] [PubMed]
27. Spear, G.T.; St John, E.; Zariffard, M.R. Bacterial vaginosis and human immunodeficiency virus infection. *AIDS Res. Ther.* **2007**, *4*, 25. [CrossRef]
28. Lamont, R.F.; Taylor-Robinson, D. The role of bacterial vaginosis, aerobic vaginitis, abnormal vaginal flora and the risk of preterm birth. *BJOG Int. J. Obstet. Gynaecol.* **2010**, *117*, 119–120. [CrossRef]
29. Larsson, P.G.; Platz-Christensen, J.J.; Dalaker, K.; Eriksson, K.; Fahraeus, L.; Irminger, K.; Jerve, F.; Stray-Pedersen, B.; Wolner-Hanssen, P. Treatment with 2% clindamycin vaginal cream prior to first trimester surgical abortion to reduce signs of postoperative infection: A prospective, double-blinded, placebo-controlled, multicenter study. *Acta Obstet. Gynecol. Scand.* **2000**, *79*, 390–396. [CrossRef]
30. Llahi-Camp, J.M.; Rai, R.; Ison, C.; Regan, L.; Taylor-Robinson, D. Association of bacterial vaginosis with a history of second trimester miscarriage. *Hum. Reprod.* **1996**, *11*, 1575–1578. [CrossRef]
31. Jacobsson, B.; Pernevi, P.; Chidekel, L.; Jorgen Platz-Christensen, J. Bacterial vaginosis in early pregnancy may predispose for preterm birth and postpartum endometritis. *Acta Obstet. Gynecol. Scand.* **2002**, *81*, 1006–1010. [CrossRef]
32. Ma, B.; Forney, L.J.; Ravel, J. Vaginal microbiome: Rethinking health and disease. *Annu. Rev. Microbiol.* **2012**, *66*, 371–389. [CrossRef] [PubMed]
33. Aldunate, M.; Srbinovski, D.; Hearps, A.C.; Latham, C.F.; Ramsland, P.A.; Gugasyan, R.; Cone, R.A.; Tachedjian, G. Antimicrobial and immune modulatory effects of lactic acid and short chain fatty acids produced by vaginal microbiota associated with eubiosis and bacterial vaginosis. *Front. Physiol.* **2015**, *6*, 164. [CrossRef] [PubMed]
34. Gajer, P.; Brotman, R.M.; Bai, G.; Sakamoto, J.; Schutte, U.M.; Zhong, X.; Koenig, S.S.; Fu, L.; Ma, Z.S.; Zhou, X.; et al. Temporal dynamics of the human vaginal microbiota. *Sci. Transl. Med.* **2012**, *4*, 132–152. [CrossRef] [PubMed]
35. Boris, S.; Suarez, J.E.; Vazquez, F.; Barbes, C. Adherence of human vaginal lactobacilli to vaginal epithelial cells and interaction with uropathogens. *Infect. Immun.* **1998**, *66*, 1985–1989. [CrossRef]
36. Martin, R.; Soberon, N.; Vazquez, F.; Suarez, J.E. Vaginal microbiota: Composition, protective role, associated pathologies, and therapeutic perspectives. *Enferm. Infecc. Microbiol. Clin.* **2008**, *26*, 160–167. [CrossRef]
37. Strus, M.; Brzychczy-Wloch, M.; Gosiewski, T.; Kochan, P.; Heczko, P.B. The in vitro effect of hydrogen peroxide on vaginal microbial communities. *FEMS Immunol. Med. Microbiol.* **2006**, *48*, 56–63. [CrossRef]
38. DiGiulio, D.B.; Callahan, B.J.; McMurdie, P.J.; Costello, E.K.; Lyell, D.J.; Robaczewska, A.; Sun, C.L.; Goltsman, D.S.; Wong, R.J.; Shaw, G.; et al. Temporal and spatial variation of the human microbiota during pregnancy. *Proc. Natl. Acad. Sci. USA* **2015**, *112*, 11060–11065. [CrossRef]
39. Bradshaw, C.S.; Brotman, R.M. Making inroads into improving treatment of bacterial vaginosis—Striving for long-term cure. *BMC Infect. Dis.* **2015**, *15*, 292. [CrossRef]
40. Srinivasan, S.; Liu, C.; Mitchell, C.M.; Fiedler, T.L.; Thomas, K.K.; Agnew, K.J.; Marrazzo, J.M.; Fredricks, D.N. Temporal variability of human vaginal bacteria and relationship with bacterial vaginosis. *PLoS ONE* **2010**, *5*, e10197. [CrossRef]
41. Pyles, R.B.; Vincent, K.L.; Baum, M.M.; Elsom, B.; Miller, A.L.; Maxwell, C.; Eaves-Pyles, T.D.; Li, G.; Popov, V.L.; Nusbaum, R.J.; et al. Cultivated vaginal microbiomes alter HIV-1 infection and antiretroviral efficacy in colonized epithelial multilayer cultures. *PLoS ONE* **2014**, *9*, e93419. [CrossRef]

42. Hanff, P.A.; Rosol-Donoghue, J.A.; Spiegel, C.A.; Wilson, K.H.; Moore, L.H. Leptotrichia sanguinegens sp. nov., a new agent of postpartum and neonatal bacteremia. *Clin. Infect. Dis.* **1995**, *20* (Suppl. 2), S237–S239. [CrossRef] [PubMed]
43. Geissdorfer, W.; Bohmer, C.; Pelz, K.; Schoerner, C.; Frobenius, W.; Bogdan, C. Tuboovarian abscess caused by Atopobium vaginae following transvaginal oocyte recovery. *J. Clin. Microbiol.* **2003**, *41*, 2788–2790. [CrossRef] [PubMed]
44. Yamagishi, Y.; Mikamo, H.; Tanaka, K.; Watanabe, K. A case of uterine endometritis caused by Atopobium vaginae. *J. Infect. Chemother.* **2011**, *17*, 119–121. [CrossRef] [PubMed]
45. Knoester, M.; Lashley, L.E.; Wessels, E.; Oepkes, D.; Kuijper, E.J. First report of Atopobium vaginae bacteremia with fetal loss after chorionic villus sampling. *J. Clin. Microbiol.* **2011**, *49*, 1684–1686. [CrossRef]
46. Adimora, A.A.; Schoenbach, V.J.; Doherty, I.A. Concurrent sexual partnerships among men in the United States. *Am. J. Public Health* **2007**, *97*, 2230–2237. [CrossRef]
47. Nagy, E.; Froman, G.; Mardh, P.A. Fibronectin binding of Lactobacillus species isolated from women with and without bacterial vaginosis. *J. Med. Microbiol.* **1992**, *37*, 38–42. [CrossRef]
48. Smith, S.B.; Ravel, J. The vaginal microbiota, host defence and reproductive physiology. *J. Physiol.* **2017**, *595*, 451–463. [CrossRef]
49. Boskey, E.R.; Cone, R.A.; Whaley, K.J.; Moench, T.R. Origins of vaginal acidity: High D/L lactate ratio is consistent with bacteria being the primary source. *Hum. Reprod.* **2001**, *16*, 1809–1813. [CrossRef]
50. Mossop, H.; Linhares, I.M.; Bongiovanni, A.M.; Ledger, W.J.; Witkin, S.S. Influence of lactic acid on endogenous and viral RNA-induced immune mediator production by vaginal epithelial cells. *Obstet. Gynecol.* **2011**, *118*, 840–846. [CrossRef]
51. Gartner, L.; Hiatt, J. *Atlas de Histología*; McGraw-Hill: New York, NY, USA, 2008.
52. Amabebe, E.; Anumba, D.O.C. The vaginal microenvironment: The physiologic role of lactobacilli. *Front. Med.* **2018**, *5*, 181. [CrossRef]
53. Lewis, F.M.; Bernstein, K.T.; Aral, S.O. Vaginal microbiome and its relationship to behavior, sexual health, and sexually transmitted diseases. *Obstet. Gynecol.* **2017**, *129*, 643–654. [CrossRef] [PubMed]
54. Borgogna, J.-L.C.; Yeoman, C.J. The application of molecular methods towards an understanding of the role of the vaginal microbiome in health and disease. In *Methods in Microbiology*; Elsevier: Amsterdam, The Netherlands, 2017; Volume 44, pp. 37–91.
55. Yamamoto, T.; Zhou, X.; Williams, C.J.; Hochwalt, A.; Forney, L.J. Bacterial populations in the vaginas of healthy adolescent women. *J. Pediatric Adolesc. Gynecol.* **2009**, *22*, 11–18. [CrossRef] [PubMed]
56. Hickey, R.J.; Zhou, X.; Pierson, J.D.; Ravel, J.; Forney, L.J. Understanding vaginal microbiome complexity from an ecological perspective. *Transl. Res. J. Lab. Clin. Med.* **2012**, *160*, 267–282. [CrossRef] [PubMed]
57. Hyman, R.W.; Herndon, C.N.; Jiang, H.; Palm, C.; Fukushima, M.; Bernstein, D.; Vo, K.C.; Zelenko, Z.; Davis, R.W.; Giudice, L.C. The dynamics of the vaginal microbiome during infertility therapy with in vitro fertilization-embryo transfer. *J. Assist. Reprod. Genet.* **2012**, *29*, 105–115. [CrossRef]
58. Brotman, R.M.; Ravel, J.; Bavoil, P.M.; Gravitt, P.E.; Ghanem, K.G. Microbiome, sex hormones, and immune responses in the reproductive tract: Challenges for vaccine development against sexually transmitted infections. *Vaccine* **2014**, *32*, 1543–1552. [CrossRef]
59. Srinivasan, S.; Hoffman, N.G.; Morgan, M.T.; Matsen, F.A.; Fiedler, T.L.; Hall, R.W.; Ross, F.J.; McCoy, C.O.; Bumgarner, R.; Marrazzo, J.M.; et al. Bacterial communities in women with bacterial vaginosis: High resolution phylogenetic analyses reveal relationships of microbiota to clinical criteria. *PLoS ONE* **2012**, *7*, e37818. [CrossRef]
60. Petrova, M.I.; van den Broek, M.; Balzarini, J.; Vanderleyden, J.; Lebeer, S. Vaginal microbiota and its role in HIV transmission and infection. *FEMS Microbiol. Rev.* **2013**, *37*, 762–792. [CrossRef]
61. Beghini, J.; Linhares, I.M.; Giraldo, P.C.; Ledger, W.J.; Witkin, S.S. Differential expression of lactic acid isomers, extracellular matrix metalloproteinase inducer, and matrix metalloproteinase-8 in vaginal fluid from women with vaginal disorders. *BJOG Int. J. Obstet. Gynaecol.* **2015**, *122*, 1580–1585. [CrossRef]
62. Verstraelen, H.; Verhelst, R.; Claeys, G.; De Backer, E.; Temmerman, M.; Vaneechoutte, M. Longitudinal analysis of the vaginal microflora in pregnancy suggests that L. crispatus promotes the stability of the normal vaginal microflora and that L. gasseri and/or L. iners are more conducive to the occurrence of abnormal vaginal microflora. *BMC Microbiol.* **2009**, *9*, 116. [CrossRef]

63. Aagaard, K.; Riehle, K.; Ma, J.; Segata, N.; Mistretta, T.A.; Coarfa, C.; Raza, S.; Rosenbaum, S.; Van den Veyver, I.; Milosavljevic, A.; et al. A metagenomic approach to characterization of the vaginal microbiome signature in pregnancy. *PLoS ONE* **2012**, *7*, e36466. [CrossRef]
64. Romero, R.; Hassan, S.S.; Gajer, P.; Tarca, A.L.; Fadrosh, D.W.; Bieda, J.; Chaemsaithong, P.; Miranda, J.; Chaiworapongsa, T.; Ravel, J. The vaginal microbiota of pregnant women who subsequently have spontaneous preterm labor and delivery and those with a normal delivery at term. *Microbiome* **2014**, *2*, 18. [CrossRef]
65. DiGiulio, D.B. Diversity of microbes in amniotic fluid. *Semin. Fetal Neonatal Med.* **2012**, *17*, 2–11. [CrossRef]
66. Bradley, F.; Birse, K.; Hasselrot, K.; Noel-Romas, L.; Introini, A.; Wefer, H.; Seifert, M.; Engstrand, L.; Tjernlund, A.; Broliden, K.; et al. The vaginal microbiome amplifies sex hormone-associated cyclic changes in cervicovaginal inflammation and epithelial barrier disruption. *Am. J. Reprod. Immunol.* **2018**, *80*, e12863. [CrossRef]
67. Brotman, R.; Gajer, P.; Holm, J.; Robinson, C.; Ma, B.; Humphrys, M.; Tuddenham, S.; Ravel, J.; Ghanem, K. 4: Hormonal contraception is associated with stability and lactobacillus-dominance of the vaginal microbiota in a two-year observational study. *Am. J. Obstet. Gynecol.* **2016**, *215*, S828–S829. [CrossRef]
68. Borgdorff, H.; Tsivtsivadze, E.; Verhelst, R.; Marzorati, M.; Jurriaans, S.; Ndayisaba, G.F.; Schuren, F.H.; van de Wijgert, J.H. Lactobacillus-dominated cervicovaginal microbiota associated with reduced HIV/STI prevalence and genital HIV viral load in African women. *ISME J.* **2014**, *8*, 1781–1793. [CrossRef]
69. Miller, E.A.; Beasley, D.E.; Dunn, R.R.; Archie, E.A. Lactobacilli dominance and vaginal pH: Why is the human vaginal microbiome unique? *Front. Microbiol.* **2016**, *7*, 1936. [CrossRef]
70. Brotman, R.M.; He, X.; Gajer, P.; Fadrosh, D.; Sharma, E.; Mongodin, E.F.; Ravel, J.; Glover, E.D.; Rath, J.M. Association between cigarette smoking and the vaginal microbiota: A pilot study. *BMC Infect. Dis.* **2014**, *14*, 471. [CrossRef]
71. Shiraishi, T.; Fukuda, K.; Morotomi, N.; Imamura, Y.; Mishima, J.; Imai, S.; Miyazawa, K.; Taniguchi, H. Influence of menstruation on the microbiota of healthy women's labia minora as analyzed using a 16S rRNA gene-based clone library method. *Jpn. J. Infect. Dis.* **2011**, *64*, 76–80.
72. Mitchell, C.M.; Fredricks, D.N.; Winer, R.L.; Koutsky, L. Effect of sexual debut on vaginal microbiota in a cohort of young women. *Obstet. Gynecol.* **2012**, *120*, 1306–1313. [CrossRef]
73. Romero, R.; Dey, S.K.; Fisher, S.J. Preterm labor: One syndrome, many causes. *Science* **2014**, *345*, 760–765. [CrossRef]
74. Kenyon, C.; Colebunders, R.; Crucitti, T. The global epidemiology of bacterial vaginosis: A systematic review. *Am. J. Obstet. Gynecol.* **2013**, *209*, 505–523. [CrossRef]
75. Neggers, Y.H.; Nansel, T.R.; Andrews, W.W.; Schwebke, J.R.; Yu, K.F.; Goldenberg, R.L.; Klebanoff, M.A. Dietary intake of selected nutrients affects bacterial vaginosis in women. *J. Nutr.* **2007**, *137*, 2128–2133. [CrossRef]
76. Hickey, R.J.; Abdo, Z.; Zhou, X.; Nemeth, K.; Hansmann, M.; Osborn, T.W., III; Wang, F.; Forney, L.J. Effects of tampons and menses on the composition and diversity of vaginal microbial communities over time. *BJOG Int. J. Obstet. Gynaecol.* **2013**, *120*, 695–704, discussion 704–696. [CrossRef]
77. Peipert, J.F.; Lapane, K.L.; Allsworth, J.E.; Redding, C.A.; Blume, J.D.; Stein, M.D. Bacterial vaginosis, race, and sexually transmitted infections: Does race modify the association? *Sex. Transm. Dis.* **2008**, *35*, 363–367. [CrossRef]
78. Koumans, E.H.; Sternberg, M.; Bruce, C.; McQuillan, G.; Kendrick, J.; Sutton, M.; Markowitz, L.E. The prevalence of bacterial vaginosis in the United States, 2001-2004; associations with symptoms, sexual behaviors, and reproductive health. *Sex. Transm. Dis.* **2007**, *34*, 864–869. [CrossRef]
79. Fettweis, J.M.; Brooks, J.P.; Serrano, M.G.; Sheth, N.U.; Girerd, P.H.; Edwards, D.J.; Strauss, J.F.; the Vaginal Microbiome, C.; Jefferson, K.K.; Buck, G.A. Differences in vaginal microbiome in African American women versus women of European ancestry. *Microbiology* **2014**, *160*, 2272–2282. [CrossRef]
80. Gautam, R.; Borgdorff, H.; Jespers, V.; Francis, S.C.; Verhelst, R.; Mwaura, M.; Delany-Moretlwe, S.; Ndayisaba, G.; Kyongo, J.K.; Hardy, L.; et al. Correlates of the molecular vaginal microbiota composition of African women. *BMC Infect. Dis.* **2015**, *15*, 86. [CrossRef]
81. Jespers, V.; van de Wijgert, J.; Cools, P.; Verhelst, R.; Verstraelen, H.; Delany-Moretlwe, S.; Mwaura, M.; Ndayisaba, G.F.; Mandaliya, K.; Menten, J.; et al. The significance of Lactobacillus crispatus and L. vaginalis for vaginal health and the negative effect of recent sex: A cross-sectional descriptive study across groups of African women. *BMC Infect. Dis.* **2015**, *15*, 115. [CrossRef]

82. Borgdorff, H.; van der Veer, C.; van Houdt, R.; Alberts, C.J.; de Vries, H.J.; Bruisten, S.M.; Snijder, M.B.; Prins, M.; Geerlings, S.E.; Schim van der Loeff, M.F.; et al. The association between ethnicity and vaginal microbiota composition in Amsterdam, the Netherlands. *PLoS ONE* **2017**, *12*, e0181135. [CrossRef]
83. Blekhman, R.; Goodrich, J.K.; Huang, K.; Sun, Q.; Bukowski, R.; Bell, J.T.; Spector, T.D.; Keinan, A.; Ley, R.E.; Gevers, D.; et al. Host genetic variation impacts microbiome composition across human body sites. *Genome Biol.* **2015**, *16*, 191. [CrossRef]
84. Onywera, H.; Williamson, A.L.; Mbulawa, Z.Z.A.; Coetzee, D.; Meiring, T.L. Factors associated with the composition and diversity of the cervical microbiota of reproductive-age Black South African women: A retrospective cross-sectional study. *PeerJ* **2019**, *7*, e7488. [CrossRef] [PubMed]
85. Cherpes, T.L.; Hillier, S.L.; Meyn, L.A.; Busch, J.L.; Krohn, M.A. A delicate balance: Risk factors for acquisition of bacterial vaginosis include sexual activity, absence of hydrogen peroxide-producing lactobacilli, black race, and positive herpes simplex virus type 2 serology. *Sex. Transm. Dis.* **2008**, *35*, 78–83. [CrossRef] [PubMed]
86. Nelson, T.M.; Borgogna, J.C.; Michalek, R.D.; Roberts, D.W.; Rath, J.M.; Glover, E.D.; Ravel, J.; Shardell, M.D.; Yeoman, C.J.; Brotman, R.M. Cigarette smoking is associated with an altered vaginal tract metabolomic profile. *Sci. Rep.* **2018**, *8*, 852. [CrossRef]
87. Chrousos, G.P. Regulation and dysregulation of the hypothalamic-pituitary-adrenal axis. The corticotropin-releasing hormone perspective. *Endocrinol. Metab. Clin. N. Am.* **1992**, *21*, 833–858. [CrossRef]
88. Wadhwa, P.D.; Culhane, J.F.; Rauh, V.; Barve, S.S. Stress and preterm birth: Neuroendocrine, immune/inflammatory, and vascular mechanisms. *Matern. Child Health J.* **2001**, *5*, 119–125. [CrossRef]
89. Wadhwa, P.D.; Culhane, J.F.; Rauh, V.; Barve, S.S.; Hogan, V.; Sandman, C.A.; Hobel, C.J.; Chicz-DeMet, A.; Dunkel-Schetter, C.; Garite, T.J.; et al. Stress, infection and preterm birth: A biobehavioural perspective. *Paediatr. Perinat. Epidemiol.* **2001**, *15* (Suppl. 2), 17–29. [CrossRef]
90. Amabebe, E.; Anumba, D.O.C. Psychosocial stress, cortisol levels, and maintenance of vaginal health. *Front. Endocrinol.* **2018**, *9*, 568. [CrossRef]
91. Fethers, K.A.; Fairley, C.K.; Hocking, J.S.; Gurrin, L.C.; Bradshaw, C.S. Sexual risk factors and bacterial vaginosis: A systematic review and meta-analysis. *Clin. Infect. Dis.* **2008**, *47*, 1426–1435. [CrossRef]
92. Jespers, V.; Crucitti, T.; Menten, J.; Verhelst, R.; Mwaura, M.; Mandaliya, K.; Ndayisaba, G.F.; Delany-Moretlwe, S.; Verstraelen, H.; Hardy, L.; et al. Prevalence and correlates of bacterial vaginosis in different sub-populations of women in sub-Saharan Africa: A cross-sectional study. *PLoS ONE* **2014**, *9*, e109670. [CrossRef]
93. Forcey, D.S.; Vodstrcil, L.A.; Hocking, J.S.; Fairley, C.K.; Law, M.; McNair, R.P.; Bradshaw, C.S. Factors associated with bacterial vaginosis among women who have sex with women: A systematic review. *PLoS ONE* **2015**, *10*, e0141905. [CrossRef]
94. Marrazzo, J.M.; Fiedler, T.L.; Srinivasan, S.; Thomas, K.K.; Liu, C.; Ko, D.; Xie, H.; Saracino, M.; Fredricks, D.N. Extravaginal reservoirs of vaginal bacteria as risk factors for incident bacterial vaginosis. *J. Infect. Dis.* **2012**, *205*, 1580–1588. [CrossRef] [PubMed]
95. Simon, C. Introduction: Do microbes in the female reproductive function matter? *Fertil. Steril.* **2018**, *110*, 325–326. [CrossRef] [PubMed]
96. Brotman, R.M.; Klebanoff, M.A.; Nansel, T.R.; Andrews, W.W.; Schwebke, J.R.; Zhang, J.; Yu, K.F.; Zenilman, J.M.; Scharfstein, D.O. A longitudinal study of vaginal douching and bacterial vaginosis—A marginal structural modeling analysis. *Am. J. Epidemiol.* **2008**, *168*, 188–196. [CrossRef] [PubMed]
97. Low, N.; Chersich, M.F.; Schmidlin, K.; Egger, M.; Francis, S.C.; van de Wijgert, J.H.; Hayes, R.J.; Baeten, J.M.; Brown, J.; Delany-Moretlwe, S.; et al. Intravaginal practices, bacterial vaginosis, and HIV infection in women: Individual participant data meta-analysis. *PLoS Med.* **2011**, *8*, e1000416. [CrossRef]
98. Fashemi, B.; Delaney, M.L.; Onderdonk, A.B.; Fichorova, R.N. Effects of feminine hygiene products on the vaginal mucosal biome. *Microb. Ecol. Health Dis.* **2013**, *24*. [CrossRef] [PubMed]
99. Hillier, S.L.M.J.; Holmes, K.K. Bacterial vaginosis. In *Sexually Transmitted Diseases*, 4th ed.; Holmes, K.K., Sparling, P.F., Mardh, P.A., Eds.; McGraw-Hill: New York, NY, USA, 2008; pp. 737–768.
100. Schwebke, J.R. Abnormal vaginal flora as a biological risk factor for acquisition of HIV infection and sexually transmitted diseases. *J. Infect. Dis.* **2005**, *192*, 1315–1317. [CrossRef] [PubMed]
101. St Amant, D.C.; Valentin-Bon, I.E.; Jerse, A.E. Inhibition of Neisseria gonorrhoeae by Lactobacillus species that are commonly isolated from the female genital tract. *Infect. Immun.* **2002**, *70*, 7169–7171. [CrossRef]

102. Kalyoussef, S.; Nieves, E.; Dinerman, E.; Carpenter, C.; Shankar, V.; Oh, J.; Burd, B.; Angeletti, R.H.; Buckheit, K.W.; Fredricks, D.N.; et al. Lactobacillus proteins are associated with the bactericidal activity against E. coli of female genital tract secretions. *PLoS ONE* **2012**, *7*, e49506. [CrossRef]
103. Brotman, R.M.; Klebanoff, M.A.; Nansel, T.R.; Yu, K.F.; Andrews, W.W.; Zhang, J.; Schwebke, J.R. Bacterial vaginosis assessed by gram stain and diminished colonization resistance to incident gonococcal, chlamydial, and trichomonal genital infection. *J. Infect. Dis.* **2010**, *202*, 1907–1915. [CrossRef]
104. Schwebke, J.R.; Desmond, R. A randomized trial of metronidazole in asymptomatic bacterial vaginosis to prevent the acquisition of sexually transmitted diseases. *Am. J. Obstet. Gynecol.* **2007**, *196*, 517.e511–e516. [CrossRef]
105. Schwebke, J.R.; Lee, J.Y.; Lensing, S.; Philip, S.S.; Wiesenfeld, H.C.; Sena, A.C.; Trainor, N.; Acevado, N.; Saylor, L.; Rompalo, A.M.: et al. Home screening for bacterial vaginosis to prevent sexually transmitted diseases. *Clin. Infect. Dis.* **2016**, *62*, 531–536. [CrossRef] [PubMed]
106. Sutton, M.; Sternberg, M.; Koumans, E.H.; McQuillan, G.; Berman, S.; Markowitz, L. The prevalence of Trichomonas vaginalis infection among reproductive-age women in the United States, 2001–2004. *Clin. Infect. Dis.* **2007**, *45*, 1319–1326. [CrossRef] [PubMed]
107. Mirmonsef, P.; Krass, L.; Landay, A.; Spear, G.T. The role of bacterial vaginosis and trichomonas in HIV transmission across the female genital tract. *Curr. HIV Res.* **2012**, *10*, 202–210. [CrossRef] [PubMed]
108. Fichorova, R.N.; Buck, O.R.; Yamamoto, H.S.; Fashemi, T.; Dawood, H.Y.; Fashemi, B.; Hayes, G.R.; Beach, D.H.; Takagi, Y.; Delaney, M.L.; et al. The villain team-up or how Trichomonas vaginalis and bacterial vaginosis alter innate immunity in concert. *Sex. Transm. Infect.* **2013**, *89*, 460–466. [CrossRef]
109. Balkus, J.E.; Richardson, B.A.; Rabe, L.K.; Taha, T.E.; Mgodi, N.; Kasaro, M.P.; Ramjee, G.; Hoffman, I.F.; Abdool Karim, S.S. Bacterial vaginosis and the risk of trichomonas vaginalis acquisition among HIV-1-negative women. *Sex. Transm. Dis.* **2014**, *41*, 123–128. [CrossRef]
110. Brotman, R.M.; Bradford, L.L.; Conrad, M.; Gajer, P.; Ault, K.; Peralta, L.; Forney, L.J.; Carlton, J.M.; Abdo, Z.; Ravel, J. Association between Trichomonas vaginalis and vaginal bacterial community composition among reproductive-age women. *Sex. Transm. Dis.* **2012**, *39*, 807–812. [CrossRef]
111. Fastring, D.R.; Amedee, A.; Gatski, M.; Clark, R.A.; Mena, L.A.; Levison, J.; Schmidt, N.; Rice, J.; Gustat, J.; Kissinger, P. Co-occurrence of Trichomonas vaginalis and bacterial vaginosis and vaginal shedding of HIV-1 RNA. *Sex. Transm. Dis.* **2014**, *41*, 173–179. [CrossRef]
112. Nagot, N.; Ouedraogo, A.; Defer, M.C.; Vallo, R.; Mayaud, P.; Van de Perre, P. Association between bacterial vaginosis and Herpes simplex virus type-2 infection: Implications for HIV acquisition studies. *Sex. Transm. Infect.* **2007**, *83*, 365–368. [CrossRef]
113. Allsworth, J.E.; Lewis, V.A.; Peipert, J.F. Viral sexually transmitted infections and bacterial vaginosis: 2001–2004 National health and nutrition examination survey data. *Sex. Transm. Dis.* **2008**, *35*, 791–796. [CrossRef]
114. Esber, A.; Vicetti Miguel, R.D.; Cherpes, T.L.; Klebanoff, M.A.; Gallo, M.F.; Turner, A.N. Risk of bacterial vaginosis among women with herpes simplex virus type 2 infection: A systematic review and meta-analysis. *J. Infect. Dis.* **2015**, *212*, 8–17. [CrossRef]
115. Oh, J.E.; Kim, B.C.; Chang, D.H.; Kwon, M.; Lee, S.Y.; Kang, D.; Kim, J.Y.; Hwang, I.; Yu, J.W.; Nakae, S.; et al. Dysbiosis-induced IL-33 contributes to impaired antiviral immunity in the genital mucosa. *Proc. Natl. Acad. Sci. USA* **2016**, *113*, E762–E771. [CrossRef] [PubMed]
116. Abbai, N.S.; Nyirenda, M.; Naidoo, S.; Ramjee, G. Prevalent herpes simplex virus-2 increases the risk of incident bacterial vaginosis in women from South Africa. *AIDS Behav.* **2018**, *22*, 2172–2180. [CrossRef] [PubMed]
117. King, C.C.; Jamieson, D.J.; Wiener, J.; Cu-Uvin, S.; Klein, R.S.; Rompalo, A.M.; Shah, K.V.; Sobel, J.D. Bacterial vaginosis and the natural history of human papillomavirus. *Infect. Dis. Obstet. Gynecol.* **2011**, *2011*, 319460. [CrossRef] [PubMed]
118. Lee, J.E.; Lee, S.; Lee, H.; Song, Y.M.; Lee, K.; Han, M.J.; Sung, J.; Ko, G. Association of the vaginal microbiota with human papillomavirus infection in a Korean twin cohort. *PLoS ONE* **2013**, *8*, e63514. [CrossRef] [PubMed]
119. Brotman, R.M.; Shardell, M.D.; Gajer, P.; Tracy, J.K.; Zenilman, J.M.; Ravel, J.; Gravitt, P.E. Interplay between the temporal dynamics of the vaginal microbiota and human papillomavirus detection. *J. Infect. Dis.* **2014**, *210*, 1723–1733. [CrossRef]

120. Mitra, A.; MacIntyre, D.A.; Lee, Y.S.; Smith, A.; Marchesi, J.R.; Lehne, B.; Bhatia, R.; Lyons, D.; Paraskevaidis, E.; Li, J.V.; et al. Cervical intraepithelial neoplasia disease progression is associated with increased vaginal microbiome diversity. *Sci. Rep.* **2015**, *5*, 16865. [CrossRef]
121. Dahoud, W.; Michael, C.W.; Gokozan, H.; Nakanishi, A.K.; Harbhajanka, A. Association of bacterial vaginosis and human papilloma virus infection with cervical squamous intraepithelial lesions. *Am. J. Clin. Pathol.* **2019**, *152*, 185–189. [CrossRef]
122. Sanchez-Garcia, E.K.; Contreras-Paredes, A.; Martinez-Abundis, E.; Garcia-Chan, D.; Lizano, M.; de la Cruz-Hernandez, E. Molecular epidemiology of bacterial vaginosis and its association with genital micro-organisms in asymptomatic women. *J. Med. Microbiol.* **2019**, *68*, 1373–1382. [CrossRef]
123. Atashili, J.; Poole, C.; Ndumbe, P.M.; Adimora, A.A.; Smith, J.S. Bacterial vaginosis and HIV acquisition: A meta-analysis of published studies. *AIDS* **2008**, *22*, 1493–1501. [CrossRef]
124. Nunn, K.L.; Wang, Y.Y.; Harit, D.; Humphrys, M.S.; Ma, B.; Cone, R.; Ravel, J.; Lai, S.K. Enhanced trapping of HIV-1 by human cervicovaginal mucus is associated with lactobacillus crispatus-dominant microbiota. *mBio* **2015**, *6*. [CrossRef]
125. Cone, R.A. Vaginal microbiota and sexually transmitted infections that may influence transmission of cell-associated HIV. *J. Infect. Dis.* **2014**, *210* (Suppl. 3), S616–S621. [CrossRef]
126. Sabo, M.C.; Lehman, D.A.; Wang, B.; Richardson, B.A.; Srinivasan, S.; Osborn, L.; Matemo, D.; Kinuthia, J.; Fiedler, T.L.; Munch, M.M.; et al. Associations between vaginal bacteria implicated in HIV acquisition risk and proinflammatory cytokines and chemokines. *Sex. Transm. Infect.* **2019**. [CrossRef] [PubMed]
127. Westrom, L.; Joesoef, R.; Reynolds, G.; Hagdu, A.; Thompson, S.E. Pelvic inflammatory disease and fertility. A cohort study of 1844 women with laparoscopically verified disease and 657 control women with normal laparoscopic results. *Sex. Transm. Dis.* **1992**, *19*, 185–192. [CrossRef]
128. Ness, R.B.; Soper, D.E.; Holley, R.L.; Peipert, J.; Randall, H.; Sweet, R.L.; Sondheimer, S.J.; Hendrix, S.L.; Amortegui, A.; Trucco, G.; et al. Effectiveness of inpatient and outpatient treatment strategies for women with pelvic inflammatory disease: Results from the Pelvic Inflammatory Disease Evaluation and Clinical Health (PEACH) randomized trial. *Am. J. Obstet. Gynecol.* **2002**, *186*, 929–937. [CrossRef] [PubMed]
129. Kiviat, N.B.; Wolner-Hanssen, P.; Eschenbach, D.A.; Wasserheit, J.N.; Paavonen, J.A.; Bell, T.A.; Critchlow, C.W.; Stamm, W.E.; Moore, D.E.; Holmes, K.K. Endometrial histopathology in patients with culture-proved upper genital tract infection and laparoscopically diagnosed acute salpingitis. *Am. J. Surg. Pathol.* **1990**, *14*, 167–175. [CrossRef]
130. Haggerty, C.L.; Totten, P.A.; Astete, S.G.; Lee, S.; Hoferka, S.L.; Kelsey, S.F.; Ness, R.B. Failure of cefoxitin and doxycycline to eradicate endometrial Mycoplasma genitalium and the consequence for clinical cure of pelvic inflammatory disease. *Sex. Transm. Infect.* **2008**, *84*, 338–342. [CrossRef] [PubMed]
131. Cohen, C.R.; Manhart, L.E.; Bukusi, E.A.; Astete, S.; Brunham, R.C.; Holmes, K.K.; Sinei, S.K.; Bwayo, J.J.; Totten, P.A. Association between Mycoplasma genitalium and acute endometritis. *Lancet* **2002**, *359*, 765–766. [CrossRef]
132. Soper, D.E.; Brockwell, N.J.; Dalton, H.P.; Johnson, D. Observations concerning the microbial etiology of acute salpingitis. *Am. J. Obstet. Gynecol.* **1994**, *170*, 1008–1014. [CrossRef]
133. Jossens, M.O.; Schachter, J.; Sweet, R.L. Risk factors associated with pelvic inflammatory disease of differing microbial etiologies. *Obstet. Gynecol.* **1994**, *83*, 989–997. [CrossRef]
134. Haggerty, C.L.; Hillier, S.L.; Bass, D.C.; Ness, R.B.; PID Evaluation and Clinical Health Study Investigators. Bacterial vaginosis and anaerobic bacteria are associated with endometritis. *Clin. Infect. Dis.* **2004**, *39*, 990–995. [CrossRef]
135. Ness, R.B.; Kip, K.E.; Hillier, S.L.; Soper, D.E.; Stamm, C.A.; Sweet, R.L.; Rice, P.; Richter, H.E. A cluster analysis of bacterial vaginosis-associated microflora and pelvic inflammatory disease. *Am. J. Epidemiol.* **2005**, *162*, 585–590. [CrossRef] [PubMed]
136. Gundi, V.A.; Desbriere, R.; La Scola, B. Leptotrichia amnionii and the female reproductive tract. *Emerg. Infect. Dis.* **2004**, *10*, 2056–2057. [CrossRef] [PubMed]
137. Gardella, C.; Riley, D.E.; Hitti, J.; Agnew, K.; Krieger, J.N.; Eschenbach, D. Identification and sequencing of bacterial rDNAs in culture-negative amniotic fluid from women in premature labor. *Am. J. Perinatol.* **2004**, *21*, 319–323. [CrossRef] [PubMed]
138. Hebb, J.K.; Cohen, C.R.; Astete, S.G.; Bukusi, E.A.; Totten, P.A. Detection of novel organisms associated with salpingitis, by use of 16S rDNA polymerase chain reaction. *J. Infect. Dis.* **2004**, *190*, 2109–2120. [CrossRef]

139. Haggerty, C.L.; Totten, P.A.; Tang, G.; Astete, S.G.; Ferris, M.J.; Norori, J.; Bass, D.C.; Martin, D.H.; Taylor, B.D.; Ness, R.B. Identification of novel microbes associated with pelvic inflammatory disease and infertility. *Sex. Transm. Infect.* **2016**, *92*, 441–446. [CrossRef]
140. Taylor, B.D.; Totten, P.A.; Astete, S.G.; Ferris, M.J.; Martin, D.H.; Ness, R.B.; Haggerty, C.L. Toll-like receptor variants and cervical Atopobium vaginae infection in women with pelvic inflammatory disease. *Am. J. Reprod. Immunol.* **2018**, *79*. [CrossRef]
141. Behrman, R.E.; Butler, A.S. (Eds.) *Preterm Birth: Causes, Consequences, and Prevention*; National Academies Press: Washington, DC, USA, 2007. [CrossRef]
142. Goldenberg, R.L.; Culhane, J.F.; Iams, J.D.; Romero, R. Epidemiology and causes of preterm birth. *Lancet* **2008**, *371*, 75–84. [CrossRef]
143. Romero, R.; Mazor, M.; Munoz, H.; Gomez, R.; Galasso, M.; Sherer, D.M. The preterm labor syndrome. *Ann. N. Y. Acad. Sci.* **1994**, *734*, 414–429. [CrossRef]
144. Goncalves, L.F.; Chaiworapongsa, T.; Romero, R. Intrauterine infection and prematurity. *Ment. Retard. Dev. Disabil. Res. Rev.* **2002**, *8*, 3–13. [CrossRef]
145. Goldenberg, R.L.; Culhane, J.F. Infection as a cause of preterm birth. *Clin. Perinatol.* **2003**, *30*, 677–700. [CrossRef]
146. Wen, S.W.; Smith, G.; Yang, Q.; Walker, M. Epidemiology of preterm birth and neonatal outcome. *Semin. Fetal Neonatal Med.* **2004**, *9*, 429–435. [CrossRef] [PubMed]
147. Donati, L.; Di Vico, A.; Nucci, M.; Quagliozzi, L.; Spagnuolo, T.; Labianca, A.; Bracaglia, M.; Ianniello, F.; Caruso, A.; Paradisi, G. Vaginal microbial flora and outcome of pregnancy. *Arch. Gynecol. Obstet.* **2010**, *281*, 589–600. [CrossRef] [PubMed]
148. Nelson, D.B.; Hanlon, A.L.; Wu, G.; Liu, C.; Fredricks, D.N. First trimester levels of BV-associated bacteria and risk of miscarriage among women early in pregnancy. *Matern. Child Health J.* **2015**, *19*, 2682–2687. [CrossRef] [PubMed]
149. Hay, P.E.; Lamont, R.F.; Taylor-Robinson, D.; Morgan, D.J.; Ison, C.; Pearson, J. Abnormal bacterial colonisation of the genital tract and subsequent preterm delivery and late miscarriage. *BMJ* **1994**, *308*, 295–298. [CrossRef] [PubMed]
150. Kurki, T.; Sivonen, A.; Renkonen, O.V.; Savia, E.; Ylikorkala, O. Bacterial vaginosis in early pregnancy and pregnancy outcome. *Obstet. Gynecol.* **1992**, *80*, 173–177. [CrossRef]
151. Lamont, R.F.; Nhan-Chang, C.L.; Sobel, J.D.; Workowski, K.; Conde-Agudelo, A.; Romero, R. Treatment of abnormal vaginal flora in early pregnancy with clindamycin for the prevention of spontaneous preterm birth: A systematic review and metaanalysis. *Am. J. Obstet. Gynecol.* **2011**, *205*, 177–190. [CrossRef]
152. Lamont, R.F. Advances in the prevention of infection-related preterm birth. *Front. Immunol.* **2015**, *6*, 566. [CrossRef]
153. Lamont, R.F.; Sobel, J.D.; Akins, R.A.; Hassan, S.S.; Chaiworapongsa, T.; Kusanovic, J.P.; Romero, R. The vaginal microbiome: New information about genital tract flora using molecular based techniques. *BJOG Int. J. Obstet. Gynaecol.* **2011**, *118*, 533–549. [CrossRef]
154. Callahan, B.J.; DiGiulio, D.B.; Goltsman, D.S.A.; Sun, C.L.; Costello, E.K.; Jeganathan, P.; Biggio, J.R.; Wong, R.J.; Druzin, M.L.; Shaw, G.M.; et al. Replication and refinement of a vaginal microbial signature of preterm birth in two racially distinct cohorts of US women. *Proc. Natl. Acad. Sci. USA* **2017**, *114*, 9966–9971. [CrossRef]
155. Hyman, R.W.; Fukushima, M.; Jiang, H.; Fung, E.; Rand, L.; Johnson, B.; Vo, K.C.; Caughey, A.B.; Hilton, J.F.; Davis, R.W.; et al. Diversity of the vaginal microbiome correlates with preterm birth. *Reprod. Sci.* **2014**, *21*, 32–40. [CrossRef]
156. Nelson, D.B.; Shin, H.; Wu, J.; Dominguez-Bello, M.G. The gestational vaginal microbiome and spontaneous preterm birth among nulliparous African American women. *Am. J. Perinatol.* **2016**, *33*, 887–893. [CrossRef] [PubMed]
157. Stafford, G.P.; Parker, J.L.; Amabebe, E.; Kistler, J.; Reynolds, S.; Stern, V.; Paley, M.; Anumba, D.O.C. Spontaneous preterm birth is associated with differential expression of vaginal metabolites by lactobacilli-dominated microflora. *Front. Physiol.* **2017**, *8*, 615. [CrossRef] [PubMed]
158. Stout, M.J.; Zhou, Y.; Wylie, K.M.; Tarr, P.I.; Macones, G.A.; Tuuli, M.G. Early pregnancy vaginal microbiome trends and preterm birth. *Am. J. Obstet. Gynecol.* **2017**, *217*, 356.e351–e355. [CrossRef] [PubMed]

159. Farr, A.; Kiss, H.; Hagmann, M.; Machal, S.; Holzer, I.; Kueronya, V.; Husslein, P.W.; Petricevic, L. Role of lactobacillus species in the intermediate vaginal flora in early pregnancy: A retrospective cohort study. *PLoS ONE* **2015**, *10*, e0144181. [CrossRef] [PubMed]
160. Kindinger, L.M.; Bennett, P.R.; Lee, Y.S.; Marchesi, J.R.; Smith, A.; Cacciatore, S.; Holmes, E.; Nicholson, J.K.; Teoh, T.G.; MacIntyre, D.A. The interaction between vaginal microbiota, cervical length, and vaginal progesterone treatment for preterm birth risk. *Microbiome* **2017**, *5*, 6. [CrossRef] [PubMed]
161. Tachedjian, G.; Aldunate, M.; Bradshaw, C.S.; Cone, R.A. The role of lactic acid production by probiotic Lactobacillus species in vaginal health. *Res. Microbiol.* **2017**, *168*, 782–792. [CrossRef]
162. Lamont, R.F.; Taylor-Robinson, D.; Wigglesworth, J.S.; Furr, P.M.; Evans, R.T.; Elder, M.G. The role of mycoplasmas, ureaplasmas and chlamydiae in the genital tract of women presenting in spontaneous early preterm labour. *J. Med. Microbiol.* **1987**, *24*, 253–257. [CrossRef]
163. Taylor-Robinson, D.; Lamont, R.F. Mycoplasmas in pregnancy. *BJOG Int. J. Obstet. Gynaecol.* **2011**, *118*, 164–174. [CrossRef]
164. Marchesi, J.R.; Adams, D.H.; Fava, F.; Hermes, G.D.; Hirschfield, G.M.; Hold, G.; Quraishi, M.N.; Kinross, J.; Smidt, H.; Tuohy, K.M.; et al. The gut microbiota and host health: A new clinical frontier. *Gut* **2016**, *65*, 330–339. [CrossRef]
165. Verstraelen, H.; Delanghe, J.; Roelens, K.; Blot, S.; Claeys, G.; Temmerman, M. Subclinical iron deficiency is a strong predictor of bacterial vaginosis in early pregnancy. *BMC Infect. Dis.* **2005**, *5*, 55. [CrossRef]
166. Tohill, B.C.; Heilig, C.M.; Klein, R.S.; Rompalo, A.; Cu-Uvin, S.; Piwoz, E.G.; Jamieson, D.J.; Duerr, A. Nutritional biomarkers associated with gynecological conditions among US women with or at risk of HIV infection. *Am. J. Clin. Nutr.* **2007**, *85*, 1327–1334. [CrossRef] [PubMed]
167. Bodnar, L.M.; Krohn, M.A.; Simhan, H.N. Maternal vitamin D deficiency is associated with bacterial vaginosis in the first trimester of pregnancy. *J. Nutr.* **2009**, *139*, 1157–1161. [CrossRef] [PubMed]
168. Klebanoff, M.A.; Turner, A.N. Bacterial vaginosis and season, a proxy for vitamin D status. *Sex. Transm. Dis.* **2014**, *41*, 295–299. [CrossRef] [PubMed]
169. Akoh, C.C.; Pressman, E.K.; Cooper, E.; Queenan, R.A.; Pillittere, J.; O'Brien, K.O. Low vitamin D is associated with infections and proinflammatory cytokines during pregnancy. *Reprod. Sci.* **2018**, *25*, 414–423. [CrossRef]
170. Thoma, M.E.; Klebanoff, M.A.; Rovner, A.J.; Nansel, T.R.; Neggers, Y.; Andrews, W.W.; Schwebke, J.R. Bacterial vaginosis is associated with variation in dietary indices. *J. Nutr.* **2011**, *141*, 1698–1704. [CrossRef]
171. Reid, G. Probiotic lactobacilli for urogenital health in women. *J. Clin. Gastroenterol.* **2008**, *42* (Suppl. 3), S234–S236. [CrossRef]
172. Chenoll, E.; Moreno, I.; Sanchez, M.; Garcia-Grau, I.; Silva, A.; Gonzalez-Monfort, M.; Genoves, S.; Vilella, F.; Seco-Durban, C.; Simon, C.; et al. Selection of new probiotics for endometrial health. *Front. Cell. Infect. Microbiol.* **2019**, *9*, 114. [CrossRef]
173. Machado, D.; Castro, J.; Palmeira-de-Oliveira, A.; Martinez-de-Oliveira, J.; Cerca, N. Bacterial vaginosis biofilms: Challenges to current therapies and emerging solutions. *Front. Microbiol.* **2015**, *6*, 1528. [CrossRef]
174. Chetwin, E.; Manhanzva, M.T.; Abrahams, A.G.; Froissart, R.; Gamieldien, H.; Jaspan, H.; Jaumdally, S.Z.; Barnabas, S.L.; Dabee, S.; Happel, A.U.; et al. Antimicrobial and inflammatory properties of South African clinical Lactobacillus isolates and vaginal probiotics. *Sci. Rep.* **2019**, *9*, 1917. [CrossRef]
175. Al-Ghazzewi, F.H.; Tester, R.F. Biotherapeutic agents and vaginal health. *J. Appl. Microbiol.* **2016**, *121*, 18–27. [CrossRef]
176. Buggio, L.; Somigliana, E.; Borghi, A.; Vercellini, P. Probiotics and vaginal microecology: Fact or fancy? *BMC Women's Health* **2019**, *19*, 25. [CrossRef] [PubMed]
177. Tan, H.; Fu, Y.; Yang, C.; Ma, J. Effects of metronidazole combined probiotics over metronidazole alone for the treatment of bacterial vaginosis: A meta-analysis of randomized clinical trials. *Arch. Gynecol. Obstet.* **2017**, *295*, 1331–1339. [CrossRef] [PubMed]
178. Oduyebo, O.O.; Anorlu, R.I.; Ogunsola, F.T. The effects of antimicrobial therapy on bacterial vaginosis in non-pregnant women. *Cochrane Database Syst. Rev.* **2009**. [CrossRef] [PubMed]
179. Huang, H.; Song, L.; Zhao, W. Effects of probiotics for the treatment of bacterial vaginosis in adult women: A meta-analysis of randomized clinical trials. *Arch. Gynecol. Obstet.* **2014**, *289*, 1225–1234. [CrossRef] [PubMed]
180. Xie, H.Y.; Feng, D.; Wei, D.M.; Mei, L.; Chen, H.; Wang, X.; Fang, F. Probiotics for vulvovaginal candidiasis in non-pregnant women. *Cochrane Database Syst. Rev.* **2017**, *11*, CD010496. [CrossRef]
181. Sobel, J.D. Vulvovaginal candidosis. *Lancet* **2007**, *369*, 1961–1971. [CrossRef]

182. Sobel, J.D. Recurrent vulvovaginal candidiasis. *Am. J. Obstet. Gynecol.* **2016**, *214*, 15–21. [CrossRef]
183. Blostein, F.; Levin-Sparenberg, E.; Wagner, J.; Foxman, B. Recurrent vulvovaginal candidiasis. *Ann. Epidemiol.* **2017**, *27*, 575–582. [CrossRef]
184. De Seta, F.; Parazzini, F.; De Leo, R.; Banco, R.; Maso, G.P.; De Santo, D.; Sartore, A.; Stabile, G.; Inglese, S.; Tonon, M.; et al. Lactobacillus plantarum P17630 for preventing Candida vaginitis recurrence: A retrospective comparative study. *Eur. J. Obstet. Gynecol. Reprod. Biol.* **2014**, *182*, 136–139. [CrossRef]
185. Jarde, A.; Lewis-Mikhael, A.M.; Moayyedi, P.; Stearns, J.C.; Collins, S.M.; Beyene, J.; McDonald, S.D. Pregnancy outcomes in women taking probiotics or prebiotics: A systematic review and meta-analysis. *BMC Pregnancy Childbirth* **2018**, *18*, 14. [CrossRef]
186. Reid, G.; Bocking, A. The potential for probiotics to prevent bacterial vaginosis and preterm labor. *Am. J. Obstet. Gynecol.* **2003**, *189*, 1202–1208. [CrossRef]
187. Kirihara, N.; Kamitomo, M.; Tabira, T.; Hashimoto, T.; Taniguchi, H.; Maeda, T. Effect of probiotics on perinatal outcome in patients at high risk of preterm birth. *J. Obstet. Gynaecol. Res.* **2018**, *44*, 241–247. [CrossRef] [PubMed]
188. Reid, G.; Younes, J.A.; Van der Mei, H.C.; Gloor, G.B.; Knight, R.; Busscher, H.J. Microbiota restoration: Natural and supplemented recovery of human microbial communities. *Nat. Rev. Microbiol.* **2011**, *9*, 27–38. [CrossRef] [PubMed]
189. Lindsay, K.L.; Brennan, L.; Kennelly, M.A.; Maguire, O.C.; Smith, T.; Curran, S.; Coffey, M.; Foley, M.E.; Hatunic, M.; Shanahan, F.; et al. Impact of probiotics in women with gestational diabetes mellitus on metabolic health: A randomized controlled trial. *Am. J. Obstet. Gynecol.* **2015**, *212*, 496. [CrossRef] [PubMed]
190. Gille, C.; Boer, B.; Marschal, M.; Urschitz, M.S.; Heinecke, V.; Hund, V.; Speidel, S.; Tarnow, I.; Mylonas, I.; Franz, A.; et al. Effect of probiotics on vaginal health in pregnancy. EFFPRO, a randomized controlled trial. *Am. J. Obstet. Gynecol.* **2016**, *215*, e601–e608. [CrossRef]
191. Daskalakis, G.J.; Karambelas, A.K. Vaginal probiotic administration in the management of preterm premature rupture of membranes. *Fetal Diagn. Ther.* **2017**, *42*, 92–98. [CrossRef]
192. Nordqvist, M.; Jacobsson, B.; Brantsaeter, A.L.; Myhre, R.; Nilsson, S.; Sengpiel, V. Timing of probiotic milk consumption during pregnancy and effects on the incidence of preeclampsia and preterm delivery: A prospective observational cohort study in Norway. *BMJ Open* **2018**, *8*, e018021. [CrossRef]
193. Yeganegi, M.; Watson, C.S.; Martins, A.; Kim, S.O.; Reid, G.; Challis, J.R.; Bocking, A.D. Effect of Lactobacillus rhamnosus GR-1 supernatant and fetal sex on lipopolysaccharide-induced cytokine and prostaglandin-regulating enzymes in human placental trophoblast cells: Implications for treatment of bacterial vaginosis and prevention of preterm labor. *Am. J. Obstet. Gynecol.* **2009**, *200*, 532. [CrossRef]
194. Bloise, E.; Torricelli, M.; Novembri, R.; Borges, L.E.; Carrarelli, P.; Reis, F.M.; Petraglia, F. Heat-killed lactobacillus rhamnosus GG modulates urocortin and cytokine release in primary trophoblast cells. *Placenta* **2010**, *31*, 867–872. [CrossRef]
195. Haahr, T.; Ersboll, A.S.; Karlsen, M.A.; Svare, J.; Sneider, K.; Hee, L.; Weile, L.K.; Ziobrowska-Bech, A.; Ostergaard, C.; Jensen, J.S.; et al. Treatment of bacterial vaginosis in pregnancy in order to reduce the risk of spontaneous preterm delivery—A clinical recommendation. *Acta Obstet. Gynecol. Scand.* **2016**, *95*, 850–860. [CrossRef]
196. Olsen, P.; Williamson, M.; Traynor, V.; Georgiou, C. The impact of oral probiotics on vaginal Group B streptococcal colonisation rates in pregnant women: A pilot randomised control study. *Women Birth. J. Aust. Coll. Midwives* **2018**, *31*, 31–37. [CrossRef] [PubMed]
197. Mirmonsef, P.; Hotton, A.L.; Gilbert, D.; Burgad, D.; Landay, A.; Weber, K.M.; Cohen, M.; Ravel, J.; Spear, G.T. Free glycogen in vaginal fluids is associated with Lactobacillus colonization and low vaginal pH. *PLoS ONE* **2014**, *9*, e102467. [CrossRef] [PubMed]
198. Romero, R.; Hassan, S.S.; Gajer, P.; Tarca, A.L.; Fadrosh, D.W.; Nikita, L.; Galuppi, M.; Lamont, R.F.; Chaemsaithong, P.; Miranda, J.; et al. The composition and stability of the vaginal microbiota of normal pregnant women is different from that of non-pregnant women. *Microbiome* **2014**, *2*, 4. [CrossRef] [PubMed]
199. Muhleisen, A.L.; Herbst-Kralovetz, M.M. Menopause and the vaginal microbiome. *Maturitas* **2016**, *91*, 42–50. [CrossRef]
200. Nelson, H.D. Commonly used types of postmenopausal estrogen for treatment of hot flashes: Scientific review. *JAMA* **2004**, *291*, 1610–1620. [CrossRef]

201. Devillard, E.; Burton, J.P.; Hammond, J.-A.; Lam, D.; Reid, G. Novel insight into the vaginal microflora in postmenopausal women under hormone replacement therapy as analyzed by PCR-denaturing gradient gel electrophoresis. *Eur. J. Obstet. Gynecol. Reprod. Biol.* **2004**, *117*, 76–81. [CrossRef]
202. Ginkel, P.D.; Soper, D.E.; Bump, R.C.; Dalton, H.P. Vaginal flora in postmenopausal women: The effect of estrogen replacement. *Infect. Dis. Obstet. Gynecol.* **1993**, *1*, 94–97. [CrossRef]
203. Heinemann, C.; Reid, G. Vaginal microbial diversity among postmenopausal women with and without hormone replacement therapy. *Can. J. Microbiol.* **2005**, *51*, 777–781. [CrossRef]
204. Galhardo, C.; Soares, J.J.; Simões, R.; Haidar, M.; de Lima Rodrigues, G.; Baracat, E. Estrogen effects on the vaginal pH, flora and cytology in late postmenopause after a long period without hormone therapy. *Clin. Exp. Obstet. Gynecol.* **2006**, *33*, 85–89.
205. Setty, P.; Redekal, L.; Warren, M.P. Vaginal estrogen use and effects on quality of life and urogenital morbidity in postmenopausal women after publication of the Women's Health Initiative in New York City. *Menopause* **2016**, *23*, 7–10. [CrossRef]
206. Achilles, S.L.; Hillier, S.L. The complexity of contraceptives: Understanding their impact on genital immune cells and vaginal microbiota. *AIDS* **2013**, *27* (Suppl. 1), S5–S15. [CrossRef]
207. Riggs, M.; Klebanoff, M.; Nansel, T.; Zhang, J.; Schwebke, J.; Andrews, W. Longitudinal association between hormonal contraceptives and bacterial vaginosis in women of reproductive age. *Sex. Transm. Dis.* **2007**, *34*, 954–959. [CrossRef] [PubMed]
208. Vodstrcil, L.A.; Hocking, J.S.; Law, M.; Walker, S.; Tabrizi, S.N.; Fairley, C.K.; Bradshaw, C.S. Hormonal contraception is associated with a reduced risk of bacterial vaginosis: A systematic review and meta-analysis. *PLoS ONE* **2013**, *8*, e73055. [CrossRef] [PubMed]
209. Mitchell, C.M.; McLemore, L.; Westerberg, K.; Astronomo, R.; Smythe, K.; Gardella, C.; Mack, M.; Magaret, A.; Patton, D.; Agnew, K.; et al. Long-term effect of depot medroxyprogesterone acetate on vaginal microbiota, epithelial thickness and HIV target cells. *J. Infect. Dis.* **2014**, *210*, 651–655. [CrossRef] [PubMed]
210. Fichorova, R.N.; Chen, P.L.; Morrison, C.S.; Doncel, G.F.; Mendonca, K.; Kwok, C.; Chipato, T.; Salata, R.; Mauck, C. The contribution of cervicovaginal infections to the immunomodulatory effects of hormonal contraception. *mBio* **2015**, *6*. [CrossRef] [PubMed]
211. Brooks, J.P.; Edwards, D.J.; Blithe, D.L.; Fettweis, J.M.; Serrano, M.G.; Sheth, N.U.; Strauss, J.F., III; Buck, G.A.; Jefferson, K.K. Effects of combined oral contraceptives, depot medroxyprogesterone acetate and the levonorgestrel-releasing intrauterine system on the vaginal microbiome. *Contraception* **2017**, *95*, 405–413. [CrossRef] [PubMed]
212. Bassis, C.M.; Allsworth, J.E.; Wahl, H.N.; Sack, D.E.; Young, V.B.; Bell, J.D. Effects of intrauterine contraception on the vaginal microbiota. *Contraception* **2017**, *96*, 189–195. [CrossRef] [PubMed]
213. Achilles, S.L.; Austin, M.N.; Meyn, L.A.; Mhlanga, F.; Chirenje, Z.M.; Hillier, S.L. Impact of contraceptive initiation on vaginal microbiota. *Am. J. Obstet. Gynecol.* **2018**, *218*, e1–e10. [CrossRef]
214. Srinivasan, S.; Morgan, M.T.; Fiedler, T.L.; Djukovic, D.; Hoffman, N.G.; Raftery, D.; Marrazzo, J.M.; Fredricks, D.N. Metabolic signatures of bacterial vaginosis. *mBio* **2015**, *6*. [CrossRef]
215. Eschenbach, D.A. Bacterial vaginosis: Resistance, recurrence, and/or reinfection? *Clin. Infect. Dis.* **2007**, *44*, 220–221. [CrossRef]
216. Hardy, L.; Jespers, V.; Dahchour, N.; Mwambarangwe, L.; Musengamana, V.; Vaneechoutte, M.; Crucitti, T. Unravelling the bacterial vaginosis-associated biofilm: A multiplex gardnerella vaginalis and atopobium vaginae fluorescence in situ hybridization assay using peptide nucleic acid probes. *PLoS ONE* **2015**, *10*, e0136658. [CrossRef] [PubMed]
217. Mastromarino, P.; Vitali, B.; Mosca, L. Bacterial vaginosis: A review on clinical trials with probiotics. *New Microbiol.* **2013**, *36*, 229–238. [PubMed]
218. Workowski, K.A. Centers for disease control and prevention sexually transmitted diseases treatment guidelines. *Clin. Infect. Dis.* **2015**, *61*, S759–S762. [CrossRef] [PubMed]
219. de Simone, C. The unregulated probiotic market. *Clin. Gastroenterol. Hepatol.* **2018**. [CrossRef]
220. Kolacek, S.; Hojsak, I.; Canani, R.B.; Guarino, A.; Indrio, F.; Pot, B.; Shamir, R.; Szajewska, H.; Vandenplas, Y.; Van Goudoever, J. Commercial probiotic products: A call for improved quality control. A Position Paper by the ESPGHAN Working Group for Probiotics and Prebiotics. *J. Pediatr. Gastroenterol. Nutr.* **2017**, *65*, 117–124. [CrossRef]

221. Hamilton, D.T.; Morris, M. The racial disparities in STI in the U.S.: Concurrency, STI prevalence, and heterogeneity in partner selection. *Epidemics* **2015**, *11*, 56–61. [CrossRef]
222. Dominguez-Bello, M.G.; De Jesus-Laboy, K.M.; Shen, N.; Cox, L.M.; Amir, A.; Gonzalez, A.; Bokulich, N.A.; Song, S.J.; Hoashi, M.; Rivera-Vinas, J.I.; et al. Partial restoration of the microbiota of cesarean-born infants via vaginal microbial transfer. *Nat. Med.* **2016**, *22*, 250–253. [CrossRef]
223. Onderdonk, A.B.; Delaney, M.L.; Fichorova, R.N. The human microbiome during bacterial vaginosis. *Am. Soc. Microbiol.* **2016**, *29*, 2. [CrossRef]
224. Goldstein, I.; Alexander, J.L. Practical aspects in the management of vaginal atrophy and sexual dysfunction in perimenopausal and postmenopausal women. *J. Sex. Med.* **2005**, *2*, 154–165. [CrossRef]

Review

Current Knowledge about the Effect of Nutritional Status, Supplemented Nutrition Diet, and Gut Microbiota on Hepatic Ischemia-Reperfusion and Regeneration in Liver Surgery

María Eugenia Cornide-Petronio [1,†], Ana Isabel Álvarez-Mercado [2,3,4,†], Mónica B. Jiménez-Castro [1,‡] and Carmen Peralta [1,5,*,‡]

1 Institut d'Investigacions Biomèdiques August Pi I Sunyer (IDIBAPS), 08036 Barcelona, Spain; cornide@clinic.cat (M.E.C.-P.); monicabjimenez@hotmail.com (M.B.J.-C.)
2 Department of Biochemistry and Molecular Biology II, School of Pharmacy, University of Granada, 18071 Granada, Spain; alvarezmercado@ugr.es
3 Institute of Nutrition and Food Technology "José Mataix," Center of Biomedical Research, University of Granada, Avda. del Conocimiento s/n, 18016 Armilla, Granada, Spain
4 Instituto de Investigación Biosanitaria ibs, GRANADA, Complejo Hospitalario Universitario de Granada, 18014 Granada, Spain
5 Centro de Investigación Biomédica en Red de Enfermedades Hepáticas y Digestivas (CIBERehd), 08036 Barcelona, Spain
* Correspondence: cperalta@clinic.cat; Tel.: +34-932275400
† These authors equally contributed to this work.
‡ These authors equally contributed to this work.

Received: 19 November 2019; Accepted: 15 January 2020; Published: 21 January 2020

Abstract: Ischemia-reperfusion (I/R) injury is an unresolved problem in liver resection and transplantation. The preexisting nutritional status related to the gut microbial profile might contribute to primary non-function after surgery. Clinical studies evaluating artificial nutrition in liver resection are limited. The optimal nutritional regimen to support regeneration has not yet been exactly defined. However, overnutrition and specific diet factors are crucial for the nonalcoholic or nonalcoholic steatohepatitis liver diseases. Gut-derived microbial products and the activation of innate immunity system and inflammatory response, leading to exacerbation of I/R injury or impaired regeneration after resection. This review summarizes the role of starvation, supplemented nutrition diet, nutritional status, and alterations in microbiota on hepatic I/R and regeneration. We discuss the most updated effects of nutritional interventions, their ability to alter microbiota, some of the controversies, and the suitability of these interventions as potential therapeutic strategies in hepatic resection and transplantation, overall highlighting the relevance of considering the extended criteria liver grafts in the translational liver surgery.

Keywords: ischemia-reperfusion injury; nutritional status; supplemented nutrition; gut microbiota; partial hepatectomy; liver transplantation

---

1. Introduction

An ischemic period is commonly required during hepatectomy or transplantation to avoid possible bleeding or blood transfusions. However, reduction of blood flow damages the liver and impairs liver regeneration [1]. Although ischemia-reperfusion (I/R) injury is commonly associated with poor post-operative results after liver surgery [2], no effective strategies are currently available to resolve this clinical problem. The mechanisms responsible for I/R injury are extremely complex, different depending on the liver type (steatotic versus non-steatotic), and involve a wide range of different cells and pro-inflammatory mediators [1–6]. Warm ischemia is associated with hepatic resections, and warm and cold ischemia is associated with liver transplantation (LT). The type of ischemia must be distinguished due to existing debate about the specific pathophysiological mechanisms of each surgical procedure. Other factors to be characterized in I/R injury are the percentage and duration of hepatic ischemia applied and the presence of regeneration (associated with hepatic resections) [7,8]. Steatotic livers have been demonstrated to be less tolerant of I/R injury than non-steatotic livers; therefore, the presence of fatty infiltration in the liver is associated with poor outcome following surgery [9–12]. Steatotic LT shows increased rates of graft failure compared with the post-operative outcomes of non-steatotic LT [9,13,14]. Similarly, complication rates following resection are two–three-fold higher in patients with hepatic steatosis [10,15]. Given the increasing prevalence of steatosis, and consequently the increase in the number of steatotic livers subjected to surgical conditions [16], the development of protective strategies in liver surgery are required.

Recent advances suggest new concerns about the pathophysiology of hepatic I/R injury. Preexisting nutritional status might affect the post-operative metabolism, liver function, inflammation, and regenerative capacity [17,18]. Starvation exacerbates warm ischemic injury due to the amount of glycogen stored in the liver [19–22]. Adenosine-5′-triphosphate (ATP) depletion during ischemia induces an acceleration of glycolysis [23]. Although glycolysis is essential for cell survival, its effects may also be detrimental due to lactate accumulation [23]. Overnutrition and specific diet factors are crucial for the pathogenesis and progression of nonalcoholic fatty liver disease (NAFLD) or nonalcoholic steatohepatitis [24]. Although there have been a wide variety of experimental studies on factors and nutritional substrates supporting or inhibiting liver regeneration after resection, a limited number of clinical studies have been addressed [25]. The intestinal microbiota is important to regulate liver functions [26,27] and is crucial in the pathogenesis of NAFLD [28–30]. Dietary components, host-intrinsic factors of the gastrointestinal tract affect microbial composition [27,31]. The activation of innate immunity and inflammation caused by gut-derived microbial compounds can exacerbate I/R injury or impair regeneration after liver resections.

The aim of the present review was to summarize the current knowledge from 2014 to 2019 about the effect of starvation, nutritional interventions, and gut microbiota alterations on morbidity and mortality in both experimental and clinical studies of liver surgery. A clear distinction between warm and cold I/R injury (associated with liver resections and LT, respectively) is discussed. The complicated differentiation on experimental models using steatotic and non-steatotic livers is addressed to elucidate the mechanisms responsible of liver I/R injury and for the establishment of new targets and protective strategies. The different results regarding the potential benefits of starvation, nutritional diets, and gut microbiota alterations in different studies (experimental, translational, and clinical studies) in hepatic surgery are discussed. All of this might be useful for the design of appropriate experimental models and treatments in clinical liver surgery.

2. Starvation Effects on I/R Injury Associated with Liver Surgery

Experimental studies have shown that liver I/R injury is influenced by different nutrients. For instance, protein restriction improved hepatic I/R injury by up-regulating hydrogen sulfide [32]. The supplementation of vitamins C and E in the diet protected against hepatic I/R injury. This effect was exerted by the up-regulation of antioxidant enzymes as well as the down-regulation of cell adhesion molecules [33]. However, although these experimental studies have demonstrated some beneficial effects of pre-operative diet restriction/fasting in liver I/R injury, the underlying mechanisms remain to be clarified. Other findings are contradictory [34–36]. Experimental studies have shown that fasting exacerbates normothermic ischemic injury [19–22]. Therefore, to support the clinical translation of starvation, the mechanisms behind the fasting-induced protection against I/R injury need to be elucidated [37]. Nil per os (NPO) status in patients undergoing hepatectomy to avoid potential problems, potentially associated with the general anesthesia, may be associated with immunomodulation risks to patients [38,39]. The NPO-associated fasting induces inflammatory responses in surgery [40]. The fasting state results in hyperglycemia, post-surgical infections, and increased length of stay [41–44]. Similarly, in clinical transplantation, donor starvation because the prolonged hospitalization or lack of an appropriate nutritional support would favor hepatic damage and primary nonfunction [45].

2.1. Studies of Short-Term Starvation (12–24 h)

The most recent preclinical studies investigating the effects of short-term starvation (12–24 h) on experimental models of normothermic I/R injury are summarized in Table 1. Twelve hours' fasting protected against apoptosis and necrosis associated with I/R injury [46]. Higher levels of serum β-hydroxybutyric acid (BHB) and, consequently, forkhead box protein O1 (FOXO1) over-expression were detected following the 12 h fast, thereby increasing antioxidant mechanisms including heme oxygenase 1 (HO-1) and autophagy activity. BHB inhibited the nucleotide oligomerization domain-like receptor family, pyrin domain containing 3 (NLRP3) inflammasome activity, the high-mobility group box 1 (HMGB1) release, and nuclear factor κ-light-chain-enhancer of activated B cells (NF-κB) activation [46]. In an ex vivo perfused rat liver model based on 60 min of ischemia and 60 min of reperfusion, the authors reported that starvation for 18 h fails to provide protection against liver I/R injury. The benefits of feeding were explained, at least partially, by increased energy metabolism (availability of energetic substrates) such as glycogen and high ATP levels [47]. These contradictory results [46,47] could be explained by the use of different experimental models of I/R (in vivo and ex vivo, respectively).

Table 1. Starvation approach in the setting of ischemia-reperfusion (I/R) injury in studies from 2014 to 2019.

Starvation Time	Model	Specie	Main Therapeutic Effects
Short-term: 12 h	Ischemia WIT: 60 min RT: 0, 1, 3, 6, 12 h [46]	Mice	↓ Liver injury, inflammation, apoptosis ↑ BHB, FOXO1 and HO-1
Short-term: 18 h	Ex vivo Ischemia WIT: 60 min RT: 60 min [47]	Rats	↑ Liver injury, inflammation, apoptosis ↓ Energetic substrates (ATP, glycogen)
Short-term: 24 h	Ischemia WIT: 60 min RT: 6 h [37]	Mice	↓ Liver injury, inflammation, HMGB1 ↑ Sirt1 activity, autophagy
	Ischemia WIT: 90 min RT: 6 h [48]	Mice	↓ Liver injury, inflammation, caspase-3 ↑ Sirt1 activity, autophagy, anti-apoptotic proteins
	Ischemia WIT: 60 min RT: 6 h [49]	Humans	↓ Liver injury, inflammation, oxidative stress ↑ Nrf2, HO-1 and Nqo1
Long-term: 2–3 days	Ischemia WIT: 60 min RT: 6 h [37]	Mice	↑ Liver injury, inflammation, HMGB1
	Ischemia WIT: 90 min RT: 6 h [48]	Mice	↓ Liver injury, inflammation, caspase-3 ↑ Sirt1 activity, autophagy, anti-apoptotic proteins
Long-term: 3–7 days	Ischemia WIT: 30 min RT: 24 h [50]	Mice	↓ Liver injury

Note: ATP, adenosine triphosphate; BHB, β-hydroxybutyric acid; FOXO1, forkhead box protein O1; h, hour; HMGB1, high-mobility group box 1; HO-1, heme oxigenase 1; min, minute; NF-κB, nuclear factor kappa-light-chain-enhancer of activated B cells; Nqo1, NAD(P)H quinone dehydrogenase 1; Nrf2, nuclear factor erythroid-derived 2-related factor 2; RT, reperfusion time; Sirt1, sirtuin 1; and WIT, warm ischemia time.

Short-term fasting for 24 h protected against hepatic I/R injury by regulating the response of innate immune cells [37]. Authors have shown that such benefits might be explained by the reduction in the circulating HMGB1 levels, which induces changes in sirtuin 1 (Sirt1) and autophagy, resulting in the anti-inflammatory regulation of short-term fasting [37]. In contrast with the results obtained in the ex vivo perfused rat liver model after 18 h fasting [47], the authors failed to find a correlation between the energy parameters, such as hepatic glycogen stores and fasting-induced protection. Altogether this suggests the relevance of using in vivo I/R models that simulate the clinical conditions as much as possible.

Qin et al. showed that starvation for 24 h inhibited hepatic I/R damage [48]. The authors suggested that starvation had anti-apoptotic effects in I/R by increasing the expression of anti-apoptotic protein such as B-cell lymphoma (BCL)-2/BCL-xl/phospho-protein kinase B (P-Akt) and decreased caspase-3 activity [48]. Similar to Rickenbacher et al. [37], the authors also concluded that starvation induced autophagy in the liver via the Sirt1 pathway [48]. Therefore, the results obtained in preclinical studies of fasting for 24 h suggest that starvation reduces cell death during hepatic I/R. Fasting-activated Sirt1 induced autophagy and promoted anti-apoptosis [48].

In the clinical context, liver resection is usually carried out under vascular occlusion to regulate bleeding [51]. Regeneration affects the mechanisms responsible of I/R injury, and I/R negatively affects liver regeneration. Thus, the beneficial effects of starvation reported to date might not be extrapolated to surgical conditions requiring partial hepatectomy (PH) under I/R.

To the best of our knowledge, only Zhan et al. [49] recently analyzed the effects of short-term fasting on PH under I/R in humans (Table 1). Thus, in a prospective, single-blinded, randomized study of 30 patients per group, 24 h fasting reduced damage, inflammation, and oxidative stress through regulation of nuclear factor erythroid-derived 2-related factor 2 (Nrf2), HO-1, and NAD(P)H quinone dehydrogenase 1 (Nqo1) signaling pathways [49]. However, postsurgical complications of control and fasting groups were similar [49]. Further clinical studies are required to confirm the benefits of 24 h of fasting in PH.

2.2. *Studies of Long-Term Starvation (Two to Seven Days)*

In addition to the investigations on the effects of short-term fasting for 24 h, Rickenbacher et al. [37] and Qin et al. [48] studied the effects of long-term starvation for two and three days (Table 1). Rickenbacher et al. showed that fasting for 24 h, but not two or three days, can reduce I/R injury via the Sirt1-mediated down-regulation of HMGB1 in circulation [37]. However, Qin et al. [48] found even more protective effects against I/R injury at two and three days of fasting than 24 h of fasting in mice. The reasons for these different findings may be related to the different experimental model used, such as duration of ischemia (60 min versus 90 min of ischemia). Three days of fasting or one week of preoperative protein/energy restriction decreased transaminases and hemorrhagic necrosis after 30 min of ischemia [50].

Further experimental investigations and clinical trials are needed to determine the effects of starvation and the exact fasting duration (one, two, or three days) to produce the greatest advantages in patients. Long-term diet restriction (more than 24 h) may be difficult to apply for human preoperative management. Experimental models that reproduce the clinical conditions might be useful for the implementation of protective treatments in clinical conditions in the short-term [52]. The studies mentioned above have been reported in non-steatotic livers. The prevalence of obesity ranges from 24% to 45% of the population; therefore, increases in the number of steatotic livers subjected to liver surgery are expected. Steatotic livers show poor regenerative response and increased vulnerability to I/R injury, and the mechanisms involved in the I/R pathology and protective strategies are different depending on the type of the liver (presence or absence of steatosis) submitted to surgery. Thus, future research in experimental models of PH with I/R and LT are required to understand the underlying mechanisms of starvation, especially in sub-optimal livers in order to ameliorate the viability of livers subjected to surgery and reduce consequently the post-operative problems.

3. Nutritional Support by Nutraceuticals and Functional Foods on Liver Surgery under Hepatic Ischemia-Reperfusion

The preoperative nutritional state considerably affects postoperative metabolism, organ function, and inflammatory responses [17], and nutritional status affects the liver regenerative capacity [18]. Therefore, the basal alimentary condition of the patient plays an important role in predicting postoperative complications. Patients with end-stage liver diseases who undergo LT usually present with malnutrition, which directly impacts the deterioration of the patient's clinical condition, affecting post-transplantation survival [24]. The post-transplantation survival is even more relevant in the case of liver steatosis (the main feature of NAFLD) as these organs show high vulnerability to I/R injury and regenerative failure in comparison with non-steatotic livers [53].

As mentioned above, coinciding with the progressive adoption of the Western lifestyle and changes in nutritional habits, many studies have evidenced the increased incidence and prevalence of NAFLD and other related disorders [54]. Also, malnutrition induces dysbiosis with translocation of bacteria- and/or pathogen-derived components from the gut to the liver [55].

Conversely, several dietary components significantly benefit health [56], presenting antioxidant or anti-inflammatory properties as well as contributing to modifying the gut microbiome [18]. As a result, the re-establishment and maintenance of the correct nutritional status by these nutraceuticals and functional foods before, during, and/or after surgery could lead to improvements in complications related to I/R injury, representing a potential approach alone or in combination with other therapies to improve patient outcomes. Eventually, strategies based on nutrition support could become a major adjunct to the conventional management of I/R injury.

Combination of different nutrition tools like anthropometry, and body composition analysis, have been reported to formulate a composite score for malnutrition assessment [57]. The goals of nutritional therapy are mainly focused on improving protein malnutrition and regulate nutrient deficiencies. Studies to address I/R injury complications by dietary supplementation and functional foods in liver surgery covering 2014 to 2019 are summarized in Table 2.

3.1. Plant-Derived Supplements and Other Food Additives

Three studies focusing on nutrition support based on plant-derived supplements and other food additives were reported from 2014 to 2019 [58–60]. All of them targeted oxidative stress and inflammatory responses related to I/R injury in murine models. The more remarkable findings were strengths of the antioxidant defense systems and anti-inflammatory properties after the intervention. For instance, ankaflavin, a traditional food additive used in Eastern Asia and China, significantly decreased the proliferation of Kupffer cells and the protein expression of inflammatory cytokines (tumor necrosis factor α (TNF-α), interleukin (IL)-6, and IL-1β) and reduced apoptosis and liver steatosis in high-fat-diet-fed mice [58].

A similar plant-derived strategy tested the potential benefits of apocynin (4-hydroxy-3-methoxyacetophenone) in rats under I/R injury. In this case, a single dose of apocynin 30 min before surgery induced the production of superoxide dismutase (SOD), reduced lipid peroxidation, and decreased glutathione (GSH) limiting the cellular stress triggered by ischemia [59]. Also, Korean red ginseng extract, which contains ginsenosides, phenolic compounds, polysaccharides, and polyacetylenes, showed a chemopreventive effect through antioxidant, apoptotic, and anti-cell proliferation in various cancers. In concordance with these findings, a study conducted in rats in which hepatic cancer had previously been induced, supplementation starting two weeks before surgery and eight weeks after PH revealed chemopreventive effects by prevention of oxidative stress and regulation of redox-enzymes [60]. The potential limitation of all these studies is related to the limited specificity of the different plant-derived supplements and additives. The relevance of the changes on oxidative stress, TNF-α, IL-6, and/or IL-1β induced by such treatment requires further investigation. Studies aimed at evaluating if such benefits can be extrapolated in steatotic liver undergoing surgery might be of clinical and scientific relevance. The potential toxicity and side effects of these components, dependent on the concentrations, required to confer protection should be investigated.

Table 2. Studies to address hepatic I/R injury by dietary supplementation and functional foods.

Drug	Administration	Model	Specie	Main Therapeutic Effects
Ankaflavin (food additive) [58]	Gavage (orally) 0.624 mg/kg daily for 1 week	Ischemia, fatty liver WIT: 60 min RT: 3 h	Mice	↓ Liver injury, steatosis, oxidative stress, apoptosis, inflammatory cytokines (TNF-α, IL-6, IL-1β)
Apocynin (organic compound related to vanillin) [59]	Intraperitoneally 20 mg/kg 30 min before surgery	Ischemia WIT: 60 min RT: 60 min	Rats	↓ Oxidative stress (MPO) ↑ Antioxidant levels (SOD)
Korean red ginseng extract [60]	Orally 0.5%, 1%, or 2% for 10 weeks	PH RT: 7 weeks	Rats	↓ Lipid peroxidation, cytochrome P450 signaling pathway ↑Antioxidant levels (tGSH, GST, GPx),
Antioxidative nutrient-rich enteral diet (Polyphenols, Vitamin C and E) [33]	Orally ad libitum for 7 days	Ischemia WIT: 60 min RT: 6 h	Mice	↓ Liver injury, necrosis, inflammatory cytokines (IL-6, CXCL1), MDA, cell adhesion molecules, neutrophils and macrophage infiltration ↑ Antioxidant levels (SOD1, SOD2)
Dexpanthenol (analogue of provitamin B5) [61]	Intraperitoneally 500 mg/kg during the ischemic period	Ischemia WIT: 60 min RT: 60 min	Rats	↓ Oxidative stress (MPO), histologic tissue damage ↑ Antioxidant levels (SOD, tGSH)
Vitamin C [62]	Intravenous 50–200 mg/kg after surgery	Ischemia WIT: 3 × 15 min pringle maneuver with 5 min between occlusion RT: 4 h	Swine	↓ Inflammatory cytokines (IL-1β, IL-8, TNF-α), procoagulant response (PAI-1, tissue factor)
Rosa mosqueta oil [63]	Orally 0.4 mL/g/day for 21 days	Ischemia WIT: 60 min RT: 20 h	Rats	↓ Liver injury, inflammation, oxidative stress ↑ α-linolenic acid, EPA and DHA fatty acids levels
Tilapia fish oil [17]	Gavage (orally) 0.4% body weight for 3 weeks	Ischemia WIT: 30 min RT: 1, 12, and 24 h	Rats	↓ Liver injury, antioxidant levels (CAT, SOD, GPx), tissue TBARS, histological tissue damage
Fish oil [64]	Gavage (orally) 12 mL/kg daily	PH RT: 1, 2, 3, and 5 days	Mice	↓ Liver injury, total bilirubin ↑ Proliferation, AMPK activation, liver-to-body weight ratio, tight junction, and BSEP protein expression
L-arginine [65]	Gavage (orally) 10% in 1 mL/100g of solution 15 min before surgery and 24 h until date of death	PH RT: 24 h, 72 h, and 7 days	Rats	↑ Alkaline phosphatase No effect in regeneration
L-glutamine [66]	Gavage (orally) 1 mL/100g body weight 6 h and 15 min before surgery	PH RT: 24 h, 72 h, and 7 days	Rats	↑ Regeneration, albumin No effect in liver function
Omega-3 fatty acids [67]	Orally 10 mg/kg/day for 28 days	PH RT: 7 days	Rats	↓ Inflammatory cellular infiltrate No effect in regeneration
Omega-3 fatty acids [18]	Gavage (orally) 1 mL/100g (10% v/v) 15 min and 24 h before surgery	PH RT: 24 h, 72 h, and 7 days	Rats	↓ GGT No effect in regeneration
Immunonutrients (EPA, arginine, and nucleotides) [68]	Orally 1000 kcal/day for 5 days before surgery	PH RT: 1, 3, 7, and 14 days	Humans	↓ Inflammatory response (IL-6), infection, severe complications ↑ Resolving E1
Immunonutrientes (EPA, arginine, and nucleotides) [69]	Orally 3 × 237 mL 1020 kcal, 54 g protein, 12.6 g arginine, 1.3 g nucleotides, 3.3 g EPA/day × 5 days before surgery	PH RT: 1, 3, 5, 7, 10, and 30 days	Humans	No benefits
Immunomodulating diet enriched with HWP [70]	Intravenous 20 mL/h 24 h after surgery	LDLT CIT: 132 ± 100 min RT: 0, 1, 2, 3, and 4 weeks	Humans	↓ Incidence of bacteremia
Hydrolyzed whey peptide (HWP) [71]	Orally 4 mL every 6 h after reperfusion	Ischemia, steatotic liver WIT: 30 min RT: 6 and 12 h	Rats	↓ Liver injury, inflammatory cytokines (TNF-α, IL-6), iNOS, oxidative stress (UCP-2), necrosis ↑ Survival
Lipid emulsion [72]	Intravenous 5 mL 4 h after surgery	PH + I/R, steatotic liver WIT: 60 min RT: 12, 24, and 48 h	Rats	↓ Liver injury, TGF-β ↑ Regeneration (HGF, cyclin A and E), IL-6, ATP, phospholipid levels
BCAA [73]	Orally 1000 mg valine, 2000 mg leucine, 1000 mg isoleucine in 500 mL until 2 h before surgery	PH RT: 0 day	Humans	↓ Lactate levels No effect in morbidity rates
BCAA [74]	Orally 4 g BCAA granules with: 952 mg L-isoleuciene, 1904 mg L-leucine, 1144 mg L-valine twice daily for 6 months	PH RT: 1–2 weeks until 1, 3, and 6 months	Humans	↑ Functional regeneration No effect in infectious, nutritional and immunologic status

Note: AMPK, AMP-activated protein kinase; ATP, adenosine triphosphate; BCAA, branched chain amino acids; BSEP, bile salt export pump; CAT, catalase; CXCL1, chemokine ligand 1; DHA, docosahexaenoic acid; EPA, eicosapentaenoic acid; GGT, gamma glutamyltransferase; GPx, glutathione peroxidase; GST, glutathione s-transferases; HGF, hepatic growth factor; HWP, hydrolyzed whey peptide; I/R, ischemia reperfusion; IL, interleukin; iNOS, nitric oxide synthase; LDLT, living donor liver transplantation; mg, milligram; min, minutes; MPO, myeloperoxidase; PH, partial hepatectomy; PAI-1, plasminogen activation inhibitor-1; RT, reperfusion time; S1P, sphingosine-1-phosphate; SOD, superoxide dismutase, TBARS, thiobarbituric acid reactive substances; TGF-β, tumor growth factor β; tGSH, total glutathione; TNF-α, tumor necrosis factor α; UCP2, uncoupling protein 2; and WIT, warm ischemia time.

3.2. Vitamins

Various vitamins deficiencies have been reported in receptors submitted to LT. Folate deficiency is caused by a decreased intake and absorption, dysregulation in renal excretion and limited hepatic storage. Folate and B12 supplementation is crucial to protect liver against alcoholic hepatitis [75]. Hypovitaminosis A is associated with impairment in immune function and increased risk of fibrosis, which are risk factors in liver surgery [76]. An anti-oxidative nutrient-rich enteral ordinary diet enhanced with vitamins C and E and supplemented with polyphenols (a combination of catechin and proanthocyanidin) for seven days before ischemic insult in mice was able to mitigate liver I/R injury, improving antioxidant and inflammatory parameters that reduced hepatocellular damage [33].

Dexpanthenol, also known as pro-vitamin B5, is oxidized to pantothenic acid (PA), which increases GSH content, coenzyme A (Co A), and ATP synthesis, thus playing a crucial role against oxidative stress and inflammation. In an experimental model of hepatic I/R in rats, a single dose of dexpanthenol before I/R induced the suppression of oxidative stress and increased antioxidant levels [61]. In a swine model of multiple injuries including I/R injury and hemorrhage, the authors observed a moderate improvement in coagulation dysfunction after intravenous provision of high-dose vitamin C and a reduction in proinflammatory/procoagulant response [62].

All these studies indicate the potential importance of vitamins in reducing the inflammation and damage in surgical conditions of I/R. The usefulness of vitamins in the presence of steatosis and in surgical conditions requiring ischemia and regeneration, such as liver resection or liver-related LT, remains to be elucidated.

3.3. Fish and Rosa Mosqueta Oils

Based on the well-established protective components of rosa mosqueta oil (i.e., α-linolenic acid (ALA) and tocopherols), Dossi et al. reported that rosa mosqueta oil supplementation before the induction of I/R in rats increased liver ALA and its derived eicosapentaenoic acid (EPA) and docosahexaenoic acid (DHA) fatty acid contents, with increases in α- and γ-tocopherols, normalized liver oxidative stress parameters, and ameliorated liver and serum inflammation indexes [63].

Fish-oil-supplemented diets have been shown to reduce I/R injury. In this sense, a study conducted to identify the effect of tilapia fish oil, which is rich in unsaturated fatty acids, administrated to rats by gavage during three weeks before I/R revealed that after ischemia and 1, 12, and 24 h of reperfusion, antioxidant enzyme activities of catalase (CAT), SOD, and glutathione peroxidase (GPx) decreased in the intervention group. Lipid peroxidation and liver damage decreased in this group [17]. Similarly, daily oral supplementation for 12 days with fish oil, comprising 40% DHA and 40% EPA, induced AMP-activated protein kinase (AMPK) activation and promoted the recovery of liver function during PH [64]. The role of each component included in either rosa-mosqueta- or fish-oil-supplemented diets on the mechanisms responsible for hepatic I/R remains unknown. The main mechanism involved in the effects of such treatments on I/R damage remain to be elucidated. This is a potential problem due to difficulties for the establishment of target signaling pathways in liver surgery. The effect of rosa mosqueta and fish oil supplementation in steatotic liver undergoing PH under vascular occlusion as well as in LT should be investigated.

3.4. Fatty Acids, Arginine, and Nucleotides

Polyunsaturated fatty acids (PUFAs) are fatty acids with two or more double bonds in their carbon chain. PUFAs can be further categorized according to the location of the first double bond relative to the terminal methyl group: Omega-3 and omega-6 and are characterized by the presence of a double bond three and six atoms away from the methyl terminus, respectively [77]. Long-chain PUFAs (LC-PUFAs), particularly omega-3 LC-PUFAs EPA and DHA, are associated with beneficial health effects [78].

In experimental and clinical studies performed in animals and humans, fatty acids, arginine, and nucleotides have shown the ability to modulate immune and inflammatory responses [18,69]. These nutrients, among others, have been labeled as pharmaconutrients [18].

Supplementation with amino acids, such as arginine, affects urea genesis, gluconeogenesis, and protein synthesis. Diets enriched with these amino acids increases the hepatic catabolism functions [79]. Enteral immunonutrition with arginine reduces the risk of infections in patients submitted to major operations [80]. The supplementation with L-arginine diet in rats hepatectomized was unable to confirm benefits in liver regeneration [65]. Conversely, a similar study using supplementation of L-glutamine in the diet of rats after PH revealed an increase in the amount of albumin and beneficial effects for liver regeneration [66]. Glutamine favors liver regeneration [66].

Omega-3 fatty acids affect the production of pro-inflammatory mediators, such as growth factors, chemokines, and matrix proteases, showing anti-inflammatory and immunomodulatory effects due to their rapid incorporation into cell membranes [67,68]. However, their effect on regeneration in livers undergoing resection has not been widely reported. Two studies evaluated whether omega-3 fatty acids protect against regeneration failure in PH in rats. Neither long-term supplementation before surgery [67] nor a preoperative supplementation plus the same dose every 24 h during the seven days post-surgery [18] showed any influence on the liver regeneration.

Concerning EPA, a study conducted in patients who underwent major hepatobiliary resection reported that preoperative immunonutrition decreased inflammation and protected against post-surgery infections and complications [68]. However, these benefits cannot be exclusively attributed to EPA because the oral supplementation was also enriched with arginine and nucleotides. A similar approach but with controversial results was conducted by Russell et al. Indeed, any benefit of preoperative immunonutrition was reported with arginine and n-3 fatty acids [69]. In a retrospective study reported by Kamo et al., liver recipients suffering from infection after LT were submitted to enteral immunonutrition enriched with nucleotides, arginine and omega-3 fatty acids, and hydrolyzed whey peptide (HWP) (an immunonutritional liquid). The main finding was a lower incidence of bacteremia in the intervention group compared with the control group [70].

For steatotic livers, Nii et al. tested the effects of HWP on hepatic I/R injury in rats with steatotic livers administered immediately after reperfusion and every six hours thereafter. This treatment ameliorated liver damage, improving function, histology, and survival following I/R [71]. In conditions of PH under I/R, a lipid emulsion comprising 52% linoleic acid, 22% oleic acid, 3% palmitic acid, 8% linolenic acid, 4% stearic acid, 1% other fatty acids, 8.184 g/L egg phospholipids, and 15 g/L glycerine infused in rats immediately after surgery for four hours protected against damage and regenerative failure [72].

3.5. Branched-Chain Amino Acid

A branched-chain amino acid (BCAA) is an amino acid with an aliphatic side-chain with a branch. BCAAs promote protein synthesis and glucose metabolism and are involved in fatty acid oxidation [81]. BCAAs favor liver regeneration, nutrition status, and hepatic encephalopathy. BCAAs have the ability to reduce oxidative stress and liver inflammation as well as lactate production [73].

A randomized controlled trial conducted in patients submitted to hepatectomy showed that supplementation with BCAAs administered two times a day for six months after surgery improved liver functionality and regenerative capacity [74]. Similarly, in patients submitted to liver resection, the preoperative BCAA supplementation decreased blood lactate, which is exacerbated by surgical stress patients [73].

3.6. Probiotics

Probiotics are cultures of single or multiple microbes that can regulate the properties of the existing gut microbiota. Probiotics can promote anti-inflammatory effects in gut, thereby preventing bacterial translocation and endotoxin generation [82] and are involved in the synthesis of antimicrobial agents that inhibit the invasion of pathogenic bacteria [83]. Probiotics might regulate the immune system, inhibiting the release of cytokines like TNF-α [84] and inducing the release of anti-inflammatory cytokines like IL-10 and tumor growth factor β (TGF-β) [85].

Current evidence has indicated the advantages resulting from the use of probiotics to prevent the infections after LT, as well as to improve the circulatory diseases associated with cirrhosis, hepatic encephalopathy, and Child–Pugh class [86,87]. The improvement in the neutrophil phagocytic capacity induced by probiotics regulated the infections, preventing bacterial translocation. These effects resulted in the restoration of the immune system [88–90].

In addition to the different types of nutritional support, the routes of administration should be considered. Oral intake is the first line therapy used to treat malnutrition and decrease the complications (hepatic encephalopathy, infections, and ascites among others) in liver diseases. However, the impact on survival remain to be elucidated [91,92]. It has been described that an increased dietary intake by oral nutrition improved liver function and lowered mortality compared with the enteral and parenteral nutrition [93,94]. Hasse et al. [95] demonstrated early enteral feeding beneficial effects like improved nitrogen balance and fewer viral infections associated with LT. Parenteral nutrition might be used as a second line approach in those who cannot be fed adequately by the oral or enteral route for instance in patients with unprotected airways and advanced hepatic encephalopathy [96,97]. All these data are not conclusive for selecting the most appropriate administration route of nutritional support. In a comparison between parenteral and early enteral nutrition, both strategies were equally effective to the maintenance of nutritional state [97]. The European Society for Parenteral and Enteral Nutrition (ESPEN) guidelines for organ transplantation recommend enteral nutrition or oral nutritional supplementation to improve nutritional status and liver function [93,98–101]. Enteral nutrition reduces the incidence of viral and bacterial infections. For enteral nutrition, the ESPEN guidelines recommend the use of more concentrated high-energy formulas in patients with ascites and BCAA-enriched formulas in hepatic encephalopathy patients [95].

4. Gut Microbiota and Hepatic Ischemia Reperfusion in Liver Surgery

The gut microbiota is crucial to the effects of diet, drugs, and disease [102]. The microorganisms that exist within the gastrointestinal ecosystem are termed gut microbiota, playing an essential role in the stimulation of immune response [103], the maintenance of intestinal barrier integrity [104], modulation of host–cell proliferation and vascularization [105,106], and regulation of neurological [107] and endocrine [108] functions. The human gut microbiota provides an energy source [109], is involved in the synthesis of vitamins and neurotransmitters [110], metabolizes bile salts [111], and eliminates toxins [112].

Disequilibrium in the microbiota composition, commonly referred to as dysbiosis, may lead to several diseases [113,114]. The gut and liver (the gut–liver axis) (Figure 1) communicate bidirectionally through the biliary tract, the portal vein, and the systemic circulation [115]. The translocation of bacterial products from the intestine to the liver induces inflammation in different cell types, such as Kupffer cells and a fibrotic response in hepatic stellate cells, resulting in deleterious effects on hepatocytes [116]. Bacterial translocation and fungal cell wall components are increased in experimental models of ethanol-induced liver disease [117].

Alterations in gut microbiota are important for determining the occurrence and progression of alcoholic liver disease (ALD) [118–120], NAFLD [121,122], nonalcoholic steatohepatitis (NASH) [123,124], cirrhosis [125,126], and hepatocellular carcinoma (HCC) [127]. Fecal microbiota transplantation could induce hepatitis B virus e-antigen (HBeAg) clearance in patients with persistent positive HBeAg, even after long-term antiviral treatment [128]. Ferrere et al. [129] observed that ALD in mice were reduced by fecal transplantation from alcohol-fed mice resistant to ALD or with prebiotics.

Evidence points to the involvement of the gut microbiota in the pathogenesis of NAFLD [130,131]. Cogger et al. showed that liver sinusoidal endothelial cells (LSECs) fenestrae are inversely and positively correlated with the gut abundance of Bacteroidetes and Firmicutes, respectively [132]. The gut microbiota also has an emerging role in NASH as a source of inflammatory stimuli [130,133].

Increased intestinal permeability and elevated plasma lipopolysaccharide (LPS) [134,135] observed in NASH may also contribute to LSECs' pro-inflammatory function [136].

Gut microbiota shifts the influence of hepatic metabolism through regulation of hepatic gene expression without direct contact with the liver [137,138].

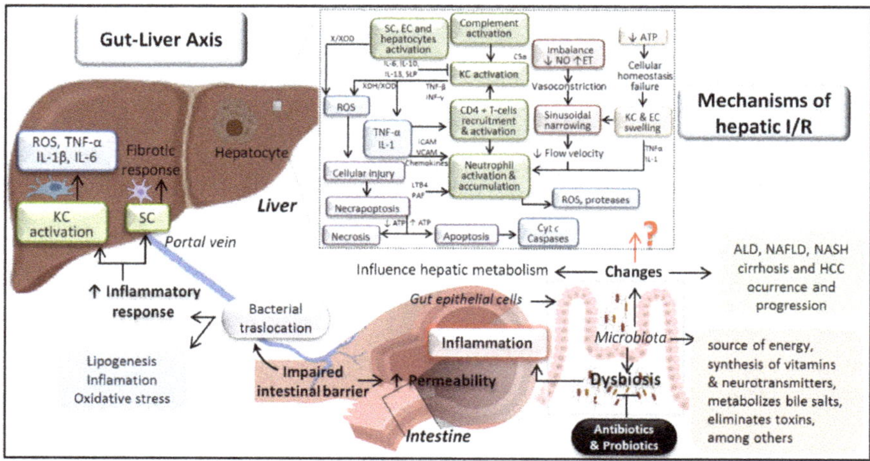

Figure 1. Gut microbiota and hepatic I/R. The dotted box summarizes the mechanisms involved in hepatic I/R injury and how some of these have been altered in the liver by changes in the gut microbiota. ALD, alcoholic liver disease; ATP, adenosine triphosphate; Cyt c, cytochrome c; EC, endothelial cell; ET, endothelin; HCC, hepatocellular carcinoma; ICAM, intracellular cell adhesion molecule; IL, interleukin; INF, interferon; KC, Kupffer cell; LTB4, leucotriene B4; NAFLD, nonalcoholic fatty liver disease; NASH, nonalcoholic steatohepatitis; NO, nitric oxide; PAF, platelet activating factor; ROS, reactive oxygen species; SC, stellate cell; TNF, tumor necrosis factor; VCAM, vascular cell adhesion molecule; and X/XOD, xanthine/xanthine oxidase.

As a result, ischemia produced during liver surgery (i.e., LT or liver resection) is expected to alter the microbiota profile, potentially affecting inflammation, the immune response, and even regeneration. The gut–liver axis is widely implicated in the pathogenesis of liver diseases such as NAFLD, NASH, HCC, and acute liver failure [139]. The gut microbiota may also contribute to the generation of memory alloreactive T cells. T cells were reported to be important in transplant rejection and many experimental and clinical studies have shown that the intestinal microbiota is altered after allogeneic transplantation [140].

In the context of I/R injury, hepatic steatosis is a key factor to consider due to negative influences on patients' outcomes [141]. Gut microbiota fundamentally influences processes such as lipogenesis, which is affected by the absorption of monosaccharides in the intestinal lumen by the microbiota [142], and bile acids, since they are able to de-conjugate them and turning them into secondary bile acids, which are capable of interacting with a nuclear receptor of the farnesoid receptor X [143]. Changes in gut microbiota promote the development of NAFLD since affect inflammation, insulin resistance, bile acids, and choline metabolism. The Western diet is associated with intestinal microbial dysbiosis [144] and the development and prevalence of NAFLD [145]. I/R injury is a common cause of rejection when grafts are sourced from NAFLD donors; the prevalence of the problem is increasing [141].

The gut microbiota alterations in NAFLD patients remain to be characterized [114]. Several reviews have highlighted studies focused on strategies to prevent and target gut microbiota (probiotics, prebiotics, diet or fecal microbiota transplantation, among others) in NAFLD [114,115,140,146]. Others have addressed the management of nutrition in patients with end-stage liver disease undergoing

LT [146,147]. However, studies evaluating changes in gut microbial populations and diversity caused by hepatic I/R and their consequences in liver function and regeneration are limited. From 2014 to 2019, authors only examined the effect of therapeutic approaches on intestinal microbiota and hepatic injury and such strategies were mainly based in the use of antibiotics. Despite this, the effects of antibiotics on hepatic damage being caused by regulation of the intestinal microbiota remain to be clarified. None of these studies aimed to improve damage induced by I/R in steatotic livers.

Intestinal microbial characterization and alteration in early phase and subsequent intestinal barrier dysfunction during acute rejection after LT have been reported [148–153]. Due to the high sensitivity of microbial changes during acute rejection after LT, intestinal microbial variation has been suggested to predict acute rejection in the early phase after LT [148]. Therefore, gut microbial profiles have been suggested as predictive injury biomarkers in LT [153].

Gut microbiota might affect immune mediators such as IL-6 and regulate liver regeneration. Following the administration of antibiotics (Table 3), the number of CD1d-dependent natural killer T (NKT) cells was reduced after partial hepatectomy (PH) [154]. NKT cells and activated Kupffer cells produced high levels of interferon-γ (IFNγ) and IL-12. Thus, antibiotic administration after PH could negatively affect regenerative response [154]. It has been reported that PH resulted in an upregulation of more than 6000 bacterial genes, some of them involved in regeneration and was also accompanied by changes in the gut microbiota (e.g., an increase in *Bacteroidetes* and *Rikenellaceae,* and decreases in *Clostridiales, Lachnospiraceae,* and *Ruminococcaceae*) [155,156].

Table 3. Therapeutic strategies in modulation of gut microbiota in liver surgery from 2014 to 2019.

Drug	Administration	Model	Specie	Main Therapeutic Effects
Ampicillin, neomycin sulfate, metronidazole and vancomycin [154]	Orally 1 g/L ampicillin, neomycin sulfate, metronidazole, and 500 mg/L vancomycin for 4 weeks	PH	Mice	↓ Liver regeneration ↑ IFNγ, IL-12
Gentamicin [157]	Gavage 2 mL daily for 3 weeks	LT CIT: Not indicated RT: 1 week and 2 weeks	Rats	↓ Liver injury, necrosis, inflammation
Rifaximin [158]	Orally 550 mg twice daily for 28 days	LT CIT: 440 min RT: not indicated	Humans	↓ Liver injury, inflammation, early allograft dysfunction
Amoxicillin [159]	Gavage 50 mg/mL for 10 days before LT	LT CIT: 18 h RT: 6 h	Mice	↓ Liver injury, inflammation, CHOP, mTORC1 activity ↑ PGE2, EP4, autophagy
Neomycin, erythromycin and ampicillin-sulbactam [159]	Orally 1 g neomycin, erythromycin 4× and 3 g ampicillin-sulbactam before or on day of LT	LT CIT: not indicated RT: not indicated	Humans	↓ Liver injury, inflammation, CHOP, early allograft dysfunction ↑ EP4, LC3B, autophagy
Cyclosporine A [160]	Intragastrically 2 mg/kg twice daily for 28 days after LT	LT CIT: not indicated RT: 28 days	Rats	↓ Liver injury, inflammation
Tacrolimus [161]	Subcutaneously, 1.0, 0.5, or 0.1 mg/kg every 12 h for 7 days and intragastrically once daily for 8–29 days after LT	LT CIT: not indicated RT: 30 days	Rats	↓ Liver injury
Retinoic acid [162]	Gavage 25 μg/g body weight 48 h before surgery	PH	Mice	↑ Liver regeneration, FGF21
Probiotics [163]	Orally 2 g/day LP, LA-11, and BL-88, total of 2.6×10^{14} CFU daily for 6 days before surgery and 10 days after surgery	PH RT: 10 days	Humans	↓ Infectious complications, septicemia, plasma endotoxin, serum zonulin concentration ↑ Liver barrier
Time-restricted feeding [164]	Food restriction: 8–10 h/day, 12 weeks before surgery	Ischemia WIT: 60 min RT: 6, 12, 24 h	Mice	↓ Liver injury, inflammation, oxidative stress, apoptosis

Note: BL-88, *Bifido-bacterium longum* 88; CFU, colony forming units; CHOP, CCAAT/enhancer-binding protein homologous protein; CIT, cold ischemia time; EP, prostaglandin E2 receptor; FGF21, fibroblast growth factor 21; IFNγ, interferon-gamma, IL, interleukin; LA-11, *Lactobaciullus acidophilus* 11; LC3B, Light Chain 3 isoform B; LP, *Lactobacillus plantarum*; LT, liver transplantation; mTORC1, mammalian target of rapamycin complex 1; PGE2, prostaglandin E2; PH, partial hepatectomy; RT, reperfusion time; and WIT, warm ischemia time.

The administration of antibiotics reduces hepatic injury in rats submitted to LT with acute rejection, but the microvilli of the ileum epithelial cells were destroyed, inducing alterations in microbiota [157]. Further studies are required for a more understanding of the immunity interactions between gut microbiota and the rejection after LT [157]. Two retrospective studies support the notion that antibiotics (rifaximin, neomycin, erythromycin, and ampicillin-sulbactam) administration prior to LT reduce infections associated with LT, thus reducing the liver injury, inflammation, and early allograft dysfunction [158,159]. However, further randomized controlled clinical trials are required to elucidate the exact mechanisms of action of such antibiotics, their target signaling pathways, and the optimal duration of treatment. Further experiments in animal LT models will be required to elucidate the specific molecular signaling pathways through which antibiotics may exert their actions, as well as to investigate whether the protection on hepatic damage induced by the treatment with antibiotics is exerted throughout changes in the gut microbiome.

Survival outcomes after LT have constantly improved using upgraded immunosuppressive agents [165]. However, the inadequate or excessive immunosuppression is associated with a higher risk of rejection, higher incidence of infection, drug toxicity, and increased mortality [166–170]. Experimental studies in rats have investigated the effect of immunosuppressive agents on the intestinal microbiota in LT. The results showed that cyclosporine A ameliorated hepatic injury and partially restore the intestinal microbiota after LT [160]. An optimal dosage of tacrolimus (FK506) induced normal graft function, and stable gut microbiota after LT in rats. This resulted in increased probiotics, including *Faecalibacterium prausnitzii* and *Bifidobacterium* spp. and decreased pathogenic endotoxin-producing bacteria, such as the *Bacteroides–Prevotella* group and *Enterobacteriaceae*. Thus, the use of the gut microbiota might be a novel strategy for the assessment of the dosage of immunosuppressive medications and its effects in receptors submitted to LT [161].

Retinoic acid, naturally present in the gastrointestinal tract, has a relevant effect in regulating lipid homeostasis [171,172] and can facilitate PH-induced liver regeneration [173,174]. Given the intimate relationship between gut-derived signaling and liver regeneration, authors hypothesized that retinoic acid may regulate gut microbiota thereby promoting liver regeneration [162]. Retinoic-acid-accelerated liver regeneration was associated with a reduction in the ratio of Firmicutes to Bacteroidetes. Retinoic acid had benefits on lipid circulation and regulated the FGF21-LKB1-AMPK pathway, which promoted energy metabolism and consequently the regenerative process in the liver [162]. Further studies will be required to elucidate the interaction between the modulation of microbiota and the improvement in proliferation induced by the retinoic acid. This will allow the development of clinical therapeutic strategies to promote liver regeneration.

In line with the results described above, the evidence suggests that probiotics play an important role in the stability of the intestinal microbiological environment and regulate intestinal microbiota. A double-center and double-blind randomized clinical trial conducted in colorectal liver metastases patients showed that the incidence of infectious complications after preoperative and postoperative supplementation with probiotics decreased blood *Escherichia coli*, *Staphylococcus aureus*m, and *Aeruginosin* populations, improved intestinal barrier function, and reduced postoperative infection rate [163].

As time-restricted feeding (TRF) is a promising intervention against the worldwide trend of obesity and other metabolic diseases [175], a study conducted in mice investigated whether alteration in gut microbiota caused by TRF could alleviate hepatic I/R injury [164]. The results confirmed the adverse effect of I/R on the gut microbial population. However, TRF prior to surgery reduced the damage, oxidative stress, and inflammatory biomarkers associated with I/R, likely due to intestinal increases in Firmicutes phylum, Clostridia and Bacilli classes, Clostridiales and Lactobacillales orders, and Lachnospiraceae and Ruminococcaceae families, which could be hallmarks of a healthy gut [164].

5. Future Perspectives and Conclusions

The temporary occlusion of hepatic inflow is commonly used during liver resection or LT, creating an unsolved problem in clinical practice associated with post-operative morbidity and mortality.

Experimental studies have shown that liver I/R injury is influenced by various nutrients, suggesting the importance of dietary control for preventing I/R injury.

Today, starvation is not a feasible strategy in clinical practice. Future clinical and preclinical studies on PH with I/R and LT are required to understand the underlying mechanisms of starvation to increase the quality of livers subjected to surgery and reduce the post-operative disorders. Controversial results have been reported in experimental models of starvation under I/R conditions [37,48], which might be explained by the use of different times of ischemia (60 or 90 min). The literature draws upon research data that support the duration of ischemia differentially affects hepatic I/R injury [176–178]. This is of clinical interest since, in clinical practice, the timing of ischemia dependent on the complications associated with surgery cannot be predicted, whereas the effects resulting from starvation are dependent on the duration of ischemia and the duration of starvation. In clinical practice, long-term diet restriction of more than 24 h is difficult to apply for preoperative management in LT. Liver donors are often kept in the intensive care unit for periods no longer than six hours after diagnosis of brain death. The time frame between the declaration of brain death and organ procurement provides a shorter window for the starvation intervention. The effects of starvation on steatotic livers undergoing surgery should be evaluated since the mechanisms responsible for I/R and consequently the useful therapeutic strategies in clinical practice might be different in steatotic and non-steatotic livers submitted to surgery. The number of steatotic livers submitted to surgery is expected to increase, though steatotic livers show regenerative failure responses and reduced tolerance to I/R injury compared with non-steatotic livers. Therefore, research in experimental models of PH with I/R and LT that closely reproduce the clinical conditions is required to understand the underlying mechanisms of starvation, especially in sub-optimal livers.

To summarize, several nutrients and dietary supplements have antioxidant or anti-inflammatory properties and contribute to modifying the gut microbiome. These properties might warrant investigations using them as potential strategies to counteract I/R injury complications and promote regeneration from a nutritional point of view. The diagnosis of nutritional status and its re-establishment and maintenance, as well as providing adequate nutritional support during all phases of the surgery, could be considered the first step to formulating adequate I/R injury therapy. From our view, studies using this approach are insufficient, with only 20 studies from 2014 to 2019, with considerable variability in models, time, and administration. This suggests that the effects of such approaches on hepatic I/R injury are specific for each surgical procedure (for instance, warm ischemia associated with hepatic resections versus LT, times of ischemia, and type of treatment: Short or prolonged fasting).

Most studies based on nutrients and dietary supplements reported benefits on liver function and oxidative stress parameters, but we did not find many studies aimed to improve liver regeneration (six of 20) and only three reported improvements in this parameter. As steatotic grafts show increased vulnerability to I/R when they are transplanted and pre-existing steatosis is related with impairment of liver regeneration following PH [53,141], more than the only three studies performed in steatotic liver seems to be warranted. We only found one study reporting the use of probiotics as a strategy. As a dysbiotic microbiota induces the translocation of several bacterial components into the portal vein and favors the activation of innate immunity and inflammation [114], modulation of gut microbiota from a nutritional point of view is mandatory for evaluating and modifying alterations associated with I/R injury and, in consequence, further studies in this area are needed.

In our view, a strategy more appropriate for clinical practice is the re-establishment and maintenance of the correct nutrient deficiencies using nutraceuticals and functional foods before, during, and/or after surgery, dependent on the patient's requirements. In hepatic resections, this strategy is suitable for the treatment of patients before during or after surgery, whereas in the case of LT, this strategy was only possible after LT with considerable difficulties during liver surgery.

For us, the use of plant-derived supplements, fish, and rosa mosqueta oils show limitations and are inadvisable due their limited specificity and the potential toxicity and side effects of these components. Vitamins, branched-chain amino acid, fatty acids, arginine, and nucleotides can be administered in clinical practice only if deficiencies exist in the patients. Thus, exhaustive studies in patients are required since, for instance, hypervitaminosis is associated with toxic effects. Given the limited studies on the effect of administering vitamins in surgery, conclusions about their efficacy cannot be drawn. Before the administration of fatty acid, the deficiencies in specific types of fatty acid in the patient must be determined. In some cases, for instance EPA supplementation, benefits have been reported but whether the potential benefits are exclusively attributed to EPA is unknown because oral supplementation was also enriched with arginine and nucleotides. Only through exhaustive studies of the patient's deficiencies can we select the most effective treatment for the patient. Unfortunately, these studies are not performed routinely in clinical practice since, in many cases, surgery is performed an emergency situation but the techniques that evaluate such components are complex, time consuming, and expensive.

Although I/R is known to have detrimental effects on the gut microbial population, studies reporting interventions targeting gut microbiota in the I/R setting are limited. A more accurate characterization of the gut microbiome and host responses using different liver surgery models, stages of liver disease, and larger cohorts of patients is required. A comprehensive understanding of the intestine microbiota's role during hepatic surgery is lacking. Maintaining the stability and/or restauration of the intestinal microbiological environment could be a safe and sustainable tool for mitigating I/R injury, which could even effect regeneration. Although regulation of the gut microbiota has been primarily achieved through the use of probiotics, as well as through dietary intervention, studies recently reported using mainly antibiotics and mostly focused on avoiding graft rejection and infectious complications post-surgery [148,158,159,163]. Further investigations are required to elucidate whether personalized and precision medicine approaches based on gut microbiota are necessary dependent on the type of surgical procedure. Dose, frequency, and route of modulation of gut microbiota should be addressed.

Probiotics supplementation requires special consideration. This is associated with the regulation of infections by altering gut microbiota and improvements in inflammation and immunological problems associated with liver surgery. Of clinical interest, gut microbial profiles have been suggested as predictive injury biomarkers in LT. However, before the application of probiotics, an exhaustive examination of the alterations in the intestinal microbiota must be performed for the administration of specific probiotics that counteract such deficiencies in the patients. An alternative to the use of probiotics would be the administration of antibiotics. However, the specificity and the appropriate dose must be determined to prevent harmful effects to ileum epithelial cells and the mucosal barrier. Rapid techniques that routinely evaluate intestinal microflora would be necessary if the aim is to establish probiotics as a useful strategy in clinical of liver surgery, especially in LT. Consequently, nutritional support must be personalized based on the patient's deficiencies. To date, I/R injury is a common complication for patients undergoing liver surgery and its relationship with changes in the gut microbiota is not totally understood. The understanding of such changes and mechanisms involved could help with restoring unhealthy microbial diversity and the richness of species, providing a potential therapeutic tool for treating I/R damage.

Author Contributions: Conceptualization: M.B.J.-C., and C.P.; Writing—Original Draft Preparation: M.E.C.-P., A.I.Á.-M., M.B.J.-C., and C.P.; Writing—Review and Editing: M.E.C.-P., A.I.Á.-M., M.B.J.-C., and C.P.; Funding Acquisition: C.P. All authors have read and agreed to the published version of the manuscript.

Funding: This research was supported by the Ministerio de Ciencia, Innovacion y Universidades (RTI2018-095114-B-I00) Madrid, Spain; European Union (Fondos Feder, "una manera de hacer Europa"); CERCA Program/Generalitat de Catalunya and Secretaria d'Universitats I Recerca del Departament d'Economia I Coneixement (2017 SGR-551) Barcelona, Spain.

Conflicts of Interest: The authors declare no conflict of interest.

Abbreviations

AKT	Protein kinase B
ALA	α-linolenic acid
ALD	Alcoholic liver disease
AMPK	AMP-activated protein kinase
ATP	Adenosine triphosphate
BCAA	Branched-chain amino acid
BCL	B-cell lymphoma
BHB	β-hydroxybutyric acid
CAT	Catalase
Co A	Coenzyme A
DHA	Docosahexaenoic acid
EPA	Eicosapentaenoic acid
ESPEN	European Society for Parenteral and Enteral Nutrition
FOXO1	Forkhead box protein O1
GSH	Glutathione
HBeAg	Hepatitis B virus e-antigen
HCC	Hepatocellular carcinoma
HMGB1	High mobility group box 1
HO-1	Heme oxygenase 1
HWP	Hydrolyzed whey peptide
I/R	Ischemia-reperfusion
IFNγ	Interferon-gamma
IL	Interleukin
LC-PUFAs	Long-chain PUFAs
LSEC	Liver sinusoidal endothelial cells
LT	Liver transplantation
NAFLD	Nonalcoholic fatty liver disease
NASH	Nonalcoholic steatohepatitis
NF-κB	Nuclear factor kappa-light-chain-enhancer of activated B cells
NKT	Natural killer T
NLRP3	Nucleotide oligomerization domain-like receptor family, pyrin domain containing protein 3
NPO	*Nil per os*
Nqo1	NAD(P)H quinone dehydrogenase 1
Nrf2	Nuclear factor erythroid-derived 2-related factor 2
PA	Pantothenic acid
PH	Partial hepatectomy
PUFAs	Polyunsaturated fatty acids
Sirt1	Sirtuin 1
SOD	Superoxide dismutase
TGF-β	Tumor growth factor beta
TNF-α	Tumor necrosis factor alpha
TRF	Time restricted feeding

References

1. Peralta, C.; Jiménez-Castro, M.B.; Gracia-Sancho, J. Hepatic ischemia and reperfusion injury: Effects on the liver sinusoidal milieu. *J. Hepatol.* **2013**, *59*, 1094–1106. [CrossRef]
2. Fu, P.; Li, W. Nitric oxide in liver ischemia-reperfusion injury. In *Liver Pathophysiology*; Muriel, P., Ed.; Elsevier Inc.: London, UK, 2017; Volume 8, pp. 125–127.
3. Selzner, N.; Rudiger, H.; Graf, R.; Clavien, P. Protective strategies against ischemic injury of the liver. *Gastroenterology* **2003**, *125*, 917–936. [CrossRef]
4. Jaeschke, H. Molecular mechanisms of hepatic ischemia-reperfusion injury and preconditioning. *Am. J. Physiol. Gastrointest. Liver Physiol.* **2003**, *284*, G15–G26. [CrossRef]

5. Montalvo-Jave, E.E.; Escalante-Tattersfield, T.; Ortega-Salgado, J.A.; Piña, E.; Geller, D.A. Factors in the pathophysiology of the liver ischemia-reperfusion injury. *J. Surg. Res.* **2008**, *147*, 153–159. [CrossRef] [PubMed]
6. Gracia-Sancho, J.; Villarreal, G., Jr.; Zhang, Y.; Yu, J.X.; Liu, Y.; Tullius, S.G.; García-Cardeña, G. Flow cessation triggers endothelial dysfunction during organ cold storage conditions: Strategies for pharmacologic intervention. *Transplantation* **2010**, *90*, 142–149. [CrossRef] [PubMed]
7. Gracia-Sancho, J.; Casillas-Ramírez, A.; Peralta, C. Molecular pathways in protecting the liver from ischaemia/reperfusion injury: A 2015 update. *Clin. Sci.* **2015**, *129*, 345–362. [CrossRef] [PubMed]
8. Ramalho, F.; Alfany-Fernandez, I.; Casillas-Ramírez, A.; Massip-Salcedo, M.; Serafín, A.; Rimola, A.; Arroyo, V.; Rodes, J.; Rosello-Catafau, J.; Peralta, C. Are angiotensin II receptor antagonists useful strategies in steatotic and nonsteatotic livers in conditions of partial hepatectomy under ischemia-reperfusion? *J. Pharmacol. Exp. Ther.* **2009**, *329*, 130–140. [CrossRef]
9. Ploeg, R.J.; D'Alessandro, A.M.; Knechtle, S.J.; Stegall, M.D.; Pirsch, J.D.; Hoffmann, R.M.; Sasaki, T.; Sollinger, H.W.; Belzer, F.O.; Kalayoglu, M. Risk factors for primary dysfunction after liver transplantation—A multivariate analysis. *Transplantation* **1993**, *55*, 807–813. [CrossRef]
10. Behrns, K.E.; Tsiotos, G.G.; DeSouza, N.F.; Krishna, M.K.; Ludwig, J.; Nagorney, D.M. Hepatic steatosis as a potential risk factor for major hepatic resection. *J. Gastrointest. Surg.* **1998**, *2*, 292–298. [CrossRef]
11. Selzner, M.; Clavien, P.A. Fatty liver in liver transplantation and surgery. *Semin. Liver Dis.* **2001**, *21*, 105–113. [CrossRef]
12. D'Alessandro, A.M.; Kalayoglu, M.; Sollinger, H.W.; Hoffmann, R.M.; Reed, A.; Knechtle, S.J.; Pirsch, J.D.; Hafez, G.R.; Lorentzen, D.; Belzer, F.O. The predictive value of donor liver biopsies for the development of primary nonfunction after orthotopic liver transplantation. *Transplantation* **1991**, *51*, 157–163. [CrossRef] [PubMed]
13. Adam, R.; Reynes, M.; Johann, M.; Morino, M.; Astarcioglu, I.; Kafetzis, I.; Castaing, D.; Bismuth, H. The outcome of steatotic grafts in liver transplantation. *Transplant Proc.* **1991**, *23*, 1538–1540. [PubMed]
14. Todo, S.; Demetris, A.J.; Makowka, L.; Teperman, L.; Podesta, L.; Shaver, T.; Tzakis, A.; Starzl, T.E. Primary nonfunction of hepatic allografts with preexisting fatty infiltration. *Transplantation* **1989**, *47*, 903–905. [CrossRef] [PubMed]
15. Belghiti, J.; Hiramatsu, K.; Benoist, S.; Massault, P.; Sauvanet, A.; Farges, O. Seven hundred forty-seven hepatectomies in the 1990s: An update to evaluate the actual risk of liver resection. *J. Am. Coll. Surg.* **2000**, *191*, 38–46. [CrossRef]
16. Safwan, M.; Collins, K.M.; Abouljoud, M.S.; Salgia, R. Outcome of liver transplantation in patients with prior bariatric surgery. *Liver Transplant.* **2017**, *23*, 1415–1421. [CrossRef] [PubMed]
17. Saïdi, S.A.; Abdelkafi, S.; Jbahi, S.; Van Pelt, J.; El-Feki, A. Temporal changes in hepatic antioxidant enzyme activities after ischemia and reperfusion in a rat liver ischemia model: Effect of dietary fish oil. *Hum. Exp. Toxicol.* **2015**, *34*, 249–259. [CrossRef]
18. Silva, R.M.; Malafaia, O.; Torres, O.J.; Czeczko, N.G.; Marinho Junior, C.H.; Kozlowski, R.K. Evaluation of liver regeneration diet supplemented with omega-3 fatty acids: Experimental study in rats. *Rev. Col. Bras. Cir.* **2015**, *42*, 393–397. [CrossRef]
19. Caraceni, P.; Nardo, B.; Domenicali, M.; Turi, P.; Vici, M.; Simoncini, M.; De Maria, N.; Trevisani, F.; Van Thiel, D.H.; Derenzini, M.; et al. Ischemia-reperfusion injury in rat fatty liver: Role of nutritional status. *Hepatology* **1999**, *29*, 1139–1146. [CrossRef]
20. Gasbarrini, A.; Borle, A.B.; Farghali, H.; Caraceni, P.; Van Thiel, D. Fasting enhances the effects of anoxia on ATP, Cai2+ and cell injury in isolated rat hepatocytes. *Biochim. Biophys. Acta* **1993**, *1178*, 9–19. [CrossRef]
21. Bradford, B.U.; Marotto, M.; Lemasters, J.J.; Thurman, R.G. New, simple models to evaluate zone-specific damage due to hypoxia in the perfused rat liver: Time course and effect of nutritional state. *J. Pharmacol. Exp. Ther.* **1986**, *236*, 263–268.
22. Tanigawa, K.; Kim, Y.M.; Lancaster, J.R., Jr.; Zar, H.A. Fasting augments lipid peroxidation during reperfusion after ischemia in the perfused rat liver. *Crit. Care Med.* **1999**, *27*, 401–406. [CrossRef] [PubMed]
23. Jimenez-Castro, M.B.; Casillas-Ramirez, A.; Massip-Salcedo, M.; Elias-Miro, M.; Serafin, A.; Rimola, A.; Rodes, J.; Peralta, C. Cyclic adenosine 3′,5′-monophosphate in rat steatotic liver transplantation. *Liver Transplant.* **2011**, *17*, 1099–1110.

24. Hammad, A.; Kaido, T.; Uemoto, S. Perioperative nutritional therapy in liver transplantation. *Surg. Today* **2015**, *45*, 271–283. [CrossRef] [PubMed]
25. Chiarla, C.; Giovannini, I.; Giuliante, F.; Ardito, F.; Vellone, M.; De Rose, A.M.; Nuzzo, G. Parenteral nutrition in liver resection. *J. Nutr. Metab.* **2012**, *2012*, 508103. [CrossRef] [PubMed]
26. Sonnenburg, J.L.; Bäckhed, F. Diet-microbiota interactions as moderators of human metabolism. *Nature* **2016**, *535*, 56–64. [CrossRef] [PubMed]
27. Wahlström, A.; Sayin, S.I.; Marschall, H.U.; Bäckhed, F. Intestinal Crosstalk between Bile Acids and Microbiota and Its Impact on Host Metabolism. *Cell Metab.* **2016**, *24*, 41–50. [CrossRef]
28. Moschen, A.R.; Kaser, S.; Tilg, H. Non-alcoholic steatohepatitis: A microbiota-driven disease. *Trends Endocrinol. Metab.* **2013**, *24*, 537–545. [CrossRef]
29. Acharya, C.; Sahingur, S.E.; Bajaj, J.S. Microbiota, cirrhosis, and the emerging oral-gut-liver axis. *JCI Insight* **2017**, *2*, 94416. [CrossRef]
30. Abu-Shanab, A.; Quigley, E.M. The role of the gut microbiota in nonalcoholic fatty liver disease. *Nat. Rev. Gastroenterol. Hepatol.* **2010**, *7*, 691–701. [CrossRef]
31. Ahuja, M.; Schwartz, D.M.; Tandon, M.; Son, A.; Zeng, M.; Swaim, W.; Eckhaus, M.; Hoffman, V.; Cui, Y.; Xiao, B.; et al. Orai1-Mediated Antimicrobial Secretion from Pancreatic Acini Shapes the Gut Microbiome and Regulates Gut Innate Immunity. *Cell Metab.* **2017**, *25*, 635–646. [CrossRef]
32. Hine, C.; Harputlugil, E.; Zhang, Y.; Ruckenstuhl, C.; Lee, B.C.; Brace, L.; Longchamp, A.; Treviño-Villarreal, J.H.; Mejia, P.; Ozaki, C.K.; et al. Endogenous hydrogen sulfide production is essential for dietary restriction benefits. *Cell* **2015**, *160*, 132–144. [CrossRef] [PubMed]
33. Miyauchi, T.; Uchida, Y.; Kadono, K.; Hirao, H.; Kawasoe, J.; Watanabe, T.; Ueda, S.; Jobara, K.; Kaido, T.; Okajima, H.; et al. Preventive Effect of Antioxidative Nutrient-Rich Enteral Diet Against Liver Ischemia and Reperfusion Injury. *JPEN J. Parenter. Enteral Nutr.* **2019**, *43*, 133–144. [CrossRef] [PubMed]
34. Mitchell, J.R.; Verweij, M.; Brand, K.; van de Ven, M.; Goemaere, N.; van den Engel, S.; Chu, T.; Forrer, F.; Müller, C.; de Jong, M.; et al. Short-term dietary restriction and fasting precondition against ischemia reperfusion injury in mice. *Aging Cell* **2010**, *9*, 40–53. [CrossRef] [PubMed]
35. Domenicali, M.; Caraceni, P.; Vendemiale, G.; Grattagliano, I.; Nardo, B.; Dall'Agata, M.; Santoni, B.; Trevisani, F.; Cavallari, A.; Altomare, E.; et al. Food deprivation exacerbates mitochondrial oxidative stress in rat liver exposed to ischemia-reperfusion injury. *J. Nutr.* **2001**, *131*, 105–110. [CrossRef] [PubMed]
36. Van Ginhoven, T.M.; Mitchell, J.R.; Verweij, M.; Hoeijmakers, J.H.; Ijzermans, J.N.; de Bruin, R.W. The use of preoperative nutritional interventions to protect against hepatic ischemia-reperfusion injury. *Liver Transplant.* **2009**, *15*, 1183–1191. [CrossRef]
37. Rickenbacher, A.; Jang, J.H.; Limani, P.; Ungethüm, U.; Lehmann, K.; Oberkofler, C.E.; Weber, A.; Graf, R.; Humar, B.; Clavien, P.A. Fasting protects liver from ischemic injury through Sirt1-mediated downregulation of circulating HMGB1 in mice. *J. Hepatol.* **2014**, *61*, 301–308. [CrossRef]
38. Awad, S.; Varadhan, K.K.; Ljungqvist, O.; Lobo, D.N. A meta-analysis of randomized controlled trials on preoperative oral carbohydrate treatment in elective surgery. *Clin. Nutr.* **2013**, *32*, 34–44. [CrossRef]
39. Maltby, J.R.; Sutherland, A.D.; Sale, J.P.; Shaffer, E.A. Preoperative oral fluids: Is a five-hour fast justified prior to elective surgery? *Anesth. Analg.* **1986**, *65*, 1112–1116. [CrossRef]
40. Page, A.J.; Ejaz, A.; Spolverato, G.; Zavadsky, T.; Grant, M.C.; Galante, D.J.; Wick, E.C.; Weiss, M.; Makary, M.A.; Wu, C.L.; et al. Enhanced recovery after surgery protocols for open hepatectomy—Physiology, immunomodulation, and implementation. *J. Gastrointest. Surg.* **2015**, *19*, 387–399. [CrossRef]
41. Ljungqvist, O.; Nygren, J.; Thorell, A. Modulation of post-operative insulin resistance by pre-operative carbohydrate loading. *Proc. Nutr. Soc.* **2002**, *61*, 329–336. [CrossRef]
42. Soop, M.; Nygren, J.; Myrenfors, P.; Thorell, A.; Ljungqvist, O. Preoperative oral carbohydrate treatment attenuates immediate postoperative insulin resistance. *Am. J. Physiol. Endocrinol. Metab.* **2001**, *280*, E576–E583. [CrossRef] [PubMed]
43. Hausel, J.; Nygren, J.; Lagerkranser, M.; Hellström, P.M.; Hammarqvist, F.; Almström, C.; Lindh, A.; Thorell, A.; Ljungqvist, O. A carbohydrate-rich drink reduces preoperative discomfort in elective surgery patients. *Anesth. Analg.* **2001**, *93*, 1344–1350. [CrossRef] [PubMed]
44. Eshuis, W.J.; Hermanides, J.; van Dalen, J.W.; van Samkar, G.; Busch, O.R.; van Gulik, T.M.; DeVries, J.H.; Hoekstra, J.B.; Gouma, D.J. Early postoperative hyperglycemia is associated with postoperative complications after pancreatoduodenectomy. *Ann. Surg.* **2011**, *253*, 739–744. [CrossRef] [PubMed]

45. Pruim, J.; van Woerden, W.F.; Knol, E.; Klompmaker, I.J.; de Bruijn, K.M.; Persijn, G.G.; Slooff, M.J. Donor data in liver grafts with primary non-function–a preliminary analysis by the European Liver Registry. *Transplant Proc.* **1989**, *21*, 2383–2384. [PubMed]
46. Miyauchi, T.; Uchida, Y.; Kadono, K.; Hirao, H.; Kawasoe, J.; Watanabe, T.; Ueda, S.; Okajima, H.; Terajima, H.; Uemoto, S. Up-regulation of FOXO1 and reduced inflammation by β-hydroxybutyric acid are essential diet restriction benefits against liver injury. *Proc. Natl. Acad. Sci. USA* **2019**, *116*, 13533–13542. [CrossRef]
47. Papegay, B.; Stadler, M.; Nuyens, V.; Kruys, V.; Boogaerts, J.G.; Vamecq, J. Short fasting does not protect perfused ex vivo rat liver against ischemia-reperfusion. On the importance of a minimal cell energy charge. *Nutrition* **2017**, *35*, 21–27. [CrossRef]
48. Qin, J.; Zhou, J.; Dai, X.; Zhou, H.; Pan, X.; Wang, X.; Zhang, F.; Rao, J.; Lu, L. Short-term starvation attenuates liver ischemia-reperfusion injury (IRI) by Sirt1-autophagy signaling in mice. *Am. J. Transl. Res.* **2016**, *8*, 3364–3375.
49. Zhan, C.; Dai, X.; Shen, G.; Lu, X.; Wang, X.; Lu, L.; Qian, X.; Rao, J. Preoperative short-term fasting protects liver injury in patients undergoing hepatectomy. *Ann. Transl. Med.* **2018**, *6*, 449. [CrossRef]
50. Mauro, C.R.; Tao, M.; Yu, P.; Treviño-Villerreal, J.H.; Longchamp, A.; Kristal, B.S.; Ozaki, C.K.; Mitchell, J.R. Preoperative dietary restriction reduces intimal hyperplasia and protects from ischemia-reperfusion injury. *J. Vasc. Surg.* **2016**, *63*, 500–509. [CrossRef]
51. De Meijer, V.E.; Kalish, B.T.; Puder, M.; Ijzermans, J.N. Systematic review and meta-analysis of steatosis as a risk factor in major hepatic resection. *Br. J. Surg.* **2010**, *97*, 1331–1339. [CrossRef]
52. Bachellier, P.; Rosso, E.; Pessaux, P.; Oussoultzoglou, E.; Nobili, C.; Panaro, F.; Jaeck, D. Risk factors for liver failure and mortality after hepatectomy associated with portal vein resection. *Ann. Surg.* **2011**, *253*, 173–179. [CrossRef] [PubMed]
53. Álvarez-Mercado, A.I.; Bujaldon, E.; Gracia-Sancho, J.; Peralta, C. The Role of Adipokines in Surgical Procedures Requiring Both Liver Regeneration and Vascular Occlusion. *Int. J. Mol. Sci.* **2018**, *19*, 3395. [CrossRef] [PubMed]
54. López-Velázquez, J.A.; Silva-Vidal, K.V.; Ponciano-Rodríguez, G.; Chávez-Tapia, N.C.; Arrese, M.; Uribe, M.; Méndez-Sánchez, N. The prevalence of nonalcoholic fatty liver disease in the Americas. *Ann. Hepatol.* **2014**, *13*, 166–178. [CrossRef]
55. Vasco, M.; Paolillo, R.; Schiano, C.; Sommese, L.; Cuomo, O.; Napoli, C. Compromised nutritional status in patients with end-stage liver disease: Role of gut microbiota. *Hepatobiliary Pancreat. Dis. Int.* **2018**, *17*, 290–300. [CrossRef] [PubMed]
56. Chandrasekara, A.; Josheph Kumar, T. Roots and Tuber Crops as Functional Foods: A Review on Phytochemical Constituents and Their Potential Health Benefits. *Int. J. Food Sci.* **2016**, *2016*, 3631647. [CrossRef] [PubMed]
57. Bakshi, N.; Singh, K. Nutrition assessment in patients undergoing liver transplant. *Indian J. Crit. Care Med.* **2014**, *18*, 672–681. [CrossRef]
58. Yang, H.J.; Tang, L.M.; Zhou, X.J.; Qian, J.; Zhu, J.; Lu, L.; Wang, X.H. Ankaflavin ameliorates steatotic liver ischemia-reperfusion injury in mice. *Hepatobiliary Pancreat. Dis. Int.* **2015**, *14*, 619–625. [CrossRef]
59. Yücel, A.; Aydogan, M.S.; Ucar, M.; Sarıcı, K.B.; Karaaslan, M.G. Effects of Apocynin on Liver Ischemia-Reperfusion Injury in Rats. *Transplant Proc.* **2019**, *51*, 1180–1183. [CrossRef]
60. Kim, H.; Hong, M.K.; Choi, H.; Moon, H.S.; Lee, H.J. Chemopreventive effects of korean red ginseng extract on rat hepatocarcinogenesis. *J. Cancer* **2015**, *6*, 1–8. [CrossRef]
61. Ucar, M.; Aydogan, M.S.; Vardı, N.; Parlakpınar, H. Protective Effect of Dexpanthenol on Ischemia-Reperfusion-Induced Liver Injury. *Transplant Proc.* **2018**, *50*, 3135–3143. [CrossRef]
62. Reynolds, P.S.; Fisher, B.J.; McCarter, J.; Sweeney, C.; Martin, E.J.; Middleton, P.; Ellenberg, M.; Fowler, E.; Brophy, D.F.; Fowler, A.A., 3rd; et al. Interventional vitamin C: A strategy for attenuation of coagulopathy and inflammation in a swine multiple injuries model. *J. Trauma Acute Care Surg.* **2018**, *85*, S57–S67. [CrossRef] [PubMed]
63. Dossi, C.G.; González-Mañán, D.; Romero, N.; Silva, D.; Videla, L.A.; Tapia, G.S. Anti-oxidative and anti-inflammatory effects of Rosa Mosqueta oil supplementation in rat liver ischemia-reperfusion. *Food Funct.* **2018**, *9*, 4847–4857. [CrossRef] [PubMed]

64. Yao, H.; Fu, X.; Zi, X.; Jia, W.; Qiu, Y. Perioperative oral supplementation with fish oil promotes liver regeneration following partial hepatectomy in mice via AMPK activation. *Mol. Med. Rep.* **2018**, *17*, 3905–3911. [CrossRef] [PubMed]
65. Montenegro, W.S.; Malafaia, O.; Nassif, P.A.; Moreira, L.B.; Prestes, M.A.; Kume, M.H.; Jurkonis, L.B.; Cella, I.F. Evaluation of liver regeneration with use of diet supplemented with L-arginine. *Acta Cir. Bras.* **2014**, *29*, 603–607. [CrossRef] [PubMed]
66. Magalhães, C.R.; Malafaia, O.; Torres, O.J.; Moreira, L.B.; Tefil, S.C.; Pinherio Mda, R.; Harada, B.A. Liver regeneration with l-glutamine supplemented diet: Experimental study in rats. *Rev. Col. Bras. Cir.* **2014**, *41*, 117–121. [CrossRef]
67. Akbari, M.; Celik, S.U.; Kocaay, A.F.; Cetinkaya, O.A.; Demirer, S. Omega-3 fatty acid supplementation does not influence liver regeneration in rats after partial hepatectomy. *Clin. Exp. Hepatol.* **2018**, *4*, 253–259. [CrossRef]
68. Uno, H.; Furukawa, K.; Suzuki, D.; Shimizu, H.; Ohtsuka, M.; Kato, A.; Yoshitomi, H.; Miyazaki, M. Immunonutrition suppresses acute inflammatory responses through modulation of resolvin E1 in patients undergoing major hepatobiliary resection. *Surgery* **2016**, *160*, 228–236. [CrossRef]
69. Russell, K.; Zhang, H.G.; Gillanders, L.K.; Bartlett, A.S.; Fisk, H.L.; Calder, P.C.; Swan, P.J.; Plank, L.D. Preoperative immunonutrition in patients undergoing liver resection: A prospective randomized trial. *World J. Hepatol.* **2019**, *11*, 305–317. [CrossRef]
70. Kamo, N.; Kaido, T.; Hamaguchi, Y.; Uozumi, R.; Okumura, S.; Kobayashi, A.; Shirai, H.; Yagi, S.; Okajima, H.; Uemoto, S. Impact of Enteral Nutrition with an Immunomodulating Diet Enriched with Hydrolyzed Whey Peptide on Infection After Liver Transplantation. *World J. Surg.* **2018**, *42*, 3715–3725. [CrossRef]
71. Nii, A.; Utsunomiya, T.; Shimada, M.; Ikegami, T.; Ishibashi, H.; Imura, S.; Morine, Y.; Ikemoto, T.; Sasaki, H.; Kawashima, A. Hydrolyzed whey peptide-based diet ameliorates hepatic ischemia-reperfusion injury in the rat nonalcoholic fatty liver. *Surg. Today* **2014**, *44*, 2354–2360. [CrossRef]
72. Mendes-Braz, M.; Elias-Miró, M.; Kleuser, B.; Fayyaz, S.; Jiménez-Castro, M.B.; Massip-Salcedo, M.; Gracia-Sancho, J.; Ramalho, F.S.; Rodes, J.; Peralta, C. The effects of glucose and lipids in steatotic and non-steatotic livers in conditions of partial hepatectomy under ischaemia-reperfusion. *Liver Int.* **2014**, *34*, e271–e289. [CrossRef]
73. Nanno, Y.; Toyama, H.; Terai, S.; Mizumoto, T.; Tanaka, M.; Kido, M.; Ajiki, T.; Fukumoto, T. Preoperative Oral Branched-Chain Amino Acid Supplementation Suppresses Intraoperative and Postoperative Blood Lactate Levels in Patients Undergoing Major Hepatectomy. *JPEN J. Parenter. Enteral Nutr.* **2019**, *43*, 220–225. [CrossRef]
74. Beppu, T.; Nitta, H.; Hayashi, H.; Imai, K.; Okabe, H.; Nakagawa, S.; Hashimoto, D.; Chikamoto, A.; Ishiko, T.; Yoshida, M.; et al. Effect of branched-chain amino acid supplementation on functional liver regeneration in patients undergoing portal vein embolization and sequential hepatectomy: A randomized controlled trial. *J. Gastroenterol.* **2015**, *50*, 1197–1205. [CrossRef] [PubMed]
75. Müller, M.J. Malnutrition in cirrhosis. *J. Hepatol.* **1995**, *23* (Suppl. S1), 31–35.
76. Riggio, O.; Ariosto, F.; Merli, M.; Caschera, M.; Zullo, A.; Balducci, G.; Ziparo, V.; Pedretti, G.; Fiaccadori, F.; Bottari, E.; et al. Short-term oral zinc supplementation does not improve chronic hepatic encephalopathy. Results of a double-blind crossover trial. *Dig. Dis. Sci.* **1991**, *36*, 1204–1208. [CrossRef]
77. Gorelick, P.B.; Counts, S.E.; Nyenhuis, D. Vascular cognitive impairment and dementia. *Biochim. Biophys. Acta* **2016**, *1862*, 860–868. [CrossRef]
78. Gemperlein, K.; Dietrich, D.; Kohlstedt, M.; Zipf, G.; Bernauer, H.S.; Wittmann, C.; Wenzel, S.C.; Müller, R. Polyunsaturated fatty acid production by Yarrowia lipolytica employing designed myxobacterial PUFA synthases. *Nat. Commun.* **2019**, *10*, 4055. [CrossRef]
79. Senkal, M.; Mumme, A.; Eickhoff, U.; Geier, B.; Späth, G.; Wulfert, D.; Joosten, U.; Frei, A.; Kemen, M. Early postoperative enteral immunonutrition: Clinical outcome and cost-comparison analysis in surgical patients. *Crit. Care Med.* **1997**, *25*, 1489–1496. [CrossRef]
80. Barbul, A.; Fishel, R.S.; Shimazu, S.; Wasserkrug, H.L.; Yoshimura, N.N.; Tao, R.C.; Efron, G. Intravenous hyperalimentation with high arginine levels improves wound healing and immune function. *J. Surg. Res.* **1985**, *38*, 328–334. [CrossRef]
81. Bifari, F.; Nisoli, E. Branched-chain amino acids differently modulate catabolic and anabolic states in mammals: A pharmacological point of view. *Br. J. Pharmacol.* **2017**, *174*, 1366–1377. [CrossRef]

82. Malaguarnera, G.; Giordano, M.; Nunnari, G.; Bertino, G.; Malaguarnera, M. Gut microbiota in alcoholic liver disease: Pathogenetic role and therapeutic perspectives. *World J. Gastroenterol.* **2014**, *20*, 16639–16648. [CrossRef] [PubMed]
83. Jones, S.E.; Versalovic, J. Probiotic Lactobacillus reuteri biofilms produce antimicrobial and anti-inflammatory factors. *BMC Microbiol.* **2009**, *9*, 35. [CrossRef] [PubMed]
84. Borruel, N.; Carol, M.; Casellas, F.; Antolín, M.; de Lara, F.; Espín, E.; Naval, J.; Guarner, F.; Malagelada, J.R. Increased mucosal tumour necrosis factor alpha production in Crohn's disease can be downregulated ex vivo by probiotic bacteria. *Gut* **2002**, *51*, 659–664. [CrossRef] [PubMed]
85. McCarthy, J.; O'Mahony, L.; O'Callaghan, L.; Sheil, B.; Vaughan, E.E.; Fitzsimons, N.; Fitzgibbon, J.; O'Sullivan, G.C.; Kiely, B.; Collins, J.K.; et al. Double blind, placebo controlled trial of two probiotic strains in interleukin 10 knockout mice and mechanistic link with cytokine balance. *Gut* **2003**, *52*, 975–980. [CrossRef] [PubMed]
86. Sheth, A.A.; Garcia-Tsao, G. Probiotics and liver disease. *J. Clin. Gastroenterol.* **2008**, *42* (Suppl. S2), S80–S84. [CrossRef] [PubMed]
87. Rayes, N.; Seehofer, D.; Theruvath, T.; Schiller, R.A.; Langrehr, J.M.; Jonas, S.; Bengmark, S.; Neuhaus, P. Supply of pre- and probiotics reduces bacterial infection rates after liver transplantation—A randomized, double-blind trial. *Am. J. Transplant* **2005**, *5*, 125–130. [CrossRef]
88. Stadlbauer, V.; Mookerjee, R.P.; Hodges, S.; Wright, G.A.; Davies, N.A.; Jalan, R. Effect of probiotic treatment on deranged neutrophil function and cytokine responses in patients with compensated alcoholic cirrhosis. *J. Hepatol.* **2008**, *48*, 945–951. [CrossRef]
89. Sharma, P.; Sharma, B.C.; Puri, V.; Sarin, S.K. An open-label randomized controlled trial of lactulose and probiotics in the treatment of minimal hepatic encephalopathy. *Eur. J. Gastroenterol. Hepatol.* **2008**, *20*, 506–511. [CrossRef]
90. Bajaj, J.S.; Saeian, K.; Christensen, K.M.; Hafeezullah, M.; Varma, R.R.; Franco, J.; Pleuss, J.A.; Krakower, G.; Hoffmann, R.G.; Binion, D.G. Probiotic yogurt for the treatment of minimal hepatic encephalopathy. *Am. J. Gastroenterol.* **2008**, *103*, 1707–1715. [CrossRef]
91. Stickel, F.; Hoehn, B.; Schuppan, D.; Seitz, H.K. Review article: Nutritional therapy in alcoholic liver disease. *Aliment. Pharmacol. Ther.* **2003**, *18*, 357–373. [CrossRef]
92. Halsted, C.H. Nutrition and alcoholic liver disease. *Semin. Liver Dis.* **2004**, *24*, 289–304. [CrossRef] [PubMed]
93. Elwyn, D.H.; Kinney, J.M.; Askanazi, J. Energy expenditure in surgical patients. *Surg. Clin. N. Am.* **1981**, *61*, 545–556. [CrossRef]
94. Cabré, E.; Periago, J.L.; Abad-Lacruz, A.; González-Huix, F.; González, J.; Esteve-Comas, M.; Fernández-Bañares, F.; Planas, R.; Gil, A.; Sánchez-Medina, F.; et al. Plasma fatty acid profile in advanced cirrhosis: Unsaturation deficit of lipid fractions. *Am. J. Gastroenterol.* **1990**, *85*, 1597–1604. [CrossRef]
95. Hasse, J.M.; Blue, L.S.; Liepa, G.U.; Goldstein, R.M.; Jennings, L.W.; Mor, E.; Husberg, B.S.; Levy, M.F.; Gonwa, T.A.; Klintmalm, G.B. Early enteral nutrition support in patients undergoing liver transplantation. *JPEN J. Parenter. Enteral Nutr.* **1995**, *19*, 437–443. [CrossRef] [PubMed]
96. Lochs, H.; Plauth, M. Liver cirrhosis: Rationale and modalities for nutritional support—The European Society of Parenteral and Enteral Nutrition consensus and beyond. *Curr. Opin. Clin. Nutr. Metab. Care* **1999**, *2*, 345–349. [CrossRef]
97. Fan, S.T.; Lo, C.M.; Lai, E.C.; Chu, K.M.; Liu, C.L.; Wong, J. Perioperative nutritional support in patients undergoing hepatectomy for hepatocellular carcinoma. *N. Engl. J. Med.* **1994**, *331*, 1547–1552. [CrossRef]
98. Morgan, M.Y.; Madden, A.M.; Jennings, G.; Elia, M.; Fuller, N.J. Two-component models are of limited value for the assessment of body composition in patients with cirrhosis. *Am. J. Clin. Nutr.* **2006**, *84*, 1151–1162. [CrossRef]
99. Plauth, M.; Cabré, E.; Riggio, O.; Assis-Camilo, M.; Pirlich, M.; Kondrup, J.; Ferenci, P.; Holm, E.; Vom Dahl, S.; DGEM (German Society for Nutritional Medicine); et al. ESPEN Guidelines on Enteral Nutrition: Liver disease. *Clin. Nutr.* **2006**, *25*, 285–294. [CrossRef]
100. DiCecco, S.R.; Francisco-Ziller, N. Nutrition in alcoholic liver disease. *Nutr. Clin. Pract.* **2006**, *21*, 245–254. [CrossRef]
101. Weimann, A.; Braga, M.; Harsanyi, L.; Laviano, A.; Ljungqvist, O.; Soeters, P.; Jauch, K.W.; Kemen, M.; Hiesmayr, J.M.; DGEM (German Society for Nutritional Medicine); et al. ESPEN Guidelines on Enteral Nutrition: Surgery including organ transplantation. *Clin. Nutr.* **2006**, *25*, 224–244. [CrossRef]

102. Sarin, S.K.; Pande, A.; Schnabl, B. Microbiome as a therapeutic target in alcohol-related liver disease. *J. Hepatol.* **2019**, *70*, 260–272. [CrossRef] [PubMed]
103. Fulde, M.; Hornef, M.W. Maturation of the enteric mucosal innate immune system during the postnatal period. *Immunol. Rev.* **2014**, *260*, 21–34. [CrossRef] [PubMed]
104. Kamada, N.; Chen, G.Y.; Inohara, N.; Núñez, G. Control of pathogens and pathobionts by the gut microbiota. *Nat. Immunol.* **2013**, *14*, 685–690. [CrossRef] [PubMed]
105. Ijssennagger, N.; Belzer, C.; Hooiveld, G.J.; Dekker, J.; van Mil, S.W.; Müller, M.; Kleerebezem, M.; van der Meer, R. Gut microbiota facilitates dietary heme-induced epithelial hyperproliferation by opening the mucus barrier in colon. *Proc. Natl. Acad. Sci. USA* **2015**, *112*, 10038–10043. [CrossRef]
106. Reinhardt, C.; Bergentall, M.; Greiner, T.U.; Schaffner, F.; Ostergren-Lundén, G.; Petersen, L.C.; Ruf, W.; Bäckhed, F. Tissue factor and PAR1 promote microbiota-induced intestinal vascular remodelling. *Nature* **2012**, *483*, 627–631. [CrossRef]
107. Yano, J.M.; Yu, K.; Donaldson, G.P.; Shastri, G.G.; Ann, P.; Ma, L.; Nagler, C.R.; Ismagilov, R.F.; Mazmanian, S.K.; Hsiao, E.Y. Indigenous bacteria from the gut microbiota regulate host serotonin biosynthesis. *Cell* **2015**, *161*, 264–276. [CrossRef]
108. Neuman, H.; Debelius, J.W.; Knight, R.; Koren, O. Microbial endocrinology: The interplay between the microbiota and the endocrine system. *FEMS Microbiol. Rev.* **2015**, *39*, 509–521. [CrossRef]
109. Canfora, E.E.; Jocken, J.W.; Blaak, E.E. Short-chain fatty acids in control of body weight and insulin sensitivity. *Nat. Rev. Endocrinol.* **2015**, *11*, 577–591. [CrossRef]
110. Yatsunenko, T.; Rey, F.E.; Manary, M.J.; Trehan, I.; Dominguez-Bello, M.G.; Contreras, M.; Magris, M.; Hidalgo, G.; Baldassano, R.N.; Anokhin, A.P.; et al. Human gut microbiome viewed across age and geography. *Nature* **2012**, *486*, 222–227. [CrossRef]
111. Devlin, A.S.; Fischbach, M.A. A biosynthetic pathway for a prominent class of microbiota-derived bile acids. *Nat. Chem. Biol.* **2015**, *11*, 685–690. [CrossRef]
112. Haiser, H.J.; Gootenberg, D.B.; Chatman, K.; Sirasani, G.; Balskus, E.P.; Turnbaugh, P.J. Predicting and manipulating cardiac drug inactivation by the human gut bacterium *Eggerthella lenta*. *Science* **2013**, *341*, 295–298. [CrossRef]
113. Plaza-Díaz, J.; Álvarez-Mercado, A.I.; Ruiz-Marín, C.M.; Reina-Pérez, I.; Pérez-Alonso, A.J.; Sánchez-Andujar, M.B.; Torné, P.; Gallart-Aragón, T.; Sánchez-Barrón, M.T.; Reyes Lartategui, S.; et al. Association of breast and gut microbiota dysbiosis and the risk of breast cancer: A case-control clinical study. *BMC Cancer* **2019**, *19*, 495. [CrossRef] [PubMed]
114. Álvarez-Mercado, A.I.; Navarro-Oliveros, M.; Robles-Sánchez, C.; Plaza-Díaz, J.; Sáez-Lara, M.J.; Muñoz-Quezada, S.; Fontana, L.; Abadía-Molina, F. Microbial Population Changes and Their Relationship with Human Health and Disease. *Microorganisms* **2019**, *7*, 68. [CrossRef] [PubMed]
115. Tripathi, A.; Debelius, J.; Brenner, D.A.; Karin, M.; Loomba, R.; Schnabl, B.; Knight, R. The gut-liver axis and the intersection with the microbiome. *Nat. Rev. Gastroenterol. Hepatol.* **2018**, *15*, 397–411. [CrossRef] [PubMed]
116. Chen, P.; Schnabl, B. Host-microbiome interactions in alcoholic liver disease. *Gut Liver* **2014**, *8*, 237–241. [CrossRef] [PubMed]
117. Yang, A.M.; Inamine, T.; Hochrath, K.; Chen, P.; Wang, L.; Llorente, C.; Bluemel, S.; Hartmann, P.; Xu, J.; Koyama, Y.; et al. Intestinal fungi contribute to development of alcoholic liver disease. *J. Clin. Investig.* **2017**, *127*, 2829–2841. [CrossRef] [PubMed]
118. Gabbard, S.L.; Lacy, B.E.; Levine, G.M.; Crowell, M.D. The impact of alcohol consumption and cholecystectomy on small intestinal bacterial overgrowth. *Dig. Dis. Sci.* **2014**, *59*, 638–644. [CrossRef]
119. Kirpich, I.A.; Solovieva, N.V.; Leikhter, S.N.; Shidakova, N.A.; Lebedeva, O.V.; Sidorov, P.I.; Bazhukova, T.A.; Soloviev, A.G.; Barve, S.S.; McClain, C.J.; et al. Probiotics restore bowel flora and improve liver enzymes in human alcohol-induced liver injury: A pilot study. *Alcohol* **2008**, *42*, 675–682. [CrossRef]
120. Casafont Morencos, F.; de las Heras Castaño, G.; Martín Ramos, L.; López Arias, M.J.; Ledesma, F.; Pons Romero, F. Small bowel bacterial overgrowth in patients with alcoholic cirrhosis. *Dig. Dis. Sci.* **1996**, *41*, 552–556. [CrossRef]
121. Michail, S.; Lin, M.; Frey, M.R.; Fanter, R.; Paliy, O.; Hilbush, B.; Reo, N.V. Altered gut microbial energy and metabolism in children with non-alcoholic fatty liver disease. *FEMS Microbiol. Ecol.* **2015**, *91*, 1–9. [CrossRef]

122. Raman, M.; Ahmed, I.; Gillevet, P.M.; Probert, C.S.; Ratcliffe, N.M.; Smith, S.; Greenwood, R.; Sikaroodi, M.; Lam, V.; Crotty, P.; et al. Fecal microbiome and volatile organic compound metabolome in obese humans with nonalcoholic fatty liver disease. *Clin. Gastroenterol. Hepatol.* **2013**, *11*, 868–875. [CrossRef] [PubMed]
123. Boursier, J.; Mueller, O.; Barret, M.; Machado, M.; Fizanne, L.; Araujo-Perez, F.; Guy, C.D.; Seed, P.C.; Rawls, J.F.; David, L.A.; et al. The severity of nonalcoholic fatty liver disease is associated with gut dysbiosis and shift in the metabolic function of the gut microbiota. *Hepatology* **2016**, *63*, 764–775. [CrossRef] [PubMed]
124. Del Chierico, F.; Nobili, V.; Vernocchi, P.; Russo, A.; De Stefanis, C.; Gnani, D.; Furlanello, C.; Zandonà, A.; Paci, P.; Capuani, G.; et al. Gut microbiota profiling of pediatric nonalcoholic fatty liver disease and obese patients unveiled by an integrated meta-omics-based approach. *Hepatology* **2017**, *65*, 451–464. [CrossRef] [PubMed]
125. Qin, N.; Yang, F.; Li, A.; Prifti, E.; Chen, Y.; Shao, L.; Guo, J.; Le Chatelier, E.; Yao, J.; Wu, L.; et al. Alterations of the human gut microbiome in liver cirrhosis. *Nature* **2014**, *513*, 59–64. [CrossRef]
126. Chen, Y.; Ji, F.; Guo, J.; Shi, D.; Fang, D.; Li, L. Dysbiosis of small intestinal microbiota in liver cirrhosis and its association with etiology. *Sci. Rep.* **2016**, *6*, 34055. [CrossRef]
127. Yu, L.X.; Schwabe, R.F. The gut microbiome and liver cancer: Mechanisms and clinical translation. *Nat. Rev. Gastroenterol. Hepatol.* **2017**, *14*, 527–539. [CrossRef]
128. Ren, Y.D.; Ye, Z.S.; Yang, L.Z.; Jin, L.X.; Wei, W.J.; Deng, Y.Y.; Chen, X.X.; Xiao, C.X.; Yu, X.F.; Xu, H.Z.; et al. Fecal microbiota transplantation induces hepatitis B virus e-antigen (HBeAg) clearance in patients with positive HBeAg after long-term antiviral therapy. *Hepatology* **2017**, *65*, 1765–1768. [CrossRef]
129. Ferrere, G.; Wrzosek, L.; Cailleux, F.; Turpin, W.; Puchois, V.; Spatz, M.; Ciocan, D.; Rainteau, D.; Humbert, L.; Hugot, C.; et al. Fecal microbiota manipulation prevents dysbiosis and alcohol-induced liver injury in mice. *J. Hepatol.* **2017**, *66*, 806–815. [CrossRef]
130. Marra, F.; Svegliati-Baroni, G. Lipotoxicity and the gut-liver axis in NASH pathogenesis. *J. Hepatol.* **2018**, *68*, 280–295. [CrossRef]
131. Soderborg, T.K.; Clark, S.E.; Mulligan, C.E.; Janssen, R.C.; Babcock, L.; Ir, D.; Young, B.; Krebs, N.; Lemas, D.J.; Johnson, L.K.; et al. The gut microbiota in infants of obese mothers increases inflammation and susceptibility to NAFLD. *Nat. Commun.* **2018**, *9*, 4462. [CrossRef]
132. Cogger, V.C.; Mohamad, M.; Solon-Biet, S.M.; Senior, A.M.; Warren, A.; O'Reilly, J.N.; Tung, B.T.; Svistounov, D.; McMahon, A.C.; Fraser, R.; et al. Dietary macronutrients and the aging liver sinusoidal endothelial cell. *Am. J. Physiol. Heart Circ. Physiol.* **2016**, *310*, H1064–H1070. [CrossRef] [PubMed]
133. Friedman, S.L.; Neuschwander-Tetri, B.A.; Rinella, M.; Sanyal, A.J. Mechanisms of NAFLD development and therapeutic strategies. *Nat. Med.* **2018**, *24*, 908–922. [CrossRef] [PubMed]
134. Harte, A.L.; da Silva, N.F.; Creely, S.J.; McGee, K.C.; Billyard, T.; Youssef-Elabd, E.M.; Tripathi, G.; Ashour, E.; Abdalla, M.S.; Sharada, H.M.; et al. Elevated endotoxin levels in non-alcoholic fatty liver disease. *J. Inflamm.* **2010**, *7*, 15. [CrossRef] [PubMed]
135. Brun, P.; Castagliuolo, I.; Di Leo, V.; Buda, A.; Pinzani, M.; Palù, G.; Martines, D. Increased intestinal permeability in obese mice: New evidence in the pathogenesis of nonalcoholic steatohepatitis. *Am. J. Physiol. Gastrointest. Liver Physiol.* **2007**, *292*, G518–G525. [CrossRef]
136. Wu, J.; Meng, Z.; Jiang, M.; Zhang, E.; Trippler, M.; Broering, R.; Bucchi, A.; Krux, F.; Dittmer, U.; Yang, D.; et al. Toll-like receptor-induced innate immune responses in non-parenchymal liver cells are cell type-specific. *Immunology* **2010**, *129*, 363–374. [CrossRef]
137. Björkholm, B.; Bok, C.M.; Lundin, A.; Rafter, J.; Hibberd, M.L.; Pettersson, S. Intestinal microbiota regulate xenobiotic metabolism in the liver. *PLoS ONE* **2009**, *4*, e6958. [CrossRef]
138. Chuang, H.L.; Huang, Y.T.; Chiu, C.C.; Liao, C.D.; Hsu, F.L.; Huang, C.C.; Hou, C.C. Metabolomics characterization of energy metabolism reveals glycogen accumulation in gut-microbiota-lacking mice. *J. Nutr. Biochem.* **2012**, *23*, 752–758. [CrossRef]
139. Wiest, R.; Albillos, A.; Trauner, M.; Bajaj, J.S.; Jalan, R. Targeting the gut-liver axis in liver disease. *J. Hepatol.* **2017**, *67*, 1084–1103. [CrossRef]
140. Wang, W.; Xu, S.; Ren, Z.; Jiang, J.; Zheng, S. Gut microbiota and allogeneic transplantation. *J. Transl. Med.* **2015**, *13*, 275. [CrossRef]
141. Álvarez-Mercado, A.I.; Gulfo, J.; Romero Gómez, M.; Jiménez-Castro, M.B.; Gracia-Sancho, J.; Peralta, C. Use of Steatotic Grafts in Liver Transplantation: Current Status. *Liver Transplant.* **2019**, *25*, 771–786. [CrossRef]

142. Bäckhed, F.; Ding, H.; Wang, T.; Hooper, L.V.; Koh, G.Y.; Nagy, A.; Semenkovich, C.F.; Gordon, J.I. The gut microbiota as an environmental factor that regulates fat storage. *Proc. Natl. Acad. Sci. USA* **2004**, *101*, 15718–15723. [CrossRef] [PubMed]
143. Schnabl, B.; Brenner, D.A. Interactions between the intestinal microbiome and liver diseases. *Gastroenterology* **2014**, *146*, 1513–1524. [CrossRef] [PubMed]
144. Martinez, K.B.; Leone, V.; Chang, E.B. Western diets, gut dysbiosis, and metabolic diseases: Are they linked? *Gut Microbes* **2017**, *8*, 130–142. [CrossRef]
145. Fan, J.G.; Kim, S.U.; Wong, V.W. New trends on obesity and NAFLD in Asia. *J. Hepatol.* **2017**, *67*, 862–873. [CrossRef] [PubMed]
146. Hammad, A.; Kaido, T.; Aliyev, V.; Mandato, C.; Uemoto, S. Nutritional Therapy in Liver Transplantation. *Nutrients* **2017**, *9*, 1126. [CrossRef]
147. Zhang, Q.K.; Wang, M.L. The management of perioperative nutrition in patients with end stage liver disease undergoing liver transplantation. *Hepatobiliary Surg. Nutr.* **2015**, *4*, 336–344.
148. Ren, Z.; Jiang, J.; Lu, H.; Chen, X.; He, Y.; Zhang, H.; Xie, H.; Wang, W.; Zheng, S.; Zhou, L. Intestinal microbial variation may predict early acute rejection after liver transplantation in rats. *Transplantation* **2014**, *98*, 844–852. [CrossRef]
149. Grąt, M.; Hołówko, W.; Wronka, K.M.; Grąt, K.; Lewandowski, Z.; Kosińska, I.; Krasnodębski, M.; Wasilewicz, M.; Gałęcka, M.; Szachta, P.; et al. The relevance of intestinal dysbiosis in liver transplant candidates. *Transpl. Infect. Dis.* **2015**, *17*, 174–184. [CrossRef]
150. Bajaj, J.S.; Fagan, A.; Sikaroodi, M.; White, M.B.; Sterling, R.K.; Gilles, H.; Heuman, D.; Stravitz, R.T.; Matherly, S.C.; Siddiqui, M.S.; et al. Liver transplant modulates gut microbial dysbiosis and cognitive function in cirrhosis. *Liver Transplant.* **2017**, *23*, 907–914. [CrossRef]
151. Sun, L.Y.; Yang, Y.S.; Qu, W.; Zhu, Z.J.; Wei, L.; Ye, Z.S.; Zhang, J.R.; Sun, X.Y.; Zeng, Z.G. Gut microbiota of liver transplantation recipients. *Sci. Rep.* **2017**, *7*, 3762. [CrossRef]
152. Bajaj, J.S.; Kakiyama, G.; Cox, I.J.; Nittono, H.; Takei, H.; White, M.; Fagan, A.; Gavis, E.A.; Heuman, D.M.; Gilles, H.C.; et al. Alterations in gut microbial function following liver transplant. *Liver Transplant.* **2018**, *24*, 752–761. [CrossRef] [PubMed]
153. Tian, X.; Yang, Z.; Luo, F.; Zheng, S. Gut microbial balance and liver transplantation: Alteration, management, and prediction. *Front. Med.* **2018**, *12*, 123–129. [CrossRef] [PubMed]
154. Wu, X.; Sun, R.; Chen, Y.; Zheng, X.; Bai, L.; Lian, Z.; Wei, H.; Tian, Z. Oral ampicillin inhibits liver regeneration by breaking hepatic innate immune tolerance normally maintained by gut commensal bacteria. *Hepatology* **2015**, *62*, 253–264. [CrossRef] [PubMed]
155. Liu, H.X.; Rocha, C.S.; Dandekar, S.; Wan, Y.J. Functional analysis of the relationship between intestinal microbiota and the expression of hepatic genes and pathways during the course of liver regeneration. *J. Hepatol.* **2016**, *64*, 641–650. [CrossRef] [PubMed]
156. Adolph, T.E.; Grander, C.; Moschen, A.R.; Tilg, H. Liver-Microbiome Axis in Health and Disease. *Trends Immunol.* **2018**, *39*, 712–723. [CrossRef] [PubMed]
157. Xie, Y.; Chen, H.; Zhu, B.; Qin, N.; Chen, Y.; Li, Z.; Deng, M.; Jiang, H.; Xu, X.; Yang, J.; et al. Effect of intestinal microbiota alteration on hepatic damage in rats with acute rejection after liver transplantation. *Microb. Ecol.* **2014**, *68*, 871–880. [CrossRef]
158. Ito, T.; Nakamura, K.; Kageyama, S.; Korayem, I.M.; Hirao, H.; Kadono, K.; Aziz, J.; Younan, S.; DiNorcia, J., 3rd; Agopian, V.G.; et al. Impact of Rifaximin Therapy on Ischemia/Reperfusion Injury in Liver Transplantation: A Propensity Score-Matched Analysis. *Liver Transplant.* **2019**, in press. [CrossRef]
159. Nakamura, K.; Kageyama, S.; Ito, T.; Hirao, H.; Kadono, K.; Aziz, A.; Dery, K.J.; Everly, M.J.; Taura, K.; Uemoto, S.; et al. Antibiotic pretreatment alleviates liver transplant damage in mice and humans. *J. Clin. Investig.* **2019**, *129*, 3420–3434. [CrossRef]
160. Jia, J.; Tian, X.; Jiang, J.; Ren, Z.; Lu, H.; He, N.; Xie, H.; Zhou, L.; Zheng, S. Structural shifts in the intestinal microbiota of rats treated with cyclosporine A after orthotropic liver transplantation. *Front. Med.* **2019**, *13*, 451–460. [CrossRef]
161. Jiang, J.W.; Ren, Z.G.; Lu, H.F.; Zhang, H.; Li, A.; Cui, G.Y.; Jia, J.J.; Xie, H.Y.; Chen, X.H.; He, Y.; et al. Optimal immunosuppressor induces stable gut microbiota after liver transplantation. *World J. Gastroenterol.* **2018**, *24*, 3871–3883. [CrossRef]

162. Liu, H.X.; Hu, Y.; Wan, Y.J. Microbiota and bile acid profiles in retinoic acid-primed mice that exhibit accelerated liver regeneration. *Oncotarget* **2016**, *7*, 1096–1106. [CrossRef]
163. Liu, Z.; Li, C.; Huang, M.; Tong, C.; Zhang, X.; Wang, L.; Peng, H.; Lan, P.; Zhang, P.; Huang, N.; et al. Positive regulatory effects of perioperative probiotic treatment on postoperative liver complications after colorectal liver metastases surgery: A double-center and double-blind randomized clinical trial. *BMC Gastroenterol.* **2015**, *15*, 34. [CrossRef] [PubMed]
164. Ren, J.; Hu, D.; Mao, Y.; Yang, H.; Liao, W.; Xu, W.; Ge, P.; Zhang, H.; Sang, X.; Lu, X.; et al. Alteration in gut microbiota caused by time-restricted feeding alleviate hepatic ischaemia reperfusion injury in mice. *J. Cell. Mol. Med.* **2019**, *23*, 1714–1722. [CrossRef] [PubMed]
165. Cheng, E.Y.; Everly, M.J. Trends of Immunosuppression and Outcomes Following Liver Transplantation: An Analysis of the United Network for Organ Sharing Registry. In *Clinical Transplants 2014*; Everly, M.J., Terasaki, P.I., Eds.; UCLA Immunogenetics Center: Los Angeles, CA, USA, 2015; Volume 2, pp. 13–26.
166. Ravaioli, M.; Neri, F.; Lazzarotto, T.; Bertuzzo, V.R.; Di Gioia, P.; Stacchini, G.; Morelli, M.C.; Ercolani, G.; Cescon, M.; Chiereghin, A.; et al. Immunosuppression Modifications Based on an Immune Response Assay: Results of a Randomized, Controlled Trial. *Transplantation* **2015**, *99*, 1625–1632. [CrossRef] [PubMed]
167. Jiang, J.W.; Ren, Z.G.; Cui, G.Y.; Zhang, Z.; Xie, H.Y.; Zhou, L. Chronic bile duct hyperplasia is a chronic graft dysfunction following liver transplantation. *World J. Gastroenterol.* **2012**, *18*, 1038–1047. [CrossRef]
168. Zhang, W.; Fung, J. Limitations of current liver transplant immunosuppressive regimens: Renal considerations. *Hepatobiliary Pancreat. Dis. Int.* **2017**, *16*, 27–32. [CrossRef]
169. Humar, A.; Michaels, M.; AST ID Working Group on Infectious Disease Monitoring. American Society of Transplantation recommendations for screening, monitoring and reporting of infectious complications in immunosuppression trials in recipients of organ transplantation. *Am. J. Transplant.* **2006**, *6*, 262–274. [CrossRef]
170. Jia, J.J.; Lin, B.Y.; He, J.J.; Geng, L.; Kadel, D.; Wang, L.; Yu, D.D.; Shen, T.; Yang, Z.; Ye, Y.F.; et al. "Minimizing tacrolimus" strategy and long-term survival after liver transplantation. *World J. Gastroenterol.* **2014**, *20*, 11363–11369. [CrossRef]
171. He, Y.Q.; Gong, L.; Fang, Y.P.; Zhan, Q.; Liu, H.X.; Lu, Y.L.; Guo, G.L.; Lehman-McKeeman, L.; Fang, J.W.; Wan, Y.J. The role of retinoic acid in hepatic lipid homeostasis defined by genomic binding and transcriptome profiling. *BMC Genom.* **2013**, *14*, 575. [CrossRef]
172. Yang, F.; He, Y.Q.; Liu, H.X.; Tsuei, J.; Jiang, X.Y.; Yang, L.; Wang, Z.T.; Wan, Y.J. All-trans retinoic acid regulates hepatic bile acid homeostasis. *Biochem. Pharmacol.* **2014**, *91*, 483–489. [CrossRef]
173. Huang, W.D.; Ma, K.; Zhang, J.; Qatanani, M.; Cuvillier, J.; Liu, J.; Dong, B.N.; Huang, X.F.; Moore, D.D. Nuclear receptordependent bile acid signaling is required for normal liver regeneration. *Science* **2006**, *312*, 233–236. [CrossRef] [PubMed]
174. Liu, H.X.; Ly, I.; Hu, Y.; Wan, Y.J. Retinoic acid regulates cell cycle genes and accelerates normal mouse liver regeneration. *Biochem. Pharmacol.* **2014**, *91*, 256–265. [CrossRef] [PubMed]
175. Melkani, G.C.; Panda, S. Time-restricted feeding for prevention and treatment of cardiometabolic disorders. *J. Physiol.* **2017**, *595*, 3691–3700. [CrossRef] [PubMed]
176. Gonzalez-Flecha, B.; Cutrin, J.; Boveris, A. Time course and mechanism of oxidative stress and tissue damage in rat liver subjected to in vivo ischemia-reperfusion. *J. Clin. Investig.* **1993**, *91*, 456–464. [CrossRef] [PubMed]
177. Kawachi, S.; Hines, I.N.; Laroux, F.S.; Hoffman, J.; Bharwani, S.; Gray, L.; Leffer, D.; Grisham, M.B. Nitric oxide synthase and postischemic liver injury. *Biochem. Biophys. Res. Commun.* **2000**, *276*, 851–854. [CrossRef]
178. Kuboki, S.; Shin, T.; Huber, N.; Eismann, T.; Galloway, E.; Schuster, R.; Blanchard, J.; Edwards, M.J.; Lentsch, A.B. Hepatocyte signaling through CXC chemokine receptor-2 is detrimental to liver recovery after ischemia/reperfusion in mice. *Hepatology* **2008**, *48*, 1213–1223. [CrossRef] [PubMed]

Review

The Evolving Microbiome from Pregnancy to Early Infancy: A Comprehensive Review

María Dolores Mesa [1,2,*,†], Begoña Loureiro [3,†], Iris Iglesia [4,5,†], Sergi Fernandez Gonzalez [6,7,†], Elisa Llurba Olivé [8,9,10,†], Oscar García Algar [6,11,†], María José Solana [12,†], Mª Jesús Cabero Perez [13,†], Talia Sainz [14,15,16,†], Leopoldo Martinez [15,17,†], Diana Escuder-Vieco [18,†], Anna Parra-Llorca [19,†], María Sánchez-Campillo [20,†], Gerardo Rodriguez Martinez [5,21,†], Dolores Gómez Roig [6,7,†], Myriam Perez Gruz [6,7,†], Vicente Andreu-Fernández [6,11,†], Jordi Clotet [6,11,†], Sebastian Sailer [6,11,†], Isabel Iglesias-Platas [6,7,22,†], Jesús López-Herce [12,†], Rosa Aras [15,†], Carmen Pallás-Alonso [18,†], Miguel Saenz de Pipaon [23,†], Máximo Vento [19,†], María Gormaz [19,†], Elvira Larqué Daza [20,†], Cristina Calvo [14,15,16,24,†] and Fernando Cabañas [25,†]

1. Department of Biochemistry and Molecular Biology II, Institute of Nutrition and Food Technology "José Mataix", Biomedical Research Center, University of Granada, Parque Tecnológico de la Salud, Avenida del Conocimiento s/n, Armilla, 18100 Granada, Spain
2. ibs.GRANADA, Instituto de Investigación Biosanitaria, Complejo Hospitalario Universitario de Granada, 18014 Granada, Spain
3. Neonatology Unit, University Hospital Cruces, Biocruces Bizkaia Health Research Institute, 48903 Barakaldo, Spain; begona.loureirogonzalez@osakidetza.eus
4. Growth, Exercise, Nutrition and Development (GENUD) Research Group, Universidad de Zaragoza, 50009 Zaragoza, Spain; iglesia@unizar.es
5. Instituto de Investigación Sanitaria Aragón (IIS Aragón), 50009 Zaragoza, Spain; gerard@unizar.es
6. BCNatal-Barcelona Center for Maternal-Fetal and Neonatal Medicine (Hospital Sant Joan de Deu and Hospital Clínic), 08028 Barcelona, Spain; sfernandezgo@sjdhospitalbarcelona.org (S.F.G.); ogarciaa@clinic.cat (O.G.A.); lgomezroig@hsjdbcn.org (D.G.R.); mperezc@sjdhospitalbarcelona.org (M.P.G.); viandreu@clinic.cat (V.A.-F.); jclotet@clinic.cat (J.C.); sebastiansailer34@gmail.com (S.S.); iiglesias@sjdhospitalbarcelona.org (I.I.-P.)
7. Institut de Recerca Sant Joan de Déu (IR-SJD), 08028 Barcelona, Spain
8. Obstetrics and Gynecology Department, High Risk Unit, Sant Pau University Hospital, 08025 Barcelona, Spain; llurba@yahoo.es
9. Women and Perinatal Health Research Group, Biomedical Research Institute Sant Pau (IIB-Sant Pau), Sant Pau University Hospital, 08041 Barcelona, Spain
10. School of Medicine, Universitat Autònoma de Barcelona, 08193 Barcelona, Spain
11. Neonatology Unit, Hospital Clinic-Maternitat, ICGON, BCNatal, Gran Via de les Corts Catalanes, 585, 08007 Barcelona, Spain
12. Servicio de Cuidados Intensivos Pediátricos, Hospital General Universitario Gregorio Marañón, Departamento de Salud Pública y Materno infantil, Universidad Complutense de Madrid, 28040 Madrid, Spain; mjsolana@hotmail.com (M.J.S.); pielvi@hotmail.com (J.L.-H.)
13. Hospital Universitario Marqués de Valdecilla, Santander, 39008 Cantabria, Spain; mariajesuscabero@gmail.com
14. Servicio de Pediatría, Enfermedades Infecciosas y Tropicales, Hospital La Paz, 28046 Madrid, Spain; tsainzcosta@gmail.com (T.S.); ccalvorey@gmail.com (C.C.)
15. Instituto de Investigación Hospital la Paz (IdiPAZ), 28029 Madrid, Spain; leopoldo.martinez@salud.madrid.org (L.M.); rosa.aras@hotmail.com (R.A.)
16. Red de investigación Traslacional en Infectología Pediátrica (RITIP), 28046 Madrid, Spain
17. Servicio de Cirugía Pediátrica, Hospital La Paz, 28046 Madrid, Spain
18. Donated Milk Bank, Health Research Institute i + 12, University Hospital 12 de Octubre, Universidad Complutense, 28040 Madrid, Spain; diana.e.vieco@gmail.com (D.E.-V.); kpallas.hdoc@gmail.com (C.P.-A.)
19. Neonatal Research Group, Health Research Institute La Fe, University and Polytechnic Hospital La Fe, 46026 Valencia, Spain; annaparrallorca@gmail.com (A.P.-L.); Maximo.Vento@uv.es (M.V.); mgormi@yahoo.es (M.G.)

[20] Department of Physiology, Faculty of Biology, University of Murcia, 30100 Murcia, Spain; medit2011@gmail.com (M.S.-C.); elvirada@um.es (E.L.D.)
[21] Department of Pediatrics, Faculty of Medicine, Hospital Clínico Universitario Lozano Blesa, 50009 Zaragoza, Spain
[22] Neonatology Unit, Hospital Sant Joan de Déu, Institut de Recerca Sant Joan de Déu, BCNatal, 08028 Barcelona, Spain
[23] Department of Neonatology La Paz University Hospital, 28046 Madrid, Spain; miguel.saenz@salud.madrid.org
[24] European Network of Excellence for Pediatric Clinical Research, 27100 Bari, Italy
[25] Department of Paediatrics-Neonatology Quironsalud, Madrid University Hospital and Biomedical Research Foundation-IDIPAZ, La Paz University Hospital, 28046 Madrid, Spain; fernando.cabanas@quironsalud.es
* Correspondence: mdmesa@ugr.es; Tel.: +34-958-246187
† All the authors are the members of SAMID Network. RETICS funded by the PN I+D+I 2018–2021 (Spain), ISCIII-Sub-Directorate General for Research Assessment and Promotion and the European Regional Development Fund (FEDER), ref. RD16/0022.

Received: 28 November 2019; Accepted: 20 December 2019; Published: 2 January 2020

Abstract: Pregnancy induces a number of immunological, hormonal, and metabolic changes that are necessary for the mother to adapt her body to this new physiological situation. The microbiome of the mother, the placenta and the fetus influence the fetus growth and undoubtedly plays a major role in the adequate development of the newborn infant. Hence, the microbiome modulates the inflammatory mechanisms related to physiological and pathological processes that are involved in the perinatal progress through different mechanisms. The present review summarizes the actual knowledge related to physiological changes in the microbiota occurring in the mother, the fetus, and the child, both during neonatal period and beyond. In addition, we approach some specific pathological situations during the perinatal periods, as well as the influence of the type of delivery and feeding.

Keywords: microbiome; pregnancy; fetus; placenta; newborn; infancy; critical illness; sepsis; allergy

---

1. Introduction

Pregnancy induces a number of immunological, hormonal, and metabolic changes necessary for the normal development of the fetus and for a timely onset of labor and successful delivery [1]. It has been described that maternal microbiota influences prenatal and early postnatal offspring development and health outcomes [2,3]. There is a lack of consensus about the real nature of microbiome changes during pregnancy, since discrepant and unpredictable findings have been described [4–6]. These differences could be explained by the difference in gestational age, genetics, ethnicity, and environmental factors surrounding the participants included in those studies. Indeed, it has been described that maternal microbiota composition during pregnancy is related to maternal diet [7–9], and by pre-pregnancy weight and weight gain over the course of pregnancy [10–13]. Koren et al. described that the amounts of anti-inflammatory butyrate-producer commensal bacteria present in non-pregnant women gut microbiota decrease while bacteria associated with pro-inflammatory responses, such as *Proteobacteria*, increase during pregnancy [4]. Similarly, bacterial diversity tends to be reduced in vaginal microbiota during pregnancy while increasing vaginal *Streptococci* along with several specific *Lactobacilli* strains, which are thought to prevent the growth of pathogenic bacteria, as well as to help human digestion, and influence host innate and adaptive immune system responses [4,14]. Furthermore, the classical paradigm of the fetus as a sterile organism is under discussion, since a characteristic microbiome has been identified in the placenta, the amniotic fluid, and the fetus in healthy pregnancies [15,16]. However, this issue is under discussion. Perez-Muñoz et al. argued the weakness of evidence supporting the "in utero colonization hypothesis", due to methodological difficulties, and concluded that current scientific evidence does not support the existence of microbiome within the healthy fetal milieu [17].

Gut microbiota influences the immune function [18], and thus may modulate the response through different microbial-derived metabolites, especially short-chain fatty acids (SCFAs) such as butyrate, acetate, or propionate [19]. These are the key drivers of T-cell subset proliferation and activity [19,20]. Gastrointestinal bacteria generate SCFAs after fermentation of complex dietary carbohydrates. These metabolites may have an influence both in the mother and in the newborn by down-regulation of pro-inflammatory responses at the specific sites where the allergens are located, which typically precedes asthma in childhood [21]. In addition, the may also influence bone marrow stimulation by reprogramming the immunological tone of the mammalian ecosystem [22].

Finally, it is important to consider that the discrepancies of the data obtained to date could be influenced by a number of factors such as the dietary pattern, the ethnicity, the geographic location, and the research methodology. The limitations of classical culturable methods have been improved with new molecular methods used to characterize the microbiota. However, these new methods have their own limitations, as reagent, laboratory contamination, and the inability to differentiate living and dead microorganisms. Indeed, recent research complements the study of microbiome with metabolomics and proteomic analysis in order to complete the whole metabolic picture of the microbiota and its metabolic status. Therefore, further studies are needed to confirm the evolution of microbiota during pregnancy and its influence in healthy and complicated labors and the newborn [23].

The present review summarizes the actual knowledge related to changes in maternal and fetal microbiota occurring during pregnancy, which may influence the newborn and infant development. In addition, changes in specific pathological infancy situations have also been revised.

2. Changes in the Microbiome during Pregnancy

During pregnancy, the female body undergoes hormonal, metabolic, and immunological changes to preserve the health of both the mother and the offspring [1]. These changes alter the mother microbiota at different sites such as the gut, the vagina, and the oral cavity. However, published data are not consistent, since a number of factors might influence the microbiota profile such as the diet, antibiotic, or other supplement intakes, as well as the methodology of research. Therefore, a holistic approach is needed to understand all this information.

2.1. Gut Microbiota

The gut microbiota shifts substantially throughout the progression of the pregnancy and is characterized by reduced individual richness (alpha-diversity) (Figure 1), and increased inter-subject beta-diversity [4]. These changes are not related to, although they may be influenced by, the diet, antibiotic treatments, gestational diabetes, or pre-pregnancy body mass index, but are vital for a healthy pregnancy [4]. It has been suggested that other factors, such as the state of the host immune and endocrine systems, may actively contribute to the observed modifications [24]. During the first trimester, the gut microbiota pattern is similar in many aspects to that of healthy non-pregnant women, showing a predominance of *Firmicutes*, mainly *Clostridiales*, over *Bacteroidetes* [25]. Then, maternal gut microbiota declines in butyrate-producing bacteria, while *Bifidobacteria*, *Proteobacteria*, and lactic acid-producing bacteria increase from the first to the third trimester, when the microbiota resembles an unpredictably disease-associated dysbiosis that differs greatly among normal pregnancies [4]. Changes in the host immune system of the gastrointestinal mucosa together with metabolic hormonal changes may trigger a low-grade pro-inflammatory status that could facilitate an increased diffusion of glucose from the gut epithelium towards the lumen, and thus may induce weight gain while modifying the gut microbiota during normal pregnancies [26]. Indeed, changes in the microbiota may contribute to the evolution of this process. In addition, disruption of maternal gut microbiota during the third trimester [27] may affect host metabolism in order to provide an energy supply for the fetus [4,26]. Moreover, it has been reported that the gut microbiota during pregnancy is a critical determinant of offspring health [13,28], and that potentially determines the development of atopy and autoimmune phenotypes in the offspring [28]. However, the relationship among the immune system, the gut microbiota, and metabolism in pregnancy is unclear, and more research is needed to stablish final conclusions.

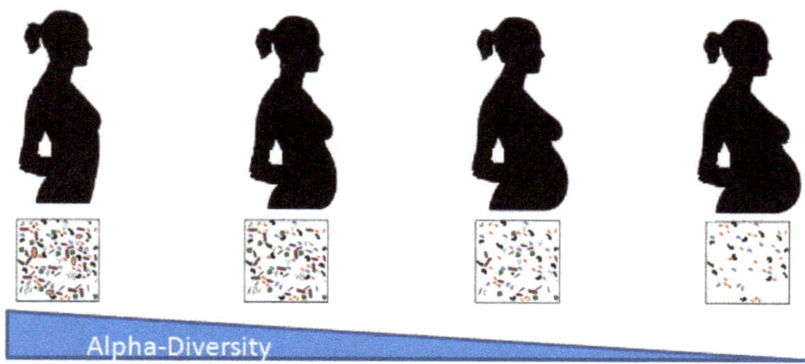

Figure 1. Alpha-diversity changes in gut microbiota during pregnancy.

2.2. Vaginal Microbiota

The composition of the vaginal microbiota is dynamic, corresponding with hormonal fluctuations throughout the woman's reproductive life, and also during pregnancy. A number of protective lactic acid-producing *Lactobacillus* species dominates the healthy vaginal microbiota in most reproductive-age women. These bacteria protect against vaginal dysbiosis and inhibit opportunistic infections through the direct and indirect protective effects of *Lactobacillus* products, such as lactic acid and bacteriocin among others. Lactic acid decreases vaginal pH and thus inhibits a broad range of infections [29], can directly affect host immune functions, by inhibiting pro-inflammatory responses, and also help to release mediators from vaginal epithelial cells and stimulate antiviral response [30]. In addition, *Lactobacillus*-derived bacteriocins may inhibit pathogen growth [31]. The degree of protection varies according to the predominant *Lactobacillus* specie [30]. Vaginal dysbiosis is comprised of a wide array of strict and facultative anaerobes that correlate to increased risk of infection, diseases, and poor reproductive and obstetric outcomes [32].

During normal pregnancy, the composition of the vaginal microbiota changes as a function of gestational age, with an increase in the relative abundance for *Lactobacillus* spp., such as *L. crispatus*, *L. jensenii*, *L. gasserii*, *L. vaginalis*, and a decrease in anaerobe or strict anaerobe microbial species, such as *Atopobium*, *Prevotella*, *Sneathia*, *Gardenerella*, *Ruminococcaceae*, *Parvimonas*, *Mobilincus* [33]. Those authors reported for the first time, that the composition and stability of the vaginal microbiota of normal pregnant women is different from that of non-pregnant women. In fact, low risk pregnant women have more stable vaginal flora throughout the pregnancy than non-pregnant women. Normal changes in the vaginal flora during pregnancy are transitions to another *Lactobacillus* community, and this stability would protect against ascending infections through the genital tract. In addition, they reported that *Lactobacillus* communities vary depending on the ethnicity of the women [33]. Stout et al. [34] confirmed that vaginal microbiota richness and diversity remained stable during the first and second trimesters of gestation in pregnancies ended at term, whereas in woman with preterm born, the richness and diversity decreased early in pregnancy. Therefore, early pregnancy may be an important environment, modulating preterm delivery. A meta-analysis reported significant diversity differences in vaginal microbiomes in the first trimester, between women with term and preterm outcomes, indicating a potential diagnostic utility of microbiome-related biomarkers [35]. In addition, the increase of pathogens in the vagina is associated with complications of pregnancy, in particular with an increased risk of preterm birth and spontaneous abortion [6].

2.3. Oral Microbiota

An increase in the microbial load in the oral cavity during pregnancy has been described. It has been hypothesized that pregnancy creates a nutrient environment that is more favorable to some sensitive strains [36]. The presence of pathogenic bacteria *Porphyromonas gingivalis* and *Aggregatibacter actinomycotemcomitans* in gingival sulcus were significantly higher during early and middle stages of pregnancy compared to non-pregnant women [37]. The oral alpha-diversity index was higher in the third trimester compared to non-pregnant women, and this may be related to the increase of progesterone and estradiol. [38]. One underlying mechanism refers to estrogens being substituted for vitamin K in bacterial anaerobic respiration, especially for black-pigmented *Bacteroides* such as *Bacteroides melaninogenicus* and *Prevotella intermedia* [38].

2.4. Placental Microbiota and Fetal Colonization

The classical paradigm of fetal environment as a sterile harbor has traditionally explained that microbes, and thus microbiome, are acquired both vertically (from the mother) and horizontally (from other humans or from the environment) during and after birth. However, recent data have questioned the traditional accepted dogma of human microbiome acquisition, proposing that neither the placenta, the amniotic fluid, nor the fetus are sterile.

Several findings using both culture and metagenomic techniques have suggested the presence of a low biomass microbial community in the healthy placenta [39–43]. The abundance of different species of *Lactobacillus*, *Propionibacterium*, and members of the *Enterobacteriaceae* family have been detected by DNA-based studies in placental tissue of pregnant women at term and it is under debate [16]. In addition, other authors have confirmed a distinct microbiota in both the placenta and amniotic fluid of healthy women at the time of elective C-section, characterized by low richness, low diversity, and the predominance of *Proteobacteria* [44]. Similarly, other studies have found microbes in amniotic fluid and umbilical cord blood in healthy asymptomatic women, as well as in those with pregnancy complications [45–47].

However, it is unclear where the fetal microbiota comes from, and when is the first fetal exposition. The presence of a different placental microbiota compared to the vagina raises the possibility that the infant may be first seeded in utero from other sources. Microorganism may pass through the placenta and colonize the fetus ascending from the vagina, from the oral cavity, from the urinary track, or from the intestinal lumen of the mother. These microorganisms may reach via the hematogenous route, the placenta, and then be transmitted to the fetus [48]. Some of those oral bacteria, such as *Fusobacterium nucleatum*, may be transmitted hematogenously during placentation by binding to the vascular endothelium, and modifying its permeability and the translation of other common commensals, such as *Escherichia coli* [49]. In addition, Franasiak et al. observed that *Flavobacterium* and *Lactobacillus* represent the majority of endometrial bacterium at the time of embryo transfer, supporting a new hypothesis of the endometrial environment participation [50].

Different studies have also detected microbiome in the first baby fecal sample, the meconium, supporting the in utero exposure to bacteria [51,52]. *Staphylococcus* has been reported as the most prevalent bacteria in meconium samples, followed by *Enterobacteriaceae*, *Enterococcus*, *Lactobacillus*, and *Bifidobacterium* even in infants born by C-section [52,53]. Modification in placental microbiota may be related with adverse pregnancy outcomes of pregnancy or symptoms of clinical infection [40].

On the contrary, Perez-Muñoz et al. [17] critically revised scientific evidence supporting both the "sterile womb" and "in utero colonization" hypotheses. These authors concluded that there is more evidence supporting a sterile womb environment. They suggest that methodological approaches, in which contamination is very easy at different steps and does not use appropriate controls, are responsible for the microorganism colonization described in utero. One well-controlled study compared oral, vaginal, and placenta samples with paired contamination controls. This study reported that when using molecular methods, placental samples were undistinguishable from their paired-contaminated samples. They concluded that while there were distinctive microbial signatures in oral and vaginal samples, they did not find a characteristic placental microbiota, evidencing a sterile environment [54]. Therefore, conclusions remain unachievable, and more studies are needed in this area.

3. Changes in the Microbiome Related to the Type of Delivery

There is great controversy in the scientific community about the relationship of the meconium and infant gut microbiota profile, and the type of delivery. Microbiome studies on early infancy have demonstrated a significant influence of the mode of delivery on the microbiome composition, suggesting the likely association of the infant gut bacteria with maternal vaginal or skin microbiome habitats. A systematic review has concluded that the diversity and colonization pattern of the gut microbiota were significantly associated to the mode of delivery during the first three months of life, which is a critical period of life for immunological programming [55]. However, the observed differences disappear after 6 months of infants' life, when solid foods are included in the diet [56]. It is important to clarify the influence of factors commonly accompanying C-section delivery on the microbiome, due to the potential influence on some non-communicable diseases, such as neonatal skin infection, asthma, allergies, obesity, inflammatory bowel disease, or type I diabetes mellitus [56,57].

Vaginally delivered newborn have shown bacterial communities resembling their own mother's vaginal microbiota, dominated by *Lactobacillus*, *Prevotella*, or *Sneathia* spp. In contrast, C-section-born infants harbored bacterial communities similar to those found on the skin surface niche, dominated by *Staphylococcus*, *Corynebacterium*, and *Propionibacterium* spp. [58] or potentially pathogenic microbial communities such as *Klebsiella*, *Enterococcus*, and *Clostridium* [57]. Other authors have reported that *Bifidobacterium* [59] and *Bacteroides* [55] seem to be significantly more frequent in vaginally compared with C-section delivered infants, which were mainly colonized by *Clostridium* and *Lactobacillus* [55]. The high abundance of *Bifidobacterium* species in infants is considered to promote the maturation of the healthy immune system, while high presence of *Clostridium difficile* is considered as one of the major intra-hospital hazards of severe gastrointestinal infections during infancy [55]. Another study proposed that some species of *Propionibacterium* were most abundant in the meconium of vaginally delivered Chinese infants, whereas C-section-born children had higher amounts of *Bacillus licheniformis*. In addition, the diversity of the microbial composition was also higher in vaginal than in C-section deliveries, although no correlation with maternal microbiome was reported [60]. Similarly, a metagenomic analysis found a *Propionibacterium*-enriched meconium in vaginal delivery mothers, which may proceed from skin or fecal microbes through direct contact during the natural labor [61]. Therefore, there is no consensus regarding the most colonizable pattern of the first microbiota community in the first three days after birth, although it seems that according to phyla, vaginal deliveries are more related to *Actinobacteria* and *Bacteroidetes*, while C-section deliveries are more related to *Firmicutes*. In addition, it has also been suggested that the transfer of maternal vaginal microbes plays a minor role in seeding infant stool microbiota since the overlap of maternal vaginal microbiota and infant faecal microbiota is minimal, while the similarity between maternal rectal microbiota and infant microbiota was more pronounced [62].

The discrepances of the results obtained could be due to different factors associated to C-section delivery such as antibiotic administration, but also to breastfeeding, maternal obesity, gestational diabetes mellitus, and even the analytical methodology. In addition, the diversity from *Firmicutes* and *Bacteroides* colonization levels on infants gut microbiota may be influenced by geographical variation such as the latitude [63].

Some authors have proposed that the lower presence of *Bifidobacteria* and *Bacteroides*, and the abundance of *Clostridia* and *Lactobacillus*, in infants delivered by C-section could be explained by perinatal antibiotics administration [55]. Mothers delivering by C-section receive antibiotic prophylaxis before the beginning of surgery or, in some countries, after the cord clamping to minimize the direct exposure of the neonate to antibiotics [64]. In addition, Azad et al. determined that intrapartum antibiotics both in C-section and vaginal deliveries are associated with infant gut microbiota dysbiosis, although breastfeeding modifies some of these effects [65]. Nevertheless, Martinez et al. [66] performed antibiotic-free C-section delivery in mice and determined that these mice did not have the dynamic developmental gut microbiota changes observed in control natural born mice, evidencing the involvement of maternal vaginal bacteria in a proper metabolic development even in absence of antibiotics supporting the hypothesis of the antibiotic-modulated dysbiosis. It is worth to take into account that perinatal antibiotic administration may be associated with increased risk of developing morbidities such as asthma, allergies and obesity, which may be influenced by dysbiosis. In accordance, epidemiological data show that atopic diseases appear more often in infants born by C-section than after vaginal delivery [67,68].

Furthermore, bacterial richness and diversity were lower in the infant gut of babies born after elective C-section and higher in emergency C-section, suggesting that colonization may be affected differently in both situations. It is important to highlight that emergency C-section and vaginal delivery labor are frequently accompanied by rupture of fetal membranes, and exposing the fetus to maternal vaginal bacteria [65].

Importantly, C-section may decrease the colonization of milk-digested bacteria including the genus *Lactobacillus* in newborns during the first months of life [58]. In addition, the mode of delivery has a relevant impact on the microbiota composition of colostrums and milk [69,70], which also may be influenced by antibiotics administrated during C-section. It has been proposed that infants born by C-section lacked the early provision of breast milk essential to attain a proper gut microbiota that contains microbes such as *Lactobacilli* and *Bifidobacteria.* This could explain the higher colonization rates of these genera in vaginal compared to C-section-delivered infants [71]. In fact, Sakwinska et al. reported that only vaginal delivered and fully breastfed infants had gut microbiota dominated by *Bifidobacteria* [62].

Finally, there are several potential preventive intervention strategies to restore the gut microbiota after C-section [72]. The intervention could be focused on maternal administration of probiotics and prebiotics during gestation. There is a great interest about "seeding approaches" as "vaginal seeding" to reverse the effects of C-section delivery mode on the microbiome in early life, but at the same time there are critical voices concerned about safety and efficacy of this practice [56,72]. In addition, the intervention could concentrate on the neonate using "seeding" methods such as encouraging breastfeeding instead of formula feeding, or the use of infant enriched formulas. In this sense, supplementation with symbiotic, the combination of synergistic pre- and probiotics, might offer an innovative strategy to re-establish the delayed colonization of *Bifidobacterium* spp. in C-section-delivered children [73].

4. Microbiome and the Type of Feeding

Maternal diet establishes long-lasting effects on offspring gut microbial composition, which may have important clinical implications [74,75]. Complex interactions between breast milk cytokines and microbiota guide the microbiological, immunological, and metabolic programming of infants' health, which may explain the higher risk of obesity in infants with overweight and excessive weight gain mothers [76]. In addition, data supporting the notion of bacterial translocation from the maternal gut to extra-intestinal sites during pregnancy are emerging and potentially explain the presence of bacteria in breast milk [28].

Some authors have reported changes in meconium microbiota when delaying the collection of meconium samples by one day, supporting that the type of feeding or the environment has an influence after the birth, which may be more determinant to establish the intestinal microbiome during childhood [53]. Breast milk has been recognized as the gold standard for human nutrition [77]. The type of feeding has an important impact on gut microbial composition in preterm infants. In preterm infants, breast milk has been associated with improved growth and cognitive development [78] and a reduced risk of necrotizing enterocolitis and late sepsis onset [76,79,80]. Occasionally, the absence of mother's own milk (MOM) requires the use of donated human milk (DHM). A prospective cohort study has been launched to determine the impact of DHM upon preterm gut microbiota admitted in a neonatal intensive care unit. Despite the high variability of DHMs, no differences in microbial diversity and richness were found, although feeding type significantly influenced the preterm microbiota composition and predictive functional profiles. Inferred metagenomic analyses showed higher presence of *Bifidobacterium* in the MOM, a genus related to enrichment in the glycan biosynthesis and metabolism pathway, as well as an unclassified *Enterobacteriaceae* and lower unclassified *Clostridiaceae* compared with the DHM or in the formula fed groups. After adjusting for gender, postnatal age, weight, and gestational age, the diversity of gut microbiota increased over time and was constantly higher in infants fed their MOM relative to infants with other types of feeding. In addition, DHM favors an intestinal microbiome more similar to MOM despite the differences between MOM and DHM [81]. Preterm infants are prone to develop free radical-associated conditions [82] that may be influenced by the microbiota. In a recent study, urine oxidative stress biomarkers such as 8-hydroxy-deoxyguanosine (8OHdG/2dG), orto-tyrosine, and F2 isoprostanes, neuroprostanes, neurofurans, and di-homo-isoprostanes were longitudinally measured in preterm infants fed either MOM or DHM using validated mass spectrometry techniques. No significant differences for any of the markers studied were found between preterm babies fed MOM or DHM [83]. However, exfoliated epithelial intestinal cells transcriptome of preterm infants fed their MOM or a DHM induced a differential gene expression of specific genes which may contribute to a more efficient antioxidant response in the postnatal period [84]. Therefore, using DHM could have potential long-term benefits on intestinal functionality, the immune system, and metabolism [85–87]. However, available pasteurization methods cause changes that may blunt many of the positive aspects derived from the use of MOM [88–90]. Further studies are needed to understand the complex links between microbiome and breastfeeding, its impact on health programming, and to develop sensitive methods capable of providing human milk as similar as possible to their MOM, when the latter is not available.

5. Microbiome in Pathological and Adverse Pregnancy Outcomes

Some studies have compared the fetal and mother microbiome in relation to adverse outcomes such as prematurity or low birth-weight without reaching firm conclusions. Ardissone et al. [91] compared the meconium microbiome in newborn before and after 33 weeks of gestation and concluded that *Enterococcus* and *Enterobacter* negatively correlated with gestational age, and *Lactobacillus* and *Phortorhabdus* were more abundant in newborns with less than 33 weeks of gestation. They indicated that the composition of the microbiome may be involved in the inflammatory response that leads to premature birth more than the colonization alone. Specifically, preterm subjects with severe chorioamnionitis had higher abundance of *Ureaplasma parvum*, *Fusobacterium nucleatum*, and *Streptococcus agalactiae* [16]. The placental microbiome varies as a consequence of an excess of gestational weight gain, but is not related to obesity among women with spontaneous preterm birth. Indeed, this placental dysbiosis affects different bacterially encoded metabolic pathways that may be related to pregnancy outcomes [92]. Furthermore, it has been reported high abundance of *Burkholderia*, *Actinomycetales*, and *Alphaproteobacteria* in placental samples from gravidae delivered preterm, and of *Streptococcus* and *Acinetobacter* in placental samples from patients with a history of antepartum urinary infection. In contrast, *Paenibacillus* predominated in term placental specimens [15]. Other authors have proposed that the fetal intestinal microbiota derives from swallowed amniotic fluid, and that they may trigger an inflammatory response which leads to premature birth [91]. Considering that some *Lactobacillus* strains may possess potential anti-inflammatory activities, and could regulate blood glucose levels in diabetic humans [93], the low abundance of *Lactobacillus* in placentas of low birth weight neonates reported by Zheng et al. [94] might be related to a pro-inflammatory status in these pregnancies. Thus, the higher sensitivity of fetal intestinal tissue to inflammatory stimuli may induce labor due to an immune-mediated reaction. However, as mentioned previously, the presence of placental microbiota is under discussion due to methodological doubts, and these data have to be discussed with caution.

Finally, a number of bacteria, viruses, and protozoa infections have been associated with pregnancy complications. Liu et al [95] analyzed the gut microbiome in pregnant women affected by preeclampsia. They showed an overall increase in pathogenic bacteria such as *Clostridium perfringens* and *Bulleidia moorei* and a reduction in probiotic bacteria *Coprococcus catus*. A correlation between periodontitis and the risk of spontaneous abortion or miscarriage has also been described [96]. More well-controlled studies should be carried out in order to identify interactions between pregnancy microbiome and mother and children health which might help to predict gestational and newborn complications and search for new therapeutic targets in adverse obstetrical conditions.

6. Microbiome and Obese Pregnancy

Epidemiological evidence shows that 50% of women in childbearing age and 20%–25% of pregnant women in Europe can be affected by overweight or obesity [97], increasing the cardiometabolic risk in mothers [98] and the susceptibility to metabolic diseases in offspring [99–102]. Pregnancy-associated changes are different in overweight or obese women compared to normal-weight pregnant women. Overweight pregnant women show a reduction in the number of *Bifidobacterium* and *Bacteroides*, and an increase in the number of *Staphylococcus*, *Enterobacteriaceae*, and *Escherichia coli* [11]. Additionally, higher levels of *Staphylococcus* and *Akkermansia muciniphila*, and lower levels of *Bifidobacterium* were detected in women with excessive weight gain during pregnancy as compared to normal-weight ones [76]. Consequently, this altered maternal microbiome will contribute to shape an altered composition of the offspring's microbiome [103,104] and thus influence their future health.

Vaginal-born neonates from overweight or obese mothers show increased numbers of *Bacteroides* and depleted in *Enterococcus*, *Acinetobacter*, *Pseudomonas*, and *Hydrogenophilus* [104]. When specifically examining phyla level relative taxonomic abundance among preterm women by virtue of maternal weight gain, other authors have reported an appreciable and significant increased abundance of *Firmicutes*, *Actinobacteria*, and *Cyanobacteria*, and decreased relative abundance of *Proteobacteria* [92]. Furthermore, this altered maternal microbiota composition may be transferred from mother to fetus during the prenatal period [94] and through lactation [105].

In addition, gut microbiota can induce obesity in children by several mechanisms. For example, lower amounts of *Bifidobacteria* can affect weight gain in infants through mucosal host-microbe crosstalk, and immune and inflammatory dysregulation. Moreover, higher presence of *Bacteroides*, *Clostridium*, and *Staphylococcus* can stimulate greater energy extraction from food, combined with a reduced control of inflammation during the first six months of life in infants of overweight mothers [12]. These first months of life are of great importance since rapid weight gain during this period is associated with an increased risk of obesity during childhood, and this influence is even more important than the birth weight [106].

7. Microbiome in Critical Ill Children

Critical illness itself or its treatment can influence the composition of microbiota [107,108]. Although broad-spectrum antibiotics are probably the factor which further alters its composition, other factors can alter the ecosystem in which develops the microbiota, such as enteral or parenteral feeding, drugs administration, disease co-morbidities, central venous catheters, or intubation and mechanical ventilation. These studies have shown that the intestinal microbiota of critical patients has low diversity, with a shortage of key commensal bacteria and overgrowth of pathogenic bacteria such as *Clostridium difficile*, and some species of *Enteococcus*, *Escherichia* and *Shigella* [107,109–111]. In addition, the microbiota changes throughout the stay in the intensive care units (ICU) [112], and the possibility of pathogenic colonization increases with the time of stay in the unit.

To our best knowledge, only one study has analyzed the microbiota in children in a pediatric ICU (PICU) [113]. These authors found that the skin, oral, and fecal microbiota differs sharply from critically ill children compared with healthy children and adults. They reported a PICU-associated dysbiosis with less alpha-diversity, different composition (beta-diversity), and the loss of body site-specificity, increasing the abundance of nosocomial pathogens across all body sites and reducing gut commensals such as *Faecalibacterium* [113]. A number of studies have shown an association between the microbiota and the immune function [114], the systemic inflammation [115], the metabolism of nutrients [116], the function of the central nervous system [117], the circadian rhythm [118], and the digestive system [119]. Therefore, PICU-associated dysbiosis may contribute to malnutrition, nosocomial infection, neurocognitive alteration, organ dysfunction, and sepsis associated to critical illness [113], and may also have an effect on the lung, the brain, and the kidneys [107].

Critically ill patient conditions may contribute to changes in the oropharynx microbiota, such as the increase of *Klebsiella* or *Pseudomonas* proliferation. On one hand, sedation and endotracheal intubation decrease mucociliary clearance and cough, reducing the elimination of microorganisms. On the other hand, mechanical ventilation, pneumonia, and acute respiratory distress syndrome (ARDS) favor alveolar edema, increasing the amount of nutrients available and decreasing the amount of oxygen in some areas. These facts stimulate bacterial proliferation [120], and increase the risk of nosocomial infection and ARDS [121].

In addition to the critical patients, associated dysbiosis, hypoperfusion, and reperfusion of the intestinal wall produce an intense inflammation of the digestive mucosa which alters the gradient of oxygen concentration and increases the concentration of nitrates favoring the growth of pathogenic flora. Furthermore, the slowing down of intestinal transit, frequent drugs (sedatives, opioids, catecholamines), and the alteration of the mechanisms of microbial elimination (decreased production of bile salts and IgA, pharmacological alkalinization of pH, etc.) may also influence the alteration of the digestive functions [122]. Freedberg et al. observed that colonization by some microorganisms prior to admission in ICU was associated with increased risk of infection by that same germ, and subsequently increased mortality [123]. This fact indicates that the gastrointestinal microbiome can help stratification and early identification of the risk of ICU patient complications.

8. Microbiome and Sepsis in the Newborn

The modification of the normal microbiota pattern can contribute to the development of a systemic inflammatory response with increased cytokine production, sepsis, multi-organ failure, and morbi-mortality [107,109–111]. In spite of variation in net incidence, neonatal sepsis remains one of the leading causes of preventable neonatal morbidity and mortality throughout the world. The main agents responsible for sepsis are group B *Streptococcus* (GBS), *Escherichia coli*, and coagulase-negative *Staphylococci* (CONS) [124]. However, this scenario may be modified depending on the use of antibiotics and/or the implementation of non-culture diagnostic techniques [125].

In recent years, there has been growing interest in the role of commensal bacteria in an individual´s susceptibility to infection. A few studies have evaluated the maternal vaginal microbiota in relation to GBS carrier status. Although it seems that some specific taxa might be associated with the presence of GBS [126], there is no apparent parallel reduction of the predominant commensal bacteria *Lactobacilli* [127]. Indirect evidence suggests that the neonatal gut microbiome might be of relevance in GBS infection, since different colonizing species have been found in the stool of infants from GBS positive and negative mothers, while the protective effect of pre and probiotics has also been suggested [127].

It seems that gastrointestinal microbiota might induce an increase in permeability, modulating gut and systemic immune response, and decreasing the tight junction integrity [128]. As a consequence, intestinal bacteria can promote the systemic inflammatory response syndrome, facilitate bacterial translocation, and cause late-onset sepsis and necrotizing enterocolitis, especially affecting premature neonates. Most, but not all, of the evidence suggests that premature newborns with low microbiome gut diversity, or with predominance of *Staphylococcus, Firmicutes*, and *Proteobacteria* are associated with increased risk for late-onset sepsis compared to those premature infants at lower risk [129]. Furthermore, gut colonization with *Bifidobacterium* and increased presence of prebiotic oligosaccharides in feces, has been related to less disruption of the mucosal barrier and gut epithelial translocation, providing an improved gut development and protection [130]. It remains unclear if invasion of the bloodstream during sepsis is caused by the same microorganisms identified in stool [131] or by others [129], in which case the gut microbiota would act as a facilitating mechanism by interfering with the gut barrier or intestinal immune function. Further studies are needed to tease out if the differences observed in gut colonization in ICU patients predispose to sepsis or if they respond to other factors such as the diet, site differences in initiating and advancing feeds, breastfeeding, the use of antibiotics, or interpatient transmission within the neonatal intensive care units [131].

9. Microbiome and Allergic Conditions

Allergy disorders represent an important global health burden with an increasing prevalence in infants and children, mainly as food allergies, atopic eczema [132], and respiratory pathologies such as rhinitis [133] or asthma [134]. Their causes are multifactorial and contemplate interactions between genetic, environmental, and socioeconomic factors leading to different symptoms or phenotypes [135]. Among this heterogeneity, a restricted microbial exposure at early life seems to play an important role influencing allergic diseases, and asthma onset [136].

9.1. Gut Microbiome and Atopy

Eczema or atopic dermatitis (AD) is the first typical allergic manifestation in newborns [137]. A recent study has reported a high proportion of *Faecalibacterium prausnitzii* on the gut microbiome from AD subjects. The presence of these bacteria is lower in Crohn's disease patients, as well as anti-inflammatory fecal bacteria metabolites [138]. Besides, it has been shown that infants with AD improved their symptomatology when the abundance of fecal *Coprococcus eutactus*, a butyrate-producing bacterium, is increased [139]. Consequently, it has been proposed that dysbiotic gut microbiota and subsequent dysregulation of the gut inflammation may promote an aberrant Th2-type immune response to allergens altering the epithelial barrier in AD skin [140].

9.2. Gut Microbiome and Food Allergy

Available literature on animal models suggests that gut microbiome may have an important role in the susceptibility to food sensitization and food allergy, mainly at early stages of life [141]. Chen et al. [142] recently showed both lower microbiota alpha-diversity and altered gut microbiota composition (an increased number of *Firmicutes* in detriment of *Bacteroidetes*) in children with food sensitization in early life compared with children without these conditions. Among the causes, the increasing use of antibiotics both in humans and in agriculture, and the lower intake of dietary fiber may have an impact on these situations [143].

9.3. Gut Microbiome and Asthma

Allergies are the strongest risk factors for childhood asthma in Western countries [144], but the relationship between asthma and the microbiota is not clear. Although it seems that the diversity of the gut microbiota in infancy is even more determinant for asthma onset than the prevalence of specific bacterial taxa, it has been suggested that there might be specific important bacterial species related to the prevention of asthma, and that gut microbial diversity during the first month of life may be the most important factor associated with asthma development at school age than with other allergic manifestations [136]. In addition, another study has indicated that the neonatal gut microbiota influences susceptibility to childhood allergic asthma via alterations in the gut microenvironment that modulates CD4+ T-cell proliferation and functions. These authors have observed a characteristic depletion of dihomo-γ-linoleate, a precursor of anti-inflammatory ω-3 polyunsaturated fatty acid and prostaglandins that may be related [145].

As described previously, different factors have been associated with infant microbiome and the risk of asthma, such as furry pets exposure [146], gestational age, the mode of delivery (vaginal vs. C-section), and antibiotic treatment (direct vs. indirect via mother) among others [147,148]. However, there is no doubt that a key issue is the type of feeding. A systematic review addressing the effect of breastfeeding in the development of asthma concluded that children who were breastfed for a longer time during the first two years of life had a lower risk of developing asthma, and this effect could be mediated by an adequate and early shaping of the gut microbiota [149,150], although whether the dysbiotic microbiota is the cause or the consequence of atopic and allergic diseases is still unknown [140]. Besides, interventional studies have suggested that pre- and probiotics could prevent or down-regulate the severity of some diseases, such as asthma or allergies, but the biological mechanisms, as well as the best taxa or type of intervention, require further research [151].

10. Microbiome and Infection in Infants

The role of microbiome diversity and its variations in the incidence and susceptibility to infection has also aroused great interest beyond the neonatal period. In view of the interaction between the microbiota and the immune system, the implications are probably major and remain challenging, but for some authors, is even more attractive the idea of its usefulness as a diagnostic tool, a preventive strategy, or even a therapeutic target. As described in the neonatal period, in most infectious diseases scenarios, a decrease in alpha and beta diversity of the microbiota seems to be present. Regarding respiratory infections, diversity of the oropharyngeal and nasopharyngeal microbiota in children with pneumonia was lower compared with healthy controls. Furthermore, a correlation between the presence of certain taxa in sputum and the clinical course of community acquired pneumonia has been described [152,153].

HIV infected children present reduced gastrointestinal microbial diversity [154]. Modulation of the intestinal microbiome through nutritional supplementation, with the aim of decreasing bacterial permeability, has been attempted in the context of HIV infection with scarce success [155,156]. In addition, the microbiome has been suggested to impact the risk of different infectious diseases. Both vaginal and penile microbiotas modify the risk of sexual acquisition of HIV, due to their influence on inflammatory pathways and metabolization of antiretroviral drugs [157,158]. Recent studies have shown how an altered vaginal microbioma increases the risk of vertical transmission of HIV [159]. These studies beautifully exemplify the potential influence of the microbiome on the risk of infections, as well as its implications in pharmacokinetics modulating bacterial metabolism.

Finally, based on the potential role of the gut microbiota as a modulator of the immune function, attempts of supplementation with pre and probiotics have also been carried out. Two randomized controlled trials have analyzed the impact of probiotic supplementation on children with acute gastroenteritis without proving any beneficial clinical outcome [160,161]. Supplementation with prebiotics or probiotics may also enhance vaccine response and thus becomes a new tool for the improvement of vaccine efficacy [162]. However, results have been controversial in this field and warrant further investigation. The evidence for a beneficial effect of probiotics on vaccine response was strongest for oral vaccinations and for parenteral influenza vaccination, and depended on the choice of probiotic, strain, dose, viability, purity, and the time and duration of administration [163].

11. Conclusions

There are many data confirming the interaction of microbiota in pregnancy and in the newborn period, on the establishment of labor, children growth and development, and susceptibility to infections and diseases. However, most studies are descriptive and entangling factors influencing the human microbiome such as the age, race, type of feeding, mother's diet, and antibiotics treatments is challenging. Whatever it is, what is clear is that a number of microbiota-derived substances may easily reach the bloodstream, and impact human metabolism.

Recent advances in genome sequencing technologies, metabolomics, proteomics, transcriptomics, and bioinformatics will enable researchers to explore the fascinating field of the microbiota and, in particular, its functions at a more detailed level. Therefore, larger and prospective studies are needed to characterize the evolution of the microbiota during different conditions and its influence on healthy and pathological pregnancies, on labor onset, and on the perinatal period, in order to promote the development of new preventive, diagnostic, and therapeutic tools.

Author Contributions: All authors have contributed to the writing of the work. M.D.M. had primary responsibility for the final content. All authors have read and agreed to the published version of the manuscript.

Funding: This research was funded by the PN I+D+I 2008–2011 (Spain), ISCIII- Sub-Directorate General for Research Assessment and Promotion and the European Regional Development Fund (ERDF), RETICS Maternal and Child Health and Development Network, SAMID Network, Ref. RD16/0022/0015. Anna Parra-Llorca acknowledges Rio Hortega grant CM18/00165 from the Instituto de Investigación en Salud Carlos III (Ministry of Science, Universities and Innovation; Kingdom of Spain); Talía Sainz is funded by The Instituto de Salud Carlos III- Spanish Ministry of Science and Innovation cofounded by FEDER (EU). (Grant nº JR16/00021); Cristina Calvo is a member of the IdiPAZ Research Institute, Madrid. Spain. Translational Research Network for Pediatric Infectious Diseases (RITIP), Madrid, Spain. TEDDY Network Member (European Network of Excellence for Pediatric Clinical Research, Bari, Italy).

Conflicts of Interest: The authors declare no conflict of interest.

References

1. Newbern, D.; Freemark, M. Placental hormones and the control of maternal metabolism and fetal growth. *Curr. Opin. Endocrinol. Diabetes Obes.* **2011**, *18*, 409–416. [CrossRef] [PubMed]
2. Rodríguez, J.M.; Murphy, K.; Stanton, C.; Ross, R.P.; Kober, O.; Juge, N.; Avershina, E.; Rudi, K.; Narbad, A.; Jenmalm, M.C.; et al. The composition of the gut microbiota throughout life, with an emphasis on early life. *Microb. Ecol. Health Dis.* **2015**, *26*, 26050. [CrossRef] [PubMed]
3. Jašarević, E.; Bale, T.L. Prenatal and postnatal contributions of the maternal microbiome on offspring programming. *Front. Neuroendocrinol.* **2019**, *55*, 100797. [CrossRef] [PubMed]
4. Koren, O.; Goodrich, J.K.; Cullender, T.C.; Spor, A.; Laitinen, K.; Bäckhed, H.K.; Gonzalez, A.; Werner, J.J.; Angenent, L.T.; Knight, R.; et al. Host remodeling of the gut microbiome and metabolic changes during pregnancy. *Cell* **2012**, *150*, 470–480. [CrossRef]
5. DiGiulio, D.B.; Callahan, B.J.; McMurdie, P.J.; Costello, E.K.; Lyell, D.J.; Robaczewska, A.; Sun, C.L.; Goltsman, D.S.; Wong, R.J.; Shaw, G.; et al. Temporal and spatial variation of the human microbiota during pregnancy. *Proc. Natl. Acad. Sci. USA* **2015**, *112*, 11060–11065. [CrossRef]
6. Peelen, M.J.; Luef, B.M.; Lamont, R.F.; de Milliano, I.; Jensen, J.S.; Limpens, J.; Hajenius, P.J.; Jørgensen, J.S.; Menon, R.; PREBIC Biomarker Working Group 2014–2018. The influence of the vaginal microbiota on preterm birth: A systematic review and recommendations for a minimum dataset for future research. *Placenta* **2019**, *79*, 30–39. [CrossRef]
7. Wankhade, U.D.; Zhong, Y.; Kang, P.; Alfaro, M.; Chintapalli, S.V.; Piccolo, B.D.; Mercer, K.E.; Andres, A.; Thakali, K.M.; Shankar, K. Maternal high-fat diet programs offspring liver steatosis in a sexually dimorphic manner in association with changes in gut microbial ecology in mice. *Sci. Rep.* **2018**, *8*, 16502. [CrossRef]
8. Olivier-Van Stichelen, S.; Rother, K.I.; Hanover, J.A. Maternal exposure to non-nutritive sweeteners impacts progeny's metabolism and microbiome. *Front. Microbiol.* **2019**, *10*, 1360. [CrossRef]
9. Garcia-Mantrana, I.; Selma-Royo, M.; Alcantara, C.; Collado, M.C. Shifts on Gut Microbiota Associated to Mediterranean Diet Adherence and Specific Dietary Intakes on General Adult Population. *Front. Microbiol.* **2018**, *9*, 890. [CrossRef]
10. Collado, M.C.; Isolauri, E.; Laitinen, K.; Salminen, S. Distinct composition of gut microbiota during pregnancy in overweight and normal-weight women. *Am. J. Clin. Nutr.* **2008**, *88*, 894–899. [CrossRef]
11. Santacruz, A.; Collado, M.C.; Garcia-Valdes, L.; Segura, M.T.; Martin-Lagos, J.A.; Anjos, T.; Martí-Romero, M.; Lopez, R.M.; Florido, J.; Campoy, C.; et al. Gut microbiota composition is associated with body weight, weight gain and biochemical parameters in pregnant women. *Br. J. Nutr.* **2010**, *104*, 83–92. [CrossRef] [PubMed]
12. Collado, M.C.; Isolauri, E.; Laitinen, K.; Salminen, S. Effect of mother's weight on infant's microbiota acquisition, composition, and activity during early infancy: A prospective follow-up study initiated in early pregnancy. *Am. J. Clin. Nutr.* **2010**, *92*, 1023–1230. [CrossRef] [PubMed]
13. Gomez-Arango, L.F.; Barrett, H.L.; McIntyre, H.D.; Callaway, L.K.; Morrison, M.; Dekker Nitert, M.; SPRING Trial Group. Connections Between the Gut Microbiome and Metabolic Hormones in Early Pregnancy in Overweight and Obese Women. *Diabetes* **2016**, *65*, 2214–2223. [CrossRef] [PubMed]

14. Aagaard, K.; Riehle, K.; Ma, J.; Segata, N.; Mistretta, T.A.; Coarfa, C.; Raza, S.; Rosenbaum, S.; Van den Veyver, I.; Milosavljevic, A.; et al. A metagenomic approach to characterization of the vaginal microbiome signature in pregnancy. *PLoS ONE* **2012**, *7*, e36466. [CrossRef] [PubMed]
15. Aagaard, K.; Ma, J.; Antony, K.M.; Ganu, R.; Petrosino, J.; Versalovic, J. The placenta harbors a unique microbiome. *Sci. Transl. Med.* **2014**, *21*, 237ra65. [CrossRef] [PubMed]
16. Prince, A.L.; Ma, J.; Kannan, P.S.; Alvarez, M.; Gisslen, T.; Harris, R.A.; Sweeney, E.L.; Knox, C.L.; Lambers, D.S.; Jobe, A.H.; et al. The placental membrane microbiome is altered among subjects with spontaneous preterm birth with and without chorioamnionitis. *Am. J. Obstet. Gynecol.* **2016**, *214*, 627. [CrossRef]
17. Perez-Muñoz, M.E.; Arrieta, M.C.; Ramer-Tait, A.E.; Walter, J. A critical assessment of the "sterile womb" and "in utero colonization" hypotheses: Implications for research on the pioneer infant microbiome. *Microbiome* **2017**, *28*, 48. [CrossRef]
18. Guinane, C.M.; Cotter, P.D. Role of the gut microbiota in health and chronic gastrointestinal disease: Understanding a hidden metabolic organ. *Therap. Adv. Gastroenterol.* **2013**, *6*, 295–308. [CrossRef]
19. Smith, P.M.; Howitt, M.R.; Panikov, N.; Michaud, M.; Gallini, C.A.; Bohlooly, Y.M.; Glickman, J.N.; Garrett, W.S. The microbial metabolites, short-chain fatty acids, regulate colonic Treg cell homeostasis. *Science* **2013**, *341*, 569–573. [CrossRef]
20. Furusawa, Y.; Obata, Y.; Hase, K. Commensal microbiota regulates T cell fate decision in the gut. *Semin. Immunopathol.* **2015**, *37*, 17–25. [CrossRef]
21. Havstad, S.; Johnson, C.C.; Kim, H.; Levin, A.M.; Zoratti, E.M.; Joseph, C.L.; Ownby, D.R.; Wegienka, G. Atopic phenotypes identified with latent class analyses at age 2 years. *J. Allergy Clin. Immunol.* **2014**, *134*, 722–727. [CrossRef] [PubMed]
22. Lynch, S.V. Gut Microbiota and Allergic Disease. New Insights. *Ann. Am. Thorac. Soc.* **2016**, *13* (Suppl. 1), S51–S54. [CrossRef] [PubMed]
23. Adak, A.; Khan, M.R. An insight into gut microbiota and its functionalities. *Cell. Mol. Life Sci.* **2019**, *76*, 473–493. [CrossRef] [PubMed]
24. Neuman, H.; Koren, O. The Pregnancy Microbiome. *Nestle Nutr. Inst. Workshop Ser.* **2017**, *88*, 1–9. [CrossRef]
25. Walters, W.A.; Xu, Z.; Knight, R. Meta-analyses of human gut microbes associated with obesity and IBD. *FEBS Lett.* **2014**, *588*, 4223–4233. [CrossRef]
26. Gosalbes, M.J.; Compte, J.; Moriano-Gutierrez, S.; Vallès, Y.; Jiménez-Hernández, N.; Pons, X.; Artacho, A.; Francino, M.P. Metabolic adaptation in the human gut microbiota during pregnancy and the first year of life. *EBioMedicine* **2019**, *39*, 497–509. [CrossRef]
27. Nyangahu, D.D.; Lennard, K.S.; Brown, B.P.; Darby, M.G.; Wendoh, J.M.; Havyarimana, E.; Smith, P.; Butcher, J.; Stintzi, A.; Mulder, N.; et al. Disruption of maternal gut microbiota during gestation alters offspring microbiota and immunity. *Microbiome* **2018**, *6*, 124. [CrossRef]
28. Nyangahu, D.D.; Jaspan, H.B. Influence of maternal microbiota during pregnancy on infant immunity. *Clin. Exp. Immunol.* **2019**, *23*. [CrossRef]
29. O'Hanlon, D.E.; Moench, T.R.; Cone, R.A. Vaginal pH and microbicidal lactic acid when lactobacilli dominate the microbiota. *PLoS ONE* **2013**, *8*, e80074. [CrossRef]
30. Smith, S.B.; Ravel, J. The vaginal microbiota, host defence and reproductive physiology. *J. Physiol.* **2017**, *595*, 451–463. [CrossRef]
31. Stoyancheva, G.; Marzotto, M.; Dellaglio, F.; Torriani, S. Bacteriocin production and gene sequencing analysis from vaginal Lactobacillus strains. *Arch. Microbiol.* **2014**, *196*, 645–653. [CrossRef] [PubMed]
32. Kroon, S.J.; Ravel, J.; Huston, W.M. Cervicovaginal microbiota, women's health, and reproductive outcomes. *Fertil. Steril.* **2018**, *110*, 327–336. [CrossRef] [PubMed]
33. Romero, R.; Hassan, S.S.; Gajer, P.; Tarca, A.L.; Fadrosh, D.W.; Nikita, L.; Galuppi, M.; Lamont, R.F.; Chaemsaithong, P.; Miranda, J.; et al. The composition and stability of the vaginal microbiota of normal pregnant women is different from that of non-pregnant women. *Microbiome* **2014**, *2*, 4. [CrossRef] [PubMed]
34. Stout, M.J.; Zhou, Y.; Wylie, K.M.; Tarr, P.I.; Macones, G.A.; Tuuli, M.G. Early pregnancy vaginal microbiome trends and preterm birth. *Am. J. Obstet. Gynecol.* **2017**, *217*, 356.e1–356.e18. [CrossRef]
35. Haque, M.M.; Merchant, M.; Kumar, P.N.; Dutta, A.; Mande, S.S. First-trimester vaginal microbiome diversity: A potential indicator of preterm delivery risk. *Sci. Rep.* **2017**, *7*, 16145. [CrossRef]
36. Lin, W.; Jiang, W.; Hu, X.; Gao, L.; Ai, D.; Pan, H.; Niu, C.; Yuan, K.; Zhou, X.; Xu, C.; et al. Ecological Shifts of Supragingival Microbiota in Association with Pregnancy. *Front. Cell. Infect. Microbiol.* **2018**, *8*, 24. [CrossRef]

37. Fujiwara, N.; Tsuruda, K.; Iwamoto, Y.; Kato, F.; Odaki, T.; Yamane, N.; Hori, Y.; Harashima, Y.; Sakoda, A.; Tagaya, A.; et al. Significant increase of oral bacteria in the early pregnancy period in Japanese women. *J. Investig. Clin. Dent.* **2017**, *8*, 1. [CrossRef]
38. Kornman, K.S.; Loesche, W.J. Effects of estradiol and progesteroneon *Bacteroides melaninogenicus* and *Bacteroides gingivalis*. *Infect. Immun.* **1982**, *35*, 256–263.
39. Nuriel-Ohayon, M.; Neuman, H.; Koren, O. Microbial Changes during Pregnancy, Birth, and Infancy. *Front. Microbiol.* **2016**, *7*, 1031. [CrossRef]
40. Pelzer, E.; Gomez-Arango, L.F.; Barrett, H.L.; Nitert, M.D. Review: Maternal health and the placental microbiome. *Placenta* **2017**, *54*, 30–37. [CrossRef]
41. D'Argenio, V. The Prenatal Microbiome: A New Player for Human Health. *High. Throughput* **2018**, *7*, 38. [CrossRef] [PubMed]
42. Jimenez, E.; Fernandez, L.; Marin, M.L.; Martin, R.; Odriozola, J.M.; Nueno-Palop, C.; Narbad, A.; Olivares, M.; Xaus, J.; Rodríguez, J.M. Isolation of commensal bacteria from umbilical cord blood of healthy neonates born by C-section. *Curr. Microbiol.* **2005**, *51*, 270–274. [CrossRef] [PubMed]
43. Wassenaar, T.M.; Panigrahi, P. Is a foetus developing in a sterile environment? *Lett. Appl. Microbiol.* **2014**, *59*, 572–579. [CrossRef] [PubMed]
44. Collado, M.C.; Rautava, S.; Aakko, J.; Isolauri, E.; Salminen, S. Human gut colonisation may be initiated in utero by distinct microbial communities in the placenta and amniotic fluid. *Sci. Rep.* **2016**, *22*, 23129. [CrossRef] [PubMed]
45. DiGiulio, D.B.; Gervasi, M.; Romero, R.; Mazaki-Tovi, S.; Vaisbuch, E.; Kusanovic, J.P.; Seok, K.S.; Gómez, R.; Mittal, P.; Gotsch, F.; et al. Microbial invasion of the amniotic cavity in preeclampsia as assessed by cultivation and sequence-based methods. *J. Perinat. Med.* **2010**, *38*, 503–513. [CrossRef]
46. DiGiulio, D.B.; Romero, R.; Kusanovic, J.P.; Gómez, R.; Kim, C.J.; Seok, K.S.; Gotsch, F.; Mazaki-Tovi, S.; Vaisbuch, E.; Sanders, K.; et al. Prevalence and diversity of microbes in the amniotic fluid, the fetal inflammatory response, and pregnancy outcome in women with preterm pre-labor rupture of membranes. *Am. J. Reprod. Immunol.* **2010**, *64*, 38–57. [CrossRef]
47. Romero, R.; Miranda, J.; Chaemsaithong, P.; Chaiworapongsa, T.; Kusanovic, J.P.; Dong, Z.; Ahmed, A.I.; Shaman, M.; Lannaman, K.; Yoon, B.H.; et al. Sterile and microbial-associated intra-amniotic inflammation in preterm prelabor rupture of membranes. *J. Matern.-Fetal Neonatal Med.* **2015**, *28*, 1394–1409. [CrossRef]
48. McClure, E.M.; Goldenberg, R.L. Infection and stillbirth. *Semin. Fetal Neonatal Med.* **2009**, *14*, 182–189. [CrossRef]
49. Fardini, Y.; Wang, X.; Témoin, S.; Nithianantham, S.; Lee, D.; Shoham, M.; Han, Y.W. Fusobacterium nucleatumadhesinFadA binds vascular endothelial cadherin and alters endothelial integrity. *Mol. Microbiol.* **2011**, *82*, 1468–1480. [CrossRef]
50. Franasiak, J.M.; Werner, M.D.; Juneau, C.R.; Tao, X.; Landis, J.; Zhan, Y.; Treff, N.R.; Scott, R.T. Endometrial microbiome at the time of embryo transfer: Next-generation sequencing of the 16S ribosomal subunit. *J. Assist. Reprod. Genet.* **2016**, *33*, 129–136. [CrossRef]
51. Jimenez, E.; Marín, M.L.; Martín, R.; Odriozola, J.M.; Olivares, M.; Xaus, J.; Fernández, L.; Rodríguez, J.M. Is meconium from healthy newborns actually sterile? *Res. Microbiol.* **2008**, *159*, 187–193. [CrossRef] [PubMed]
52. Hansen, R.; Scott, K.P.; Khan, S.; Martin, J.C.; Berry, S.H.; Stevenson, M.; Okpapi, A.; Munro, M.J.; Hold, G.L. First-Pass Meconium Samples from Healthy Term Vaginally-Delivered Neonates: An Analysis of the Microbiota. *PLoS ONE* **2015**, *10*, e0133320. [CrossRef] [PubMed]
53. Martin, R.; Makino, H.; Cetinyurek Yavuz, A.; Ben-Amor, K.; Roelofs, M.; Ishikawa, E.; Kubota, H.; Swinkels, S.; Sakai, T.; Oishi, K.; et al. Early-Life Events, Including Mode of Delivery and Type of Feeding, Siblings and Gender, Shape the Developing Gut Microbiota. *PLoS ONE* **2016**, *11*, e0158498. [CrossRef] [PubMed]
54. Lauder, A.P.; Roche, A.M.; Sherrill-Mix, S.; Bailey, A.; Laughlin, A.L.; Bittinger, K.; Leite, R.; Elovitz, M.A.; Parry, S.; Bushman, F.D. Comparison of placenta samples with contamination controls does not provide evidence for a distinct placenta microbiota. *Microbiome* **2016**, *23*, 29. [CrossRef]
55. Rutayisire, E.; Huang, K.; Liu, Y.; Tao, F. The mode of delivery affects the diversity and colonization pattern of the gut microbiota during the first year of infants' life: A systematic review. *BMC Gastroenterol.* **2016**, *16*, 86. [CrossRef]

56. Stinson, L.F.; Payne, M.S.; Keelan, J.A. A Critical Review of the Bacterial Baptism Hypothesis and the Impact of C-section Delivery on the Infant Microbiome. *Front. Med.* **2018**, *4*, 135. [CrossRef]
57. Montoya-Williams, D.; Lemas, D.J.; Spiryda, L.; Patel, K.; Carney, O.O.; Neu, J.; Carson, T.L. The Neonatal Microbiome and Its Partial Role in Mediating the Association between Birth by C-Section and Adverse Pediatric Outcomes. *Neonatology* **2018**, *114*, 103–111. [CrossRef]
58. Dominguez-Bello, M.G.; Costello, E.K.; Contreras, M.; Magris, M.; Hidalgo, G.; Fierer, N.; Knight, R. Delivery mode shapes the acquisition and structure of the initial microbiota across multiple body habitats in newborns. *Proc. Natl. Acad. Sci. USA* **2010**, *107*, 11971–11975. [CrossRef]
59. Biasucci, G.; Rubini, M.; Riboni, S.; Morelli, L.; Bessi, E.; Retetangos, C. Mode of delivery affects the bacterial community in the newborn gut. *Early Hum. Dev.* **2010**, *86* (Suppl. 1), 13–15. [CrossRef]
60. Shi, Y.C.; Guo, H.; Chen, J.; Sun, G.; Ren, R.R.; Guo, M.Z.; Peng, L.H.; Yang, Y.S. Initial meconium microbiome in Chinese neonates delivered naturally or by C-section. *Sci. Rep.* **2018**, *8*, 3255. [CrossRef]
61. Backhed, F.; Roswall, J.; Peng, Y.; Feng, Q.; Jia, H.; Kovatcheva-Datchary, P.; Datchary, P.; Li, Y.; Xia, Y.; Xie, H.; et al. Dynamics and Stabilization of the Human Gut Microbiome during the First Year of Life. *Cell Host. Microbe* **2015**, *17*, 690–703. [CrossRef] [PubMed]
62. Sakwinska, O.; Foata, F.; Berger, B.; Brüssow, H.; Combremont, S.; Mercenier, A.; Dogra, S.; Soh, S.E.; Yen, J.C.K.; Heong, G.Y.S.; et al. Does the maternal vaginal microbiota play a role in seeding the microbiota of neonatal gut and nose? *Benef. Microbes* **2017**, *8*, 763–778. [CrossRef] [PubMed]
63. Escobar, J.S.; Klotz, B.; Valdes, B.E.; Agudelo, G.M. The gut microbiota of Colombians differs from that of Americans, Europeans and Asians. *BMC Microbiol.* **2014**, *14*, 311. [CrossRef] [PubMed]
64. Seedat, F.; Stinton, C.; Patterson, J.; Geppert, J.; Tan, B.; Robinson, E.R.; McCarthy, N.D.; Uthman, O.A.; Freeman, K.; Johnson, S.A.; et al. Adverse events in women and children who have received intrapartum antibiotic prophylaxis treatment: A systematic review. *BMC Pregnancy Childbirth* **2017**, *17*, 247. [CrossRef]
65. Azad, M.B.; Konya, T.; Maughan, H.; Guttman, D.S.; Field, C.J.; Chari, R.S.; Sears, M.R.; Becker, A.B.; Scott, J.A.; Kozyrskyj, A.L.; et al. Gut microbiota of healthy Canadian infants: Profiles by mode of delivery and infant diet at 4 months. *CMAJ* **2013**, *185*, 385–394. [CrossRef]
66. Martinez, K.A.; Devlin, J.C.; Lacher, C.R.; Yin, Y.; Cai, Y.; Wang, J.; Dominguez-Bello, M.G. Increased weight gain by C-section: Functional significance of the primordial microbiome. *Sci. Adv.* **2017**, *3*, eaao1874. [CrossRef]
67. Laubereau, B.; Filipiak-Pittroff, B.; von Berg, A.; Grübl, A.; Reinhardt, D.; Wichmann, H.E.; Koletzko, S.; GINI Study Group. C-section and gastrointestinal symptoms, atopic dermatitis, and sensitisation during the first year of life. *Arch. Dis. Child.* **2004**, *89*, 993–997. [CrossRef]
68. Negele, K.; Heinrich, J.; Borte, M.; von Berg, A.; Schaaf, B.; Lehmann, I.; Wichmann, H.E.; Bolte, G.; LISA Study Group. Mode of delivery and development of atopic disease during the first 2 years of life. *Pediatr. Allergy Immunol.* **2004**, *15*, 48–54. [CrossRef]
69. Cabrera-Rubio, R.; Mira-Pascual, L.; Mira, A.; Collado, M.C. Impact of mode of delivery on the milk microbiota composition of healthy women. *J. Dev. Orig. Health. Dis.* **2016**, *7*, 54–60. [CrossRef]
70. Toscano, M.; De Grandi, R.; Peroni, D.G.; Grossi, E.; Facchin, V.; Comberiati, P.; Drago, L. Impact of delivery mode on the colostrum microbiota composition. *BMC Microbiol.* **2017**, *17*, 205. [CrossRef]
71. Neu, J.; Rushing, J. C-section versus vaginal delivery: Long-term infant outcomes and the hygiene hypothesis. *Clin. Perinatol.* **2011**, *38*, 321–331. [CrossRef] [PubMed]
72. Moya-Perez, A.; Luczynski, P.; Renes, I.B.; Wang, S.; Borre, Y.; Anthony Ryan, C.; Knol, J.; Stanton, C.; Dinan, T.G.; Cryan, J.F. Intervention strategies for cesarean section-induced alterations in the microbiota-gut-brain axis. *Nutr. Rev.* **2017**, *75*, 225–240. [CrossRef] [PubMed]
73. Francavilla, R.; Cristofori, F.; Tripaldi, M.E.; Indrio, F. Intervention for Dysbiosis in Children Born by C-Section. *Ann. Nutr. Metab.* **2018**, *73* (Suppl. 3), 33–39. [CrossRef]
74. Penders, J.; Thijs, C.; Vink, C.; Stelma, F.F.; Snijders, B.; Kummeling, I.; van den Brandt, P.A.; Stobberingh, E.E. Factors influencing the composition of the intestinal microbiota in early infancy. *Pediatrics* **2006**, *118*, 511–521. [CrossRef] [PubMed]
75. Clarke, G.; O'Mahony, S.M.; Dinan, T.G.; Cryan, J.F. Priming for health: Gut microbiota acquired in early life regulates physiology, brain and behaviour. *Acta Paediatr.* **2014**, *103*, 812–819. [CrossRef]
76. Collado, M.C.; Laitinen, K.; Salminen, S.; Isolauri, E. Maternal weight and excessive weight gain during pregnancy modify the immunomodulatory potential of breast milk. *Pediatr. Res.* **2012**, *72*, 77–85. [CrossRef]

77. American Academy of Pediatrics. Breastfeeding and the use of human milk. *Pediatrics* **2012**, *129*, e827–e884. [CrossRef]
78. Belfort, M.B.; Anderson, P.J.; Nowak, V.A.; Lee, K.J.; Molesworth, C.; Thompson, D.K.; Doyle, L.W.; Inder, T.E. Breast milk feeding, brain development, and neurocognitive outcomes: A 7-year longitudinal study in infants born at less than 30 weeks' gestation. *J. Pediatr.* **2016**, *177*, 133–139. [CrossRef]
79. Meinzen-Derr, J.; Poindexter, B.; Wrage, L.; Morrow, A.L.; Stoll, B.; Donovan, E.F. Role of human milk in extremely low birth weight infants' risk of necrotizing enterocolitis or death. *J. Perinatol.* **2009**, *29*, 57–62. [CrossRef]
80. Ballard, O.; Morrow, A.L. Human milk composition: Nutrients and bioactive factors. *Pediatr. Clin. N. Am.* **2013**, *60*, 49–74. [CrossRef]
81. Parra-Llorca, A.; Gormaz, M.; Alcantara, C.; Cernada, M.; Nuñez-Ramiro, A.; Vento, M.; Collado, M.C. Preterm Gut Microbiome Depending on Feeding Type: Significance of Donor Human Milk. *Front. Microbiol.* **2018**, *27*, 1376. [CrossRef] [PubMed]
82. Vento, M.; Hummler, H.; Dawson, J.A.; Escobar, J.; Kuligowski, J. Use of Oxygen in the Resuscitation of Neonates. In *Perinatal and Prenatal Disorders*; Dennery, P.A., Buonocore, G., Saugstad, O.D., Eds.; Humana Press: New York, NY, USA, 2014; pp. 213–244.
83. Parra-Llorca, A.; Gormaz, M.; Sánchez-Illana, A.; Piñeiro-Ramos, J.D.; Collado, M.C.; Serna, E.; Cernada, M.; Nuñez-Ramiro, A.; Ramón-Beltrán, A.; Oger, C.; et al. Does pasteurized donor human milk efficiently protect preterm infants against oxidative stress? *Antioxid. Redox Signal.* **2019**, *31*, 791–799. [CrossRef] [PubMed]
84. Parra-Llorca, A.; Gormaz, M.; Lorente-Pozo, S.; Cernada, M.; García-Robles, A.; Torres-Cuevas, I.; Kuligowski, J.; Collado, M.C.; Serna, E.; Vento, M. Impact of donor human milk in the preterm very low birth weight gut transcriptome profile by use of exfoliated intestinal cells. *Nutrients* **2019**, *11*, 2677. [CrossRef]
85. Bertino, E.; Giuliani, F.; Occhi, L.; Coscia, A.; Tonetto, P.; Marchino, F.; Fabris, C. Benefits of donor human milk for preterm infants: Current evidence. *Early Hum. Dev.* **2009**, *85*, S9–S10. [CrossRef] [PubMed]
86. Christen, L.; Lai, C.T.; Hartmann, B.; Hartmann, P.E.; Geddes, D.T. The effect of UV-C pasteurization on bacteriostatic properties and immunological proteins of donor human milk. *PLoS ONE* **2013**, *8*, e85867. [CrossRef] [PubMed]
87. Madore, L.S.; Bora, S.; Erdei, C.; Jumani, T.; Dengos, A.R.; Sen, S. Effects of donor breastmilk feeding on growth and early neurodevelopmental outcomes in preterm infants: An observational study. *Clin. Ther.* **2017**, *39*, 1210–1220. [CrossRef]
88. Untalan, P.B.; Keeney, S.E.; Palkowetz, K.H.; Rivera, A.; Goldman, S. Heat susceptibility of interleukin-10 and other cytokines in donor human milk. *Breastfeed. Med.* **2009**, *4*, 137–144. [CrossRef]
89. Sousa, S.G.; Santos, M.D.; Fidalgo, L.G.; Delgadillo, I.; Saraiva, J.A. Effect of thermal pasteurisation and high-pressure processing on immunoglobulin content and lysozyme and lactoperoxidase activity in human colostrum. *Food Chem.* **2014**, *151*, 79–85. [CrossRef]
90. Peila, C.; Moro, G.E.; Bertino, E.; Cavallarin, L.; Giribaldi, M.; Giuliani, F.; Cresi, F.; Coscia, A. The effect of holder pasteurization on nutrients and biologically-active components in donor human milk: A review. *Nutrients* **2016**, *8*, 477. [CrossRef]
91. Ardissone, A.N.; De La Cruz, D.M.; Davis-Richardson, A.G.; Rechcigl, K.T.; Li, N.; Drew, J.C.; Murgas-Torrazza, R.; Sharma, R.; Hudak, M.L.; Triplett, E.W.; et al. Meconium microbiome analysis identifies bacteria correlated with premature birth. *PLoS ONE* **2014**, *9*, e90784. [CrossRef]
92. Antony, K.M.; Ma, J.; Mitchell, K.B.; Racusin, D.A.; Versalovic, J.; Aagaard, K. The preterm placental microbiome varies in association with excess maternal gestational weight gain. *Am. J. Obstet. Gynecol.* **2015**, *212*, 653.e1–653.e16. [CrossRef] [PubMed]
93. Ejtahed, H.S.; Mohtadi-Nia, J.; Homayouni-Rad, A.; Niafar, M.; Asghari-Jafarabadi, M.; Mofid, V. Probiotic yogurt improves antioxidant status in type 2 diabetic patients. *Nutrition* **2012**, *28*, 539–543. [CrossRef] [PubMed]
94. Zheng, J.; Xiao, X.; Zhang, Q.; Mao, L.; Yu, M.; Xu, J. The placental microbiome varies in association with low birth weight in full-term neonates. *Nutrients* **2015**, *7*, 6924–6937. [CrossRef] [PubMed]
95. Liu, J.; Yang, H.; Yin, Z.; Jiang, X.; Zhong, H.; Qiu, D.; Zhu, F.; Li, R. Remodeling of the gut microbiota and structural shifts in Preeclampsia patients in South China. *Eur. J. Clin. Microbiol. Infect. Dis.* **2017**, *36*, 713–719. [CrossRef]

96. Chanomethaporn, A.; Chayasadom, A.; Wara-Aswapati, N.; Kongwattanakul, K.; Suwannarong, W.; Tangwanichgapong, K.; Sumanonta, G.; Matangkasombut, O.; Dasanayake, A.P.; Pitiphat, W. Association between periodontitis and spontaneous abortion: A case-control study. *J. Periodontol.* **2019**, *90*, 381–390. [CrossRef]
97. Stevens, G.A.; Singh, G.M.; Lu, Y.; Danaei, G.; Lin, J.K.; Finucane, M.M.; Bahalim, A.N.; McIntire, R.K.; Gutierrez, H.R.; Cowan, M.; et al. National, regional, and global trends in adult overweight and obesity prevalences. *Popul. Health Metr.* **2012**, *10*, 22. [CrossRef]
98. Zhu, Y.; Zhang, C. Prevalence of Gestational Diabetes and Risk of Progression to Type 2 Diabetes: A Global Perspective. *Curr. Diabates Rep.* **2016**, *16*, 7. [CrossRef]
99. Eriksson, J.G.; Sandboge, S.; Salonen, M.; Kajantie, E.; Osmond, C. Maternal weight in pregnancy and offspring body composition in late adulthood: Findings from the Helsinki Birth Cohort Study (HBCS). *Ann. Med.* **2015**, *47*, 94–99. [CrossRef]
100. Hussen, H.I.; Persson, M.; Moradi, T. Maternal overweight and obesity are associated with increased risk of type 1 diabetes in offspring of parents without diabetes regardless of ethnicity. *Diabetologia* **2015**, *58*, 1464–1473. [CrossRef]
101. Gaillard, R.; Santos, S.; Duijts, L.; Felix, J.F. Childhood Health Consequences of Maternal Obesity during Pregnancy: A Narrative Review. *Ann. Nutr. Metab.* **2016**, *69*, 171–180. [CrossRef]
102. Toemen, L.; Gishti, O.; van Osch-Gevers, L.; Steegers, E.A.; Helbing, W.A.; Felix, J.F.; Reiss, I.K.; Duijts, L.; Gaillard, R.; Jaddoe, V.W. Maternal obesity, gestational weight gain and childhood cardiac outcomes: Role of childhood body mass index. *Int. J. Obes.* **2016**, *40*, 1070–1078. [CrossRef] [PubMed]
103. Ma, J.; Prince, A.L.; Bader, D.; Hu, M.; Ganu, R.; Baquero, K.; Blundell, P.; Alan Harris, R.; Frias, A.E.; Grove, K.L.; et al. High-fat maternal diet during pregnancy persistently alters the offspring microbiome in a primate model. *Nat. Commun.* **2014**, *20*, 3889. [CrossRef] [PubMed]
104. Mueller, N.T.; Shin, H.; Pizoni, A.; Werlang, I.C.; Matte, U.; Goldani, M.Z.; Goldani, H.A.; Dominguez-Bello, M.G. Birth mode-dependent association between pre-pregnancy maternal weight status and the neonatal intestinal microbiome. *Sci. Rep.* **2016**, *1*, 23133. [CrossRef] [PubMed]
105. Garcia-Mantrana, I.; Collado, M.C. Obesity and overweight: Impact on maternal and milk microbiome and their role for infant health and nutrition. *Mol. Nutr. Food Res.* **2016**, *60*, 865–875. [CrossRef] [PubMed]
106. Taveras, E.M.; Rifas-Shiman, S.L.; Belfort, M.B.; Kleinman, K.P.; Oken, E.; Gillman, M.W. Weight status in the first 6 months of life and obesity at 3 years of age. *Pediatrics* **2009**, *123*, 1177–1183. [CrossRef]
107. Jacobs, M.C.; Haak, B.W.; Hugenholtz, F.; Wiersinga, W.J. Gut microbiota and host defense in critical illness. *Curr. Opin. Crit. Care* **2017**, *23*, 257–263. [CrossRef]
108. Williams, V.; Angurana, S.K. Probiotics do have a role to play in treating critically ill children. *Acta Paediatr.* **2019**, *108*, 180. [CrossRef]
109. Kitsios, G.D.; Morowitz, M.J.; Dickson, R.P.; Huffnagle, G.B.; McVerry, B.J.; Morris, A. Dysbiosis in the intensive care unit: Microbiome science coming to the bedside. *J. Crit. Care* **2017**, *38*, 84–91. [CrossRef]
110. Lyons, J.D.; Coopersmith, C.M. Pathophysiology of the Gut and the Microbiome in the Host Response. *Pediatr. Crit. Care Med.* **2017**, *18*, S46–S49. [CrossRef]
111. Wischmeyer, P.E.; McDonald, D.; Knight, R. Role of the microbiome, probiotics, and 'dysbiosis therapy' in critical illness. *Curr. Opin. Crit. Care* **2016**, *22*, 347–353. [CrossRef]
112. Lankelma, J.M.; van Vught, L.A.; Belzer, C.; Schultz, M.J.; van der Poll, T.; de Vos, W.M.; Wiersinga, W.J. Critically ill patients demonstrate large interpersonal variation in intestinal microbiota dysregulation: A pilot study. *Intensive Care Med.* **2017**, *43*, 59–68. [CrossRef] [PubMed]
113. Rogers, M.B.; Firek, B.; Shi, M.; Yeh, A.; Brower-Sinning, R.; Aveson, V.; Kohl, B.L.; Fabio, A.; Carcillo, J.A.; Morowitz, M.J. Disruption of the microbiota across multiple body sites in critically ill children. *Microbiome* **2016**, *4*, 66. [CrossRef] [PubMed]
114. Surana, N.K.; Kasper, D.L. Deciphering the tête-à-tête between the microbiota and the immune system. *J. Clin. Investig.* **2014**, *124*, 4197–4203. [CrossRef]
115. Trompette, A.; Gollwitzer, E.S.; Yadava, K.; Sichelstiel, A.K.; Sprenger, N.; Ngom-Bru, C.; Blanchard, C.; Junt, T.; Nicod, L.P.; Harris, N.L.; et al. Gut microbiota metabolism of dietary fiber influences allergic airway disease and hematopoiesis. *Nat. Med.* **2014**, *20*, 159–166. [CrossRef] [PubMed]
116. Tremaroli, V.; Bäckhed, F. Functional interactions between the gut microbiota and host metabolism. *Nature* **2012**, *489*, 242–249. [CrossRef] [PubMed]

117. Mayer, E.A.; Tillisch, K.; Gupta, A. Gut/brain axis and the microbiota. *J. Clin. Investig.* **2015**, *2*, 926–938. [CrossRef]
118. Leone, V.; Gibbons, S.M.; Martinez, K.; Hutchison, A.L.; Huang, E.Y.; Cham, C.M.; Pierre, J.F.; Heneghan, A.F.; Nadimpalli, A.; Hubert, N.; et al. Effects of diurnal variation of gut microbes and high-fat feeding on host circadian clock function and metabolism. *Cell Host Microbe* **2015**, *13*, 681–689. [CrossRef]
119. Wolff, N.S.; Hugenholtz, F.; Wiersinga, W.J. The emerging role of the microbiota in the ICU. *Crit. Care* **2018**, *22*, 78. [CrossRef]
120. Rajilić-Stojanović, M.; de Vos, W.M. The first 1000 cultured species of the human gastrointestinal microbiota. *FEMS Microbiol. Rev.* **2014**, *38*, 996–1047. [CrossRef]
121. Haaka, B.W.; Levi, M.; Wiersinga, W.J. Microbiota-targeted therapies on the intensive care unit. *Curr. Opin. Crit. Care* **2017**, *23*, 167–174. [CrossRef]
122. Dickson, R.P. The microbiome and critical illness. *Lancet Respir. Med.* **2016**, *4*, 59–72. [CrossRef]
123. Freedberg, D.E.; Zhou, M.J.; Cohen, M.E.; Annavajhala, M.K.; Khan, S.; Moscoso, D.I.; Brooks, C.; Whittier, S.; Chong, D.H.; Uhlemann, A.C.; et al. Pathogen colonization of the gastrointestinal microbiome at intensive care unit admission and risk for subsequent death or infection. *Intensive Care Med.* **2018**, *44*, 1203–1211. [CrossRef] [PubMed]
124. Shane, A.L.; Stoll, B.J. Recent developments and current issues in the epidemiology, diagnosis, and management of bacterial and fungal neonatal sepsis. *Am. J. Perinatol.* **2013**, *30*, 131–141. [CrossRef] [PubMed]
125. Özenci, V.; Schubert, U. Earlier and more targeted treatment of neonatal sepsis. *Acta Paediatr.* **2018**, *108*, 169–170. [CrossRef]
126. Rosen, G.H.; Randis, T.M.; Desai, P.V.; Sapra, K.J.; Ma, B.; Gajer, P.; Humphrys, M.S.; Ravel, J.; Gelbe, S.E.; Ratner, A.J. Group B *Streptococcus* and the vaginal microbiota. *J. Infect. Dis.* **2017**, *16*, 744–751. [CrossRef]
127. Kolter, J.; Henneke, P. Codevelopment of microbiota and innate immunity and the risk for Group B Streptococcal disease. *Front. Immunol.* **2017**, *8*, 1497. [CrossRef]
128. Berrington, J.E.; Stewart, C.J.; Embleton, N.D.; Cummings, S.P. Gut microbiota in preterm infants: Assessment and relevance to health and disease. *Arch. Dis. Child.-Fetal Neonatal Ed.* **2013**, *98*, F286–F290. [CrossRef]
129. Madan, J.C.; Salari, R.C.; Saxena, D.; Davidson, L.; O'Toole, G.A.; Moore, J.H.; Sogin, M.L.; Foster, J.A.; Edwards, W.H.; Palumbo, P.; et al. Gut microbial colonisation in premature neonates predicts neonatal sepsis. *Arch. Dis. Child.-Fetal Neonatal Ed.* **2012**, *97*, F456–F462. [CrossRef]
130. Stewart, C.J.; Embleton, N.D.; Marrs, E.C.L.; Smith, D.P.; Fofanova, T.; Nelson, A. Longitudinal development of the gut microbiome and metabolome in preterm neonates with late onset sepsis and healthy controls. *Microbiome* **2017**, *5*, 75. [CrossRef]
131. Carl, M.A.; Ndao, I.M.; Springman, A.C.; Manning, S.D.; Johnson, J.R.; Johnston, B.D.; Burnham, C.A.D.; Weinstock, E.S.; Weinstock, G.M.; Wylie, T.N.; et al. Sepsis from the gut: The enteric habitat of bacteria that cause late-onset neonatal bloodstream infections. *Clin. Infect. Dis.* **2014**, *58*, 1211–1218. [CrossRef]
132. Dunlop, J.H.; Keet, C.A. Epidemiology of Food Allergy. *Immunol. Allergy. Clin. N. Am.* **2018**, *38*, 13–25. [CrossRef] [PubMed]
133. Bousquet, J.; Khaltaev, N.; Cruz, A.A.; Denburg, J.; Fokkens, W.J.; Togias, A.; Burnham, C.A.; Weinstock, E.S.; Weinstock, G.M.; Wylie, T.N.; et al. Allergic Rhinitis and its Impact on Asthma (ARIA) 2008 update (in collaboration with the World Health Organization, GA(2)LEN and AllerGen). *Allergy* **2008**, *63* (Suppl. 86), 8–160. [CrossRef] [PubMed]
134. Bousquet, J.; Kaltaev, N.; World Health Organization. Global Surveillance, Prevention and Control of Chronic Respiratory Diseases: A Comprehensive Approach. Global Alliance against Chronic Respiratory Diseases: 2007. Available online: http://www.who.int/iris/handle/10665/43776 (accessed on 2 August 2019).
135. Johansson, S.G.; Bieber, T.; Dahl, R.; Friedmann, P.S.; Lanier, B.Q.; Lockey, R.F.; Motala, C.; Ortega Martell, J.A.; Platts-Mills, T.A.; Ring, J.; et al. Revised nomenclature for allergy for global use: Report of the Nomenclature Review Committee of the World Allergy Organization, October 2003. *Allergy Clin. Immunol.* **2004**, *113*, 832–836. [CrossRef] [PubMed]
136. Abrahamsson, T.R.; Jakobsson, H.E.; Andersson, A.F.; Bjorksten, B.; Engstrand, L.; Jenmalm, M.C. Low gut microbiota diversity in early infancy precedes asthma at school age. *Clin. Exp. Allergy* **2013**, *44*, 842–850. [CrossRef]

137. Wopereis, H.; Sim, K.; Shaw, A.; Warner, J.O.; Knol, J.; Kroll, J.S. Intestinal microbiota in infants at high risk for allergy: Effects of prebiotics and role in eczema development. *J. Allergy Clin. Immunol.* **2017**, *141*, 1334–1342. [CrossRef]
138. Song, H.; Yoo, Y.; Hwang, J.; Na, Y.C.; Kim, H.S. Faecalibacterium prausnitzii subspecies-level dysbiosis in the human gut microbiome underlying atopic dermatitis. *J. Allergy Clin. Immunol.* **2015**, *137*, 852–860. [CrossRef]
139. Nylund, L.; Nermes, M.; Isolauri, E.; Salminen, S.; de Vos, W.M.; Satokari, R. Severity of atopic disease inversely correlates with intestinal microbiota diversity and butyrate-producing bacteria. *Allergy* **2015**, *70*, 241–244. [CrossRef]
140. Muir, A.B.; Benitez, A.J.; Dods, K.; Spergel, J.M.; Fillon, S.A. Microbiome and its impact on gastrointestinal atopy. *Allergy* **2016**, *71*, 1256–1263. [CrossRef]
141. Stefka, A.T.; Feehley, T.; Tripathi, P.; Qiu, J.; McCoy, K.; Mazmanian, S.K.; Tjota, M.Y.; Seo, G.Y.; Cao, S.; Theriault, B.R.; et al. Commensal bacteria protect against food allergen sensitization. *Proc. Natl. Acad. Sci. USA* **2014**, *111*, 13145–13150. [CrossRef]
142. Chen, C.C.; Chen, K.J.; Kong, M.S.; Chang, H.J.; Huang, J.L. Alterations in the gut microbiotas of children with food sensitization in early life. *Pediatr. Allergy Immunol.* **2015**, *27*, 254–262. [CrossRef]
143. Berni Canani, R.; Gilbert, J.A.; Nagler, C.R. The role of the commensal microbiota in the regulation of tolerance to dietary allergens. *Curr. Opin. Allergy Clin. Immunol.* **2015**, *15*, 243–249. [CrossRef] [PubMed]
144. Simpson, A.; Tan, V.Y.; Winn, J.; Svensen, M.; Bishop, C.M.; Heckerman, D.E.; Buchan, I.; Custovic, A. Beyond atopy: Multiple patterns of sensitization in relation to asthma in a birth cohort study. *Am. J. Respir. Crit. Care Med.* **2010**, *181*, 1200–1206. [CrossRef] [PubMed]
145. Fujimura, K.E.; Sitarik, A.R.; Havstad, S.; Lin, D.L.; Levan, S.; Fadrosh, D.; Panzer, A.R.; LaMere, B.; Rackaityte, E.; Lukacs, N.W.; et al. Neonatal gut microbiota associates with childhood multisensitized atopy and T cell differentiation. *Nat. Med.* **2016**, *22*, 1187–1191. [CrossRef] [PubMed]
146. Tun, H.M.; Konya, T.; Takaro, T.K.; Brook, J.R.; Chari, R.; Field, C.J.; Guttman, D.S.; Becker, A.B.; Mandhane, P.J.; Turvey, S.E.; et al. Exposure to household furry pets influences the gut microbiota of infant at 3–4 months following various birth scenarios. *Microbiome* **2017**, *5*, 40. [CrossRef] [PubMed]
147. Marques, T.M.; Wall, R.; Ross, R.P.; Fitzgerald, G.F.; Ryan, C.A.; Stanton, C. Programming infant gut microbiota: Influence of dietary and environmental factors. *Curr. Opin. Biotechnol.* **2010**, *21*, 149–156. [CrossRef]
148. Chong, C.Y.L.; Bloomfield, F.H.; O'Sullivan, J.M. Factors Affecting Gastrointestinal Microbiome Development in Neonates. *Nutrients* **2018**, *10*, 274. [CrossRef]
149. Dogaru, C.M.; Nyffenegger, D.; Pescatore, A.M.; Spycher, B.D.; Kuehni, C.E. Breastfeeding and childhood asthma: Systematic review and meta-analysis. *Am. J. Epidemiol.* **2014**, *179*, 1153–1167. [CrossRef]
150. Groer, M.W.; Luciano, A.A.; Dishaw, L.J.; Ashmeade, T.L.; Miller, E.; Gilbert, J.A. Development of the preterm infant gut microbiome: A research priority. *Microbiome* **2014**, *2*, 38. [CrossRef]
151. Mennini, M.; Dahdah, L.; Artesani, M.C.; Fiocchi, A.; Martelli, A. Probiotics in Asthma and Allergy Prevention. *Front. Pediatr.* **2017**, *5*, 165. [CrossRef]
152. Sakwinska, O.; Bastic Schmid, V.; Berger, B.; Bruttin, A.; Keitel, K.; Lepage, M.; Moine, D.; Ngom Bru, C.; Brüssow, H.; Gervaix, A. Nasopharyngeal microbiota in healthy children and pneumonia patients. *J. Clin. Microbiol.* **2014**, *52*, 1590–1594. [CrossRef]
153. Pettigrew, M.M.; Gent, J.F.; Kong, Y.; Wade, M.; Gansebom, S.; Bramley, A.M.; Jain, S.; Arnold, S.L.; McCullers, J.A. Association of sputum microbiota profiles with severity of community-acquired pneumonia in children. *BMC Infect. Dis.* **2016**, *8*, 317. [CrossRef] [PubMed]
154. Machiavelli, A.; Duarte, R.T.D.; Pires, M.M.S.; Zárate-Bladés, C.R.; Pinto, A.R. The impact of in utero HIV exposure on gut microbiota, inflammation, and microbial translocation. *Gut Microbes* **2019**, *10*, 599–614. [CrossRef]
155. Kaur, U.S.; Shet, A.; Rajnala, N.; Gopalan, B.P.; Moar, P.D.H.; Singh, B.P.; Chaturvedi, R.; Tandon, R. High Abundance of genus Prevotella in the gut of perinatally HIV-infected children is associated with IP-10 levels despite therapy. *Sci. Rep.* **2018**, *5*, 17679. [CrossRef] [PubMed]

156. Serrano-Villar, S.; de Lagarde, M.; Vázquez-Castellanos, J.; Vallejo, A.; Bernadino, J.I.; Madrid, N.; Matarranz, M.; Díaz-Santiago, A.; Gutiérrez, C.; Cabello, A.; et al. Effects of Immunonutrition in Advanced Human Immunodeficiency Virus Disease: A Randomized Placebo-controlled Clinical Trial [Promaltia Study]. *Clin. Infect. Dis.* **2019**, *1*, 120–130. [CrossRef] [PubMed]
157. Liu, C.M.; Prodger, J.L.; Tobian, A.A.R.; Abraham, A.G.; Kigozi, G.; Hungate, B.A.; Aziz, M.; Nalugoda, F.; Sariya, S.; Serwadda, D.; et al. Penile Anaerobic Dysbiosis as a Risk Factor for HIV Infection. *MBio* **2017**, *25*, 8. [CrossRef]
158. Klatt, N.R.; Cheu, R.; Birse, K.; Zevin, A.S.; Perner, M.; Noël-Romas, L.; Grobler, A.; Westmacott, G.; Xie, I.Y.; Butler, J.; et al. Vaginal bacteria modify HIV tenofovir microbicide efficacy in African women. *Science* **2017**, *2*, 938–945. [CrossRef]
159. Frank, D.N.; Manigart, O.; Leroy, V.; Meda, N.; Valéa, D.; Zhang, W.; Dabis, F.; Pace, N.R.; Van de Perre, P.; Janoff, E. Altered vaginal microbiota are associated with perinatal mother-to-child transmission of HIV in African women from Burkina Faso. *J. Acquir. Immune Defic. Syndr.* **2012**, *1*, 299–306. [CrossRef]
160. Freedman, S.B.; Williamson-Urquhart, S.; Farion, K.J.; Gouin, S.; Willan, A.R.; Poonai, N.; Hurley, K.; Sherman, P.M.; Finkelstein, Y.; Lee, B.E.; et al. Multicenter Trial of a Combination Probiotic for Children with Gastroenteritis. *N. Engl. J. Med.* **2018**, *22*, 2015–2026. [CrossRef]
161. Schnadower, D.; Tarr, P.I.; Casper, T.C.; Gorelick, M.H.; Dean, J.M.; O'Connell, K.J.; Mahajan, P.; Levine, A.C.; Bhatt, S.R.; Roskind, C.G.; et al. Lactobacillus rhamnosus GG versus Placebo for Acute Gastroenteritis in Children. *N. Engl. J. Med.* **2018**, *22*, 2002–2014. [CrossRef]
162. Yeh, T.L.; Shih, P.C.; Liu, S.J.; Lin, C.H.; Liu, J.M.; Lei, W.T.; Lin, C.Y. The influence of prebiotic or probiotic supplementation on antibody titers after influenza vaccination: A systematic review and meta-analysis of randomized controlled trials. *Drug Des. Dev. Ther.* **2018**, *25*, 217–230. [CrossRef]
163. Zimmermann, P.; Curtis, N. The influence of probiotics on vaccine responses—A systematic review. *Vaccine* **2018**, *4*, 207–213. [CrossRef] [PubMed]

Review

Xylitol's Health Benefits beyond Dental Health: A Comprehensive Review

Krista Salli, Markus J. Lehtinen, Kirsti Tiihonen and Arthur C. Ouwehand *

Global Health & Nutrition Sciences, DuPont Nutrition & Biosciences, 02460 Kantvik, Finland
* Correspondence: arthur.ouwehand@dupont.com; Tel.: +358-40-5956-353

Received: 24 June 2019; Accepted: 31 July 2019; Published: 6 August 2019

Abstract: Xylitol has been widely documented to have dental health benefits, such as reducing the risk for dental caries. Here we report on other health benefits that have been investigated for xylitol. In skin, xylitol has been reported to improve barrier function and suppress the growth of potential skin pathogens. As a non-digestible carbohydrate, xylitol enters the colon where it is fermented by members of the colonic microbiota; species of the genus *Anaerostipes* have been reported to ferment xylitol and produce butyrate. The most common *Lactobacillus* and *Bifidobacterium* species do not appear to be able to grow on xylitol. The non-digestible but fermentable nature of xylitol also contributes to a constipation relieving effect and improved bone mineral density. Xylitol also modulates the immune system, which, together with its antimicrobial activity contribute to a reduced respiratory tract infection, sinusitis, and otitis media risk. As a low caloric sweetener, xylitol may contribute to weight management. It has been suggested that xylitol also increases satiety, but these results are not convincing yet. The benefit of xylitol on metabolic health, in addition to the benefit of the mere replacement of sucrose, remains to be determined in humans. Additional health benefits of xylitol have thus been reported and indicate further opportunities but need to be confirmed in human studies.

Keywords: sugar alcohol; prebiotic; bowel function; immune function; respiratory tract infections; otitis media; sinusitis; weight management; satiety; bone health

---

1. Introduction

Xylitol is a five-carbon sugar alcohol ($C_5H_{12}O_5$, Figure 1) with a molecular weight of 152.15 g/mol, which is commonly used as a sweetener in sugar-free confectionery. It also naturally occurs in fruits and vegetables (plums, strawberries, cauliflower, and pumpkin [1]). It is equisweet to sucrose and has a very similar sweetness-time intensity to sucrose. Xylitol is the sweetest of all polyols [2]. Xylitol is best known for its dental benefits, such as reducing the risk for dental caries [3]. This is thought to function through three mechanisms: xylitol replaces cariogenic sucrose, xylitol may stimulate salivation, and xylitol may have specific inhibitory effects on *Streptococcus mutans*—the main causative microbe of dental caries [4]. Although a recent meta-analysis concluded that there is a need for high-quality studies on the dental benefits of xylitol, the same study concluded nevertheless that xylitol is an effective strategy as a self-applied caries preventive agent [3]. Furthermore, the European Food Safety Agency has approved a health claim "xylitol chewing gum reduces the risk of caries in children" [5]. Here, however, we want to focus on other potential health benefits of xylitol, such as skincare, respiratory, digestive, immune health, and weight management.

Approximately half of the consumed xylitol is absorbed; the liver readily converts it to xylose by a non-specific cytoplasmic NAD-dependent dehydrogenase. The formed xylose is phosphorylated via a specific xylulokinase to xylulose-5-phosphate, an intermediate of the pentose-phosphate pathway before conversion to glucose, which is only slowly released into the bloodstream or stored as glycogen [6,7].

Figure 1. Chemical structure of xylitol ©DuPont Nutrition & Biosciences.

Xylitol is safe for human consumption and in general well tolerated. However, as with all sugar alcohols, overconsumption (>20 g) is associated with digestive symptoms such as bloating and loose stools [8]. When consumption is seized, the symptoms disappear.

2. Skin

2.1. Skin Introduction

The skin acts as a barrier between the body and its surrounding environment. The epidermis is made up of the stratum corneum (outermost layer of the skin, Figure 2); formed by terminally differentiated epidermal keratinocytes and lipids, which play a main role as a physical and chemical permeability barrier. Under this lies the stratum granulosum, which forms a paracellular barrier that regulates the loss of moisture through the skin, as shown in Figure 2. Below that are the stratum spinosum, basal cells, and melanocytes, which are also part of the epidermis. The epidermal barrier, which is constantly being renewed, is characterized by its capacity to adapt to changing conditions in the environment [9]. The dermis, the next layer, supports the epidermis and produces matrix proteins such as elastin and collagen, as shown in Figure 2.

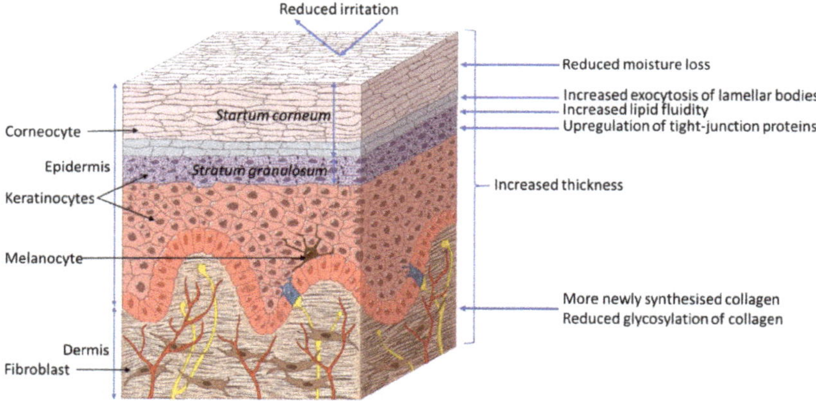

Figure 2. Proposed effects of xylitol on skin health. ©DuPont Nutrition & Biosciences.

2.2. Xylitol Benefits to Skin

Xylitol (100 mM) for 2 h has been observed, in an epidermal-equivalent skin model, to improve lipid fluidity in the uppermost layer of the stratum granulosum. The model consisted of normal human epidermal keratinocytes (NHEKs); isolated from donated skin samples; cultured *ex vivo*, and studied microscopically using lipid specific staining. The improved lipid fluidity accelerated the release of lipids and accelerates the exocytosis of lamellar bodies to the intercellular domain between stratum granulosum and stratum corneum thereby improving the lamellar structure and accelerating epidermal permeability barrier recovery [10]. Indeed, volunteers ($n = 7$) who had the inside of their forearms mechanically irritated by repeated tape stripping, were observed to have significantly less moisture loss; approximately 20%, when exposed to 100 mM xylitol for 10 min as compared to water. This was measurable both 1.5 and 2 h after exposure [10].

Further studies with NHEKs have shown that the viability and intracellular calcium concentration were not affected by 0.0045%–0.45% xylitol (calcium regulates keratinocyte differentiation) after 24 and 48 h as compared to the cell culture medium alone. However, xylitol up-regulated the expression of filaggrin, loricrin, involucrin, and occludin mRNA as measured by qPCR [11]. These proteins are involved in barrier function and tight junction (TJ) formation in the skin; occludin is the major protein in TJs, filaggrin or filament aggregating protein is a filament associated protein that binds keratin fibers in epithelial cells, loricrin is the major protein in cornified cells and contributes to barrier function of the skin, involucrin is bound to loricrin [12]. Moreover, 0.45% xylitol stimulated the mitogen-activated protein kinase (MAPK) pathway in the NHEKs and induced the activation-dependent translocation of protein kinase Cδ, after 48h as determined by Western blotting, a key promoter of epidermal differentiation [11]. The effect on the other cell types in the epidermis was not investigated in this model. Twelve healthy volunteers with dry skin received topical exposure to a combination of 5% glycerol and 5% xylitol for 14 days. This was observed to be associated with increased hydration, reduced moisture loss and increased dermal and epidermal thickness, as measured from biopsies and histological staining, compared to the untreated control arm of the same volunteer. In agreement with the above-described *ex vivo* keratinocyte studies, increased expression of filaggrin in epidermal cells in biopsies taken from the volunteers was also observed [13]. The separate contribution of xylitol and glycerol in the observed effects cannot be determined from this study.

In a study with hairless mice (23/group), skin irritation induced by 3 h topical application of 5% sodium dodecyl sulfate (SDS) was reduced with concomitant exposure to 8.26% xylitol or 5% glycerol (same osmolarity); transepidermal water loss was reduced and in the irritated area blood flow was reduced as well, as determined by videomicroscopy. Histological staining indicated that the epidermal thickness was increased in response to xylitol treatment compared to SDS alone [14]. Also in healthy adult volunteers (n = 16), the transepidermal water loss induced by experimental irritation with 0.1% SDS could be inhibited by simultaneous exposure for 24 h to 4.5% or 15% xylitol and 2.6% or 9.0% glycerol, but not 5.4% or 18% mannitol (same osmolarity) as compared to another site on the same arm with 0.1% SDS alone for 24 h [15]. These results suggest a polyol-specific response.

In a study with male rats, the inclusion of 10% xylitol to basic chow for 20 months was observed to be associated with a thicker skin and more acid-soluble collagen was observed, as determined from biopsies. Also, less collagen fluorescence was observed, which is a marker for collagen glycosylation and aging [16]. However, no difference in collagenase soluble and insoluble collagen was observed nor more total collagen as compared to control animals fed the same chow without xylitol [17]. Three months dietary supplementation with 10% xylitol in basic chow has been reported to increase the amounts of acid-soluble and total collagen (expressed as hydroxyproline) in the skin of streptozotocin-induced type 1 diabetic male rats (10 animals/group) as compared to type 1 diabetic animals fed unsupplemented chow. Also here, reduced hexose concentrations of acid-soluble collagen and reduce fluorescence of the collagenase-soluble fraction; indicating reduced glycosylation were observed. Similar observations on increased were made for non-diabetic rats (10 animals/group) after three months on 10% xylitol supplemented chow as compared to non-diabetic rats fed unsupplemented chow; for acid-soluble and total collagen, as well as reduced hexose concentrations of acid-soluble collagen and reduced fluorescence of the collagenase-soluble fraction in the skin [18].

The selective antimicrobial activity of xylitol, observed in dental health, has also been applied to wound care. *In vitro* studies with a Lubbock Chronic Wound Biofilm model have shown that the application of 2%, 10%, and 20% xylitol in water reduced growth *Pseudomonas aeruginosa*, *Staphylococcus aureus*, and *Enterococcus faecalis* compared to the water control. The highest concentration was observed to completely abolish biofilm formation [19]. Furthermore, another *in vitro* study showed that the combination of 5% xylitol and 2% lactoferrin could reduce the biofilm formation of *P. aeruginosa* and methicillin-resistant *S. aureus* after 72 h in a colony drip flow reactor, as compared to base wound dressing alone [20]. The anti-*S. aureus* potential of xylitol has also been investigated in human volunteers. Seventeen volunteers with atopic dermatitis received skin lotion with or without a combination of 5% xylitol and 0.2% farnesol on either arm for seven days. Compared to the control arm treated with unsupplemented lotion, *S. aureus* was significantly reduced, and skin moisture increased [21]. The contribution of xylitol alone cannot be deduced from this study. A further potential benefit of xylitol in wound care is the negative dissolution energy [2] which gives a cooling effect to the tissue.

2.3. Conclusions

Topical exposure of the skin with xylitol has thus been shown to reduce skin moisture loss. The mechanism appears to relate to increased tight junction and barrier formation in the skin. Also, dietary exposure to xylitol has been found to improve skin thickness. The antimicrobial activity against skin pathogens has been documented mainly in combination with other compounds and the contribution of xylitol to the observed effects needs to be determined. Furthermore, many of these results have been obtained *in vitro* and in animal models at relatively high doses (10% of the diet); their applicability to humans thus still needs to confirmed.

3. Digestive Tract

3.1. Introduction

The digestive tract can be largely divided into the stomach, small intestine, and large intestine (colon). Much of the digestion and nutrient absorption takes place in the stomach and small intestine. Although the upper digestive tract harbors a microbiota [22], it is especially the colon that is host to a diverse and extensive microbiota [23]. This colonic microbiota ferments non-digested dietary components, mainly fiber, and other components that have escaped digestion as well as sloughed-off cells and secretions. The colon absorbs the fermentation products together with water from the digesta; in particular short-chain fatty acids are an important additional energy source.

Xylitol is not digested by human enzymes and approximately 50% of the consumed xylitol is absorbed through passive diffusion in the small intestine [6]. The remaining 50% of the dietary xylitol thus enters the colon where it can serve as an energy and carbon source for the intestinal microbiota and leads to the formation of short-chain fatty acids which provide energy to the host and support immune system homeostasis [24]. These properties of xylitol are very similar to what is expected from a prebiotic; a substrate that is selectively utilized by host microorganisms conferring a health benefit [25]. The increased concentration of xylitol in the digesta leads to an increased osmotic pressure which contributes to water retention in the digesta and thus may lead to laxative effects when consumed in excess (>20 g) [8,24]. However, this property of xylitol can also be used to address constipation; which is in line with the prebiotic nature of xylitol.

3.2. Prebiotic Benefits of Xylitol

Simulations of fermentation by the colonic microbiota *in vitro* have shown that exposure of this microbiota to xylitol leads to a rapid disappearance of the xylitol, as determined by enzymatic colorimetry, indicating that it is readily fermented by the simulated intestinal microbiota. Gas chromatographic analysis of the simulated colonic digesta showed an increased formation of butyric acid compared to the non-supplemented control simulations [26]. Strains from the genus *Anaerostipes* have been observed by 16S rRNA denaturing gradient gel electrophoresis (DGGE) analyses to be associated with the increased production of butyric acid in fecal cultures [27]. Production of butyric acid is considered beneficial for colonic health as it is the preferred energy source for colonocytes and is thought to be associated with a reduced risk for colorectal cancer [28]. Furthermore, butyric acid promotes the generation of regulatory T-cells that promote immune system balance [29]. In rats (at least 5 animals/group), early fecal microscopy studies indicated that 20% of dietary xylitol caused a shift from fecal Gram-negative to Gram-positive bacteria after six weeks compared to animals fed an unsupplemented diet; the magnitude of this change was, however, not reported. Similar observations were made in humans; six volunteers, after an overnight fast, consumed in a cross-over design randomly a single 30 g dose of xylitol or a single 30 g dose of glucose (control) in 200 mL water. Fecal microscopy indicated an increase in Gram-positive bacteria from 20%–30% to 50%–55% for glucose and xylitol, respectively, and a concomitant decrease in Gram-negative bacteria was observed. Furthermore, a reduction in the fecal level of yeasts was reported, from Log10 9.2–9.4 colony forming units (CFU)/g feces during the control phase to Log10 7.2–7.5 CFU/g feces after xylitol consumption [30]. The type of yeast that was reduced was not reported, but *in vitro* studies have reported that xylitol can suppress the growth of *Candida* with a minimal inhibitory concentration of 200 mg/mL and a 99.95% reduction in colony-forming units at 400 mg/mL [31]. Recent mouse studies (5 animals/group) have reported that consumption of xylitol (40 or 194 mg/kg body weight/day) for 15 weeks was associated with an increase in the genus *Prevotella*, the phyla Eubacteria and Firmicutes and a reduction in the phylum Bacteroidetes by DGGE analysis [32]. Others have made similar observations, terminal restriction fragment length polymorphism (TRFLP) analysis indicated reduced levels of *Bacteroides* and *Clostridium* cluster XIVa and increased levels of *Prevotella* in mice (7 animals/group) fed 5% xylitol for 28 days as compared to animals fed unsupplemented chow [33]. In studies with cyclophosphamide-immune suppressed mice, 5%–10% xylitol (12 animals) was observed to lead to significantly lower fecal counts of *Candida albicans* (7.58 vs. 5.22 Log10 CFU/g, control and xylitol respectively) and significantly less and fewer cases of *C. albicans* invasion of the gastric wall as compared to animals not fed xylitol (10 animals); 80% vs. 10% of animals, control and xylitol respectively [34]. Furthermore, urinary HPLC analysis indicated an increased metabolism of daidzein to equol when mouse diet (7 animals/group) was supplemented with 0.05% daidzein (control) or 0.05% daidzein and 5% xylitol for 28 days [33]; this may contribute improved bone health.

These observations are in agreement with the definition of prebiotics [25]; furthermore, xylitol is utilized only by a limited number of organisms and changes the metabolism of the microbiota; as expected for a prebiotic, Table 1. As Table 1 also clearly shows, commercial probiotics have been shown to be unable to grow on xylitol as sole carbon and energy source.

Table 1. Non-exhaustive list of organisms that are able to grow or not to grow in the presence of xylitol, or that have the capacity to metabolize xylitol *in vitro* or not.

Organisms Reported to Grow on Xylitol	Reference	Organisms Reported Not to Grow on Xylitol	Reference
Anaerostipes hadrus (strain dependent), *A. caccae*	[27]	*Lactobacillus plantarum* 299v, *L. plantarum* 931, *L. rhamnosus* GG, *L. rhamnosus* LB21, *L. paracasei* F19, *L. reuteri* PTA5289	[35]
		Bifidobacterium lactis 1100, *B. lactis* Bb-12, *B. longum* 913, *B. lactis* 420, *L. acidophilus* NCFM, *L. casei* 921, *L. casei* Shirota, *L. bulgaricus* 365, *L. johnsonii* LA1, *L. paracasei* F19, *L. plantarum* 299v, *L. reuteri* SD2112, *L. rhamnosus* GG, *L. rhamnosus*, Lc-705, *Streptococcus mutans* Ingbritt	[36]
		L. plantarum 299v, *L. reuteri* DSM17938	[37]
		Coprococcus catus, *Eubacterium halli*, *E. limosum*, *E. rectale*, *Faecalibacterium prausnitzii*, *Megasphera elsedenii*, *Ruminococcus faecis*, *R. hominis*, *R. intestinalis*, *R. inulinivoruans*	[27]
		S. pneumoniae	
		S. mutans, *S. salivarius*, *S. sanguis*	
		Candida albicans, *S. mutans*	[38]
		Staphylococcus epidermidis, *Staphylococcus aureus*, *Pseudomonas aeruginosa*	

Even though organisms may not be able to metabolize and grow on xylitol, there may still be an opportunity for synergy with xylitol and probiotic bacteria, as was shown with the combination of *Lactobacillus plantarum* Inducia in combination with 5% xylitol which was reported to completely stop spore germination of *Clostridioides* (formerly *Clostridium*) *difficile*, *in vitro* after 48 h. In addition, prefeeding with a single dose of 0.2 g xylitol improved the survival of hamsters in a *C. difficile* challenge model (5 out of 9 survived in the xylitol test against 2 out of 15 in the unsupplemented group). Fecal colonization with *C. difficile* quantified by real-time PCR was lower in the xylitol group, 3.5 vs. 4.9 Log10 gene copy number/g in the control group. Real-time PCR *Lactobacillus* fecal counts, however, were highest in the xylitol group, 6.6 vs. 4.6 Log10 gene copy number/g in the control group [39].

3.3. Benefits of Xylitol on Bowel Function

Similar as other prebiotics [40], xylitol has been used to relieve constipation. To investigate the normalization of bowel function post-laparoscopic surgery; 60 patients were randomized to consume xylitol chewing gum (amount not reported) three times per day and 60 patients allocated to a non-chewing gum control group. The time to first flatus (−5.7 h) and first bowel sounds (3.8 h) was observed to be significantly reduced compared to the control group. There was, however, no influence on time to first bowel movement [41]. This result is very similar to what was observed with xylitol chewing gum (2.40–2.74 g xylitol/dose) every two hours until first flatus, for normalizing bowel function after Caesarian section; 40 women in xylitol chewing gum group and 40 women in non-xylitol chewing gum control group. Time to first bowel sounds (−1.1 h) and first flatus (−0.9 h) were significantly reduced, but no effect was observed for time to first bowel movement compared to the control group [42]. However, xylitol chewing gum (0.86 g xylitol/dose; 43 subjects) three times/day has been shown to contribute to earlier normalization of bowel function after elective proctectomy; time to first flatus (−6.9 h) and time to first stool (−12.3 h) were significantly reduced compared to the control group (no chewing gum; 46 subjects). Interestingly, also post-operative opioid use was reduced in the xylitol chewing group by approximately 20% as compared to the control group. No differences in post-operative complications were observed [43].

3.4. Conclusions

Xylitol has been shown to modulate intestinal microbial composition and activity *in vitro* and in animal studies. Although these data are promising, data in humans are limited. Similarly, for improving bowel function, human data exists but are limited to specific patient groups. There is thus a need for studies in, otherwise healthy, humans with constipation.

4. Nose, Throat and Ear

4.1. Introduction

As all the body sites that are exposed to the outside environment, also the respiratory tract is colonized by a microbiota. An important function of this microbiota is to hamper the establishment of exogenous microbes; in particular potential pathogens. As with the microbiota in other body sites, the respiratory microbiota evolves from birth to an 'adult-like' microbiota [44]. In contrast to viral gastrointestinal infections, it seems that during an upper respiratory tract viral infection the nasal microbiota is relatively stable as was demonstrated in an experimental rhinovirus challenge study in humans [45]. The microbiota composition also differs at different sites along the respiratory tract. The anterior nares may be colonized by *Staphylococcus* spp., *Cutibacterium* (formerly *Propionibacterium*) spp., *Streptococcus* spp. and *Corynebacterium* spp. [46]. The nasopharyngeal microbiota demonstrates considerable overlap with the anterior nares and consists of *Moraxella* spp., *Staphylococcus* spp., *Corynebacterium* spp., *Dolosigranulum* spp., *Haemophilus* spp. and *Streptococcus* spp. [46]. The microbiota of the oropharynx is characterized by *Streptococcus* spp., *Neisseria* spp., *Rothia* spp., *Veillonella* spp., *Prevotella* spp. and *Leptotrichia* spp. [46]. Some of these potential pathogens can spread from the nasopharynx into the sinus cavity during viral respiratory infection and cause sinus infection; *S. aureus*, *Staphylococcus epidermidis*, and Gram-negative bacteria such as *P. aeruginosa* and *Klebsiella pneumoniae*, predominate in chronic rhinosinusitis [47]. Acute otitis media (AOM) is defined as the presence of middle ear effusion (thick or sticky fluid behind the eardrum in the middle ear) and a rapid onset of signs or symptoms of middle-ear inflammation, such as ear pain, discharge from the ear or fever. Also here, the key step in the pathogenesis is the colonization of the upper airways with pathogenic bacteria; in particular *S. pneumoniae* and *H. influenzae*, which move from the nasopharynx through the eustachian tube to the middle ear [48].

4.2. Benefits of Xylitol in Respiratory Health

In vitro studies have shown that 1% and 5% xylitol markedly reduced the growth of alpha-hemolytic streptococci, including *S. pneumoniae* in a dose dependent manner. The inhibitory growth pattern was similar to that previously seen with *S. mutans*. Xylitol reduced slightly the growth of beta-hemolytic streptococci but not that of *H. influenzae* or *Moraxella catarrhalis* [49]. Although *in vitro* inhibition of *S. pneumoniae* was observed, nasal infection of rats (20 animals/group) with *S. pneumoniae* could not be reduced, as evaluated by PCR, with 3 day exposure to dietary xylitol (20%) or nasal spray with 5% xylitol compared to control animals not exposed to xylitol [50].

Furthermore, 250 µl of 5% xylitol sprayed for 4 days into each nostril of 21 healthy volunteers significantly decreased the number of nasal coagulase-negative *Staphylococcus* compared with saline control treatment in the same volunteers. Counts were reduced from 597 CFU/nasal swab during the control treatment to 99 CFU/nasal swab during the xylitol treatment; no other organisms were assessed [51].

A nasal spray with xylitol has been reported to improve the quality of life in patients with non-allergic nasal congestion. Subjects were randomized to either receive xylitol spray twice daily for 5 days ($n = 14$) or saline ($n = 14$). Objective rhinometry measures were not significantly different from control and baseline, and subjective measures of nasal obstruction, by questionnaire, only exhibited a trend for improvement from baseline. However, the Rhinoconjunctivitis Quality of Life Questionnaire indicated a significant improvement from baseline for the xylitol group, but not for the control group [52].

Despite some anti-pathogenic effects by xylitol on some potential pathogens of the upper respiratory tract, the consumption of 5 pieces of 15% xylitol-containing chewing gum by 106 pharyngitis patients for three months was not found to be associated with a reduction in pharyngitis and did not perform better in reducing symptoms; difficulty in swallowing and sore throat as compared to no chewing gum control subjects ($n = 110$). Data were collected by questionnaire [53]. Inhalation of xylitol aerosol has been suggested to reduce salt concentration in airway surface liquid (ASL); increased salt concentrations are associated with reduced antimicrobial activity of ASL and may partially explain the pathogenesis of cystic fibrosis [51].

As will be discussed below under immune-modulatory effects of xylitol (Section 6.2) there is substantial animal model data indicating a benefit of xylitol consumption and immune modulation which improves resistance to experimental viral infections by the human respiratory syncytial virus (hRSV) and influenza A virus (H1N1).

4.3. Benefits of Xylitol in Sinusitis

A reduction of the ionic composition of ASL by xylitol has been hypothesized to be beneficial not only for respiratory tract infections but also for the treatment of sinusitis. *In vitro*, 5% and 10% xylitol in saline significantly reduced *S. epidermidis* and *S. aureus* biofilm formation after 1 h, and after 24 h also of *P. aeruginosa* compared to saline. After 4 h 5% and 10% xylitol significantly reduced the growth of planktonic *S. epidermidis*, *S. aureus*, and *P. aeruginosa* compared to saline. There was no difference between 5% and 10% xylitol [54]. As mentioned above, 2%, 10%, and 20% xylitol in water have also been shown to inhibit the growth of *P. aeruginosa* in a biofilm model [19].

Indeed, in experimental sinusitis through *P. aeruginosa* infection of 26 rabbits, and local pre-administration (20 min) of 0.1 mL 5% xylitol for five days, reduced the number of recovered *P. aeruginosa* compared to administration with saline in the other sinus of the same rabbit (control). Culturing showed counts of 5.37×10^6 CFU in control sinuses and 1.93×10^6 CFU in xylitol pretreated sinuses. However, simultaneous or subsequent administration of xylitol and *P. aeruginosa* infection resulted only in a non-significant reduction in *P. aeruginosa* [55].

A 10-day nasal irrigation with a 5% xylitol solution by 15 subjects with chronic rhinosinusitis resulted in a significant reduction in Sino-Nasal Outcome Test 20 (SNOT-20) score compared to control irrigation with saline. The volunteers, however, did not self-report an improvement in their sino-nasal wellbeing. No adverse events were reported [56]. In a subsequent study with 30 patients with chronic rhinosinusitis, nasal irrigation with a 5% xylitol solution for 30 days has indeed been found to lead to an improvement in symptoms of chronic rhinosinusitis reported as SNOT-22 [57]. As a potential mechanism, a reduction in the viscoelasticity of mucus has been proposed [58].

4.4. Acute Otitis Media

As noted above, *S. pneumoniae* is one of the main causative agents of AOM; 1% and 5% xylitol has been shown to inhibit the growth of *S. pneumoniae in vitro* [49]. Ultrastructural analysis of the pneumococci showed that the cell wall became more diffuse, the polysaccharide capsule became ragged and the proportion of damaged pneumococci increased after exposure to 5% xylitol for 2 h, but not after exposure to other sugars or control medium [59]. In fact, exposure to 5% xylitol lowered pneumococcal capsular locus (*cpsB*) gene expression levels significantly compared with those in the control and glucose media [60]. However, in clinical trials, xylitol did not decrease nasopharyngeal carriage of pneumococci; even though AOM risk was reduced. Nevertheless, xylitol at 0.5% solution has been observed to reduce the growth of 20 pneumococcal clinical isolates *in vitro* compared to other carbon sources. Also *in vitro* pneumococcal biofilm formation was reduced and expression of genes involved in biofilm formation—capsule, competence, and autolysin—was reduced [61].

A recent Cochrane review investigated the benefit of the prophylactic administration of xylitol to healthy children up to 12 years of age on the risk for the development of AOM. In all, 5 clinical trials were identified and included in the analysis, which involved 3405 children in total. Doses used ranged from 8.4 to 10 g/day. The authors concluded that there is moderate-quality evidence that xylitol (in any form) can reduce the risk of AOM from 30% in the control group to approximately 22%. However, xylitol was not found to be effective in reducing AOM among healthy children during respiratory infection or among otitis-prone healthy children [48]. Furthermore, the authors expressed the concern that there is only a limited number of studies, mainly from the same research group. In that sense, it is interesting to see that at least two clinical trials are on the way to investigate the effect of xylitol on AOM (clinicaltrials.gov: NCT02950311 and NCT03055091 [62]).

4.5. Conclusions

Some subjective benefits for xylitol were observed in relieving congestion; overall these results are not convincing. Also for sinusitis, results are inconclusive. For AOM, however, there is quite convincing evidence on the potential benefit of xylitol in reducing its risk.

5. Bone

5.1. Introduction

Although bone may appear to be a rather static tissue, it is actually in continuous turnover. It is, therefore, important that there is a correct balance in the resorption and reconstruction of bone tissue. There is a continued risk for reduced reconstruction and especially with aging a risk for osteoporosis. Dietary means to improve mineral absorption, bone mineral density, and bone strength are thus welcome.

5.2. Effects of Xylitol on Bone Strength

In non-challenged animals (12 rats/group) on a diet supplemented with 10% or 20% (w/w) xylitol for 40 days, higher levels of both serum Ca^{2+} (double and triple that of the control group for 10% and 20% xylitol, respectively) and 25% and 80% increase in alkaline phosphatase activity (for 10% and 20% xylitol, respectively) were observed compared to the unsupplemented control group. Microfocus X-ray computed tomography did not show significant differences in the three-dimensional bone structure or trabecular bone structure of the femur. However, the histological analysis indicated an increase in trabeculae. Furthermore, both xylitol groups showed 3% and 6% higher bone density for 10% and 20% xylitol, respectively, than the control group fed an unsupplemented diet [63]. Xylitol has also been shown to reduce bone resorption by 42% in tetracyclin-challenged animals (10 rats/group) on a diet supplemented with 1 molar xylitol per kilogram dry feed for 31 days, compared to the control animals on a non-supplemented basal diet [64]. A similar study with 5%, 10% and 20% dietary xylitol in tetracyclin-challenged animals (10 rats/group) for 31 days noted a retarding effect on bone resorption of about 25% in the 10% xylitol group, about 40% in the 20% xylitol group, and undetectable in the 5% xylitol group. Furthermore, the effect was detected as early as 2 days after the beginning of xylitol-feeding and was maintained throughout the experimental period of 31 days compared to the unsupplemented control group [65]. This is in an agreement with observations in an ovariectomized rat model (10 animals/group). After three months on a 10% (w/w) xylitol diet, humeral ash, calcium and phosphorus loss was abrogated as compared to animals not supplemented with xylitol and no significant difference compared to sham operated animals. Furthermore, there was no loss of stress and strain resistance upon xylitol supplementation compared to sham operated animals; while elasticity was maintained. Diets between the groups were isocaloric [66].

In an injected type II collagen-induced arthritis model with 20 rats/ group, administration of 10% dietary xylitol for 17 days led to a significant protective effect against the imbalance in bone metabolism. This was seen in greater values of osteoid thickness, as well as in lower values of the number of osteoclasts on bone surface, trabecular separation, and eroded surface/bone surface in the xylitol-fed animals as compared to arthritic animals few the unsupplemented diet. In the case of trabecular bone volume, trabecular number and trabecular separation this was not different from the non-arthritic rats [67]. These observations can partially be explained by an increased bone formation activity induced by xylitol and a diminished bone resorption activity. Also, in a streptozotocin-induced type I diabetic osteoporosis model with ten rats/group, 3-month dietary supplementation with 10% and 20% xylitol has been shown to reduce the loss of trabecular bone volume and bone strength. Tibia density and ash weight in both xylitol groups were significantly different from diabetic rats fed the unsupplemented diet but similar to unsupplemented healthy rats. This was similar for tibia and femur stress tolerance and for histomorphometric assessed tibia trabecular bone volume; both xylitol groups were significantly different from diabetic rats fed the unsupplemented diet but similar to unsupplemented healthy rats [68].

As discussed above, in a mouse study, 28 days of 5% dietary xylitol was observed to stimulate the conversion of daidzian to equol [33]. The conversion of isoflavones to equol has been suggested to be responsible for their positive effects on bone health [69], whether dietary xylitol plus isoflavonoids exert a favorable effect on bone health remains, however, to be studied [33].

5.3. Conclusions

The ability of xylitol to positively influence bone health is in line with its prebiotic properties. Being undigestible but fermented in the colon, leads to a production of short-chain fatty acids and a reduction in pH of the digesta. This improves the solubility and absorption of minerals such as calcium. Furthermore, it has been shown in mice that butyrate stimulates bone formation via regulatory T cell-dependent mechanisms [70] thus linking the butyrogenic effect of xylitol [18] to bone health. These observations are, however, all in animals. Human studies are required to validate these benefits.

Furthermore, the levels of dietary xylitol in animal studies are high (up to 20%) and not feasible for humans.

6. Immune Function

6.1. Introduction

As the first line of defense against foreign compounds and potential pathogenic micro-organisms, the body has physicochemical barriers such as the skin and mucous membranes. As mentioned above, xylitol may beneficially affect the skin barrier function, and as will be discussed below, xylitol also improves mucous membrane function; especially in the oropharynx. Below these barriers, the body relies on the immune system which can roughly be divided into a non-specific, fast-working, innate immunity and highly specific, but slower reacting, acquired immunity [71]. Xylitol may exert its effects on the immune system indirectly by prebiotic effect as discussed above or directly by influencing host (e.g., immune) cell metabolism [72].

6.2. Immune Modulatory Effects of Xylitol

Xylitol has been found to potentiate immune responses mainly in animal models. A single 0.5 mL dose of 20% xylitol within 24 h after hatching of ten female broiler chicks was found to improve splenocyte proliferation by B-cell and T-cell mitogens (concanavalin A and pokeweed mitogen) compared to 0.5 mL of 20% glucose. Furthermore, antibody titers to keyhole limpet hemocyanin (KHL) and *Mycobacterium butyricum* injected at day 5 were higher at day 12 post-hatching compared to animals that received glucose [73]; indicating an improved acquired immune response development in chicks. The effect of xylitol on innate immunity has been studied in rats. Rats (20 animals/group) fed 20% dietary xylitol exhibited a 6.7% higher increase in the percentage of activated neutrophils from baseline than in the unsupplemented control group after 2 weeks. Likewise, the strength of the oxidative burst per neutrophil was 13.5% higher in the xylitol group as compared to the control group [74]. When rats (20 animals/group) were infected with an intraperitoneal inoculation of *Streptococcus pneumoniae* after two weeks supplementation with 10% or 20% dietary xylitol, or no supplementation (control). The mean survival time was 11 h longer in the 10% xylitol and 12 h longer in the 20% group compared to the control group [74].

Anti-bacterial effects of xylitol have been well documented especially against oral [75] and respiratory pathogens [19]; see also earlier sections. However, only a few studies have investigated its effect on viral infections. Human respiratory syncytial virus (hRSV) is the most common cause of bronchiolitis and pneumonia in infants. There is a need for prophylactic and therapeutic strategies to control hRSV infection. Mice (5/group) receiving dietary xylitol (3.3–33 mg/kg/d in phosphate-buffered saline; PBS) for 14 d prior to hRSV challenge and for a further 3 d post-challenge had significantly lower lung virus titers compared to PBS only, control mice. In line with lower viral load, also fewer CD3(+) and CD3(+)CD8(+) lymphocytes were found in bronchoalveolar lavage, indicating less need for lymphocyte recruitment to control the viral infection [76]. Similar effects were observed for the anti-viral drug ribavirin (40 mg/kg/d during the 3 days post hRSV infection) in the study [76]. The results indicate an improved innate immune response but nevertheless combined with a reduced inflammatory response to hRSV infection. Another mouse study (five mice/group) investigated the effect of xylitol consumption (3.3 or 33 mg/kg/d) during 5 days prior to influenza infection and three days post-infection. Mortality in mice infected with influenza A virus (H1N1) could not be influenced by prophylactic oral application of xylitol or red ginseng. However, combining the two remarkably reduced mortality. With a higher dose of xylitol (33 mg/kg body weight/day) being more effective than the lower xylitol dose (3.3 mg/kg body weight/day). Interestingly, dietary administration of 33 mg/kg/d xylitol significantly reduced the lung viral titer compared the PBS control [77].

6.3. Anti-Inflammatory Effects of Xylitol

The studies discussed above indicate that xylitol may have anti-inflammatory effects on skin by improving the epithelial tight junctions and thus limiting the leakage of microbial and other foreign components into the host. It has been further shown that 0.0045%–0.45% xylitol exerts direct anti-inflammatory effects after 24 and 48h on NHEKs stimulated *ex vivo* with toll-like receptor agonists lipopolysaccharide (LPS), lipoteichoic acid and polyI:C, as compared to the cell culture medium alone [11]. Although the authors noted some skin donor-dependent effect, xylitol was in general effective in suppressing inflammatory cytokine interleukin (IL)-1α and IL-1β upregulation, and also in decreasing tumor necrosis factor (TNF)-α after polyI:C induction. It can be hypothesized that this reduced inflammatory response contributes to improved skin barrier function. Further evidence on anti-inflammatory effects was observed in a hairless mouse model (23 animals/group). Inflammatory responses induced by irritation of the skin for 3h with 5% SDS were substantially reduced by concomitant topical xylitol administration (at 8.26% or 16.52%); normalizing the level of lymphocytes and reducing the expression of inflammatory cytokines IL-1β and TNF-α, but not IL-1α in skin biopsies, compared to biopsies only treated with 5% SDS [14]. On the other hand, intraperitoneal injection of *Escherichia coli* LPS caused an increase in α1 acid glycoprotein in 10 and 12-day-old male broiler chicks (16 animals/group) as expected. This acute-phase inflammatory marker protein was, however, not affected by the inclusion of 6% xylitol (+9% glucose) in the diet for 7 days [78]. Nevertheless, the LPS induced reduction in body weight gain, feed intake and feed efficiency were partially prevented by the xylitol diet as compared to the 15% dietary glucose control; suggesting a reduced physiological stress response to the immune challenge.

6.4. Conclusions

In animal models, xylitol has been observed to stimulate innate and acquired immunity; mainly against bacterial infectious agents. For viral infections, results are less conclusive. Also, the anti-inflammatory effects of xylitol are somewhat inconclusive and based on animal studies. Information on the potential effects on human inflammatory responses is lacking.

7. Weight Management

7.1. Introduction

Overweight and obesity are an increasing health risk not only in affluent countries but increasingly also in developing countries. Strategies to aid consumers with weight management are thus very welcome and xylitol may play a role here. A potential mechanism by which xylitol could contribute to weight management and reduced energy intake is through the induction of satiety. In addition to weight management, there may also be a benefit in counteracting the consequences of overweight and obesity, commonly referred to as metabolic syndrome; insulin resistance, high serum cholesterol and hyperlipidemia [79].

7.2. Effects of Xylitol on Weight Management

An obvious contribution of xylitol to weight management is through the replacement of sucrose. The caloric value of sucrose is 3.87 kcal/g and for xylitol approximately 2.4 kcal/g [2]. As xylitol is equisweet to sucrose, replacing sucrose with xylitol will reduce the caloric value of a particular food while maintaining taste. In confectionery, xylitol will also contribute the same bulk as sucrose. Whether this will contribute to long-term weight loss is uncertain.

For short-term weight management, a high-fat diet animal model (6 rats/group), reported a smaller bodyweight increase after an 8-week intervention, with less visceral (−12.9% and −15.5%) and epididymal fat (−15.5% and −17%) was observed in rats on 1 and 2 g xylitol/100 kcal of diet, respectively, as compared to animals fed an unsupplemented high-fat diet. This may be explained by the observation that adipose tissue of the xylitol-fed rats exhibited significantly higher levels of mRNAs encoding peroxisome proliferator-activated receptor (PPAR)γ, adiponectin, hormone-sensitive lipase (HSL) and adipose triglyceride lipase (ATGL). These factors regulate lipid metabolism and storage and may have caused a miniaturization of adipocytes, lipolysis, and liver fatty acid oxidation [7]. Further animal studies (12 rats/group) have also reported lower weight for animals consuming 10% or 20% dietary xylitol for 40 days; with 10% xylitol approximately 5% lower body weight and with 20% approximately 15% lower body weight [63]. In a fructose-streptozotocin-induced type 2 diabetic rat model, 7 animals/group, were fed 0 (control), 2.5%, 5% and 10% dietary xylitol for 4 weeks. A dose-dependent reduction in food and fluid intake was noted compared to diabetic control animals, where 10% xylitol was not different from non-diabetic animals. Bodyweight gain was, however, similar to the control animals but less than the healthy animals [80].

A one-year study with 91 obese subjects suggests an inverse relation between xylitol consumption and weight loss; a high intake of xylitol would predict for a small weight loss. People in the two lower quartiles had a 5.5-fold greater chance of losing more than 10% weight, while subjects in the highest quartile and a 14 chance of losing less than 10% weight [81]. Whether this is just a correlation or an actual causality remains to be determined.

7.3. Benefits of Xylitol on Satiety

Nasogastric administration of 50 g xylitol in 300 mL water to 10 obese and 10 lean volunteers after an 8 h fasting, induced an increase in cholecystokinin (CCK) and glucagon-like peptide-1 (GLP-1) compared to water alone [82]; both are indicated as satiation hormones. This was associated with an increased time to gastric emptying in both groups as compared to the control (water). However, subjective feelings of appetite were not influenced compared to the water control [82]. Similarly, an earlier study indicated that 25 g xylitol in yogurt for 10 days had no influence on reported fullness in 16 healthy lean adults. However, the combination of 12.5 g xylitol and 12.5 g polydextrose resulted in an increased subjective feeling of fullness [83]. Interestingly, clinical studies have reported that a single dose of 30 g xylitol in 200 mL water resulted in a change in gastric emptying half-time from 39.8 min during the glucose control to 77.5 min during the xylitol test with 5 healthy volunteers in a cross over design study. This delay in gastric emptying was associated with increased plasma motilin [84]. Motilin is involved in the regulation of small intestinal motility [85]. After ingestion of 25 g of xylitol in 50 mL water by ten healthy volunteers, the gastric emptying halftime was increased from 58 min to 91 min compared to the water only control as well as the 25 g glucose comparator in a crossover study. Food intake after xylitol preloading was reduced from 920 (water control) to 690 kcal [86]. Similar observations were made by King and co-workers [83] who observed that during a ten-day ingestion of yogurt containing 25 g of xylitol, 90 min prior to lunch, reduced the combined caloric intake by 11.9%. This difference did, however, not reach statistical significance compared to control.

7.4. Benefits of Xylitol on Metabolic Health

Xylitol, although having a similar sweetness as sucrose and glucose, has different molecular properties and thus does not lead to an increase in blood glucose or insulin levels [83]. Xylitol has a glycemic index of 7 ± 7 compared to a value of 100 for glucose; not surprisingly, the serum insulin and C-peptide responses to xylitol are negligible [87]. Carbohydrate and lipid oxidation were not observed to be influenced when eight healthy non-obese males consumed a single dose of 25 g xylitol after an overnight fast [87].

In animal models of type-2 diabetes (7 rats/group), induced through high-fructose feeding and injection of streptozotocin, administration of xylitol at 2.5%, 5% and 10% in drinking water during 4, respectively 5 weeks has been observed to improve serum insulin concentration at all tested xylitol concentration and glucose tolerance at 10% but not 2.5% and 5% xylitol [80,88]. In a study with 10 obese and 10 lean, non-diabetic volunteers; nasogastric administration of 50 g xylitol in 300 mL water after an 8 h fasting, resulted in a small but significant increase in serum glucose after administrations of xylitol compared with placebo. The authors hypothesized that this could be due to a decrease in plasma glucose over time after placebo intake rather than an increase in plasma glucose after xylitol intake [82]. However, the small increase is in line with earlier reports [87] and can be explained by the normal metabolism of absorbed xylitol to glucose by the liver [7].

In a non-diabetic non-high-fat diet rat model, total cholesterol and low-density lipoprotein (LDL)-cholesterol were significantly reduced (approximately 50% and 75%, respectively) after three weeks in the 10% xylitol drinking water group (6 animals) compared to water only control (5 animals) [89]. In a fructose-streptozotocin-induced type 2 diabetic rat model, 7 animals/group, were fed 0 (control), 2.5%, 5%, and 10% dietary xylitol for 4 weeks a dose-dependent reduction in serum cholesterol was observed. This was in particular driven by a dose-dependent reduction in LDL-cholesterol, where 10% xylitol reached a level lower than the non-diabetic control animals [80]. A similar trend has been reported for humans as well, but only with high doses (40–100 g/day) of xylitol [90].

In a high-fructose streptozotocin-induced, diabetes animal model (7 rats/group), administration of 10% xylitol in drinking water was not found to improve serum triglycerides after 5 weeks as compared to diabetic animals in the unsupplemented control group [88]. However, a fructose-streptozotocin-induced type 2 diabetic rat model, 7 animals/group, were fed 0 (control), 2.5%, 5%, and 10% dietary xylitol for 4 weeks observed a dose-dependent increase in serum triglycerides [80]. A differential lipidemic response between healthy and type 2 diabetic animal models and humans has been suggested [91].

7.5. Conclusions

While there is some indication for improved short-term weight loss in animal models, the long-term data in humans is inconclusive. There is some indication that xylitol may influence satiety hormones and gastric emptying in humans. Whether this translates into an effect on weight management remains to be determined. The benefit of xylitol on metabolic health; in addition to the benefit of the mere replacement of sucrose, remains to be determined in humans. Although there are indications for reduced LDL-cholesterol with xylitol consumption, this would need to be confirmed with lower dietary doses in humans as well as the effect of xylitol on serum triglycerides.

8. Discussion

The dental health benefits of xylitol are well established [3]. Here, we have highlighted that xylitol also has other potential health benefits, Figure 3. Many of these are related to oral-pharyngeal health. Changes in the respiratory microbiota are associated with positive effects on respiratory infections, sinusitis, and acute otitis media. Also, the immune function modulating effects of xylitol may contribute to the reduction in respiratory-related infections. Furthermore, topical or oral administration of xylitol seems to have anti-inflammatory effects on immune function and could be beneficial in controlling for example skin inflammation. As a non-digestible, non-absorbed, selectively fermentable carbohydrate, xylitol also exhibits the characteristics of prebiotics. Xylitol consumption is associated with changes in microbiota composition and metabolic activity, and influences bowel and immune function, and positively influences bone health. Being a low caloric sweetener, xylitol may contribute to weight management; but also by stimulating satiety and contributing to improved serum cholesterol levels. Finally, the topical application of xylitol is associated with improved skin moisture and improved skin barrier.

There are thus many opportunities for additional health benefits of xylitol. However, a limitation is that many of these novel health end-points are mainly based on *in vitro* and animal studies, and limited human intervention studies. This is helpful for the exploration of new health targets and for their mechanistic understanding. Furthermore, it should be observed that animal studies often used 6%–20% of xylitol in the diet, which obviously is beyond what is feasible for human consumption. There is, therefore, a rationale and especially a need to investigate the feasibility of these potential health benefits in humans.

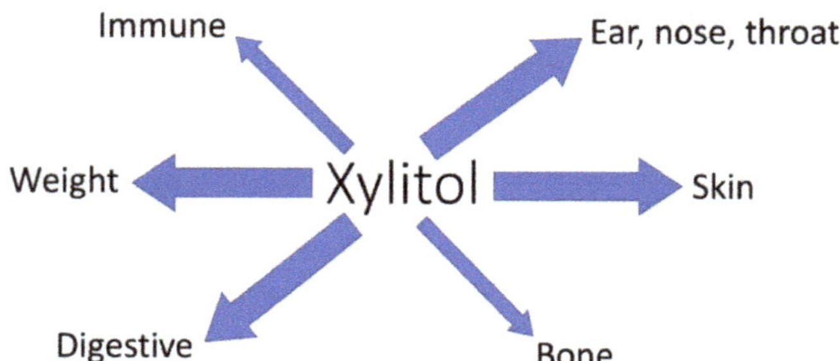

Figure 3. Summary of non-dental health benefits of xylitol. Arrow thickness indicates the level of documentation. Thin arrows indicate only *in vitro* or animal data, while thick arrows indicate some level of human data.

The purpose of the current review was to focus on xylitol. However, it may be relevant to place this into the perspective of other sugar alcohols; without embarking on an in-depth review. In addition to 4 g/day xylitol, one month of 4 g/day sorbitol and to a lesser degree 4 g/day mannitol but not 3 g/day erythritol reduced tetracycline induced bone resorption in rats [64]. Inhaled mannitol may improve some lung functions in cystic fibrosis patients as indicated in a recent Cochrane review [92]. Some polyols, such e.g., lactitol [93] and sorbitol [94], have been suggested to have prebiotic potential. For improving bowel function, lactitol appears to be the sugar alcohol of choice [95]. Mannitol can work as an antioxidant and protect hyaluronic acid in the skin [96]. Lactitol has been reported to stimulate secretory IgA production [97]. Erythritol causes no increase in blood serum glucose level [82]. While sorbitol and erythritol have been shown to reduce glucose absorption from the intestine and improve muscular glucose absorption *ex vivo* [98–100]. Thus, while other sugar alcohols have multiple potential beneficial health effects, xylitol seems to be the more versatile or more investigated one.

Author Contributions: Conceptualization, A.C.O.; resources, K.S.; writing—original draft preparation, A.C.O.; writing—review and editing, K.S., M.J.L., K.T., and A.C.O.

Funding: This research received no external funding.

Conflicts of Interest: All authors are employees of DuPont Nutrition & Biosciences. DuPont Nutrition & Biosciences manufactures and markets xylitol. The authors declare no other conflict of interest.

References

1. Ur-Rehman, S.; Mushtaq, Z.; Zahoor, T.; Jamil, A.; Murtaza, M.A. Xylitol: A review on bioproduction, application, health benefits, and related safety issues. *Crit. Rev. Food Sci. Nutr.* **2015**, *55*, 1514–1528. [CrossRef] [PubMed]
2. Bond, M.; Dunning, N. Xylitol. In *Sweeteners and Sugar Alternatives in Food Technology*; Mitchell, H., Ed.; Blackwell Publishing: Oxford, UK, 2006; pp. 295–324.
3. Janakiram, C.; Deepan Kumar, C.V.; Joseph, J. Xylitol in preventing dental caries: A systematic review and meta-analyses. *J. Nat. Sci. Biol. Med.* **2017**, *8*, 16–21. [CrossRef] [PubMed]
4. Mäkinen, K.K. Gastrointestinal Disturbances Associated with the Consumption of Sugar Alcohols with Special Consideration of Xylitol: Scientific Review and Instructions for Dentists and Other Health-Care Professionals. *Int. J. Dent.* **2016**, *2016*, 5967907. [CrossRef] [PubMed]
5. European Food Safety Authority. Xylitol chewing gum/pastilles and reduction of the risk of tooth decay. *EFSA J.* **2008**, *6*, 852. [CrossRef]
6. Livesey, G. Health potential of polyols as sugar replacers, with emphasis on low glycaemic properties. *Nutr. Res. Rev.* **2003**, *16*, 163–191. [CrossRef] [PubMed]
7. Amo, K.; Arai, H.; Uebanso, T.; Fukaya, M.; Koganei, M.; Sasaki, H.; Yamamoto, H.; Taketani, Y.; Takeda, E. Effects of xylitol on metabolic parameters and visceral fat accumulation. *J. Clin. Biochem. Nutr.* **2011**, *49*, 1–7. [CrossRef] [PubMed]

8. Storey, D.; Lee, A.; Bornet, F.; Brouns, F. Gastrointestinal tolerance of erythritol and xylitol ingested in a liquid. *Eur. J. Clin. Nutr.* **2007**, *61*, 349–354. [CrossRef]
9. Abdayem, R.; Haftek, M. [The epidermal barrier]. *Ann. Derm. Venereol.* **2018**, *145*, 293–301. [CrossRef]
10. Umino, Y.; Ipponjima, S.; Denda, M. Modulation of lipid fluidity likely contributes to the fructose/xylitol-induced acceleration of epidermal permeability barrier recovery. *Arch. Dermatol. Res.* **2019**, *311*, 317–324. [CrossRef]
11. Payér, E.; Szabó-Papp, J.; Ambrus, L.; Szöllösi, A.G.; Andrási, M.; Dikstein, S.; Kemény, L.; Juhász, I.; Szegedi, A.; Bíró, T.; et al. Beyond the physico-chemical barrier: Glycerol and xylitol markedly yet differentially alter gene expression profiles and modify signalling pathways in human epidermal keratinocytes. *Exp. Dermatol.* **2018**, *27*, 280–284. [CrossRef] [PubMed]
12. Kirschner, N.; Rosenthal, R.; Gunzel, D.; Moll, I.; Brandner, J.M. Tight junctions and differentiation–A chicken or the egg question? *Exp. Dermatol.* **2012**, *21*, 171–175. [CrossRef] [PubMed]
13. Korponyai, C.; Szél, E.; Behány, Z.; Varga, E.; Mohos, G.; Dura, A.; Dikstein, S.; Kemény, L.; Erös, G. Effects of Locally Applied Glycerol and Xylitol on the Hydration, Barrier Function and Morphological Parameters of the Skin. *Acta Derm. Venereol.* **2017**, *97*, 182–187. [CrossRef] [PubMed]
14. Szél, E.; Polyánka, H.; Szabó, K.; Hartmann, P.; Degovics, D.; Balázs, B.; Németh, I.B.; Korponyai, C.; Csányi, E.; Kaszaki, J.; et al. Anti-irritant and anti-inflammatory effects of glycerol and xylitol in sodium lauryl sulphate-induced acute irritation. *J. Eur. Acad. Dermatol. Venereol.* **2015**, *29*, 2333–2341. [CrossRef] [PubMed]
15. Korponyai, C.; Kovács, R.K.; Erös, G.; Dikstein, S.; Kemény, L. Antiirritant properties of polyols and amino acids. *Dermatitis* **2011**, *22*, 141–146. [PubMed]
16. Odetti, P.R.; Borgoglio, A.; Rolandi, R. Age-related increase of collagen fluorescence in human subcutaneous tissue. *Metabolism* **1992**, *41*, 655–658. [CrossRef]
17. Mattila, P.T.; Pelkonen, P.; Knuuttila, M.L. Effects of a long-term dietary xylitol supplementation on collagen content and fluorescence of the skin in aged rats. *Gerontology* **2005**, *51*, 166–169. [CrossRef]
18. Knuuttila, M.L.; Kuoksa, T.H.; Svanberg, M.J.; Mattila, P.T.; Karjalainen, K.M.; Kolehmainen, E. Effects of dietary xylitol on collagen content and glycosylation in healthy and diabetic rats. *Life Sci.* **2000**, *67*, 283–290. [CrossRef]
19. Dowd, S.E.; Sun, Y.; Smith, E.; Kennedy, J.P.; Jones, C.E.; Wolcott, R. Effects of biofilm treatments on the multi-species Lubbock chronic wound biofilm model. *J. Wound Care* **2009**, *18*, 510–512. [CrossRef]
20. Ammons, M.C.; Ward, L.S.; James, G.A. Anti-biofilm efficacy of a lactoferrin/xylitol wound hydrogel used in combination with silver wound dressings. *Int. Wound J.* **2011**, *8*, 268–273. [CrossRef]
21. Katsuyama, M.; Kobayashi, Y.; Ichikawa, H.; Mizuno, A.; Miyachi, Y.; Matsunaga, K.; Kawashima, M. A novel method to control the balance of skin microflora Part 2. A study to assess the effect of a cream containing farnesol and xylitol on atopic dry skin. *J. Dermatol. Sci.* **2005**, *38*, 207–213. [CrossRef]
22. El Aidy, S.; Van Den Bogert, B.; Kleerebezem, M. The small intestine microbiota, nutritional modulation and relevance for health. *Curr. Opin. Biotechnol.* **2015**, *32*, 14–20. [CrossRef] [PubMed]
23. Hibberd, A.A.; Lyra, A.; Ouwehand, A.C.; Rolny, P.; Lindegren, H.; Cedgard, L.; Wettergren, Y. Intestinal microbiota is altered in patients with colon cancer and modified by probiotic intervention. *BMJ Open Gastroenterol.* **2017**, *4*, e000145. [CrossRef] [PubMed]
24. Lenhart, A.; Chey, W.D. A Systematic Review of the Effects of Polyols on Gastrointestinal Health and Irritable Bowel Syndrome. *Adv. Nutr.* **2017**, *8*, 587–596. [CrossRef] [PubMed]
25. Gibson, G.R.; Hutkins, R.; Sanders, M.E.; Prescott, S.L.; Reimer, R.A.; Salminen, S.J.; Scott, K.; Stanton, C.; Swanson, K.S.; Cani, P.D.; et al. Expert consensus document: The International Scientific Association for Probiotics and Prebiotics (ISAPP) consensus statement on the definition and scope of prebiotics. *Nat. Rev. Gastroenterol. Hepatol.* **2017**, *14*, 491. [CrossRef] [PubMed]
26. Mäkeläinen, H.S.; Mäkivuokko, H.A.; Salminen, S.J.; Rautonen, N.E.; Ouwehand, A.C. The effects of polydextrose and xylitol on microbial community and activity in a 4-stage colon simulator. *J. Food Sci.* **2007**, *72*, M153–M159. [CrossRef] [PubMed]
27. Sato, T.; Kusuhara, S.; Yokoi, W.; Ito, M.; Miyazaki, K. Prebiotic potential of L-sorbose and xylitol in promoting the growth and metabolic activity of specific butyrate-producing bacteria in human fecal culture. *FEMS Microbiol. Ecol.* **2017**, *93*. [CrossRef] [PubMed]
28. Scheppach, W.; Weiler, F. The butyrate story: Old wine in new bottles. *Curr. Opin. Clin. Nutr. Metab. Care* **2004**, *7*, 563–567. [CrossRef]

29. Arpaia, N.; Campbell, C.; Fan, X.; Dikiy, S.; van der Veeken, J.; deRoos, P.; Liu, H.; Cross, J.R.; Pfeffer, K.; Coffer, P.J.; et al. Metabolites produced by commensal bacteria promote peripheral regulatory T-cell generation. *Nature* **2013**, *504*, 451–455. [CrossRef]
30. Salminen, S.; Salminen, E.; Koivistoinen, P.; Bridges, J.; Marks, V. Gut microflora interactions with xylitol in the mouse, rat and man. *Food Chem. Toxicol.* **1985**, *23*, 985–990. [CrossRef]
31. Talattof, Z.; Azad, A.; Zahed, M.; Shahradnia, N. Antifungal Activity of Xylitol against Candida albicans: An in vitro Study. *J. Contemp. Dent. Pract.* **2018**, *19*, 125–129. [CrossRef]
32. Uebanso, T.; Kano, S.; Yoshimoto, A.; Naito, C.; Shimohata, T.; Mawatari, K.; Takahashi, A. Effects of Consuming Xylitol on Gut Microbiota and Lipid Metabolism in Mice. *Nutrients* **2017**, *9*, 756. [CrossRef] [PubMed]
33. Tamura, M.; Hoshi, C.; Hori, S. Xylitol affects the intestinal microbiota and metabolism of daidzein in adult male mice. *Int. J. Mol. Sci.* **2013**, *14*, 23993–24007. [CrossRef] [PubMed]
34. Vargas, S.L.; Patrick, C.C.; Ayers, G.D.; Hughes, W.T. Modulating effect of dietary carbohydrate supplementation on Candida albicans colonization and invasion in a neutropenic mouse model. *Infect. Immun.* **1993**, *61*, 619–626. [PubMed]
35. Hedberg, M.; Hasslöf, P.; Sjöström, I.; Twetman, S.; Stecksén-Blicks, C. Sugar fermentation in probiotic bacteria-an in vitro study. *Oral Microbiol. Immunol.* **2008**, *23*, 482–485. [CrossRef] [PubMed]
36. Haukioja, A.; Söderling, E.; Tenovuo, J. Acid production from sugars and sugar alcohols by probiotic lactobacilli and bifidobacteria in vitro. *Caries Res.* **2008**, *42*, 449–453. [CrossRef] [PubMed]
37. Keller, M.K.; Twetman, S. Acid production in dental plaque after exposure to probiotic bacteria. *BMC Oral Health* **2012**, *12*, 44. [CrossRef] [PubMed]
38. Brambilla, E.; Ionescu, A.C.; Cazzaniga, G.; Ottobelli, M.; Samaranayake, L.P. Levorotatory carbohydrates and xylitol subdue Streptococcus mutans and Candida albicans adhesion and biofilm formation. *J. Basic Microbiol.* **2016**, *56*, 480–492. [CrossRef]
39. Rätsep, M.; Kõljalg, S.; Sepp, E.; Smidt, I.; Truusalu, K.; Songisepp, E.; Stsepetova, J.; Naaber, P.; Mikelsaar, R.H.; Mikelsaar, M. A combination of the probiotic and prebiotic product can prevent the germination of Clostridium difficile spores and infection. *Anaerobe* **2017**, *47*, 94–103. [CrossRef]
40. Yu, T.; Zheng, Y.P.; Tan, J.C.; Xiong, W.J.; Wang, Y.; Lin, L. Effects of Prebiotics and Synbiotics on Functional Constipation. *Am. J. Med Sci.* **2017**, *353*, 282–292. [CrossRef]
41. Gong, Y.; Zhang, Q.; Qiao, L.; Lv, D.; Ruan, J.; Chen, H.; Gong, J.; Shi, G. Xylitol Gum Chewing to Achieve Early Postoperative Restoration of Bowel Motility After Laparoscopic Surgery. *Surg. Laparosc. Endosc. Percutan. Tech.* **2015**, *25*, 303–306. [CrossRef]
42. Lee, J.T.; Hsieh, M.H.; Cheng, P.J.; Lin, J.R. The Role of Xylitol Gum Chewing in Restoring Postoperative Bowel Activity After Cesarean Section. *Biol. Res. Nurs.* **2016**, *18*, 167–172. [CrossRef] [PubMed]
43. Yang, P.; Long, W.J.; Wei, L. Chewing Xylitol Gum could Accelerate Bowel motility Recovery after Elective Open Proctectomy for Rectal Cancer. *Rev. Investig. Clin.* **2018**, *70*, 53–58. [CrossRef] [PubMed]
44. Ottman, N.; Smidt, H.; de Vos, W.M.; Belzer, C. The function of our microbiota: Who is out there and what do they do? *Front. Cell. Infect. Microbiol.* **2012**, *2*, 104. [CrossRef] [PubMed]
45. Lehtinen, M.J.; Hibberd, A.A.; Mannikko, S.; Yeung, N.; Kauko, T.; Forssten, S.; Lehtoranta, L.; Lahtinen, S.J.; Stahl, B.; Lyra, A.; et al. Nasal microbiota clusters associate with inflammatory response, viral load, and symptom severity in experimental rhinovirus challenge. *Sci. Rep.* **2018**, *8*, 11411. [CrossRef] [PubMed]
46. Man, W.H.; de Steenhuijsen Piters, W.A.; Bogaert, D. The microbiota of the respiratory tract: Gatekeeper to respiratory health. *Nat. Rev. Microbiol.* **2017**, *15*, 259–270. [CrossRef] [PubMed]
47. Brook, I. Microbiology of chronic rhinosinusitis. *Eur. J. Clin. Microbiol. Infect. Dis.* **2016**, *35*, 1059–1068. [CrossRef] [PubMed]
48. Azarpazhooh, A.; Lawrence, H.P.; Shah, P.S. Xylitol for preventing acute otitis media in children up to 12 years of age. *Cochrane Database Syst. Rev.* **2016**, *8*, CD007095. [CrossRef] [PubMed]
49. Kontiokari, T.; Uhari, M.; Koskela, M. Effect of xylitol on growth of nasopharyngeal bacteria in vitro. *Antimicrob. Agents Chemother.* **1995**, *39*, 1820–1823. [CrossRef]
50. Kontiokari, T.; Svanberg, M.; Mattila, P.; Leinonen, M.; Uhari, M. Quantitative analysis of the effect of xylitol on pneumococcal nasal colonisation in rats. *FEMS Microbiol. Lett.* **1999**, *178*, 313–317. [CrossRef]

51. Zabner, J.; Seiler, M.P.; Launspach, J.L.; Karp, P.H.; Kearney, W.R.; Look, D.C.; Smith, J.J.; Welsh, M.J. The osmolyte xylitol reduces the salt concentration of airway surface liquid and may enhance bacterial killing. *Proc. Natl. Acad. Sci. USA* **2000**, *97*, 11614–11619. [CrossRef] [PubMed]
52. Cingi, C.; Birdane, L.; Ural, A.; Oghan, F.; Bal, C. Comparison of nasal hyperosmolar xylitol and xylometazoline solutions on quality of life in patients with inferior turbinate hypertrophy secondary to nonallergic rhinitis. *Int. Forum Allergy Rhinol.* **2014**, *4*, 475–479. [CrossRef] [PubMed]
53. Little, P.; Stuart, B.; Wingrove, Z.; Mullee, M.; Thomas, T.; Johnson, S.; Leydon, G.; Richards-Hall, S.; Williamson, I.; Yao, L.; et al. Probiotic capsules and xylitol chewing gum to manage symptoms of pharyngitis: A randomized controlled factorial trial. *CMAJ: Can. Med Assoc. J.* **2017**, *189*, E1543–E1550. [CrossRef] [PubMed]
54. Jain, R.; Lee, T.; Hardcastle, T.; Biswas, K.; Radcliff, F.; Douglas, R. The in vitro effect of xylitol on chronic rhinosinusitis biofilms. *Rhinology* **2016**, *54*, 323–328. [CrossRef]
55. Brown, C.L.; Graham, S.M.; Cable, B.B.; Ozer, E.A.; Taft, P.J.; Zabner, J. Xylitol enhances bacterial killing in the rabbit maxillary sinus. *Laryngoscope* **2004**, *114*, 2021–2024. [CrossRef] [PubMed]
56. Weissman, J.D.; Fernandez, F.; Hwang, P.H. Xylitol nasal irrigation in the management of chronic rhinosinusitis: A pilot study. *Laryngoscope* **2011**, *121*, 2468–2472. [CrossRef] [PubMed]
57. Lin, L.; Tang, X.; Wei, J.; Dai, F.; Sun, G. Xylitol nasal irrigation in the treatment of chronic rhinosinusitis. *Am. J. Otolaryngol.* **2017**, *38*, 383–389. [CrossRef] [PubMed]
58. Hardcastle, T.; Jain, R.; Radcliff, F.; Waldvogel-Thurlow, S.; Zoing, M.; Biswas, K.; Douglas, R. The in vitro mucolytic effect of xylitol and dornase alfa on chronic rhinosinusitis mucus. *Int. Forum Allergy Rhinol.* **2017**, *7*, 889–896. [CrossRef]
59. Tapiainen, T.; Sormunen, R.; Kaijalainen, T.; Kontiokari, T.; Ikäheimo, I.; Uhari, M. Ultrastructure of *Streptococcus pneumoniae* after exposure to xylitol. *J. Antimicrob. Chemother.* **2004**, *54*, 225–228. [CrossRef]
60. Kurola, P.; Tapiainen, T.; Kaijalainen, T.; Uhari, M.; Saukkoriipi, A. Xylitol and capsular gene expression in *Streptococcus pneumoniae*. *J. Med. Microbiol.* **2009**, *58*, 1470–1473. [CrossRef]
61. Kurola, P.; Tapiainen, T.; Sevander, J.; Kaijalainen, T.; Leinonen, M.; Uhari, M.; Saukkoriipi, A. Effect of xylitol and other carbon sources on *Streptococcus pneumoniae* biofilm formation and gene expression in vitro. *Apmis* **2011**, *119*, 135–142. [CrossRef]
62. Persaud, N.; Laupacis, A.; Azarpazhooh, A.; Birken, C.; Hoch, J.S.; Isaranuwatchai, W.; Maguire, J.L.; Mamdani, M.M.; Thorpe, K.; Allen, C.; et al. Xylitol for the prevention of acute otitis media episodes in children aged 2–4 years: Protocol for a pragmatic randomised controlled trial. *BMJ Open* **2018**, *8*, e020941. [CrossRef] [PubMed]
63. Sato, H.; Ide, Y.; Nasu, M.; Numabe, Y. The effects of oral xylitol administration on bone density in rat femur. *Odontology* **2011**, *99*, 28–33. [CrossRef] [PubMed]
64. Mattila, P.T.; Svanberg, M.J.; Mäkinen, K.K.; Knuuttila, M.L. Dietary xylitol, sorbitol and D-mannitol but not erythritol retard bone resorption in rats. *J. Nutr.* **1996**, *126*, 1865–1870. [CrossRef] [PubMed]
65. Mattila, P.; Svanberg, M.; Knuuttila, M. Diminished bone resorption in rats after oral xylitol administration: A dose-response study. *Calcif. Tissue Int.* **1995**, *56*, 232–235. [CrossRef] [PubMed]
66. Mattila, P.T.; Svanberg, M.J.; Pökkä, P.; Knuuttila, M.L. Dietary xylitol protects against weakening of bone biomechanical properties in ovariectomized rats. *J. Nutr.* **1998**, *128*, 1811–1814. [CrossRef] [PubMed]
67. Kaivosoja, S.M.; Mattila, P.T.; Knuuttila, M.L. Dietary xylitol protects against the imbalance in bone metabolism during the early phase of collagen type II-induced arthritis in dark agouti rats. *Metabolism* **2008**, *57*, 1052–1055. [CrossRef] [PubMed]
68. Mattila, P.T.; Knuuttila, M.L.; Svanberg, M.J. Dietary xylitol supplementation prevents osteoporotic changes in streptozotocin-diabetic rats. *Metabolism* **1998**, *47*, 578–583. [CrossRef]
69. Castelo-Branco, C.; Soveral, I. Phytoestrogens and bone health at different reproductive stages. *Gynecol. Endocrinol.* **2013**, *29*, 735–743. [CrossRef] [PubMed]
70. Tyagi, A.M.; Yu, M.; Darby, T.M.; Vaccaro, C.; Li, J.Y.; Owens, J.A.; Hsu, E.; Adams, J.; Weitzmann, M.N.; Jones, R.M.; et al. The Microbial Metabolite Butyrate Stimulates Bone Formation via T Regulatory Cell-Mediated Regulation of WNT10B Expression. *Immunity* **2018**, *49*, 1116–1131.e7. [CrossRef] [PubMed]
71. Gleeson, M.; Bosch, J. The human immune system. In *Excercise Immunology*; Gleeson, M., Bishop, N., Walsh, N., Eds.; Routledge: London, UK, 2013.

72. Jakob, A.; Williamson, J.R.; Asakura, T. Xylitol metabolism in perfused rat liver. Interactions with gluconeogenesis and ketogenesis. *J. Biol. Chem.* **1971**, *246*, 7623–7631. [PubMed]
73. Takahashi, K.; Akiba, Y. Single administration of xylitol to newly hatched chicks enhances growth, digestive enzyme activity and immune responses by 12 d of age. *Br. Poult. Sci.* **2005**, *46*, 635–640. [CrossRef] [PubMed]
74. Renko, M.; Valkonen, P.; Tapiainen, T.; Kontiokari, T.; Mattila, P.; Knuuttila, M.; Svanberg, M.; Leinonen, M.; Karttunen, R.; Uhari, M. Xylitol-supplemented nutrition enhances bacterial killing and prolongs survival of rats in experimental pneumococcal sepsis. *BMC Microbiol.* **2008**, *8*, 45. [CrossRef] [PubMed]
75. Salli, K.M.; Gürsoy, U.K.; Söderling, E.M.; Ouwehand, A.C. Effects of Xylitol and Sucrose Mint Products on Streptococcus mutans Colonization in a Dental Simulator Model. *Curr. Microbiol.* **2017**, *74*, 1153–1159. [CrossRef] [PubMed]
76. Xu, M.L.; Wi, G.R.; Kim, H.J.; Kim, H.J. Ameliorating Effect of Dietary Xylitol on Human Respiratory Syncytial Virus (hRSV) Infection. *Biol. Pharm. Bull.* **2016**, *39*, 540–546. [CrossRef] [PubMed]
77. Yin, S.Y.; Kim, H.J.; Kim, H.J. Protective effect of dietary xylitol on influenza A virus infection. *PloS ONE* **2014**, *9*, e84633. [CrossRef] [PubMed]
78. Takahashi, K.; Mashiko, T.; Akiba, Y. Effect of dietary concentration of xylitol on growth in male broiler chicks during immunological stress. *Poult. Sci.* **2000**, *79*, 743–747. [CrossRef] [PubMed]
79. Grundy, S.M. Metabolic syndrome update. *Trends Cardiovasc. Med.* **2016**, *26*, 364–373. [CrossRef] [PubMed]
80. Rahman, A.; Islam, S. Xylitol improves pancreatic islets morphology to ameliorate type 2 diabetes in rats: A dose response study. *J. Food Sci.* **2014**, *79*, H1436–H1442. [CrossRef] [PubMed]
81. Geidenstam, N.; Al-Majdoub, M.; Ekman, M.; Spegel, P.; Ridderstrale, M. Metabolite profiling of obese individuals before and after a one year weight loss program. *Int. J. Obes. (Lond.)* **2017**, *41*, 1369–1378. [CrossRef]
82. Wölnerhanssen, B.K.; Cajacob, L.; Keller, N.; Doody, A.; Rehfeld, J.F.; Drewe, J.; Peterli, R.; Beglinger, C.; Meyer-Gerspach, A.C. Gut hormone secretion, gastric emptying, and glycemic responses to erythritol and xylitol in lean and obese subjects. *Am. J. Physiol. Endocrinol. Metab.* **2016**, *310*, E1053–E1061. [CrossRef]
83. King, N.A.; Craig, S.A.S.; Pepper, T.; Blundell, J.E. Evaluation of the independent and combined effects of xylitol and polydextrose consumed as a snack on hunger and energy intake over 10 d. *Br. J. Nutr.* **2005**, *93*, 911–915. [CrossRef] [PubMed]
84. Salminen, E.K.; Salminen, S.J.; Porkka, L.; Kwasowski, P.; Marks, V.; Koivistoinen, P.E. Xylitol vs glucose: Effect on the rate of gastric emptying and motilin, insulin, and gastric inhibitory polypeptide release. *Am. J. Clin. Nutr.* **1989**, *49*, 1228–1232. [CrossRef] [PubMed]
85. Ohno, T.; Mochiki, E.; Kuwano, H. The roles of motilin and ghrelin in gastrointestinal motility. *Int. J. Pept.* **2010**, *2010*. [CrossRef] [PubMed]
86. Shafer, R.B.; Levine, A.S.; Marlette, J.M.; Morley, J.E. Effects of xylitol on gastric emptying and food intake. *Am. J. Clin. Nutr.* **1987**, *45*, 744–747. [CrossRef] [PubMed]
87. Natah, S.S.; Hussien, K.R.; Tuominen, J.A.; Koivisto, V.A. Metabolic response to lactitol and xylitol in healthy men. *Am. J. Clin. Nutr.* **1997**, *65*, 947–950. [CrossRef]
88. Islam, S.; Indrajit, M. Effects of xylitol on blood glucose, glucose tolerance, serum insulin and lipid profile in a type 2 diabetes model of rats. *Ann. Nutr. Metab.* **2012**, *61*, 57–64. [CrossRef] [PubMed]
89. Islam, M.S. Effects of xylitol as a sugar substitute on diabetes-related parameters in nondiabetic rats. *J. Med. Food* **2011**, *14*, 505–511. [CrossRef]
90. Förster, H.; Quadbeck, R.; Gottstein, U. Metabolic tolerance to high doses of oral xylitol in human volunteers not previously adapted to xylitol. *Int. J. Vitam. Nutr. Res. Suppl.* **1982**, *22*, 67–88.
91. Islam, M.S. Xylitol increases serum triglyceride in normal but not in a type 2 diabetes model of rats. *J. Med. Food* **2012**, *15*, 677. [CrossRef]
92. Nevitt, S.J.; Thornton, J.; Murray, C.S.; Dwyer, T. Inhaled mannitol for cystic fibrosis. *Cochrane Database Syst. Rev.* **2018**, *2*, CD008649. [CrossRef]
93. Drakoularakou, A.; Hasselwander, O.; Edinburgh, M.; Ouwehand, A.C. Lactitol, an emerging prebiotic: Functional properties with a focus on digestive health. *Food Sci. Technol. Bull.* **2007**, *3*, 71–80. [CrossRef]
94. Sarmiento-Rubiano, L.A.; Zuniga, M.; Perez-Martinez, G.; Yebra, M.J. Dietary supplementation with sorbitol results in selective enrichment of lactobacilli in rat intestine. *Res. Microbiol.* **2007**, *158*, 694–701. [CrossRef] [PubMed]
95. Mueller-Lissner, S.A.; Wald, A. Constipation in adults. *BMJ Clin. Evid.* **2010**, *7*, 413.

96. Andre, P.; Villain, F. Free radical scavenging properties of mannitol and its role as a constituent of hyaluronic acid fillers: A literature review. *Int. J. Cosmet. Sci.* **2017**, *39*, 355–360. [CrossRef] [PubMed]
97. Yamamoto, Y.; Kubota, N.; Takahashi, T.; To, M.; Hayashi, T.; Shimizu, T.; Kamata, Y.; Saruta, J.; Tsukinoki, K. Continuous combined intake of polydextrose and lactitol stimulates cecal fermentation and salivary IgA secretion in rats. *J. Oral. Sci.* **2017**, *59*, 603–610. [CrossRef]
98. Chukwuma, C.I.; Ibrahim, M.A.; Islam, M.S. Maltitol inhibits small intestinal glucose absorption and increases insulin mediated muscle glucose uptake ex vivo but not in normal and type 2 diabetic rats. *Int. J. Food Sci. Nutr.* **2017**, *68*, 73–81. [CrossRef]
99. Chukwuma, C.I.; Mopuri, R.; Nagiah, S.; Chuturgoon, A.A.; Islam, M.S. Erythritol reduces small intestinal glucose absorption, increases muscle glucose uptake, improves glucose metabolic enzymes activities and increases expression of Glut-4 and IRS-1 in type 2 diabetic rats. *Eur. J. Nutr.* **2018**, *57*, 2431–2444. [CrossRef]
100. Chukwuma, C.I.; Islam, M.S. Sorbitol increases muscle glucose uptake ex vivo and inhibits intestinal glucose absorption ex vivo and in normal and type 2 diabetic rats. *Appl. Physiol. Nutr. Metab.* **2017**, *42*, 377–383. [CrossRef]

Review

Gut Microbiome: Profound Implications for Diet and Disease

Ronald D. Hills Jr. [1,*], Benjamin A. Pontefract [2,3], Hillary R. Mishcon [1], Cody A. Black [1,4], Steven C. Sutton [1] and Cory R. Theberge [1]

1 Department of Pharmaceutical Sciences, College of Pharmacy, University of New England, Portland, ME 04103, USA
2 Pharmacy Service, Boise Veterans Affairs Medical Center, Boise, ID 83702, USA
3 College of Pharmacy, Ferris State University, Big Rapids, MI 49307, USA
4 College of Pharmacy, University of Texas at Austin, San Antonio, TX 78229, USA
* Correspondence: rhills@une.edu; Tel.: +1-207-221-4049

Received: 30 May 2019; Accepted: 11 July 2019; Published: 16 July 2019

Abstract: The gut microbiome plays an important role in human health and influences the development of chronic diseases ranging from metabolic disease to gastrointestinal disorders and colorectal cancer. Of increasing prevalence in Western societies, these conditions carry a high burden of care. Dietary patterns and environmental factors have a profound effect on shaping gut microbiota in real time. Diverse populations of intestinal bacteria mediate their beneficial effects through the fermentation of dietary fiber to produce short-chain fatty acids, endogenous signals with important roles in lipid homeostasis and reducing inflammation. Recent progress shows that an individual's starting microbial profile is a key determinant in predicting their response to intervention with live probiotics. The gut microbiota is complex and challenging to characterize. Enterotypes have been proposed using metrics such as alpha species diversity, the ratio of Firmicutes to Bacteroidetes phyla, and the relative abundance of beneficial genera (e.g., *Bifidobacterium, Akkermansia*) versus facultative anaerobes (*E. coli*), pro-inflammatory *Ruminococcus*, or nonbacterial microbes. Microbiota composition and relative populations of bacterial species are linked to physiologic health along different axes. We review the role of diet quality, carbohydrate intake, fermentable FODMAPs, and prebiotic fiber in maintaining healthy gut flora. The implications are discussed for various conditions including obesity, diabetes, irritable bowel syndrome, inflammatory bowel disease, depression, and cardiovascular disease.

Keywords: gut microbiota; nutrition; habitual diets; Western diet; obesity; cardiometabolic risk factors; chronic health conditions; gastrointestinal disorders; prebiotics and probiotics

---

1. Introduction to Gut Microbiota and Disease

The intestinal microbiome has recently been implicated in a host of chronic diseases ranging from inflammatory bowel disease (IBD), type 2 diabetes (T2D), and cardiovascular disease (CVD) to colorectal cancer [1–3]. The community of ~200 prevalent bacteria, virus, and fungi inhabiting the human gastrointestinal (GI) tract provide unique metabolic functions to the host and are fundamentally important in health and disease [4,5]. Microbiome refers to the collective genomes of all microorganisms inhabiting an environment. While isolating and culturing each individual species is an intractable task, a cutting-edge method of sequence analysis, metagenomics, has enabled the reconstruction of microbial species and their function from the collective nucleotide contents contained in a stool sample. Shotgun metagenomic sequencing analysis discovered 1952 unclassified bacteria species in the human gut microbiome in addition to the 553 bacteria previously cultured from the gut [6]. A central question in medicine concerns the nature of the relationship between human health and the gut microbiota, which refers to the community of microorganisms themselves, the relative abundance of individual species populations, and their function.

Metagenomics and analysis of twins data has revealed that environmental factors such as diet and household cohabitation greatly outweigh heritable genetic contributions to the composition and function of gut microbiota [7]. Analogous to the genetic heritability statistic, Rothschild et al. constructed a microbiome-association index. Significant associations are observed between the gut microbiome and host phenotypes for body mass index (BMI) (25%), waist-to-hip ratio (24%), fasting glucose levels (22%), glycemic status (25%), high-density lipoprotein (HDL) cholesterol levels (36%), and monthly lactose consumption (36%) [7]. Compared to BMI, waist-to-hip ratio is an anthropometric measurement of central obesity and stronger predictor of diastolic and systolic blood pressure, total cholesterol/HDL, and triglycerides [8] as well as death from CVD [9].

The Western diet has profound effects on the diversity and populations of microbial species that make up gut flora [10]. The U.S. is home to the largest number of immigrants in the world, many of whom develop metabolic diseases post immigration. Earlier epidemiological evidence revealed a fourfold increase in obesity risk is possible within 15 years of emigrating to the U.S. compared to populations remaining in their birth country [11]. In a recent cross-sectional and longitudinal study of a multi-generational Asian American cohort, emigrating to the U.S. was shown to reduce gut microbial diversity and function [12]. Alpha diversity was measured using the Shannon entropy, a quantitative index that accounts for the abundance and evenness of species residing in the host, as opposed to species richness, which is the number of species present. Within the gram-negative Bacteroidetes phylum, bacterial strains from the genus *Prevotella*, whose enzymes degrade plant fiber, became displaced by dominant strains from the genus *Bacteroides* according to an individual's time spent in the U.S. The ratio of *Bacteroides* to *Prevotella* increased by factors of 10, correlating with the time in decades spent in the U.S. Prior to this study, metagenomics had identified three clusters of variation in the human gut, referred to as enterotypes [13]. The first enterotype, high in *Bacteroides* and low in *Prevotella*, is found in individuals on a long-term Western diet high in animal protein, the nutrient choline, and saturated fat [14]. The second enterotype is high in *Prevotella*, low in *Bacteroides*, and associated with a plant-based diet rich in fiber, simple sugars, and plant-derived compounds. While less distinct, a third potential enterotype was found with a slightly higher population of genus *Ruminococcus* within the phylum Firmicutes. Enriched *Ruminococcus* is associated with irritable bowel syndrome (IBS) [15], and transient blooms of pro-inflammatory *Ruminococcus* have been associated with active flare-ups in IBD [16]. *R. gnavus*, a prevalent gut microbe that proliferates in IBD, has been found to secrete a unique L-rhamnose oligosaccharide that induces tumor necrosis factor alpha (TNFα), a major pro-inflammatory cytokine [17].

2. Microbiota, the Immune Response, and Diet in IBD

IBD is a chronic GI disorder characterized by an overactive immune response to the gut microbiome. A serious, debilitating condition, IBD affects growth and development in children, increases the risk of colorectal cancer, and can lead to life-threatening complications [18]. There are two forms of IBD, Crohn's disease and ulcerative colitis, that differ in the inflamed areas of the intestine. Normally, anaerobic microbes in the gut derive their nutrients from fermentation of indigestible oligosaccharides and other carbohydrates escaping proximal digestion [19]. In IBD, respiratory electron acceptors generated as a byproduct of the inflammatory host response become environmental stressors that support bacterial growth [20]. The disorder results in oxidative stress for the host and the microbiome, leading to gut dysbiosis in the form of decreased community richness and proliferation of facultative anaerobic Enterobacteriaceae and adherent invasive strains of *Escherichia coli* [16,20,21]. Drug therapies for IBD have traditionally included immunosuppressants in the form of corticosteroids, antimetabolite agents, or anti-TNF antibodies, often with ancillary administration of antibiotics [22]. An alternative treatment, given predominantly to children, is a defined enteral nutrition formula. Dietary therapy has the advantage of obviating the need for immunosuppression and is thought to work by altering the composition of gut microbiota.

A longitudinal study involving metagenomic analysis was conducted of 90 children initiating treatment for Crohn's disease [22]. GI symptoms, mucosal inflammation, and microbial communities were compared for dietary and anti-TNF therapy and antibiotic use relative to healthy children. Microbial communities separated into two clusters based on composition. The dysbiotic community associated with active disease was characterized by increased fungal representation, increased lactose-fermenting bacteria *(Streptococcus, Lactobacillus, Klebsiella)*, and the presence of human DNA in the stool (from epithelial cells and white blood cells). Crohn's patients also had reduced relative abundance of *Prevotella* and increased *Escherichia* compared to healthy children. Treatment with antibiotics in the last six months was strongly associated with microbial dysbiosis [22], consistent with earlier findings that oral antibiotics for acne are a risk factor for new onset Crohn's disease [23]. Antibiotic-treatment was observed to enrich fungi such as *Candida* and *Saccharomyces* [22]. Treatment with the enteral nutrition [24] or antibody therapies, on the other hand, reduced inflammation and markedly improved gut microbiota. The relative populations of fungi were reduced within one week of receiving the defined dietary formula, which lacked fiber [22]. Since a defined formula was effective for restoring healthy microbiota, it is conceivable that a more general oral diet with the proper nutrition can restore the intraluminal environment [25–27].

3. Microbial Metabolites and Short-Chain Fatty Acids

3.1. SCFA Receptor Activation

Short-chain fatty acids (SCFAs) have attracted considerable attention for their role in human health [28]. Obligate anaerobic bacteria (phyla Firmicutes and Bacteroidetes) encode a variety of enzymes for hydrolyzing complex carbohydrates (chains of sugar molecules) not digestible by the host such as resistant starch and fiber. Certain genera such as *Lactobacillus* and *Bifidobacterium* specialize in oligosaccharide fermentation, utilizing galactooligosaccharides (GOS), fructooligosaccharides (FOS), and polysaccharide inulin [29]. Carbohydrate fermentation by anaerobes provides the host with important SCFAs such as acetate, propionate, and butyrate [30]. Several receptors have been identified for SCFAs such as free fatty acid receptor 3 (FFAR3 or GPR41) and niacin receptor 1 (GPR109A) [31]. GPR41 and GPR109A are G-protein coupled receptors (GPCRs) found on intestinal epithelial cells, immune cells, and adipocytes. As endogenous agonists in GPCR signal transduction, SCFAs have a profound effect on physiological processes [32,33] independent of delivering calories to the host as carbon molecules [34]. GPR41 is associated with increased energy expenditure, leptin hormone expression, and decreased food intake [31,35]. Analogous to the activity of niacin, butyrate activates GPR109A to suppress colonic inflammation and colon cancer development [36]. Niacin is a known lipid-lowering agent: GPR109A inhibits triglyceride hydrolysis (lipolysis) in adipocytes, lowering blood levels of triglyceride and low-density lipoprotein (LDL) to reduce atherogenic activity. Acetate and propionate activate cell surface receptor GPR43 to induce neutrophil chemotaxis. GPR43 is anti-lipolysis and implicated in IBD, but contradictory results in mouse models leave doubt as to whether an agonist or antagonist will best treat colitis [35]. There is a growing interest in pursuing GPR41 and GPR43 as drug targets for the chronic inflammatory disorders asthma, arthritis, and obesity [37]. Much work remains to be done to establish the appropriate disease models needed to study these conditions.

Colonic epithelial cells (colonocytes) are the control switch separating microbial homeostasis from gut dysbiosis [38]. It is known that antibiotics deplete microbes that ferment essential SCFAs such as butyrate, which are normally responsible for maintaining microbial homeostasis [24,39]. The lack of butyrate silences metabolic signaling in the gut. Mitochondrial beta-oxidation in colonocytes becomes disabled, resulting in a transfer of oxygen, which freely diffuses across cell membranes from the blood to the GI lumen. Oxygen in the colon then allows for pathogenic facultative anaerobes such as *E. coli* [40] to outcompete the benign obligate anaerobes that characterize a healthy gut [41,42]. Microbial homeostasis is normally maintained by peroxisome proliferator-activated receptor gamma (PPAR-γ). PPAR-γ is a nuclear receptor activated by butyrate and other ligands, is found in adipocytes and colonocytes, and is responsible for activating genes involved in glucose and lipid metabolism. Lack of butyrate signaling results in nitrate electron acceptors being released into the colon, which facultative anaerobes can also use for cell respiration, breaking down carbohydrates into carbon dioxide rather than fermenting them [20]. Facultative anaerobes, including Proteobacteria, could further affect nutrition by catabolizing SCFAs present in the lumen [38]. The metabolic reprogramming of colonocytes is analogous to that of macrophages, which become polarized toward anaerobic glycolysis in response to proinflammatory signals. In ulcerative colitis, excessive epithelial repair results in lower PPAR-γ synthesis, which reduces beta-oxidation and increases oxygenation of colonocytes. Inflamed mucosae in colitis patients are increased in Proteobacteria, a major phylum of gram-negative bacteria, but decreased in gram-positive Firmicutes. Treatment with PPAR-γ agonist, however, can improve the microbial balance [43].

3.2. Fecal Biomarkers and IBS

Fecal biomarkers such as inflammatory proteins, antimicrobial peptides, and SCFA levels are emerging as a non-invasive screening tool for assessing and diagnosing various health conditions [44]. Patients with IBD have lower fecal levels of acetate, propionate and butyrate, and higher levels of lactic and pyruvic acids than healthy individuals [45]. Given the relationship between bacterial fermentation products and atherosclerosis, ongoing research aims to characterize the fecal microbiota and SCFA signatures of individuals with high blood lipid levels [46]. High levels of isobutyric acid could be one such biomarker for hypercholesterolemia. Colonoscopy is an invasive exam relied on in the United States as a periodic screen for colorectal cancer [18], but annual screening is performed in many countries using a non-invasive fecal immunochemical test, which looks for microscopic blood in the stool [47]. Current efforts are underway to identify novel microbial biomarkers for colorectal cancer given that it is associated with increased fecal levels of *F. nucleatum*, a promoter of tumorigenesis [3].

Unlike structural disorders such as IBD, IBS is a functional disorder and collection of GI symptoms observed in the absence of macroscopic signs of inflammation. Despite affecting 10–15% of the population and the potential for low quality of life, its etiology is unclear and current drug treatments are largely ineffective [48]. Diagnosis has traditionally relied on symptom criteria, stool characteristics, and questionnaires, once all other pathologies are ruled out [49]. The Rome criteria sets classifications for four subtypes: IBS with predominant diarrhea (IBS-D), IBS with predominant constipation (IBS-C), and IBS with mixed or alternating-type bowel habits (IBS-M) depending on whether >25% of bowel movements belong to soft or hard type stool categories or both, respectively, followed by IBS unclassified (IBS-U) [50]. It has been known for some time that IBS patients have reduced microbial diversity compared to healthy subjects [51], see also References 6–9 in [51]. Inflammatory proteins such as human β-defensin 2, a bactericide, have been identified as a useful fecal biomarker in IBS and IBD [48,51,52]. Lastly, the concentration difference in two SCFAs, propionic minus butyric acid, has been shown to be positive for all four IBS subtypes but negative in healthy subjects [53].

3.3. Leaky Gut

Elevated levels of interleukin 6, a pro-inflammatory peptide cytokine, and plasma levels of lipopolysaccharide (LPS) endotoxin, a marker of gram-negative bacterial translocation, were found to be elevated in a subpopulation of IBS-D patients with small intestinal permeability, analogous to that observed in celiac disease [54]. It is hypothesized that psychological stress can exacerbate the inflammatory condition by allowing translocation of harmful bacterial products across the intestinal epithelium. Known as "leaky gut", a compromised epithelial barrier allows toxins and antigens in the GI lumen to enter the bloodstream. A healthy gut flora is important in maintaining the intestinal barrier. By increasing the expression of tight cell junction proteins, beneficial probiotics such as *Lactobacillus* and *Bifidobacterium* can limit the development of autoimmune diseases in genetically susceptible individuals [55] and fatty liver disease in obese individuals [56]. In alcoholic liver disease, alcohol consumption causes gut permeability by reducing the expression of REG3, a bactericidal protein normally responsible for restricting the mucosal colonization of luminal bacteria [57].

3.4. Gut-Brain Interactions

In the last decade, it has been discovered that the enteric and central nervous systems are linked via a bidirectional communication network termed the gut-brain axis. Gut-brain communication is disrupted in the cases of IBS and microbial dysbiosis [58], in the former leading to changes in intestinal motility and secretion and causing visceral hypersensitivity (hyperalgesia) [59]. Recurrent abdominal pain is a hallmark characteristic of IBS. Autism spectrum disorder, which is often associated with constipation, has been connected to gut dysbiosis in the form of an increased Firmicutes/Bacteroidetes ratio and high levels of facultative anaerobes *Escherichia/Shigella* and the fungal genus *Candida* [60,61]. It is suggested that leaky gut contributes to the pathogenesis of autism by increasing systemic metabolites that alter the neuroimmune and neuroendocrine systems, thus affecting the brain and neurodevelopment [61–63].

For the last century, the ketogenic diet (KD) has been used to treat refractory epilepsy in children's hospitals [64], achieving a 50% reduction in seizure rates [65]. KD restricts the proportion of carbohydrate intake to create a state of ketosis in which the body relies on ketone bodies for energy rather than glucose. Clinical studies are now investigating the use of KD for treating neurological conditions including autism, Alzheimer's, and Parkinson's disease, with promising results obtained for small cohorts [66]. The mechanism of action was initially thought to result from the normalization of aberrant energy metabolism associated with these disorders, but the role of the gut microbiota is now coming into focus. A recent comparison of KD-fed conventionally raised mice versus mice treated with antibiotics or reared germ-free revealed that alterations in the gut microbiota are required to reproduce the anti-seizure effects of KD [67]. Following KD was observed to enrich the populations of the anaerobic genera *Akkermansia* and *Parabacteroides*. Moreover, increased levels of the inhibitory neurotransmitter γ-aminobutyric acid (GABA) were detected in metabolite profiles of the brain hippocampus of KD-fed mice and were observed to be microbiota-dependent. GABA is a principal means of reducing communication between brain cells, and neuronal excitability is enhanced in neurological conditions such as epilepsy, anxiety, and Alzheimer's disease [66,68]. Besides dietary intervention, these and other observations suggest that supplementation with prebiotics or probiotics could be used to improve cognitive symptoms associated with neurological conditions ranging from autism to Alzheimer's and Parkinson's [69,70], giving rise to the notion of "psychobiotics" [71,72].

Fecal microbiota transplantation (FMT) is yet another therapeutic option, which involves the engraftment of microbes from a healthy donor [73]. In a study of 18 autistic children, an eight-week course of FMT resulted in behavioral improvement and an 80% reduction in GI symptoms and abdominal pain associated with autism [74]. Outcomes remained improved when assessed eight weeks after treatment had ended, lending support to the hypothesis that gut microbiota are at least partially responsible for autism symptoms. Analysis of microbiota composition showed that FMT increased overall bacterial diversity and the abundance of fermentative *Bifidobacterium* and *Prevotella* in autistic individuals even after treatment cessation. In other clinical studies, FMT has demonstrated a 90% success rate for treating recurrent *Clostridioides difficile* infection, clinical remission rates of up to 78% in treating IBD, and symptom resolution or improvement in up to 70% of IBS patients [75]. Interest is now growing for the application of FMT in other disorders ranging from Parkinson's to metabolic syndrome [75,76]. In patients with metabolic syndrome, FMT was shown to improve insulin sensitivity for those with decreased baseline microbial diversity, but the effects did not persist in the long-term [77].

Other lines of clinical evidence on the gut-brain interaction show that gut microbiota influences the central nervous system by alterations in the release of neuroendocrine hormones and neurotransmitter activity. Dysfunctions in GABA receptor signaling are implicated in anxiety and depression, and beneficial bacteria *Lactobacillus* and *Bifidobacterium* convert the amino acid glutamate into GABA [78,79]. Metagenomic analysis of a 1054-person Flemish cohort revealed that butyrate-producing *Faecalibacterium* and *Coprococcus* associate with higher quality of life and improved mental health, while *Dialister* and *Coprococcus* are depleted in cases of depression [79]. To improve cognitive symptoms associated with clinical depression and anxiety, beneficial probiotic strains of *B. longum* and *L. helveticus* have been administered clinically with promising results [72,80]. In a study comparing young and middle-aged mice, dietary supplementation with prebiotic inulin was observed to increase *Bifidobacterium* and *Akkermansia*, reduce neuroinflammation and anxiety, and improve cognition in middle-aged mice [81]. The fact that alterations in gut microbiota can provide cognitive symptom relief could offer one basis for the relationship observed between quality of diet and one's mental health status [82].

4. Gut Microbiota and Metabolic Syndrome

4.1. Obesity, Microbial Diversity, and SCFA Supplementation

Clear links are emerging between the microbiome and its effects on host metabolism, with profound implications for human health given the rise of obesity and metabolic syndrome in Western society [83]. A study of four twin pairs discordant for obesity by Ridaura et al. revealed differences in their microbiota [84], with the lean individuals exhibiting an increase in bacterial SCFA fermentation and transformation of bile acids. To show that SCFA production was transmissible, the human fecal microbiota was transplanted into lean and obese mice. Obese mice were also cohoused with lean mice for 10 days, which countered weight gain due to an invasion of their microbiome by specific members of Bacteroidetes when a low-fat diet was administered. Such findings highlight the role of environmental factors in shaping gut microbiota and the development of obesity.

A study of human and mouse microbiota correlated obesity with differences in the relative abundance of two dominant bacterial divisions and showed that obese individuals have an increased capacity to harvest energy from the diet [85]. Relative to lean mice and humans, obese individuals have an increased relative abundance of Firmicutes, and reduced abundance of Bacteroidetes. The observation that reduced microbial diversity enhances calorie harvesting is also supported by a metagenomic analysis comparing microbiotas belonging to identical and fraternal twins and their mothers [86]. More recent work demonstrated that individuals with low microbial gene count have more systemic inflammation, adiposity, insulin resistance, and dyslipidemia [87]. Low gene count individuals gained more weight over time and were dominant in *Bacteroides* and *Ruminococcus* genera, while 36 genera including *Faecalibacterium*, *Bifidobacterium*, *Lactobacillus*, and *Akkermansia* were significantly associated with high gene count, lean individuals. In an analogous study involving 49 overweight or obese individuals, following an energy-restricted diet for six weeks was observed to partially restore microbial gene richness [88].

In human and rodent studies, one species of the Verrucomicrobia phylum inversely correlates with obesity and T2D, *Akkermansia muciniphila*, a mucus colonizer that can use mucin as its sole carbon and nitrogen source in times of caloric restriction. Treatment in mice with a probiotic strain of *A. muciniphila* or its prebiotic FOS was shown to reverse high fat diet-induced weight gain and insulin resistance, increase intestinal endocannabinoids controlling inflammation and the gut barrier, and counteract diet-induced decreases in mucus layer thickness [89]. In mouse fed a high-fat/high-sucrose diet, polyphenol-rich cranberry extract was found to protect against metabolic syndrome and intestinal inflammation by increasing the relative abundance of *Akkermansia* [90]. In humans, *A. muciniphila* levels at baseline and after a six-week calorie restriction diet were observed to correlate inversely with fasting glucose, waist-to-hip ratio, and plasma triglycerides [91]. A recent pilot study was conducted in overweight or obese insulin-resistant volunteers. Daily oral supplementation with 10^{10} *A. muciniphila* cells was found to improve insulin sensitivity, reduce insulinemia, and decrease body weight over a three-month period [92]. Such successful studies suggest that *A. muciniphila* could find use as a next generation probiotic to combat metabolic syndrome [93].

Roux-en-Y gastric bypass (RYGB) surgery is one of the most effective treatments for morbid obesity and T2D. RYGB reduces adiposity, improves glucose metabolism, increases resting energy expenditure, and results in rapid and sustained weight loss, but these effects cannot simply be attributed to decreased food intake and absorption [94,95]. In patients post-gastric-bypass, the abundance of Firmicutes has been found to decrease [96]. *Prevotella* is observed to increase three months after surgery relative to obese individuals, while *Faecalibacterium prausnitzii* is lower in diabetic subjects and correlates negatively with low-grade inflammatory markers [97]. In a mouse model, RYGB has been shown to restructure microbiota via a rapid and sustained increase in the relative abundance of *Akkermansia* downstream of the site of surgery in the gut [94].

Jiao et al. examined the effects of orally administering doses of the SCFAs acetic, propionic, and butyric acid to weaned pigs [98]. SCFA administration was observed to decrease serum levels of triglycerides, total cholesterol, and insulin, while increasing serum concentrations of the leptin hormone. Remarkably, the study demonstrated that SCFAs attenuate fat deposition by inhibiting feed intake, reducing lipogenesis, and enhancing lipolysis. Another study of 12 men undergoing colonic infusions showed that receiving an enema containing SCFAs can increase fasting fat oxidation and resting energy expenditure [99]. In a healthy diet, the bacterial fermentation of fiber into SCFAs promotes microbial diversity and is one mechanism by which high fiber intake inhibits weight gain [100,101], even outweighing heritable contributions to obesity [102].

The metabolic effects of butyrate were measured in a study of mice fed a high-fat diet (60% of calories from lard) [103]. Oral but not intravenous administration of butyrate was shown to act on the gut-brain circuitry via the vagus nerve, decreasing food intake and preventing diet-induced obesity, hyperinsulinemia, hypertriglyceridemia, and fatty liver disease. Interestingly, butyrate also promoted fat oxidation and activated brown adipose tissue. The finding that butyrate improves energy metabolism without eliciting any ill effects suggests that oral supplementation might be a promising strategy for combatting cardiometabolic disease [104]. Butyrate was further shown to alter the gut microbiota independent of the vagus nerve [103]. Specific genera within the subclass Erysipelotrichia were significantly increased, bringing the relative abundance of the Firmicutes phylum from 26% to 32% relative to controls, while the Bacteroidetes phylum decreased from 71% to 66%. The ratio of Firmicutes to Bacteroidetes increased by 21% upon butyrate administration. Given that Firmicutes generally correlate with a less beneficial metabolic profile [105], it appears that specific species of Erysipelotrichia are beneficial to host energy metabolism.

4.2. Microbiota in Diabetes

Both obesity and diabetes are characterized by insulin resistance and low-grade inflammation. A mouse study by Cani et al. points to bacterial LPS as a causative factor of insulin resistance, obesity, and diabetes [106]. Feeding and fasting cycles increased or decreased plasma levels of LPS, respectively, and metabolic endotoxemia was observed in mice fed a four-week high-fat diet that increased the proportion of gram-negative bacteria in the gut, raising plasma LPS concentration by a factor of two to three. Endotoxemia could also be induced via subcutaneous infusion of LPS for four weeks, resulting in weight gain and increased fasting hyperglycemia and hyperinsulinemia. LPS produces inflammation in adipocytes through the activation of toll-like receptor 4 signaling [107]. Thus, prebiotics that improve intestinal microbiota and reduce intestinal permeability are of potential clinical use for the treatment of diabetes [108,109]. Randomized controlled trials have reported improvements in glycemia and cardiovascular markers in T2D patients taking resistant starch, resistant dextrin, or inulin [110].

Consumption of dietary fiber has positive metabolic health effects including increased satiety, decreased weight gain, and lowered blood glucose and cholesterol levels, serving to reduce the risk of CVD and T2D [111–113]. Fiber has historically been classified as either soluble or insoluble, but plant cell walls often contain both and this distinction does not always predict physiological function [114].

It can be more useful to classify fibers into four categories based on whether they are readily fermented and whether they form a viscous cross-linked gel [115]. Insoluble fiber (wheat bran) is poorly fermented and does not alter viscosity. Soluble, nonviscous fiber (inulin, wheat dextrin, resistant starch) is readily fermented. Conversely, viscous gel-forming fibers can be fermentable (β-glucan) or not (psyllium). Improvements in metabolism can arise from three factors: microbial fermentation of soluble fiber into SCFAs [33,95,100], delayed nutrient absorption and improved cholesterol/glucose due to viscous gel formation [115,116], and the ability of insoluble fiber to reduce insulin resistance by interfering with protein absorption [112]. In conventional rats, a high-fat diet was found to reduce butyrate formation and increase liver cholesterol and triglyceride content compared to rats fed a low-fat diet, but these effects could be partially reversed by adding fermentable dietary fiber to the high-fat diet [117]. In a 12-week mouse study, supplementing a high-fat diet with 10% fermentable flaxseed fiber dramatically increased butyrate production, energy expenditure, and *Bifidobacterium* and *Akkermansia* levels, while countering weight gain [118]. In contrast to the Western diet, consuming daily servings of fiber, fruit, and vegetables promotes the alpha diversity of bacterial species in the gut [12,102,119–121].

Suez et al. investigated the impact of non-caloric artificial sweeteners (NAS) on glucose tolerance [122]. Commercial formulations of saccharin, sucralose, or aspartame were added to the drinking water of lean mice for 11 weeks. The 10% NAS solutions were well below the known toxic doses given per kg body weight. While mice drinking water, glucose, or sucrose had similar glucose tolerance curves, all three NAS-consuming groups developed glucose intolerance, which could be reversed upon antibiotic treatment. NAS was also shown to induce changes in gut microbiota previously observed in T2D; notably, the over-representation of gram-negative *Bacteroides* and under-representation of gram-positive Clostridiales. Bacterial taxa were enriched in the metabolic pathways involved in glycan degradation, contributing to enhanced capacity for energy harvest [85]. Lastly, Suez et al. assessed long-term NAS consumption in a clinical nutrition study using a food frequency questionnaire given to 381 non-diabetic individuals. Significant positive correlations were found between NAS consumption and measures of metabolic syndrome including increased weight, waist-to-hip ratio, fasting blood glucose, and hemoglobin A1c [122].

The link between NAS consumption in mice and alterations in gut microbiota lends support to the notion that individuals can have a personalized response to dietary components based on existing or acquired differences in their microbiota. A study of 800 healthy and prediabetic Israelis revealed high interpersonal variability in their postprandial glucose responses to the same foods, which could be attributed to differences in gut microbiota and other factors [123]. A machine learning algorithm was developed by Zeevi et al. and found to accurately predict personalized glycemic responses to real-life meals using information on blood parameters, dietary habits, anthropometric measures, physical activity, and gut microbiota. Twenty-six new participants were then recruited for a randomized controlled trial. The algorithm was found to be capable of choosing a personalized diet that successfully lowered the post-meal glycemic responses for each individual [123]. An analogous study of Midwestern Americans predicted glycemic responses once the abundances of *Prevotella* and *Bacteroides* were taken into account [124]. Such studies highlight the significance of individual microbial profiles in constructing therapeutic interventions, of great potential relevance to the emerging field of personalized nutrition [125].

Finally, diabetes medications have been connected to positive changes in gut microbiota. Metagenomic analysis of 345 Chinese volunteers revealed that diabetics have a decrease in butyrate-producing bacteria and an increase in opportunistic pathogens relative to healthy subjects [126]. A four-month placebo-controlled study was recently performed on 40 newly diagnosed T2D patients [127]. In individuals given the gold standard T2D drug, metformin, rapid alterations were observed in the composition of the gut microbiome. In the entire cohort, a negative association was observed between hemoglobin A1c blood levels and *B. adolescentis*, a species whose replication rate was increased by metformin. Transfer of fecal samples before and after metformin treatment to germ-free mice showed that improved glucose tolerance can arise solely from the metformin-altered microbiota. At the chemical level, the antidiabetic effects were attributed to increased microbial production of SCFAs and changes observed in the bacterial expression of metal-binding proteins [127].

In a rodent study, mice fed a high-fat diet containing lard oil had reduced expression of sodium glucose cotransporter-1 (SGLT1) [128]. SGLT1 is normally required for healthy glucose sensing in the upper small intestine in order to lower endogenous glucose production by the liver. Treatment with metformin was observed to restore SGLT1 expression and enhance intestinal glucose uptake. Metformin also increased the abundance of *Lactobacillus* bacteria in the upper small intestine. The antidiabetic effect was transmissible upon fecal transplantation, showing that the intestinal microbiota restores SGLT1 expression and glucose sensing in untreated obese rats. Before treatment, mice consuming the high-fat diet had a decreased abundance of gram-positive phylum Actinobacteria, while phylum Proteobacteria and genus *Escherichia* were increased relative to the control group consuming regular chow. The molecular link to SGLT1 expression is unknown, but it is likely that microbial metabolites such as SCFAs activate glucose sensing. Metagenomic analysis of a Dutch cohort corroborated that SCFA concentrations are higher in metformin users compared to diabetics not taking metformin [120]. Analysis of a Colombian community found that metformin users had higher levels of SCFA-producing *A. muciniphila*, *B. bifidum*, and *Prevotella* [129].

A subset of patients cannot tolerate metformin due to adverse GI effects including abdominal pain, bloating, nausea, and diarrhea. A small clinical trial was recently conducted in nondiabetic individuals, confirming that metformin alters gut microbiota independent of glycemic status [130]. Interestingly, the bacterial abundance of 12 genera at baseline predicted whether healthy individuals would experience adverse GI effects upon treatment with metformin. This observation provides a glimpse at how gut microbiota, which are shaped by diet, can mediate individualized therapeutic responses to a medication. Lastly, diabetes medication acarbose is a minimally absorbed glucoamylase inhibitor that prevents starch digestion by humans. A mouse study monitored acarbose-treated mice fed either a Western-style high-starch diet or a high-fiber diet rich in plant polysaccharides [131]. Analogous to metformin treatment, high doses of acarbose were sufficient to alter gut bacterial taxa and increase butyrate production even in those consuming a high-starch diet, but the bacterial composition quickly reverted upon cessation of acarbose treatment. Altogether, these studies suggest that alterations in the gut microbial community are prominent contributors to the mechanism of action in antihyperglycemic agents.

4.3. Dietary Choline and Atherosclerosis

Metabolomic analysis was used to monitor 2000 metabolites present in the blood plasma of patients undergoing cardiac evaluation in order to identify potential predictors of CVD events [132]. Three small molecules were found to predict CVD risk: choline, trimethylamine *N*-oxide (TMAO), and betaine. Each are metabolites of phosphatidylcholine, a dietary lipid found in high quantities in egg yolk, liver, and other high-fat animal products. Choline, also called lecithin, is an essential nutrient that is marketed as a dietary supplement. Hydrolysis of phosphatidylcholine liberates choline, which is metabolized by gut microbes into trimethylamine (TMA) gas, which the liver in turn converts into TMAO. In mice fed radiolabeled phosphatidylcholine, increased blood levels of TMAO were revealed to contribute to greater arterial plaque development [132]. In another study, atherosclerosis susceptibility could be transmitted from atherogenic-prone mouse strains to atherogenic-resistant strains via cecal microbial transplantation [133].

The National Institutes of Health funded two prospective clinical studies on TMAO [134]. In the first study, the phosphatidylcholine challenge, plasma levels of TMAO were observed to rise after consumption of two eggs traced with isotope-labeled phosphatidylcholine. TMAO generation could be suppressed by administering a weeklong course of antibiotics to reduce gut bacteria. One month after withdrawal of antibiotics, TMAO generation returned in a follow-up choline challenge test. In a second cohort of 4007 adults undergoing cardiac evaluation, participants with the highest quartile of fasting plasma TMAO levels had a significantly increased risk of experiencing a major adverse CVD event within the three-year follow-up period (hazard ratio, 2.5, relative to lowest quartile). Another study of patients with stable coronary artery disease found a four-fold increase in all-cause five-year mortality risk for those in the highest TMAO quartile [135]. The atherogenicity of choline metabolite TMAO helps explain the correlation that exists between CVD and excessive consumption of animal products [136]. A causal link between dietary cholesterol and CVD, on the other hand, has not been demonstrated and would be difficult to prove given the fact that cholesterol-containing foods are also high in saturated fat, with the exception of eggs and shrimp [137]. A long-term study of 29,615 participants recently showed that consuming eggs with yolk elevates one's CVD risk in a dose-dependent fashion [138], with each half an egg consumed per day elevating absolute risk by 1.1% and all-cause mortality by 1.9%. One egg yolk contains 120 mg choline.

A structural analog of choline and natural product found in some foods, 3,3-dimethyl-1-butanol (DMB), has been shown in mice to reduce TMAO levels by non-lethal inhibition of TMA lyase [139], giving credence to the notion of "drugging the microbiome." In a study of mice fed a Western diet, DMB reduced plasma TMAO and prevented cardiac dysfunction, inflammation, and fibrosis, but had no effect on body weight and dyslipidemia [140]. Efforts are underway to determine the TMA-forming potential of different bacterial species and develop new treatment strategies for restraining the proliferation of TMA producers [141]. L-carnitine is another trimethylamine abundant in red meat that is also sold as a dietary supplement. Similar to choline, studies in rodents and humans show that carnitine increases plasma TMAO levels, accelerates atherosclerosis, and increases CVD risk [142]. Interestingly, comparison of carnitine challenge tests in habitual omnivores versus vegans/vegetarians reveals that omnivores harbor a microbiota capable of generating 20-fold higher levels of TMAO [142,143].

The connection between TMAO and CVD has important implications for meat consumption given that beef and pork contain 100 mg choline per 100-g serving (veal: 400 mg). Fish and chicken are not far behind with 70–80 mg choline per serving. Some studies have observed a modest increase in relative risk of CVD mortality (between 26% and 34%) for the highest quantile consumption of unprocessed red meat or both processed and unprocessed red meat [144,145]. Comparative risk assessment using a national survey, however, did not find a significant contribution for unprocessed red meat alone [146], and an earlier meta-analysis calculated its relative risk ratio per 100-g serving to be 1.00 (95% confidence interval: 0.81–1.23) [147]. It is likely that the quality of the comparison diet is a confounding variable contributing to disparate findings on the contribution of meat to CVD [148].

Improved cardiovascular health has been associated with one's degree of adherence to a Mediterranean-style diet, which limits consumption of red meat and dairy while emphasizing plant-based foods and healthy fats [149–151]. The relative reduction in CVD morbidity risk obtained for those in the highest quantile of adherence to the Mediterranean diet, considering all dietary components combined, is observed from meta-analyses to be in the vicinity of 30%, or even up to 45% for high risk populations [152]. The microbiome was recently assessed by De Filippis et al. in 123 Italian individuals habitually following omnivore, vegetarian, or vegan diets [153]. To score their adherence to the Mediterranean diet, individuals were stratified along an 11-food unit dietary index. Individuals consuming vegetable-based diets had higher adherence to the Mediterranean diet, were increased in *Prevotella* and fiber-degrading bacteria, and had higher fecal levels of SCFAs. Omnivores on the other hand had a higher ratio of Firmicutes to Bacteroidetes in the gut and elevated TMAO in the urine [153].

The scientific community has also debated the extent to which red meat elevates the risk of colorectal cancer, another condition prominent in Western society [154,155]. Gut microbiota associated with colorectal cancer were recently shown to have an increase in genes associated with TMA lyase and protein catabolism, while microbe carbohydrate degradation pathways were depleted [156,157]. Dietary choline is not observed to correlate with cancer incidence, while betaine, a methyl group donor, is associated with reduced colorectal cancer risk [158]. Again, overall diet quality is likely a significant factor. A study using a polyposis cancer model in mice showed that a high-fiber diet increases SCFA-producing bacteria as well as the expression of butyrate receptor GPR109A, serving to suppress colon carcinogenesis [159]. A case-control study conducted in China found an inverse association between vegetable fiber intake and colorectal cancer (Q4 versus Q1 odds ratio: 0.51; 95% confidence interval: 0.31–0.85) [160]. Strong associations were also observed for total, soluble, and insoluble fiber intakes, but not fruit, soy, or grain fiber. A comparative risk assessment estimated that suboptimal food group intake levels account for 38% of new colorectal cancer cases [161]. Microbial overgrowth was recently shown to fulfill the ecological Koch's postulates [162] of disease causation in colorectal cancer. Rather than a specific pathogen, a matrix-enclosed ecosystem of bacteria, or biofilm, extracted from tumor patients was found to induce tumorigenesis in mice [163].

5. Microbial Interventions

5.1. Probiotics

Probiotics are defined as "live microorganisms that, when administered in adequate amounts, confer a health benefit on the host" [164]. Probiotics are available over-the-counter or by prescription containing microorganisms similar to the commensal bacteria found in the gut, most commonly lactic acid-producing *Bifidobacterium* and *Lactobacillus* spp. As a whole, there is clinical evidence to support the use of probiotics for treating acute infectious diarrhea, antibiotic-associated diarrhea, *C. difficile*-associated diarrhea, ulcerative colitis, and irritable bowel syndrome, but not for acute pancreatitis or Crohn's disease [165–170]. Commonly prescribed antibiotics carry a risk of *C. difficile* infection, which can cause severe complications and has an estimated treatment cost of $24,205 USD per patient. Co-administration of probiotics, which lower the risk of *C. difficile* infection, has therefore been proposed as a prophylactic whenever antibiotics are prescribed [171]. Clinical research into probiotics is species- and often strain-specific, with particular bacteria investigated for separate disease states [172]. Probiotic bacteria can potentially provide various health benefits through normalizing perturbed microbiota and intestinal motility, competitively excluding pathogens, and increasing SCFA production [173–175].

Different probiotic species have been studied for ameliorating GI symptoms, though it is not always clear which species or strains are most beneficial [176]. Earlier work observed that the ratio of Firmicutes to Bacteroidetes was elevated in 62 IBS patients relative to 46 control subjects in Helsinki, Finland [177]. Surprisingly, both groups were dominant in the relative abundance of Firmicutes (90% and 83%, respectively), leaving doubt as to the representativeness or overall health of the small cohort (64% was estimated for an 1135-person Dutch cohort [120]). *Bifidobacterium* was one genus of strictly anaerobic gram-positive Actinobacteria whose numbers were markedly decreased (16–47%) in patients diagnosed with IBS-M, IBS-D, or IBS-C relative to healthy controls [177]. Other studies have confirmed that probiotic supplementation with bifidobacteria results in modest improvement of GI symptoms experienced in IBS-C and IBD patients [167,178]. Correlating microbial profiles to gut health is more complicated for other species. Within the Firmicutes phyla, *Streptococcus* are found to be decreased in IBS-C but increased in IBS-D, while *Allisonella* are decreased in IBS-C and IBS-D but increased in IBS-M [15]. Genera within Bacteroidetes such as *Prevotella* and *Bacteroides* may be increased or decreased in IBS [15,177]. It has been noted that there is a strong positive association between IBS and small intestinal bacterial overgrowth (SIBO) [179]. This gave rise to the initial idea of treating the condition with antibiotics, but patient response varies widely and GI symptoms may even worsen. Recent antibiotic exposure actually correlates positively with the development of SIBO [180]. SIBO and GI symptoms have been shown to be exacerbated in healthy individuals who switch to a high-sugar, low-fiber diet for only seven days, leading to a decrease in small intestinal microbial diversity and an increase in epithelial permeability [180].

One challenge with the probiotic market is that, unless specific disease-related claims are made, commercial products are poorly regulated. Probiotics are trademarked by brand rather than by bacterial strain, and formulations or manufacturing protocols can change over time, having a dramatic impact on efficacy [181]. It has been shown in particular that strains within the same genus or species can have substantially different effects on the host, differing in their ability to grow and survive the intestinal environment, adhere to intestinal epithelial cells, and inhibit pathogen invasion [182,183]. After the isolation of *E. coli* Nissle 1917 from the stool of a World War I soldier who did not catch dysentery, nonpathogenic strains of *E. coli* gained some acceptance as probiotics. *E. coli* is unique in that it relies on monosaccharide and disaccharide nutrients broken down from complex carbohydrates by strict anaerobe species of bacteria [184]. Beneficial *E. coli* strains have been used to treat patients suffering from infectious diseases, likely due to their ability to outcompete enteric pathogens for nutrients [40]. Recent mouse studies give cause for caution, however. Cocolonization of *E. coli* O157:H7, a notorious foodborne pathogen, with a nonpathogenic strain of *E. coli* in germ-free mice actually increased the pathogen's virulence and production of Shiga toxins, which are encoded by viral prophage genes, by up to 12-fold [185]. In another study, probiotic *E. coli* Nissle 1917 was observed to undergo genomic adaptation in response to selective and diet-dependent host pressures within a transit period of five weeks [186] To gain advantage especially in low-diversity guts, competitive adaptations in genes were acquired that affected intestinal adhesion and the utilization of carbohydrates and mucin components as carbon energy sources. In mice that were previously exposed to antibiotics, the *E. coli* strains acquired mutations responsible for antibiotic resistance [186]. Such studies underscore the centrally important role that horizontal gene exchange plays in the evolution of gut bacteria [187].

Several species of *Lactobacillus* and *Bifidobacterium* have now become the staples in the field of probiotics. Notable commercial multi-strain formulations have been subjected to clinical studies including Visbiome® (formerly VSL#3) [188], BIO-25 [189], and Ther-Biotic® Complete [190]. Visbiome® contains several strains from well-known probiotic species *L. plantarum* DSM24730, *Streptococcus thermophilus* DSM24731, *B. breve* DSM24732, *L. paracasei* DSM24733, *L. delbrueckii subsp. bulgaricus* DSM24734, *L. acidophilus* DSM24735, *B. longum* DSM24736, and *B. infantis* DSM24737. Lactobacilli and bifidobacteria such as these have been extensively tested for their anti-inflammatory effects in colitis as well as their beneficial effects on gut motility, particularly for the treatment of constipation [173,191–193]. While *E. coli* is LPS-producing, *B. breve* has been shown to reduce LPS-induced epithelial cell shedding, which is observed in relapsing IBD patients [194]. Populations of *Lactobacillus* are reduced in alcohol consumption and in high fat diet-induced obesity [55,195]. Supplementation with probiotic strain *L. rhamnosus* GG has been shown to decrease microbial overgrowth, restore mucosal integrity, reduce microbial translocation, and ameliorate alcohol-induced liver injury [55,196]. Lastly, the use of probiotics has been proposed as an alternative or adjuvant to antibiotic treatment [197]. In the case of enterohemorrhagic *E. coli* O157:H7, antibiotics are not effective due to the release of additional toxin. Probiotics *L. acidophilus* R0052 and *L. rhamnosus* R0011 have been observed to prevent epithelial injury by reducing adhesion of *E. coli* O157:H7 and also enteropathogenic *E. coli* O127:H6 [198].

A clinical study of healthy adults given the probiotic *L. paracasei* DG revealed that the changes observed in the underlying gut microbiota can depend on an individual's starting microbial profile [199]. Study participants with low initial fecal butyrate levels experienced a four-fold increase in butyrate production and a 55% decrease in *Ruminococcus*, a member of the Clostridia class responsible for degrading resistant starch. On the other hand, individuals with high starting butyrate levels experienced a 49% decrease in butyrate production and a decrease in six Clostridia genera including *Faecalibacterium*, an anti-inflammatory butyrate producer beneficial to mental health [79]. Other studies corroborate that a patient's initial fecal microbial pattern can help predict their response to a probiotic intervention [189], suggesting it will one day be possible to optimize the dose of bacterial strains administered for an individual [200]. An individual's microbiome has also been shown to influence the production of butyrate upon dietary supplementation with fermentable resistant starch according to which bacterial taxa become amplified [201]. Given the relation between the microbiome and metabolic disease, current research is now exploring probiotic interventions as an adjuvant therapy for improving cardiometabolic profiles [202,203]. Positive results have been obtained using the multi-strain formulation Ecologic® Barrier for T2D [204]. In rats, Ecologic® Barrier was previously shown to improve depression-related behavior independent of consumption of a high-fat Western-style diet [205]. Ecologic® Barrier contains the following strains: *Bifidobacterium bifidum* W23, *B. lactis* W52, *Lactobacillus acidophilus* W37, *L. brevis* W63, *L. casei* W56, *L. salivarius* W24, *Lactococcus lactis* W19, and *Lc. lactis* W58. Lastly, two strains of *L. gasseri* isolated from human intestine and breast milk were found to reduce visceral fat mass in obese adults, but the effects diminished once treatment with SBT2055 was ceased, indicating that the probiotic needs to be continually supplied [206,207].

5.2. Prebiotics

In some clinical studies, a probiotic is administered in combination with a prebiotic compound that promotes bacterial growth, together termed a synbiotic. The requirements of a prebiotic are that it is not digested in the upper GI tract, can be fermented by intestinal microbiota, selectively stimulates beneficial bacteria growth and diversity, and has a positive effect on host health [208,209]. Prebiotics include FOS, GOS, and polyol sugar alcohols used as nutritive sweeteners [193,210]. Inulin is a soluble fiber and fructan, or variable length polymer of fructose, that is indigestible to humans and has minimal impact on blood glucose levels [211]. Believed to be most effective in nurturing the growth of many species of probiotic [193], inulin has been tested in successful synbiotic treatments for ulcerative colitis [191,211]. More recently, supplementation with butyrate and inulin was found to lower diastolic blood pressure, fasting blood sugar, and waist-to-hip ratio in T2D patients [104].

Numerous studies reveal that significant health benefits can be obtained from prebiotic administration alone [110,193,211]. Prebiotics such as GOS and FOS have been shown to improve microbial profiles by increasing bifidobacteria and decreasing *E. coli* [193,212]. See Table 5 in Reference [193] for a summary of prebiotic clinical trials. In a double-blind, randomized controlled trial of two separate cohorts in Canada, 16 weeks of FOS-enriched inulin supplementation (8 g/day) decreased body fat, serum triglycerides, and interleukin 6 in overweight or obese children compared to those given an isocaloric dose of maltodextrin placebo [213]. Bifidobacteria in fecal samples increased from 6% to 10% of mean bacterial abundance with prebiotic treatment, while Firmicutes decreased from 69% to 63% and *Ruminococcus* from 2.3% to 1.4%. In an animal study, rats fed a high-fat/high-sucrose diet along with FOS experienced a normalization in insulin resistance, leptin levels, dyslipidemia, and gut microbiota [214]. Moreover, prebiotic FOS was observed to limit knee joint damage in this diet-induced model of osteoarthritis, to levels approaching that obtained with moderate aerobic exercise. The effects of prebiotic therapy also depend on individual's starting microbial profile. In a study comparing FOS, sorghum and arabinoxylan, equally high SCFA production was observed in volunteers whose microbiota was dominant in fiber-utilizing *Prevotella*, but *Bacteroides*-dominated individuals showed different SCFA levels in response to each fiber [215].

Given the relationship between gut microbiota and inflammation, research is underway to examine the effects of anti-inflammatory omega-3 polyunsaturated fatty acids (PUFAs) on microbial diversity. Consuming a Western diet high in animal protein is known to elevate the ratio of omega-6 to omega-3 PUFAs by up to a factor of 10, producing an inflammatory response mediated by hormone-like eicosanoids in the body [149,216]. The omega-3 PUFAs docosahexaenoic acid (DHA) and eicosapentaenoic acid (EPA), however, are inflammation-resolving and have anti-colorectal cancer activity, see References 4–6 in [217]. Human studies show that dietary supplementation with EPA and DHA increases the intestinal abundance of *Bifidobacterium* and *Lactobacillus*, while decreasing *Faecalibacterium* [217,218]. Conflicting results were reported for the effect of omega-3 fatty acids on the ratio of Firmicutes to Bacteroidetes phyla. Lastly, a metabolomic analysis was recently conducted of 876 adult female twins. After adjusting for dietary fiber intake, the consumption and circulating levels of omega-3 fatty acids were found to be significantly correlated with microbial alpha diversity as measured by the Shannon index [219].

6. Implications for Diet and Nutrition

6.1. Dietary and other Microbiome Covariates

A metagenomic analysis was conducted of 1135 participants from a Dutch population using deep sequencing [120]. The sequencing data enabled the detection of associations in microbiota with 126 different environmental factors including diet, disease, and medication use. Higher intakes of total carbohydrates were most strongly associated with decreased microbiome diversity: bifidobacteria increased while *Lactobacillus, Streptococcus,* and *Roseburia* genera decreased. The Shannon diversity index decreased according to intake levels of total carbohydrates, followed by sugar-sweetened beverages, bread, beer, savory snacks, and, to a lesser extent, total fats, pulses, and legumes. Diversity was also reduced in individuals self-reporting IBS, and antibiotic use was associated with decreases in two species of *Bifidobacterium*. On the other hand, microbial diversity increased with fruit, coffee, vegetable, and red wine intake and to a smaller extent eating breakfast and drinking tea. Red wine consumption was associated with an increased abundance of *F. prausnitzii* [120], an anti-inflammatory species implicated in lean-type, high-richness microbiota [87]. Coffee, tea, and red wine are high in polyphenols, compounds associated with prebiotic and bifidogenic activity, see References 19–21 in Reference [120]. In a recent meta-analysis, consuming up to three cups/day in coffee was found to decrease all-cause and CVD mortality in a dose-dependent fashion irrespective of caffeine content [220].

A similar population-level analysis of an 1106-person Belgian cohort across 69 covariates [221] showed that the Bristol stool scale, an indicator of gut transit time, and the use of medications have the largest explanatory value for microbiome variation. A total bacterial richness of 664 genera was found, but variance between individuals arose primarily from differences in the relative abundance of 14 core genera. Consistent with previously characterized enterotypes [13], bacterial taxa with the largest variation in abundance were *Prevotella, Bacteroides,* and Ruminococcaceae. *Prevotella* correlated with softer type stools, while Ruminococcaceae was the dominant family in hard type stools. Overall species richness declines with shorter gut transit times and the abundance of core species increases, likely because specific bacteria are selected for with a fast growth potential or high degree of mucosal adherence to avoid washout [221,222]. Other factors that turned out to be microbiome covariates were recent smoking history as well as the use of antibiotics, osmotic laxatives, IBD drugs, and antidepressants [221]. In a recent mouse study, six days of treatment with over-the-counter laxative polyethylene glycol had long-term effects on the gut [223]. Bacterial family S24-7 went from 50% of total microbial abundance to apparent extinction, while family Bacteroidaceae, also in order Bacteroidales, experienced an expansion from 20% to 60% microbial abundance. Osmotic stress was observed to decimate the mucus barrier and cause the immune system to generate a lasting antibody response against commensal bacteria [223]. Fecal samples were recently collected from 758 Korean men to examine the effects of cigarette smoking on the microbiome [224]. While no differences were observed between former smokers and those who never smoked, current smokers had an increased proportion of Bacteroidetes and decreased levels of Firmicutes and Proteobacteria.

Notable dietary covariates in the Belgian population study included consumption of fruits, alcohol, meat, soy products, and soda as well as one's preference for dark chocolate [221]. Surprisingly, mode of birth and history of breastfeeding were not associated with one's adult microbiota composition, and household pets only predicted a minimal fraction of microbiome variation [221]. An earlier study showed that household dogs primarily alter their owner's skin microbiota rather than the gut microbiota [225]. More dominant influencers of the microbiome are the urbanization of outdoor areas, increased building confinement, and cleaning, each of which diminish overall microbial diversity, shifting from gram-positive (e.g., Actinobacteria) to gram-negative and potentially pathogenic species [226–228].

Consistent with the Belgian [221] and other studies [7,120], earlier analysis of the Dutch population cohort revealed that bacterial taxa could explain BMI and blood lipids independent of age, gender, and host genetics [229]. Species richness was negatively correlated with both BMI and triglycerides and positively correlated with protective levels of HDL cholesterol [120,229]. A significant correlation is not observed, however, between gut microbiota and LDL or total cholesterol levels [7,120,221,229]. The absence of correlation between plasma LDL and the microbiome is notable given that the latter is associated with metabolic disease. Despite plasma LDL being used as the principal target in lipid-lowering therapy for the last three decades, recent evidence suggests that triglyceride, HDL, and apolipoprotein B blood levels may be more useful CVD predictors [230–235]. Many factors confound the relationship between plasma LDL concentration and CVD. While one in three individuals are hyper-responders to dietary cholesterol, the ratio of LDL to HDL is minimally affected when others, particularly the elderly, consume an additional 100 mg/day [236]. For individuals with similar LDL concentrations, a predominance of small dense LDL particles (sdLDL) increases one's CVD risk [236], as does a higher proportion of covalently modified LDL particles, known as lipoprotein(a) [237]. Widely prescribed statin drugs are effective at lowering LDL and to some extent apolipoprotein B concentration, but they do not decrease the proportion of sdLDL and have been found to raise plasma lipoprotein (a) by up to 20%, contributing to what has been termed "residual" CVD risk [238,239]. The lack of an association between plasma LDL concentration and the microbiome is not surprising given these confounding factors.

6.2. FODMAPs and Gut Health

Fermentable oligosaccharides (fructans, GOS), disaccharides (lactose), monosaccharides (fructose), and polyols (sorbitol, xylitol) are termed FODMAPs [240]. Consumption of dietary FODMAPs pulls water into the small intestine and colon, causing luminal distension. Fermentation of FODMAPs by gut bacteria and yeast then produces hydrogen or methane gas. Restricting FODMAPs in one's diet has been shown to help alleviate functional GI symptoms in IBS patients (bloating, abdominal pain, diarrhea), but no effects have been reported for intestinal inflammation in IBD [27,240]. Wheat, rye and barley contain fructans and supply much of the FODMAPs contained in the Western diet. A double-blind crossover challenge was conducted of 59 adults self-reporting non-celiac gluten sensitivity (NCGS), who had previously followed a gluten-free diet for at least six months [241]. Participants completed three seven-day challenges in which a muesli bar was consumed containing either FOS, wheat gluten, or placebo, with the amounts of fructan/gluten equal to that contained in four slices of wheat sandwich bread. IBS symptom scores worsened in the fructan challenge ($P = 0.04$), while symptoms were actually slightly improved relative to placebo upon consumption of gluten ($P = 0.55$). The finding that fructans are responsible for GI symptoms in self-identified NCGS patients, and not gluten, is also supported by a crossover trial in which 37 subjects with NCGS and IBS followed a low-FODMAP diet before switching to a high- or low-gluten diet [242]. Regardless of the source of symptoms, NCGS and IBS at least have overlapping features and are not entirely separate entities [243].

Long-term implementation of a low-FODMAP diet is problematic due to the restriction of healthy plant foods and the fact that FODMAPs are prebiotics that support gut microbiota. Apples, pears, and stone fruits are high in fructose and other FODMAPs. Legumes and pulses are also high FODMAP, as are several vegetables including onion, garlic, and cauliflower. When administered properly by a trained dietitian, the FODMAP elimination diet is intended to be a process rather than a rigid exclusion diet. The initial elimination phase lasts 2–6 weeks in order to get GI symptoms under control. In the challenge phase, specific foods or types of FODMAPs are reintroduced one at a time and in increasing amounts. The patient is instructed to keep a detailed food diary so they can learn what FODMAPs are best tolerated and can eventually be incorporated into the final integration phase of the diet. Two clinical challenges can occur during this process: a patient's symptoms may not respond, or they do respond and then the patient becomes reluctant to reintroduce FODMAPs [244]. While long-term studies are lacking, following a low-FODMAP diet reduces the diversity and quality of dietary components being consumed [245], and healthy diet diversity has been linked to more diverse microbiota and better health outcomes [246]. Short-term FODMAP restriction has been shown to disturb the gut microbiota in as little as 2–3 weeks, reducing total bacterial abundance and the population of *Bifidobacterium*, while increasing the ratio of Firmicutes to Bacteroidetes [247,248].

6.3. Ketogenic Diet

KD and low-carbohydrate diets have become a popular and effective tool for losing weight and can improve blood CVD parameters in the short-term [249,250]. However, 20-year studies involving a large prospective cohort reveal that diet quality and the source of protein and fat can ultimately determine health outcomes in low (40% of caloric intake) carbohydrate diets [251,252]. In research by de Koning et al., it was found that high plant-based intake of protein and fat reduces the hazard ratio (HR) for T2D to 0.78, whereas high intake of animal protein and fat maximizes the risk (HR: 1.37) [251]. Adjusting for red and processed meat intake was observed to lower the association with animal sources (HR: 1.11). In strict KD, below ground vegetables and legumes high in net carbs, and most fruits, are restricted in order limit total carbohydrate intake to 50 g/day. Restricting plant-based carbohydrates can have considerable effects on gut microbiota given that fiber and prebiotics are required for bacterial diversity [65,119,208]. The reduction in fiber can also contribute to constipation, a common side effect of KD.

In an anti-seizure mouse model, KD was shown to reduce gut bacterial alpha diversity, while elevating the relative abundance of *A. muciniphila*, but KD was only followed for three weeks [67]. A much longer study of 10 multiple sclerosis patients found that total bacterial abundance and diversity decreased in the short-term but recovered during weeks 12–24 of KD treatment [253]. *Akkermansia* was observed to increase initially but then declined during long-term KD and pioneer bacteria steadily declined [253]. Pioneer bacteria such as *Bifidobacterium* and *Clostridium* are the first to colonize newborns and patients recovering from a course of antibiotic treatment. Twenty children with refractory epilepsy were recently treated with KD for six months [254]. Treatment lowered alpha diversity and decreased the Firmicutes/Bacteroidetes ratio. In 10 of the children who were non-responsive to treatment (<50% seizure reduction), the relative abundance of Ruminococcaceae and Clostridia became enriched, suggesting specific bacteria may serve as an efficacy biomarker or potential therapeutic target [254]. Such alterations in gut microbiota associated with long-term KD suggest the importance of a properly balanced, high quality diet [65].

6.4. Role of Carbohydrate Intake

Consuming excess carbohydrates as part of a Western diet high in refined grains, starch, and added sugar negatively impacts gut microbiota. The first connection between the microbiome and metabolic health was noted in 1970, when the International Sugar Research Foundation found that a high-sugar diet led to high serum triglycerides in conventional rats but not germ-free rats [255]. In a modern Dutch population study, the largest dietary predictor of low gut bacterial diversity was the total intake of carbohydrates, followed by consumption levels of beer, bread, and soda [120]. A study of 178 elderly subjects by Claesson et al. found that patients in long-term residential care consumed a diet higher in fat and lower in fiber than seniors living in their community [246]. Diet diversity was scored using the healthy food diversity index, which differentiates between healthy and unhealthy foods across all food groups, and found to positively correlate with gut bacterial diversity. Individual microbiota clustered based on long-term care or community living status, and microbiota composition significantly correlated with frailty, co-morbidity, and inflammation markers [246]. While obesity research has traditionally compared low versus high fat diets, a rat study found that a low-fat/high-sucrose diet led to reduced bacterial diversity, increased Firmicutes: Bacteroidetes, a bloom in Ruminococcaceae, gut inflammation, altered vagal gut-brain communication, and obesity, similar to an isocaloric high-fat/high-sucrose diet [105].

Diets high in total carbohydrates and sugar correlate with increased fungus *Candida* and methanogen *Methanobrevibacter*, genera from different domains of life that correlate negatively with consumption of amino acids, protein, and fatty acids [256]. *Methanobrevibacter smithii* is the most prevalent archaeon in the human gut and can comprise up to 10% of all anaerobes in healthy adults. In a mouse model, *M. smithii* has been shown to increase host adiposity by directing *Bacteroides thetaiotaomicron* to ferment plant polysaccharides (fructans) in the diet to the SCFA acetate [257]. Bacterial fermentation of undigested dietary polysaccharides into SCFAs is estimated to account for 5 to 10% of daily caloric intake in the typical diet [258]. Elevated *M. smithii* has also been identified in IBS patients, especially those with IBS-C, in whom methane gas delays gut transit [259]. *M. smithii* copy number was observed to correlate inversely with stool frequency ($R = -0.42$).

Candida are the predominant fungal species capable of colonizing the gut. Overall the mycobiome is less stable than the microbiome [260]. While bacterial population structure primarily associates with long-term diet [14,246], *Candida* can vary extensively in time in response to recent carbohydrate consumption, antibiotic use, and environmental sources [22]. In a study of 98 healthy volunteers by Hoffmann et al., *Candida* correlated positively with long-term intake of total carbohydrates and sugar, and was strongly associated with recent carbohydrate intake [256]. Unlike *Candida* and *Methanobrevibacter*, bacterial populations were observed to associate more strongly with long-term dietary habits than with recent food consumption. *Prevotella* and *Ruminococcus* increased with carbohydrate intake and decreased with animal products, while the reverse effect was observed for *Bacteroides* [256]. A model of syntrophy was proposed in which methanogenesis supports *Ruminococcus* metabolism and *Candida* degrades starch into simple sugars, allowing for substrate fermentation by *Prevotella*.

Stool sample studies have found *Candida* in 63% of individuals, with 11% showing *Candida* overgrowth [261]. Overgrowth can lead to invasive, systemic fungal infection in cancer patients or immunocompromised individuals, resulting in a high mortality rate. In a mouse chemotherapy model, *C. albicans* infection was observed to drive mucosal dysbiosis, allowing *Stenotrophomonas*, *Alphaproteobacteria*, and lactic acid-fermenting *Enterococcus* to proliferate while bacterial diversity declined [262]. Antibiotic treatment is also a strong risk factor for systemic candidiasis. In cell growth assays, SCFAs and lactic acid are shown to have a fungistatic but not fungicidal effect, suggesting that a healthy microbiome prevents *Candida* overgrowth [263]. Lactic acid is responsible for the antimicrobial activity of lactobacilli towards pathogens. Beneficial probiotic strain *L. rhamnosus* GG was additionally shown to bear an exopolysaccharide that interferes with *Candida* growth, hypha formation, and intestinal adhesion [264].

Excessive sugar or starch consumption can lead to *Candida* dysbiosis. Candidiasis is mostly attributed to *C. albicans*, a species which has intrinsic resistance to the fungistatic effect of SCFAs. Interestingly, SCFA resistance is dependent on monomeric glucose being present in the growth media; growth rates are attenuated when the disaccharide maltose is used as a nutrient source [263]. In a study of 120 individuals with chronic intestinal *Candida* overgrowth, diet therapy cured 85% of patients three months after conventional antifungal therapy, compared to 42% of subjects receiving nystatin alone [261]. Patients in the diet group avoided foods high in simple sugars and starch, cured and fatty meats, milk and dairy products, and alcohol.

The notion of cutting starch and sugar to promote intestinal health can be traced to the 1920s, when gastroenterologist Sydney Haas began treating celiac patients using the specific carbohydrate diet (SCD) [265]. SCD was later popularized as a diet for reducing microbial overgrowth by biochemist Elaine Gottschall, who created a dictionary of legal/illegal foods and ingredients [266,267]. The diet prohibits grains (wheat, barley, oats, rice, corn), potatoes, processed meats, added sugars, and disaccharides (lactose, sucrose), while allowing fresh (not canned) fruit, vegetables, and juices not from concentrate [268]. SCD limits dairy to butter, eggs, and aged cheeses containing minimal lactose. Beer, sweet wine, liqueurs, and mucilaginous fibers are restricted as are additives and preservatives like maltodextrin, pectin, guar/gums, and FOS. Sugar alcohols are prohibited, and honey is the recommended sweetener in SCD. A strict three-month period is first observed to starve off overgrowing bacteria and yeast, after which legumes may be selectively introduced. Unlike a low-FODMAP dietary strategy, SCD is intended to be a long-term exclusion diet. While avoiding FODMAPs can improve IBS symptoms in the short-term, cases of drug-free clinical remission have been reported in IBD patients following SCD, with complete resolution of mucosal inflammation in some Crohn's patients [27,269].

Artificial food ingredients are specifically being linked to gut dysbiosis. Maltodextrin, a polysaccharide derived from starch hydrolysis, is a common food additive that enables adherent invasive strains of *E. coli* to adhere to intestinal epithelial cells and grow into biofilm, contributing to gut dysbiosis and intestinal inflammation [21]. Polysorbate-80, an emulsifier used in processed foods, has been shown to enhance translocation of pathogenic *E. coli* strains across colonocytes [21]. In a mouse study by Chassaing et al., low (0.1–1.0%) mass concentrations of emulsifiers polysorbate-80 and carboxymethylcellulose induced low-grade inflammation, obesity, and dysglycemia in wild-type mice and promoted robust colitis in mice predisposed to the disorder [270]. Fecal transplants to germ-free mice demonstrated that changes in microbiota were responsible. The emulsifiers reduced microbial diversity and levels of health-promoting Bacteroidales, while increasing mucolytic *Ruminococcus gnavus* and pro-inflammatory Proteobacteria. Reduced mucus thickness was also observed in the emulsifier-treated mice, along with bacterial encroachment into the normally sterile inner mucus layer [270]. Microbiota encroachment has been implicated in IBD and metabolic syndrome. In humans, the average bacterial-epithelial distance of closest bacteria correlates inversely with BMI, fasting glucose levels, and hemoglobin A1c [271]. Such observations point to the consumption of processed foods as one potentiator of the global obesity epidemic [272].

6.5. Intermittent Fasting

Excessive caloric intake results in fat being stored in white adipose tissue, while energy expenditure by fat oxidation predominantly occurs from thermogenesis of brown adipose tissue. Conversion of white adipocytes, known as beiging, is thus a promising strategy for treatment of metabolic disease. Recently, Li et al. were able to selectively induce the beiging of white adipose tissue in mice using the natural strategy of intermittent fasting [273]. Mice placed on an every-other-day fasting regimen had the same cumulative food intake as the ad libitum control group, but experienced a shift in gut microbiota, increase in fermentation products acetate and lactate, and a reversal of diet-induced obesity. Transport of acetate and lactate across the adipocyte membrane is driven by monocarboxylate transporter 1, whose expression was found to be upregulated in beige cells. Beiging was not observed in germ-free mice, but could be restored upon fecal transplantation of gut microbiota [273]. A previous study in mice demonstrated that cold exposure activates white fat beiging and increases insulin sensitivity via changes in the microbiome [274]. These observations reveal the existence of a microbiota-beige fat axis. In other work, Panda et al. found that diet-induced obesity dampens daily cyclical fluctuations in mice microbiota [275]. Restricting feeding to an eight-hour window each day partially restored circadian fluctuations, including a decrease in the abundance of *Lactobacillus* observed during the feeding phase. Intermittent fasting, longer multiday fasts, and fasting-mimicking diets have been shown to improve gut barrier function, increase microbial diversity, enhance antioxidative microbial pathways, and even reverse intestinal inflammation in models of IBD [276–278].

7. Other Considerations

7.1. Endocannabinoid System

In addition to altered microbiota and low-grade inflammation, obesity is characterized by increased endocannabinoid (eCB) system tone. A study of the eCB system in lean and obese mice was performed by blocking or activating cannabinoid receptor 1 (CB_1) [279]. SR141716A, a CB_1 antagonist that reduces food intake, significantly reduced gut permeability and plasma LPS levels in obese mice, decreasing both adiposity and blood glucose levels. In contrast, agonist HU-210 increased eCB system tone in lean mice and raised plasma LPS. Increased gut permeability with HU-210 was attributed to a decrease in the expression of two epithelial tight junction proteins. By comparing diet-induced obesity and intervention with antibiotics or prebiotics, microbiota associated with obesity were shown to be responsible for increasing the expression levels of CB_1 in colonocytes and adipose tissue [279].

Endocannabinoids are an appealing therapeutic strategy for many conditions such as treating inflammation in IBD [280]. Cannabinoid antagonist cannabidiol has been shown to counteract the inflammatory environment induced by LPS in mice and in human colonic cultures derived from ulcerative colitis patients, at least in part due to PPAR-γ activation [281]. The use of CB_1 agonists has been proposed for increasing GI transit time in IBS-D, while antagonists could prove useful for IBS-C [282]. Partial agonist tetrahydrocannabinol (THC) increases food intake in the short-term, but in epidemiological surveys, obesity is observed to be less prevalent among cannabis users [283]. In mice fed a high-fat diet, chronic treatment with THC was recently shown to stave off increases in the ratio of Firmicutes to Bacteroidetes, increase the abundance of *A. muciniphila*, and prevent diet-induced obesity [284].

7.2. Medication Dysbiosis

Oral administration of high dose antibiotics can result in rapid changes to gut microbiota and is implicated in dysbiosis [22,285–287]. Over-the-counter and prescription non-antibiotic medicines also influence the gut microbiome. Proton pump inhibitors (PPIs) are a widely used class of drugs that function by raising gastric pH. PPIs are an effective short-term indicated therapy for gastroesophageal reflux, peptic ulcers, and *H. pylori* infection, but many chronically afflicted patients take long-term or off-label dosing. Meta-analyses have shown that PPI use increases the risk of developing SIBO and *C. difficile* infection (odds ratios: 1.71 and 1.99; 95% confidence intervals: 1.20–2.43 and 1.73–2.30, respectively) [288,289]. Antibiotics, PPIs, and atypical antipsychotics have each been implicated in reducing alpha microbial diversity [286,290,291]. Second-generation antipsychotic medications, which contribute to weight gain and metabolic syndrome, gradually increase the ratio of Firmicutes to Bacteroidetes in association with BMI and decrease the abundance of *Akkermansia* [292,293]. Efforts are now underway to examine how bacterial taxa each respond to treatment with drugs from other common therapeutic classes [290,294]. Opioids can cause severe constipation and at high doses in mice enable bacterial translocation through disruption of the gut barrier [290,295]. Changes in microbiota have been implicated in the creation of intestinal lesions by nonsteroidal anti-inflammatory drugs, which reduce blood flow to the gut and weaken the hydrophobic mucosal barrier. Lastly, GI symptoms are a common side effect of statins, which affect bile acid metabolism and have been shown to increase the abundance of five bacterial families including Enterobacteriaceae [290].

The interrelationships discussed in this article between diet, environmental factors, gut microbiota, and their physiological outcomes are summarized in Table 1.

Table 1. Summary of diet-microbiota interactions in health and disease.

Healthy Microbiota	Gut Dysbiosis	Other Cause/Consequence
High dietary fiber intake [115]	Western diet; low core diversity [10,83]	High in choline/fat/added sugar [105,117]
Plant foods low in choline [151]	High [TMAO] in blood [134]	Arterial plaque formation [135]
Fruits and vegetables; prebiotic-containing foods [4]	Low fiber intake/low FODMAP carbs [244]	Beer, bread, sugar/artificially-sweetened beverages [120,122]
High α species diversity; butyrate-producing [4,105,120]	Low short-chain fatty acid fermentation [100]	Intestinal inflammation [25,117]
Anti-inflammatory omega-3 [217]	Diet high in omega-6 fatty acids	Pro-inflammatory [149]
Lean body mass, increased lipolysis [84]	Obesity, vagal remodeling, increased energy harvest [85,105]	Increased appetite/lipogenesis [103]
High *Prevotella*/low *Bacteroides*; abundance of *A. muciniphila* [12,14,91]	Abundance of *Ruminococcus* [16,105]	High Firmicutes:Bacteroidetes ratio [85,105]
Glucose and lipid homeostasis [100]	Insulin resistance, bacterial encroachment [76,106,271]	Cardiovascular disease [111,151]
Beneficial bacteria/probiotics: *Bifidobacterium, Lactobacillus* [192,206]	Oxidative stress; facultative anaerobes; *E. coli* [38]	Broad-spectrum antibiotics [22,39,287]; medication dysbiosis [290]
Gut-brain interactions [78]	Mental health issues or visceral pain [72,296]	Leaky gut, plasma endotoxin, psychological stress; emulsifiers [54,272]
Regular intestinal motility [222,259]	Structural or functional bowel disorders [22,50]	Colorectal cancer [3]
Healthy fecal biomarkers [53]	Need butyrate/inulin supplementation [81,104,213]	Potential for fecal transplant [73,76]
Intermittent fasting; adipose beiging [273]	Excess starch/sugar consumption [120]	*Candida* overgrowth; gluten sensitivity [241,256]

8. Conclusions and Future Directions for Research

The past decade of research has begun to reveal the overarching roles the gut microbiome plays in human health. Particular species of *Bifidobacterium*, *Akkermansia*, and *Lactobacillus* are beneficial to the human host and are included in many probiotic preparations, but genera such as *Bacteroides* and *Ruminococcus* are implicated in negative health outcomes. Antibiotic use and modern sanitation have contributed to a decrease in the diversity of the human microbiome [287]. Core microbial diversity and the ratio of Firmicutes to Bacteroidetes are general indicators of health and may change with age, though inter-individual variation is large and quality of diet and environmental factors play a dominant role [246,297–299]. Future research will need to characterize the changes in bacterial composition accompanying different disease states and the corresponding expression patterns in genes of both microbe and host [296,300]. Increased age is associated with oxidative stress and a pro-inflammatory state, and improvements in microbiota have been shown to extend life span in animal models of aging, though human aging studies are lacking [81,278,301,302]. Prebiotics and dietary fiber increase the relative abundance of beneficial anaerobic bacteria, increase butyrate fermentation, and have favorable metabolic effects. Propionate, on the other hand, is an SCFA used as a food preservative that has recently been linked to insulin resistance when consumed in typical concentrations [303]. Lastly, negative results are being reported for gut microbiota-produced acetate. In rats fed a high-fat diet, increased acetate production was found to promote obesity and metabolic syndrome [304]. In an analogous rat model, colonic infusion with resistant starch plus exogenous acetate delayed the development of obesity and insulin resistance and protected the mucosal barrier [305]. Genera such as *Faecalibacterium* and *Roseburia* were observed to enable the conversion of acetate into butyrate, increasing serum and fecal butyrate levels.

While our knowledge of commensal and pathogenic bacteria has grown considerably, future research will need to further address the role of nonbacterial microbes in the human gut, including viruses, eukaryotes, yeasts, and archaea [256,306,307]. Viruses parasitic to bacteria, known as bacteriophages, have been shown to coexist over time with the bacterial species they prey on. Phage predation can also lead to cascading effects on other species, including blooms in non-targeted bacteria [308]. An abnormal enteric virome has been found in IBD patients, in whom an increase in bacteriophage richness contributes to decreasing bacterial diversity and gut dysbiosis [309]. The most prevalent eukaryote in the human intestine, *Blastocystis*, is a single-celled heterokont protist that colonizes a considerable fraction of individuals in industrialized (0.5–30%) and developing (30–76%) nations [310]. It has been hypothesized that *Blastocystis* can prey on bacterial species in the gut in its ameboid form [306] and can contribute to the pathogenesis of IBS [311]. In a mouse study by Yason et al., infection with a pathogenic subtype of *Blastocystis* (ST7) was observed to decrease intestinal levels of beneficial *Bifidobacterium* and *Lactobacillus* while increasing *E. coli* content, seeming to fulfill Koch's postulate that infection of a healthy individual leads to disease [312]. In asymptomatic individuals, however, nonpathogenic *Blastocystis* correlates positively with microbial diversity and inversely with BMI, fecal calprotectin levels, Crohn's disease, and colorectal cancer [313]. As a genus, species of *Blastocystis* have incredibly divergent genomes. The percentage of proteins unique to each subtype ranges from 6% to 20%, and orthologous proteins have a median amino acid sequence identity of only 60% [314].

Diet and nutritional status are important determinants in human health. Efforts to characterize the relationship between diet and health have pivoted from studying the effects of individual nutrients to examining the roles of dietary patterns and specific diets [149–151]. The role of diet in shaping gut microbiota, host metabolism, and lipid homeostasis is changing our view of the steps a person can take to make improvements in their systemic health [10,315]. Correlations between microbial diversity across as many as 60 different dietary covariates reveal the importance of a high quality, balanced diet [120], supporting the view that dietary supplementation of individual nutrients does not take the place of a sound diet [316]. Observations that individual foods stimulate the growth of specific bacterial taxa suggest that intestinal bacteria could actually be serving to guide our food preferences, appetite, and feelings of satiety [221,317]. By influencing metabolism and inflammation, diet and nutrition can outweigh genetic and environmental factors in determining health outcomes for chronic Western conditions such as diabetes, obesity, IBS, IBD, colorectal cancer, and depression [1,2,318].

One research question that remains is what constitutes an optimal health-promoting microbiome, and how individuals with different starting microbiota can achieve such microflora. In characterizing gut eubiosis and dysbiosis, the effects of particular microbial species cannot be considered simply in isolation, giving rise to the notion of ecological Koch's postulates of disease causation [162]. Changes in stool consistency and water content have hampered quantification of absolute microbial loads, and new methods are needed to identify pathological markers [319]. While fecal samples are generally thought to be representative of colonic microbial communities, further research is needed to characterize the different microbial communities that occur along the length of the GI tract [320]. A study of five gut sections taken from pigs found a predominance of *Lactobacillus* in the small intestine and *Prevotella* in the colon, suggesting that rapid utilization of simple carbohydrates drives microbial competition in the upper intestine, while polysaccharide fermentation is left mainly to the colon [321].

Inter-individual variation in gut microbiota could explain the disparity in outcomes often observed with lifestyle interventions and why one-size-fits-all diets are not always effective [83,125,201]. The influence of diet type on the relative abundances of microbial populations can be complex and difficult to reproduce across different clinical studies, in part due to the number of individual species involved in each phylum and genus [142]. Individuals have been shown to have highly personalized microbiome responses to different foods depending on their prior history of dietary diversity [322]. Rapid modifications in gut microbiota are possible when adopting a new dietary strategy, such as following an exclusively plant- or animal-based diet [323]. Microbial markers have even been proposed as an objective means of measuring adherence to a given dietary pattern in order to more accurately correlate resultant health outcomes [150]. Microbes collectively encode 150-fold more genes than the human genome [5]. Enzymes in gut bacteria across the main taxonomic groupings have been shown to metabolize 176 common oral drugs, suggesting that differences in gut microbiota may shape individual responses to drug therapy [324]. Ultimately, determining the full landscape of host-microbiota interactions will enable advances in personalized medicine, precision nutrition [125,325], and the development of next-generation probiotics tailored to the individual [326].

Author Contributions: Conceptualization, R.D.H., B.A.P., S.C.S., H.R.M., C.A.B., and C.R.T.; study design, R.D.H.; writing—Original draft preparation, R.D.H.; writing—Review and editing, B.A.P., R.D.H., S.C.S., H.R.M., C.A.B., and C.R.T.

Funding: C.A.B. received funding from the American Foundation for Pharmaceutical Education—Gateway to Research Award. R.D.H. acknowledges financial support from the National Science Foundation, grant MCB 1516826. This paper is dedicated to the memory of Dan Veilleux.

Acknowledgments: The authors thank Dan MicKool and John Redwanski for helpful discussion.

Conflicts of Interest: C.R.T. has an ownership stake in Noble Wellness, LLC. S.C.S. is the founder of Simulation Consultation Services. The funders had no role in the design of the study; in the collection, analyses, or interpretation of data; in the writing of the manuscript, or in the decision to publish the results.

Abbreviations

BMI	body mass index
CB_1	cannabinoid receptor 1
CVD	cardiovascular disease
DHA	docosahexaenoic acid
DMB	3,3-dimethyl-1-butanol
eCB	endocannabinoid
EPA	eicosapentaenoic acid
FMT	fecal microbial transplantation
FODMAP	fermentable oligo-, di-, mono-saccharides and polyols
FOS	fructo-oligosaccharide
GABA	γ-aminobutyric acid
GI	gastrointestinal
GOS	galacto-oligosaccharide
GPCR	G-protein coupled receptor
GPR109A	niacin receptor 1
GPR41	free fatty acid receptor 3
HR	hazard ratio
HDL	high-density lipoprotein
IBD	inflammatory bowel disease
IBS	irritable bowel syndrome
IBS-C	IBS with predominant constipation
IBS-D	IBS with predominant diarrhea
IBS-M	IBS with alternating bowel habits
KD	ketogenic diet
LDL	low-density lipoprotein
LPS	lipopolysaccharide (endotoxin)
NAS	non-caloric artificial sweetener
NCGS	non-celiac gluten sensitivity
P	probability value
PPAR-γ	peroxisome proliferator-activated receptor gamma

PPI	proton pump inhibitor
PUFA	polyunsaturated fatty acid
R	Pearson correlation coefficient
RYGB	Roux-en-Y gastric bypass
SCD	specific carbohydrate diet
SCFA	short-chain fatty acid
sdLDL	small dense low-density lipoprotein particle
SGLT1	sodium glucose cotransporter-1
SIBO	small intestinal bacterial overgrowth
THC	tetrahydrocannabinol
TMA	trimethylamine
TMAO	trimethylamine *N*-oxide
TNF	tumor necrosis factor
T2D	type 2 diabetes

References

1. Conlon, M.A.; Bird, A.R. The impact of diet and lifestyle on gut microbiota and human health. *Nutrients* **2015**, *7*, 17–44. [CrossRef] [PubMed]
2. Singh, R.K.; Chang, H.W.; Yan, D.; Lee, K.M.; Ucmak, D.; Wong, K.; Abrouk, M.; Farahnik, B.; Nakamura, M.; Zhu, T.H.; et al. Influence of diet on the gut microbiome and implications for human health. *J. Transl. Med.* **2017**, *15*, 73. [CrossRef]
3. Brennan, C.A.; Garrett, W.S. Gut microbiota, inflammation, and colorectal cancer. *Annu. Rev. Microbiol.* **2016**, *70*, 395–411. [CrossRef] [PubMed]
4. Valdes, A.M.; Walter, J.; Segal, E.; Spector, T.D. Role of the gut microbiota in nutrition and health. *BMJ* **2018**, *361*, k2179. [CrossRef] [PubMed]
5. Qin, J.; Li, R.; Raes, J.; Arumugam, M.; Burgdorf, K.S.; Manichanh, C.; Nielsen, T.; Pons, N.; Levenez, F.; Yamada, T.; et al. A human gut microbial gene catalogue established by metagenomic sequencing. *Nature* **2010**, *464*, 59–65. [CrossRef] [PubMed]
6. Almeida, A.; Mitchell, A.L.; Boland, M.; Forster, S.C.; Gloor, G.B.; Tarkowska, A.; Lawley, T.D.; Finn, R.D. A new genomic blueprint of the human gut microbiota. *Nature* **2019**, *568*, 499–504. [CrossRef] [PubMed]
7. Rothschild, D.; Weissbrod, O.; Barkan, E.; Kurilshikov, A.; Korem, T.; Zeevi, D.; Costea, P.I.; Godneva, A.; Kalka, I.N.; Bar, N.; et al. Environment dominates over host genetics in shaping human gut microbiota. *Nature* **2018**, *555*, 210–215. [CrossRef] [PubMed]
8. Noble, R.E. Waist-to-hip ratio versus BMI as predictors of cardiac risk in obese adult women. *West. J. Med.* **2001**, *174*, 240–241. [CrossRef] [PubMed]
9. Hamer, M.; O'Donovan, G.; Stensel, D.; Stamatakis, E. Normal-Weight central obesity and risk for mortality. *Ann. Intern. Med.* **2017**, *166*, 917–918. [CrossRef] [PubMed]
10. Gentile, C.L.; Weir, T.L. The gut microbiota at the intersection of diet and human health. *Science* **2018**, *362*, 776–780. [CrossRef] [PubMed]
11. Lauderdale, D.S.; Rathouz, P.J. Body mass index in a US national sample of Asian Americans: Effects of nativity, years since immigration and socioeconomic status. *Int. J. Obes.* **2000**, *24*, 1188–1194. [CrossRef]
12. Vangay, P.; Johnson, A.J.; Ward, T.L.; Al-Ghalith, G.A.; Shields-Cutler, R.R.; Hillmann, B.M.; Lucas, S.K.; Beura, L.K.; Thompson, E.A.; Till, L.M.; et al. US immigration westernizes the human gut microbiome. *Cell* **2018**, *175*, 962–972. [CrossRef] [PubMed]
13. Arumugam, M.; Raes, J.; Pelletier, E.; Le Paslier, D.; Yamada, T.; Mende, D.R.; Fernandes, G.R.; Tap, J.; Bruls, T.; Batto, J.M.; et al. Enterotypes of the human gut microbiome. *Nature* **2011**, *473*, 174–180. [CrossRef] [PubMed]
14. Wu, G.D.; Chen, J.; Hoffmann, C.; Bittinger, K.; Chen, Y.Y.; Keilbaugh, S.A.; Bewtra, M.; Knights, D.; Walters, W.A.; Knight, R.; et al. Linking long-term dietary patterns with gut microbial enterotypes. *Science* **2011**, *334*, 105–108. [CrossRef] [PubMed]
15. Rajilic-Stojanovic, M.; Jonkers, D.M.; Salonen, A.; Hanevik, K.; Raes, J.; Jalanka, J.; De Vos, W.M.; Manichanh, C.; Golic, N.; Enck, P.; et al. Intestinal microbiota and diet in IBS: Causes, consequences, or epiphenomena? *Am. J. Gastroenterol.* **2015**, *110*, 278–287. [CrossRef] [PubMed]

16. Hall, A.B.; Yassour, M.; Sauk, J.; Garner, A.; Jiang, X.; Arthur, T.; Lagoudas, G.K.; Vatanen, T.; Fornelos, N.; Wilson, R.; et al. A novel *Ruminococcus gnavus* clade enriched in inflammatory bowel disease patients. *Genome Med.* **2017**, *9*, 103. [CrossRef] [PubMed]
17. Henke, M.T.; Kenny, D.J.; Cassilly, C.D.; Vlamakis, H.; Xavier, R.J.; Clardy, J. *Ruminococcus gnavus*, a member of the human gut microbiome associated with Crohn's disease, produces an inflammatory polysaccharide. *Proc. Natl. Acad. Sci. USA* **2019**, *116*, 12672–12677. [CrossRef]
18. Kim, E.R.; Chang, D.K. Colorectal cancer in inflammatory bowel disease: The risk, pathogenesis, prevention and diagnosis. *World J. Gastroenterol.* **2014**, *20*, 9872–9881. [CrossRef] [PubMed]
19. Jandhyala, S.M.; Talukdar, R.; Subramanyam, C.; Vuyyuru, H.; Sasikala, M.; Nageshwar Reddy, D. Role of the normal gut microbiota. *World J. Gastroenterol.* **2015**, *21*, 8787–8803. [CrossRef]
20. Winter, S.E.; Baumler, A.J. Dysbiosis in the inflamed intestine: Chance favors the prepared microbe. *Gut Microbes* **2014**, *5*, 71–73. [CrossRef]
21. Martinez-Medina, M.; Garcia-Gil, L.J. Escherichia coli in chronic inflammatory bowel diseases: An update on adherent invasive Escherichia coli pathogenicity. *World J. Gastrointest. Pathophysiol.* **2014**, *5*, 213–227. [CrossRef] [PubMed]
22. Lewis, J.D.; Chen, E.Z.; Baldassano, R.N.; Otley, A.R.; Griffiths, A.M.; Lee, D.; Bittinger, K.; Bailey, A.; Friedman, E.S.; Hoffmann, C.; et al. Inflammation, antibiotics, and diet as environmental stressors of the gut microbiome in pediatric Crohn's disease. *Cell Host Microbe* **2015**, *18*, 489–500. [CrossRef] [PubMed]
23. Margolis, D.J.; Fanelli, M.; Hoffstad, O.; Lewis, J.D. Potential association between the oral tetracycline class of antimicrobials used to treat acne and inflammatory bowel disease. *Am. J. Gastroenterol.* **2010**, *105*, 2610–2616. [CrossRef] [PubMed]
24. Lloyd, D.A.; Powell-Tuck, J. Artificial nutrition: Principles and practice of enteral feeding. *Clin. Colon Rectal Surg.* **2004**, *17*, 107–118. [CrossRef] [PubMed]
25. Llewellyn, S.R.; Britton, G.J.; Contijoch, E.J.; Vennaro, O.H.; Mortha, A.; Colombel, J.F.; Grinspan, A.; Clemente, J.C.; Merad, M.; Faith, J.J. Interactions between diet and the intestinal microbiota alter intestinal permeability and colitis severity in mice. *Gastroenterology* **2018**, *154*, 1037–1046. [CrossRef] [PubMed]
26. Chiba, M.; Abe, T.; Tsuda, H.; Sugawara, T.; Tsuda, S.; Tozawa, H.; Fujiwara, K.; Imai, H. Lifestyle-related disease in Crohn's disease: Relapse prevention by a semi-vegetarian diet. *World J. Gastroenterol.* **2010**, *16*, 2484–2495. [CrossRef] [PubMed]
27. Lewis, J.D.; Abreu, M.T. Diet as a trigger or therapy for inflammatory bowel diseases. *Gastroenterology* **2017**, *152*, 398–414. [CrossRef] [PubMed]
28. Roy, C.C.; Kien, C.L.; Bouthillier, L.; Levy, E. Short-chain fatty acids: Ready for prime time? *Nutr. Clin. Pract.* **2006**, *21*, 351–366. [CrossRef] [PubMed]
29. Sims, I.M.; Ryan, J.L.; Kim, S.H. In vitro fermentation of prebiotic oligosaccharides by *Bifidobacterium lactis* HN019 and *Lactobacillus* spp. *Anaerobe* **2014**, *25*, 11–17. [CrossRef]
30. Andoh, A. Physiological role of gut microbiota for maintaining human health. *Digestion* **2016**, *93*, 176–181. [CrossRef]
31. Ohira, H.; Tsutsui, W.; Fujioka, Y. Are short chain fatty acids in gut microbiota defensive players for inflammation and atherosclerosis? *J. Atheroscler. Thromb.* **2017**, *24*, 660–672. [CrossRef] [PubMed]
32. Sivaprakasam, S.; Prasad, P.D.; Singh, N. Benefits of short-chain fatty acids and their receptors in inflammation and carcinogenesis. *Pharmacol. Ther.* **2016**, *164*, 144–151. [CrossRef] [PubMed]
33. Morrison, D.J.; Preston, T. Formation of short chain fatty acids by the gut microbiota and their impact on human metabolism. *Gut Microbes* **2016**, *7*, 189–200. [CrossRef] [PubMed]
34. Bergman, E.N. Energy contributions of volatile fatty acids from the gastrointestinal tract in various species. *Physiol. Rev.* **1990**, *70*, 567–590. [CrossRef] [PubMed]
35. Kim, S.; Kim, J.H.; Park, B.O.; Kwak, Y.S. Perspectives on the therapeutic potential of short-chain fatty acid receptors. *BMB Rep.* **2014**, *47*, 173–178. [CrossRef] [PubMed]
36. Singh, N.; Gurav, A.; Sivaprakasam, S.; Brady, E.; Padia, R.; Shi, H.; Thangaraju, M.; Prasad, P.D.; Manicassamy, S.; Munn, D.H.; et al. Activation of Gpr109a, receptor for niacin and the commensal metabolite butyrate, suppresses colonic inflammation and carcinogenesis. *Immunity* **2014**, *40*, 128–139. [CrossRef] [PubMed]
37. Ang, Z.; Ding, J.L. GPR41 and GPR43 in obesity and inflammation—Protective or causative? *Front. Immunol.* **2016**, *7*, 28. [CrossRef] [PubMed]

38. Litvak, Y.; Byndloss, M.X.; Baumler, A.J. Colonocyte metabolism shapes the gut microbiota. *Science* **2018**, *362*, eaat9076. [CrossRef]
39. Romick-Rosendale, L.E.; Haslam, D.B.; Lane, A.; Denson, L.; Lake, K.; Wilkey, A.; Watanabe, M.; Bauer, S.; Litts, B.; Luebbering, N.; et al. Antibiotic exposure and reduced short chain fatty acid production after hematopoietic stem cell transplant. *Biol. Blood Marrow Transpl.* **2018**, *24*, 2418–2424. [CrossRef]
40. Wassenaar, T.M. Insights from 100 years of research with probiotic *E. coli*. *Eur. J. Microbiol. Immunol.* **2016**, *6*, 147–161. [CrossRef]
41. Byndloss, M.X.; Olsan, E.E.; Rivera-Chavez, F.; Tiffany, C.R.; Cevallos, S.A.; Lokken, K.L.; Torres, T.P.; Byndloss, A.J.; Faber, F.; Gao, Y.; et al. Microbiota-activated PPAR-gamma signaling inhibits dysbiotic Enterobacteriaceae expansion. *Science* **2017**, *357*, 570–575. [CrossRef] [PubMed]
42. Winter, S.E.; Winter, M.G.; Xavier, M.N.; Thiennimitr, P.; Poon, V.; Keestra, A.M.; Laughlin, R.C.; Gomez, G.; Wu, J.; Lawhon, S.D.; et al. Host-derived nitrate boosts growth of *E. coli* in the inflamed gut. *Science* **2013**, *339*, 708–711. [CrossRef] [PubMed]
43. Xu, J.; Chen, N.; Wu, Z.; Song, Y.; Zhang, Y.; Wu, N.; Zhang, F.; Ren, X.; Liu, Y. 5-Aminosalicylic acid alters the gut bacterial microbiota in patients with ulcerative colitis. *Front. Microbiol.* **2018**, *9*, 1274. [CrossRef] [PubMed]
44. Pang, T.; Leach, S.T.; Katz, T.; Day, A.S.; Ooi, C.Y. Fecal biomarkers of intestinal health and disease in children. *Front. Pediatr.* **2014**, *2*, 6. [CrossRef] [PubMed]
45. Huda-Faujan, N.; Abdulamir, A.S.; Fatimah, A.B.; Anas, O.M.; Shuhaimi, M.; Yazid, A.M.; Loong, Y.Y. The impact of the level of the intestinal short chain fatty acids in inflammatory bowel disease patients versus healthy subjects. *Open Biochem. J.* **2010**, *4*, 53–58. [CrossRef] [PubMed]
46. Granado-Serrano, A.B.; Martin-Gari, M.; Sanchez, V.; Riart Solans, M.; Berdun, R.; Ludwig, I.A.; Rubio, L.; Vilaprinyo, E.; Portero-Otin, M.; Serrano, J.C.E. Faecal bacterial and short-chain fatty acids signature in hypercholesterolemia. *Sci. Rep.* **2019**, *9*, 1772. [CrossRef] [PubMed]
47. Imperiale, T.F.; Gruber, R.N.; Stump, T.E.; Emmett, T.W.; Monahan, P.O. Performance characteristics of fecal immunochemical tests for colorectal cancer and advanced adenomatous polyps: A systematic review and meta-analysis. *Ann. Intern. Med.* **2019**, *170*, 319–329. [CrossRef]
48. Lazaridis, N.; Germanidis, G. Current insights into the innate immune system dysfunction in irritable bowel syndrome. *Ann. Gastroenterol.* **2018**, *31*, 171–187. [CrossRef]
49. Yao, X.; Yang, Y.S.; Cui, L.H.; Zhao, K.B.; Zhang, Z.H.; Peng, L.H.; Guo, X.; Sun, G.; Shang, J.; Wang, W.F.; et al. Subtypes of irritable bowel syndrome on Rome III criteria: A multicenter study. *J. Gastroenterol. Hepatol.* **2012**, *27*, 760–765. [CrossRef]
50. Lacy, B.E.; Mearin, F.; Chang, L.; Chey, W.D.; Lembo, A.J.; Simren, M.; Spiller, R. Bowel Disorders. *Gastroenterology* **2016**, *150*, 1393–1407. [CrossRef]
51. Hughes, P.A.; Zola, H.; Penttila, I.A.; Blackshaw, L.A.; Andrews, J.M.; Krumbiegel, D. Immune activation in irritable bowel syndrome: Can neuroimmune interactions explain symptoms? *Am. J. Gastroenterol.* **2013**, *108*, 1066–1074. [CrossRef]
52. Langhorst, J.; Junge, A.; Rueffer, A.; Wehkamp, J.; Foell, D.; Michalsen, A.; Musial, F.; Dobos, G.J. Elevated human beta-defensin-2 levels indicate an activation of the innate immune system in patients with irritable bowel syndrome. *Am. J. Gastroenterol.* **2009**, *104*, 404–410. [CrossRef] [PubMed]
53. Farup, P.G.; Rudi, K.; Hestad, K. Faecal short-chain fatty acids - a diagnostic biomarker for irritable bowel syndrome? *BMC Gastroenterol.* **2016**, *16*, 51. [CrossRef] [PubMed]
54. Linsalata, M.; Riezzo, G.; D'Attoma, B.; Clemente, C.; Orlando, A.; Russo, F. Noninvasive biomarkers of gut barrier function identify two subtypes of patients suffering from diarrhoea predominant-IBS: A case-control study. *BMC Gastroenterol.* **2018**, *18*, 167. [CrossRef]
55. Mu, Q.; Kirby, J.; Reilly, C.M.; Luo, X.M. Leaky gut as a danger signal for autoimmune diseases. *Front. Immunol.* **2017**, *8*, 598. [CrossRef] [PubMed]
56. Briskey, D.; Heritage, M.; Jaskowski, L.A.; Peake, J.; Gobe, G.; Subramaniam, V.N.; Crawford, D.; Campbell, C.; Vitetta, L. Probiotics modify tight-junction proteins in an animal model of nonalcoholic fatty liver disease. *Ther. Adv. Gastroenterol.* **2016**, *9*, 463–472. [CrossRef] [PubMed]

57. Wang, L.; Fouts, D.E.; Starkel, P.; Hartmann, P.; Chen, P.; Llorente, C.; DePew, J.; Moncera, K.; Ho, S.B.; Brenner, D.A.; et al. Intestinal REG3 lectins protect against alcoholic steatohepatitis by reducing mucosa-associated microbiota and preventing bacterial translocation. *Cell Host Microbe* **2016**, *19*, 227–239. [CrossRef] [PubMed]
58. Carabotti, M.; Scirocco, A.; Maselli, M.A.; Severi, C. The gut-brain axis: Interactions between enteric microbiota, central and enteric nervous systems. *Ann. Gastroenterol.* **2015**, *28*, 203–209. [PubMed]
59. Quigley, E.M.M. The gut-brain axis and the microbiome: Clues to pathophysiology and opportunities for novel management strategies in irritable bowel syndrome (IBS). *J. Clin. Med.* **2018**, *7*, 6. [CrossRef]
60. Strati, F.; Cavalieri, D.; Albanese, D.; De Felice, C.; Donati, C.; Hayek, J.; Jousson, O.; Leoncini, S.; Renzi, D.; Calabro, A.; et al. New evidences on the altered gut microbiota in autism spectrum disorders. *Microbiome* **2017**, *5*, 24. [CrossRef]
61. Fowlie, G.; Cohen, N.; Ming, X. The perturbance of microbiome and gut-brain axis in autism spectrum disorders. *Int. J. Mol. Sci.* **2018**, *19*, 2251. [CrossRef] [PubMed]
62. De Angelis, M.; Francavilla, R.; Piccolo, M.; De Giacomo, A.; Gobbetti, M. Autism spectrum disorders and intestinal microbiota. *Gut Microbes* **2015**, *6*, 207–213. [CrossRef] [PubMed]
63. Sinaiscalco, D.; Brigida, A.L.; Antonucci, N. Autism and neuro-immune-gut link. *AIMS Mol. Sci.* **2018**, *5*, 166–172. [CrossRef]
64. Wheless, J.W. History of the ketogenic diet. *Epilepsia* **2008**, *49*, 3–5. [CrossRef]
65. Reddel, S.; Putignani, L.; Del Chierico, F. The impact of low-FODMAPs, gluten-free, and ketogenic diets on gut microbiota modulation in pathological conditions. *Nutrients* **2019**, *11*, 373. [CrossRef] [PubMed]
66. Stafstrom, C.E.; Rho, J.M. The ketogenic diet as a treatment paradigm for diverse neurological disorders. *Front. Pharmacol.* **2012**, *3*, 59. [CrossRef] [PubMed]
67. Olson, C.A.; Vuong, H.E.; Yano, J.M.; Liang, Q.Y.; Nusbaum, D.J.; Hsiao, E.Y. The gut microbiota mediates the anti-seizure effects of the ketogenic diet. *Cell* **2018**, *173*, 1728–1741. [CrossRef]
68. Wong, C.G.; Bottiglieri, T.; Snead, O.C., III. GABA, gamma-hydroxybutyric acid, and neurological disease. *Ann. Neurol.* **2003**, *54*, S3–S12. [CrossRef]
69. Doenyas, C. Dietary interventions for autism spectrum disorder: New perspectives from the gut-brain axis. *Physiol. Behav.* **2018**, *194*, 577–582. [CrossRef]
70. Sampson, T.R.; Debelius, J.W.; Thron, T.; Janssen, S.; Shastri, G.G.; Ilhan, Z.E.; Challis, C.; Schretter, C.E.; Rocha, S.; Gradinaru, V.; et al. Gut microbiota regulate motor deficits and neuroinflammation in a model of Parkinson's disease. *Cell* **2016**, *167*, 1469–1480. [CrossRef]
71. Sarkar, A.; Lehto, S.M.; Harty, S.; Dinan, T.G.; Cryan, J.F.; Burnet, P.W.J. Psychobiotics and the manipulation of bacteria-gut-brain signals. *Trends Neurosci.* **2016**, *39*, 763–781. [CrossRef] [PubMed]
72. Skonieczna-Zydecka, K.; Marlicz, W.; Misera, A.; Koulaouzidis, A.; Loniewski, I. Microbiome-the missing link in the gut-brain axis: Focus on its role in gastrointestinal and mental health. *J. Clin. Med.* **2018**, *7*, 521. [CrossRef] [PubMed]
73. Cammarota, G.; Ianiro, G.; Tilg, H.; Rajilic-Stojanovic, M.; Kump, P.; Satokari, R.; Sokol, H.; Arkkila, P.; Pintus, C.; Hart, A.; et al. European consensus conference on faecal microbiota transplantation in clinical practice. *Gut* **2017**, *66*, 569–580. [CrossRef] [PubMed]
74. Kang, D.W.; Adams, J.B.; Gregory, A.C.; Borody, T.; Chittick, L.; Fasano, A.; Khoruts, A.; Geis, E.; Maldonado, J.; McDonough-Means, S.; et al. Microbiota Transfer Therapy alters gut ecosystem and improves gastrointestinal and autism symptoms: An open-label study. *Microbiome* **2017**, *5*, 10. [CrossRef] [PubMed]
75. Choi, H.H.; Cho, Y.S. Fecal microbiota transplantation: Current applications, effectiveness, and future perspectives. *Clin. Endosc.* **2016**, *49*, 257–265. [CrossRef] [PubMed]
76. De Groot, P.F.; Frissen, M.N.; De Clercq, N.C.; Nieuwdorp, M. Fecal microbiota transplantation in metabolic syndrome: History, present and future. *Gut Microbes* **2017**, *8*, 253–267. [CrossRef] [PubMed]
77. Kootte, R.S.; Levin, E.; Salojarvi, J.; Smits, L.P.; Hartstra, A.V.; Udayappan, S.D.; Hermes, G.; Bouter, K.E.; Koopen, A.M.; Holst, J.J.; et al. Improvement of insulin sensitivity after lean donor feces in metabolic syndrome Is driven by baseline intestinal microbiota composition. *Cell Metab.* **2017**, *26*, 611–619. [CrossRef] [PubMed]
78. Foster, J.A.; Neufeld, K.A.M. Gut-brain axis: How the microbiome influences anxiety and depression. *Trends Neurosci.* **2013**, *36*, 305–312. [CrossRef] [PubMed]

79. Valles-Colomer, M.; Falony, G.; Darzi, Y.; Tigchelaar, E.F.; Wang, J.; Tito, R.Y.; Schiweck, C.; Kurilshikov, A.; Joossens, M.; Wijmenga, C.; et al. The neuroactive potential of the human gut microbiota in quality of life and depression. *Nat. Microbiol.* **2019**, *4*, 623–632. [CrossRef] [PubMed]
80. Wallace, C.J.K.; Milev, R. The effects of probiotics on depressive symptoms in humans: A systematic review. *Ann. Gen. Psychiatry* **2017**, *16*, 14. [CrossRef] [PubMed]
81. Boehme, M.; Van De Wouw, M.; Bastiaanssen, T.F.S.; Olavarria-Ramirez, L.; Lyons, K.; Fouhy, F.; Golubeva, A.V.; Moloney, G.M.; Minuto, C.; Sandhu, K.V.; et al. Mid-life microbiota crises: Middle age is associated with pervasive neuroimmune alterations that are reversed by targeting the gut microbiome. *Mol. Psychiatry* **2019**, in press. [CrossRef] [PubMed]
82. Banta, J.E.; Segovia-Siapco, G.; Crocker, C.B.; Montoya, D.; Alhusseini, N. Mental health status and dietary intake among California adults: A population-based survey. *Int. J. Food Sci. Nutr.* **2019**, *70*, 759–770. [CrossRef] [PubMed]
83. Sonnenburg, J.L.; Backhed, F. Diet-microbiota interactions as moderators of human metabolism. *Nature* **2016**, *535*, 56–64. [CrossRef] [PubMed]
84. Ridaura, V.K.; Faith, J.J.; Rey, F.E.; Cheng, J.; Duncan, A.E.; Kau, A.L.; Griffin, N.W.; Lombard, V.; Henrissat, B.; Bain, J.R.; et al. Gut microbiota from twins discordant for obesity modulate metabolism in mice. *Science* **2013**, *341*, 1241214. [CrossRef] [PubMed]
85. Turnbaugh, P.J.; Ley, R.E.; Mahowald, M.A.; Magrini, V.; Mardis, E.R.; Gordon, J.I. An obesity-associated gut microbiome with increased capacity for energy harvest. *Nature* **2006**, *444*, 1027–1031. [CrossRef] [PubMed]
86. Turnbaugh, P.J.; Hamady, M.; Yatsunenko, T.; Cantarel, B.L.; Duncan, A.; Ley, R.E.; Sogin, M.L.; Jones, W.J.; Roe, B.A.; Affourtit, J.P.; et al. A core gut microbiome in obese and lean twins. *Nature* **2009**, *457*, 480–484. [CrossRef]
87. Le Chatelier, E.; Nielsen, T.; Qin, J.; Prifti, E.; Hildebrand, F.; Falony, G.; Almeida, M.; Arumugam, M.; Batto, J.M.; Kennedy, S.; et al. Richness of human gut microbiome correlates with metabolic markers. *Nature* **2013**, *500*, 541–546. [CrossRef]
88. Cotillard, A.; Kennedy, S.P.; Kong, L.C.; Prifti, E.; Pons, N.; Le Chatelier, E.; Almeida, M.; Quinquis, B.; Levenez, F.; Galleron, N.; et al. Dietary intervention impact on gut microbial gene richness. *Nature* **2013**, *500*, 585–588. [CrossRef]
89. Everard, A.; Belzer, C.; Geurts, L.; Ouwerkerk, J.P.; Druart, C.; Bindels, L.B.; Guiot, Y.; Derrien, M.; Muccioli, G.G.; Delzenne, N.M.; et al. Cross-talk between *Akkermansia muciniphila* and intestinal epithelium controls diet-induced obesity. *Proc. Natl. Acad. Sci. USA* **2013**, *110*, 9066–9071. [CrossRef]
90. Anhe, F.F.; Roy, D.; Pilon, G.; Dudonne, S.; Matamoros, S.; Varin, T.V.; Garofalo, C.; Moine, Q.; Desjardins, Y.; Levy, E.; et al. A polyphenol-rich cranberry extract protects from diet-induced obesity, insulin resistance and intestinal inflammation in association with increased *Akkermansia* spp. population in the gut microbiota of mice. *Gut* **2015**, *64*, 872–883. [CrossRef]
91. Dao, M.C.; Everard, A.; Aron-Wisnewsky, J.; Sokolovska, N.; Prifti, E.; Verger, E.O.; Kayser, B.D.; Levenez, F.; Chilloux, J.; Hoyles, L.; et al. *Akkermansia muciniphila* and improved metabolic health during a dietary intervention in obesity: Relationship with gut microbiome richness and ecology. *Gut* **2016**, *65*, 426–436. [CrossRef] [PubMed]
92. Depommier, C.; Everard, A.; Druart, C.; Plovier, H.; Van Hul, M.; Vieira-Silva, S.; Falony, G.; Raes, J.; Maiter, D.; Delzenne, N.M.; et al. Supplementation with *Akkermansia muciniphila* in overweight and obese human volunteers: A proof-of-concept exploratory study. *Nat. Med.* **2019**, *25*, 1096–1103. [CrossRef] [PubMed]
93. Zhai, Q.; Feng, S.; Arjan, N.; Chen, W. A next generation probiotic, *Akkermansia muciniphila*. *Crit. Rev. Food Sci. Nutr.* **2018**, in press. [CrossRef] [PubMed]
94. Liou, A.P.; Paziuk, M.; Luevano, J.M., Jr.; Machineni, S.; Turnbaugh, P.J.; Kaplan, L.M. Conserved shifts in the gut microbiota due to gastric bypass reduce host weight and adiposity. *Sci. Transl. Med.* **2013**, *5*, 178ra41. [CrossRef] [PubMed]
95. Kasubuchi, M.; Hasegawa, S.; Hiramatsu, T.; Ichimura, A.; Kimura, I. Dietary gut microbial metabolites, short-chain fatty acids, and host metabolic regulation. *Nutrients* **2015**, *7*, 2839–2849. [CrossRef] [PubMed]
96. Zhang, H.; DiBaise, J.K.; Zuccolo, A.; Kudrna, D.; Braidotti, M.; Yu, Y.; Parameswaran, P.; Crowell, M.D.; Wing, R.; Rittmann, B.E.; et al. Human gut microbiota in obesity and after gastric bypass. *Proc. Natl. Acad. Sci. USA* **2009**, *106*, 2365–2370. [CrossRef] [PubMed]

97. Furet, J.P.; Kong, L.C.; Tap, J.; Poitou, C.; Basdevant, A.; Bouillot, J.L.; Mariat, D.; Corthier, G.; Dore, J.; Henegar, C.; et al. Differential adaptation of human gut microbiota to bariatric surgery-induced weight loss: Links with metabolic and low-grade inflammation markers. *Diabetes* **2010**, *59*, 3049–3057. [CrossRef] [PubMed]
98. Jiao, A.R.; Diao, H.; Yu, B.; He, J.; Yu, J.; Zheng, P.; Huang, Z.Q.; Luo, Y.H.; Luo, J.Q.; Mao, X.B.; et al. Oral administration of short chain fatty acids could attenuate fat deposition of pigs. *PLoS ONE* **2018**, *13*, e0196867. [CrossRef] [PubMed]
99. Canfora, E.E.; Van Der Beek, C.M.; Jocken, J.W.E.; Goossens, G.H.; Holst, J.J.; Olde Damink, S.W.M.; Lenaerts, K.; Dejong, C.H.C.; Blaak, E.E. Colonic infusions of short-chain fatty acid mixtures promote energy metabolism in overweight/obese men: A randomized crossover trial. *Sci. Rep.* **2017**, *7*, 2360. [CrossRef]
100. Chambers, E.S.; Preston, T.; Frost, G.; Morrison, D.J. Role of gut microbiota-generated short-chain fatty acids in metabolic and cardiovascular health. *Curr. Nutr. Rep.* **2018**, *7*, 198–206. [CrossRef]
101. Fernandes, J.; Su, W.; Rahat-Rozenbloom, S.; Wolever, T.M.; Comelli, E.M. Adiposity, gut microbiota and faecal short chain fatty acids are linked in adult humans. *Nutr. Diabetes* **2014**, *4*, e121. [CrossRef] [PubMed]
102. Menni, C.; Jackson, M.A.; Pallister, T.; Steves, C.J.; Spector, T.D.; Valdes, A.M. Gut microbiome diversity and high-fibre intake are related to lower long-term weight gain. *Int. J. Obes.* **2017**, *41*, 1099–1105. [CrossRef]
103. Li, Z.; Yi, C.X.; Katiraei, S.; Kooijman, S.; Zhou, E.; Chung, C.K.; Gao, Y.; Van Den Heuvel, J.K.; Meijer, O.C.; Berbee, J.F.P.; et al. Butyrate reduces appetite and activates brown adipose tissue via the gut-brain neural circuit. *Gut* **2018**, *67*, 1269–1279. [CrossRef] [PubMed]
104. Roshanravan, N.; Mahdavi, R.; Alizadeh, E.; Jafarabadi, M.A.; Hedayati, M.; Ghavami, A.; Alipour, S.; Alamdari, N.M.; Barati, M.; Ostadrahimi, A. Effect of butyrate and inulin supplementation on glycemic status, lipid profile and glucagon-like peptide 1 level in patients with type 2 diabetes: A randomized double-blind, placebo-controlled trial. *Horm. Metab. Res.* **2017**, *49*, 886–891. [CrossRef] [PubMed]
105. Sen, T.; Cawthon, C.R.; Ihde, B.T.; Hajnal, A.; DiLorenzo, P.M.; De La Serre, C.B.; Czaja, K. Diet-driven microbiota dysbiosis is associated with vagal remodeling and obesity. *Physiol. Behav.* **2017**, *173*, 305–317. [CrossRef] [PubMed]
106. Cani, P.D.; Amar, J.; Iglesias, M.A.; Poggi, M.; Knauf, C.; Bastelica, D.; Neyrinck, A.M.; Fava, F.; Tuohy, K.M.; Chabo, C.; et al. Metabolic endotoxemia initiates obesity and insulin resistance. *Diabetes* **2007**, *56*, 1761–1772. [CrossRef] [PubMed]
107. Boutagy, N.E.; McMillan, R.P.; Frisard, M.I.; Hulver, M.W. Metabolic endotoxemia with obesity: Is it real and is it relevant? *Biochimie* **2016**, *124*, 11–20. [CrossRef] [PubMed]
108. Pedersen, C.; Gallagher, E.; Horton, F.; Ellis, R.J.; Ijaz, U.Z.; Wu, H.; Jaiyeola, E.; Diribe, O.; Duparc, T.; Cani, P.D.; et al. Host-microbiome interactions in human type 2 diabetes following prebiotic fibre (galacto-oligosaccharide) intake. *Br. J. Nutr.* **2016**, *116*, 1869–1877. [CrossRef]
109. Cox, A.J.; Zhang, P.; Bowden, D.W.; Devereaux, B.; Davoren, P.M.; Cripps, A.W.; West, N.P. Increased intestinal permeability as a risk factor for type 2 diabetes. *Diabetes Metab.* **2017**, *43*, 163–166. [CrossRef]
110. Colantonio, A.G.; Werner, S.L.; Brown, M. The effects of prebiotics and substances with prebiotic properties on metabolic and inflammatory biomarkers in individuals with type 2 diabetes mellitus: A systematic review. *J. Acad. Nutr. Diet.* **2019**, in press. [CrossRef]
111. Dahl, W.J.; Agro, N.C.; Eliasson, A.M.; Mialki, K.L.; Olivera, J.D.; Rusch, C.T.; Young, C.N. Health benefits of fiber fermentation. *J. Am. Coll. Nutr.* **2017**, *36*, 127–136. [CrossRef] [PubMed]
112. Weickert, M.O.; Pfeiffer, A.F.H. Impact of dietary fiber consumption on insulin resistance and the prevention of type 2 diabetes. *J. Nutr.* **2018**, *148*, 7–12. [CrossRef] [PubMed]
113. Reynolds, A.; Mann, J.; Cummings, J.; Winter, N.; Mete, E.; Te Morenga, L. Carbohydrate quality and human health: A series of systematic reviews and meta-analyses. *Lancet* **2019**, *393*, 434–445. [CrossRef]
114. McKenzie, Y.A.; Bowyer, R.K.; Leach, H.; Gulia, P.; Horobin, J.; O'Sullivan, N.A.; Pettitt, C.; Reeves, L.B.; Seamark, L.; Williams, M.; et al. British Dietetic Association systematic review and evidence-based practice guidelines for the dietary management of irritable bowel syndrome in adults (2016 update). *J. Hum. Nutr. Diet.* **2016**, *29*, 549–575. [CrossRef] [PubMed]
115. McRorie, J.W., Jr. Evidence-based approach to fiber supplements and clinically meaningful health benefits, Part 1: What to look for and how to recommend an effective fiber therapy. *Nutr. Today* **2015**, *50*, 82–89. [CrossRef] [PubMed]

116. Lambeau, K.V.; McRorie, J.W., Jr. Fiber supplements and clinically proven health benefits: How to recognize and recommend an effective fiber therapy. *J. Am. Assoc. Nurse Pract.* **2017**, *29*, 216–223. [CrossRef] [PubMed]
117. Jakobsdottir, G.; Xu, J.; Molin, G.; Ahrne, S.; Nyman, M. High-fat diet reduces the formation of butyrate, but increases succinate, inflammation, liver fat and cholesterol in rats, while dietary fibre counteracts these effects. *PLoS ONE* **2013**, *8*, e80476. [CrossRef]
118. Arora, T.; Rudenko, O.; Egerod, K.L.; Husted, A.S.; Kovatcheva-Datchary, P.; Akrami, R.; Kristensen, M.; Schwartz, T.W.; Backhed, F. Microbial fermentation of flaxseed fibers modulates the transcriptome of GPR41-expressing enteroendocrine cells and protects mice against diet-induced obesity. *Am. J. Physiol. Endocrinol. Metab.* **2019**, *316*, E453–E463. [CrossRef]
119. Simpson, H.L.; Campbell, B.J. Review article: Dietary fibre-microbiota interactions. *Aliment. Pharmacol. Ther.* **2015**, *42*, 158–179. [CrossRef]
120. Zhernakova, A.; Kurilshikov, A.; Bonder, M.J.; Tigchelaar, E.F.; Schirmer, M.; Vatanen, T.; Mujagic, Z.; Vila, A.V.; Falony, G.; Vieira-Silva, S.; et al. Population-based metagenomics analysis reveals markers for gut microbiome composition and diversity. *Science* **2016**, *352*, 565–569. [CrossRef]
121. Trompette, A.; Gollwitzer, E.S.; Yadava, K.; Sichelstiel, A.K.; Sprenger, N.; Ngom-Bru, C.; Blanchard, C.; Junt, T.; Nicod, L.P.; Harris, N.L.; et al. Gut microbiota metabolism of dietary fiber influences allergic airway disease and hematopoiesis. *Nat. Med.* **2014**, *20*, 159–166. [CrossRef] [PubMed]
122. Suez, J.; Korem, T.; Zeevi, D.; Zilberman-Schapira, G.; Thaiss, C.A.; Maza, O.; Israeli, D.; Zmora, N.; Gilad, S.; Weinberger, A.; et al. Artificial sweeteners induce glucose intolerance by altering the gut microbiota. *Nature* **2014**, *514*, 181–186. [CrossRef] [PubMed]
123. Zeevi, D.; Korem, T.; Zmora, N.; Israeli, D.; Rothschild, D.; Weinberger, A.; Ben-Yacov, O.; Lador, D.; Avnit-Sagi, T.; Lotan-Pompan, M.; et al. Personalized nutrition by prediction of glycemic responses. *Cell* **2015**, *163*, 1079–1094. [CrossRef] [PubMed]
124. Mendes-Soares, H.; Raveh-Sadka, T.; Azulay, S.; Ben-Shlomo, Y.; Cohen, Y.; Ofek, T.; Stevens, J.; Bachrach, D.; Kashyap, P.; Segal, L.; et al. Model of personalized postprandial glycemic response to food developed for an Israeli cohort predicts responses in Midwestern American individuals. *Am. J. Clin. Nutr.* **2019**, *110*, 63–75. [CrossRef] [PubMed]
125. Bashiardes, S.; Godneva, A.; Elinav, E.; Segal, E. Towards utilization of the human genome and microbiome for personalized nutrition. *Curr. Opin. Biotechnol.* **2018**, *51*, 57–63. [CrossRef] [PubMed]
126. Qin, J.; Li, Y.; Cai, Z.; Li, S.; Zhu, J.; Zhang, F.; Liang, S.; Zhang, W.; Guan, Y.; Shen, D.; et al. A metagenome-wide association study of gut microbiota in type 2 diabetes. *Nature* **2012**, *490*, 55–60. [CrossRef] [PubMed]
127. Wu, H.; Esteve, E.; Tremaroli, V.; Khan, M.T.; Caesar, R.; Manneras-Holm, L.; Stahlman, M.; Olsson, L.M.; Serino, M.; Planas-Felix, M.; et al. Metformin alters the gut microbiome of individuals with treatment-naive type 2 diabetes, contributing to the therapeutic effects of the drug. *Nat. Med.* **2017**, *23*, 850–858. [CrossRef]
128. Bauer, P.V.; Duca, F.A.; Waise, T.M.Z.; Rasmussen, B.A.; Abraham, M.A.; Dranse, H.J.; Puri, A.; O'Brien, C.A.; Lam, T.K.T. Metformin alters upper small intestinal microbiota that impact a glucose-SGLT1-sensing glucoregulatory pathway. *Cell Metab.* **2018**, *27*, 101–117. [CrossRef]
129. De La Cuesta-Zuluaga, J.; Mueller, N.T.; Corrales-Agudelo, V.; Velasquez-Mejia, E.P.; Carmona, J.A.; Abad, J.M.; Escobar, J.S. Metformin Is associated with higher relative abundance of mucin-degrading *Akkermansia muciniphila* and several short-chain fatty acid-producing microbiota in the gut. *Diabetes Care* **2017**, *40*, 54–62. [CrossRef]
130. Bryrup, T.; Thomsen, C.W.; Kern, T.; Allin, K.H.; Brandslund, I.; Jorgensen, N.R.; Vestergaard, H.; Hansen, T.; Hansen, T.H.; Pedersen, O.; et al. Metformin-induced changes of the gut microbiota in healthy young men: Results of a non-blinded, one-armed intervention study. *Diabetologia* **2019**, *62*, 1024–1035. [CrossRef]
131. Baxter, N.T.; Lesniak, N.A.; Sinani, H.; Schloss, P.D.; Koropatkin, N.M. The glucoamylase inhibitor acarbose has a diet-dependent and reversible effect on the murine gut microbiome. *mSphere* **2019**, *4*, e00528-18. [CrossRef] [PubMed]
132. Wang, Z.; Klipfell, E.; Bennett, B.J.; Koeth, R.; Levison, B.S.; Dugar, B.; Feldstein, A.E.; Britt, E.B.; Fu, X.; Chung, Y.M.; et al. Gut flora metabolism of phosphatidylcholine promotes cardiovascular disease. *Nature* **2011**, *472*, 57–63. [CrossRef] [PubMed]

133. Gregory, J.C.; Buffa, J.A.; Org, E.; Wang, Z.; Levison, B.S.; Zhu, W.; Wagner, M.A.; Bennett, B.J.; Li, L.; DiDonato, J.A.; et al. Transmission of atherosclerosis susceptibility with gut microbial transplantation. *J. Biol. Chem.* **2015**, *290*, 5647–5660. [CrossRef] [PubMed]
134. Tang, W.H.; Wang, Z.; Levison, B.S.; Koeth, R.A.; Britt, E.B.; Fu, X.; Wu, Y.; Hazen, S.L. Intestinal microbial metabolism of phosphatidylcholine and cardiovascular risk. *N. Engl. J. Med.* **2013**, *368*, 1575–1584. [CrossRef] [PubMed]
135. Senthong, V.; Wang, Z.; Li, X.S.; Fan, Y.; Wu, Y.; Tang, W.H.; Hazen, S.L. Intestinal microbiota-generated metabolite trimethylamine-N-oxide and 5-year mortality risk in stable coronary artery disease: The contributory role of intestinal microbiota in a COURAGE-like patient cohort. *J. Am. Heart Assoc.* **2016**, *5*, e002816. [CrossRef] [PubMed]
136. Petersen, K.S.; Flock, M.R.; Richter, C.K.; Mukherjea, R.; Slavin, J.L.; Kris-Etherton, P.M. Healthy dietary patterns for preventing cardiometabolic disease: The role of plant-based foods and animal products. *Curr. Dev. Nutr.* **2017**, *1*, e001289. [CrossRef] [PubMed]
137. Soliman, G.A. Dietary cholesterol and the lack of evidence in cardiovascular disease. *Nutrients* **2018**, *10*, 780. [CrossRef]
138. Zhong, V.W.; Van Horn, L.; Cornelis, M.C.; Wilkins, J.T.; Ning, H.; Carnethon, M.R.; Greenland, P.; Mentz, R.J.; Tucker, K.L.; Zhao, L.; et al. Associations of dietary cholesterol or egg consumption with incident cardiovascular disease and mortality. *JAMA* **2019**, *321*, 1081–1095. [CrossRef]
139. Wang, Z.; Roberts, A.B.; Buffa, J.A.; Levison, B.S.; Zhu, W.; Org, E.; Gu, X.; Huang, Y.; Zamanian-Daryoush, M.; Culley, M.K.; et al. Non-lethal inhibition of gut microbial trimethylamine production for the treatment of atherosclerosis. *Cell* **2015**, *163*, 1585–1595. [CrossRef]
140. Chen, K.; Zheng, X.; Feng, M.; Li, D.; Zhang, H. Gut microbiota-dependent metabolite trimethylamine N-oxide contributes to cardiac dysfunction in Western diet-induced obese mice. *Front. Physiol.* **2017**, *8*, 139. [CrossRef]
141. Rath, S.; Heidrich, B.; Pieper, D.H.; Vital, M. Uncovering the trimethylamine-producing bacteria of the human gut microbiota. *Microbiome* **2017**, *5*, 54. [CrossRef] [PubMed]
142. Koeth, R.A.; Wang, Z.; Levison, B.S.; Buffa, J.A.; Org, E.; Sheehy, B.T.; Britt, E.B.; Fu, X.; Wu, Y.; Li, L.; et al. Intestinal microbiota metabolism of L-carnitine, a nutrient in red meat, promotes atherosclerosis. *Nat. Med.* **2013**, *19*, 576–585. [CrossRef] [PubMed]
143. Koeth, R.A.; Lam-Galvez, B.R.; Kirsop, J.; Wang, Z.; Levison, B.S.; Gu, X.; Copeland, M.F.; Bartlett, D.; Cody, D.B.; Dai, H.J.; et al. L-Carnitine in omnivorous diets induces an atherogenic gut microbial pathway in humans. *J. Clin. Investig.* **2019**, *129*, 373–387. [CrossRef] [PubMed]
144. Alshahrani, S.M.; Fraser, G.E.; Sabate, J.; Knutsen, R.; Shavlik, D.; Mashchak, A.; Lloren, J.I.; Orlich, M.J. Red and processed meat and mortality in a low meat intake population. *Nutrients* **2019**, *11*, 622. [CrossRef] [PubMed]
145. Bellavia, A.; Stilling, F.; Wolk, A. High red meat intake and all-cause cardiovascular and cancer mortality: Is the risk modified by fruit and vegetable intake? *Am. J. Clin. Nutr.* **2016**, *104*, 1137–1143. [CrossRef] [PubMed]
146. Micha, R.; Penalvo, J.L.; Cudhea, F.; Imamura, F.; Rehm, C.D.; Mozaffarian, D. Association between dietary factors and mortality from heart disease, stroke, and type 2 diabetes in the United States. *JAMA* **2017**, *317*, 912–924. [CrossRef] [PubMed]
147. Micha, R.; Wallace, S.K.; Mozaffarian, D. Red and processed meat consumption and risk of incident coronary heart disease, stroke, and diabetes mellitus: A systematic review and meta-analysis. *Circulation* **2010**, *121*, 2271–2283. [CrossRef]
148. Guasch-Ferre, M.; Satija, A.; Blondin, S.A.; Janiszewski, M.; Emlen, E.; O'Connor, L.E.; Campbell, W.W.; Hu, F.B.; Willett, W.C.; Stampfer, M.J. Meta-analysis of randomized controlled trials of red meat consumption in comparison with various comparison diets on cardiovascular risk factors. *Circulation* **2019**, *139*, 1828–1845. [CrossRef]
149. Hills, R.D., Jr.; Erpenbeck, E. Guide to popular diets, food choices, and their health outcome. *Health Care Curr. Rev.* **2018**, *6*, 223. [CrossRef]
150. Jin, Q.; Black, A.; Kales, S.N.; Vattem, D.; Ruiz-Canela, M.; Sotos-Prieto, M. Metabolomics and microbiomes as potential tools to evaluate the effects of the Mediterranean diet. *Nutrients* **2019**, *11*, 207. [CrossRef]

151. Lara, K.M.; Levitan, E.B.; Gutierrez, O.M.; Shikany, J.M.; Safford, M.M.; Judd, S.E.; Rosenson, R.S. Dietary patterns and incident heart failure in U.S. adults without known coronary disease. *J. Am. Coll. Cardiol.* **2019**, *73*, 2036–2045. [CrossRef] [PubMed]
152. Grosso, G.; Marventano, S.; Yang, J.; Micek, A.; Pajak, A.; Scalfi, L.; Galvano, F.; Kales, S.N. A comprehensive meta-analysis on evidence of Mediterranean diet and cardiovascular disease: Are individual components equal? *Crit. Rev. Food Sci. Nutr.* **2017**, *57*, 3218–3232. [CrossRef] [PubMed]
153. De Filippis, F.; Pellegrini, N.; Vannini, L.; Jeffery, I.B.; La Storia, A.; Laghi, L.; Serrazanetti, D.I.; Di Cagno, R.; Ferrocino, I.; Lazzi, C.; et al. High-level adherence to a Mediterranean diet beneficially impacts the gut microbiota and associated metabolome. *Gut* **2016**, *65*, 1812–1821. [CrossRef] [PubMed]
154. Alexander, D.D.; Weed, D.L.; Miller, P.E.; Mohamed, M.A. Red meat and colorectal cancer: A quantitative update on the state of the epidemiologic science. *J. Am. Coll. Nutr.* **2015**, *34*, 521–543. [CrossRef] [PubMed]
155. Schwingshackl, L.; Schwedhelm, C.; Galbete, C.; Hoffmann, G. Adherence to Mediterranean Diet and risk of cancer: An updated systematic review and meta-analysis. *Nutrients* **2017**, *9*, 1063. [CrossRef] [PubMed]
156. Thomas, A.M.; Manghi, P.; Asnicar, F.; Pasolli, E.; Armanini, F.; Zolfo, M.; Beghini, F.; Manara, S.; Karcher, N.; Pozzi, C.; et al. Metagenomic analysis of colorectal cancer datasets identifies cross-cohort microbial diagnostic signatures and a link with choline degradation. *Nat. Med.* **2019**, *25*, 667–678. [CrossRef]
157. Wirbel, J.; Pyl, P.T.; Kartal, E.; Zych, K.; Kashani, A.; Milanese, A.; Fleck, J.S.; Voigt, A.Y.; Palleja, A.; Ponnudurai, R.; et al. Meta-analysis of fecal metagenomes reveals global microbial signatures that are specific for colorectal cancer. *Nat. Med.* **2019**, *25*, 679–689. [CrossRef]
158. Youn, J.; Cho, E.; Lee, J.E. Association of choline and betaine levels with cancer incidence and survival: A meta-analysis. *Clin. Nutr.* **2019**, *38*, 100–109. [CrossRef]
159. Bishehsari, F.; Engen, P.A.; Preite, N.Z.; Tuncil, Y.E.; Naqib, A.; Shaikh, M.; Rossi, M.; Wilber, S.; Green, S.J.; Hamaker, B.R.; et al. Dietary fiber treatment corrects the composition of gut microbiota, promotes SCFA production, and suppresses colon carcinogenesis. *Genes* **2018**, *9*, 102. [CrossRef]
160. Song, Y.; Liu, M.; Yang, F.G.; Cui, L.H.; Lu, X.Y.; Chen, C. Dietary fibre and the risk of colorectal cancer: A case- control study. *Asian Pac. J. Cancer Prev.* **2015**, *16*, 3747–3752. [CrossRef]
161. Zhang, F.F.; Cudhea, F.; Shan, Z.; Michaud, D.S.; Imamura, F.; Eom, H.; Ruan, M.; Rehm, C.D.; Liu, J.; Du, M.; et al. Preventable cancer burden associated with poor diet in the United States. *JNCI Cancer Spectr.* **2019**, *3*, pkz034. [CrossRef]
162. Vonaesch, P.; Anderson, M.; Sansonetti, P.J. Pathogens, microbiome and the host: Emergence of the ecological Koch's postulates. *FEMS Microbiol. Rev.* **2018**, *42*, 273–292. [CrossRef]
163. Tomkovich, S.; Dejea, C.M.; Winglee, K.; Drewes, J.L.; Chung, L.; Housseau, F.; Pope, J.L.; Gauthier, J.; Sun, X.; Muhlbauer, M.; et al. Human colon mucosal biofilms from healthy or colon cancer hosts are carcinogenic. *J. Clin. Investig.* **2019**, *130*, 1699–1712. [CrossRef] [PubMed]
164. Hill, C.; Guarner, F.; Reid, G.; Gibson, G.R.; Merenstein, D.J.; Pot, B.; Morelli, L.; Canani, R.B.; Flint, H.J.; Salminen, S.; et al. Expert consensus document. The International Scientific Association for Probiotics and Prebiotics consensus statement on the scope and appropriate use of the term probiotic. *Nat. Rev. Gastroenterol. Hepatol.* **2014**, *11*, 506–514. [CrossRef] [PubMed]
165. Shen, J.; Zuo, Z.X.; Mao, A.P. Effect of probiotics on inducing remission and maintaining therapy in ulcerative colitis, Crohn's disease, and pouchitis: Meta-analysis of randomized controlled trials. *Inflamm. Bowel Dis.* **2014**, *20*, 21–35. [CrossRef]
166. Wilkins, T.; Sequoia, J. Probiotics for gastrointestinal conditions: A summary of the evidence. *Am. Fam. Physician* **2017**, *96*, 170–178. [PubMed]
167. Tojo, R.; Suarez, A.; Clemente, M.G.; De Los Reyes-Gavilan, C.G.; Margolles, A.; Gueimonde, M.; Ruas-Madiedo, P. Intestinal microbiota in health and disease: Role of bifidobacteria in gut homeostasis. *World J. Gastroenterol.* **2014**, *20*, 15163–15176. [CrossRef]
168. Sanchez, B.; Delgado, S.; Blanco-Miguez, A.; Lourenco, A.; Gueimonde, M.; Margolles, A. Probiotics, gut microbiota, and their influence on host health and disease. *Mol. Nutr. Food Res.* **2017**, *61*, 1600240. [CrossRef]
169. Allen, S.J.; Martinez, E.G.; Gregorio, G.V.; Dans, L.F. Probiotics for treating acute infectious diarrhoea. *Cochrane Database Syst. Rev.* **2010**, *11*, CD003048. [CrossRef] [PubMed]
170. Goldenberg, J.Z.; Yap, C.; Lytvyn, L.; Lo, C.K.; Beardsley, J.; Mertz, D.; Johnston, B.C. Probiotics for the prevention of *Clostridium difficile*-associated diarrhea in adults and children. *Cochrane Database Syst. Rev.* **2017**, *12*, CD006095. [CrossRef]

171. Goldenberg, J.Z.; Mertz, D.; Johnston, B.C. Probiotics to prevent *Clostridium difficile* infection in patients receiving antibiotics. *JAMA* **2018**, *320*, 499–500. [CrossRef] [PubMed]
172. Fijan, S. Microorganisms with claimed probiotic properties: An overview of recent literature. *Int. J. Env. Res. Public Health* **2014**, *11*, 4745–4767. [CrossRef] [PubMed]
173. Dimidi, E.; Christodoulides, S.; Scott, S.M.; Whelan, K. Mechanisms of action of probiotics and the gastrointestinal microbiota on gut motility and constipation. *Adv. Nutr.* **2017**, *8*, 484–494. [CrossRef] [PubMed]
174. Plaza-Diaz, J.; Ruiz-Ojeda, F.J.; Gil-Campos, M.; Gil, A. Mechanisms of action of probiotics. *Adv. Nutr.* **2019**, *10*, S49–S66. [CrossRef] [PubMed]
175. LeBlanc, J.G.; Chain, F.; Martin, R.; Bermudez-Humaran, L.G.; Courau, S.; Langella, P. Beneficial effects on host energy metabolism of short-chain fatty acids and vitamins produced by commensal and probiotic bacteria. *Microb. Cell Fact.* **2017**, *16*, 79. [CrossRef] [PubMed]
176. Ford, A.C.; Quigley, E.M.; Lacy, B.E.; Lembo, A.J.; Saito, Y.A.; Schiller, L.R.; Soffer, E.E.; Spiegel, B.M.; Moayyedi, P. Efficacy of prebiotics, probiotics, and synbiotics in irritable bowel syndrome and chronic idiopathic constipation: Systematic review and meta-analysis. *Am. J. Gastroenterol.* **2014**, *109*, 1547–1561. [CrossRef] [PubMed]
177. Rajilic-Stojanovic, M.; Biagi, E.; Heilig, H.G.; Kajander, K.; Kekkonen, R.A.; Tims, S.; De Vos, W.M. Global and deep molecular analysis of microbiota signatures in fecal samples from patients with irritable bowel syndrome. *Gastroenterology* **2011**, *141*, 1792–1801. [CrossRef] [PubMed]
178. O'Callaghan, A.; Van Sinderen, D. Bifidobacteria and their role as members of the human gut microbiota. *Front. Microbiol.* **2016**, *7*, 925. [CrossRef]
179. Giamarellos-Bourboulis, E.J.; Pyleris, E.; Barbatzas, C.; Pistiki, A.; Pimentel, M. Small intestinal bacterial overgrowth is associated with irritable bowel syndrome and is independent of proton pump inhibitor usage. *BMC Gastroenterol.* **2016**, *16*, 67. [CrossRef]
180. Saffouri, G.B.; Shields-Cutler, R.R.; Chen, J.; Yang, Y.; Lekatz, H.R.; Hale, V.L.; Cho, J.M.; Battaglioli, E.J.; Bhattarai, Y.; Thompson, K.J.; et al. Small intestinal microbial dysbiosis underlies symptoms associated with functional gastrointestinal disorders. *Nat. Commun.* **2019**, *10*, 2012. [CrossRef]
181. De Simone, C. The unregulated probiotic market. *Clin. Gastroenterol. Hepatol.* **2018**, *17*, 809–817. [CrossRef] [PubMed]
182. Campana, R.; Van Hemert, S.; Baffone, W. Strain-specific probiotic properties of lactic acid bacteria and their interference with human intestinal pathogens invasion. *Gut Pathog.* **2017**, *9*, 12. [CrossRef] [PubMed]
183. Wang, L.; Hu, L.; Xu, Q.; Yin, B.; Fang, D.; Wang, G.; Zhao, J.; Zhang, H.; Chen, W. *Bifidobacterium adolescentis* exerts strain-specific effects on constipation induced by loperamide in BALB/c mice. *Int. J. Mol. Sci.* **2017**, *18*, 318. [CrossRef] [PubMed]
184. Conway, T.; Cohen, P.S. Commensal and pathogenic *Escherichia coli* metabolism in the gut. *Microbiol. Spectr.* **2015**, *3*. [CrossRef] [PubMed]
185. Goswami, K.; Chen, C.; Xiaoli, L.; Eaton, K.A.; Dudley, E.G. Coculture of *Escherichia coli* O157:H7 with a nonpathogenic E. coli strain increases toxin production and virulence in a germfree mouse model. *Infect. Immun.* **2015**, *83*, 4185–4193. [CrossRef] [PubMed]
186. Crook, N.; Ferreiro, A.; Gasparrini, A.J.; Pesesky, M.W.; Gibson, M.K.; Wang, B.; Sun, X.; Condiotte, Z.; Dobrowolski, S.; Peterson, D.; et al. Adaptive strategies of the candidate probiotic *E. coli* Nissle in the mammalian gut. *Cell Host Microbe* **2019**, *25*, 499–512. [CrossRef] [PubMed]
187. Lerner, A.; Matthias, T.; Aminov, R. Potential effects of horizontal gene exchange in the human gut. *Front. Immunol.* **2017**, *8*, 1630. [CrossRef] [PubMed]
188. Trinchieri, V.; Laghi, L.; Vitali, B.; Parolin, C.; Giusti, I.; Capobianco, D.; Mastromarino, P.; De Simone, C. Efficacy and safety of a multistrain probiotic formulation depends from manufacturing. *Front. Immunol.* **2017**, *8*, 1474. [CrossRef]
189. Hod, K.; Dekel, R.; Aviv Cohen, N.; Sperber, A.; Ron, Y.; Boaz, M.; Berliner, S.; Maharshak, N. The effect of a multispecies probiotic on microbiota composition in a clinical trial of patients with diarrhea-predominant irritable bowel syndrome. *Neurogastroenterol. Motil.* **2018**, *30*, e13456. [CrossRef]
190. Lee, S.H.; Joo, N.S.; Kim, K.M.; Kim, K.N. The therapeutic effect of a multistrain probiotic on diarrhea-predominant irritable bowel syndrome: A pilot study. *Gastroenterol. Res. Pract.* **2018**, *2018*, 8791916. [CrossRef]

191. Plaza-Diaz, J.; Ruiz-Ojeda, F.J.; Vilchez-Padial, L.M.; Gil, A. Evidence of the anti-inflammatory effects of probiotics and synbiotics in intestinal chronic diseases. *Nutrients* **2017**, *9*, 555. [CrossRef]
192. Floch, M.H.; Walker, W.A.; Sanders, M.E.; Nieuwdorp, M.; Kim, A.S.; Brenner, D.A.; Qamar, A.A.; Miloh, T.A.; Guarino, A.; Guslandi, M.; et al. Recommendations for probiotic use–2015 Update: Proceedings and consensus opinion. *J. Clin. Gastroenterol.* **2015**, *49*, S69–S73. [CrossRef] [PubMed]
193. Markowiak, P.; Slizewska, K. Effects of probiotics, prebiotics, and synbiotics on human health. *Nutrients* **2017**, *9*, 1021. [CrossRef] [PubMed]
194. Hughes, K.R.; Harnisch, L.C.; Alcon-Giner, C.; Mitra, S.; Wright, C.J.; Ketskemety, J.; Van Sinderen, D.; Watson, A.J.; Hall, L.J. *Bifidobacterium breve* reduces apoptotic epithelial cell shedding in an exopolysaccharide and MyD88-dependent manner. *Open Biol.* **2017**, *7*, 160155. [CrossRef] [PubMed]
195. Lam, Y.Y.; Ha, C.W.; Campbell, C.R.; Mitchell, A.J.; Dinudom, A.; Oscarsson, J.; Cook, D.I.; Hunt, N.H.; Caterson, I.D.; Holmes, A.J.; et al. Increased gut permeability and microbiota change associate with mesenteric fat inflammation and metabolic dysfunction in diet-induced obese mice. *PLoS ONE* **2012**, *7*, e34233. [CrossRef] [PubMed]
196. Wang, Y.; Kirpich, I.; Liu, Y.; Ma, Z.; Barve, S.; McClain, C.J.; Feng, W. *Lactobacillus rhamnosus* GG treatment potentiates intestinal hypoxia-inducible factor, promotes intestinal integrity and ameliorates alcohol-induced liver injury. *Am. J. Pathol.* **2011**, *179*, 2866–2875. [CrossRef] [PubMed]
197. Reid, G. Probiotics to prevent the need for, and augment the use of, antibiotics. *Can. J. Infect. Dis. Med. Microbiol.* **2006**, *17*, 219–295. [CrossRef]
198. Sherman, P.M.; Johnson-Henry, K.C.; Yeung, H.P.; Ngo, P.S.; Goulet, J.; Tompkins, T.A. Probiotics reduce enterohemorrhagic *Escherichia coli* O157:H7- and enteropathogenic *E. coli* O127:H6-induced changes in polarized T84 epithelial cell monolayers by reducing bacterial adhesion and cytoskeletal rearrangements. *Infect. Immun.* **2005**, *73*, 5183–5188. [CrossRef]
199. Ferrario, C.; Taverniti, V.; Milani, C.; Fiore, W.; Laureati, M.; De Noni, I.; Stuknyte, M.; Chouaia, B.; Riso, P.; Guglielmetti, S. Modulation of fecal Clostridiales bacteria and butyrate by probiotic intervention with *Lactobacillus paracasei* DG varies among healthy adults. *J. Nutr.* **2014**, *144*, 1787–1796. [CrossRef]
200. Barbara, G.; Cremon, C.; Azpiroz, F. Probiotics in irritable bowel syndrome: Where are we? *Neurogastroenterol. Motil.* **2018**, *30*, e13513. [CrossRef]
201. Baxter, N.T.; Schmidt, A.W.; Venkataraman, A.; Kim, K.S.; Waldron, C.; Schmidt, T.M. Dynamics of human gut microbiota and short-chain fatty acids in response to dietary interventions with three fermentable fibers. *mBio* **2019**, *10*, e02566-18. [CrossRef] [PubMed]
202. Bordalo Tonucci, L.; Dos Santos, K.M.; De Luces Fortes Ferreira, C.L.; Ribeiro, S.M.; De Oliveira, L.L.; Martino, H.S. Gut microbiota and probiotics: Focus on diabetes mellitus. *Crit. Rev. Food Sci. Nutr.* **2017**, *57*, 2296–2309. [CrossRef] [PubMed]
203. Sanchez, M.; Darimont, C.; Drapeau, V.; Emady-Azar, S.; Lepage, M.; Rezzonico, E.; Ngom-Bru, C.; Berger, B.; Philippe, L.; Ammon-Zuffrey, C.; et al. Effect of *Lactobacillus rhamnosus* CGMCC1.3724 supplementation on weight loss and maintenance in obese men and women. *Br. J. Nutr.* **2014**, *111*, 1507–1519. [CrossRef] [PubMed]
204. Sabico, S.; Al-Mashharawi, A.; Al-Daghri, N.M.; Wani, K.; Amer, O.E.; Hussain, D.S.; Ahmed Ansari, M.G.; Masoud, M.S.; Alokail, M.S.; McTernan, P.G. Effects of a 6-month multi-strain probiotics supplementation in endotoxemic, inflammatory and cardiometabolic status of T2DM patients: A randomized, double-blind, placebo-controlled trial. *Clin. Nutr.* **2018**, *38*, 1561–1569. [CrossRef] [PubMed]
205. Abildgaard, A.; Elfving, B.; Hokland, M.; Wegener, G.; Lund, S. Probiotic treatment reduces depressive-like behaviour in rats independently of diet. *Psychoneuroendocrinology* **2017**, *79*, 40–48. [CrossRef]
206. Kadooka, Y.; Sato, M.; Ogawa, A.; Miyoshi, M.; Uenishi, H.; Ogawa, H.; Ikuyama, K.; Kagoshima, M.; Tsuchida, T. Effect of *Lactobacillus gasseri* SBT2055 in fermented milk on abdominal adiposity in adults in a randomised controlled trial. *Br. J. Nutr.* **2013**, *110*, 1696–1703. [CrossRef] [PubMed]
207. Kim, J.; Yun, J.M.; Kim, M.K.; Kwon, O.; Cho, B. *Lactobacillus gasseri* BNR17 supplementation reduces the visceral fat accumulation and waist circumference in obese adults: A randomized, double-blind, placebo-controlled trial. *J. Med. Food* **2018**, *21*, 454–461. [CrossRef]
208. Valcheva, R.; Dieleman, L.A. Prebiotics: Definition and protective mechanisms. *Best Pract. Res. Clin. Gastroenterol.* **2016**, *30*, 27–37. [CrossRef]

209. Nie, Y.; Lin, Q.; Luo, F. Effects of non-starch polysaccharides on inflammatory bowel disease. *Int. J. Mol. Sci.* **2017**, *18*, 1372. [CrossRef]
210. Ruiz-Ojeda, F.J.; Plaza-Diaz, J.; Saez-Lara, M.J.; Gil, A. Effects of sweeteners on the gut microbiota: A review of experimental studies and clinical trials. *Adv. Nutr.* **2019**, *10*, S31–S48. [CrossRef]
211. Leenen, C.H.; Dieleman, L.A. Inulin and oligofructose in chronic inflammatory bowel disease. *J. Nutr.* **2007**, *137*, 2572S–2575S. [CrossRef] [PubMed]
212. Musilova, S.; Rada, V.; Marounek, M.; Nevoral, J.; Duskova, D.; Bunesova, V.; Vlkova, E.; Zelenka, R. Prebiotic effects of a novel combination of galactooligosaccharides and maltodextrins. *J. Med. Food* **2015**, *18*, 685–689. [CrossRef]
213. Nicolucci, A.C.; Hume, M.P.; Martinez, I.; Mayengbam, S.; Walter, J.; Reimer, R.A. Prebiotics reduce body fat and alter intestinal microbiota in children who are overweight or with obesity. *Gastroenterology* **2017**, *153*, 711–722. [CrossRef] [PubMed]
214. Rios, J.L.; Bomhof, M.R.; Reimer, R.A.; Hart, D.A.; Collins, K.H.; Herzog, W. Protective effect of prebiotic and exercise intervention on knee health in a rat model of diet-induced obesity. *Sci. Rep.* **2019**, *9*, 3893. [CrossRef] [PubMed]
215. Chen, T.; Long, W.; Zhang, C.; Liu, S.; Zhao, L.; Hamaker, B.R. Fiber-utilizing capacity varies in *Prevotella*- versus *Bacteroides*-dominated gut microbiota. *Sci. Rep.* **2017**, *7*, 2594. [CrossRef] [PubMed]
216. Thomas, S.; Browne, H.; Mobasheri, A.; Rayman, M.P. What is the evidence for a role for diet and nutrition in osteoarthritis? *Rheumatology* **2018**, *57*, iv61–iv74. [CrossRef] [PubMed]
217. Watson, H.; Mitra, S.; Croden, F.C.; Taylor, M.; Wood, H.M.; Perry, S.L.; Spencer, J.A.; Quirke, P.; Toogood, G.J.; Lawton, C.L.; et al. A randomised trial of the effect of omega-3 polyunsaturated fatty acid supplements on the human intestinal microbiota. *Gut* **2018**, *67*, 1974–1983. [CrossRef] [PubMed]
218. Costantini, L.; Molinari, R.; Farinon, B.; Merendino, N. Impact of omega-3 fatty acids on the gut microbiota. *Int. J. Mol. Sci.* **2017**, *18*, 2645. [CrossRef]
219. Menni, C.; Zierer, J.; Pallister, T.; Jackson, M.A.; Long, T.; Mohney, R.P.; Steves, C.J.; Spector, T.D.; Valdes, A.M. Omega-3 fatty acids correlate with gut microbiome diversity and production of N-carbamylglutamate in middle aged and elderly women. *Sci. Rep.* **2017**, *7*, 11079. [CrossRef]
220. Kim, Y.; Je, Y.; Giovannucci, E. Coffee consumption and all-cause and cause-specific mortality: A meta-analysis by potential modifiers. *Eur. J. Epidemiol.* **2019**, *34*, 731–752. [CrossRef]
221. Falony, G.; Joossens, M.; Vieira-Silva, S.; Wang, J.; Darzi, Y.; Faust, K.; Kurilshikov, A.; Bonder, M.J.; Valles-Colomer, M.; Vandeputte, D.; et al. Population-level analysis of gut microbiome variation. *Science* **2016**, *352*, 560–564. [CrossRef] [PubMed]
222. Vandeputte, D.; Falony, G.; Vieira-Silva, S.; Tito, R.Y.; Joossens, M.; Raes, J. Stool consistency is strongly associated with gut microbiota richness and composition, enterotypes and bacterial growth rates. *Gut* **2016**, *65*, 57–62. [CrossRef] [PubMed]
223. Tropini, C.; Moss, E.L.; Merrill, B.D.; Ng, K.M.; Higginbottom, S.K.; Casavant, E.P.; Gonzalez, C.G.; Fremin, B.; Bouley, D.M.; Elias, J.E.; et al. Transient osmotic perturbation causes long-term alteration to the gut microbiota. *Cell* **2018**, *173*, 1742–1754. [CrossRef] [PubMed]
224. Lee, S.H.; Yun, Y.; Kim, S.J.; Lee, E.J.; Chang, Y.; Ryu, S.; Shin, H.; Kim, H.L.; Kim, H.N.; Lee, J.H. Association between cigarette smoking status and composition of gut microbiota: Population-based cross-sectional study. *J. Clin. Med.* **2018**, *7*, 282. [CrossRef] [PubMed]
225. Song, S.J.; Lauber, C.; Costello, E.K.; Lozupone, C.A.; Humphrey, G.; Berg-Lyons, D.; Caporaso, J.G.; Knights, D.; Clemente, J.C.; Nakielny, S.; et al. Cohabiting family members share microbiota with one another and with their dogs. *eLife* **2013**, *2*, e00458. [CrossRef] [PubMed]
226. Parajuli, A.; Gronroos, M.; Siter, N.; Puhakka, R.; Vari, H.K.; Roslund, M.I.; Jumpponen, A.; Nurminen, N.; Laitinen, O.H.; Hyoty, H.; et al. Urbanization reduces transfer of diverse environmental microbiota indoors. *Front. Microbiol.* **2018**, *9*, 84. [CrossRef] [PubMed]
227. Mahnert, A.; Moissl-Eichinger, C.; Zojer, M.; Bogumil, D.; Mizrahi, I.; Rattei, T.; Martinez, J.L.; Berg, G. Man-made microbial resistances in built environments. *Nat. Commun.* **2019**, *10*, 968. [CrossRef]
228. Fragiadakis, G.K.; Smits, S.A.; Sonnenburg, E.D.; Van Treuren, W.; Reid, G.; Knight, R.; Manjurano, A.; Changalucha, J.; Dominguez-Bello, M.G.; Leach, J.; et al. Links between environment, diet, and the hunter-gatherer microbiome. *Gut Microbes* **2019**, *10*, 216–227. [CrossRef]

229. Fu, J.; Bonder, M.J.; Cenit, M.C.; Tigchelaar, E.F.; Maatman, A.; Dekens, J.A.; Brandsma, E.; Marczynska, J.; Imhann, F.; Weersma, R.K.; et al. The gut microbiome contributes to a substantial proportion of the variation in blood lipids. *Circ. Res.* **2015**, *117*, 817–824. [CrossRef]
230. Elshazly, M.B.; Martin, S.S.; Blaha, M.J.; Joshi, P.H.; Toth, P.P.; McEvoy, J.W.; Al-Hijji, M.A.; Kulkarni, K.R.; Kwiterovich, P.O.; Blumenthal, R.S.; et al. Non-high-density lipoprotein cholesterol, guideline targets, and population percentiles for secondary prevention in 1.3 million adults: The VLDL-2 study (very large database of lipids). *J. Am. Coll. Cardiol.* **2013**, *62*, 1960–1965. [CrossRef]
231. Ravnskov, U.; De Lorgeril, M.; Diamond, D.M.; Hama, R.; Hamazaki, T.; Hammarskjold, B.; Hynes, N.; Kendrick, M.; Langsjoen, P.H.; Mascitelli, L.; et al. LDL-C does not cause cardiovascular disease: A comprehensive review of the current literature. *Expert Rev. Clin. Pharmacol.* **2018**, *11*, 959–970. [CrossRef] [PubMed]
232. Wu, T.T.; Zheng, Y.Y.; Yang, Y.N.; Li, X.M.; Ma, Y.T.; Xie, X. Age, sex, and cardiovascular risk attributable to lipoprotein cholesterol among chinese individuals with coronary artery disease: A case-control study. *Metab. Syndr. Relat. Disord.* **2019**, *17*, 223–231. [CrossRef] [PubMed]
233. Ference, B.A.; Ginsberg, H.N.; Graham, I.; Ray, K.K.; Packard, C.J.; Bruckert, E.; Hegele, R.A.; Krauss, R.M.; Raal, F.J.; Schunkert, H.; et al. Low-density lipoproteins cause atherosclerotic cardiovascular disease. 1. Evidence from genetic, epidemiologic, and clinical studies. A consensus statement from the European Atherosclerosis Society Consensus Panel. *Eur. Heart J.* **2017**, *38*, 2459–2472. [CrossRef] [PubMed]
234. Kastelein, J.J.; Van Der Steeg, W.A.; Holme, I.; Gaffney, M.; Cater, N.B.; Barter, P.; Deedwania, P.; Olsson, A.G.; Boekholdt, S.M.; Demicco, D.A.; et al. Lipids, apolipoproteins, and their ratios in relation to cardiovascular events with statin treatment. *Circulation* **2008**, *117*, 3002–3009. [CrossRef] [PubMed]
235. Meeusen, J.W.; Donato, L.J.; Jaffe, A.S. Should apolipoprotein B replace LDL cholesterol as therapeutic targets are lowered? *Curr. Opin. Lipidol.* **2016**, *27*, 359–366. [CrossRef] [PubMed]
236. Greene, C.M.; Waters, D.; Clark, R.M.; Contois, J.H.; Fernandez, M.L. Plasma LDL and HDL characteristics and carotenoid content are positively influenced by egg consumption in an elderly population. *Nutr. Metab.* **2006**, *3*, 6. [CrossRef] [PubMed]
237. Willeit, P.; Ridker, P.M.; Nestel, P.J.; Simes, J.; Tonkin, A.M.; Pedersen, T.R.; Schwartz, G.G.; Olsson, A.G.; Colhoun, H.M.; Kronenberg, F.; et al. Baseline and on-statin treatment lipoprotein(a) levels for prediction of cardiovascular events: Individual patient-data meta-analysis of statin outcome trials. *Lancet* **2018**, *392*, 1311–1320. [CrossRef]
238. Tsimikas, S.; Gordts, P.; Nora, C.; Yeang, C.; Witztum, J.L. Statin therapy increases lipoprotein(a) levels. *Eur. Heart J.* **2019**, in press. [CrossRef]
239. Choi, C.U.; Seo, H.S.; Lee, E.M.; Shin, S.Y.; Choi, U.J.; Na, J.O.; Lim, H.E.; Kim, J.W.; Kim, E.J.; Rha, S.W.; et al. Statins do not decrease small, dense low-density lipoprotein. *Tex. Heart Inst. J.* **2010**, *37*, 421–428.
240. Magge, S.; Lembo, A. Low-FODMAP diet for treatment of irritable bowel syndrome. *Gastroenterol. Hepatol.* **2012**, *8*, 739–745.
241. Skodje, G.I.; Sarna, V.K.; Minelle, I.H.; Rolfsen, K.L.; Muir, J.G.; Gibson, P.R.; Veierod, M.B.; Henriksen, C.; Lundin, K.E.A. Fructan, rather than gluten, induces symptoms in patients with self-reported non-Celiac gluten sensitivity. *Gastroenterology* **2018**, *154*, 529–539. [CrossRef] [PubMed]
242. Biesiekierski, J.R.; Peters, S.L.; Newnham, E.D.; Rosella, O.; Muir, J.G.; Gibson, P.R. No effects of gluten in patients with self-reported non-celiac gluten sensitivity after dietary reduction of fermentable, poorly absorbed, short-chain carbohydrates. *Gastroenterology* **2013**, *145*, 320–328. [CrossRef] [PubMed]
243. Fasano, A.; Sapone, A.; Zevallos, V.; Schuppan, D. Nonceliac gluten sensitivity. *Gastroenterology* **2015**, *148*, 1195–1204. [CrossRef] [PubMed]
244. Halmos, E.P.; Gibson, P.R. Controversies and reality of the FODMAP diet for patients with irritable bowel syndrome. *J. Gastroenterol. Hepatol.* **2019**, in press. [CrossRef] [PubMed]
245. Staudacher, H.M.; Ralph, F.S.E.; Irving, P.M.; Whelan, K.; Lomer, M.C.E. Nutrient intake, diet quality, and diet diversity in irritable bowel syndrome and the impact of the low FODMAP diet. *J. Acad. Nutr. Diet.* **2019**, in press. [CrossRef] [PubMed]
246. Claesson, M.J.; Jeffery, I.B.; Conde, S.; Power, S.E.; O'Connor, E.M.; Cusack, S.; Harris, H.M.; Coakley, M.; Lakshminarayanan, B.; O'Sullivan, O.; et al. Gut microbiota composition correlates with diet and health in the elderly. *Nature* **2012**, *488*, 178–184. [CrossRef] [PubMed]

247. Halmos, E.P.; Christophersen, C.T.; Bird, A.R.; Shepherd, S.J.; Gibson, P.R.; Muir, J.G. Diets that differ in their FODMAP content alter the colonic luminal microenvironment. *Gut* **2015**, *64*, 93–100. [CrossRef]
248. Dieterich, W.; Schuppan, D.; Schink, M.; Schwappacher, R.; Wirtz, S.; Agaimy, A.; Neurath, M.F.; Zopf, Y. Influence of low FODMAP and gluten-free diets on disease activity and intestinal microbiota in patients with non-celiac gluten sensitivity. *Clin. Nutr.* **2019**, *38*, 697–707. [CrossRef]
249. Kosinski, C.; Jornayvaz, F.R. Effects of ketogenic diets on cardiovascular risk factors: Evidence from animal and human studies. *Nutrients* **2017**, *9*, 517. [CrossRef]
250. Ebbeling, C.B.; Feldman, H.A.; Klein, G.L.; Wong, J.M.W.; Bielak, L.; Steltz, S.K.; Luoto, P.K.; Wolfe, R.R.; Wong, W.W.; Ludwig, D.S. Effects of a low carbohydrate diet on energy expenditure during weight loss maintenance: Randomized trial. *BMJ* **2018**, *363*, k4583. [CrossRef]
251. De Koning, L.; Fung, T.T.; Liao, X.; Chiuve, S.E.; Rimm, E.B.; Willett, W.C.; Spiegelman, D.; Hu, F.B. Low-carbohydrate diet scores and risk of type 2 diabetes in men. *Am. J. Clin. Nutr.* **2011**, *93*, 844–850. [CrossRef] [PubMed]
252. Seidelmann, S.B.; Claggett, B.; Cheng, S.; Henglin, M.; Shah, A.; Steffen, L.M.; Folsom, A.R.; Rimm, E.B.; Willett, W.C.; Solomon, S.D. Dietary carbohydrate intake and mortality: A prospective cohort study and meta-analysis. *Lancet Public Health* **2018**, *3*, e419–e428. [CrossRef]
253. Swidsinski, A.; Dorffel, Y.; Loening-Baucke, V.; Gille, C.; Goktas, O.; Reisshauer, A.; Neuhaus, J.; Weylandt, K.H.; Guschin, A.; Bock, M. Reduced mass and diversity of the colonic microbiome in patients with multiple sclerosis and their improvement with ketogenic diet. *Front. Microbiol.* **2017**, *8*, 1141. [CrossRef] [PubMed]
254. Zhang, Y.; Zhou, S.; Zhou, Y.; Yu, L.; Zhang, L.; Wang, Y. Altered gut microbiome composition in children with refractory epilepsy after ketogenic diet. *Epilepsy Res.* **2018**, *145*, 163–168. [CrossRef] [PubMed]
255. Kearns, C.E.; Apollonio, D.; Glantz, S.A. Sugar industry sponsorship of germ-free rodent studies linking sucrose to hyperlipidemia and cancer: An historical analysis of internal documents. *PLoS Biol.* **2017**, *15*, e2003460. [CrossRef] [PubMed]
256. Hoffmann, C.; Dollive, S.; Grunberg, S.; Chen, J.; Li, H.; Wu, G.D.; Lewis, J.D.; Bushman, F.D. Archaea and fungi of the human gut microbiome: Correlations with diet and bacterial residents. *PLoS ONE* **2013**, *8*, e66019. [CrossRef]
257. Samuel, B.S.; Gordon, J.I. A humanized gnotobiotic mouse model of host-archaeal-bacterial mutualism. *Proc. Natl. Acad. Sci. USA* **2006**, *103*, 10011–10016. [CrossRef]
258. McNeil, N.I. The contribution of the large intestine to energy supplies in man. *Am. J. Clin. Nutr.* **1984**, *39*, 338–342. [CrossRef]
259. Ghoshal, U.; Shukla, R.; Srivastava, D.; Ghoshal, U.C. Irritable bowel syndrome, particularly the constipation-predominant form, involves an increase in *Methanobrevibacter smithii*, which is associated with higher methane production. *Gut Liver* **2016**, *10*, 932–938. [CrossRef]
260. Hallen-Adams, H.E.; Suhr, M.J. Fungi in the healthy human gastrointestinal tract. *Virulence* **2017**, *8*, 352–358. [CrossRef]
261. Otasevic, S.; Momcilovic, S.; Petrovic, M.; Radulovic, O.; Stojanovic, N.M.; Arsic-Arsenijevic, V. The dietary modification and treatment of intestinal *Candida* overgrowth - a pilot study. *J. Mycol. Med.* **2018**, *28*, 623–627. [CrossRef] [PubMed]
262. Bertolini, M.; Ranjan, A.; Thompson, A.; Diaz, P.I.; Sobue, T.; Maas, K.; Dongari-Bagtzoglou, A. *Candida albicans* induces mucosal bacterial dysbiosis that promotes invasive infection. *PLoS Pathog.* **2019**, *15*, e1007717. [CrossRef] [PubMed]
263. Cottier, F.; Tan, A.S.; Xu, X.; Wang, Y.; Pavelka, N. MIG1 regulates resistance of *Candida albicans* against the fungistatic effect of weak organic acids. *Eukaryot. Cell* **2015**, *14*, 1054–1061. [CrossRef]
264. Allonsius, C.N.; Van Den Broek, M.F.L.; De Boeck, I.; Kiekens, S.; Oerlemans, E.F.M.; Kiekens, F.; Foubert, K.; Vandenheuvel, D.; Cos, P.; Delputte, P.; et al. Interplay between *Lactobacillus rhamnosus* GG and *Candida* and the involvement of exopolysaccharides. *Microb. Biotechnol.* **2017**, *10*, 1753–1763. [CrossRef] [PubMed]
265. Haas, S.V.; Haas, M.P. The treatment of celiac disease with the specific carbohydrate diet; report on 191 additional cases. *Am. J. Gastroenterol.* **1955**, *23*, 344–360. [PubMed]
266. Breaking the Vicious Cycle and the Specific Carbohydrate Diet. Available online: www.breakingtheviciouscycle.info (accessed on 14 July 2019).

267. Gottschall, E. *Breaking the Vicious Cycle: Intestinal Health Through Diet*; Kirkton Press: Baltimore, ON, Canada, 1994; p. 205.
268. Nutrition in Immune Balance (NiMBAL)—Food Table. Available online: www.nimbal.org/legalillegal-food-list (accessed on 26 June 2019).
269. Kakodkar, S.; Farooqui, A.J.; Mikolaitis, S.L.; Mutlu, E.A. The specific carbohydrate diet for inflammatory bowel disease: A case series. *J. Acad. Nutr. Diet.* **2015**, *115*, 1226–1232. [CrossRef]
270. Chassaing, B.; Koren, O.; Goodrich, J.K.; Poole, A.C.; Srinivasan, S.; Ley, R.E.; Gewirtz, A.T. Dietary emulsifiers impact the mouse gut microbiota promoting colitis and metabolic syndrome. *Nature* **2015**, *519*, 92–96. [CrossRef]
271. Chassaing, B.; Raja, S.M.; Lewis, J.D.; Srinivasan, S.; Gewirtz, A.T. Colonic microbiota encroachment correlates with dysglycemia in humans. *Cell. Mol. Gastroenterol. Hepatol.* **2017**, *4*, 205–221. [CrossRef] [PubMed]
272. Miclotte, L.; Van De Wiele, T. Food processing, gut microbiota and the globesity problem. *Crit. Rev. Food Sci. Nutr.* **2019**, 1–14, in press. [CrossRef]
273. Li, G.; Xie, C.; Lu, S.; Nichols, R.G.; Tian, Y.; Li, L.; Patel, D.; Ma, Y.; Brocker, C.N.; Yan, T.; et al. Intermittent fasting promotes white adipose browning and decreases obesity by shaping the gut microbiota. *Cell Metab.* **2017**, *26*, 672–685. [CrossRef]
274. Chevalier, C.; Stojanovic, O.; Colin, D.J.; Suarez-Zamorano, N.; Tarallo, V.; Veyrat-Durebex, C.; Rigo, D.; Fabbiano, S.; Stevanovic, A.; Hagemann, S.; et al. Gut microbiota orchestrates energy homeostasis during cold. *Cell* **2015**, *163*, 1360–1374. [CrossRef] [PubMed]
275. Zarrinpar, A.; Chaix, A.; Yooseph, S.; Panda, S. Diet and feeding pattern affect the diurnal dynamics of the gut microbiome. *Cell Metab.* **2014**, *20*, 1006–1017. [CrossRef] [PubMed]
276. Cignarella, F.; Cantoni, C.; Ghezzi, L.; Salter, A.; Dorsett, Y.; Chen, L.; Phillips, D.; Weinstock, G.M.; Fontana, L.; Cross, A.H.; et al. Intermittent fasting confers protection in CNS autoimmunity by altering the gut microbiota. *Cell Metab.* **2018**, *27*, 1222–1235. [CrossRef] [PubMed]
277. Rangan, P.; Choi, I.; Wei, M.; Navarrete, G.; Guen, E.; Brandhorst, S.; Enyati, N.; Pasia, G.; Maesincee, D.; Ocon, V.; et al. Fasting-mimicking diet modulates microbiota and promotes intestinal regeneration to reduce inflammatory bowel disease pathology. *Cell Rep.* **2019**, *26*, 2704–2719. [CrossRef]
278. Catterson, J.H.; Khericha, M.; Dyson, M.C.; Vincent, A.J.; Callard, R.; Haveron, S.M.; Rajasingam, A.; Ahmad, M.; Partridge, L. Short-term, intermittent fasting induces long-lasting gut health and TOR-independent lifespan extension. *Curr. Biol.* **2018**, *28*, 1714–1724. [CrossRef] [PubMed]
279. Muccioli, G.G.; Naslain, D.; Backhed, F.; Reigstad, C.S.; Lambert, D.M.; Delzenne, N.M.; Cani, P.D. The endocannabinoid system links gut microbiota to adipogenesis. *Mol. Syst. Biol.* **2010**, *6*, 392. [CrossRef]
280. Pesce, M.; Esposito, G.; Sarnelli, G. Endocannabinoids in the treatment of gasytrointestinal inflammation and symptoms. *Curr. Opin. Pharmacol.* **2018**, *43*, 81–86. [CrossRef]
281. De Filippis, D.; Esposito, G.; Cirillo, C.; Cipriano, M.; De Winter, B.Y.; Scuderi, C.; Sarnelli, G.; Cuomo, R.; Steardo, L.; De Man, J.G.; et al. Cannabidiol reduces intestinal inflammation through the control of neuroimmune axis. *PLoS ONE* **2011**, *6*, e28159. [CrossRef]
282. Uranga, J.A.; Vera, G.; Abalo, R. Cannabinoid pharmacology and therapy in gut disorders. *Biochem. Pharmacol.* **2018**, *157*, 134–147. [CrossRef]
283. Le Strat, Y.; Le Foll, B. Obesity and cannabis use: Results from 2 representative national surveys. *Am. J. Epidemiol.* **2011**, *174*, 929–933. [CrossRef]
284. Cluny, N.L.; Keenan, C.M.; Reimer, R.A.; Le Foll, B.; Sharkey, K.A. Prevention of diet-induced obesity effects on body weight and gut microbiota in mice treated chronically with delta9-tetrahydrocannabinol. *PLoS ONE* **2015**, *10*, e0144270. [CrossRef] [PubMed]
285. Zhang, L.; Huang, Y.; Zhou, Y.; Buckley, T.; Wang, H.H. Antibiotic administration routes significantly influence the levels of antibiotic resistance in gut microbiota. *Antimicrob. Agents Chemother.* **2013**, *57*, 3659–3666. [CrossRef] [PubMed]
286. O'Donoghue, C.; Solomon, K.; Fenelon, L.; Fitzpatrick, F.; Kyne, L. Effect of proton pump inhibitors and antibiotics on the gut microbiome of hospitalised older persons. *J. Infect.* **2016**, *72*, 498–500. [CrossRef] [PubMed]
287. Blaser, M.J. Antibiotic use and its consequences for the normal microbiome. *Science* **2016**, *352*, 544–545. [CrossRef] [PubMed]

288. Su, T.; Lai, S.; Lee, A.; He, X.; Chen, S. Meta-analysis: Proton pump inhibitors moderately increase the risk of small intestinal bacterial overgrowth. *J. Gastroenterol.* **2018**, *53*, 27–36. [CrossRef] [PubMed]
289. Trifan, A.; Stanciu, C.; Girleanu, I.; Stoica, O.C.; Singeap, A.M.; Maxim, R.; Chiriac, S.A.; Ciobica, A.; Boiculese, L. Proton pump inhibitors therapy and risk of *Clostridium difficile* infection: Systematic review and meta-analysis. *World J. Gastroenterol.* **2017**, *23*, 6500–6515. [CrossRef]
290. Le Bastard, Q.; Al-Ghalith, G.A.; Gregoire, M.; Chapelet, G.; Javaudin, F.; Dailly, E.; Batard, E.; Knights, D.; Montassier, E. Systematic review: Human gut dysbiosis induced by non-antibiotic prescription medications. *Aliment. Pharmacol. Ther.* **2018**, *47*, 332–345. [CrossRef] [PubMed]
291. Imhann, F.; Bonder, M.J.; Vich Vila, A.; Fu, J.; Mujagic, Z.; Vork, L.; Tigchelaar, E.F.; Jankipersadsing, S.A.; Cenit, M.C.; Harmsen, H.J.; et al. Proton pump inhibitors affect the gut microbiome. *Gut* **2016**, *65*, 740–748. [CrossRef]
292. Bahr, S.M.; Tyler, B.C.; Wooldridge, N.; Butcher, B.D.; Burns, T.L.; Teesch, L.M.; Oltman, C.L.; Azcarate-Peril, M.A.; Kirby, J.R.; Calarge, C.A. Use of the second-generation antipsychotic, risperidone, and secondary weight gain are associated with an altered gut microbiota in children. *Transl. Psychiatry* **2015**, *5*, e652. [CrossRef]
293. Flowers, S.A.; Evans, S.J.; Ward, K.M.; McInnis, M.G.; Ellingrod, V.L. Interaction between atypical antipsychotics and the gut microbiome in a bipolar disease cohort. *Pharmacotherapy* **2017**, *37*, 261–267. [CrossRef]
294. Rogers, M.A.M.; Aronoff, D.M. The influence of non-steroidal anti-inflammatory drugs on the gut microbiome. *Clin. Microbiol. Infect.* **2016**, *22*, 178. [CrossRef] [PubMed]
295. Meng, J.; Yu, H.; Ma, J.; Wang, J.; Banerjee, S.; Charboneau, R.; Barke, R.A.; Roy, S. Morphine induces bacterial translocation in mice by compromising intestinal barrier function in a TLR-dependent manner. *PLoS ONE* **2013**, *8*, e54040. [CrossRef] [PubMed]
296. Sharon, G.; Cruz, N.J.; Kang, D.W.; Gandal, M.J.; Wang, B.; Kim, Y.M.; Zink, E.M.; Casey, C.P.; Taylor, B.C.; Lane, C.J.; et al. Human gut microbiota from autism spectrum disorder promote behavioral symptoms in mice. *Cell* **2019**, *177*, 1600–1618. [CrossRef] [PubMed]
297. Martinez, G.P.; Bauerl, C.; Collado, M.C. Understanding gut microbiota in elderly's health will enable intervention through probiotics. *Benef. Microbes* **2014**, *5*, 235–246. [CrossRef] [PubMed]
298. Davenport, E.R.; Mizrahi-Man, O.; Michelini, K.; Barreiro, L.B.; Ober, C.; Gilad, Y. Seasonal variation in human gut microbiome composition. *PLoS ONE* **2014**, *9*, e90731. [CrossRef] [PubMed]
299. O'Toole, P.W.; Jeffery, I.B. Gut microbiota and aging. *Science* **2015**, *350*, 1214–1215. [CrossRef]
300. Armour, C.R.; Nayfach, S.; Pollard, K.S.; Sharpton, T.J. A metagenomic meta-analysis reveals functional signatures of health and disease in the human gut microbiome. *mSystems* **2019**, *4*, e00332-18. [CrossRef]
301. Brunt, V.E.; Gioscia-Ryan, R.A.; Richey, J.J.; Zigler, M.C.; Cuevas, L.M.; Gonzalez, A.; Vazquez-Baeza, Y.; Battson, M.L.; Smithson, A.T.; Gilley, A.D.; et al. Suppression of the gut microbiome ameliorates age-related arterial dysfunction and oxidative stress in mice. *J. Physiol.* **2019**, *597*, 2361–2378. [CrossRef]
302. Smith, P.; Willemsen, D.; Popkes, M.; Metge, F.; Gandiwa, E.; Reichard, M.; Valenzano, D.R. Regulation of life span by the gut microbiota in the short-lived African turquoise killifish. *eLife* **2017**, *6*, e27014. [CrossRef]
303. Tirosh, A.; Calay, E.S.; Tuncman, G.; Claiborn, K.C.; Inouye, K.E.; Eguchi, K.; Alcala, M.; Rathaus, M.; Hollander, K.S.; Ron, I.; et al. The short-chain fatty acid propionate increases glucagon and FABP4 production, impairing insulin action in mice and humans. *Sci. Transl. Med.* **2019**, *11*, eaav0120. [CrossRef]
304. Perry, R.J.; Peng, L.; Barry, N.A.; Cline, G.W.; Zhang, D.; Cardone, R.L.; Petersen, K.F.; Kibbey, R.G.; Goodman, A.L.; Shulman, G.I. Acetate mediates a microbiome-brain-beta-cell axis to promote metabolic syndrome. *Nature* **2016**, *534*, 213–217. [CrossRef] [PubMed]
305. Si, X.; Shang, W.; Zhou, Z.; Strappe, P.; Wang, B.; Bird, A.; Blanchard, C. Gut microbiome-induced shift of acetate to butyrate positively manages dysbiosis in high fat diet. *Mol. Nutr. Food Res.* **2018**, *62*, 1700670. [CrossRef] [PubMed]
306. Laforest-Lapointe, I.; Arrieta, M.C. Microbial eukaryotes: A missing link in gut microbiome studies. *mSystems* **2018**, *3*, e00201-17. [CrossRef] [PubMed]

307. Rampelli, S.; Turroni, S.; Schnorr, S.L.; Soverini, M.; Quercia, S.; Barone, M.; Castagnetti, A.; Biagi, E.; Gallinella, G.; Brigidi, P.; et al. Characterization of the human DNA gut virome across populations with different subsistence strategies and geographical origin. *Environ. Microbiol.* **2017**, *19*, 4728–4735. [CrossRef] [PubMed]
308. Hsu, B.B.; Gibson, T.E.; Yeliseyev, V.; Liu, Q.; Lyon, L.; Bry, L.; Silver, P.A.; Gerber, G.K. Dynamic modulation of the gut microbiota and metabolome by bacteriophages in a mouse model. *Cell Host Microbe* **2019**, *25*, 803–814. [CrossRef] [PubMed]
309. Norman, J.M.; Handley, S.A.; Baldridge, M.T.; Droit, L.; Liu, C.Y.; Keller, B.C.; Kambal, A.; Monaco, C.L.; Zhao, G.; Fleshner, P.; et al. Disease-specific alterations in the enteric virome in inflammatory bowel disease. *Cell* **2015**, *160*, 447–460. [CrossRef] [PubMed]
310. Audebert, C.; Even, G.; Cian, A. The Blastocystis Investigation Group; Loywick, A.; Merlin, S.; Viscogliosi, E.; Chabe, M. Colonization with the enteric protozoa *Blastocystis* is associated with increased diversity of human gut bacterial microbiota. *Sci. Rep.* **2016**, *6*, 25255. [CrossRef] [PubMed]
311. Poirier, P.; Wawrzyniak, I.; Vivares, C.P.; Delbac, F.; El Alaoui, H. New insights into *Blastocystis* spp.: A potential link with irritable bowel syndrome. *PLoS Pathog.* **2012**, *8*, e1002545. [CrossRef]
312. Yason, J.A.; Liang, Y.R.; Png, C.W.; Zhang, Y.; Tan, K.S.W. Interactions between a pathogenic *Blastocystis* subtype and gut microbiota: In vitro and in vivo studies. *Microbiome* **2019**, *7*, 30. [CrossRef]
313. Nieves-Ramirez, M.E.; Partida-Rodriguez, O.; Laforest-Lapointe, I.; Reynolds, L.A.; Brown, E.M.; Valdez-Salazar, A.; Moran-Silva, P.; Rojas-Velazquez, L.; Morien, E.; Parfrey, L.W.; et al. Asymptomatic intestinal colonization with protist *Blastocystis* is strongly associated with distinct microbiome ecological patterns. *mSystems* **2018**, *3*, e00007-18. [CrossRef]
314. Gentekaki, E.; Curtis, B.A.; Stairs, C.W.; Klimes, V.; Elias, M.; Salas-Leiva, D.E.; Herman, E.K.; Eme, L.; Arias, M.C.; Henrissat, B.; et al. Extreme genome diversity in the hyper-prevalent parasitic eukaryote *Blastocystis*. *PLoS Biol.* **2017**, *15*, e2003769. [CrossRef] [PubMed]
315. Zmora, N.; Suez, J.; Elinav, E. You are what you eat: Diet, health and the gut microbiota. *Nat. Rev. Gastroenterol. Hepatol.* **2019**, *16*, 35–56. [CrossRef] [PubMed]
316. Chen, F.; Du, M.; Blumberg, J.B.; Ho Chui, K.K.; Ruan, M.; Rogers, G.; Shan, Z.; Zeng, L.; Zhang, F.F. Association among dietary supplement use, nutrient intake, and mortality among U.S. adults: A cohort study. *Ann. Intern. Med.* **2019**, *170*, 604–613. [CrossRef] [PubMed]
317. Fetissov, S.O. Role of the gut microbiota in host appetite control: Bacterial growth to animal feeding behaviour. *Nat. Rev. Endocrinol.* **2017**, *13*, 11–25. [CrossRef] [PubMed]
318. Thorburn, A.N.; Macia, L.; Mackay, C.R. Diet, metabolites, and "Western-lifestyle" inflammatory diseases. *Immunity* **2014**, *40*, 833–842. [CrossRef] [PubMed]
319. Vieira-Silva, S.; Sabino, J.; Valles-Colomer, M.; Falony, G.; Kathagen, G.; Caenepeel, C.; Cleynen, I.; Van Der Merwe, S.; Vermeire, S.; Raes, J. Quantitative microbiome profiling disentangles inflammation- and bile duct obstruction-associated microbiota alterations across PSC/IBD diagnoses. *Nat. Microbiol.* **2019**, in press. [CrossRef] [PubMed]
320. Hillman, E.T.; Lu, H.; Yao, T.; Nakatsu, C.H. Microbial ecology along the gastrointestinal tract. *Microbes Environ.* **2017**, *32*, 300–313. [CrossRef] [PubMed]
321. Crespo-Piazuelo, D.; Estelle, J.; Revilla, M.; Criado-Mesas, L.; Ramayo-Caldas, Y.; Ovilo, C.; Fernandez, A.I.; Ballester, M.; Folch, J.M. Characterization of bacterial microbiota compositions along the intestinal tract in pigs and their interactions and functions. *Sci. Rep.* **2018**, *8*, 12727. [CrossRef]
322. Johnson, A.J.; Vangay, P.; Al-Ghalith, G.A.; Hillmann, B.M.; Ward, T.L.; Shields-Cutler, R.R.; Kim, A.D.; Shmagel, A.K.; Syed, A.N.; Walter, J.; et al. Daily sampling reveals personalized diet-microbiome associations in humans. *Cell Host Microbe* **2019**, *25*, 789–802. [CrossRef]
323. David, L.A.; Maurice, C.F.; Carmody, R.N.; Gootenberg, D.B.; Button, J.E.; Wolfe, B.E.; Ling, A.V.; Devlin, A.S.; Varma, Y.; Fischbach, M.A.; et al. Diet rapidly and reproducibly alters the human gut microbiome. *Nature* **2014**, *505*, 559–563. [CrossRef]
324. Zimmermann, M.; Zimmermann-Kogadeeva, M.; Wegmann, R.; Goodman, A.L. Mapping human microbiome drug metabolism by gut bacteria and their genes. *Nature* **2019**, *570*, 462–467. [CrossRef] [PubMed]

325. De Toro-Martin, J.; Arsenault, B.J.; Despres, J.P.; Vohl, M.C. Precision nutrition: A review of personalized nutritional approaches for the prevention and management of metabolic syndrome. *Nutrients* **2017**, *9*, 913. [CrossRef] [PubMed]
326. Zhang, N.; Ju, Z.; Zuo, T. Time for food: The impact of diet on gut microbiota and human health. *Nutrition* **2018**, *51–52*, 80–85. [CrossRef] [PubMed]

MDPI
St. Alban-Anlage 66
4052 Basel
Switzerland
Tel. +41 61 683 77 34
Fax +41 61 302 89 18
www.mdpi.com

Nutrients Editorial Office
E-mail: nutrients@mdpi.com
www.mdpi.com/journal/nutrients

www.ingramcontent.com/pod-product-compliance
Lightning Source LLC
LaVergne TN
LVHW071615170726
843515LV00009B/2399

* 9 7 8 3 0 3 9 3 6 9 1 6 4 *